DESIGNER'S HANDBOOK OF INTEGRATED CIRCUITS

OTHER McGRAW-HILL HANDBOOKS OF INTEREST

Baumeister • Marks' Standard Handbook for Mechanical Engineers
Beeman • Industrial Power Systems Handbook
Condon and Odishaw • Handbook of Physics
Considine • Energy Technology Handbook
Coombs • Basic Electronic Instrument Handbook
Coombs • Printed Circuits Handbook
Croft, Carr, and Watt • American Electricians' Handbook
Fink and Christiansen • Electronics Engineers' Handbook
Fink and Beaty • Standard Handbook for Electrical Engineers
Giacoletto • Electronics Designers' Handbook
Harper • Handbook of Components for Electronics
Harper • Handbook of Electronic Packaging
Harper • Handbook of Materials and Processes for Electronics
Harper • Handbook of Thick Film Hybrid Microelectronics
Harper • Handbook of Wiring, Cabling, and Interconnecting for Electronics
Helms • The McGraw-Hill Computer Handbook
Hicks • Standard Handbook of Engineering Calculations
Jasik • Antenna Engineering Handbook
Juran • Quality Control Handbook
Kaufman and Seidman • Handbook of Electronics Calculations
Kaufman and Seidman • Handbook for Electronics Engineering Technicians
Kurtz and Shoemaker • The Lineman's and Cableman's Handbook
Maissel and Glang • Handbook of Thin Film Technology
Markus • Electronics Dictionary
McPartland • McGraw-Hill's National Electrical Code Handbook
Perry • Engineering Manual
Skolnik • Radar Handbook
Stout • Handbook of Microprocessor Applications
Stout • Handbook of Microcircuit Design and Applications
Stout and Kaufman • Handbook of Operational Amplifier Circuit Design
Tuma • Engineering Mathematics Handbook
Tuma • Handbook of Physical Calculations
Tuma • Technology Mathematics Handbook
Williams • Electronic Filter Design Handbook

DESIGNER'S HANDBOOK OF INTEGRATED CIRCUITS

Arthur B. Williams, Editor in Chief
Vice President of Engineering,
Research, and Development

Coherent Communications Systems Corp.
Hauppauge, N.Y.

McGraw-Hill Book Company

New York   St. Louis   San Francisco   Auckland
Bogotá   Hamburg   Johannesburg   London   Madrid
Mexico   Montreal   New Delhi   Panama   Paris
São Paulo   Singapore   Sydney   Tokyo   Toronto

Library of Congress Cataloging in Publication Data
Main entry under title:

Designer's handbook of integrated circuits.

Includes index.
1.Integrated circuits—Handbooks, manuals,
etc. I. Williams, Arthur Bernard, 1940–
TK7874.D476   1984   621.381'73   82-14955
ISBN 0-07-070435-X

234567890 KGP KGP 8987654

ISBN 0-07-070435-X

The editors for this book were Harold B. Crawford and James T. Halston,
the designer was Graphic Images Ltd., and the production
supervisor was Sara L. Fliess. It was set in Times Roman
by Techna Type.

Printed and bound by the Kingsport Press

TO MY WIFE ELLEN AND CHILDREN
HOWARD, BONNIE, AND ROBIN

Contents

List of Contributors

Hamil Aldridge, Paradyne Corp., Largo, Fla. (*SSI Logic Circuits*)

Peter Alfke, Director, Applications Engineering, Advanced Micro Devices Inc., Sunnyvale Calif. (*MSI Logic Circuits*)

Don Birkley, Tektronix Corp., Beaverton, Oreg. (*Microprocessors*)

Peter D. Bradshaw, Director of Advanced Applications, Array Technology Inc., San Jose, Calif. (*A/D and D/A Conversion*)

Eric G. Breeze, Atari Corp., Sunnyvale, Calif. (*Optoelectronics*)

Brian Cayton, Marketing Manager, Standard Microsyststems Corp., Hauppauge, N.Y. (*LSI Peripheral Devices*)

Earl V. Cole, Atari Corp., Sunnyvale, Calif. (*Optoelectronics*)

Robert C. Frostholm, Account Manager, Automotive Marketing, National Semiconductor Corp., Santa Clara, Calif. (*IC Power-Mangement Circuits*)

Sid Ghosh, TRW Vidar Corp., Mountainview, Calif. (*Phase-Locked Loops*)

Randall J. Hipp, Mostek Corp., Carrollton, Tex. (*Telecommunications Circuits*)

Robert C. Jones, Mostek Corp., Carrollton, Tex. (*Telecommunications Circuits*)

Darin L. Kincaid, Mostek Corp., Carrollton, Tex. (*Telecommunications Circuits*)

Dave Kohlmeier, Tektronix Corp., Beaverton, Oreg. (*Microprocessors*)

Glen M. Masker, Mostek Corp., Carrollton, Tex. (*Telecommunications Circuits*)

William M. Otsuka, President, Optomicronix, Cupertino, Calif. (*Optoelectronics*)

H. Ilhan Refioglu, Exar Integrated Systems Inc., Sunnyvale, Calif. (*Timing Circuits*)

Joel Silverman, Marketing Manager, Siliconix Inc., Santa Clara, Calif. (*Function Circuits*)

Michael R. Sims, Mostek Corp., Carrollton, Tex. (*Telecommunications Circuits*)

Carroll Smith, Applications Engineer, Texas Instruments Corp., Dallas, Tex. (*Interface Circuits*)

Jerri L. Smith, Mostek Corp., Carrollton, Tex. (*Telecommunications Circuits*)

Dr. William R. Warner, Tektronix Corp., Beaverton, Oreg. (*Microprocessors*)

Arthur B. Williams, Vice President of Engineering, Research, and Development, Coherent Communications Systems Corp., Hauppauge, N.Y. (*Operational Amplifiers* and *Design of Active Filters Using Operational Amplifiers*)

Preface

Integrated circuits (ICs) have greatly simplified the design of complex analog and digital circuits. Over the past two decades an overwhelming variety of ICs have been produced by numerous manufacturers.

The engineer or technician, when faced with the task of IC selection and circuit design, must sort through a variety of manufacturers' IC catalogs and a limited number of application notes to try and determine the optimum IC and circuit configuration for the requirement.

Catalog data sheets are useful in defining the operating and worst-case parameters of a particular device, but cannot serve as a selection guide since ICs are not evaluated on a comparative basis. Also these catalogs and application notes are restricted to ICs of a particular manufacturer and are organized by IC type rather than application.

This book is intended to serve a twofold purpose. Equal emphasis is placed on IC applications as on device selection. Preferred IC circuit configurations are provided by experts so that proven practical solutions to frequently encountered design problems can be easily obtained. This book is not intended to replace IC catalogs, since the inclusion of detailed parameters on all the ICs covered would be totally impractical. Instead, the device selection and comparison charts provided, as well as the detailed discussions and design examples, will assist the designer in selecting the best device and circuit configuration for the application.

Operational amplifiers are covered in Chap. 1. Op amp theory is reviewed, both from a theoretical and practical standpoint. Numerous circuit configurations are illustrated and an extensive selection guide is provided.

Chap. 2 discusses selection and application of function circuits, such as multipliers, waveform generators, voltage-to-frequency and frequency-to-voltage converters, etc.

Active filter design using op amps is introduced in Chap. 3. Numerous preferred low-pass, high-pass, bandpass, and band-reject circuit configurations are illustrated along with design examples.

Chap. 4 extensively covers telecommunication circuits, such as pulse and DTMF dialers and encoders, CODECs, PCM line filters, and speech networks.

The theory, design, and selection of phase-locked-loop configurations and devices are discussed in Chap. 5.

Timer ICs are introduced in Chap. 6. The analysis, design, selection, and application of these versatile devices are extensively covered in this chapter.

Chap. 7 on IC power management circuits covers the principles of series pass and switching regulators, and the optimum selection and circuit configuration of these ICs.

The principles of A/D and D/A conversion are covered in Chap. 8. Various types of circuit configurations are discussed and preferred circuit structures are presented along with device selection guidelines.

Chap. 9 introduces SSI logic circuits. The various logic families and their limitations are extensively covered.

MSI logic circuits are covered in Chap. 10. Combinatorial and sequential logic applications of MSI devices are presented along with guidelines for device selection.

In Chap. 11 the selection process and considerations involved in determining the optimum microprocessor for a given application is discussed. Chip architecture, support software, and other major considerations are extensively covered.

Chap. 12 covers optoelectronics. The theory, application, and selection of LED lamps, bar graph displays, alphanumeric displays, and optocouplers are discussed along with many practical examples of the selection and design process.

LSI peripheral devices are described in Chap. 13. The operation, application, and selection of devices such as UARTs, CRT controllers, and floppy disc controllers are presented.

Application and selection of interface circuits are covered in Chap. 14, including devices such as peripheral drivers, line circuits, and display drivers.

I would like to thank the many contributors and their companies for their efforts to make this book as technically comprehensive as possible while placing sufficient emphasis on everyday applications of ICs.

Arthur B. Williams
Editor

DESIGNER'S HANDBOOK OF INTEGRATED CIRCUITS

Chapter 1

OPERATIONAL AMPLIFIERS

Arthur B. Williams Vice President of Engineering,
Research, and Development
Coherent Communications Systems Corp.
Hauppauge, N.Y.

1–1 FUNDAMENTALS OF OP AMP THEORY

The operational amplifier (op amp) is probably the most popular and versatile building block used in electronic circuitry. This device is capable of performing an infinite variety of tasks involving both linear and nonlinear operations upon electrical signals.

Until the revolutionary development of the monolithic μA709 in the mid-sixties, operational amplifiers were constructed using discrete components and were relatively costly. Many new generations of op amps have since evolved, offering special features and improved performance.

1–1a The Ideal Op Amp

An op amp is a high-gain amplifier with two input terminals, a single output terminal, and is direct coupled internally. The standard symbol is shown in Fig. 1-1a and the equivalent circuit is represented in Fig. 1-1b. The output voltage V_o is the difference of the input voltages applied to the two input terminals, multiplied by the amplifier gain A_o. A positive changing signal applied to the positive (|) input terminal results in a positive change at the output. Hence the (+) terminal is called the noninverting input. A positive changing signal applied to the negative (−) input terminal results in a negative change at the output, so the (−) terminal is referred to as the inverting input.

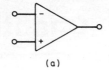

(a)

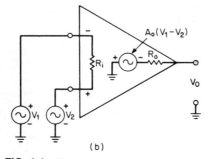

(b)

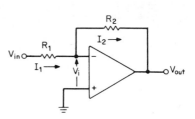

FIG. 1-1 The operational amplifier. (a) Symbol; (b) equivalent circuit.

FIG. 1-2 Inverting amplifier.

The device provides a very high gain to a differential input signal, since A_o is generally very large. A signal applied to both inputs simultaneously (common-mode input) results in no output, since the net differential input is zero.

The amplifier has an input impedance R_i and an output impedance R_o. In an ideal amplifier R_i is considered infinite and R_o zero. Also, the voltage gain A_o would be infinite.

Inverting Amplifier If negative feedback is applied from the output to the inverting input and if the amplifier is assumed ideal, the differential input voltage is forced to zero. This is called the "virtual ground effect." Therefore the inverting input can be used as a current-summing node, resulting in numerous useful amplifier configurations and a simplified approach for circuit analysis.

An inverting op amp circuit is shown in Fig. 1-2. If the amplifier is ideal, the differential input voltage V_i becomes zero, placing the inverting input at ground potential because of the negative feedback path through R_2. The resistor currents can then be given by

$$I_1 = \frac{V_{in}}{R_1} \qquad (1\text{-}1)$$

and

$$I_2 = -\frac{V_{out}}{R_2} \qquad (1\text{-}2)$$

Since the amplifier is assumed ideal, the input impedance is infinite and no current can flow into the inverting input terminal; therefore, I_1 must equal I_2. If we equate Eqs. 1-1 and 1-2 and solve for the overall closed-loop gain, we obtain

$$A_c = \frac{V_{out}}{V_{in}} = -\frac{R_2}{R_1} \qquad (1\text{-}3)$$

We can then conclude that the circuit gain of an ideal inverting amplifier is equal to the ratio of two resistors and is independent of the amplifier itself. The input impedance is R_1 and the output impedance is zero.

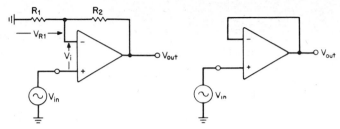

FIG. 1-3   Noninverting configuration.   **FIG. 1-4**   Voltage follower.

Noninverting Amplifier   Let us now consider an operational amplifier in the noninverting configuration of Fig. 1-3. Because of negative feedback, the differential input voltage V_i is forced to zero. V_{R_1}, the voltage across R_1, is then equal to V_{in}. Since R_1 and R_2 are in the form of a voltage divider, the output voltage must be

$$V_{out} = V_{R_1} \frac{R_1 + R_2}{R_1} = V_{in} \frac{R_1 + R_2}{R_1} \tag{1-4}$$

which can also be expressed as

$$A_c = \frac{V_{out}}{V_{in}} = 1 + \frac{R_2}{R_1} \tag{1-5}$$

The gain of a noninverting amplifier is given by the ratio of two resistors *plus 1*. The input impedance is infinite and the output impedance is zero.

If we take the special case where R_1 is infinite and R_2 is zero, the circuit reduces to the voltage-follower circuit of Fig. 1-4. Circuit gain is unity, the input impedance is infinite, and the output impedance is zero. This circuit performs the same buffering function as a cathode follower or emitter follower.

1–1b   The Feedback Equation

To derive the feedback equation for the inverting amplifier of Fig. 1-2, let us first express V_i in terms of the input and output voltages using superposition. This results in

$$V_i = \frac{R_2}{R_1 + R_2} V_{in} + \frac{R_1}{R_1 + R_2} V_{out} \tag{1-6}$$

Since the amplifier has an open-loop gain $-A_o$, we can substitute $-V_{out}/A_o$ for V_i in Eq. 1-6. By rearranging terms we obtain the overall closed-loop gain expression

$$A_c = \frac{\dfrac{R_2}{R_1 + R_2}}{\dfrac{1}{-A_o} - \dfrac{R_1}{R_1 + R_2}} \tag{1-7}$$

where

$$A_c = V_{out}/V_{in}$$

If we let

$$\beta = \frac{R_1}{R_1 + R_2}$$

then
$$1 - \beta = \frac{R_2}{R_1 + R_2}$$

We now reexpress Eq. 1-7 in terms of β. The overall gain expression then becomes

$$A_c = \frac{1 - \beta}{\dfrac{1}{-A_o} - \beta} \tag{1-8}$$

or in the preferred form

$$A_c \text{ (inverting)} = \frac{A_o (\beta - 1)}{1 + A_o \beta} \tag{1-9}$$

Let us now consider the noninverting configuration of Fig. 1-3. As a result of voltage-divider action, we can state

$$V_{R_1} = \frac{R_1}{R_1 + R_2} V_{\text{out}} = \beta V_{\text{out}} \tag{1-10}$$

It should also be evident that

$$V_i = V_{\text{in}} - V_{R_1} \tag{1-11}$$

The output voltage is then given by

$$V_{\text{out}} = A_o (V_{\text{in}} - V_{R_1}) \tag{1-12}$$

If we replace V_{R_1} by βV_{out} as given by Eq. 1-10 and then solve for the closed-loop gain, we obtain

$$A_c \text{ (noninverting)} = \frac{A_o}{1 + A_o \beta} \tag{1-13}$$

Eqs. 1-9 and 1-13 are the general feedback equations for the inverting and noninverting amplifier configurations. These expressions should be familiar to those who have studied feedback systems or servo theory. β is called the feedback factor, since it represents the fraction of the output that is fed back to the input. The product $A_o\beta$ is referred to as the "loop gain." The significance of these equations will be demonstrated in Sec. 1-2.

1–1c Glossary of Terms

The following terms define the basic parameters associated with operational amplifiers.

Channel Separation: The ratio of the change in output voltage of a driven channel to the corresponding change of another channel.

Common-Mode Input Resistance: The parallel combination of the small-signal input resistances between the two input terminals and ground.

Common-Mode Rejection Ratio: The ratio of differential to common-mode voltage gain.

Equivalent Input Noise Current: The current of an ideal current source that can be placed in shunt with the input terminals to represent an internally generated noise source.

Equivalent Input Noise Voltage: The voltage of an ideal source that can be placed in series with the input terminals to represent an internally generated noise source.

Gain Margin: The number of decibels the loop-gain is below 0 dB at the 180° phase-shift point.

Input Bias Current: The average of the two input currents with the output at zero volts.

Input Capacity: The capacity between the input terminals with either terminal grounded.

Input Offset Current: The difference of the two input currents with the output at zero volts.

Input Offset Voltage: The voltage that must be differentially introduced across the input terminals through two equal resistors to make the output zero volts.

Input Resistance: The resistance between the input terminals with either terminal grounded.

Input-Voltage Range: The range of voltage at either input terminal that cannot be exceeded for normal operation.

Large-Signal Voltage Amplification: The ratio of the peak-to-peak output voltage swing to the corresponding change in input voltage.

Maximum Peak-to-Peak Output Voltage Swing: The maximum output voltage obtainable prior to clipping with a quiescent DC output of zero volts.

Output Resistance: The small signal equivalent source impedance as seen at the output with the output voltage near zero volts.

Overshoot: The ratio of the maximum deviation of an output signal value to the final steady-state value for a step input.

Phase Margin: The number of degrees the loop-gain phase shift is below 180° at the unity loop-gain frequency.

Power Supply Rejection: (see Supply Voltage Sensitivity)

Rise Time: The time required for an output-voltage step to increase from 10% to 90% of full amplitude.

Settling Time: The time required for a step-function change of output to settle to within a specified range of the final value.

Short-Circuit Output Current: The maximum output current that will occur when the output is shorted to ground.

Slew Rate: The rate of change of the output voltage for a step input.

Supply Current: The current into the amplifier from the power supply with no load and the output at zero volts.

Supply-Voltage Sensitivity: The ratio of the change in input offset voltage to the change in supply voltages producing it.

Total Power Dissipation: The total DC power supplied to the device less the power delivered to the output load.

Unity-Gain Bandwidth: The frequency at which the open-loop gain is unity.

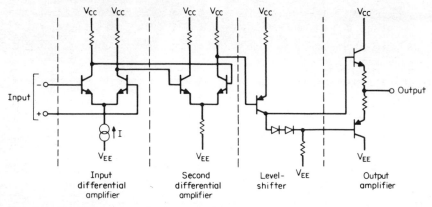

FIG. 1-5 Simplified op amp circuit.

1–2 PARAMETERS OF PRACTICAL AMPLIFIERS

1–2a Amplifier Circuit Structure

A simplified schematic of a typical op amp is shown in Fig. 1-5. The input section consists of a differential amplifier that has the major impact on the operating parameters. A second differential amplifier provides further gain and buffering. In order to obtain no DC offset at the output, a stage for level shifting is required. This stage drives the output amplifier which normally consists of a complementary PNP/NPN pair to obtain low output impedance and high current-handling capability. Current limiting may also be provided by using additional circuitry.

1–2b Errors in Operational Amplifiers

As a first-order approximation, op amps are assumed to have infinite open-loop gain, infinite input impedance, and zero output impedance. Although this assumption will simplify op amp circuit analysis and design, it may not always be valid. The designer should be aware of the limitations of op amps resulting from finite parameters and be in a position to calculate their effect.

Finite Open-Loop Gain Feedback equations for closed-loop gain were derived for both the inverting and noninverting amplifier configurations in Sec. 1-1 and are repeated for convenience.

$$A_c \text{ (inverting)} = \frac{A_o (\beta - 1)}{1 + A_o \beta} \tag{1-9}$$

$$A_c \text{ (noninverting)} = \frac{A_o}{1 + A_o \beta} \tag{1-13}$$

Both expressions have the denominator term $1 + A_o \beta$, where A_o is the open-loop gain and β is the feedback factor corresponding to the portion of the output fed back to the input. The open-loop gain of practical amplifiers is neither infinite nor real (zero phase shift), since the magnitude and phase of A_o both change with frequency. It is therefore perfectly feasible that at a high enough frequency, the $A_o \beta$ product could equal -1. The denominator of Eqs. 1-9 and 1-13 would then vanish, implying infinite closed-loop gain, which would be an oscillatory condition.

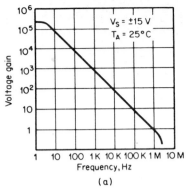
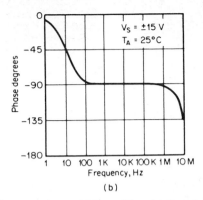

FIG. 1-6 Open-loop gain and phase shift vs. frequency for a µA741 amplifier. (*a*) Open-loop voltage gain as a function of frequency; (*b*) open-loop phase response as a function of frequency.

To prevent oscillation, the amplifier open-loop gain must be band-limited so that the product $A_o\beta$ is less than one below the frequency where the amplifier phase shift reaches 180°. This is achieved by causing a gain rolloff starting at low frequencies and continuing at a 6 decibels per octave rate. This method of ensuring stability is called "frequency compensation." For high closed-loop gains, β is very small, so less frequency compensation is required. Conversely, the voltage follower will require the most compensation, since $\beta = 1$.

A typical open-loop gain and phase-shift plot for a µA741 amplifier is shown in Fig. 1-6. The gain curve has two break points. A low-frequency breakpoint results from a frequency-compensation network deliberately introduced to maintain amplifier stability. The output phase lag increases to 45° at the breakpoint and asymptotically approaches 90° with increasing frequency. The amplitude rolls off at 6 decibels per octave in this region.

Another amplifier breakpoint occurs in the region of 5 MHz, due to amplifier parasitics. An additional 45° of phase shift occurs and the asymptotic phase limit becomes 180°. The rate of rolloff approaches 12 decibels per octave.

To determine the effect of open-loop gain on the closed-loop gain of an inverting amplifier, Eq. 1-9 can be manipulated into the following form using $R_1/(R_1 + R_2)$ for β:

$$A_c \text{ (inverting)} = \cfrac{-\dfrac{R_2}{R_1}}{\dfrac{1 + \dfrac{R_2}{R_1}}{A_o} + 1} \tag{1-14}$$

In a similar manner the following closed-loop gain expression for a noninverting amplifier can be obtained from Eq. 1-13:

$$A_c \text{ (noninverting)} = \cfrac{1 + \dfrac{R_2}{R_1}}{\dfrac{1 + \dfrac{R_2}{R_1}}{A_o} + 1} \tag{1-15}$$

TABLE 1–1 Gain Error vs. Ratio of Open-Loop
to Closed-Loop Gain

A_o/A_c	% Error (DC)	% Error (AC)
1	−50	−29
3	−25	−5
5	−16	−2
10^1	−10	−0.5
10^2	−1	−0.005
10^3	−0.1	
10^4	−0.01	

If the open-loop gain A_o is infinite, both denominators become unity, resulting in the general equations previously given

$$A_c \text{ (inverting)} = -\frac{R_2}{R_1} \tag{1-3}$$

and

$$A_c \text{ (noninverting)} = 1 + \frac{R_2}{R_1} \tag{1-5}$$

For finite values of open-loop gain, the denominator increases above one, which causes a gain reduction. A table of gain error vs. A_o/A_c can be obtained by evaluating the denominator of Eq. 1-15. These results are summarized in Table 1-1.

Clearly, the effect of open-loop gain rapidly diminishes as the ratio A_o/A_c is increased. Since the AC open-loop phase shift is usually −90° over most of the band of interest, the gain term is in quadrature with unity, thus diminishing its effect in the denominator of Eq. 1-15. Therefore the gain error decreases more rapidly for an increasing A_o/A_c ratio with AC signals as opposed to DC.

Input and Output Impedance   Bipolar-type op amps have typical input impedances of 1 MΩ, whereas FET input devices may have an input impedance of 10^{12} Ω or higher. In both cases the input capacity is typically a few picofarads. The output resistance is normally in the range of 100 Ω.

These impedances are open-loop parameters. With negative feedback, the closed-loop impedance characteristics are drastically changed. The input impedance of the noninverting amplifier can be derived as

$$R_{\text{in}} = (1 + A_o\beta) R_i \tag{1-16}$$

where R_i is the open-loop input impedance.

The output resistance with negative feedback is given by

$$R_{\text{out}} = \frac{R_o}{A_o\beta} \tag{1-17}$$

where R_o is the open-loop output impedance. For a typical case, where $R_i = 1$ MΩ, $R_o = 100$ Ω, $\beta = 0.1$, and $A_o = 10^3$, the closed-loop input and output impedances will be 100 MΩ and 1 Ω, respectively.

It is clear from Eqs. 1-16 and 1-17 that as the closed-loop gain is made larger (less negative feedback), β decreases and the impedance characteristics are degraded. Conversely, for the case of the voltage follower where $\beta = 1$, the input impedance is highest and the output impedance is minimized. For the previously discussed typical values for R_i, R_o, and A_o, the closed-loop input and output resistance will then be 10^9 and $0.1\ \Omega$, respectively.

Since A_o decreases with increasing frequency, the designer must recognize that the closed-loop impedance parameters may not be as good at the higher end of the band of operation. Also the nonideal impedances will have an effect on circuit closed-loop gain. However, except for very extreme cases, these gain errors are negligible and can be ignored. Amplifier input capacity is also neglected, except in very high impedance circuits, since it is typically only 1 or 2 pF.

DC Offsets The emitter-base voltages of the two input transistors in the input differential amplifier of Fig. 1-5 are never perfectly matched. As a result, a small differential offset voltage V_{IO} will occur at the input. This voltage is typically under 10 mV. However, when multiplied by an appreciable closed-loop gain, the resulting offset at the amplifier's output may be intolerable, especially in the case of DC coupled circuits. This offset is also temperature dependent. Even in the case of AC coupled circuits, the offset may shift the output quiescent point so that unsymmetrical clipping will occur in the case of high gains.

Fig. 1-7 illustrates some commonly used methods for adjusting the output offset. In the circuit of Fig. 1-7a, a potentiometer can be connected to the two terminals that are especially provided for this purpose with some op amps. Injecting small currents into the summing point will provide offset adjust capability for the amplifier circuits of Fig. 1-7b and c. Since V_{IO} is temperature dependent, the circuit once adjusted may still exhibit output offset with temperature. However, since the temperature coefficient of V_{IO} is typically a few microvolts per degree Celsius, this will usually not be a problem.

A small bias current I_B must flow into each base of the transistor pair in the input differential amplifiers. These currents are shown in Fig. 1-8 for the case of an inverting amplifier. The currents are nearly equal except for a small difference or offset current I_{IO}. In the circuit of Fig. 1-8, I_{IB_1} produces an error voltage at the input given by $I_{IB_1} R_{eq}$, where

$$R_{eq} = \frac{R_1 R_2}{R_1 + R_2} \tag{1-18}$$

The noninverting input bias current I_{IB_2} has no effect. To minimize the effect of I_{IB_1}, high values of R_{eq} should be avoided. Also, the circuits of Fig. 1-7 can be used.

A more frequently used approach involves introducing a resistor of R_{eq} between the noninverting input and ground, as illustrated in Fig. 1-9. This does not affect the overall

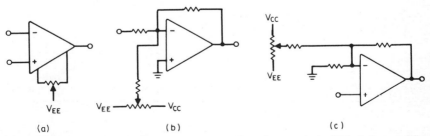

(a) (b) (c)

FIG. 1-7 Adjusting DC offset voltage. (*a*) Using offset adjust terminals; (*b*) inverting amplifier; (*c*) non-inverting amplifier.

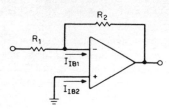

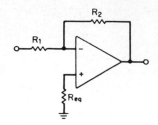

FIG. 1-8 Input bias currents in an inverting amplifier.

FIG. 1-9 Minimizing effect of DC offset current.

gain. However, a DC offset voltage of $I_{IB_2}R_{eq}$ is now introduced at the noninverting input. Since the amplifier is a differential-type device, the resulting net input voltage error due to the bias currents is $(I_{IB_1} - I_{IB_2})R_{eq}$ or $I_{IO}R_{eq}$. Since bias currents I_{IB_1} and I_{IB_2} are typically 80 nA and the offset current I_{IO} is in the range of 20 nA, a 4-to-1 reduction in offset is obtained.

In the case of the noninverting amplifier of Fig. 1-3, the gain is determined from the ratio R_2/R_1 plus 1. However, the actual values of R_1 and R_2 are nearly arbitrary and can be chosen so that the parallel combination of R_1 and R_2 is equal to the DC loading on the noninverting input; i.e., the parallel combination of all resistances connected between the noninverting input node and ground. For the voltage follower a resistor of R_{eq} can be introduced in the feedback path, as shown in Fig. 1-10.

If an amplifier can be restricted to AC operation only, the output offset can be reduced by configuring the circuit so that the DC offset gain is unity irrespective of the AC gain. This is achieved by introducing a blocking capacitor, as shown in the circuits of Fig. 1-11.

A 3-dB break point is introduced in the low-frequency response of the inverting amplifier, which can be computed by

$$f_{3\,dB} = \frac{1}{2\pi R_1 C} \tag{1-19}$$

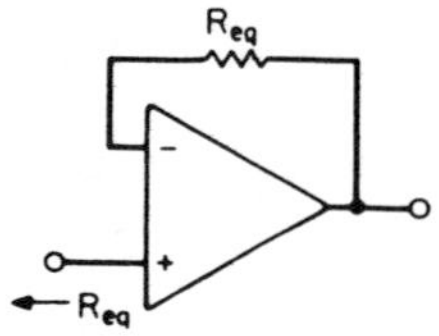

(a)

(b)

FIG. 1-10 Minimizing offset in the voltage follower.

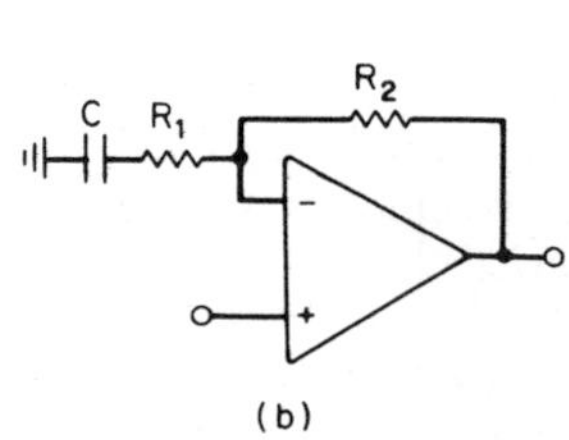

FIG. 1-11 Minimizing DC offset in AC amplifiers. (a) Inverting amplifier; (b) noninverting amplifier.

Below $f_{3\ dB}$, the response rolls off at approximately 6 decibels per octave. For AC gains in excess of three or four, the low-frequency break point of the noninverting amplifier may also be calculated by using Eq. 1-19.

Common-Mode and Power-Supply Rejection   The "common-mode rejection ratio" (CMRR) is a measure of the rejection of a common signal applied *simultaneously* to both amplifier inputs as compared to the *differential* input signal. This parameter is usually expressed in decibels and can be computed by

$$\text{CMRR (dB)} = 20 \log \frac{A_o}{A_{ocm}} \qquad (1\text{-}20)$$

where A_o is the amplifier open-loop differential gain and A_{ocm} is the open-loop common mode gain.

Fig. 1-12 shows a typical CMRR curve as a function of frequency for a μA741 op amp. This parameter remains essentially constant up to a few hundred hertz and then begins to degrade. Above a few hundred kilohertz the CMRR becomes unacceptable.

The "power-supply rejection ratio" (PSRR) is the measure of the change in input offset voltage for a change in power-supply voltage and is usually specified in microvolts per volt. This parameter is usually meaningful in determining the effect of power-supply ripple on an amplifier. The effect of the PSRR at the amplifier output can be determined by

$$V_o = \text{PSRR} \times A_c \times V_{AC} \quad (1\text{-}21)$$

where A_c is the closed-loop gain and V_{AC} is the power supply ripple. The PSRR is typically 30 $\mu V/V$ for a μA741-type op amp.

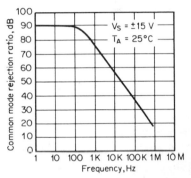

FIG. 1-12   Common-mode rejection of a μA741 amplifier.

Amplifier Noise   Op amp noise is usually specified in terms of an equivalent noise voltage at the amplifier input and is given in nanovolts/$\sqrt{\text{hertz}}$, which indicates the rms voltage within a 1-Hz bandwidth. If, for example, 500 nV of equivalent input noise is measured using a bandwidth of 25 Hz, the noise voltage is

$$V_n = \frac{500\text{ nV}}{\sqrt{25\text{ Hz}}} = \frac{500\text{ nV}}{5\sqrt{\text{Hz}}} = 100\text{ nV}/\sqrt{\text{Hz}}$$

The noise voltage V_n must be specified at a particular center frequency, since the noise distribution is not constant over the operating bandwidth of the amplifier and tends to increase below a few hundred hertz. A typical noise voltage for the μA741 is 45 nV/$\sqrt{\text{Hz}}$ measured at 1000 Hz over a 1-Hz bandwidth. As in the case of other parameters that are specified at the input, the output noise can be found in a similar fashion by multiplying V_n by A_c, the closed-loop circuit gain.

Frequently the input noise is specified in terms of volts^2/hertz, which can be converted to volts/$\sqrt{\text{hertz}}$ by taking the square root. Also an input noise current I_n may be specified in terms of amperes/$\sqrt{\text{hertz}}$.

Slewing Rate   When a large sine wave signal is applied to an op amp and the frequency is increased, at some point the amplifier will have difficulty in following the input signal. The limiting rate of change of the output voltage with respect to time is called ''slewing rate,'' and is usually expressed in volts per microsecond.

The slewing rate of an amplifier can be measured by applying a voltage step to the input and determining the slope at the output, as illustrated in Fig. 1-13. The slewing rate is then given by

$$\text{Slewing rate} = \frac{\Delta V_{out}}{\Delta t} \tag{1-22}$$

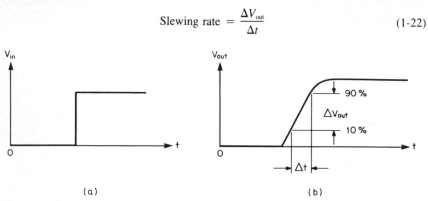

(a)   (b)

FIG. 1-13   Measurement of slewing rate. (a) Input step; (b) output waveform.

Data sheets frequently show the output response for a voltage step applied to a voltage-follower circuit. The slewing rate can be estimated by applying Eq. 1-22. For a μA741, the slewing rate is typically 0.5 V/μs. Rates as high as a few hundred volts per microsecond are obtainable using other devices.

The slewing rate is dependent upon a number of factors, such as the amplifier type, degree of frequency compensation, load capacity, circuit gain, and peak to peak output swing. The following suggestions are helpful for maximizing a slewing rate for a given amplifier type.

1. Use the minimum amount of frequency compensation that will be acceptable and avoid capacitive loads.
2. Higher circuit gain results in a slewing rate improvement.
3. Keep output signal swings as low as possible.

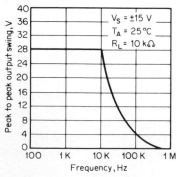

FIG. 1-14   Output voltage swing as a function of frequency for a μA741 amplifier.

Amplifier Bandwidth   Amplifier bandwidth is frequently specified in terms of the unity-gain bandwidth, that is, the frequency where the open-loop gain is reduced to one. The effect of insufficient open-loop gain on closed-loop gain was discussed earlier in the section ''Finite Open-Loop Gain,'' so the importance of maximizing the open-loop bandwidth should be evident.

Open-loop parameters are generally measured under small-signal conditions. For large output swings, the bandwidth is further reduced. The full-power bandwidth may be one or two orders of magnitude below the small-signal bandwidth. The full-power output limitation as a function of frequency for a μA741 amplifier is shown in Fig. 1-14. The output voltage swing corresponds to 5% harmonic distortion at the frequencies shown using ±15-V DC supplies. To extend the full-power bandwidth, the supply voltages should be as high as possible (without exceeding amplifier ratings). Also, if the compensation is externally provided, it should be the minimum value necessary to ensure stability.

Amplifier Stability   The closed-loop gain equations were previously given as

$$A_c \text{ (inverting) } = \frac{A_o(\beta - 1)}{1 + A_o\beta} \tag{1-9}$$

and

$$A_c \text{ (noninverting) } = \frac{A_o}{1 + A_o\beta} \tag{1-13}$$

It was also pointed out that the loop gain term $A_o\beta$ can be equal to -1 as a result of open-loop gain rolloff with frequency and amplifier phase shift. Also, if the feedback network is not purely resistive, β will introduce additional phase shift. A loop gain of -1 implies an oscillatory condition or instability, since the denominators vanish in the closed-loop gain expressions.

Phase and Gain Margin   To determine phase and gain margin, a Bode plot of loop gain vs. frequency is made. This curve can be obtained by multiplying the open-

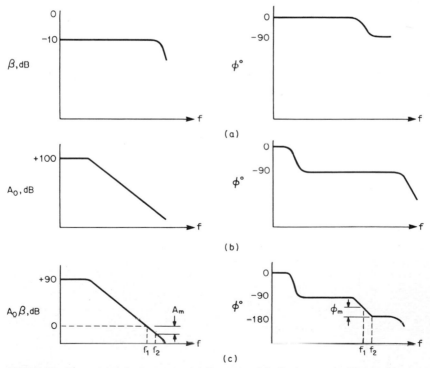

FIG. 1-15   Determining the loop-gain. (*a*) Response of feedback network; (*b*) open-loop response of amplifier; (*c*) composite loop-gain response.

loop gain and phase curves given in the data sheet by the transfer function of β, the feedback network. With the gains of A_o and β both expressed in decibels, the composite gain is the algebraic sum of the corresponding points. Likewise, the phase shifts are directly additive. Fig. 1-15 illustrates this method of point-by-point addition of these curves to obtain the loop-gain response.

Let us now consider the gain and phase curves of Fig. 1-15c. At frequency f_1, the loop-gain magnitude is unity. However, the phase shift has not yet reached 180°. (If the phase shift were −180°, a minus sign would appear in front of the A_cβ product, that is, A_cβ = −1, the oscillatory condition.) The number of degrees that the phase shift is below 180° is called the "phase margin" and is shown as ϕ_m.

Alternately, we can determine the frequency where the phase shift actually reaches 180°, which happens to be f_2. The measure of the number of decibels that the loop gain is below 0 dB (A_cβ = 1) is called the "gain margin" and is represented by A_m.

Clearly, the greater the computed phase and gain margins, the higher the likelihood of stable operation. Typical numbers for the phase and gain margins of a μA741 in a voltage-follower configuration (β = 1) are 65° and 11 dB, respectively.

Frequency Compensation An estimated nominal phase margin of less than 30° or a gain margin of less than 6 dB may result in unstable operation under worst-case conditions. By controlling the amplifier's open-loop response using external frequency compensation, stable operation can be assured.

Many op amps have built-in frequency compensation, such as a single 30-pF capacitor. However, since these amplifiers must be unconditionally stable, including in the

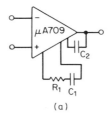

(a)

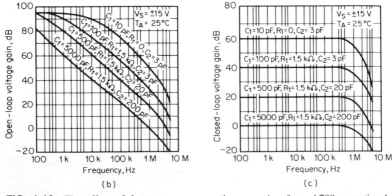

(b)   (c)

FIG. 1-16 The effect of frequency compensation on gain of a μA709 operational amplifier. (a) Amplifier configuration; (b) open-loop frequency response for various values of compensation; (c) frequency response for various closed-loop gains.

voltage-follower configuration, the compensation is larger than required for most other applications. The resulting degraded bandwidth and poorer slewing rate may not be acceptable. If the compensation is supplied externally, a smaller value can be chosen. By using only enough compensation to ensure stability, the maximum possible bandwidth and slewing rate can be obtained.

The manufacturer provides recommended values in the data sheet for fixed increments of closed-loop gains. These values have been determined to ensure stability over the range of production distributions for the device. Fig. 1-16 shows the open-loop response for a µA709-type op amp for various values of external compensation as well as the closed-loop response at fixed gains.

Stability Measurement   Open-loop measurements are generally difficult to make. Fortunately, circuit stability can be easily estimated on a closed-loop basis. If we plot the closed-loop frequency response, peaking may occur at the upper end of the band. A phase margin of 45° corresponds to the critically damped condition. With a phase margin of only 25°, approximately 6 dB of peaking will occur. As the phase margin is further reduced, the peaking will increase until the amplifier eventually breaks into oscillation. The relationship between phase margin and amplitude peaking is shown in Fig. 1-17.

The frequency-response method of determining stability is not as accurate as determining the gain and phase margins from a loop gain vs. frequency plot. However, it is certainly more convenient and is generally acceptable at least as a first cut for stability analysis.

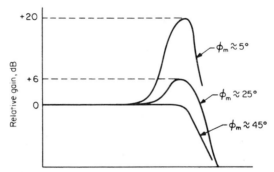

FIG. 1-17   Peaking vs. phase margin.

1–3   LINEAR AMPLIFIER CONFIGURATIONS

1–3a   Summing Amplifiers

Operation of the inverting amplifier discussed in Sec. 1-1 is based on the virtual ground effect occurring at the inverting input terminal. If we expand the input structure by introducing two additional input resistors, the circuit of Fig. 1-18 is obtained.

Three currents are now introduced flowing towards the inverting input terminal, which are given by V_1/R_1, V_2/R_2, and V_3/R_3. Since no current can flow into the op amp, these currents must flow through R_4. Since the differential input voltage is zero, the output voltage is given by

$$V_o = -\left[\frac{R_4}{R_1}V_1 + \frac{R_4}{R_2}V_2 + \frac{R_4}{R_3}V_3\right] \tag{1-23}$$

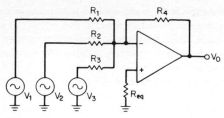

FIG. 1-18 Summing amplifier.

The composite output is the sum of all three input signals where the individual gain factors are determined by the ratio of R_4 to each summing resistor. This technique can be expanded for as many inputs as required.

All inputs are completely isolated from each other. Functions such as combining of multiple signals using arbitrary gain factors are easily implemented. To minimize DC offset, a resistor is introduced between the noninverting input and ground. Its value is determined by the parallel combination of the input summing resistors and the feedback resistor R_4. If offset is not critical, the noninverting input can be directly connected to ground.

Example 1–1 Design of Multiple Input Summing Amplifier

Design a summing amplifier to combine three signals providing gains of 1, 10, and 100, respectively. The frequencies of operation will range from DC to 100 Hz.

Solution

a) The circuit of Fig. 1-18 will be used. A μA741-type amplifier has a typical open-loop gain of 10,000 at 100 Hz, which is well in excess of the maximum closed-loop gain required ($A_o/A_c = 100$).

b) Let us assume a feedback resistor of 100 kΩ. Since individual gains of 1, 10, and 100 are specified, the summing resistors are computed from Eq. 1-23 as follows:

$$\frac{R_4}{R_1} = 1 \qquad R_1 = R_4 = 100 \text{ k}\Omega$$

$$\frac{R_4}{R_2} = 10 \qquad R_2 = \frac{R_4}{10} = 10 \text{ k}\Omega$$

$$\frac{R_4}{R_3} = 100 \qquad R_3 = \frac{R_4}{100} = 1 \text{ k}\Omega$$

where in all cases R_4 is 100 kΩ.

c) The resulting circuit is shown in Fig. 1-19. To minimize DC offset, a resistor of 1 kΩ is introduced between the noninverting input terminal and ground. Resistors should be of the 1% metal film type if circuit accuracy is critical.

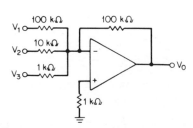

FIG. 1-19 Summing amplifier of Example 1-1.

Three-Resistor Feedback   In the inverting summing amplifier of Fig. 1-18, each gain factor is determined by the ratio of the feedback resistor R_4 to the input summing resistor. For very high gains, impractical resistor values may result; for instance, for a gain of 1000, a summing resistor of 10 kΩ results in a feedback resistor R_4 of 10 MΩ.

In the circuit of Fig. 1-20, the feedback resistor R_4 has been replaced by a T of resistors. If we pick a convenient value of R_a, resistor R_b is found by

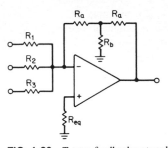

FIG. 1-20 T-type feedback network.

$$R_b = \frac{R_a^2}{R_4 - 2R_a} \qquad (1\text{-}24)$$

Resistor R_4 of 10 MΩ can now be replaced by the T where R_a is chosen at 100 kΩ and R_b is computed to be 1020 Ω.

1–3b   Differential Amplifiers

In many applications, the signal to be amplified originates from a differential-type source, such as a transducer. Frequently a common-mode voltage is simultaneously present (i.e., both sides of the source have a common signal with respect to ground, such as a DC offset or AC hum). If the amplifier operates in a differential manner, that is, the output is a function of the *difference* of the two input voltages, any common-mode signal is automatically canceled.

The circuit of Fig. 1-21 is a differential amplifier. If $R_2/R_1 = R_4/R_3$, the closed-loop differential output gain is given by

$$A_c = \frac{V_o}{V_2 - V_1} = \frac{R_2}{R_1} = \frac{R_4}{R_3} \qquad (1\text{-}25)$$

If R_2/R_1 is exactly equal to R_4/R_3, the circuit is perfectly balanced and V_{cm}, the common-mode voltage, is completely canceled. However, this is normally not the case, so a small part of the common-mode signal will appear at the output. Also, since the common-mode rejection ratio (CMRR) of the amplifier is finite, further degradation can occur.

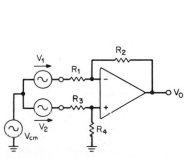

FIG. 1-21 Differential amplifier.

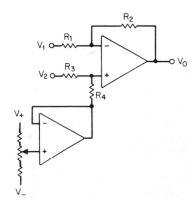

FIG. 1-22 Offset adjustment of differential amplifier.

To minimize DC offset, the parallel combination of R_1 and R_2 should equal the parallel combination of R_3 and R_4. An offset null control may also be added if the amplifier has this provision. A separate offset nulling circuit can be introduced by returning R_4 to a voltage-follower output instead of ground, as shown in Fig. 1-22.

If a differential output is required, this can be achieved by simply adding a unity gain inverter to an existing circuit, as shown in Fig. 1-23. The differential output signal is equal to $2V_1$.

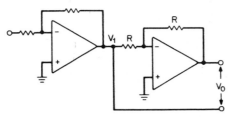

FIG. 1-23   Differential output.

Example 1–2   Design of Differential Input Amplifier

A differential input amplifier is required to have an amplification of 20 dB ($A_c = 10$) to a differential input signal. Operation is required up to 100 kHz.

Solution

Let us choose an LM101 for this application, since the μA741, which has fixed compensation, has insufficient open-loop gain at 100 kHz.

If we choose R_1 equal to 1 kΩ, then $R_1 = R_3 = 1$ kΩ and $R_2 = R_4 = 10$ kΩ using Eq. 1-25. The resulting circuit is shown in Fig. 1-24.

If a differential input and output are both required, the circuit of Fig. 1-25 can be used. The gain expression can be derived using superposition and is found to be

$$A_c = \frac{V_o}{V_i} = 2\frac{R_2}{R_1} + 1 \tag{1-26}$$

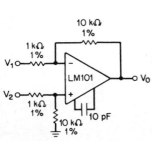

FIG. 1-24   Differential amplifier of Example 1-2.

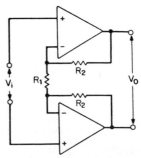

FIG. 1-25   Amplifier having both differential inputs and outputs.

By making R_1 adjustable, a gain control is introduced. Another feature of this circuit is that the input impedance is very high, since the input signal is connected directly to the noninverting input terminals rather than to summing resistors.

1–3c Instrumentation Amplifiers

The basic differential amplifier of Fig. 1-21 has a number of limitations. The input resistance is relatively low, since it is determined by R_1, R_3, and R_4. This may result in errors due to loading of the driving source. Gain cannot easily be made adjustable since all four resistors must be simultaneously varied. The CMRR of the circuit may be unacceptable.

A device that overcomes these limitations is called an "instrumentation amplifier." A popular implementation is shown in Fig. 1-26. At the expense of two additional amplifiers, superior performance is obtained in comparison to the differential amplifier of Fig. 1-21.

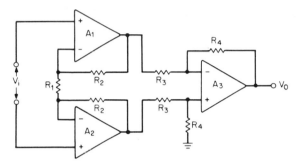

FIG. 1-26 Instrumentation amplifier.

This circuit consists of the differential input and output configuration of Fig. 1-25 followed by the differential input amplifier of Fig. 1-21. The circuit gain is given by

$$A_c = \frac{V_o}{V_i} = \left[2\frac{R_2}{R_1} + 1 \right] \frac{R_4}{R_3} \tag{1-27}$$

If we let $R_3 = R_4$, the circuit gain expression is reduced to

$$A_c = \frac{V_o}{V_i} = 2\frac{R_2}{R_1} + 1 \tag{1-28}$$

The minimum gain obtainable for this case is unity.

By making R_1 variable, the gain becomes adjustable. Resistor values should be kept low to minimize DC offset. Also, if A_1 and A_2 are comprised of a dual amplifier having inherent matching, such as an MC1458, further reduction of offset is achieved.

Example 1–3 Design of Instrumentation Amplifier

Design an amplifier having a differential input, high input impedance, and high common-mode rejection. The gain should be switchable between 3, 5, and 10.

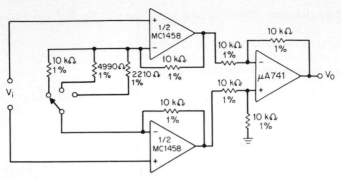

FIG. 1-27   Instrumentation amplifier of Example 1-3.

Solution

a) The instrumentation amplifier circuit of Fig. 1-26 will be used. If we let $R_2 = R_3 = R_4 = 10$ kΩ, we can then compute three different values of R_1 for the corresponding gains, using Eq. 1-28 as follows:

$$R_1 = \frac{2R_2}{A_c - 1} \qquad (1\text{-}29)$$

A_c	R_2	R_1
3	10 kΩ	10 kΩ
5	10 kΩ	5 kΩ
10	10 kΩ	2.22 kΩ

b) The resulting circuit is shown in Fig. 1-27. An MC1458 is used for A_1 and A_2 and a μA741 type is used for A_3. The resistors are of the standard 1% metal film type.

An instrumentation amplifier can also be formed using only two operational amplifiers. The circuit is shown in Fig. 1-28. The circuit gain is given by

$$A_c = 1 + R_1\left[\frac{1}{R_2} + \frac{2}{R_3}\right] \qquad (1\text{-}30)$$

Resistor R_3 may be made variable when a means of gain adjustment is required.

Although instrumentation amplifiers will reject a common-mode signal present at both input terminals, the voltage of this signal must be within the maximum ratings of the device. For example, the μA741 has a rated common-mode voltage of ± 15 V maximum (with respect to ground). Exceeding these limits may damage the device.

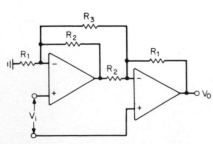

FIG. 1-28   Two op amp instrumentation amplifier.

1–3d Integrators and Differentiators

The Ideal Integrator An ideal integrator provides an output that is proportional to the integral with time of the input signal. The output signal is determined by the area under the input waveform. Mathematically, this can be expressed as

$$V_o = k \int V_i \, dt \qquad (1\text{-}31)$$

where k is a constant.

Let us now consider the ideal integrator shown in Fig. 1-29. The current in R_1 is equal to V_i/R_1, since a virtual ground exists at the inverting input. This current must then flow through capacitor C, which produces the output voltage. Since the voltage across a capacitor is given by

$$V = \frac{1}{C} \int I \, dt \qquad (1\text{-}32)$$

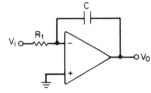

FIG. 1-29 Ideal integrator.

the output voltage can be expressed as

$$V_o = -\frac{1}{C} \int \frac{V_i}{R_1} \, dt \qquad (1\text{-}33)$$

or in the preferred form

$$V_o = -\frac{1}{R_1 C} \int V_i \, dt \qquad (1\text{-}34)$$

Working with Eq. 1-33 directly is not very desirable, since it may involve the mathematical integration of complex signals. A more desirable expression can be derived by starting with the fundamental equation for the accumulated charge on a capacitor

$$Q = CV \qquad (1\text{-}35)$$

Since the charge is the product of current and time, we can state

$$IT = CV$$

Substituting V_i/R_1 for I and V_o for V, the output voltage of Fig. 1-29 can be determined from

$$V_o = -\frac{V_i T}{R_1 C} \qquad (1\text{-}36)$$

where the negative sign occurs because of the polarity inversion. This formula can be applied in a piecewise manner to an input waveform.

Fig. 1-30 illustrates the integration of a square wave resulting in a triangular waveform. For the first half-cycle, the output charges negative by an amount determined from Eq. 1-36. For the second half-cycle, the output charges positive by exactly the same amount, since the square wave is symmetrical. This pattern is repeated in a periodic manner.

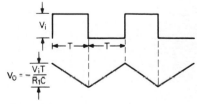

FIG. 1-30 Integration of square wave.

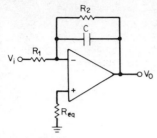

FIG. 1-31   Practical integrator.

Practical Integrators   A DC offset at the input of the integrator of Fig. 1-29 will cause a continuous ramp in one direction until the amplifier output reaches saturation. To provide DC stabilization, an additional resistor is introduced in parallel with the capacitor in the circuit of Fig. 1-31. The low-frequency gain, including the DC condition, is now limited to R_2/R_1. In order to retain the property of an integrator at the frequencies of interest, the minimum value of R_2 is computed from

$$R_2 \geqq \frac{1}{2\pi f_L C} \tag{1-37}$$

where f_L is the lowest frequency of operation. Ideally, R_2 should be at least 10 times the value given by Eq. 1-37. Resistor R_{eq} will minimize output DC offset and is equal to the parallel combination of R_1 and R_2.

Example 1–4   Design of Integrator

Design an integrator to convert a 10-V peak-to-peak square wave at 1000 Hz to a 5-V peak-to-peak triangular waveform. Use a capacitor value of 1 µF.

Solution

a) Resistor R_1 is computed by

$$R_1 = \frac{V_i T}{V_o C} = \frac{10 \times 0.5 \times 10^{-3}}{5 \times 10^{-6}} = 1 \text{ k}\Omega \tag{1-36}$$

Note that T corresponds to half the period of 1000 Hz. (The negative sign of Eq. 1-36 has been dropped, since a sign inversion is assumed.)

b) The stabilizing resistor R_2 is found from

$$R_2 \geqq \frac{1}{2\pi f_L C} \geqq \frac{1}{6.28 \times 10^3 \times 10^{-6}} \geqq 159 \ \Omega \tag{1-37}$$

Let $R_2 = 10$ kΩ. R_{eq} is then 910 Ω. The resulting circuit is illustrated in Fig. 1-32.

As a final comment on integrators, the circuit gain for a sine wave input is given by

$$A_c = j \frac{1}{2\pi f R C} \tag{1-38}$$

The gain rolls off at 6 decibels per octave with a constant phase shift of 90°.

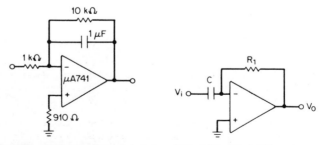

FIG. 1-32   Integrator of Example 1-4.   **FIG. 1-33**   Ideal differentiator.

Differentiators   A differentiator performs the opposite function of an integrator. Whereas the output of an integrator is the integral of the input, the differentiator performs the mathematical operation of differentiation on the input signal. An ideal differentiator is shown in Fig. 1-33. The current through the capacitor is given by $C\, dV_i/dt$, since the inverting input terminal is at virtual ground. The output voltage is then

$$V_o = -R_1C\,\frac{dV_i}{dt} \qquad\qquad (1\text{-}39)$$

If the input voltage changes in a linear fashion over the region of interest, the output voltage can be expressed as

$$V_o = -R_1C\,\frac{\Delta V_i}{\Delta t} \qquad\qquad (1\text{-}40)$$

If we apply a triangular waveform to this circuit, a square wave results at the output, as shown in Fig. 1-34. Clearly, the process of integration is reversed.

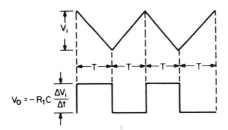

FIG. 1-34   Differentiation of a triangular waveform.   **FIG. 1-35**   Practical differentiator.

The gain of a differentiator *increases* at a rate of 6 decibels per octave, as opposed to an integrator that *decreases* at the same rate. This circuit is therefore highly susceptible to high-frequency noise. To limit the high-frequency gain, a resistor is usually introduced in series with the input capacitor, as shown in Fig. 1-35. The maximum gain is then limited to $-R_1/R_2$. The maximum value of R_2 is found from

$$R_2 \leq \frac{1}{2\pi f_h C} \qquad\qquad (1\text{-}41)$$

where f_h is the highest frequency of operation. Ideally, R_2 should be no more than one-tenth the value computed by Eq. 1-41.

1–3e   Current Sources

An ideal current source maintains a load current that is directly determined by an input voltage and independent of load impedance. This type of circuit is sometimes also known as a voltage-controlled current source (VCCS) or voltage-to-current converter.

Op amps are well suited for configuring current sources because of their nearly ideal characteristics under conditions of negative feedback. This section discusses a number of these configurations, which have various inherent features.

Unipolar Current Sources   A unipolar current source provides only a single polarity of current with respect to ground. A popular implementation is shown in Fig. 1-36a. A virtual ground appears between the two input terminals of the op amp because

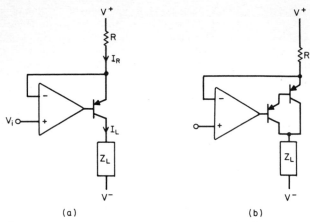

FIG. 1-36   Transistor current source. (*a*) Single transistor; (*b*) Darlington.

of negative feedback. The current through resistor R is then given by

$$I_R = \frac{V^+ - V_i}{R} \tag{1-42}$$

This current enters the emitter of the transistor and exits from the collector. However, $(1 - \alpha)$ times the emitter current flows into the base, thus reducing the current into the load. This error is usually negligible since transistor α's are typically 0.99. Replacing the single transistor by the Darlington structure of Fig. 1-36*b* further reduces this error.

Circuit output impedance is quite high and is typically tens of megohms. Higher impedances are obtainable by using an FET instead of a transistor. The output current is dependent upon the voltage difference between V^+ and V_i, so any variation in the positive supply is reflected at the output. This limitation can be overcome by referencing the input signal to V^+.

Example 1–5   Design of Current Source

A current source is needed to provide a constant DC current of 1 mA $\pm 10\%$ into a maximum resistive load of 10 kΩ. A supply voltage of ± 15 V is available.

Solution
A zener diode will be used to maintain a constant voltage across resistor R of Fig. 1-36 so that the current is maintained constant. The resulting circuit is shown in Fig. 1-37. The 7.5-V zener forces 1 mA through the 7500 Ω resistor and into the load. Observe that a total voltage of 22.5 V is available to force the 1-mA current (30 V less the 7.5-V zener voltage across the 7500 Ω resistor, neglecting the $V_{CE\,sat}$ of the transistor). Therefore, the total load resistance cannot exceed 22.5 V/1 mA or 22.5 kΩ to maintain regulation.

Since the circuit operates at DC, a μA741 will satisfy the requirements.

Bipolar Current Sources   A bipolar current source can supply a regulated current that has both positive and negative polarities with respect to ground, and is generally used for AC signals. A convenient implementation is shown in Fig. 1-38.

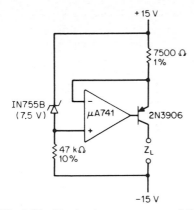

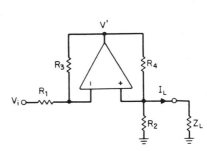

FIG. 1-37 The 1-mA current source of Example 1-5.

FIG. 1-38 Bipolar current source.

The circuit output impedance is given by

$$Z_{out} = \frac{R_4}{\dfrac{R_4}{R_2} - \dfrac{R_3}{R_1}} \tag{1-43}$$

If the ratios of R_4/R_2 and R_3/R_1 are equal, the output impedance is infinite, which corresponds to an ideal current source. The output current is given by

$$I_L = -V_i \frac{R_3}{R_1 R_4} \tag{1-44}$$

If we let $R_1 = R_2 = R_3 = R_4 = R$, then the output current expression simplifies to

$$I_L = -\frac{V_i}{R} \tag{1-45}$$

To maintain high output impedance, low values of R should be avoided. Precision values should also be used. For a tolerance of $\pm 1\%$, the minimum output impedance is

$$Z_{out} \geq 50R \tag{1-46}$$

The voltage at node V' is determined to be

$$V' = -V_i \left(2\frac{Z_L}{R} + 1 \right) \tag{1-47}$$

Excessive values of Z_L can cause amplifier clipping because of large output swings.

Floating Current Sources The previously discussed current sources provide a current into a load that is returned to either a DC voltage or ground. If the load is permitted to float, the summing properties of an op amp can be directly used to make a simple current source, as shown in Fig. 1-39. The current through the load I_L is exactly equal to V_i/R.

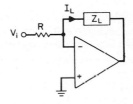

FIG. 1-39 Simple floating current source.

The output voltage is the product $-I_L Z_L$ and must remain within the available output range of the op amp.

The configuration of Fig. 1-39 has unequal impedances to ground on each side of the load. In many cases a high impedance is required on *both* sides of the load while a constant current is provided. One apparent solution would be to use two independent complementary current sources, one on each side of the load. This approach will not work, however, for unless the two current sources are *exactly* equal, the voltage at the load will start climbing until the current sources stop functioning.

A self-regulating differential current source is shown in Fig. 1-40, which provides approximately 30 mA to a balanced load for the values indicated. To understand the operation of this circuit, let us temporarily ignore the presence of R_6 and R_7. The circuit then reduces to a pair of complementary transistor current sources similar to the type shown in Fig. 1-36a. The voltage divider consisting of R_1 through R_4 produces a 3-V

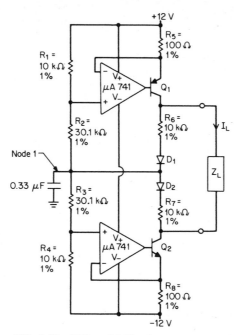

FIG. 1-40   Differential 30-mA current source.

drop across resistors R_5 and R_8. Node 1 is at ground potential. A current of 30 mA (3 V/100 Ω) flows out of current source Q_1, through load Z_L, and back into current source Q_2.

Resistors R_6 and R_7 sense the voltage across the load. If the two current sources are identical, then the voltage at the junction of D_1 and D_2 remains at ground. If the current leaving Q_1 exceeds the current entering Q_2, node 1 will move positive, decreasing current source Q_1 and increasing current source Q_2 until equilibrium is established. Conversely, if current source Q_2 is larger than Q_1, node 1 will move negative. Both current sources will again become equal.

1–4 EXTENDED OPERATION

While the op amp is a very versatile device, situations frequently arise where extension of the operating parameters becomes desirable. This section discusses some of the techniques frequently used to increase op amp capability.

1–4a Output Power Boosters

Op amps are limited in terms of output power. The µA741, for example, cannot deliver more than a few milliwatts to a resistive load. The inverting amplifier circuit of Fig. 1-41 provides a moderate power boost. This circuit offers a two-fold advantage. Output currents of 20 or 30 mA are easily obtainable. In addition, the output can swing nearly up to the power supply rails.

The circuit operates as follows. As the output swings negative, the current drawn from the negative supply increases. This results in more base drive for Q_1, which in turn provides for the increased load current requirement. In a similar manner a positive output swing results in more base drive for Q_2.

The values of R_3 and R_4 are shown for the µA741 op amp. For other devices, the values are computed from the following expression:

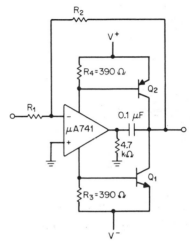

FIG. 1-41   Power booster.

$$R_3 = R_4 = \frac{0.600}{I_{cc}} \qquad (1\text{-}48)$$

where I_{cc} is the nominal power supply current for the device.

A disadvantage of this circuit is that a dead zone occurs as the output swings through zero volts, thus simultaneously turning off Q_1 and Q_2. If the open-loop gain is inadequate at the frequency of operation, excessive crossover distortion may occur.

Improved performance can be obtained using the circuit of Fig. 1-42. The output

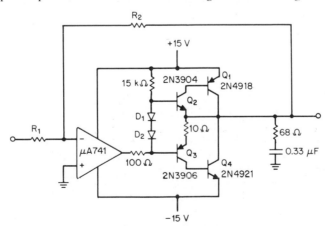

FIG. 1-42   Improved power booster.

stage operates in class AB, resulting in almost zero idle current and no crossover distortion. This circuit is capable of delivering 2 W to an 8 Ω load with less than 1% harmonic distortion.

1–4b Bandwidth Extension

Amplifier bandwidth can be extended by using an externally compensated device and as little compensation as possible for the required gain, while maintaining stability.

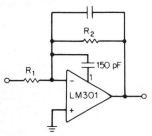

An alternate method involves "feedforward compensation." With this technique, high-frequency signals are bypassed around the input stage to drive the higher frequency secondary stages directly. This technique is applied to an LM301 in Fig. 1-43. An order of magnitude of bandwidth improvement can be obtained for the open-loop response. A small capacitor may be necessary across resistor R_2 to ensure stability. The technique of feedforward compensation is restricted to the inverting amplifier configuration.

FIG. 1-43 Feedforward compensation.

1–4c Using Single-Ended Power Supplies

Most IC operational amplifiers are specified to require dual supply voltages, typically ± 3 to ± 18 V. Frequently, two opposite polarity supplies are not available. Conventional op amps can still be used with a single supply by providing a reference voltage V_r, which replaces the circuit ground connections. This voltage is midway between ground and the single supply voltage available and should be provided from a low-impedance source.

A convenient way of generating V_r from a positive supply is shown in Fig. 1-44a. A voltage divider and bypass capacitor form an intermediate DC voltage, which is then buffered using a voltage follower. For single negative supply voltages, the circuit of Fig. 1-44b is used.

Input signals referenced to ground must first be AC coupled and then superimposed upon V_r. The output signal must also be AC coupled to remove the DC component. As a result, this technique prevents the passage of low-frequency signals, including DC.

This method is illustrated in Figure 1-45 as applied to some previously discussed circuits. The circuit of Fig. 1-44 is used to generate V_r, and all ICs are powered by a single-ended power supply.

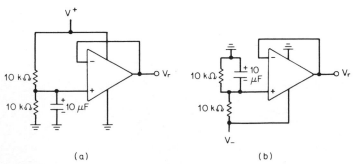

(a)   (b)

FIG. 1-44 Generating a reference voltage. (a) Single positive supply; (b) single negative supply.

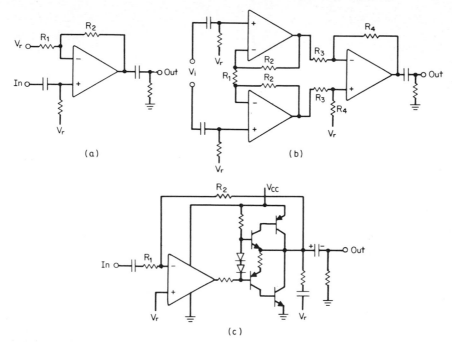

FIG. 1-45   Single-ended power-supply operation. (*a*) Noninverting amplifier; (*b*) instrument amplifier; (*c*) power booster.

1–5   OP AMP SELECTION

The success or failure of a given design is frequently determined by the parameters of the op amp used. Insufficient open-loop gain, low input impedance, poor slewing rate, or a variety of other degraded parameters can severely affect circuit performance.

To select the proper device for the application, the designer must first isolate those op amp parameters that most directly affect circuit operation. An op amp with acceptable characteristics is then chosen.

1–5a   Mechanical Data

Op amps are obtainable in a variety of standardized package styles and lead configurations. The choice of a particular package depends upon mechanical, environmental, and thermal considerations. For military-grade devices (-55 °C to $+125$ °C operating), the plastic package is not available. For commercial applications, the 8-pin plastic case is probably the most popular.

The packages shown in this section are typical for a particular manufacturer. Although devices of the same family of packages are generally interchangeable, the designer can refer to the particular manufacturer's data sheets for more detailed information.

Ceramic Dual-in-Line Package   These dual-in-line (DIP) packages consist of a ceramic base, an 8, 14, or 16 lead frame, and a ceramic cap. The circuit chip is alloy mounted to the base, the lead frame is attached to the chip, and the cap is hermetically sealed to the base using glass.

Typical package outlines are illustrated in Fig. 1-46. These packages are intended to be inserted in mounting holes on 0.300-in centers. The leads must first be compressed and then inserted into the holes provided. Tension is created by the compressed leads to hold the device in place during soldering. Solder will then readily adhere to the tin-plated leads.

This package features a hermetic seal and low thermal resistance. For operation in adverse environments, such as for military applications, a hermetic package is usually required.

Plastic Dual-in-Line Package The plastic dual-in-line (DIP) package mounts in the same holes as the ceramic version. However, the lead frame and IC chip are encapsulated in an insulating plastic compound. This package has a higher thermal resistance than the ceramic version and is usually restricted to commercial applications. It will provide reliable operation, even under conditions of high humidity. It is also lower in cost than the ceramic package. Case outline dimensions for a typical device are shown in Fig. 1-47.

Ceramic Flat Package This form of construction features a low-profile hermetically sealed package using a ceramic base and cap containing a lead frame with the attached IC chip. A glass seal ensures hermeticity. Thermal resistance is higher than either plastic or ceramic DIP packages. Use of this package is generally restricted to military applications. Typical case dimensions are shown in Fig. 1-48.

Metal Can Package A metal can package consists of a welded metal base and cap where the individual leads are held in place in the base by a glass seal. It features hermetic construction, compact size, and the facility to increase package dissipation by slipping on a heat sink. Typical 8- and 10-pin versions are illustrated in Fig. 1-49.

1–5b General-Purpose Op Amps

A large selection of op amps are available to the designer. These range from low-cost industry standards for general-purpose applications to specialized amplifiers having unique properties such as high slew rate, low power, etc. The designer is urged to use the more popular general-purpose devices wherever possible to minimize costs and ensure availability from multiple sources. The electrical properties for this type are summarized in Tables 1-2 through 1-4 for single, dual, and quad devices.

1–5c JFET Input Operational Amplifiers

This family of op amps features a JFET input stage combined with bipolar technology. The result is extremely low bias and offset current, low offset temperature coefficient, extremely high input impedance, and wide bandwidth. These devices are highly suited for applications that require good DC stability and accuracy as well as faster operation, such as high-speed D/A or A/D converters, wide-band amplifiers, and high-impedance buffers (open-loop input impedance is typically 10^{12} Ω). The electrical characteristics are summarized in Tables 1-5 through 1-7.

1–5d Operational Amplifiers with Special Features

In many cases, a general-purpose-type device cannot satisfy the critical requirements of a particular circuit application. In these cases, an op amp that is specifically designed to optimize a particular parameter may be required. The electrical properties of the more popular devices are summarized in Tables 1-8 through 1-10.

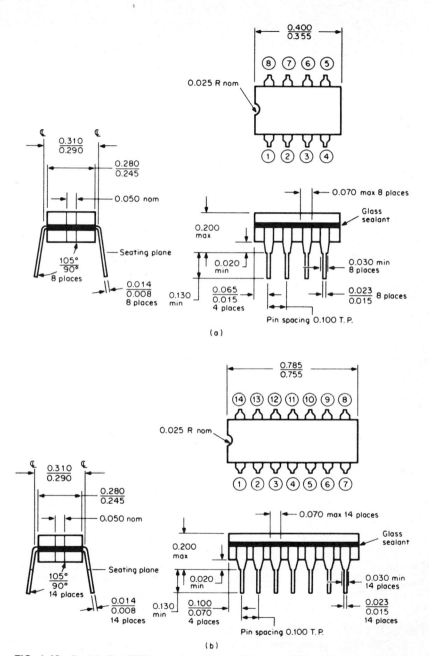

FIG. 1-46 Dual-in-line (DIP) ceramic packages. (*Courtesy of Texas Instruments Corporation.*) (*a*) 8-pin DIP ceramic; (*b*) 14-pin DIP ceramic. (*Note:* All dimensions in inches.)

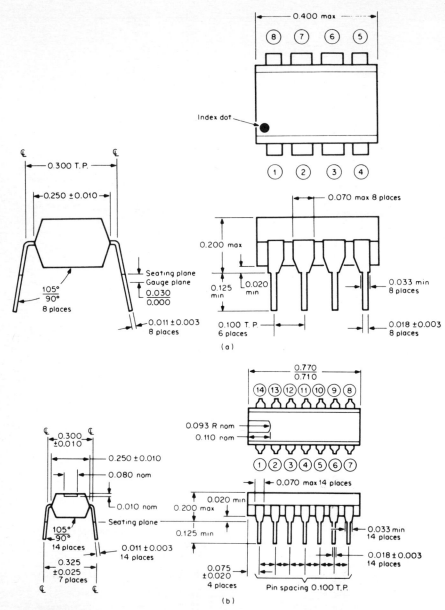

FIG. 1-47 Dual-in-line (DIP) plastic packages. (*Courtesy of Texas Instruments Corporation.*) (*a*) 8-pin DIP plastic; (*b*) 14-pin DIP plastic. (*Note:* All dimensions in inches.)

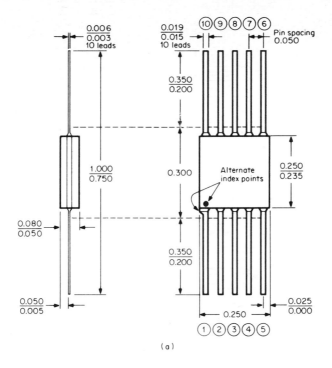

(a)

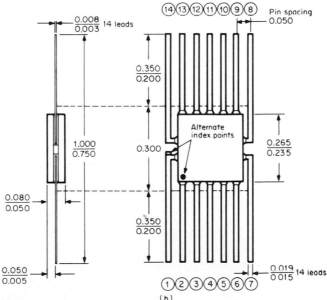

(b)

FIG. 1-48 Ceramic flat packages. (*Courtesy of Texas Instruments Corporation.*) (*a*) 10-pin flat package; (*b*) 14-pin flat package. (*Note:* All dimensions in inches.)

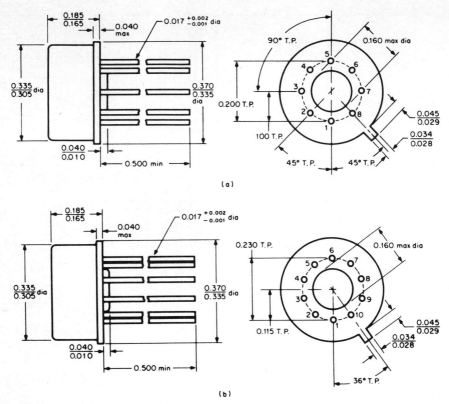

FIG. 1-49 Metal can package. (*Courtesy of Texas Instruments Corporation.*) (*a*) 8-pin construction; (*b*) 10-pin construction. (*Note:* All dimensions in inches.)

TABLE 1–2 General-Purpose Single Operational Amplifiers

Device	Input offset voltage max (mV)	Input offset voltage drift (μV/°C)	Input offset current max (nA)	Input bias current max (nA)	Volt gain min (V/V)	Unity gain BW typ (MHz)	Slew rate typ (V/μS)	Supply volt range (V)	Common-mode range (V)	Diff input range (V)	Supply current max (mA)	Features (see note 1)
Military −55°C to +125°C												
LM101	6	15	20	100	25 k	1	0.5	±3 to ±22	±12	±30	3	A, B, C
LM101A	3	15	20	100	25 k	1	0.5	±3 to ±22	±12	±30	3	A, B, C
LM107	3	15	20	100	25 k	1	0.5	±3 to ±22	±12	±30	3	B, C
TL321M	5	. . . .	30	150	50 k	1	0.5	3 to 32	−0.3 to 32	±32	1	B, C, E
μA709	6	6 typ	500	1500	25 k	1	0.3	±9 to ±18	± 8	±5	5.5	D
μA709A	3	15	250	600	25 k	1	0.3	±5 to ±22	±20	±40	3.6	D
μA741	6	15 typ	500	1500	25 k	1	0.5	±3 to ±22	±12	±30	2.8	A, B, C
μA741A	4	15	70	210	32 k	1	0.5	±3 to ±22	±12	±30	4.0	A, B, C
μA748	6	. . . .	500	1500	25 k	1	0.5	±3 to ±22	±12	±30	2.8	A, B, D
Industrial −25°C to +85°C												
LM201	10	10	750	2000	15 k	1	0.5	±3 to ±22	±12	±30	3	A, B, C
LM201A	3	15	20	100	25 k	1	0.5	±3 to ±22	±12	±30	3	A, B, C
LM207	2	20	20	100	25 k	1	0.5	±3 to ±22	±12	±30	3	B, C
TL321I	5	. . . .	30	150	50 k	1	0.5	3 to 32	−0.3 to 32	±32	1	B, C, E
Commercial 0°C to 70°C												
LM301A	10	30	70	300	15 k	1	0.5	±3 to ±18	±12	±30	3	A, B, C
LM307	10	30	50	250	15 k	1	0.5	±3 to ±18	±12	±30	3	B, C
TL321C	5	. . . .	30	150	50 k	1	0.5	3 to 32	−0.3 to 32	±32	1	B, C, E
μA709C	10	12 typ	500	1500	15 k	1	0.3	±9 to ±18	±8	±5	6.6	D
μA741C	7.5	15 typ	300	800	15 k	1	0.5	±3 to ±18	±12	±30	2.8	A, B, C
μA748C	6	6	500	1500	25 k	1	0.5	±3 to ±18	±12	±30	2.8	A, B, D

Note 1: A. Offset adjust capability; B. Output short-circuit protection; C. Internal compensation; D. External compensation; E. Single supply operation.

TABLE 1–3 General-Purpose Dual Operational Amplifiers

Device	Input offset voltage max (mV)	Input offset voltage drift (µV/°C)	Input offset current max (nA)	Input bias current max (nA)	Volt gain min (V/V)	Unity gain BW typ (MHz)	Slew rate typ (V/µS)	Supply volt range (V)	Common-mode range (V)	Diff input range (V)	Supply current max (mA)	Features (see note 1)
Military −55°C to +125°C												
LM158	5	....	30	150	25 k	1	....	±1.5 to ±16	$V^+ - 1.5$	32	1.2	B, C, E
LM158A	2	....	10	50	50 k	1	....	±1.5 to ±16	$V^+ - 1.5$	32	1.2	B, C, E
TL322M	5	10 typ	50	500	50 k	0.6	....	3 to 36	±18	±36	2.5	B, C, E
µA747	6	....	500	1500	25 k	1	0.5	±3 to ±22	±12	±30	5.6	A, B, C
µA747A	4	15	70	210	32 k	1	0.5	±3 to ±22	±12	±30	5.6	A, B, C
MC1558	6	....	500	1500	25 k	1	0.5	±3 to ±22	±12	±30	5.0	B, C
Industrial −25°C to +85°C												
LM258	7.5	7 typ	150	300	15 k	1	0.5	3 to 32	$V^+ - 1.5$	32	1.2	B, C, E
TL322I	8	10 typ	75	500	20 k	1	0.6	3 to 36	±18	32	4	B, C, E
Commercial 0°C to +70°C												
TL322C	10	10 typ	50	500	20 k	1	0.6	3 to 36	±18	32	4	B, C, E
LM358	7.5	7 typ	150	500	15 k	1	....	±1.5 to ±16	$V^+ - 1.5$	32	1.2	B, C, E
µA747C	6	....	300	800	15 k	1	0.5	±3 to ±18	±12	±30	5.6	A, B, C
MC1458	6	....	300	800	15 k	1	0.2	±3 to ±18	±15	±30	5.6	B, C

Note 1: A. Offset adjust capability; B. Output short-circuit protection; C. Internal compensation; D. External compensation; E. Single supply operation.

TABLE 1-4 General-Purpose Quad Operational Amplifiers

Device	Input offset voltage max (mV)	Input offset voltage drift (μV/°C)	Input offset current max (nA)	Input bias current max (nA)	Volt gain min (V/V)	Unity gain BW typ (MHz)	Slew rate typ (V/μS)	Supply volt range (V)	Common-mode range (V)	Diff input range (V)	Supply current max (mA)	Features (see note 1)
Military −55°C to +125°C												
LM124	7	. . . .	100	150	25 k	1	. . . .	±1.5 to ±15	V⁺ − 2	±32	2.0	B, C, E
LM124A	2	. . . .	10	50	50 k	1	. . . .	±1.5 to ±15	V⁺ − 2	±32	2.0	B, C, E
LM148	6	15 typ	75	325	25 k	1	0.6	±3 to ±22	±12	±30	3.6	B, C
MC3503	5	. . . .	50	500	50 k	1	0.6	+3 to +36	±18	±36	4	B, C, E
Industrial −25°C to +85°C												
LM224	9	. . . .	150	500	15 k	1	. . . .	±1.5 to ±15	V⁺ − 1.5	±32	2.0	B, C, E
LM248	7.5	15 typ	125	500	15 k	1	0.5	±5 to ±18	±18	±36	4.5	B, C
Commercial 0°C to +70°C												
LM324	9	7 typ	150	500	15 k	1	. . . .	±1.5 to ±15	V⁺ − 1.5	±32	2	B, C, E
LM348	7.5	15 typ	100	400	15 k	1	. . . .	±5 to ±18	±18	±36	4.5	B, C
MC3403	10	. . . .	50	500	20 k	1	0.6	+3 to +36	±18	±36	7	B, C, E

Note 1: A. Offset adjust capability; B. Output short-circuit protection; C. Internal compensation; D. External compensation; E. Single supply operation.

TABLE 1-5 JFET Input Single Operational Amplifiers

Device	Input offset voltage max (mV)	Input offset voltage drift (µV/°C)	Input offset current max (nA)	Input bias current max (nA)	Volt gain min (V/V)	Unity gain BW typ (MHz)	Slew rate typ (V/µS)	Supply volt range (V)	Common-mode range (V)	Diff input range (V)	Supply current max (mA)	Features (see note 1)
Military −55°C to +125°C												
TL080M	6	10 typ	0.1	0.2	50 k	3	13	±3.5 to ±18	±15	±30	2.8	A, B, D
TL081M	6	10 typ	0.1	0.2	50 k	3	13	±3.5 to ±18	±15	±30	2.8	A, B, C
LF155	7	20	0.05	0.1	25 k	2.5	5	±5 to ±22	±20	±40	7	A, B, C
LF155A	2.5	10	0.025	0.05	25 k	2.5	5	±5 to ±22	±20	±40	7	A, B, C
LF156	7	20	0.05	0.1	25 k	5	15	±5 to ±22	±20	±40	7	A, B, C
LF156A	2.5	10	0.025	0.05	25 k	5	15	±5 to ±22	±20	±40	7	A, B, C
LF157	7	20	0.05	0.1	25 k	25	75	±5 to ±22	±20	±40	7	A, B, C
LF157A	2.5	10	0.025	0.05	25 k	25	75	±5 to ±22	±20	±40	7	A, B, C
Industrial −25°C to +85°C												
TL080I	6	10 typ	0.1	0.2	50 k	3	13	±3.5 to ±18	±15	±30	2.8	A, B, D
TL081I	6	10 typ	0.1	0.2	50 k	3	13	±3.5 to ±18	±15	±30	2.8	A, B, C
LF255	6.5	5 typ	0.02	0.05	25 k	2.5	5	±5 to ±22	±20	±40	4	A, B, C
LF256	6.5	5 typ	0.02	0.05	25 k	5	15	±5 to ±22	±20	±40	7	A, B, C
LF257	6.5	5 typ	0.02	0.05	25 k	25	75	±5 to ±22	±20	±40	7	A, B, C
Commercial 0°C to +70°C												
TL080AC	6	10 typ	0.1	0.2	50 k	3	13	±3.5 to ±18	±15	±30	2.8	A, B, D
TL080C	15	10 typ	0.1	0.2	25 k	3	13	±3.5 to ±18	±15	±30	2.8	A, B, D
TL081AC	6	10 typ	0.1	0.2	50 k	3	13	±3.5 to ±18	±15	±30	2.8	A, B, C
TL081BC	3	10 typ	0.1	0.2	50 k	3	13	±3.5 to ±18	±15	±30	2.8	A, B, C
TL081C	15	10 typ	0.1	0.2	50 k	3	13	±3.5 to ±18	±15	±30	2.8	A, B, C
LF351	10	10 typ	0.1	0.2	25 k	4	13	±5 to ±18	±15	±30	3.4	A, B, C
LF355	13	5 typ	2	8	15 k	2.5	5	±5 to ±18	±15	±30	4	A, B, C
LF355A	2.3	5	1	5	25 k	2.5	5	±5 to ±22	±16	±40	4	A, B, C
LF356	13	5 typ	2	8	15 k	5	15	±5 to ±18	±20	±30	10	A, B, C
LF356A	2.3	5	1	5	25 k	5	15	±5 to ±22	±16	±40	10	A, B, C
LF357	13	5 typ	2	8	15 k	25	75	±5 to ±18	±20	±30	10	A, B, C
LF357A	2.3	5	1	5	25 k	25	75	±5 to ±22	±20	±40	10	A, B, C
LF13741	20	10 typ	2	8	15 k	1	0.5	±4 to ±18	±16	±30	4	A, B, C

Note 1: A. Offset adjust capability; B. Output short-circuit protection; C. Internal compensation; D. External compensation; E. Single supply operation.

TABLE 1–6 JFET Input Dual Operational Amplifier

Device	Input offset voltage max (mV)	Input offset voltage drift (μV/°C)	Input offset current max (nA)	Input bias current max (nA)	Volt gain min (V/V)	Unity gain BW typ (MHz)	Slew rate typ (V/μS)	Supply volt range (V)	Common-mode range (V)	Diff input range (V)	Supply current max (mA)	Features (see note 1)
Military — 55°C to + 125°C												
TL082M	6	10 typ	0.1	0.2	50 k	3	13	±3.5 to ±18	±15	±30	2.8	B, C
TL083M	6	10 typ	0.1	0.2	50 k	3	13	±3.5 to ±18	±15	±30	2.8	A, B, C
Industrial — 25°C to + 85°C												
TL082I	6	10 typ	0.1	0.2	50 k	3	13	±3.5 to ±18	±15	±30	2.8	B, C
TL083I	6	10 typ	0.1	0.2	50 k	3	13	±3.5 to ±18	±15	±30	2.8	A, B, C
TL288I	3	10 typ	0.1	0.4	50 k	3	13	±3.5 to ±18	±15	±30	2.8	A, B, C
Commercial 0°C to + 70°C												
TL082AC	6	10 typ	0.1	0.2	50 k	3	13	±3.5 to ±18	±15	±30	2.8	B, C
TL082BC	3	10 typ	0.1	0.2	50 k	3	13	±3.5 to ±18	±15	±30	2.8	B, C
TL082C	15	10 typ	0.2	0.4	25 k	3	13	±3.5 to ±18	±15	±30	2.8	B, C
TL083AC	6	10 typ	0.1	0.2	50 k	3	13	±3.5 to ±18	±15	±30	2.8	A, B, C
TL083C	15	10 typ	0.2	0.4	25 k	3	13	±3.5 to ±18	±15	±30	2.8	A, B, C
TL288C	3	10 typ	0.1	0.4	25 k	3	13	±4 to ±18	±15	±30	2.8	A, B, C
LF353	10	10 typ	0.1	0.2	25 k	4	13	±5 to ±18	±15	±30	2.8	B, C
LF353A	2	10 typ	0.05	0.2	50 k	4	13	±5 to ±18	±15	±30	2.8	B, C
LF354	13	10 typ	0.1	0.2	50 k	4	13	±5 to ±18	±15	±30	2.8	A, B, C
LF354A	4	10 typ	0.05	0.2	50 k	4	13	±5 to ±18	±15	±30	2.8	A, B, C

Note 1: A. Offset adjust capability; B. Output short-circuit protection; C. Internal compensation; D. External compensation; E. Single supply operation.

TABLE 1-7 JFET Input Quad Operational Amplifiers

Device	Input offset voltage max (mV)	Input offset voltage drift (μV/°C)	Input offset current max (nA)	Input bias current max (nA)	Volt gain min (V/V)	Unity gain BW typ (MHz)	Slew rate typ (V/μS)	Supply volt range (V)	Common-mode range (V)	Diff input range (V)	Supply current max (mA)	Features (see note 1)
Military −55°C to +125°C												
TL084M	9	10 typ	0.1	0.2	50 k	3	13	±3.5 to ±18	±15	±30	2.8	B, C
Industrial −25°C to +85°C												
TL084I	6	10 typ	0.1	0.2	50 k	3	13	±3.5 to ±18	±15	±30	2.8	B, C
Commercial 0°C to +70°C												
TL084AC	6	10 typ	0.1	0.2	50 k	3	13	±3.5 to ±18	±15	±30	2.8	B, C
TL084BC	3	10 typ	0.1	0.2	50 k	3	13	±3.5 to ±18	±15	±30	2.8	B, C
TL084C	15	10 typ	0.2	0.4	25 k	3	13	±3.5 to ±18	±15	±30	2.8	B, C
TL085C	15	10 typ	0.2	0.4	25 k	3	13	±3.5 to ±18	±15	±30	2.8	B, C
LF347	10	10 typ	0.01	0.2	25 k	4	13	±5 to ±18	±11	±30	3.4	B, C
LF347A	2	10 typ	0.05	0.2	50 k	4	13	±5 to ±18	±11	±30	2.8	B, C
LF347B	5	10 typ	0.1	0.1	50 k	4	13	±5 to ±18	±11	±30	2.8	B, C

Note 1: A. Offset adjust capability; B. Output short-circuit protection; C. Internal compensation; D. External compensation; E. Single supply operation.

TABLE 1-8 Special-Purpose Single Operational Amplifiers

Device	Input offset voltage max (mV)	Input offset voltage drift (µV/°C)	Input offset current max (nA)	Input bias current max (nA)	Volt gain min (V/V)	Unity gain BW typ (MHz)	Slew rate typ (V/µS)	Supply volt range (V)	Common-mode range (V)	Diff input range (V)	Supply current max (mA)	Description
Military −55°C to +125°C												
LM102	7.5	6 typ	....	100	0.999	10	10	±12 to ±18	±10	....	5.5	Voltage follower
LM108	3	15	0.4	3	25 k	1	0.3	±2 to ±20	±14	....	0.6	Precision
LM108A	1	5	0.4	3	40 k	1	0.3	±2 to ±20	±14	....	0.6	Precision
LM110	6	12	....	10	0.999	20	30	±5 to ±15	±10	....	5.5	Voltage follower
LM112	3	15	0.4	3	25 k	1	0.2	±2 to ±20	±14	....	0.6	Micro-power
LM118	4	....	50	250	20 k	15	50 min	±5 to ±18	±11.5	....	8	High speed
LM143	6	....	7	35	50 k	1	2.5	±4 to ±40	±38	±40	4	High voltage
LM144	6	....	7	35	50 k	2	30	±4 to ±40	±38	±40	4	High voltage/slew rate
µA702	5	10	2000	7500	2 k	30	100	....	....	±5	6.7	Wide bandwidth
µA715	7.5	6	250	1500	10 k	65	3.5	±6 to ±18	±10	±15	10	High speed
LM725	1.5	5	40	200	1 k	0.5	0.005	±3 to ±22	±13.5	±5	3.5	Instrumentation
µA777	7.5	50	250	....	25 k	1	0.5	±5 to ±20	±12	±30	2.8	Precision
µA791	6	200	500	....	20 k	1	0.5	±5 to ±18	±12	±30	25	High power
Industrial −25°C to +85°C												
LM202	10	15 typ	....	15	0.999	10	10	±12 to ±18	±10	....	5.5	Voltage follower
LM208	3	15	0.4	3	25 k	1	0.3	±2 to ±20	±14	....	0.6	Precision
LM208A	1	5	0.4	3	40 k	1	0.3	±2 to ±20	±14	....	0.6	Precision
LM210	4	....	....	3	0.999	20	30	±5 to ±18	±10	....	5.5	Voltage follower
LM212	2	15	0.2	2	25 k	1	0.3	±2 to ±20	±14	....	0.6	Micro-power
LM218	4	....	50	500	25 k	15	50 min	±5 to ±18	±11.5	....	1.5	High speed
Commercial 0°C to +70°C												
LM302	20	20 typ	....	3	0.9985	10	10	±12 to ±18	±10	....	5.5	Voltage follower
LM308	10	30	1.5	10	15 k	1	0.3	±2 to ±18	±14	....	0.8	Precision
LM308A	0.73	5	1.5	10	60 k	1	0.3	±2 to ±18	±14	....	0.8	Precision
LM310	10	10 typ	....	10	0.999	20	30	±5 to ±18	±10	....	5.5	Voltage follower
LM312	10	30	1.5	10	15 k	1	0.3	±2 to ±18	±14	....	0.8	Micro-power
LM318	15	....	300	750	20 k	15	50	±5 to ±18	±11.5	....	10	High speed
LM343	10	14	14	55	50 k	1	2.5	±4 to ±34	±34	....	5	High voltage
LM344	10	14	14	55	50 k	2	30	±4 to ±34	±34	....	5	High voltage/slew rate
µA702C	5	10	2000	7500	2 k	30	100	....	....	±5	6.7	Wide bandwidth
µA715C	7.5	6	250	1500	10 k	65	3.5	±6 to ±18	±10	±15	10	High speed
LM725C	3.5	2 typ	50	250	125 k	1	0.005	±3 to ±22	±13.5	±5	5	Instrumentation
µA777	7.5	50	250	....	25 k	1	0.5	±5 to ±20	±12	±30	2.8	Precision
µA791C	6	200	500	....	20 k	1	0.5	±5 to ±18	±12	±30	25	High power

TABLE 1-9 Special-Purpose Dual Operational Amplifiers

Device	Input offset voltage max (mV)	Input offset voltage drift (µV/°C)	Input offset current max (nA)	Input bias current max (nA)	Volt gain min (V/V)	Unity gain BW typ (MHz)	Slew rate typ (V/µS)	Supply volt range (V)	Common-mode range (V)	Diff input range (V)	Supply current max (mA)	Description
Military −55°C to +125°C												
TL022M	5	....	40	100	4 k	0.5	0.5	±2 to ±22	±15	±30	0.1	Low power
TL062M	6	....	0.1	0.2	4 k	1	3.5	±1.5 to ±18	±15	±30	0.2	Low power
TL072M	6	....	0.05	0.2	50 k	3	13	±3.5 to ±18	±15	±30	2.5	Low noise
RM4458	5	....	200	500	25 k	3	1.5	±3 to ±22	±15	±30	2.8	Precision
Industrial −25°C to +85°C												
TL062I	6	....	0.1	0.2	4 k	1	3.5	±1.5 to ±18	±15	±30	0.25	Low power
TL072I	6	....	0.05	0.2	50 k	3	13	±3.5 to ±18	±15	±30	2.5	Low noise
Commercial 0°C to +70°C												
TL022C	5	....	80	250	1 k	0.5	0.5	±2 to ±18	±15	±30	0.125	Low power
TL062C	6	....	0.1	0.2	4 k	1	3.5	±1.5 to ±18	±15	±30	0.25	Low power
TL072C	10	....	0.05	0.2	25 k	3	13	±3.5 to ±18	±15	±30	2.5	Low noise
TL287C	0.5	....	0.1	0.4	25 k	3	13	±4 to ±18	±15	±30	2.8	Low offset
RC4558	6	....	200	500	20 k	3	1	±3 to ±18	±15	±30	2.8	Precision
NE5532	4	....	150	800	25 k	10	9	±3 to ±20	±12	....	16	Low noise
NE5533	4	....	300	1500	25 k	10	13	±3 to ±20	±12	....	16	Low noise

TABLE 1-10 Special-Purpose Quad Operational Amplifiers

Device	Input offset voltage max (mV)	Input offset voltage drift (μV/°C)	Input offset current max (nA)	Input bias current max (nA)	Volt gain min (V/V)	Unity gain BW typ (MHz)	Slew rate typ (V/μS)	Supply volt range (V)	Common-mode range (V)	Diff input range (V)	Supply current max (mA)	Description
Military −55°C to +125°C												
TL044M	5	....	40	100	4 k	0.5	0.5	±2 to ±22	±15	±30	0.1	Low power
TL064M	9	....	0.1	0.2	4 k	1	3.5	±1.5 to ±18	±15	±30	0.2	Low power
TL074M	9	....	0.05	0.2	50 k	3	13	±3.5 to ±22	±15	±30	2.5	Low noise
LM149	6	15 typ	75	325	25 k	4	3	±3 to ±22	±12	±30	3.6	Wide band
Industrial −25°C to +85°C												
TL064I	6	....	0.1	0.2	4 k	1	3.5	±1.5 to ±18	±15	±30	0.25	Low power
TL074I	6	....	0.05	0.2	50 k	3	13	±3.5 to ±18	±15	±30	2.5	Low noise
LM249	7.5	15 typ	125	500	15 k	4	2	±5 to ±18	±18	±36	4.5	Wide band
Commercial 0°C to +70°C												
TL044C	5	....	80	250	1 k	0.5	0.5	±2 to ±18	±15	±30	0.125	Low power
TL064C	15	....	0.2	0.4	3 k	1	2.5	±1.5 to ±18	±15	±30	0.25	Low power
TL074C	10	....	0.05	0.2	25 k	3	13	±3.5 to ±18	±15	±30	2.5	Low noise
TL075C	10	....	0.05	0.2	25 k	3	13	±3.5 to ±18	±15	±30	2.5	Low noise
LM349	7.5	15 typ	100	400	15 k	4	3	±5 to ±18	±18	±36	4.5	Wide band

REFERENCES

Graeme, Jerald G.: *Applications of Operational Amplifiers,* McGraw-Hill, New York, 1973.

Hnatek, Eugene R.: *Applications of Linear Integrated Circuits,* Wiley, New York, 1975.

Stout, David: In Milton Kaufman (Ed.), *Handbook of Operational Amplifier Circuit Design,* McGraw-Hill, New York, 1976.

Chapter 2

FUNCTION CIRCUITS

Joel Silverman Marketing Manager
Siliconix Inc.
Santa Clara, Calif.

Author was with Exar Integrated Systems when this chapter was written.

INTRODUCTION

This chapter deals with a variety of ICs commonly referred to as function circuits. These circuits are usually listed in manufacturer's data books under such headings as miscellaneous and special function. These function ICs have been designed to provide a straightforward solution to problems that are frequently encountered in electronic circuit design. Since these circuits tend to be complex, their operation is usually not readily defined by a simple set of equations. This section is intended to give the designer a brief overview of some of these devices from both the applications and circuit design point of view. This knowledge will enable the designer to readily determine which device would be most suitable for a specific application. Since IC manufacturers are continuously upgrading their products, a general overview is presented that would apply to any new products that have not yet been developed.

2–1 FOUR-QUADRANT MULTIPLIERS

In a variety of analog system applications, a circuit is required that produces an output that is proportional to the product of two input signals. The circuit block that performs this function is called an analog multiplier. The need for such a multiplier extends well beyond its ability to perform arithmetic operations such as multiplication, division, squar-

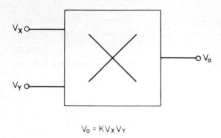

$$V_o = KV_XV_Y$$

FIG. 2-1 Block diagram of a multiplier.

ing, and square root extraction, since, as will be shown, multipliers also serve as key building blocks for AM generators and detectors, frequency translators, and phase detectors. Fig. 2-1 illustrates the conceptual block diagram of an analog multiplier. The output is defined as

$$V_o = KV_XV_Y \qquad (2\text{-}1)$$

where V_X and V_Y are the analog signals applied to the X and Y terminals, respectively, and K is the gain constant of the multiplier, which has the dimensions of (volts)$^{-1}$. Eq. 2-1 is the characteristic equation for an "ideal" four-quadrant multiplier. Thus, the magnitude and polarity of the output are solely determined by that of the input, for any combination of positive and negative input signals, within the dynamic operating range of the multiplier.

2–1a Analysis of Practical Multipliers

Eq. 2-1 illustrates the transfer function for a conceptual multiplier and as such neglects any offset terms that are inherent in any practical "nonideal" multiplier circuit. Since the output of a multiplier is a function of two independent input variables, its operating characteristics must be defined with respect to the offset voltages associated with each input, and also include any offset associated with the output stage. When these finite offset terms are considered, the actual transfer function becomes

$$V_o = K(V_X + \Phi_X)(V_Y + \Phi_Y) + \Phi_o \qquad (2\text{-}2)$$

where Φ_X and Φ_Y are the offsets associated with the X and Y inputs, respectively, and Φ_o is the output offset voltage of the multiplier. This implies that for high-precision multiplication, four separate adjustments are required, three of these to "null" out the internal offsets and the fourth to set the multiplier gain. In most applications, the multiplier gain is set to 0.1. This permits either or both inputs to have a value around the 10-V range without causing the output to exceed 10 V. Once the offsets have been trimmed and the value of K set to 0.1, the transfer function becomes

$$V_o = \frac{V_XV_Y}{10} \qquad (2\text{-}3)$$

There are several key parameters that are used to describe the operating characteristics of a multiplier. These are usually used to define and specify any deviation from the predicted transfer function. Some of the key terms are "accuracy," "linearity," and "bandwidth." A basic fundamental knowledge of these terms is necessary to accurately predict circuit performance.

"Accuracy" is specified as the deviation of the actual output from that of the ideal, for any combination of X and Y inputs within the specified operating range of the multiplier. It is usually specified in terms of a percentage of the full-scale output of the multiplier. Therefore, if a multiplier with a ± 10-V output swing is specified, with 0.5% full-scale accuracy, the output would be within ± 50 mV of its predicted value.

"Linearity" is usually defined as a deviation of the best fitting straight line and is usually expressed as a percentage of the full-scale output, since the maximum deviation occurs at the extreme ends of the multiplier dynamic range. For example, if a multiplier with a ± 10-V output was specified to have 0.5% linearity, the maximum deviation from the best fitting straight line would be ± 50 mV.

"Bandwidth" is a measure of how well the high-frequency operation of the multiplier corresponds to its low-frequency operation. Since this parameter is very dependent upon the particular application, several bandwidths are defined.

1. 1% absolute error bandwidth is the frequency where the phase vector between the actual and the ideal output is equal to 1%. This frequency is reached when the net phase shift across the multiplier is equal to 0.01 rad or 0.57°.
2. 3° phase-shift bandwidth is the frequency where the net phase shift across the multiplier is 3°.
3. 3-dB bandwidth is the frequency where the multiplier output is 3 dB below its low-frequency value for a constant amplitude input signal.
4. Transconductance bandwidth is the frequency where the transconductance of the multiplier drops 3 dB below its low-frequency value. This bandwidth defines the frequency range for operation as a phase detector or synchronous AM detector.

In order to better understand the operation of a multiplier circuit, a fundamental knowledge of the circuit is of prime importance. Fig. 2-2 illustrates the Gilbert multiplier cell, which is comprised of three emitter-coupled pairs and is suitable for four-quadrant multiplication. The following derivation assumes that all transistors are identical, and that base currents and output resistance can be neglected.

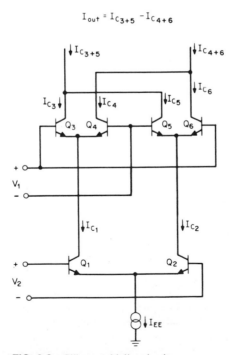

FIG. 2-2   Gilbert multiplier circuit.

The first step is to calculate the collector currents of Q_3, Q_4, Q_5, and Q_6, in terms of V_1.

Using the basic diode equation,

$$V_{BE_3} = V_T \ln \frac{I_{C_3}}{I_S}$$

$$V_{BE_4} = V_T \ln \frac{I_{C_4}}{I_S}$$

where V_T = thermal voltage = KT/q, and I_S = reverse saturation current.

$$V_1 = V_{BE_3} - V_{BE_4} = V_T \ln \frac{I_{C_3}}{I_{C_4}}$$

$$I_{C_3} = I_{C_4} e^{V_1/V_T}$$

But, $I_{C_3} + I_{C_4} = I_{C_1}$, so

$$I_{C_3} = (I_{C_1} - I_{C_3}) e^{V_1/V_T}$$

$$I_{C_3} = I_{C_1} e^{V_1/V_T} - I_{C_3} e^{V_1/V_T} \tag{2-4}$$

$$I_{C_3}(1 + e^{V_1/V_T}) = I_{C_1} e^{V_1/V_T}$$

$$I_{C_3} = \frac{I_{C_1}}{1 + e^{-V_1/V_T}}$$

The collector current of Q_4 is then solved by $I_{C_4} = I_{C_1} - I_{C_3}$:

$$I_{C_4} = I_{C_1} - \frac{I_{C_1}}{1 + e^{-V_1/V_T}}$$

$$= \frac{I_{C_1}}{1 + e^{V_1/V_T}} \tag{2-5}$$

Using the same method, the collector currents of Q_5 and Q_6 are also derived,

$$I_{C_5} = \frac{I_{C_2}}{1 + e^{V_1/V_T}} \tag{2-6}$$

$$I_{C_6} = \frac{I_{C_2}}{1 + e^{-V_1/V_T}} \tag{2-7}$$

along with the currents of Q_1 and Q_2,

$$I_{C_1} = \frac{I_{EE}}{1 + e^{-V_2/V_T}} \tag{2-8}$$

$$I_{C_2} = \frac{I_{EE}}{1 + e^{V_2/V_T}} \tag{2-9}$$

Eqs. 2-4 through 2-9 are then combined to obtain equations for the collector currents in terms of input voltages V_1 and V_2 and current I_{EE}. We can then obtain expressions for collector currents I_{C_3}, I_{C_4}, I_{C_5}, and I_{C_6} in terms of input voltages V_1 and V_2 as follows:

$$I_{C_3} = \frac{I_{EE}}{(1 + e^{-V_1/V_T})(1 + e^{-V_2/V_T})} \tag{2-10}$$

$$I_{C_4} = \frac{I_{EE}}{(1 + e^{-V_2/V_T})(1 + e^{V_1/V_T})} \tag{2-11}$$

$$I_{C_5} = \frac{I_{EE}}{(1 + e^{V_1/V_T})(1 + e^{V_2/V_T})} \tag{2-12}$$

$$I_{C_6} = \frac{I_{EE}}{(1 + e^{V_2/V_T})(1 + e^{-V_1/V_T})} \tag{2-13}$$

The differential output current is then given by

$$I_o = I_{C_{3+5}} - I_{C_{4+6}} = I_{C_3} + I_{C_5} - (I_{C_6} + I_{C_4})$$
$$= (I_{C_3} - I_{C_6}) - (I_{C_4} - I_{C_5}) \tag{2-14}$$
$$I_o = I_{EE}(\tanh^{V_1/2V_T})(\tanh^{V_2/2V_T}) \tag{2-15}$$

Thus, the final transfer function is the product of the hyperbolic tangents of the two input signals, and as such is only linear for input signals that are small with respect to V_T. In order to improve the linearity and allow operation over a wider dynamic range, the exponential transfer characteristics of the circuit must be reduced to a linear function. This is accomplished by preconditioning the X input signal, and linearizing the Y input signal as illustrated in Fig. 2-3. The operation of this circuit may be explained as follows.

The V_Y input is linearized by splitting the current source I_{EE} and adding an emitter degeneration resistor, R_Y. The effect of this resistor is to linearize the transconductance of the emitter-coupled pair over a wide operating range. However, this method cannot be used to linearize the V_1 input due to the cross-coupled transistor pair used in this section. The preconditioning for this input is accomplished via the circuit consisting of D_1, D_2, Q_7, and Q_8.

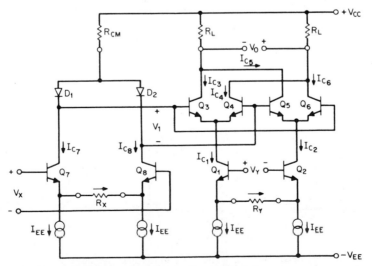

FIG. 2-3   Circuit diagram of a four-quadrant multiplier.

The complete derivation can be found in the references (see Grey and Meyer, and Grebene). The result is the derivation of

$$V_1 = 2V_T \tanh^{-1}\left(\frac{KV_x}{I_{EE}}\right) \tag{2-16}$$

Thus, causing the overall transfer function to become

$$V_o = \frac{2V_XV_YR_L}{I_{EE}R_XR_Y} \tag{2-17}$$

When using this basic analog building block, all the standard analog functions, such as multiplication, division, squaring, and square root functions, may be implemented. While the multiplication and squaring function are readily implemented (the squaring function is obtained by connecting the X and Y inputs together), the division and square root function may require some explanation.

2–1b Division

Fig. 2-4a shows the block diagram for the division function. Here it is seen that the multiplier is connected in the feedback loop of an operational amplifier. The operation of the circuit can be briefly described as follows. The denominator is applied to the X input of the multiplier and the numerator is applied to one summing input of the op amp, with the other input coming from the output of the multiplier. In closed-loop operation, the output of the op amp is forced to some voltage, such that the output of the multiplier V_Z is equal to $-V_N$. But since V_Z must be equal to $V_XV_Y/10$, the output of the operational

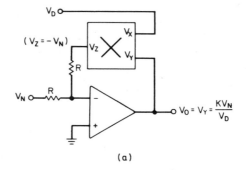

(a)

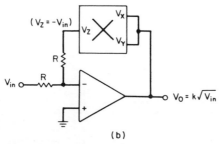

(b)

FIG. 2-4   The division and square root functions.
(a) Division; (b) square root.

amplifier must then be equal to

$$V_o = \frac{-10V_Z}{V_X} = \frac{10V_N}{V_D} \tag{2-18}$$

It is important to note that in such applications, the sign of the denominator must be negative or the polarity of the feedback will reverse itself and cause the circuit to latch up. This latch up is not destructive and is common to all division circuits.

2–1c   The Square Root Circuit

The square root circuit of Fig. 2-4b has a multiplier in the feedback loop connected as a squaring circuit. In terms of actual circuit operation, the voltage on the Z output is V_o^2 times a constant and is also equal to $-V_{in}$. Therefore the amplifier output V_o is proportional to the square root of the input. A diode is usually placed in series with the output to prevent a latch up condition, which would result if V_Z were allowed to go negative.

2–1d   Trimming of Multipliers

As discussed earlier and illustrated in Eq. 2-2, for maximum precision there are four separate adjustments that must be made in a multiplier circuit. In order to determine the best method to adjust these offsets, Eq. 2-2 must be considered further.

$$V_o = K(V_X + \Phi_X)(V_Y + \Phi_Y) + \Phi_o \tag{2-2}$$

The first step in the trimming procedure would be to adjust the output offset Φ_o to be equal to zero. This is accomplished by setting both the X and Y inputs to zero and adjusting for an output of zero volts. The second step is to adjust for the Y offset. This is accomplished by applying an AC input to the X terminal and adjusting the Y offset adjustment until a minimum output signal is observed. The same procedure is then followed to adjust for the X offset, by setting the X input equal to zero, applying an AC input to the Y terminal and adjusting for a minimum output swing. The third step in the adjustment procedure is to adjust the scale factor. This is usually set at the point where the two input signals are at their maximum; however, it may be set with different amplitudes and polarities of input signals to optimize accuracy over the entire input dynamic range.

Example 2–1   Design of a Four-Quadrant Multiplier

Design a four-quadrant multiplier circuit that has an input dynamic range of ±10 V with an output swing of ±10 V using the XR-2208. The gain equations for the multiplier and amplifier are

$$K_M \approx \frac{20}{R_X R_Y} \text{ V}^{-1} \tag{2-19}$$

$$K_A \approx \frac{R_F}{5 + R_1} \tag{2-20}$$

where $R_Y \approx 2R_X$ and all resistors are in kilohms.

Solution
The first step in the design is to consult the manufacturer's data sheet for the specific type of application, then generate a general schematic (Fig. 2-5) and determine the gain constant of the multiplier. For the device spec-

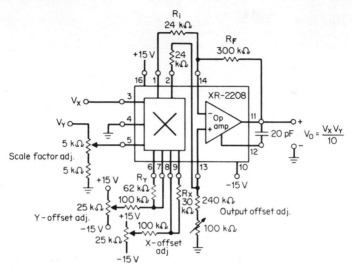

FIG. 2-5 Circuit connection for multiplication, using XR 2208.

ified, the gain constant is equal to the multiplier gain constant times the amplifier gain and is defined as

$$K = (K_M)(K_A) = \frac{V_Z}{V_X V_Y} \times \frac{V_o}{V_Z} \tag{2-21}$$

A K value of 0.1 is desired when the maximum input and output values are substituted into the above equation. The next step is to determine the actual component values required to set the gain constant. The gain is usually set slightly higher than 0.1 and then finely tuned by the scale factor adjust resistor. The desired values for K_M and K_A are then substituted into Eqs. 2-19 and 2-20 and the resistor values are computed. Let $R_X = 30$ kΩ, $K_M \approx 0.01$, and $K_A \approx 10$; then

$$R_Y = 62 \text{ k}\Omega$$
$$R_F = 300 \text{ k}\Omega$$
$$R_1 = 24 \text{ k}\Omega$$

The circuit is then constructed and trimmed as follows:

1. Apply 0 V to both inputs and adjust the output offset to 0 V using the output offset control.
2. Apply 20 V p-p at 50 Hz to the X input and 0 V to the Y input. Trim the Y offset adjust for minimum peak-to-peak output.
3. Apply 20 V p-p to the Y input and 0 V to the X input. Trim the X offset adjust for minimum peak-to-peak output.
4. Repeat step 1.
5. Apply $+10$ V to both inputs and adjust the scale factor for $V_o = +10$ V. This step may be repeated with different amplitudes and polarities of input voltages to optimize accuracy over the entire range of input voltages, or over any specific portion of the input voltage range.

It should be noted that the circuit of Fig. 2-5 can easily be converted to a squaring function, or frequency doubler, by simply connecting the X and Y inputs together. In this configuration, the squaring function is easily seen and the frequency doubling is accomplished by considering the trigometric identity

$$\cos A \cos B = \frac{\cos(A+B) + \cos(A-B)}{2} \qquad (2\text{-}22)$$

Thus, when the two inputs are tied together and $A = B$, the output is comprised of a two times frequency component along with a DC component.

Example 2–2   Wattmeter Design Using Multiplier

Design a wattmeter that is capable of measuring the power delivered to an 8 Ω speaker. Assume the amplifier has a peak output of 100 W, and the multiplier has a fixed gain constant of 0.1, with an input dynamic range of ±10 V.

Solution

The block diagram of the wattmeter is illustrated in Fig. 2-6. Its function is to multiply the voltage across the load by the current through the load. The sense resistor R_S is used to measure the current through the load and is chosen to be much smaller than R_L. Thus, the voltage drop across R_S and the power dissipated in R_S becomes negligible. The first step in the realization of the circuit is to scale the inputs to be within the dynamic range of the multiplier. Since

$$E^2 = P \times R$$
$$= 800$$
$$E \approx 28 \text{ V}$$

and
$$I^2 = P/R$$
$$= 12.5$$
$$I = 3.5 \text{ A}$$

The output voltage of the amplifier must first be scaled by a factor determined by dividing the maximum input of the multiplier by the maximum amplifier output voltage. Then resistors for the divider that are large with respect to R_L, should be chosen, so the power dissipated in them becomes

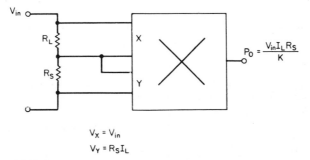

$$V_X = V_{in}$$
$$V_Y = R_S I_L$$

FIG. 2-6   Block diagram of a wattmeter.

negligible. Thus

$$\frac{R_1}{(R_1 + R_2)} = \frac{10}{28}$$

If R_1 is chosen to be 10 kΩ and R_2, 18 kΩ, the power dissipated in them is approximately 0.03% of that dissipated in the load.

The next step is to convert the current-sensing voltage to a compatible input level. The scale factor is determined by dividing the maximum multiplier input by the maximum voltage across the sense resistor (where $R_S = 0.1$ Ω). Thus, $10/0.35 = 28.6$ = scale factor. Since this scale factor is greater than one, an amplifier must be used.

The actual circuit implementation is shown in Fig. 2-7. The sense resistor of 0.1 Ω causes a decrease of 1.25% in the voltage delivered.

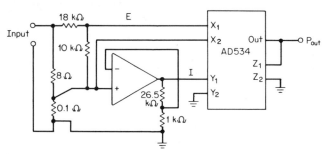

FIG. 2-7 Circuit implementation for wattmeter of Example 2-2.

2–1e Modulators

When a four-quadrant multiplier is used for arithmetic applications, much emphasis is placed on linear operation with respect to both inputs. However, there are many applications, such as modulators, or mixers, where linear operation for only one of the inputs is required. In such applications, one input is referred to as the carrier input and the other is referred to as the modulation input. A linear response is only required for the modulating input, since the carrier is usually a constant amplitude AC signal and frequently a square wave. An example is shown in Fig. 2-8. The top waveform corresponds to the modulation input, the center waveform corresponds to the carrier input, and the lower waveform corresponds to the output signal. The large-signal "carrier input" is basically used to alternately multiply the modulation input by $+1$ and -1. The spectrum of the output can easily be derived by considering the modulation signal to be

$$V_M(t) = V_M \cos(\omega_m t) \tag{2-23}$$

and the carrier signal to be

$$V_C(t) = (4/\pi)[\cos(\omega_c t) + (1/3)\cos(3\omega_c t) + (1/5)\cos(5\omega_c t) + \cdots + (1/n)\cos(n\omega_c t)] \tag{2-24}$$

which is the Fourier series for a square wave, with a swing of ± 1 for $n = 1, 3, 5, 7 \ldots$.

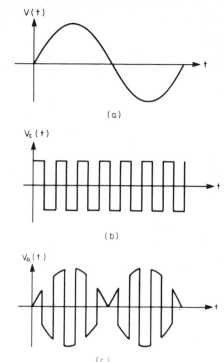

FIG. 2-8 Modulator waveforms. (*a*) Small-signal input; (*b*) large-signal modulating input; (*c*) output.

Then

$$V_o(t) = KV_M(t) \, V_C(t) \tag{2-25}$$
$$= (4K/\pi)V_M \cos(\omega_m t)[\cos(\omega_c t) + (1/3)\cos(3\omega_c t) + \cdots + (1/n)\cos(n\omega_c t)]$$
$$= (4K/\pi)V_M[\cos(\omega_c + \omega_m)t + \cos(\omega_c - \omega_m)t + \cdots + (1/n)\cos(n\omega_c + \omega_m)t$$
$$+ (1/n)\cos(n\omega_c - \omega_m)t] \tag{2-26}$$

This output spectrum is illustrated in Fig. 2-9.

An interesting feature of a balanced modulator is that the output spectrum contains no component at either the carrier or modulating frequency. That is, the output signal is a suppressed carrier AM signal, providing that the response of the modulation input is linear. If a DC component is added to the modulation input, the carrier component is no longer suppressed and the output signal becomes a conventional AM signal. This can easily be seen by considering the modulation signal to be

$$V_M(t) = V_M(1 + M \cos(\omega_m t)) \tag{2-27}$$

where M is called the modulation index. The output signal can then be calculated by substituting Eqs. 2-27 and 2-24 into 2-25, which yields

$$V_o(t) = \frac{4KV_M}{\pi} [\cos(\omega_c t) + M \cos(\omega_c + \omega_m)t + M \cos(\omega_c - \omega_m)t$$

$$+ \cdots + \frac{1}{n}\cos(n\omega_c t) + \frac{M}{n}\cos(n\omega_c + \omega_m)t + \frac{M}{N}\cos(n\omega_c - \omega_m)t] \tag{2-28}$$

where $n = 1, 3, 5, 7 \ldots$

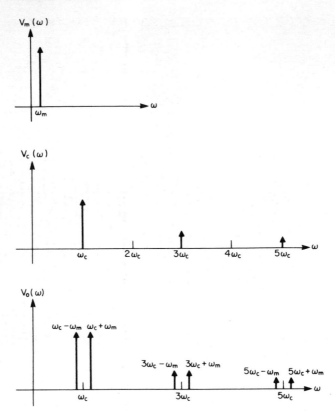

FIG. 2-9   Output spectrum for suppressed carrier AM.

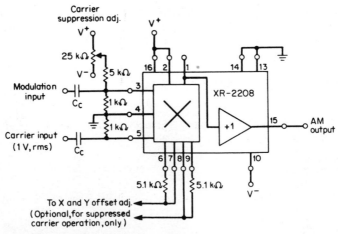

FIG. 2-10   Circuit connection for AM generation.

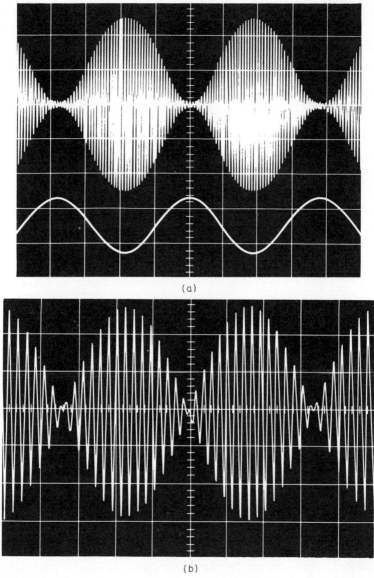

(a)

(b)

FIG. 2-11 Waveforms for AM generation. (*a*) 95% AM modulation; (*b*) suppressed carrier AM.

The effect of adding a DC component in the modulating signal is to increase the carrier amplitude in the output spectrum. This is intentionally added in conventional AM systems; however, in suppressed carrier systems it is usually the result of internal offsets within the modulator and results in what is commonly called "carrier feedthrough," an undesirable effect. Fig. 2-10 illustrates a circuit suitable for use as an AM generator, for either suppressed carrier or conventional AM generation. Fig. 2-11 shows the input and output signals for both modes of operation.

The primary difference between the two modes of operation is that for suppressed carrier AM generation, the output is at a minimum when the modulation is at its zero crossing points and a 180° phase reversal occurs in the output when the modulation input goes below this zero crossing point. For conventional AM generation, the output phase does not reverse and the minimum output signal occurs when the modulation input is at its minimum.

The following section will illustrate how multipliers can also be used to demodulate AM signals.

2–1f Demodulators

Multipliers are often utilized as synchronous AM detectors, or demodulators. This application is similar to that of the modulator and requires only one input to have linearity with respect to the input signal amplitude. Fig. 2-12 illustrates a block diagram of a typical AM demodulator. The operation of the circuit can briefly be described as follows. The input is simultaneously applied to both the X and Y inputs. However, before the signal is applied to the Y input, it is amplified and limited to convert it to a constant amplitude (carrier) signal. This carrier signal is then multiplied or mixed with the input signal to produce an output signal that is comprised of the demodulated AM signal along with a two-times frequency component. The two-times carrier component is then filtered out via the low-pass filter, yielding the demodulated output signal.

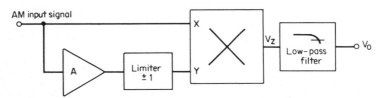

FIG. 2-12 Block diagram of a syncronous AM detector.

The circuit operation can easily be seen by considering the input signal to be of the form

$$V_{in} = V_x = V_m(t)\cos \omega_0 t \qquad (2\text{-}29)$$

where $V_m(t)$ is the modulation signal and ω_0 is the carrier frequency.

When this signal is amplified and limited, a constant amplitude carrier signal is generated of the form

$$V_Y = A_1 \cos \omega_0 t + A_2 \cos 3\omega_0 t + A_3 \cos 5\omega_0 t \cdots$$

Since the output will be low-pass filtered, all higher orders of this term may be eliminated at this time to yield an input signal

$$V_Y = A_1 \cos \omega_0 t \qquad (2\text{-}30)$$

When these two input signals V_{in} and V_Y are multiplied together, the output signal may be described as

$$V_Z = K[V_m(t)\cos \omega_0 t](A_1 \cos \omega_0 t)$$
$$= KV_m(t)[1 + \cos(2\omega_0 t)]$$

where K is the gain of the multiplier. If this signal is then passed through a low-pass filter to remove the $\cos(2\omega_0 t)$ term, the resultant output signal becomes

$$V_o = KV_m(t)$$

which corresponds to the detected input signal.

Example 2–3   Design of AM Detector

Design a 500-kHz AM detector using the XR-2208. The bandwidth of the modulating signal is 20 Hz to 20 kHz. The output impedance of the multiplier is 10 kΩ between pins 1 and 2. The block diagram of the XR-2208 is illustrated in Fig. 2-13.

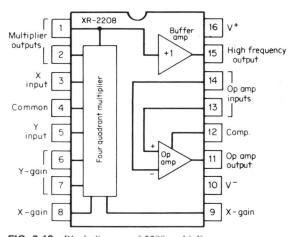

FIG. 2-13   Block diagram of 2208 multiplier.

Solution

The general schematic of an AM detector is illustrated in Fig. 2-14. The Y input gain terminals have been shorted to supply maximum gain. This will accomplish both the amplification and limiting to the Y input signal. The output filtering is accomplished by connecting a capacitor between the multiplier output terminals. The cutoff frequency is determined by

$$f_c = \frac{1}{2\pi R_o C} \tag{2-31}$$

where R_o is the output resistance of the multiplier and C is the filter capacitor value. Since $R_o = 10$ kΩ, the capacitor may be calculated by rearranging Eq. 2-31 to yield

$$C = \frac{1}{6.28 \times 10^4 f_c} \tag{2-32}$$

The cutoff frequency is chosen to be 25 kHz, which assures that the two-times carrier component will be attenuated by approximately 32 dB. The capacitor value is calculated using Eq. 2-32

$$C = \frac{1}{(6.28 \times 10^4)(25 \times 10^3)}$$

$$C \approx 620 \text{ pF}$$

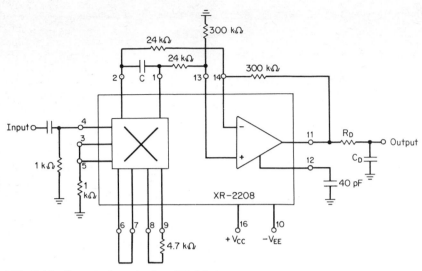

FIG. 2-14   General schematic of an AM detector.

The operational amplifier of the XR-2208 is used to buffer the output signal and to perform the differential-to-single-ended conversion of the multiplier output. R_D and C_D form a postdetection filter, which further reduces the two-times frequency component present in the output. The cutoff frequency for this filter is also chosen to be approximately 25 kHz. Assuming that the output resistance of the op amp is low, the 3-dB cutoff frequency is determined by

$$f_c = \frac{1}{2\pi R_D C_D}$$

The AM detector illustrated in Fig. 2-14 is suitable for carrier frequencies up to 100 MHz, since the usable operational range is determined by the transconductance bandwidth of the multiplier.

2–1g   Phase Detectors

Another application for multipliers where the usable bandwidth is determined by the transconductance bandwidth is phase detection. When used as a phase detector, the multiplier generates an output that is proportional to the phase difference of the input signals. This function is especially useful for phase meters, phase-locked loops, and FM demodulators. The operation of a phase detector can be described in a manner similar to that of the AM detector, where the two input signals are the same frequency; however, a finite phase difference exists between them. The input signals may then be defined as

$$V_X = A \cos \omega_0 t$$
$$V_Y = B \cos(\omega_0 t + \phi)$$

When these two inputs are multiplied together, the resultant output signal becomes

$$V_Z = V_X V_Y = KC \cos \phi + KD \cos(2\omega_0 t + \phi)$$

The two-times frequency component is then filtered out to yield

$$V_o = KC \cos \phi$$

Thus, the output is proportional to the cosine of the phase difference between the two input signals, providing that the multiplier is operating in its linear range. In many applications, the input signals are square waves or the multiplier gain is set high enough to cause limiting, and the multiplier output becomes

$$V_o = K\left(\frac{1 - 2\phi}{\pi}\right)$$

where ϕ is the phase difference in radians.

This yields the characteristic transfer function for a PLL, where the output error voltage is directly proportional to the phase difference between the input signal and the phase-locked oscillator, and is thus proportional to the frequency. The basic PLL illustrated in Fig. 2-15 is comprised of three key blocks: the phase detector, oscillator, and low-pass filter. Once the loop is locked, an error voltage is generated that is proportional to the difference between the nominal or free-running frequency of the oscillator and the input signal. This is illustrated in the characteristic transfer function for a PLL (Fig. 2-16). It is this particular transfer function that allows a PLL to be used for performing FM demodulation. This is easily seen by

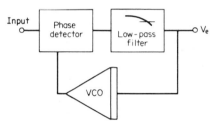

FIG. 2-15   Block diagram of a PLL.

considering FM transmission to be a form of voltage-to-frequency conversion, where the signal to be transmitted is used to frequency modulate a carrier. The phase locked loop then recovers the transmitted signal by performing a frequency-to-voltage conversion on the received carrier signal.

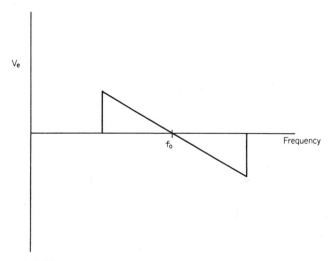

FIG. 2-16   Transfer function for a locked PLL.

TABLE 2-1 Multipliers

Type	Function	Differential inputs	Scale factor SF	Small signal BW	Nonlinearity	Manufacturer
532	$(X_1 - X_2)(Y_1 - Y_2)/10$	Yes	Set internally (10)	1 MHz	0.3%	Analog Devices
533	XY/K	No	Set externally	1 MHz	0.8%	Analog Devices
534	$A[(X_1 - X_2)(Y_1 - Y_2)/SF - (Z_1 - Z_2)]$	Yes	Set internally (10)	1 MHz	0.4%	Analog Devices
535	$A[(X_1 - X_2)(Y_1 - Y_2)/SF - (Z_1 - Z_2)]$	Yes	Set internally (10)	1 MHz	1%	Analog Devices
1494	XY/K	Yes	Set externally	800 kHz	1.3%	Motorola
1495	XY/K	Yes	Set externally	3 MHz	2%	Motorola
2208	XY/K	No	Set externally	3 MHz	0.5%	Exar
2228	XY/K	Yes	Set externally	3 MHz	0.5%	Exar
4200	$I_1 I_2/I_4$	No	Set externally	4 MHz	0.1%	Raytheon

Note: 4200 operates with a current input.

The linearity of the transfer function of Fig. 2-16 is of prime importance for FM demodulation, since any nonlinearity introduces distortion to the demodulated output. Sec. 2-2 describes oscillators and waveform generators which, when combined with multiplier ICs, are suitable for phase-locked loop applications. Table 2-1 lists several multiplier ICs along with some key performance characteristics. When designing with multiplier ICs, it is important to consider the particular type of application in order to determine the most cost effective solution.

2–2 WAVEFORM GENERATORS

Waveform generators have a wide range of applications in communication, telemetry, and process control. In addition, they are often found in the laboratory where they are used for testing and calibration. In many cases, IC function generators can provide a low-cost alternative to conventional discrete units.

The basic waveform generator is a device that generates a stable, well-defined periodic output signal that may be controlled externally. Fig. 2-17 illustrates a block diagram of a typical waveform generator, consisting of 3 sections: (1) an oscillator that generates a periodic waveform; (2) a wave shaper that converts the output of the

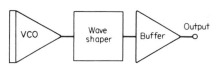

FIG. 2-17 Block diagram of a waveform generator.

oscillator into the desired waveform (usually sine or triangle); and (3) an output buffer amplifier to enable the generator to drive the required load.

The general performance characteristics of a waveform generator are determined by the performance of each of the sections that comprise the complete system. The oscillator determines the stability and linearity of a triangle output signal, as will be discussed later, and the wave shaper usually determines the distortion of the sine wave output.

2–2a Oscillators

Since the primary performance characteristics are determined by the oscillator, much emphasis has been placed on the design of an integrated oscillator that is suitable for the wide range of potential applications. For maximum versatility it is necessary that the oscillator have the following characteristics:

1. External sweep control with linear voltage-to-frequency conversion characteristics over a wide sweep range.
2. Stability with respect to:
   a. Temperature
   b. Power supply
   c. Short term (cycle to cycle).
3. Minimum external components to set frequency.

Constant Current Type   At the present time, two basic oscillator configurations are used that satisfy most of the above requirements. The first is the so-called $I/2I$ oscillator, where a capacitor is alternately charged and discharged by a constant current I, and the second is the emitter-coupled multivibrator. The first type is illustrated in Fig. 2-18 and consists of two comparators, one flip-flop, a constant-current source I, and

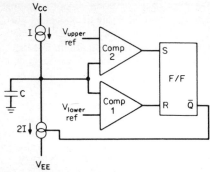

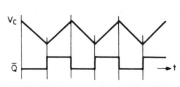

FIG. 2-18   Block diagram of an IC oscillator.

FIG. 2-19   Timing waveforms of oscillator of Fig. 2-18.

current sink $2I$, which is alternately switched on and off. The operation of the circuit can be described as follows: Assuming initially that the $2I$ current sink is off, the capacitor is therefore being charged by the current source I, and the voltage is charging at a rate of

$$\frac{dv}{dt} = \frac{I}{C}$$

This linear charging continues until the upper threshold is reached. At this time, comparator two switches, which sets the flip-flop, thus turning on the $2I$ current sink. This cancels out the I charging current and discharges the capacitor at a rate of

$$\frac{dv}{dt} = -\frac{I}{C}$$

until the lower threshold is reached and the flip-flop is reset, thus completing the timing cycle. The timing waveforms are illustrated in Fig. 2-19. The timing period is determined by summing the two individual timing cycles.

$$T = T_1 + T_2 \qquad \Delta V = V_{upper} - V_{lower}$$

$$T = \frac{\Delta VC}{I} + \frac{\Delta VC}{I}$$

$$T = \frac{2\Delta VC}{I}$$

and the frequency is defined as

$$f = \frac{1}{T} = \frac{I}{2\Delta VC} \tag{2-33}$$

The preceding derivation was based on the premise that the discharge current is precisely twice the charge current. Therefore, the triangle waveform is symmetrical and the duty cycle of the square wave is 50%. For most applications, a symmetry adjustment is provided to cancel out any offset component in the current sink. Thus, the output is a triangular waveform with a peak-to-peak swing of ΔV.

This particular oscillator is well suited for low-frequency applications since a polarized capacitor may be used for the timing function.

Example 2–4 Computation of Output Frequency

Derive an expression for the output frequency for the oscillator of Fig. 2-18. Assume that the threshold voltages are set at 1/3 and 2/3 V_{CC} and the charge current is defined as

$$I = \frac{2V_{CC}}{3R}$$

where R is an external resistor.

Solution

From Eq. 2-33

$$f = \frac{I}{2\Delta VC}$$

Substituting the equation for I into the above equation yields

$$f = \frac{V_{CC}}{3R\Delta VC}$$

Substituting 1/3 V_{CC} for ΔV

$$f = \frac{1}{RC}$$

Thus, the final transfer function is independent of both the timing current and the supply voltage. This type of oscillator depends on the tracking of the charge and discharge current in order to maintain a symmetrical triangle range and thus is only useful for sweep ranges of approximately 100:1. This particular type of problem is eliminated with an emitter-coupled multivibrator. A 1% variation between the two currents will result in a 2.4% change in the output duty cycle. Due to this need for precise tracking of the current source and current sink, this particular type of oscillator is only useful for sweep ranges of approximately 100:1.

The Emitter-Coupled Multivibrator The emitter-coupled multivibrator illustrated in Fig. 2-20 circumvents the need to precisely match a current source and a current sink by utilizing a matched pair of current sinks to alternately discharge each side of the

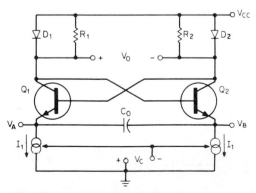

FIG. 2-20 Emitter-coupled multivibrator.

timing capacitor. Since these current sinks are operating at identical currents, the matching characteristics are excellent and the circuit is capable of maintaining a symmetrical triangle over a wide sweep range, typically 1000:1. The operation of the circuit may briefly be described as follows: Assume initially that both Q_1 and D_1 are conducting and Q_2 and D_2 are nonconducting. With these conditions, and neglecting base current effects, the emitter of Q_1 is held at $1V_{BE}$ below V_{CC} and the collector of Q_1, which is also the base of Q_2, is held at $1V_{BE}$ below V_{CC}. The emitter of Q_2 is discharging at a rate of

$$\frac{dv}{dt} = \frac{I}{C_o} \tag{2-34}$$

This discharging continues until the emitter of Q_2 reaches $2V_{BE}$ below V_{CC}, at which time Q_2 turns on. This causes D_2 to begin conducting, which turns off both Q_1 and D_1, causing the emitter of Q_2 to be pulled up to $1V_{BE}$ below V_{CC}. Since the voltage across a capacitor cannot change instantaneously, the $1V_{BE}$ voltage step is coupled to the other side of the timing capacitor C_o, changing the voltage on the emitter of Q_1 from $1V_{BE}$ below V_{CC} to V_{CC}. At this time, the Q_2, D_2 combination is conducting and the Q_1, D_1 is nonconducting, and the emitter of Q_1 is being discharged at a rate of

$$\frac{dv}{dt} = \frac{I}{C_o}$$

thus completing one-half of the timing cycle.

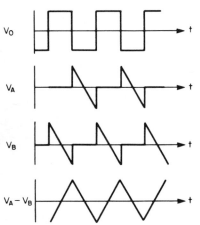

The waveforms of this oscillator are illustrated in Fig. 2-21. It should be noted that several output waveforms are available: a square wave with a peak-to-peak voltage swing of $2V_{BE}$ across diodes D_1 and D_2, and two linear ramps on the emitters of Q_1 and Q_2. If the waveform on the emitter of Q_2 is inverted and summed with the waveform on Q_1, the resultant waveform is a symmetrical triangle with a peak-to-peak voltage swing of $2V_{BE}$. As illustrated in Fig. 2-21, the square and triangular waveforms are 90° out of phase. This is extremely useful for phase-lock loop applications that will be discussed in Chap. 5.

The frequency of the oscillator is determined by taking the reciprocal of the sum of the two half-cycles

FIG. 2-21   Waveforms of emitter-coupled multivibrator of Fig. 2-20.

$$f = \frac{1}{(T_1 + T_2)}$$

But since

$$T_1 = T_2 = T$$

and rearranging Eq. 2-34, $T = dvC_o/I$. Substituting $2V_{BE}$ for dv, multiplying by 2, and taking the reciprocal, yields the output frequency

$$f_o = \frac{I}{4V_{BE}C_o} \tag{2-35}$$

Two potential problems exist with the oscillator of Fig. 2-20. First, the duty cycle of the square wave and symmetry of the triangle wave are dependent upon the tracking of two current sources. In order to eliminate the need for matching, a single current source may be used that is alternately switched between both sides of the timing capacitor. The second problem is the sensitivity of the frequency with respect to temperature, since the period is dependent upon V_{BE}. The temperature sensitivity may be calculated by using the simple formula

$$\frac{\dfrac{df}{f}}{dT} = \frac{\dfrac{-dV_{BE}}{V_{BE}}}{dT} = +2 \text{ mV}/(600 \text{ mV})°C = 3300 \text{ ppm}/°C$$

Thus, the frequency of the oscillator has a positive temperature coefficient of 3300 ppm/°C. This temperature effect may be compensated by giving it an equal and opposite temperature effect. This usually provides an improvement of an order of magnitude and therefore a temperature drift of approximately 300 ppm/°C would be expected. To further improve the temperature performance of this type of oscillator, precision references are used to accurately set the swing of the oscillator and control the charging current. This improved oscillator is illustrated in Fig. 2-22. The frequency is determined by

$$f = 1/RC$$

The temperature drift is approximately 20 ppm/°C. A thorough derivation is given in the references (see Gilbert).

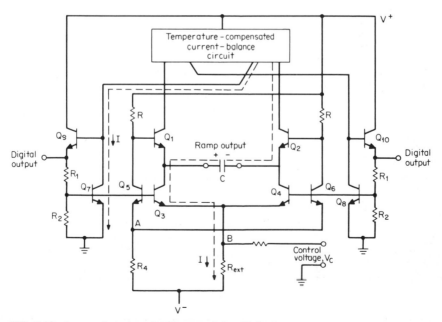

FIG. 2-22   Improved version of emitter-coupled multivibrator.

2–2b Sine Shapers

Once a good stable oscillator has been designed, a method must be determined to convert the output to a sine wave. Both oscillators in the preceding section have triangular output waveforms that greatly simplify the conversion. At the present time two sine shapers are predominantly used in IC fabrication.

Diode Breakpoint Method This method involves setting up breakpoints to form a nonlinear attenuator to shape the triangle wave into a sine wave. A circuit that performs this function is illustrated in Fig. 2-23. This method generates a sine wave in a piecewise fashion and, as such, the distortion is a function of the number of breakpoints used. Extreme care must be utilized in the design of such a sine shaper, since any nonsymmetry in the breakpoints will generate unwanted, even harmonics. When 16 breakpoints are used as illustrated in Fig. 2-23, and the circuit is trimmed, distortion of 0.5% can be obtained. Since the sine shaper contains *pnp* transistors, it is only useful at relatively low frequencies, as shown in the graph of Fig. 2-24, which illustrates distortion as a function of frequency.

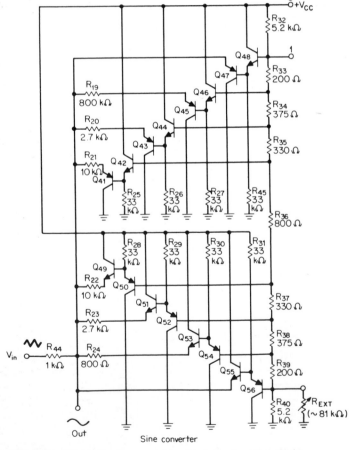

FIG. 2-23 Triangle-to-sine converter. (Diode breakpoint method.)

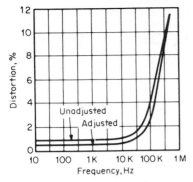

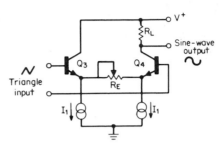

FIG. 2-24 Distortion vs. frequency for triangle-to-sine converter of Fig. 2-23.

FIG. 2-25 Emitter-coupled pair as a triangle-to-sine converter.

Emitter-Coupled Pair An alternate method of performing the triangle-to-sine conversion utilizes a differential pair with emitter degeneration, which is illustrated in Fig. 2-25. This circuit is suitable for operation up to approximately 10 MHz. The circuit operation can briefly be described as follows: Emitter degeneration resistor R_E is adjusted such that either Q_3 or Q_4 is brought near its cutoff point when the input triangle waveform reaches its peak. The transfer characteristics of the peaks then become logarithmic rather than linear, thus rounding out the peaks of the triangle, as illustrated in Fig. 2-25. When the current source, input voltage, and emitter resistor are adjusted for their optimum values, the total harmonic distortion of the output is typically 0.6%.

When these sine shapers are combined with the oscillators of the preceding section, the result is an integrated waveform generator. These waveform generators may readily be converted to AM generators by simply connecting the output of the generator to the input of a four-quadrant multiplier.

2–2c Performance Parameters

The performance of a function generator is rated by the following parameters.

Frequency Range—The range of frequencies over which functional operation is guaranteed.

Sweep Range—The ratio of the maximum to minimum output frequency that can be obtained by applying a sweep input voltage.

FM Linearity—The deviation of the best fit straight line of the control voltage vs. the output frequency.

Stability—The change of the output frequency with respect to both temperature and power-supply variations.

Triangle Wave Linearity—The deviation of the triangle output from the best fit straight line.

THD—The total harmonic distortion of the sine wave output.

There are currently several readily available integrated waveform generators. Table 2-2 summarizes these along with their basic operating characteristics. Further information is available through manufacturers' data sheets, or application notes.

TABLE 2-2 Waveform Generators

P/N	Sine	Triangle	Upper freq	Stability (PPM/°C)	Sweep range	AM input	FSK input	Manufacturer
555	No	Yes	4 MHz	300 (typ)	7:1	Yes	No	Exar
566	No	Yes	1 MHz	200 (typ)	10:1	No	No	National, Signetics
2206	Yes	Yes	1 MHz	50 max	2000:1	Yes	Yes	Exar
2207	No	Yes	1 MHz	50 max	1000:1	No	Yes	Exar, Raytheon
2209	No	Yes	1 MHz	50 max	1000:1	No	No	Exar
8038	Yes	Yes	1 MHz	50 max	1000:1	No	No	Exar, Intersil

2–2d   Applications

Most integrated waveform generators have been designed as general-purpose building blocks and, as such, are suited for a wide range of applications.

The primary applications of function generators are:

Sine wave generation
AM generation
FM generation
FSK generation
Laboratory function generator

All these applications require specific features from the waveform generator. The designer must be aware of what features are required to determine which integrated circuit would be the most cost-effective solution. For example, if a fixed frequency oscillator is required to produce a 1-kHz triangle waveform, all the circuits listed in Table 2-2 would be suitable. The IC that is chosen would simply be determined by which oscillator has the lowest cost and fewest external components. The best choice would be the 566 (the schematic of such an oscillator is illustrated in Fig. 2-26). The center frequency is determined by

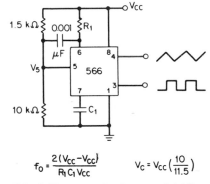

$$f_0 = \frac{2(V_{CC} - V_{CC})}{R_1 C_1 V_{CC}} \qquad V_C = V_{CC}\left(\frac{10}{11.5}\right)$$

FIG. 2-26   Schematic diagram of 1-kHz triangle wave oscillator.

$$f_0 = \frac{2(V_{CC} - V_C)}{R_1 C_1 V_{CC}}$$

where V_C is the potential at pin 5 set by the 10 kΩ and 1.5 kΩ resistor voltage divider. Frequency f_0 is then determined by substituting $0.870 V_{CC}$ for V_C, which yields

$$f_0 \approx \frac{1}{3.8 R_1 C_1}$$

If C_1 is chosen to be 0.1 μF, R_1 is then calculated to be

$$R_1 = \frac{1}{3.8 f_0 C_1}$$

$$= \frac{1}{(3.8 \times 10^3)(0.1 \times 10^{-6})}$$

$$= 2.6 \text{ k}\Omega$$

In most applications, this 2.6-kΩ value would consist of a 2.2-kΩ fixed resistor in series with a 1-kΩ potentiometer to precisely tune the oscillator to 1 kHz.

Example 2–5   Sine Wave Sweeping Oscillator

Design a sine wave oscillator with a sweep range of 20 Hz to 20 kHz, using the 2206 function generator. The equation governing the operation of this circuit is

$$f_0 = \frac{320I_T(\text{mA})}{C_0(\mu\text{F})} \tag{2-36}$$

where I_T is the current flowing through the timing terminal in milliamperes, and C_0 is the timing capacitor in microfarads. The timing terminal is internally biased at 3.125 V and the maximum allowable timing current is 3 mA. The sweep signal input varies from 0 to 10 V.

Solution

The sweep function is accomplished by connecting the timing terminal, as illustrated in Fig. 2-27. The timing current is defined as

$$I_T = I_B + I_C$$

But, $\qquad I_B = 3.125/R$

and $\qquad I_C = (3.125 - V_C)/R_C$

Substituting these into the Eq. 2-36 yields

$$f = \frac{0.320}{C_0}\left[\frac{3.125}{R} + \frac{3.125 - V_C}{R_C}\right] = \frac{1}{RC_0} + \frac{1}{R_C C_0} - \frac{V_C}{3.125 R_C C_0}$$

$$f = \frac{1}{RC_0}\left[1 + \frac{R}{R_C}(1 - V_C/3.125)\right] \tag{2-37}$$

By differentiating Eq. 2-37, the voltage-to-frequency conversion gain K may be determined.

$$K = \frac{dF}{dV_C} = -\frac{0.32}{R_C C_0} \; (\text{Hz/V})$$

For the required sweep range

$$K \approx 20 \text{ kHz/10 V} = \frac{0.32}{R_C C_0} \tag{2-38}$$

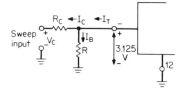

FIG. 2-27 Circuit connection for frequency sweep operation of 2206 function generator.

Thus,

$$R_C C_0 = 160 \times 10^{-6} \tag{2-39}$$

The maximum frequency of 20 kHz is obtained when $V_C = 0$. Therefore, using Eq. 2-37

$$20{,}000 = \frac{1}{RC_0}\left[1 + \frac{R}{R_C}\right] \tag{2-40}$$

$$= \frac{1}{RC_0} + \frac{1}{R_C C_0} = \frac{1}{C_0(R//R_C)}$$

The maximum timing current is 3 mA, and

$$I_T = 3.125/R_P, \qquad \text{where } R_P = R//R_C$$

If the maximum I_T is chosen to be ≈ 2.6 mA, R_P is calculated to be 1.2 kΩ. The timing capacitor value is then calculated from Eq. 2-40. Thus

$$20{,}000 = \frac{1}{1200 \times C_0}$$

$$C_0 = 0.04 \ \mu\text{F}$$

R_C is then calculated from Eq. 2-39.

$$R_C = \frac{160 \times 10^{-6}}{0.04 \times 10^{-6}}$$

$$= 4 \text{ k}\Omega$$

and R is calculated by

$$\frac{1}{R_P} = \frac{1}{R} + \frac{1}{R_c}$$

$$\frac{1}{1200} = \frac{1}{4000} + \frac{1}{R}$$

so

$$R \approx 1.7 \text{ k}\Omega$$

The complete schematic is illustrated in Fig. 2-28. Resistor R_3 sets the output amplitude. The total harmonic distortion of the output is adjusted to a minimum via R_A and R_B. It is important to note that the R and R_C values are critical in determining the sweep range of this circuit. Precision matching is required and separate trim adjustments are necessary. An alternate ap-

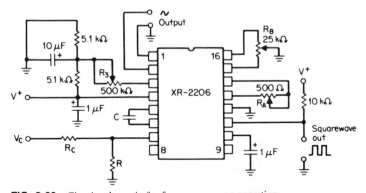

FIG. 2-28 Circuit schematic for frequency sweep operation.

proach that eliminates the need for this precision matching requires an external op amp to operate as a voltage-to-current converter. This current output may then drive the timing terminal directly, as illustrated in Fig. 2-29. A logarithmic sweep could also be generated by replacing the linear current source with one that is logarithmic.

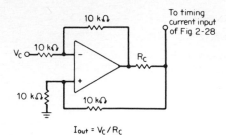

$$I_{out} = V_C/R_C$$

FIG. 2-29   Using external current source for frequency sweep operation.

FM Generation   FM generation is similar to that of sweep generation; however, the frequency deviation is usually small, typically less than 10% of the normal operating frequency. Terms used for FM generators are as follows.

Carrier frequency—The output frequency with no input present.
Deviation—The percent of change of the output frequency from no input signal to the maximum input signal.
Modulation rate—The rate of change of the output frequency.
FM nonlinearity—The maximum deviation of frequency output to voltage input, from the ideal straight-line transfer function.

The circuit for FM modulation is identical to that of the sweep generator, except the sweep range is typically less than $\pm 10\%$ and the modulation input is capacitively coupled.

Frequency-Shift Keying Generation   Frequency-shift keying (FSK) is a method used in the transmission of digital data through voice-grade channels such as telephone lines. In these applications, the data to be transmitted must first be converted to a signal that is compatible with the bandwidth of the transmission medium. The data is then transmitted in this form to the receiver, where it is demodulated and converted back to its original digital form. The device that performs this function is called a "modem," formed by the contraction of the two key elements that describe its function, MOdulator and DEModulator.

The modulation portion generates the FSK signal and the demodulator decodes the signal. This is basically a specific case of FM transmission where the output signal is one of two discrete frequencies determined by the state of the digital input. A block diagram of an FSK generator is illustrated in Fig. 2-30 along with input and output waveforms. It is important to note that in FSK generators the output signal must be phase

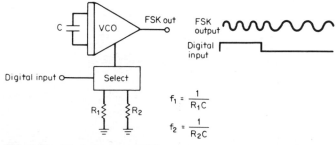

$$f_1 = \frac{1}{R_1 C}$$

$$f_2 = \frac{1}{R_2 C}$$

FIG. 2-30   Block diagram of FSK generator.

continuous. This will eliminate any unwanted sidebands from being generated
due to a step change in the output voltage.

Virtually all oscillators can be frequency-shift keyed by connecting an external transistor and a resistor to the timing
terminal, as illustrated in Fig. 2-31. However, some function generators listed in
Table 2-2 have this feature built in to the
device, which greatly simplifies circuit
design and eliminates any variations in
output frequency due to changes in the
saturation voltage of the external transistor.

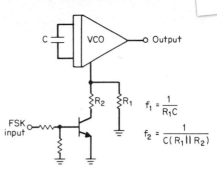

FIG. 2-31   Method of converting a fixed-frequency oscillator to FSK generator.

Example 2–6   Bell 103-Type Transmit Frequency Generation

Design an FSK generator for a 103-type modem. The output sine wave
distortion should be less than 2.5%. The output frequencies are 1070 and
1270 Hz for a space and mark, or 1 and 0, respectively. Choose the device
from Table 2-2.

Solution
The only device listed in Table 2-2 with an FSK input and a sine wave
output is the 2206. The sine wave distortion spec of 2.5% is achieved by
connecting a 200 Ω resistor between pins 13 and 14. The timing resistor
values are determined using the following formulas:

$$f_s = \frac{1}{R_7 C}$$

$$f_m = \frac{1}{R_8 C}$$

The resistors are chosen to be in the range of 10 to 100 kΩ for optimum
stability. If the timing capacitor is chosen to be 0.039 μF

$$R_7 = 20.19 \text{ k}\Omega$$
$$R_8 = 23.96 \text{ k}\Omega$$

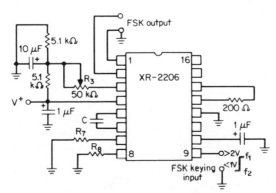

FIG. 2-32   Schematic diagram of an FSK generator.

These values are normally obtained using a fixed resistor and a variable to accurately tune in each frequency. The complete schematic is illustrated in Fig. 2-32. The output amplitude is adjusted by means of potentiometer R_3.

Sawtooth Generators   Most function generators are also capable of generating both pulse and sawtooth waveforms. This is illustrated in Fig. 2-33, where the data is fed into the FSK input. The FSK input is a digital input terminal, which is used to select either one of two timing resistors. The duty cycle of the output is given by

$$\text{Duty cycle} = \frac{R_1}{R_1 + R_2}$$

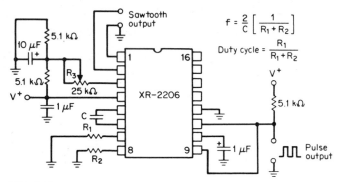

FIG. 2-33   Circuit connection for pulse and sawtooth generator.

The duty cycle of this circuit can be adjusted from approximately 0.1 to 99.9%. The variable-duty cycle range is usually approximated by the sweep range or ratio of maximum timing current to minimum timing current.

AM Generation   As discussed earlier, when the output of an oscillator is applied to one input of a four-quadrant multiplier (carrier), and a modulating signal is applied to the other input, the output of the multiplier is an AM signal. Some waveform generators have been specifically designed for such functions and have internal four-quadrant multipliers, as shown in Table 2-2. The key performance characteristics of an AM generator are the output frequency stability and linearity of the output level, with respect to the modulation input or AM input. Fig. 2-34 illustrates a block diagram of an AM generator. This circuit is capable of producing either suppressed carrier or conventional AM. The

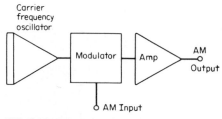

FIG. 2-34   Block diagram of an AM generator.

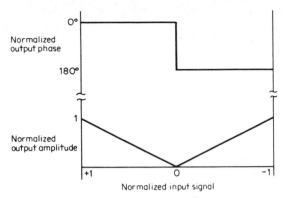

FIG. 2-35 Carrier output amplitude and phase characteristics of an AM generator.

transfer characteristics are illustrated in Fig. 2-35 for both the amplitude and phase of the output. It should be noted that if the DC level at the modulation input is zero volts, the carrier signal is suppressed; thus, no power is wasted transmitting the carrier signal.

The actual circuit implementation of an AM generator may be accomplished by either a single IC with an internal multiplier, or two ICs, one oscillator to generate the carrier frequency and a multiplier to serve as the AM modulator. The decision as to which approach is most effective is dependent upon the required stability, the frequency of the carrier, and the required linearity of the AM output.

2–3 VOLTAGE-TO-FREQUENCY CONVERTERS

There are many instances when it is necessary to convert an analog voltage to a frequency, which is then either applied to a counter to obtain a digital readout or transmitted to some remote point where it is then converted back to an analog signal by means of a frequency-to-voltage converter. The normalized characteristics of an ideal voltage-to-frequency converter are illustrated in Fig. 2-36. The equation for such a function is given by

$$f_{out} = KV_{in} \qquad (2-41)$$

where K is the conversion gain of the voltage-to-frequency converter and has the dimensions of hertz/volt.

The key performance feature of such a function is the linearity of the output frequency with respect to the input voltage. Thus, the key element in this circuit is a voltage-controlled oscillator. The previous section dealt with several VCOs that are suitable for such applications when operated in the sweep mode.

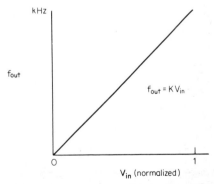

FIG. 2-36 Transfer characteristics for an ideal voltage-to-frequency converter.

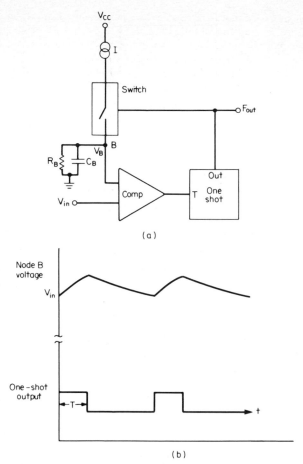

FIG. 2-37   Voltage-to-frequency converter. (*a*) Block diagram; (*b*) timing waveforms.

An alternate approach for such a converter is illustrated in Fig. 2-37*a*. The basic operation of this circuit may be described as follows: The comparator compares the input voltage to the voltage at node *B*. If the input voltage is higher, the comparator output triggers the one shot. The output of the one shot then closes the switch connecting the current source to node *B* for the duration of the one-shot period. The switch is then opened when the one-shot times out. At this time, current source *I* has injected a charge Q into the $R_B C_B$ network equal to

$$Q = I \times T$$

where T is equal to the one-shot period. If at this time the voltage on node *B* is not greater than the input voltage, the comparator will retrigger the one shot and inject another charge into the $R_B C_B$ network. This continues until the node *B* voltage is greater than the input voltage. At this point, the one shot remains off and C_B is discharged through resistor R_B until it equals V_{in} and triggers the one shot. At this point a steady-state condition has been reached, where the rate that the charge injected by current source *I* is sufficient to keep $V_B \geq V_{in}$. Since the discharge rate of C_B is proportional to V_B/R_B, the system runs at a frequency that is proportional to V_{in}. Fig. 2-37b illustrates the waveforms present at node B and the one-shot output pulse. It should be noted that the one-shot output pulse remains constant, only the repetition rate changes with V_{in}.

Fig. 2-38 illustrates the block diagram of an industry-standard voltage-to-frequency converter, the 4151. The conversion gain K is defined as

$$K = \frac{0.486\ R_S}{R_B R_0 C_0}\ \text{kHz/V} \tag{2-42}$$

Note that C_B does not occur in the conversion gain equation. Its function is to integrate the one-shot output and it also serves to keep the current source from saturating, which would occur if C_B was too small.

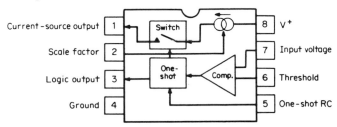

FIG. 2-38   Functional block diagram of 4151 voltage-to-frequency converter.

The actual circuit implementation of a voltage-to-frequency converter with an input dynamic range of 0 to 10 V and an output frequency range of 0 to 10 kHz is illustrated in Fig. 2-39. This circuit has the advantages of low cost and simplicity; however, linearity is only approximately 1% and has a relatively slow response time (approximately 135 ms for a 0- to 10-V step in V_{in}). This is primarily due to the passive integration of the $R_B C_B$ network.

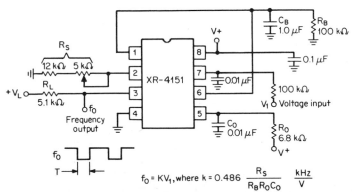

FIG. 2-39   Simple voltage-to-frequency converter.

A precision voltage-to-frequency converter may be realized with the addition of an external operational amplifier, as illustrated in Fig. 2-40. This circuit improves the linearity to approximately 0.05% and the response time to 10 μs. The circuit maintains its linearity over the full input dynamic range, all the way to zero volts input. The op amp improves the linearity by holding the output current source, pin 1, at zero volts, thus eliminating

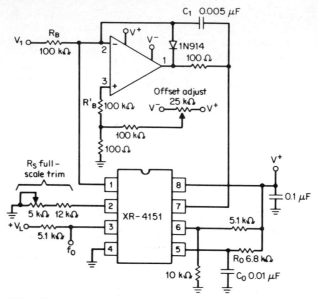

FIG. 2-40 Precision voltage-to-frequency converter.

any error due to the output conductance of the current source. The diode prevents the voltage at pin 7 from going below zero volts. A low-leakage diode is required, since any leakage at this point will degrade the accuracy.

Trimming is accomplished by setting the input signal to -10 V and adjusting R_s for a 10-kHz output, then setting the input to -10 mV and adjusting the offset for 10-Hz output.

Example 2–7 Analog-to-Digital Converter Design

Utilizing the 4151, design an analog-to-digital converter capable of accepting a 0- to 10-V input signal. The required accuracy is 0.1%.
Solution
The block diagram of an analog-to-digital converter, using a voltage-to-frequency converter, is illustrated in Fig. 2-41. The voltage-to-frequency

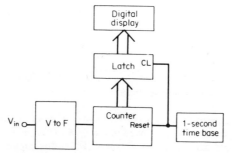

FIG. 2-41 Block diagram of analog-to-digital converter.

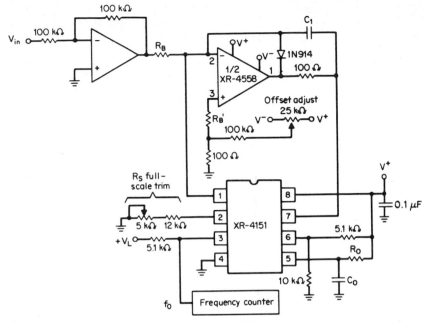

FIG. 2-42   Circuit implementation of an analog-to-digital converter.

converter is used to convert the input voltage to a frequency that is directly proportional to the input voltage. This frequency is then applied to a frequency counter with a 1-s sample time. Thus, the output frequency is counted for a 1-s time period, then latched to the output. The actual circuit implementation illustrated in Fig. 2-42 replaces the counter stage with a frequency counter to simplify the design. The first op amp is used to convert the 0- to 10-V input signal to a 0- to -10-V signal and also to buffer the input signal.

This particular technique is ideally suited for transmitting analog data over fiber-optic links. In such applications, the frequency counter would be located at the receiving end of the channel. If the frequency counter were to be replaced with a frequency-to-voltage converter, the original analog signal could be restored.

2–4 FREQUENCY-TO-VOLTAGE CONVERTERS

A frequency-to-voltage converter is a device that accepts an input frequency and converts it to an output DC signal that is proportional to the input frequency. Thus

$$V_{out} = Kf_{in} \qquad (2\text{-}43)$$

where K is the conversion gain of the frequency-to-voltage converter and has the dimensions of volts/hertz. There are several techniques that are suitable for performing the frequency-to-voltage conversion. One method utilizes a phase-locked loop and will be discussed in a later chapter. Another technique uses a 4151 voltage-to-frequency converter.

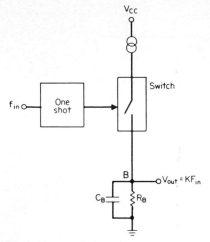

FIG. 2-43 Functional block diagram of a frequency-to-voltage converter.

Fig. 2-43 illustrates the block diagram of a circuit suitable for performing the frequency-to-voltage conversion. The operation of this circuit may be described as follows: the one shot is triggered on the rising edge of the input signal. (The one shot could also be made to trigger on the falling edge with the same results.) This causes the switch to close for the one-shot time, during which a charge equal to it is injected into the R_BC_B network, increasing the potential at node B. After the one shot times out, the voltage at node B begins to discharge through R_B. This discharge continues until the next positive edge of the input signal, which triggers the one shot for another time T. As the input frequency is increased, the time between triggering the one shot decreases; thus, the voltage at node B increases. Capacitor C_B serves as an integrator and thus filters the voltage at node B. The value of C_B can be increased to further reduce the ripple; however, the tradeoff is a slower response time to input frequency changes.

The actual circuit implementation of a frequency-to-voltage converter is illustrated in Fig. 2-44. The transfer function is given as:

$$V_o = kf_{in}$$

where $K = (2.058\ R_BR_0C_0/R_S)$ V/Hz.

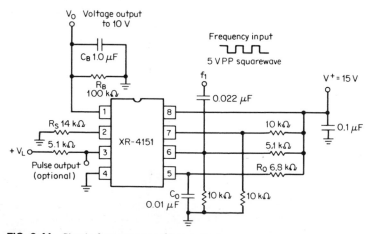

FIG. 2-44 Simple frequency-to-voltage converter.

As previously discussed, the output conductance of the current-source at pin 1 degrades linearity. This degradation can be compensated for with the addition of an op amp, as shown in Fig. 2-45, which improves the linearity of the frequency-to-voltage

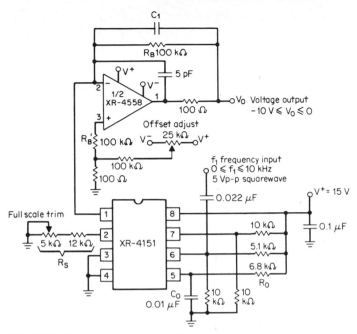

FIG. 2-45 Precision frequency-to-voltage converter.

conversion to approximately 0.05%. Unlike the precision voltage-to-frequency conversion, however, there is no significant improvement in the response time. The only simple method to reduce the response time is to increase the cutoff frequency of the low-pass filter. However, this will cause a higher output ripple; for example, with

$$C_1 = 0.1 \ \mu\text{F}$$
$$R_B = 100 \ \text{k}\Omega$$

output ripple will be approximately 100 mV and the response time approximately 10 ms. This output voltage may be filtered further to reduce output ripple, but the response time will be even slower.

2–4a General Guidelines for the 4151

The 4151 can be programmed to operate with a full-scale frequency anywhere from 1.0 Hz to 100 kHz. In the case of the voltage-to-frequency configuration, nearly any full-scale input voltage from 1.0 V and up can be tolerated, if proper scaling is employed. The following guidelines are useful in determining the external component values for any desired full-scale frequency.

1. Set $R_S = 14 \ \text{k}\Omega$ or use a 12 kΩ resistor and 5 kΩ pot.
2. Set $T = 1.1R_0C_0 = 0.75[1/f_0]$, where f_0 is the desired full-scale frequency. For optimum performance, make 6.8 k$\Omega < R_0 < 680$ kΩ and 0.001 μF $< C_0 < 1.0$ μF.
3. *a.* For the circuit of Fig. 2-39, make $C_B = 10^{-2}(1/f_0)$ F. Smaller values of C_B will give a faster response time, but will also increase the frequency offset and nonlinearity.

b. For the active integrator circuit, make

$$C_1 = 5 \times 10^{-5} [1/f_0] \text{ F}$$

The op-amp integrator must have a slew rate of at least $135 \times 10^{-6} [1/C_1]$ volts/s, where the value of C_1 is in farads.

4. a. For the circuit of Fig. 2-40, keep the values of R_B and R_B' as shown and use an input attenuator to give the desired full-scale input voltage. ($R_B' = R_B$ for minimum offset.)

b. For the precision mode circuit of Figure 2-40, set

$$R_B = V_{IO}/100 \text{ } \mu\text{A}$$

where V_{IO} is the full-scale input voltage.

Alternately, the op-amp inverting input (summing node) can be used as a current input with the full-scale input current $I_{IO} = 100 \text{ } \mu\text{A}$.

5. For the FVCs, pick the value of C_B or C_1 to give the optimum tradeoff between the response time and output ripple for the particular application.

Example 2–8   Frequency-to-Voltage Converter Design

Design a frequency-to-voltage converter operating with a 8-V power supply and a full-scale input of 750 Hz. The output must reach 63% of its final value within 200 ms. Determine the output ripple.

Solution

Use the general schematic shown in Fig. 2-45. Set $R_S = 14 \text{ k}\Omega$. Determine the one-shot time constant via rule 2 of the programming equations:

$$T = 1.1 R_0 C_0 = 0.75 [1/f_0]$$
$$T = 1 \text{ ms}$$

Set $R_0 = 9.1 \text{ k}\Omega$ and $C_0 = 0.1 \text{ } \mu\text{F}$. Since this circuit must operate from an 8-V power supply, set the full-scale output equal to 5 V. From general guideline step 4b, select

$$R_B = 5 \text{ V}/100 \text{ } \mu\text{A} = 50 \text{ k}\Omega$$

The output time constant $\tau_R \leqslant 200$ ms; therefore, $C_B \leqslant \tau_R/R_B = 200 \times 10^{-3}/50 \times 10^3 = 4 \text{ } \mu\text{F}$. The worst-case ripple is calculated by

$$V_R = T\frac{I}{C}$$
$$= (1 \text{ ms}) (135 \text{ } \mu\text{A})/4 \text{ } \mu\text{F}$$
$$\approx 34 \text{ mV}$$

2–5   OP-AMP CIRCUIT FUNCTIONS

There are a number of circuit functions that may be implemented using op amps and a few external components. Since many of the functions are currently only available as hybrid modules, and not in completely integrated form, a fundamental understanding of their operation is necessary. This will assist the designer in determining whether the discrete or the hybrid version would present the most cost effective solution, by illustrating what tradeoffs exist for some specific applications.

2–5a Precision Rectifiers

There are many instances where a diode that operates according to its ideal transfer function is required. Thus, it must appear as either an open circuit or a short circuit, depending upon the polarity of the applied voltage. This is particularly important for applications such as precision rectifiers and absolute value circuits.

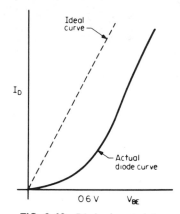

FIG. 2-46 Diode characteristics.

When diodes are used as half-wave or full-wave rectifiers, the output that is obtained deviates from that of the ideal diode. This deviation is primarily due to the diode characteristics that are illustrated in Fig. 2-46. Thus, when this nonideal diode is used as a half-wave rectifier, the output signal is offset by the diode "on" voltage, yielding an output as shown in Fig. 2-47. This effect can be eliminated by using an op amp connected as a precision rectifier. The circuit that performs this function is illustrated in Fig. 2-48. The operation of this circuit is such that when $V_{in} < 0$, and $V_{D1} > 0$, all the input current flows through R_F, since D_2 is reverse biased, generating an output voltage

$$V_o = -\frac{R_F}{R_C} V_{in}$$

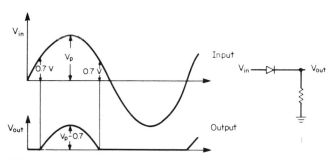

FIG. 2-47 Waveforms for a half-wave rectifier.

When $V_{in} > 0$ and $V_{D1} < 0$, diode D_1 is no longer conducting and all the input current flows through diode D_2, and the output is held virtually at zero volts.

The precision rectifier has several advantages over a standard diode; for example, the breakpoint is sharp and well defined, as well as insensitive to temperature variations. This type of rectifier also allows for full-wave rectification of very low-level input signals.

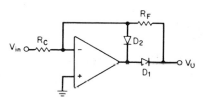

FIG. 2-48 Precision half-wave rectifier.

Example 2–9   Ideal Rectifier Design

Design a circuit that produces an output voltage that is the absolute value of the input voltage; therefore

$$V_o = |V_{in}|$$

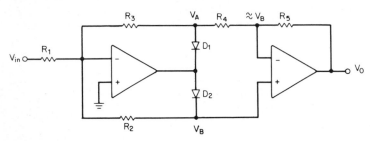

FIG. 2-49   Absolute value amplifier.

Solution

The absolute value circuit consists of two ideal diodes connected as a full-wave rectifier. This function can be realized by summing the outputs of two precision rectifier circuits identical to that of Fig. 2-48 and inverting one of the inputs. A more practical circuit is illustrated in Fig. 2-49, which consists of an ideal diode and a differencing circuit. When V_{in} is positive, D_1 conducts and D_2 is open; thus

$$V_B = 0$$

$$V_A = -V_{in}\frac{R_3}{R_1}$$

and the output is equal to

$$V_o = \frac{V_{in}\,R_3 R_5}{R_1 R_4}$$

If all resistances are equal

$$V_o = V_{in}$$

When V_{in} is negative, D_2 conducts and D_1 is open; thus

$$-\frac{V_{in}}{R_1} = \frac{V_B}{R_2} + \frac{V_B}{R_3 + R_4}$$

and

$$V_o = V_B\left[1 + \frac{R_5}{R_3 + R_4}\right]$$

If all resistances are equal

$$-V_{in} = V_B + \frac{V_B}{2} = 1.5V_B$$

and

$$V_o = -\frac{V_{in}}{1.5}[1 + 0.5]$$

thus
$$V_o = -V_{in}$$
$$V_o = |V_{in}|$$

2-5b   Logarithmic Converters

In a wide variety of applications, a system must be capable of processing input signals that vary over a wide dynamic range. This presents problems to the circuit designer, since the dynamic operational range of a circuit is dependent upon the operating range of each circuit element, the noise of each component, and the power-supply voltage. Many of these problems may be avoided with the use of logarithmic converters that basically compress the input signal, process it in this compressed form, and then expand it. Arithmetic operations may also be performed utilizing logarithmic circuits and op amps using the following identities:

$$X \times Y = e^{(\ln x + \ln y)}$$
$$\frac{X}{Y} = e^{(\ln x - \ln y)}$$

Therefore, an analog multiplier could be constructed with three logarithmic converters and an operational amplifier. Two of the converters are utilized to take the log of the input signals, and the outputs of these converters are then summed in the op amp. The output of the op amp is then applied to the input of the third converter, which takes the antilog. Thus, the output of the third converter is equal to the product of the two input signals. The functional block diagram of such a circuit is illustrated in Fig. 2-50.

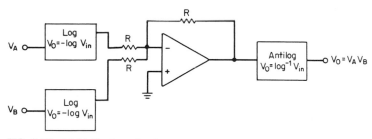

FIG. 2-50   Multiplication using log converters.

Currently, most logarithmic circuits are fabricated using a diode in the feedback path of an operational amplifier. For purposes of simplicity, all diodes used in the derivations will be considered to be well matched and defined by the following equation:

$$I = I_0 \left(e^{(qV_{BE}/KT)} - 1 \right) \qquad (2\text{-}44)$$

where q = a charge constant 1.60219×10^{-19} C
K = Boltzman's constant 1.38062×10^{-23} J/°K
T = temperature in K = °C + 273
I_0 = the reverse saturation current
V_{BE} = the voltage across the diode
I = the forward diode current.

This equation may be rearranged and solved for V_{BE} by dividing both sides by I_0 and then taking the log of both sides to yield

$$\ln\left(\frac{I}{I_0} + 1\right) = \frac{qV_{BE}}{KT}$$

Since

$$\frac{I}{I_0} \gg 1$$

$$\ln\left(\frac{I}{I_0} + 1\right) \approx \ln\left(\frac{I}{I_0}\right)$$

(2-45)

and,

$$V_{BE} = \frac{KT}{q}\ln\left(\frac{I}{I_0}\right)$$

When such a diode is connected in the feedback path of an op amp, as shown in Fig. 2-51, the output voltage may be defined as

$$V_o = \frac{KT}{q}\ln\left(\frac{V_{in}}{R_{in} I_0}\right)$$

At room temperature

$$\frac{KT}{q} \approx 25.7 \times 10^{-3} \text{ mV}$$

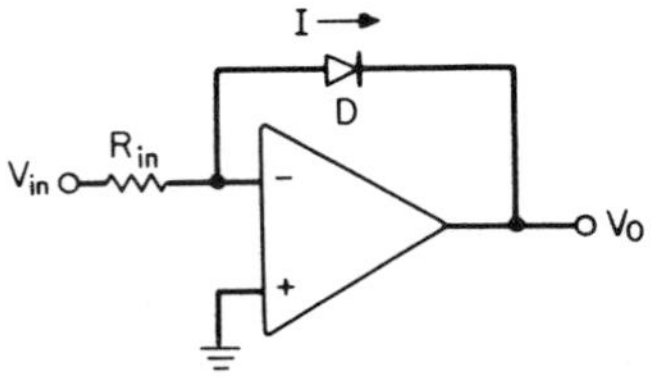

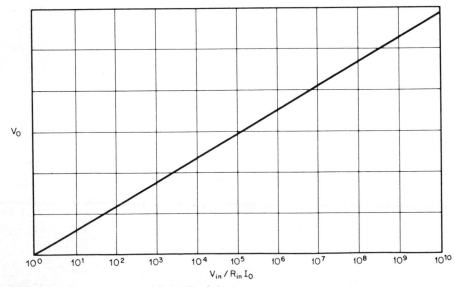

FIG. 2-51   Simple logarithmic converter.

If we assume that I_0 is a constant and plot V_o as a function of $V_{in}/R_{in}I_0$, the output is a logarithmic function as illustrated in Fig. 2-52. It is important to note that the above derivation considered the diode to have ideal behavior over its full dynamic range. However, the operating range of practical diodes is limited on both the high and low current ranges.

FIG. 2-52   Ideal transfer function for Fig. 2-51.

At the high end, the ohmic bulk resistance produces an additional voltage drop; thus, at high current levels Eq. 2-45 becomes

$$V_{BE} = \frac{KT}{q} \ln(I/I_0) + IR_B$$

where R_B is the ohmic bulk resistance.

At low current ranges, the slope undergoes some changes due to surface inversion layers and the generation recombination mechanisms in the space-charge regions.

To overcome these limitations, two approaches have been utilized; the transdiode configuration of Fig. 2-53a and the transistor connected diode of Fig. 2-53b. Each of these approaches serves to substantially increase the dynamic operating range of a logarithmic converter. Several other modifications are also made to the logarithmic converter of Fig. 2-51 to compensate for all the temperature-dependent variables in the commercially available hybrid modules.

Antilog amplifiers may be made in a similar manner to the log amplifier, except the input resistor and the log element are interchanged.

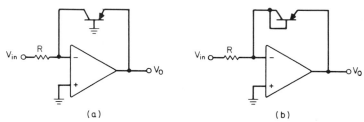

(a)   (b)

FIG. 2-53   Alternate method of generating a diode. (a) Transdiode; (b) diode connected transistor.

Example 2–10   Analysis of Antilog Amplifier

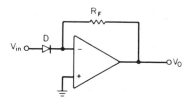

FIG. 2-54   Simple antilog amplifier.

For the antilog amplifier of Fig. 2-54 determine the transfer function, assuming the diode behavior is ideal. Then graph the transfer function at 25°C, −55°C and +125°C temperature. What constraints must be placed on V_{in} to keep the diode current from exceeding 10 mA? Assume $q/KT = 38.9$ at 25°C, 29.2 at 125°C, and 53.2 at −55°C and $I_0 = 7.3 \times 10^{-15}$ A.

Solution

Since the inverting input of the operational amplifier must remain at a virtual ground potential, the input voltage appears across the diode. The diode current is determined by

$$I = I_0 \left(e^{qV_{BE}/KT} - 1\right) \tag{2-44}$$

This current must all flow through R_F; thus, the output voltage becomes

$$V_o = R_F I$$
$$V_o = R_F I_0 (e^{qV_{BE}/KT} - 1)$$

at $T_A = 25$ °C:

$$V_o = R_F I_0 \left(e^{(38.9)V_{BE}} - 1\right)$$

Fig. 2-55 is a graph of the transfer function at 25°C, -55°C, and $+125$°C for the antilog circuit of Example 2-10. This graphically illustrates the need for temperature compensation for diode-based log circuits. The maximum allowable input signal for 10-mA diode current is readily calculated from

$$V_{BE} = \frac{KT}{q} \ln I/I_0$$

Substituting 10 mA for I yields a maximum allowable input signal of 0.718 V at 25°C.

The transfer characteristics of Fig. 2-55 illustrate the need for temperature compensation when absolute accuracy is required. The two primary contributors to the temperature drift are the KT/q and I_0 terms of the diode equation. The circuit of Fig. 2-56 can be used to compensate for these terms. The saturation current (I_0) term is reduced by adding diode D_2 and current source I_R. The current source forces a constant current through D_2, which generates the voltage V_F. If D_1 and D_2 are perfectly matched, the KT/q and I_0 terms will be equal; thus, the I_0 term will not appear in the equation for V_3, which becomes

$$V_3 = V_2 + V_F = -\frac{KT}{q}\left(\ln\frac{V_{in}}{R_1} - \ln I_0 - \ln I_R + \ln I_0\right)$$

Thus, $\qquad V_3 = -\frac{KT}{q}\ln\frac{V_{in}}{R_1 I_R}$

At this point, the only remaining temperature-dependent component is due to the T in the KT/q term. This may be removed by making the gain of the output amplifier tem-

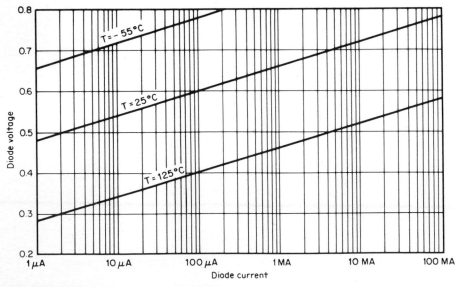

FIG. 2-55 Diode transfer function vs. temperature.

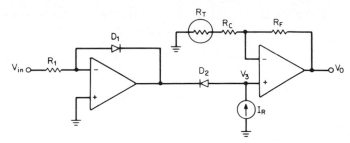

FIG. 2-56 Temperature-compensated log amplifier.

perature dependent and thus cancel the KT/q factor. This is accomplished by considering the output V_o to be

$$V_o = V_3 \left[\frac{R_F}{R_T + R_C} + 1 \right]$$

$$= \left(\frac{R_F + R_C + R_T}{R_C + R_T} \right) \left(\frac{KT}{q} \ln \frac{V_{in}}{R_1 I_R} \right)$$

If $(R_F + R_C + R_T)/(R_C + R_T)$ is chosen to have a temperature coefficient that is equal and opposite to KT/q, the temperature dependence of the diode will be compensated, yielding a final transfer function of

$$V_o = K_1 \ln(K_2 V_{in})$$

where K_1 is a constant. It is normally convenient to assign a value of 1.0 to the K_1 term, and also convert to the log base 10. This is accomplished using the logarithmic formula:

$$\frac{\ln X}{\ln 10} = \log 10 X$$

The op amp gain is set such that

$$G = 1 + \frac{R_F}{R_C + R_T} = q(KT \ln 10) = 16.9 \text{ at } 25°C$$

with the gain temperature sensitivity set to $-0.33\%/°C$. The values of R_F, R_C, and R_T may then be calculated by

$$\frac{R_F}{R_C + R_T} = 15.9$$

with the series combination of R_C and R_T chosen to have a temperature coefficient (T_R) of approximately $0.33\%/°C$. For example, if R_F was chosen to be 15.9 kΩ, the series combination of R_C and R_T would be chosen to equal 1 kΩ. The ratio of R_C to R_T may then be determined by considering R_C to have a zero temperature coefficient and R_T to have a temperature coefficient of $1\%/°C$. Thus

$$R = R_C + R_T$$

$$T_R = \frac{dR_C}{dT} \times \frac{R_C}{R} + \frac{dR_T}{dT} \times \frac{R_T}{R}$$

$$T_R = \frac{R_T}{R}$$

$$R_T = (1 \text{ k}\Omega)(0.33)$$

Therefore, $R_T = 330\ \Omega$ and $R_C = 670\ \Omega$. The resultant transfer function then becomes

$$V_o = 1 \log 10\ K_2 V_{in}$$

If $R_1 I_R$ is also set to one, the resultant transfer function becomes

$$V_o = 1 \log 10(V_{in})$$

A temperature-independent antilog amplifier may also be generated in a similar manner; however, when designing with logarithmic converters, it is important to consider whether absolute precision is required. For example, if multiplication of two input signals is to be accomplished by performing a log conversion, sum these two signals and take the antilog of the sum. The accuracy of this function is dependent solely on the tracking of the temperature coefficients of the two converters rather than the absolute accuracy of each. However, when absolute accuracy is required, log modules that are internally temperature compensated are available.

REFERENCES

Analog Devices: *Multiplier Application Guide,* Analog Devices, Norwood, Mass., 1978.

Exar: *Function Generator Data Book,* Exar Integrated Systems, Inc., Sunnyvale, Calif., 1979.

Gilbert, B.: "A Versatile Monolithic Voltage-to-Frequency Converter," *IEEE Journal of Solid-State Circuits,* SC- Dec. 1976, pp. 852–864.

Gray, P., and R. Meyer: *Analysis and Design of Integrated Circuits,* Wiley, New York, 1977.

Grebene, A.: *Analog Integrated Circuit Design,* Van Nostrand-Reinhold, New York, 1972.

Graeme, J., G. Tobey, and L. Huelsman: *Operational Amplifiers Design and Applications,* McGraw-Hill, New York, 1971.

Sheingold, D. H.: *Nonlinear Circuit Handbook,* Analog Devices, Inc., Norwood, Mass., 1976.

Chapter 3

DESIGN OF ACTIVE FILTERS USING OPERATIONAL AMPLIFIERS

Arthur B. Williams Vice President of Engineering, Research, and Development
Coherent Communications Systems Corp.
Hauppauge, N.Y.

3–1 SELECTING THE FILTER TYPE

Modern network theory has provided us with standard types of low-pass filters. These filters are characterized by a transfer function, which in the general form is expressed as

$$T(s) = \frac{e_{\text{out}}}{e_{\text{in}}} = \frac{N(s)}{D(s)} \tag{3-1}$$

The numerator $N(s)$ and denominator $D(s)$ are polynomials in s, where $s = j\omega$ and $\omega = 2\pi f$.

A transfer function can also be defined in terms of the roots of the numerator and denominator polynomials. The roots of the numerator are called zeros and the denominator roots are called poles. The filter is described as an nth order, where n is the order of the denominator polynomial.

Many types of low-pass filter families are available to the designer. Each filter within a family has a unique transfer function. The complexity and location of the poles and zeros completely define the response.

Most requirements can be satisfied by a filter chosen from one of the following families:

Butterworth	Bessel
Chebyshev	Elliptic-Function

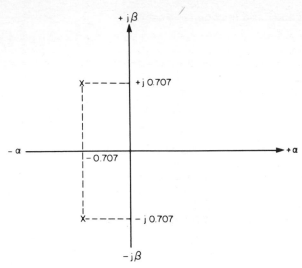

FIG. 3-1   Complex plane representation of Eq. 3-2.

A normalized $n = 2$ Butterworth transfer function is given as

$$T(s) = \frac{1}{s^2 + 1.414s + 1} \tag{3-2}$$

The two roots of the denominator polynomial are

$$s = -0.707 + j\,0.707$$

and

$$s = -0.707 - j\,0.707$$

These complex numbers can be represented in the complex frequency plane where the abscissa is α, the real part of the root, and the ordinate is β, the imaginary portion. Numerator roots (zeros) are represented as 0 and denominator roots (poles) are shown as X. The complex plane representation for Eq. 3-2 is shown in Fig. 3-1.

If we evaluate Eq. 3-2 at different values of ω, the following table can be obtained:

ω	$T\,(j\omega)$	Attenuation (dB)
0	1	0
1	$0.707 \angle -90°$	-3
2	$0.242 \angle -137°$	-12.3

The attenuation is always 3 dB at 1 rad/s for all normalized filters, except for the elliptic-function type, where the 1-rad/s attenuation is equal to the passband ripple.

When the designer is given a frequency-response specification, this requirement must also be normalized and then compared to normalized filter response curves. A satisfactory filter type and complexity can then be selected. The circuit values corresponding to the chosen filter must then be denormalized to the frequency range of operation. This denormalization procedure may also involve transformations if the ultimate requirement is other than a low-pass filter.

3–1a Frequency and Impedance Scaling

If the reactive elements of a filter are all divided by a frequency scaling factor (FSF), the resulting frequency response will occur over a new frequency range related to the original response by the same FSF. This characteristic is the basis that allows us to utilize normalized filters and their associated response curves.

The FSF is a dimensionless number defined as follows:

$$\text{FSF} = \frac{\text{desired reference frequency}}{\text{existing reference frequency}} \tag{3-3}$$

Both numerator and denominator must be expressed in identical units, generally radians/second. The reference frequency is usually the 3-dB cutoff for low-pass and high-pass filters, and the center frequency for bandpass and band-reject networks.

The following example illustrates the frequency scaling of an active low-pass filter:

Example 3–1 Frequency Scaling of Low-Pass Filter

A low-pass active filter corresponding to an $n = 2$ Butterworth type is shown in Fig. 3-2a, having the response of Fig. 3-2b. Scale this filter to a 3-dB cutoff of 100 Hz.

Solution

a) Compute FSF

$$\text{FSF} = \frac{2\,\pi\,100 \text{ rad/s}}{1 \text{ rad/s}} = 628 \tag{3-3}$$

b) Dividing all the normalized values by the FSF results in the circuit of Fig. 3-2c, having the frequency response shown in Fig. 3-2d.

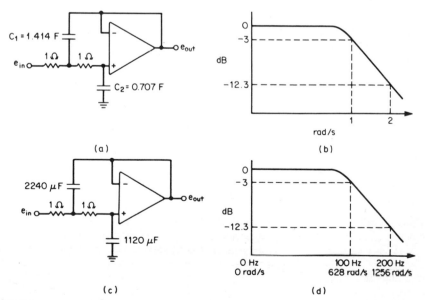

FIG. 3-2 Denormalization of Example 3-1. (*a*) Normalized low-pass active filter $n = 3$ Butterworth; (*b*) normalized frequency response; (*c*) denormalized filter; (*d*) denormalized frequency response.

The component values of Fig. 3-2c are clearly impractical, since the resistors are too low and the capacitor values are excessively large. This situation can be resolved by impedance scaling.

Any linear network, either passive or active, retains its response characteristics if all frequency-response determining elements are raised (or lowered) in impedance by the factor Z. In the case of active filters, impedance scaling can be mathematically expressed as

$$R' = ZR \qquad (3-4)$$

and

$$C' = \frac{C}{Z} \qquad (3-5)$$

where the primes denote the impedance-scaled values.

Let us now impedance scale the denormalized filter of Example 3–1. Using a Z of 10,000 we obtain the circuit of Fig. 3-3. The resulting values are certainly more practical and therefore better suited for production.

Frequency and impedance scaling are usually combined into a single operation. The combined formulas are

$$R' = R \times Z \qquad (3-6)$$

$$C' = \frac{C}{FSF \times Z} \qquad (3-7)$$

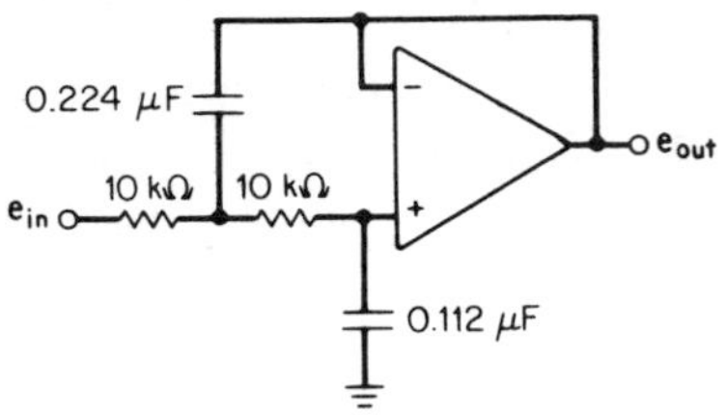

FIG. 3-3   Impedance-scaled filter of Example 3-1.

Low-Pass Normalization   A given low-pass frequency-response requirement must first be normalized so that the normalized curves can be entered to find a satisfactory filter type and complexity. The corresponding network is then scaled to the desired cutoff using the combined frequency- and impedance-scaling formulas.

To normalize a low-pass filter requirement, compute the steepness factor A_s, which is given by

$$A_s = \frac{f_s}{f_c} \qquad (3-8)$$

where f_s is the frequency corresponding to the beginning of the stopband and f_c is the passband cutoff frequency. If the attenuation corresponding to f_c is 3 dB, the steepness factor can be directly looked up on the curves (where A_s is interpreted in radians/second) to select a design that meets or exceeds the attenuation requirements.

Example 3–2   Normalization of Low-Pass Requirement

Normalize the following specification:

> Low-pass Filter
> 3 dB at 100 Hz
> 35 dB minimum at 300 Hz

Solution
Compute A_s as follows:

$$A_s = \frac{f_s}{f_c} = \frac{300 \text{ Hz}}{100 \text{ Hz}} = 3 \qquad (3-8)$$

The normalized requirement is

3 dB at 1 rad/s
35 dB at 3 rad/s

A filter can now be selected from the normalized curves that exceed 35 dB of attenuation at 3 rad/s. The corresponding values are then denormalized to a 3-dB cutoff at 100 Hz.

High-Pass Normalization   A normalized $n = 2$ Butterworth low-pass transfer function was previously given as

$$T(s) = \frac{1}{s^2 + 1.414s + 1}$$ (3-2)

A high-pass transformation can be performed by replacing s by $1/s$ in the low-pass transfer function, resulting in

$$T(s) = \frac{s^2}{s^2 + 1.414s + 1}$$ (3-9)

If we evaluate Eq. 3-9 at 0.5, 1, and ∞ rad/s we obtain

ω	$T(j\omega)$	Attenuation (dB)
0.5	$0.242 \angle 137°$	-12.3
1	$0.707 \angle 90°$	-3
∞	1	0

The response is clearly that of a high-pass filter. It should also be apparent that the low-pass attenuation determined by evaluating Eq. 3-2 now occurs at reciprocal frequencies. The high-pass transformation of a normalized low-pass transposes the low-pass attenuation to reciprocal frequencies and retains the 3-dB cutoff at 1 rads. This relationship is evident in Fig. 3-4, where both responses are compared.

The design procedure normally involves converting the high-pass requirement into a normalized low-pass specification and then directly entering the normalized curves to select an appropriate filter type.

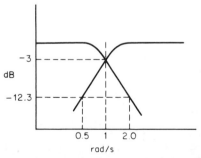

FIG. 3-4   Low-pass/high-pass relationship for normalized $n = 2$ Butterworth.

The corresponding normalized low-pass filter is then transformed into a normalized high-pass filter by replacing resistors by capacitors and capacitors by resistors using reciprocal element values. The high-pass filter is then frequency and impedance scaled to the operating frequencies.

The steepness factor for high-pass filters is given by

$$A_s = \frac{f_c}{f_s} \qquad\qquad (3\text{-}10)$$

The following example illustrates the normalization of a high-pass specification:

Example 3–3   Normalization of High-Pass Requirement

Normalize the following specification:

> High-pass filter
> 3 dB at 400 Hz
> 30 dB minimum at 100 Hz

Solution
Compute A_s

$$A_s = \frac{f_c}{f_s} = \frac{400 \text{ Hz}}{100 \text{ Hz}} = 4 \qquad\qquad (3\text{-}10)$$

The corresponding normalized low-pass filter should have the following requirement:

> 3 dB at 1 rad/s
> 30 dB minimum at 4 rad/s

Bandpass Normalization   Bandpass filters can be classified into two categories: narrow band and wide band. Wide band filters are comprised of a cascade of low-pass and high-pass filters. This approach is practical when the ratio of upper to lower passband cutoff frequencies is in excess of a frequency octave (2:1).

To normalize a wide-band requirement, separate the specification into independent low-pass and high-pass requirements. Normalize the low-pass and high-pass specifications individually, design both filters independently, and cascade the low-pass and high-pass networks to obtain the composite response.

Example 3–4   Normalization of Wide-Band Bandpass Specification

Normalize the following bandpass filter requirement:

> 3 dB at 50 and 100 Hz
> 40 dB minimum at 20 and 200 Hz

Solution
a) Determine the ratio of upper to lower cutoff:

$$\frac{100 \text{ Hz}}{50 \text{ Hz}} = 2$$

The filter should be a wide-band type.
b) Separate the requirement into low-pass and high-pass specifications and compute the steepness factors:

High-pass filter	Low-pass filter
3 dB at 50 Hz	3 dB at 100 Hz
40 dB minimum at 20 Hz	40 dB minimum at 200 Hz
$A_s = 2.5$ (3-10)	$A_s = 2$ (3-8)

Low-pass and high-pass filters are then individually designed and cascaded.

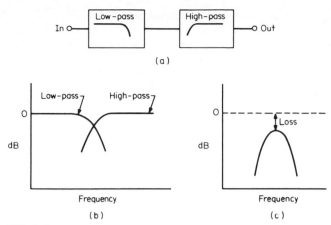

FIG. 3-5 Limitation of wide-band approach for bandpass filters. (*a*) Cascade of low-pass and high-pass filters; (*b*) frequency response of individual filters; (*c*) composite response.

If the ratio of upper to lower cutoff is less than approximately 2, the filter cannot be designed as a cascade of separate low-pass and high-pass filters. The reason for this is evident from Fig. 3-5. As the ratio of upper to lower cutoff decreases, the loss at the center frequency increases and can become prohibitive for near-unity ratios.

If $s + 1/s$ is substituted for s in the normalized low-pass transfer function, the result would be a bandpass filter having a center frequency of 1 rad/s. The frequency response of the low-pass filter will be transformed into the *bandwidth* of the bandpass filter. Each pole and zero of the low-pass filter will be transformed into a *pair* of poles and zeros in the bandpass case. Fig. 3-6 shows the relationship between a low-pass filter and the transformed bandpass equivalent.

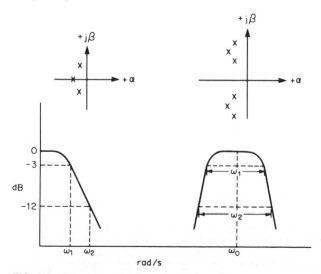

FIG. 3-6 Low-pass-to-bandpass transformation.

To design a bandpass filter the following sequence is involved:

1. Convert the bandpass requirement into a normalized low-pass specification.
2. Select a normalized low-pass filter from the normalized frequency-response curves.
3. Transform the normalized low-pass parameters into a bandpass filter at the required center frequency and bandwidth.

The response shape of a general band-pass filter is shown in Fig. 3-7. The center frequency is normally defined as

$$f_0 = \sqrt{f_L f_U} \qquad (3\text{-}11)$$

where f_L and f_U are the lower and upper passband limits, respectively, usually the 3-dB points. For the more general case

$$f_0 = \sqrt{f_1 f_2} \qquad (3\text{-}12)$$

where f_1 and f_2 correspond to any two frequencies having equal attenuation.

These equations imply geometric symmetry; that is, the entire curve below f_0 is the mirror image of the portion above f_0 when the frequency axis is *logarithmic*.

A useful parameter of bandpass filters is the frequency selectivity factor Q, which is given by

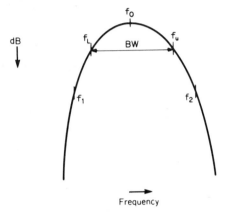

FIG. 3-7 General bandpass-filter response shape.

$$Q = \frac{f_0}{\text{BW}} \qquad (3\text{-}13)$$

where BW is the passband bandwidth. The higher the Q, the narrower the filter.

For Q's of 10 or more the filter response approaches arithmetic symmetry. The response is then symmetrical about f_0 when plotted on a *linear* frequency axis. Therefore

$$f_0 = \frac{f_1 + f_2}{2} \qquad (3\text{-}14)$$

To normalize a bandpass requirement, the specification must first be manipulated to make it geometrically symmetrical. At equal attenuation points the corresponding frequencies must satisfy

$$f_1 f_2 = f_0^2 \qquad (3\text{-}15)$$

which is an alternate form of Eq. 3-12. (For Q's well in excess of 10, Eq. 3-14 would apply.)

The given specification is first modified by calculating the corresponding opposite geometric frequency for each specified stopband frequency. Two frequency pairs will then result from each original pair specified at a particular attenuation. The pair with the lesser separation represents the more severe requirement and should be retained. All response requirements must be converted to bandwidths by subtracting the lower frequency from the upper. A steepness factor A_s can now be defined in terms of bandwidth as follows:

$$A_s = \frac{\text{stopband bandwidth}}{\text{passband bandwidth}} \qquad (3\text{-}16)$$

The normalized low-pass curves can now be entered, using the A_s directly in radians/second to select a satisfactory low-pass filter. The corresponding values are then transformed into the bandpass network.

The following example illustrates the normalization of a bandpass filter requirement.

Example 3–5   Normalization of Bandpass Requirement

Normalize the following bandpass parameters:

> Bandpass Filter
> Center Frequency of 100 Hz
> 3 dB at ± 15 Hz (85 Hz, 115 Hz)
> 40 dB at ± 30 Hz (70 Hz, 130 Hz)

Solution

a) Compute geometric center frequency

$$f_0 = \sqrt{f_L f_U} = \sqrt{85 \times 115} = 98.9 \text{ Hz} \qquad (3\text{-}11)$$

b) Compute two geometrically related stopband frequency pairs for each pair specified at a given attenuation. Let

$$f_1 = 70 \text{ Hz}$$

$$f_2 - \frac{f_0^2}{f_1} = 139.7 \text{ Hz} \qquad (3\text{-}15)$$

$$f_2 - f_1 = 139.7 \text{ Hz} - 70 \text{ Hz} = 69.7 \text{ Hz}$$

Let

$$f_2 = 130 \text{ Hz}$$

$$f_1 = \frac{f_0^2}{f_2} = 75.2 \text{ Hz} \qquad (3\text{-}15)$$

$$f_2 - f_1 = 130 \text{ Hz} - 75.2 \text{ Hz} = 54.8 \text{ Hz}$$

The second pair of frequencies has the lesser separation and therefore represents the more severe requirement.

c) Geometrically symmetrical bandpass filter specification:

$$f_0 = 98.9 \text{ Hz}$$

$$\text{BW}_{3 \text{ dB}} = 30 \text{ Hz}$$

$$\text{BW}_{40 \text{ dB}} = 54.8 \text{ Hz}$$

Compute steepness factor

$$A_s = \frac{\text{stopband bandwidth}}{\text{passband bandwidth}} = \frac{54.8 \text{ Hz}}{30 \text{ Hz}} = 1.83 \qquad (3\text{-}16)$$

The corresponding response curves are shown in Fig. 3-8.

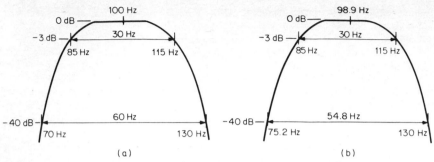

FIG. 3-8 Frequency-response requirements of Example 3-5. (*a*) Given filter requirement; (*b*) geometrically symmetrical requirement.

Band-Reject Filters Band-reject filters with a ratio of upper to lower cutoff of an octave or more can be designed as a combination of a low-pass and high-pass filter, where both inputs are paralleled and the outputs are summed. This configuration is illustrated in Fig. 3-9.

The basic assumption of this technique is that each filter has sufficient rejection in the band of the other to prevent interaction when the outputs are combined. If inadequate separation exists, the effect shown in Fig. 3-9*c* will occur.

Another category of band-reject filters are null networks. These are discussed in Sec. 3-5.

Transient Characteristics The frequency response of a filter does not completely define the network's transmission properties. A major factor in the operation of a filter is its transient characteristics.

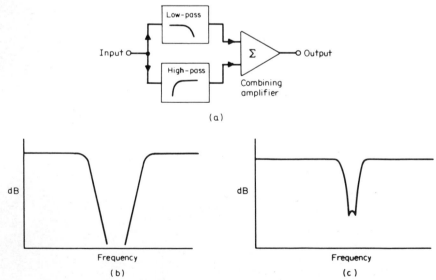

FIG. 3-9 Wide-band band-reject filters. (*a*) Configuration; (*b*) frequency response; (*c*) limitation of wide-band technique.

A complex waveform applied to a filter will be distorted if the various spectral components that comprise this waveform undergo unequal amounts of delay. This distortion may at times be unacceptable, especially if the filter is used for data transmission.

The delay incurred by the sidebands of a modulated signal in passing through a filter is called ''group delay.'' The classical definition for this parameter is

$$T_{gd} = \frac{d\phi(\omega)}{d\omega} \tag{3-17}$$

Group delay is the derivative of phase ϕ versus frequency ω; so for a constant delay, a linear phase shift is required.

Group delay curves for the standard response families are presented in the next section. To denormalize these curves divide the delay axis by $2\pi f_c$, where f_c is the 3-dB cutoff frequency. The frequency axis is multiplied by f_c.

It is important to recognize that the absolute delay is generally of little significance. It is the *delay variation* over the band of interest that disperses the various spectral components of the signal.

A more direct indication of the distortion introduced to a modulated signal by a low-pass filter is the network's step response. These curves indicate the time response to a voltage step of unity amplitude. For optimum performance, especially when the input consists of an abrupt amplitude step, the overshoot should be minimized and the transient part of the step response should rapidly decay. To denormalize these curves, which are presented in the next section, the time axis is divided by $2\pi f_c$. When a low-pass filter is transformed to any other type, these characteristics are not retained.

3–1b Characteristics of Standard Responses

Filter responses can be defined in terms of transfer functions. These transfer functions have been standardized into Butterworth, Chebyshev, Bessel, and Elliptic-Function categories. Each type has mathematical properties that are unique. These characteristics are briefly discussed in this section.

Butterworth   The Butterworth family of low-pass filters are considered *all-pole* networks, since the transfer function has either unity or a constant multiplier in the numerator, which indicates the absence of zeros.

The unique property of this family is that the roots (poles) of the transfer function denominator all fall on a circle having a radius of unity when plotted in the $j\omega$ plane.

These filters have moderately good transient characteristics. The frequency response is extremely flat near DC and asymptotically approaches a rate of $n \times 6$ decibels per octave rolloff in the stopband region. In the vicinity of cutoff the response is somewhat rounded. Nevertheless, the Butterworth family is widely used, since these designs usually result in more practical component values and less critical tolerances than most other filter types.

The attenuation of normalized Butterworth low-pass filters up to $n = 10$ is illustrated in Fig. 3-10. The corresponding delay and step response are given in Figs. 3-11 and 3-12, respectively. Butterworth pole locations and active low-pass values are tabulated in Tables 3-1 and 3-2 in Sec. 3-7.

Chebyshev   A steeper rate of descent to the stopband than Butterworth filters is obtainable from the Chebyshev family. This is obtained at the expense of passband ripples, as shown in Fig. 3-13.

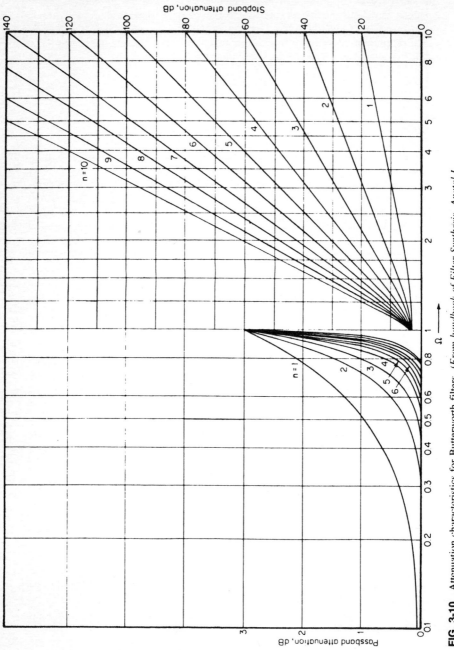

FIG. 3-10 Attenuation characteristics for Butterworth filters. (*From handbook of Filter Synthesis. Anatol I. Zverev. John Wiley & Sons, Inc.. New York, 1967. Used by permission.*)

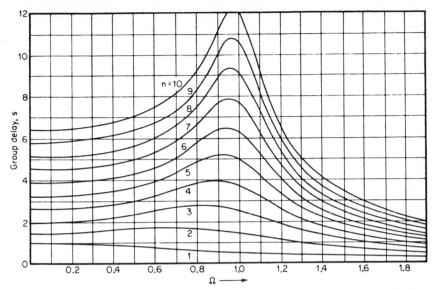

FIG. 3-11 Group-delay characteristics for Butterworth filters. (*From Handbook of Filter Synthesis, Anatol I. Zverev, John Wiley & Sons, Inc., New York, 1967. Used by permission.*)

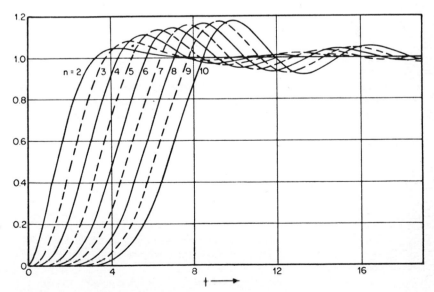

FIG. 3-12 Step response for Butterworth filters. (*From Handbook of Filter Synthesis, Anatol I. Zverev, John Wiley & Sons, Inc., New York, 1967. Used by permission.*)

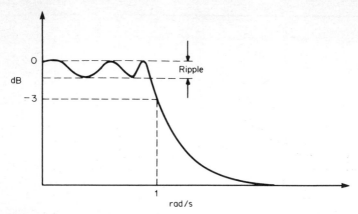

FIG. 3-13 Normalized Chebyshev low-pass filter response.

For a particular filter order n, increasing the ripple magnitude increases the rate of descent from passband to stopband and makes the response in the cutoff region more rectangular. However, the transient properties are poorer than the Butterworth family and rapidly deteriorate with increasing ripple.

The poles are located on an ellipse. In contrast to Butterworth filters, the stopband rate of descent exceeds $n \times 6$ decibels per octave for the first few octaves and eventually attains $n \times 6$ decibels per octave well into the stopband.

The normalized frequency response, delay, and step response for Chebyshev filters having ripples of 0.1 dB and 0.5 dB are shown in Figs. 3-14 through 3-19. Tables 3-3 through 3-6 provide pole locations and active low-pass values.

Bessel When the major consideration is the faithful reproduction of a pulsed waveform, the Bessel family of low-pass networks is desirable. These filters provide an excellent approximation to a constant delay, especially for high-order n's. The step response exhibits no overshoot.

The frequency-response characteristics are extremely poor in comparison with the Butterworth and Chebyshev types. The passband is rounded and the rate of rolloff in the first few octaves is inferior. Nevertheless, the superior transient properties make this family extremely useful.

The normalized frequency-response, delay, and step-response characteristics are given in Figs. 3-20 through 3-22. Pole locations and active low-pass values are provided in Tables 3-7 and 3-8, respectively.

Elliptic-Function All of the previous network families discussed are all-pole types, that is, the numerator of the transfer function is unity or a constant multiplier. The Elliptic-Function filter type contains zeros as well as poles. The introduction of zeros results in finite frequencies of infinite attenuation. These poles and zeros are located so that the passband exhibits equal ripples similar to the Chebyshev family and the stopband has equal return lobes.

The introduction of transmission zeros results in the steepest possible transition from the passband to the stopband for a given number of poles. Fig. 3-23 compares an $n = 3$ Butterworth, Chebyshev, and Elliptic-Function low-pass filter. The Elliptic-Function filter is obviously steeper than the other types.

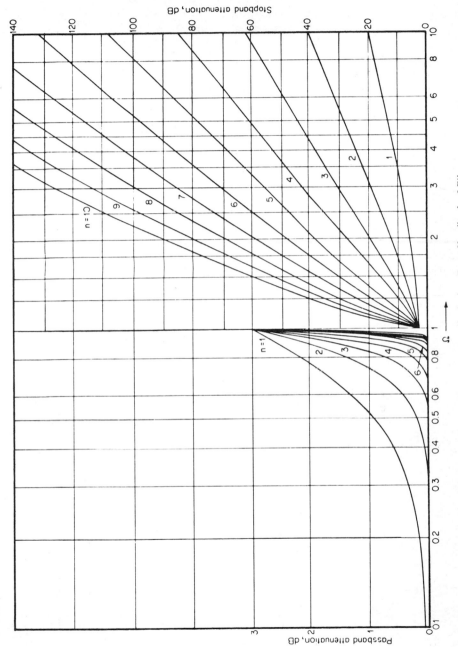

FIG. 3-14 Attenuation characteristics for Chebyshev filters with 0.1-dB ripple. *(From Handbook of Filter Synthesis, Anatol I. Zverev, John Wiley & Sons, Inc., New York, 1967. Used by permission.)*

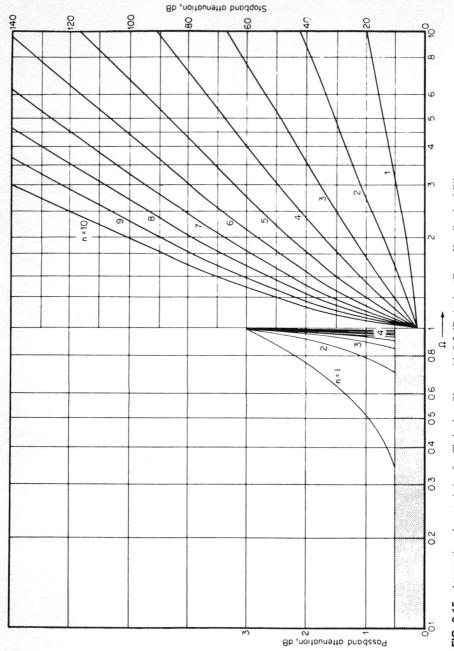

FIG. 3-15 Attenuation characteristics for Chebyshev filters with 0.5-dB ripple. (*From Handbook of Filter Synthesis, Anatol I. Zverev, John Wiley & Sons, Inc., New York, 1967. Used by permission.*)

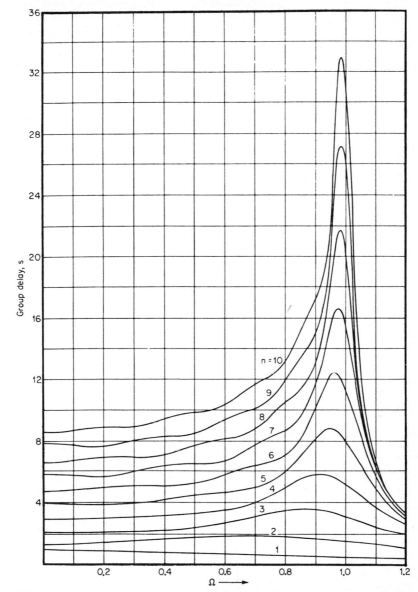

FIG. 3-16 Group-delay characteristics for Chebyshev filters with 0.1-dB ripple. (*From Handbook of Filter Synthesis, Anatol I. Zverev, John Wiley & Sons, Inc., New York, 1967. Used by permission.*)

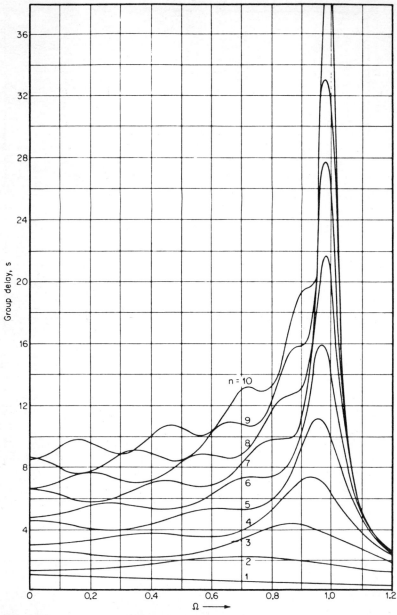

FIG. 3-17 Group-delay characteristics for Chebyshev filters with 0.5-dB ripple. (*From Handbook of Filter Synthesis, Anatol I. Zverev, John Wiley & Sons, Inc., New York, 1967. Used by permission.*)

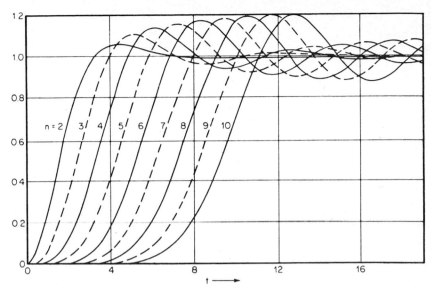

FIG. 3-18 Step response for Chebyshev filters with 0.1-dB ripple. (*From Handbook of Filter Synthesis, Anatol I. Zverev, John Wiley & Sons, Inc., New York, 1967. Used by permission.*)

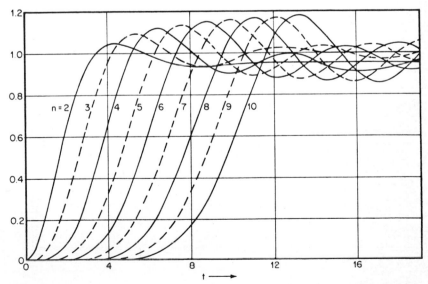

FIG. 3-19 Step response for Chebyshev filters with 0.5-dB ripple. (*From Handbook of Filter Synthesis, Anatol I. Zverev, John Wiley & Sons, Inc., New York, 1967. Used by permission.*)

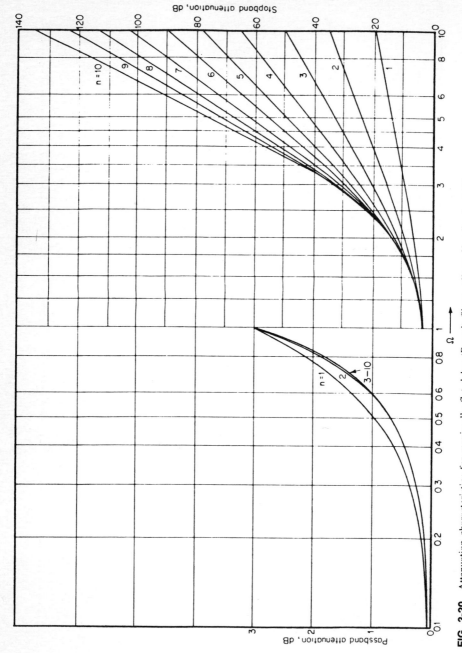

FIG. 3-20 Attenuation characteristics for maximally flat delay (Bessel) filters. *(From Handbook of Filter Synthesis, Anatol I. Zverev, John Wiley & Sons, Inc., New York, 1967. Used by permission.)*

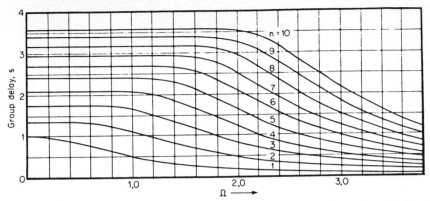

FIG. 3-21 Group-delay characteristics for maximally flat delay (Bessel) filters. (*From Handbook of Filter Synthesis, Anatol I. Zverev, John Wiley & Sons, Inc., New York, 1967. Used by permission.*)

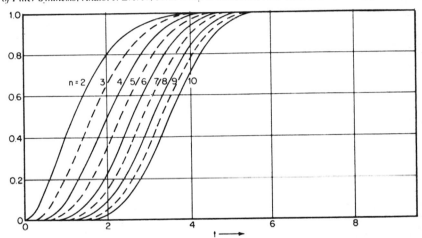

FIG. 3-22 Step response for maximally flat delay (Bessel) filters. (*From Handbook of Filter Synthesis, Anatol I. Zverev, John Wiley & Sons, Inc., New York, 1967. Used by permission.*)

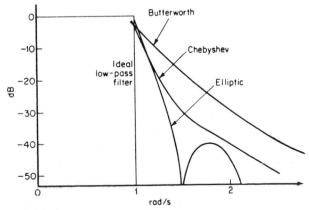

FIG. 3-23 Comparison of $n = 3$ Butterworth, Chebyshev, and Elliptic-Function filters.

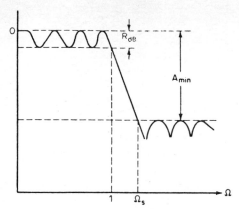

FIG. 3-24   Normalized Elliptic-Function low-pass filter response.

The faster rate of rolloff is achieved at the cost of return lobes and more complex circuitry. The return lobes are normally acceptable, provided that the minimum stopband attenuation requirements are met. Although each filter section is more complex than the other types, fewer sections will be required for the same magnitude of attenuation. For steep filter requirements, this family is a necessity.

A normalized Elliptic-Function low-pass filter response is shown in Fig. 3-24. The following definitions apply:

$$R_{dB} = \text{passband ripple in decibels}$$
$$A_{min} = \text{minimum stopband attenuation in decibels}$$
$$\Omega_s = \text{lowest stopband frequency at which } A_{min} \text{ occurs}$$

The passband attenuation at the 1-rad/s cutoff is equal to the ripple rather than 3 dB, as in the other families.

The Elliptic-Function filters tabulated in Table 3-9 are classified by n, R_{dB}, A_{min}, and Ω_s. Since the last three parameters completely define the passband and stopband limits, normalized response curves are not required.

3–2   DESIGN OF LOW-PASS FILTERS

Active low-pass filters are designed by first normalizing the design requirement and then selecting the filter type from the normalized response curves given in Sec. 3-1. The corresponding filter values given in Sec. 3-7 are then denormalized by frequency and impedance scaling.

The design of active filters can also be accomplished directly from the poles (and zeros) using closed-form formulas. This method offers some additional degrees of freedom and will also be illustrated.

3–2a   All-Pole Filters

A passive RC network has poles that are restricted to the negative real axis of the complex frequency plane. In order to obtain the complex poles required by all-pole transfer functions, active elements must be introduced. Op amps configured as voltage followers or voltage amplifiers are used to provide feedback to achieve the complex poles required.

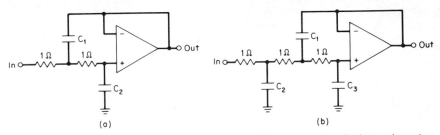

FIG. 3-25 Unity-gain active low-pass configurations. (*a*) Two-pole section; (*b*) three-pole section.

Unity-Gain Configuration Two active low-pass filter sections are shown in Fig. 3-25. These configurations consist of voltage followers imbedded in second- and third-order RC networks. The two-pole section provides a pair of complex poles, whereas the three-pole section realizes a pair of complex poles along with a single real pole.

For odd-order filters, $(n-3)/2$ two-pole sections are required and a single three-pole section is needed, since odd-order transfer functions require a single real pole. In the case of even-order filters, $n/2$ two-pole sections are used. Since each section has a low output impedance because of the voltage follower, filter sections can be directly cascaded with no interaction.

Each section has unity gain at DC. The response of individual sections may have gains in excess of unity within the passband where these sections provide sharp peaks, but the composite response will always correspond to the associated filter type.

In the normalized active low-pass values tabulated in Sec. 3-7, all resistors are assumed to be 1 Ω and the magnitudes of C_1, C_2, and C_3 are given. To design a low-pass filter, the following sequence is followed:

1. Normalize the low-pass requirement as illustrated in Sec. 3-1a.
2. Select a standard response type and complexity from the curves of Sec. 3-1b.
3. Denormalize the corresponding values given in Sec. 3-7 to the required cutoff frequency and a convenient impedance level by the denormalization formula

$$C' = \frac{C}{\text{FSF} \times Z} \tag{3-7}$$

where FSF is the frequency scaling factor $2\pi f_c$ and Z is the impedance scaling factor Z. All resistors are multiplied by Z, resulting in equal resistors throughout of Z ohms. Independent Z's can be selected for each section, since individual circuits are isolated. Also, the sequence of sections can be rearranged.

The following example illustrates the design of an active low-pass filter using the unity-gain structure.

Example 3–6   Unity-Gain Low-Pass Filter

Design an active low-pass filter having the following response:

    100 Hz   3 dB
    350 Hz   70 dB minimum
    Unity gain

Solution

a) Compute low-pass steepness factor

$$A_s = \frac{f_s}{f_c} = \frac{350 \text{ Hz}}{100 \text{ Hz}} = 3.5 \qquad (3\text{-}8)$$

b) Choose a Chebyshev 0.5-dB filter type. The curves of Fig. 3-15 indicate that for this family, an $n = 5$ design meets the normalized requirement of 70-dB minimum attenuation at 3.5 rad/s.

c) The corresponding normalized values are given in Table 3-6. A three-pole section followed by a two-pole section is shown in Fig. 3-26a.

d) Selecting an arbitrary impedance scaling factor Z of 5×10^4 and using an FSF of $2\pi f_c$, or 628, the resulting element values are as follows.

Three-Pole Section

$$C'_1 = \frac{C}{\text{FSF} \times Z} = \frac{6.842}{628 \times 5 \times 10^4} = 0.218 \text{ }\mu\text{F} \qquad (3\text{-}7)$$

$$C'_2 = 0.106 \text{ }\mu\text{F}$$

$$C'_3 = 0.00966 \text{ }\mu\text{F}$$

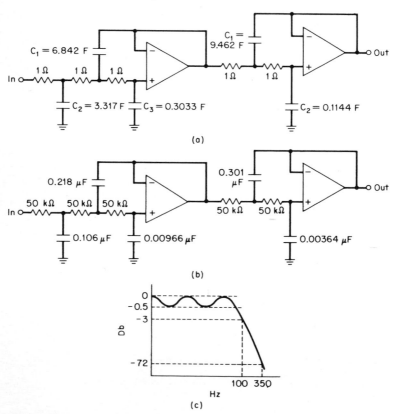

(a)

(b)

(c)

FIG. 3-26 Low-pass filter of Example 3-6. (*a*) Normalized fifth-order 0.5-dB Chebyshev low-pass filter; (*b*) denormalized filter; (*c*) frequency response.

Two-Pole Section

$$C_1' = 0.301 \ \mu F$$
$$C_2' = 0.00364 \ \mu F$$

All 1 Ω resistors are multiplied by Z, resulting in equal resistors throughout of 50 kΩ. The denormalized circuit is shown in Fig. 3-26b.

VCVS Uniform Capacitor Structure   The unity-gain low-pass filter configuration of the previous section suffers from unequal capacitor values. This results in some inconvenience, since either special nonstandard values must be ordered or two or more standard values are to be paralleled.

The voltage-controlled voltage-source (VCVS) circuit of Fig. 3-27 features equal capacitors that can be chosen as standard values. Instead of the voltage follower of the unity-gain low-pass configuration, the amplifier is configured for a gain of two. Each section is a second-order type, so $n/2$ sections are required for a nth-order filter with even n's. The element values are computed as follows: select C, then

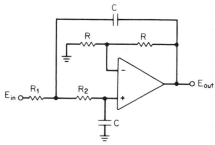

FIG. 3-27   VCVS uniform capacitor all-pole low-pass filter.

$$R_1 = \frac{1}{4\pi f_c \, \alpha \, C} \qquad (3\text{-}18)$$

and

$$R_2 = \frac{\alpha}{\pi f_c C(\alpha^2 + \beta^2)} \qquad (3\text{-}19)$$

where α and β are the normalized real and imaginary coordinates of the pole pair, and R may be any convenient value.

Example 3–7   VCVS All-Pole Low-Pass Filter

Design an active low-pass filter to meet the following requirements:

$$
\begin{array}{ll}
100 \ \text{Hz} & 3 \ \text{dB} \\
300 \ \text{Hz} & 40 \ \text{dB minimum}
\end{array}
$$

In addition, all capacitors should be 0.01 μF.
Solution
a) Compute the steepness factor

$$A_s = \frac{f_s}{f_c} = \frac{300 \ \text{Hz}}{100 \ \text{Hz}} = 3 \qquad (3\text{-}8)$$

Using Fig. 3-14, a fourth-order 0.1-dB Chebyshev low-pass filter will satisfy the normalized attenuation requirements.
b) The corresponding pole positions are given in Table 3–3 and are as follows:

$$
\begin{array}{ll}
\alpha = 0.2177 & \beta = 0.9254 \\
\text{and} \qquad \alpha = 0.5257 & \beta = 0.3833
\end{array}
$$

c) The element values are found from

Section 1 ($\alpha = 0.2177$, $\beta = 0.9254$)

$$R_1 = \frac{1}{4\pi f_c \alpha C} = \frac{1}{4\pi \times 100 \times 0.2177 \times 10^{-8}} = 365 \text{ k}\Omega \qquad (3\text{-}18)$$

$$R_2 = \frac{\alpha}{\pi f_c C(\alpha^2 + \beta^2)} = \frac{0.2177}{\pi \times 100 \times 10^{-8}(0.2177^2 + 0.9254^2)} \qquad (3\text{-}19)$$

$$= 76.7 \text{ k}\Omega$$

Section 2 ($\alpha = 0.5257$, $\beta = 0.3833$)

$$R_1 = 151 \text{ k}\Omega \qquad (3\text{-}18)$$
$$R_2 = 395 \text{ k}\Omega \qquad (3\text{-}19)$$

The resulting filter is shown in Fig. 3-28 using standard 1% resistor values. The overall circuit gain is 2^2 or 4.

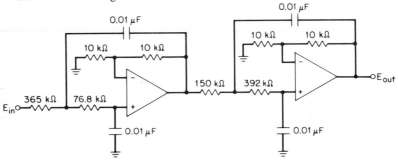

FIG. 3-28   Uniform capacitor filter of Example 3-7.

For odd-order low-pass filters, a real pole must also be implemented. This can be accomplished using the circuit of Fig. 3-29. For a normalized real pole α_0 and a preferred capacitance C, the value of R_1 is given by

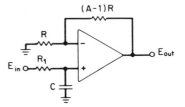

$$R_1 = \frac{1}{2\pi f_c \alpha_0 C} \qquad (3\text{-}20)$$

The magnitude of R can be any convenient value and A corresponds to the desired circuit gain. For unity gain the amplifier configuration reduces to a voltage follower.

FIG. 3-29   Real pole configuration.

The VCVS uniform capacitor structure offers the convenience of nearly arbitrary capacitor values at the expense of additional resistors. The circuit is somewhat more sensitive to component tolerances when compared to the unity-gain configuration, but is useful for most general filtering requirements.

3–2b   Elliptic-Function Filters

Elliptic-Function low-pass filters were first discussed in Sec. 3-1. They contain zeros as well as poles. These zeros begin just outside the passband and force the response to rapidly descend into the stopband beyond cutoff.

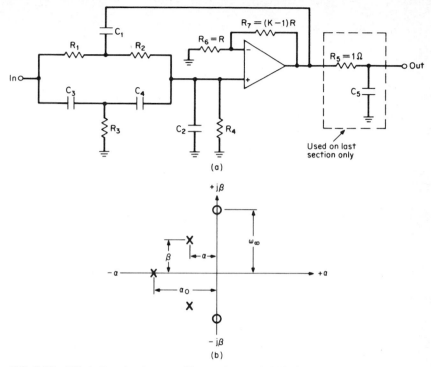

FIG. 3-30 Elliptic-Function low-pass filter section. (*a*) VCVS circuit configuration; (*b*) pole-zero pattern.

Design Method Using Tabulated Values   The schematic of an Elliptic-Function low-pass filter section is illustrated in Fig. 3-30*a*. This section provides a pair of complex conjugate poles and a pair of imaginary zeros, as shown in Fig. 3-30*b*. The R_5C_5 network introduces a single real pole at α_0 and is required only after the last section. The total number of sections for a complete filter is $(n-1)/2$, where n defines the order of the filter. Only odd-order normalized networks are tabulated in Table 3-9 to provide the most efficient utilization of a fixed number of operational amplifiers.

The values in Table 3-9 correspond to Fig. 3-30 and are followed by the magnitude of overall circuit gain and the pole-zero locations. The circuit is normalized for a cutoff of 1 rad/s, as illustrated in Fig. 3-24. The circuit can be denormalized in the same manner as the previous low-pass section types; that is, by dividing all capacitors by $Z \times$ FSF and multiplying all resistors by Z, except for R_6 and R_7, where K is determined from Table 3-9, and R may be any convenient value.

The following example demonstrates the design of a VCVS Elliptic-Function low-pass filter.

Example 3–8   VCVS Elliptic-Function Low-Pass Filter

Design an active low-pass filter to meet the following parameters:

1-dB maximum ripple below 1000 Hz
38-dB minimum attenuation above 2950 Hz

Solution

a) Compute steepness factor

$$A_s = \frac{f_s}{f_c} = \frac{2950 \text{ Hz}}{1000 \text{ Hz}} = 2.95 \tag{3-8}$$

b) Select an Elliptic-Function low-pass filter from Table 3-9 that makes the transition from less than 1 dB to more than 38 dB within a frequency ratio of 2.95.

The following network will meet these parameters:

$$n = 3$$
$$R_{dB} = 0.28 \text{ dB}$$
$$\Omega_s = 2.924$$
$$A_{min} = 39.48 \text{ dB}$$

The normalized circuit from Table 3-9 is shown in Fig. 3-31a, where

$$R_7 = (K - 1)R = 0.410R \text{ with } K = 1.410$$

c) Frequency and impedance scale the normalized filter where FSF $= 2\pi f_c = 6280$ and Z is arbitrarily chosen at 5×10^3. The resulting denormalized values are

$$R_1' = R_1 \times Z = 1860 \text{ }\Omega$$
$$R_2' = 3719 \text{ }\Omega$$
$$R_3' = 10.71 \text{ k}\Omega$$
$$R_4' = 48.19 \text{ k}\Omega$$
$$R_5' = 5 \text{ k}\Omega$$
$$C_1' = \frac{C_1}{\text{FSF} \times Z} = 0.113 \text{ }\mu\text{F} \tag{3-7}$$
$$C_2' = 0.0250 \text{ }\mu\text{F}$$
$$C_3' = 8690 \text{ pF}$$
$$C_4' = 4345 \text{ pF}$$
$$C_5' = 0.0408 \text{ }\mu\text{F}$$

Selecting R at 10 kΩ the feedback resistors are

$$R_6 = 10 \text{ k}\Omega$$
$$R_7 = 4100 \text{ }\Omega$$

The section zero is found from

$$f_\infty = \omega_\infty \times f_c = 3.35 \times 1000 \text{ Hz} = 3350 \text{ Hz}$$

The resulting network is shown in Fig. 3-31b, with the response of Fig. 3-31c.

Design Method Using Standard Capacitances   The element values for the VCVS low-pass Elliptic-Function filter configuration of Fig. 3-30 can be directly computed using a design method that allows additional degrees of freedom in capacitor selection. In comparison to using Table 3-9, this circuit is more sensitive to component tolerances, but the convenience of using standard capacitor values is a significant advantage in many instances.

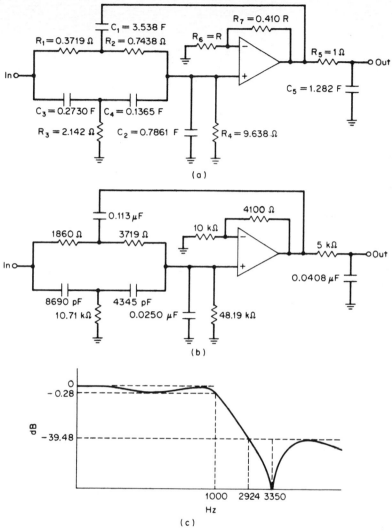

FIG. 3-31 Design of Elliptic-Function filter of Example 3-8. (*a*) Normalized circuit from Table 3-9; (*b*) denormalized filter; (*c*) frequency response.

The design proceeds as follows: First compute

$$a = \frac{2\alpha}{\sqrt{\alpha^2 + \beta^2}} \tag{3-21}$$

$$b = \frac{\omega_x^{\,2}}{\alpha^2 + \beta^2} \tag{3-22}$$

$$c = \text{FSF}\,\sqrt{\alpha^2 + \beta^2} \tag{3-23}$$

where α, β, and ω_x are obtained from Table 3-9 and FSF is $2\pi f_c$.

The element values are computed as follows: Select C, then

$$C_1 = C \tag{3-24}$$

$$C_3 = C_4 = \frac{C_1}{2} \tag{3-25}$$

$$C_2 \geqq \frac{C_1(b - 1)}{4} \tag{3-26}$$

$$R_3 = \frac{1}{cC_1\sqrt{b}} \tag{3-27}$$

$$R_1 = R_2 = 2R_3 \tag{3-28}$$

$$R_4 = \frac{4\sqrt{b}}{cC_1(1 - b) + 4cC_2} \tag{3-29}$$

$$K = 2 + \frac{2C_2}{C_1} - \frac{a}{2\sqrt{b}} + \frac{2}{C_1\sqrt{b}}\left(\frac{1}{cR_4} - aC_2\right) \tag{3-30}$$

$$\text{Section gain} = \frac{bKC_1}{4C_2 + C_1} \tag{3-31}$$

Capacitor C_5 is determined from the denormalized real pole by

$$C_5 = \frac{1}{\text{FSF} \times R_5 \times \alpha_0} \tag{3-32}$$

where both R and R_5 can be arbitrarily chosen.

Example 3–9   Elliptic-Function Low-Pass Filter Using Standard Capacitor Values

Design an active Elliptic-Function low-pass filter corresponding to $n = 3$, 0.28-dB ripple, and $\Omega_s = 2.924$ from Table 3-9, using standard capacitor values. Use a design cutoff frequency of 100 Hz.

Solution
a) The normalized poles and zeros are given as

$$\alpha = 0.3449$$
$$\beta = 1.0860$$
$$\omega_x = 3.350$$
$$\alpha_0 = 0.7801$$

b) The element values are computed as follows, where $\text{FSF} = 2\pi f_c = 628.3$:

$a = 0.6053$	(3-21)
$b = 8.6437$	(3-22)
$c = 716$	(3-23)
Select $C_1 = C = 0.1\ \mu\text{F}$	(3-24)
$C_3 = C_4 = 0.05\ \mu\text{F}$	(3-25)
$C_2 \geqq 0.191\ \mu\text{F}$	(3-26)

Let $C_2 = 0.22\ \mu F$

$R_3 = 4751\ \Omega$ (3-27)

$R_1 = R_2 = 9502\ \Omega$ (3-28)

$R_4 = 142\ k\Omega$ (3-29)

$K = 5.458$ (3-30)

Let $R = R_5 = 10\ k\Omega$

Then $C_5 = 0.204\ \mu F$ (3-32)

The resulting circuit is shown in Fig. 3-32.

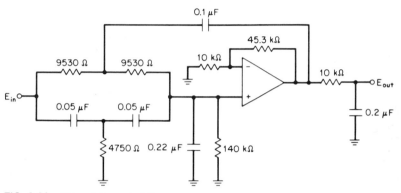

FIG. 3-32   Filter of Example 3-9.

3–3   HIGH-PASS FILTERS

After a high-pass filter requirement is normalized using the procedure of Sec. 3-1, a suitable low-pass filter type must be chosen from the normalized data and then transformed into a high-pass filter. This is accomplished by simply replacing each resistor R by a capacitor of $1/R$ F. Each capacitor C is replaced by a resistor of $1/C\ \Omega$. The normalized high-pass filter is then frequency scaled to the desired cutoff f_c and simultaneously imped-ance scaled to a convenient impedance level.

3–3a   All-Pole High-Pass Filters

Sec. 3-2 discussed the design of two types of all-pole low-pass filter configurations that were the unity-gain circuit and the VCVS uniform capacitor structure. The unity-gain low-pass configuration is best suited for transformation into a high-pass type because the equal resistors will result in a high-pass filter that has equal capacitors after the trans-formation. By selecting an appropriate impedance scaling factor the capacitor can be forced to a standard value.

The design of an all-pole high-pass filter is illustrated in the following example.

Example 3–10   Design of an All-Pole High-Pass Filter

A high-pass filter is required having the following parameters:

$$
\begin{array}{ll}
100 \text{ Hz} & 3 \text{ dB} \\
33.3 \text{ Hz} & 60 \text{ dB minimum}
\end{array}
$$

Also it is desirable to use $0.015 \ \mu\text{F}$ capacitors throughout.

Solution

a) Normalize the specification by computing the steepness factor

$$
A_s = \frac{f_c}{f_s} = \frac{100 \text{ Hz}}{33.3 \text{ Hz}} = 3 \tag{3-10}
$$

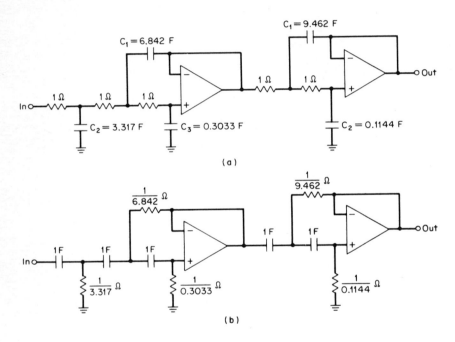

(a)

(b)

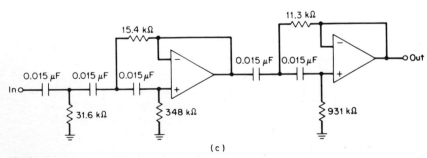

(c)

FIG. 3-33 All-pole high-pass filter of Example 3-10. (*a*) Normalized low-pass filter; (*b*) high-pass transformation; (*c*) frequency- and impedance-scaled high-pass filter.

b) The curves of Fig. 3-15 indicate that a fifth-order 0.5-dB Chebyshev low-pass filter will meet the normalized requirement. Using the unity-gain configuration, the circuit values can be obtained from Table 3-6 and are shown in Fig. 3-33a. To transform the circuit into a normalized high-pass configuration, replace each resistor with a capacitor and all capacitors with resistors using reciprocal values. The transformed high-pass filter is illustrated in Fig. 3-33b.

c) To denormalize the high-pass filter each capacitor is divided by $Z \times$ FSF and each resistor is multiplied by Z. Since capacitors of 0.015 μF are desired, the impedance scaling factor Z is found from

$$Z = \frac{C}{\text{FSF} \times C'} = \frac{1 \text{ F}}{2\pi f_c \times 0.015 \times 10^{-6}} = 106 \times 10^{3}$$

where C is the normalized capacitor and C' is the denormalized capacitance value.

If we frequency and impedance scale the high-pass filter of Fig. 3-33b by multiplying all resistors by Z and dividing all capacitors by $Z \times$ FSF, the final configuration of Fig. 3-33c is obtained. The resistors have all been rounded off to standard 1% values.

3–3b   Elliptic-Function High-Pass Filters

Elliptic-Function high-pass filters can be designed directly from the normalized low-pass values given in Table 3-9. Replacing each capacitor by a resistor and each resistor by a capacitor, using reciprocal values, results in a normalized Elliptic-Function high-pass filter. This network is then frequency and impedance scaled to the design cutoff frequency.

An Elliptic-Function high-pass filter corresponding to Table 3-9 is illustrated in Fig. 3-34. All low-pass elements except for R_6 and R_7 have been replaced by reciprocal values, since these resistors only determine the gain and do not get transformed.

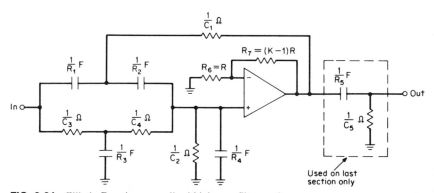

FIG. 3-34   Elliptic-Function normalized high-pass filter section.

Example 3–11   Design of Elliptic-Function High-Pass Filter

Design an Elliptic-Function high-pass filter to meet the following specifications:

3000 Hz	0.5 dB maximum
1000 Hz	35-dB minimum attenuation

Solution

a) Compute high-pass steepness factor

$$A_s = \frac{f_c}{f_s} = \frac{3000 \text{ Hz}}{1000 \text{ Hz}} = 3 \tag{3-10}$$

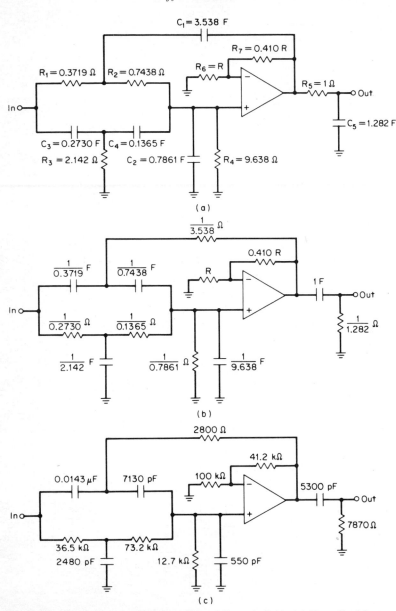

FIG. 3-35   Elliptic-Function high-pass filter of Example 3-11. (*a*) Normalized low-pass filter from Table 3-9; (*b*) transformed high-pass filter; (*c*) denormalized filter.

b) Choose a normalized Elliptic-Function low-pass filter that has over 35 dB of attenuation within a frequency ratio of 3 and a passband ripple of less than 0.5 dB.

A design is chosen from Table 3-9 that has the following parameters:

$$n = 3$$
$$R_{dB} = 0.28 \text{ dB}$$
$$\Omega_s = 2.9238$$
$$A_{min} = 39.48 \text{ dB}$$

The normalized low-pass circuit from Table 3-9 is shown in Fig. 3-35a.

c) The normalized circuit is transformed directly into the high-pass equivalent by replacing resistors with capacitors and vice versa using reciprocal values. The normalized high-pass filter is illustrated in Fig. 3-35b.

d) The circuit is denormalized by frequency and impedance scaling. The resistors are multiplied by Z and the capacitors are divided by Z × FSF, where Z is arbitrarily chosen at 10^5. The final circuit is shown in Fig. 3-35c.

3–4 BANDPASS FILTERS

When the ratio of upper to lower cutoff frequencies exceeds an octave, the bandpass requirement is partitioned into low-pass and high-pass requirements. The separate low-pass and high-pass filters are then designed as previously discussed and then cascaded. For narrower requirements, active filters are designed by transforming the low-pass transfer function to a bandpass transfer function and then realizing the bandpass poles by a cascade of active bandpass sections.

3–4a Bandpass Transformation of Low-Pass Poles

Poles corresponding to normalized standard low-pass transfer functions are tabulated in Sec. 3-7. After a bandpass requirement has been normalized and an appropriate filter type and order is chosen, the normalized low-pass poles must be transformed to bandpass poles, which are then implemented using active sections. Each complex low-pass pole results in *two* bandpass pole-pairs that require two active sections. A real low-pass pole is transformed into a single bandpass pole-pair requiring only one section.

Each bandpass section is specified in terms of center frequency and selectivity factor Q, where

$$Q = \frac{\text{section center frequency}}{\text{section 3-dB bandwidth}} \tag{3-33}$$

To make the transformation, first compute

$$Q_{bp} = \frac{f_0}{\text{BW}} \tag{3-34}$$

where f_0 is the geometric center frequency determined from the bandpass requirement and BW is the filter design bandwidth. The bandpass transformation proceeds as follows.

Complex Poles   The complex poles tabulated in Sec. 3-7 have the form

$$-\alpha \pm j\beta$$

where α is the real coordinate and β is the imaginary part. Given α, β, Q_{bp}, and f_0, the following sequence of calculations results in two sets of center frequencies and Q's corresponding to a pair of bandpass sections.

$$C = \alpha^2 + \beta^2 \tag{3-35}$$

$$D = \frac{2\alpha}{Q_{bp}} \tag{3-36}$$

$$E = \frac{C}{Q_{bp}^2} + 4 \tag{3-37}$$

$$G = \sqrt{E^2 - 4D^2} \tag{3-38}$$

$$Q = \sqrt{\frac{E + G}{2D^2}} \tag{3-39}$$

$$M = \frac{\alpha Q}{Q_{bp}} \tag{3-40}$$

$$W = M + \sqrt{M^2 - 1} \tag{3-41}$$

$$f_{ra} = \frac{f_0}{W} \tag{3-42}$$

$$f_{rb} = W f_0 \tag{3-43}$$

The resulting two bandpass sections have resonant frequencies of f_{ra} and f_{rb} in hertz and a Q defined by Eq. 3-39.

Real Poles   A normalized low-pass real pole with a tabulated value of α_0 results in a single bandpass section tuned to f_0, the filter geometric center frequency. The section Q is found from

$$Q = \frac{Q_{bp}}{\alpha_0} = 10 \tag{3-44}$$

Section Gain   The gain of a single bandpass section at the filter geometric center frequency f_0 is given by

$$A_0 = \frac{A_r}{\sqrt{1 + Q^2\left(\dfrac{f_0}{f_r} - \dfrac{f_r}{f_0}\right)^2}} \tag{3-45}$$

where A_r is the section gain at its resonant frequency f_r. The section gain will always be less at f_0 than at f_r, since the circuit peaks at f_r (except in the case of real poles where $f_r = f_0$, reducing the equation to $A_0 = A_r$). The composite circuit gain is determined by the product of the A_0 values of all sections used.

If the section Q is greater than 20, the gain equation can be simplified to

$$A_0 = \frac{A_r}{\sqrt{1 + \left(\dfrac{2Q\,\Delta f}{f_r}\right)^2}} \tag{3-46}$$

where Δf is the frequency separation between f_0 and f_r.

The following example illustrates the bandpass transformation of a set of normalized low-pass poles.

Example 3–12   Low-Pass to Bandpass Pole Transformation

Compute the pole locations and section gains for an $n = 3$ Butterworth bandpass filter with a center frequency (f_0) of 1000 Hz, a 3-dB bandwidth of 100 Hz, and a midband gain of $+30$ dB.

Solution

a) The normalized pole locations for an $n = 3$ Butterworth low-pass filter are obtained from Table 3-1 and are

$$-0.500 \pm j\,0.8660$$
$$-1.000$$

b) To obtain the bandpass poles first compute

$$Q_{bp} = \frac{f_0}{BW_{3\ dB}} = \frac{1000\ Hz}{100\ Hz} = 10 \tag{3-34}$$

The low-pass to bandpass pole transformation is performed as follows:

Complex Pole

$$\alpha = 0.5000 \qquad \beta = 0.8660$$

$$C = \alpha^2 + \beta^2 = 1.000000 \tag{3-35}$$

$$D = \frac{2\alpha}{Q_{bp}} = 0.1000000 \tag{3-36}$$

$$E = \frac{C}{Q_{bp}^2} + 4 = 4.010000 \tag{3-37}$$

$$G = \sqrt{E^2 - 4D^2} = 4.005010 \tag{3-38}$$

$$Q = \sqrt{\frac{E + G}{2D^2}} = 20.018754 \tag{3-39}$$

$$M = \frac{\alpha Q}{Q_{bp}} = 1.000938 \tag{3-40}$$

$$W = M + \sqrt{M^2 - 1} = 1.044261 \tag{3-41}$$

$$f_{ra} = \frac{f_0}{W} = 957.6\ Hz \tag{3-42}$$

$$f_{rb} = Wf_0 = 1044.3\ Hz \tag{3-43}$$

Real Pole

$$\alpha_0 = 1.0000$$

$$Q = \frac{Q_{bp}}{\alpha_0} = 10 \qquad (3\text{-}44)$$

$$f_r = f_0 = 1000 \text{ Hz}$$

c) Since a composite midband gain of $+30$ dB is required, let us distribute the gain equally among the three sections. Each section should then contribute $+10$ dB of gain ($A_0 = 3.162$). The gain at section resonant frequency f_r is obtained from the following form of Eq. 3-45:

$$A_r = A_0 \sqrt{1 + Q^2 \left(\frac{f_0}{f_r} - \frac{f_r}{f_0} \right)^2} \qquad (3\text{-}45)$$

The resulting values are

Section 1

$$f_r = 957.6 \text{ Hz}$$
$$Q = 20.02$$
$$A_r = 6.333$$

Section 2

$$f_r = 1044.3 \text{ Hz}$$
$$Q = 20.02$$
$$A_r = 6.335$$

Section 3

$$f_r = 1000.0 \text{ Hz}$$
$$Q = 10.00$$
$$A_r = 3.162$$

A block diagram of the sections is illustrated in Fig. 3-36.

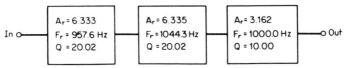

FIG. 3-36 Block realization of Example 3-12.

3–4b Sensitivity

"Sensitivity" is defined as the degree of change of a dependent variable resulting from the variation of an independent variable. Expressed mathematically, the sensitivity of y with respect to x is given by

$$S_x^y = \frac{\dfrac{dy}{y}}{\dfrac{dx}{x}} \qquad (3\text{-}47)$$

Sensitivity is a parameter used as a figure of merit to evaluate the change in a filter's characteristic, such as resonant frequency, or Q for a given change in a component value or amplifier parameter. Component value deviations will occur from tolerances, drift, etc., and will cause variations in a filter's parameters from their design values.

Let us take the case where we are given the parameter $S_{R_3}^Q = -5$. This convention means that for a 1% increment of R_3, the circuit Q will change 5% in the opposite direction.

Sensitivity is sometimes expressed as an equation such as $S_A^Q = 2Q^2$. This equation states that the sensitivity of Q with respect to amplifier gain A will increase with Q^2, so for large values of Q, the sensitivity becomes unacceptable. In the extreme case where the Q can increase dramatically for a small change in a component value, an unstable (oscillatory) condition may occur. However, filter configurations prone to instability have been avoided in this chapter.

3–4c   Bandpass Configurations

Multiple Feedback Bandpass (MFBP)   A useful implementation of a bandpass pole-pair is shown in Fig. 3-37. An op amp is used in the inverting configuration with multiple feedback paths. This structure is extremely simple and features low sensitivity to component tolerances. The element values for the circuit of Fig. 3-37a are computed as follows:

$$R_2 = \frac{Q}{\pi f_r C} \tag{3-48}$$

$$R_1 = \frac{R_2}{4Q^2} \tag{3-49}$$

where C is arbitrary.

The circuit gain at resonance is

$$A_r = 2Q^2 \tag{3-50}$$

Therefore, the amplifier open-loop gain in the region of f_r should be well in excess of $2Q^2$, so that the circuits' operation is governed mainly by the passive elements. This places a limitation on the circuit to Q's typically below 20.

Because of large closed-loop gains, even with moderate values of Q, it is highly desirable to control the closed-loop gain to prevent amplifier overdrive, etc. This results in the modified circuit of Fig. 3-37b. The input resistor has been split into two resistors, R_{1a} and R_{1b}, to form a voltage divider so that overall circuit gain can be controlled. The parallel combination of R_{1a} and R_{1b} remains equal to R_1, so the resonant frequency is retained.

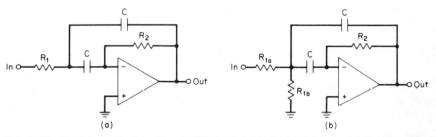

FIG. 3-37   Multiple-feedback bandpass (MFBP) ($Q < 20$). (a) MFBP basic circuit; (b) modified configuration.

The values of R_{1a} and R_{1b} are computed from

$$R_{1a} = \frac{R_2}{2A_r} \tag{3-51}$$

and

$$R_{1b} = \frac{\dfrac{R_2}{2}}{2Q^2 - A_r} \tag{3-52}$$

where A_r is the design gain at resonant frequency f_r and cannot exceed $2Q^2$.

The sensitivities of this configuration are

$$S_{R_{1a}}^{Q} = S_{R_{1a}}^{f_r} \frac{A_r}{4Q^2} \tag{3-53}$$

$$S_{R_{1b}}^{Q} = S_{R_{1b}}^{f_r} = \frac{1}{2}\left(1 + \frac{A_r}{2Q^2}\right) \tag{3-54}$$

$$S_{R_2}^{f_r} = S_{C}^{f_r} = -\frac{1}{2} \tag{3-55}$$

$$S_{R_2}^{Q} = \frac{1}{2} \tag{3-56}$$

Eq. 3-54 implies that for $Q^2/A_r \gg 1$, the resonant frequency can be directly controlled by R_{1b}, so this element can be made adjustable if frequency adjustment is desired.

Although this circuit will result in a large spread of resistor values, it is highly recommended for low Q requirements because of its low sensitivity, ease of adjustment, and circuit simplicity.

The following example illustrates the design of a bandpass filter using the MFBP configuration.

Example 3–13   Bandpass Filter Design

Design a bandpass filter to implement the computed resonant frequencies, Q's, and A_r of Example 3-12.

Solution

a) The requirements of the previous example were:

Section 1

$f_r = 957.6$ Hz
$Q = 20.02$
$A_r = 6.333$

Section 2

$f_r = 1044.3$ Hz
$Q = 20.02$
$A_r = 6.335$

Section 3

$f_r = 1000$ Hz
$Q = 10.00$
$A_r = 3.162$

b) If we choose a capacitance of 0.01 μF, the elements are computed as follows.

Section 1

$$R_2 = \frac{Q}{\pi f_r C} = \frac{20.02}{\pi \times 957.6 \times 10^{-8}} = 665.5 \text{ k}\Omega \qquad (3\text{-}48)$$

$$R_{1a} = \frac{R_2}{2A_r} = \frac{665.5 \times 10^3}{2 \times 6.333} = 52.54 \text{ k}\Omega \qquad (3\text{-}51)$$

$$R_{1b} = \frac{\dfrac{R_2}{2}}{2Q^2 - A_r} = \frac{332.8 \times 10^3}{2 \times 20.02^2 - 6.333} = 418 \ \Omega \qquad (3\text{-}52)$$

Section 2

$$R_2 = 610.2 \text{ k}\Omega$$
$$R_{1a} = 48.16 \text{ k}\Omega$$
$$R_{1b} = 384 \ \Omega$$

Section 3

$$R_2 = 318.3 \text{ k}\Omega$$
$$R_{1a} = 50.3 \text{ k}\Omega$$
$$R_{1b} = 809 \ \Omega$$

c) The final circuit is shown in Fig. 3-38. Standard 1% resistor values are used and each section has been made adjustable. Since the amplifier open-loop gain in the range of 1000 Hz should exceed $2Q^2$, or 800, a TL080 type op amp is well suited for this application.

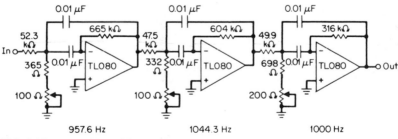

FIG. 3-38   MFBP filter of Example 3-13.

Dual Amplifier Bandpass (DABP)   The bandpass circuit of Fig. 3-39 exhibits excellent performance in terms of obtainable Q, low sensitivity, and flexibility when compared to other configurations involving two op amps.

The element values are computed as follows:

$$R = \frac{1}{2\pi f_r C} \qquad (3\text{-}57)$$

$$R_1 = QR \qquad (3\text{-}58)$$

$$R_2 = R_3 = R \qquad (3\text{-}59)$$

where C is arbitrary (R' in Fig. 3-39 can also be arbitrarily chosen at any convenient value), and section gain at f_r is fixed at two.

The circuit sensitivities are

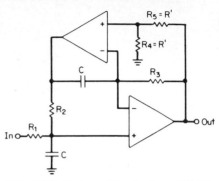

$$S_{R_1}^Q = 1 \qquad (3\text{-}60)$$

$$S_{R_2}^{f_r} = S_{R_3}^{f_r} = S_{R_4}^{f_r} = S_C^{f_r} = -\tfrac{1}{2} \quad (3\text{-}61)$$

$$S_{R_5}^{f_r} = \tfrac{1}{2} \qquad (3\text{-}62)$$

Sensitivity studies have shown that if both amplifiers are reasonably matched, small deviations of Q from the design values will occur. It is recommended that dual-type op amps be used, since both amplifier halves will then be closely matched to each other.

FIG. 3-39 Dual-amplifier bandpass (DABP) configuration ($Q < 150$).

An interesting feature of this circuit is that the Q is directly proportional to R_1 and that R_1 has no effect on resonant frequency. Therefore, by making R_1 adjustable, a variable-Q filter section is obtainable. However, it is important to recognize that a bandpass filter's response is much more affected by a tuning error than by a comparable percentage error in Q, so that Q need not be adjusted in most cases. However, it is frequently necessary to adjust section resonances, especially in very narrow bandpass filters, to avoid distortion of the passband shape resulting from tuning errors. Resonant frequency of this circuit can be adjusted by making R_2 variable.

Since each section provides a fixed gain of two at f_r, a composite filter may require an additional amplification stage if higher gains are required. If a gain reduction is desired, resistor R_1 can be split into two resistors to form a voltage divider in the same manner as in Fig. 3-37b. The resulting values are

$$R_{1a} = \frac{2R_1}{A_r} \qquad (3\text{-}63)$$

and

$$R_{1b} = \frac{R_{1a}A_r}{2 - A_r} \qquad (3\text{-}64)$$

where A_r is the desired gain at resonance.

The spread of element values of the MFBP section previously discussed is equal to $4Q^2$. In comparison, this circuit has a ratio of resistances determined by Q, so the spread is much less.

The DABP configuration has been found very useful for designs covering a wide range of Q's and frequencies. Component sensitivity is small, resonant frequency and Q are easily adjustable, and the element spread is low.

State-Variable Configuration The state-variable configuration shown in Fig. 3-40 features excellent sensitivity properties and the capability to control resonant frequency and Q independently. It is especially suited for constructing precision-active filters in a standard form. A variety of manufacturers offer hybrid biquad filter sections where frequency and Q are determined by a few external resistors.

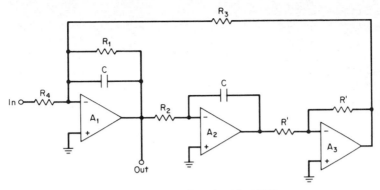

FIG. 3-40 State-variable bandpass configuration ($Q < 200$).

The component values are computed by

$$R_1 = \frac{Q}{2\pi f_r C} \tag{3-65}$$

$$R_2 = R_3 = \frac{R_1}{Q} \tag{3-66}$$

$$R_4 = \frac{R_1}{A_r} \tag{3-67}$$

where A_r is the desired gain at resonant frequency f_r. R_1 controls Q and R_3 can be made adjustable to vary resonant frequency. R' may be any convenient value.

The circuit sensitivities are:

$$S_{R_2}^{f_r} = S_{R_3}^{f_r} = S_C^{f_r} = -\tfrac{1}{2} \tag{3-68}$$

$$S_{R_1}^{Q} = 1 \tag{3-69}$$

$$S_{A_0}^{Q} = \frac{2Q}{A_0} \tag{3-70}$$

where A_0 is the amplifier open-loop gain at f_r. Therefore, according to Eq. 3-70, the section Q is limited by the finite gain of the op amp.

For high Q designs, the phase shift of the amplifiers will create an effect called "Q enhancement," whereby the actual circuit Q will be in excess of the design value. A practical solution is to make R_1 adjustable. By adjusting the element until the circuit gain becomes equal to the design value of A_r at resonance, the Q will be restored to its correct value.

The state-variable circuit is a low-sensitivity filter configuration useful for precision applications. Circuit Q's of up to 200 can be obtained over a wide range of frequencies.

3–5 BAND-REJECT FILTERS

If the ratio of upper to lower cutoff frequency of a band-reject requirement is in excess of an octave, the specification should be partitioned into low-pass and high-pass requirements. Separate filters are then designed using the techniques previously discussed and then paralleled at the input and combined at the output as shown in Fig. 3-9.

3–5a Twin-T Notch Network

A class of band-reject filters are "notch networks." These consist of single sections that provide a notch, or transmission zero, at a specified frequency or narrow band of frequencies.

The "twin-T" is a very popular form of notch network. However, in the passive case, this circuit exhibits a Q of only $\frac{1}{4}$, so it is somewhat limited when more selectivity is desired. If positive feedback is introduced, as shown in Fig. 3-41, the Q can be dramatically increased. The circuit element values are computed from

$$k = 1 - \frac{1}{4Q} \tag{3-71}$$

$$R_1 = \frac{1}{2\pi f_0 C} \tag{3-72}$$

where R and C are arbitrary.

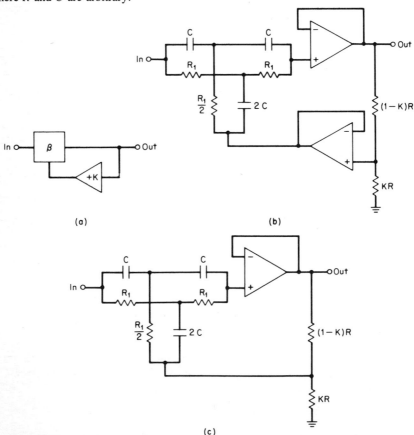

(a)

(b)

(c)

FIG. 3-41   Twin-T with positive feedback. (*a*) Block diagram; (*b*) circuit realization; (*c*) simplified configuration $R_1 \gg (1-K)R$.

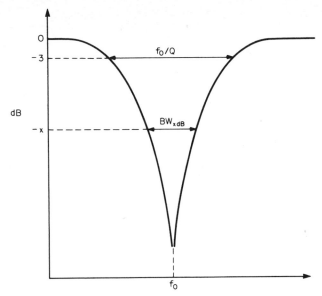

FIG. 3-42 Frequency response of twin-T with positive feedback.

The frequency response is shown in Fig. 3-42. The attenuation at any bandwidth is given by

$$A_{dB} = 10 \log \left[1 + \left(\frac{f_0}{Q \times BW_{xdB}} \right)^2 \right] \tag{3-73}$$

The circuit of Fig. 4-41c uses one less op amp than the configuration shown in Fig. 3-41b, but R_1 must be much greater than $(1 - k)R$, which may not always result in convenient values.

Eq. 3-71 can also be expressed in the form

$$Q = \frac{1}{4(1 - k)} \tag{3-74}$$

Since k is very close to unity when high Q's are required, small deviations in k can result in large Q errors, so a potentiometer is sometimes used to adjust k.

3-5b Tunable Notch Networks

The twin-T configuration does not lend itself to easy adjustment. When narrow notch networks are required, the component tolerances can result in excessive detuning, which may make the circuit useless. An adjustable form of notch network is then desired.

If we implement the configuration illustrated in the block diagram of Fig. 3-43, where $T(s)$ is a bandpass section having a gain of exactly $+1$ at center frequency, a subtraction of the input and bandpass output signals will occur. The combining amplifier's output will be zero at f_0 and exactly

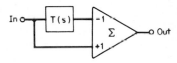

FIG. 3-43 Converting a bandpass section into a notch network.

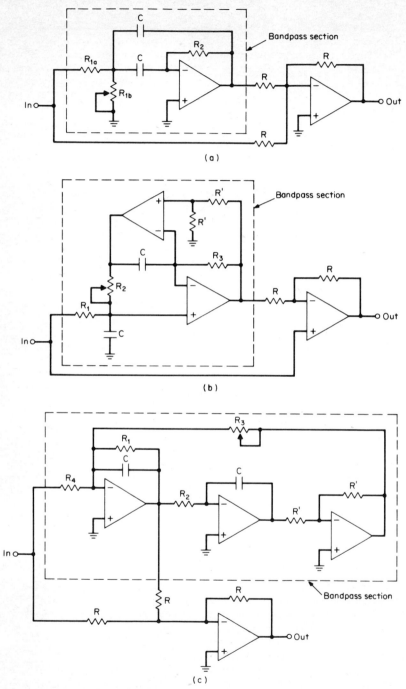

FIG. 3-44 Adjustable notch circuits. (*a*) MFBP section ($Q < 20$); (*b*) DABP section ($Q < 150$); (*c*) state-variable section ($Q < 200$).

the input signal at frequencies far removed from resonance where the contribution of the bandpass section is negligible. As a result, a notch network is obtained.

The three bandpass sections previously discussed are shown in Fig. 3-44, where by the addition of a summing amplifier, a notch network is obtained. The adjustable characteristics of the bandpass sections will now control the tuning of the notch network. In addition to the adjustable notch frequency, the Q can also be controlled by making the appropriate element variable.

The design equations for these bandpass sections were previously given and need not be repeated. The Q and resonant frequency parameters of the bandpass section directly determine the Q and resonant frequency of the notch.

3–6 CONSIDERATIONS FOR OP AMP SELECTION

Choice of the type of op amp is governed by the filter topology and circuit design parameters. In many cases the op amp will behave as an ideal device, especially at low frequencies where open-loop gain is extremely high. To ensure successful designs, the engineer must carefully evaluate circuit requirements and choose an op amp that will not result in excessive deviations of frequency response from theoretical values.

Amplifier parameters are important in the passband region. Once the minimum stopband attenuation becomes established, the amplifier parameters are of minor consequence.

3–6a Open-Loop Gain

The op amp open-loop gain requirements for an active filter are determined from the closed-loop design gain. The effects of open-loop gain on closed-loop gain can be evaluated from the formula for a noninverting amplifier.

$$A_c = \frac{A_0}{1 + A_0\beta} \tag{3-75}$$

where A_0 is the open-loop gain at the highest passband frequency and β is the feedback factor.

Fig. 3-45 illustrates the effect of open-loop gain on gain error for closed-loop gains of 1, 10, and 100. As the open-loop gain is increased, there is a diminishing return of increased accuracy. An open- to closed-loop gain ratio of 100:1 is more than adequate for most filters. In many cases a 10:1 ratio will result in acceptable accuracy.

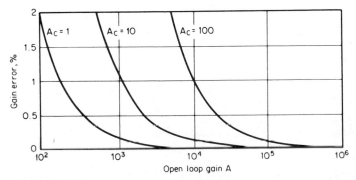

FIG. 3-45   Effect of open-loop gain on gain error.

In the case of the unity gain and VCVS filter configurations, high open- to closed-loop gain ratios are easily achieved. For the MFBP bandpass configuration the open-loop gain should exceed $10Q^2$. For the DABP and state-variable bandpass configurations, precision filters can be obtained with open-loop gains of $100Q$. Where tuning adjustments are provided much lower open-loop gains can be tolerated.

3–6b   Amplifier Phase Shift

As the open-loop gain rolls off with frequency, amplifier phase shift will increase. This results in Q enhancement in state-variable filters, where lagging phase shift will cause an increase in circuit Q. The increased Q can be estimated by

$$Q' = \frac{Q}{1 + \dfrac{2Q\,(f_{3\ dB} - 2f)}{A_0 F_{3\ dB}}} \tag{3-76}$$

where Q is the design Q, $f_{3\ dB}$ is the open-loop 3-dB breakpoint, f is the highest passband frequency of interest, and Q' is the resulting Q. If more than a 5% enhancement is calculated, a Q adjustment control may be desirable.

The closed-loop phase shift of an amplifier can be estimated from

$$\phi_c = \tan^{-1} \frac{f}{\dfrac{f_{3\ dB}}{1 + A_0\beta}} \tag{3-77}$$

For the VCVS low-pass and high-pass filter structures of Secs. 3-2 and 3-3, ϕ_c should be less than 4° for the second-order sections and less than 6° for the three-pole section where f is the filter 3-dB cutoff. The Elliptic-Function filters should have less than 4° of phase shift at cutoff.

3–6c   Input and Output Impedance

The input and output impedances of op amp are normally not of major concern in active realizations. The basic consideration is that the input impedance be much higher than the largest value of resistor connected to the input terminal and that the output impedance be much lower than the lowest resistor connected to the output terminal.

If the active filter is designed at a very high impedance level to minimize capacitor values, the input impedance of the op amp must be high enough to avoid loading errors. In the case of the noninverting VCVS configurations, the negative feedback greatly enhances input impedance, since

$$R_{in} = (1 + A_0\beta)\,R_i \tag{3-78}$$

where A_0 is the open-loop gain, β is the feedback factor, and R_i is the open-loop input impedance of the amplifier.

For maximum input impedance a JFET-type input should be used. Open-loop input impedances in the range of $10^{12}\ \Omega$ are easily obtained, so for all practical purposes this parameter can be considered ideal.

Amplifier output impedance is calculated from

$$R_{out} = \frac{R_o}{1 + A_0\beta} \tag{3-79}$$

Since R_o, the open-loop output impedance, is typically 150 Ω, the closed-loop values are in the range of a few hundred milliohms for most applications, which, of course, is insignificant.

3–6d Dynamic Range

The output swing of an active filter is limited to within a few volts of the power-supply voltage. The section gain or input signal must be limited to avoid overdriving.

Another consideration, especially in narrow bandpass filters, is a detuning of the resonant frequency for large output swings. This results from "slew rate limiting," which is the inability of the output stage to provide the steep output slope necessary for large output swings at higher frequencies.

At the other extreme the dynamic range is limited by amplifier noise. For the highest output signal-to-noise ratio, bandpass filters should have maximum design gain. This is especially important for the input filter section.

3-7 FILTER DESIGN TABLES

Butterworth pole locations and active low-pass values, Chebyshev pole locations and active low-pass values, Bessel pole locations and active low-pass values, and Elliptic-Function low-pass filter values are tabulated in this section.

TABLE 3–1 Butterworth Pole Locations

Order n	Real part $-\alpha$	Imaginary part $\pm j\beta$
2	0.7071	0.7071
3	0.5000	0.8660
	1.0000	
4	0.9239	0.3827
	0.3827	0.9239
5	0.8090	0.5878
	0.3090	0.9511
	1.0000	
6	0.9659	0.2588
	0.7071	0.7071
	0.2588	0.9659
7	0.9010	0.4339
	0.6235	0.7818
	0.2225	0.9749
	1.0000	
8	0.9808	0.1951
	0.8315	0.5556
	0.5556	0.8315
	0.1951	0.9808
9	0.9397	0.3420
	0.7660	0.6428
	0.5000	0.8660
	0.1737	0.9848
	1.0000	
10	0.9877	0.1564
	0.8910	0.4540
	0.7071	0.7071
	0.4540	0.8910
	0.1564	0.9877

TABLE 3–2 Butterworth Active Low-Pass Values*

Order n	C_1	C_2	C_3
2	1.414	0.7071	
3	3.546	1.392	0.2024
4	1.082	0.9241	
	2.613	0.3825	
5	1.753	1.354	0.4214
	3.235	0.3090	
6	1.035	0.9660	
	1.414	0.7071	
	3.863	0.2588	
7	1.531	1.336	0.4885
	1.604	0.6235	
	4.493	0.2225	
8	1.020	0.9809	
	1.202	0.8313	
	1.800	0.5557	
	5.125	0.1950	
9	1.455	1.327	0.5170
	1.305	0.7661	
	2.000	0.5000	
	5.758	0.1736	
10	1.012	0.9874	
	1.122	0.8908	
	1.414	0.7071	
	2.202	0.4540	
	6.390	0.1563	

*Reprinted from "Electronics," August 18, 1969, McGraw-Hill, Inc., 1969.

TABLE 3-3 0.1-dB Chebyshev Pole Locations

Order n	Real part $-\alpha$	Imaginary part $\pm j\beta$
2	0.6104	0.7106
3	0.3490	0.8684
	0.6979	
4	0.2177	0.9254
	0.5257	0.3833
5	0.3842	0.5884
	0.1468	0.9521
	0.4749	
6	0.3916	0.2590
	0.2867	0.7077
	0.1049	0.9667
7	0.3178	0.4341
	0.2200	0.7823
	0.0785	0.9755
	0.3528	
8	0.3058	0.1952
	0.2592	0.5558
	0.1732	0.8319
	0.06082	0.9812
9	0.2622	0.3421
	0.2137	0.6430
	0.1395	0.8663
	0.04845	0.9852
	0.2790	

TABLE 3-4 0.5-dB Chebyshev Pole Locations

Order n	Real part $-\alpha$	Imaginary part $\pm j\beta$
2	0.5129	0.7225
3	0.2683	0.8753
	0.5366	
4	0.3872	0.3850
	0.1605	0.9297
5	0.2767	0.5902
	0.1057	0.9550
	0.3420	
6	0.2784	0.2596
	0.2037	0.7091
	0.07459	0.9687
7	0.2241	0.4349
	0.1550	0.7836
	0.05534	0.9771
	0.2487	
8	0.2144	0.1955
	0.1817	0.5565
	0.1214	0.8328
	0.04264	0.9824
9	0.1831	0.3425
	0.1493	0.6436
	0.09743	0.8671
	0.03383	0.9861
	0.1949	

TABLE 3–5 0.1-dB Chebyshev Active Low-Pass Values*

Order n	C_1	C_2	C_3
2	1.638	0.6955	
3	6.653	1.825	0.1345
4	1.900	1.241	
	4.592	0.2410	
5	4.446	2.520	0.3804
	6.810	0.1580	
6	2.553	1.776	
	3.487	0.4917	
	9.531	0.1110	
7	5.175	3.322	0.5693
	4.546	0.3331	
	12.73	0.08194	
8	3.270	2.323	
	3.857	0.6890	
	5.773	0.2398	
	16.44	0.06292	
9	6.194	4.161	0.7483
	4.678	0.4655	
	7.170	0.1812	
	20.64	0.04980	
10	4.011	2.877	
	4.447	0.8756	
	5.603	0.3353	
	8.727	0.1419	
	25.32	0.04037	

*Reprinted from "Electronics," August 18, 1969, McGraw-Hill, Inc., 1969.

TABLE 3–6 0.5-dB Chebyshev Active Low-Pass Values*

Order n	C_1	C_2	C_3
2	1.950	0.6533	
3	11.23	2.250	0.0895
4	2.582	1.300	
	6.233	0.1802	
5	6.842	3.317	0.3033
	9.462	0.1144	
6	3.592	1.921	
	4.907	0.3743	
	13.40	0.07902	
7	7.973	4.483	0.4700
	6.446	0.2429	
	18.07	0.05778	
8	4.665	2.547	
	5.502	0.5303	
	8.237	0.1714	
	23.45	0.04409	
9	9.563	5.680	0.6260
	6.697	0.3419	
	10.26	0.1279	
	29.54	0.03475	
10	5.760	3.175	
	6.383	0.6773	
	8.048	0.2406	
	12.53	0.09952	
	36.36	0.02810	

*Reprinted from "Electronics," August 18, 1969, McGraw-Hill, Inc., 1969.

TABLE 3–7 Bessel Pole Locations

Order n	Real part $-\alpha$	Imaginary part $\pm j\beta$
2	1.1030	0.6368
3	1.0509	1.0025
	1.3270	
4	1.3596	0.4071
	0.9877	1.2476
5	1.3851	0.7201
	0.9606	1.4756
	1.5069	
6	1.5735	0.3213
	1.3836	0.9727
	0.9318	1.6640
7	1.6130	0.5896
	1.3797	1.1923
	0.9104	1.8375
	1.6853	
8	1.7627	0.2737
	0.8955	2.0044
	1.3780	1.3926
	1.6419	0.8253
9	1.8081	0.5126
	1.6532	1.0319
	1.3683	1.5685
	0.8788	2.1509
	1.8575	

TABLE 3–8 Bessel Active Low-Pass Values

Order n	C_1	C_2	C_3
2	0.9066	0.6800	
3	1.423	0.9880	0.2538
4	0.7351	0.6746	
	1.012	0.3900	
5	1.010	0.8712	0.3095
	1.041	0.3100	
6	0.6352	0.6100	
	0.7225	0.4835	
	1.073	0.2561	
7	0.8532	0.7792	0.3027
	0.7250	0.4151	
	1.100	0.2164	
8	0.5673	0.5540	
	0.6090	0.4861	
	0.7257	0.3590	
	1.116	0.1857	
9	0.7564	0.7070	0.2851
	0.6048	0.4352	
	0.7307	0.3157	
	1.137	0.1628	
10	0.5172	0.5092	
	0.5412	0.4682	
	0.6000	0.3896	
	0.7326	0.2792	
	1.151	0.1437	

TABLE 3-9 Elliptic-Function Low-Pass Filter Values

	R_1	R_2	R_3	R_4	R_5	C_1	C_2	C_3	C_4	C_5	K	Gain	α	β	ω_∞
						$N = 3$	$R_{dB} = 0.28$ dB								
$\Omega_s = 9.5668$	0.3369	0.6738	21.02	94.57	1.000	3.9010	0.8670	0.0278	0.0139		1.2480	1.2350	0.3693	1.0790	11.03
$A_{min} = 71.10$										1.3400			0.7465		
$\Omega_s = 6.3925$	0.3413	0.6826	9.4920	42.72		3.8510	0.8558	0.0616	0.0308		1.2680	1.2380	0.3663	1.0800	7.3700
$A_{min} = 60.50$					1.0000					1.3320			0.7505		
$\Omega_s = 4.4454$	0.3499	0.6998	4.6920	21.11		3.7580	0.8350	0.1245	0.0623		1.3070	1.2450	0.3604	1.0820	5.1170
$A_{min} = 50.86$					1.0000					1.3180			0.7585		
$\Omega_s = 2.9238$	0.3719	0.7438	2.1420	9.6380		3.5380	0.7861	0.2730	0.1365		1.4100	1.2630	0.3449	1.0860	3.3500
$A_{min} = 39.48$					1.0000					1.2820			0.7801		
$\Omega_s = 2.1301$	0.4069	0.8137	1.2290	5.5330		3.2390	0.7198	0.4764	0.2382		1.5780	1.2930	0.3191	1.0920	2.4230
$A_{min} = 30.44$					1.0000					1.2210			0.8188		
$\Omega_s = 1.5557$	0.4739	0.9479	0.7490	3.3710		2.7970	0.6215	0.7865	0.3933		1.9270	1.3550	0.2643	1.1000	1.7420
$A_{min} = 20.58$					1.0000					1.0930			0.9142		
						$N = 5$	$R_{dB} = 0.28$ dB								
$\Omega_s = 2.3662$	0.3457	0.6913	6.2320	28.04		5.7850	1.2860	0.1426	0.0713		1.0990	1.0600	0.3410	0.6681	3.9010
$A_{min} = 71.71$	0.3920	0.7840	1.4850	6.6830		3.6830	0.8185	0.4321	0.2160		1.7820	1.5160	0.1176	1.0320	2.4770
					1.0000					2.2240			0.4497		
$\Omega_s = 1.8871$	0.3544	0.7089	3.7330	16.80		5.5190	1.2260	0.2329	0.1164		1.1460	1.0770	0.3386	0.6879	3.0480
$A_{min} = 60.51$	0.4258	0.8515	1.0230	4.6060		3.3970	0.7548	0.6280	0.3140		1.9500	1.5260	0.1091	1.0310	1.9690
					1.0000					2.1400			0.4673		

	C1	C2	C3	C4	C5	C6	C7	C8	C9	C10	C11	C12	C13	C14	C15
$\Omega_s = 1.5557$	0.3685	0.7370	2.3300	10.48		5.1420	1.1420	0.3615	0.1807		1.2200	1.1040	0.3339	0.7177	2.4380
$A_{min} = 50.10$	0.4698	0.9396	0.7651	3.4430		3.0860	0.6858	0.8422	0.4211	2.0190	2.1730	1.5420	0.0970	1.0300	1.6170
					1.0000								0.4952		
$\Omega_s = 1.3054$	0.3942	0.7884	1.4390	6.4770		4.5720	1.0150	0.5564	0.2782		1.3600	1.1490	0.3226	0.7673	1.9480
$A_{min} = 39.17$	0.5280	1.0560	0.6027	2.7120		2.7570	0.6127	1.0730	0.5368	1.8310	2.4790	1.5650	0.0786	1.0270	1.3480
					1.0000								0.5462		
$\Omega_s = 1.2361$	0.4078	0.8157	1.2160	5.4740		4.3170	0.9592	0.6433	0.3216		1.4350	1.1730	0.3154	0.7915	1.8020
$A_{min} = 35.21$	0.5509	1.1010	0.5627	2.5320		2.6480	0.5884	1.1520	0.5761	1.7430	2.6040	1.5750	0.0704	1.0250	1.2730
					1.0000								0.5737		
$\Omega_s = 1.1666$	0.4284	0.8568	1.0010	4.5070		3.9820	0.8848	0.7570	0.3785		1.5510	1.2070	0.3031	0.8255	1.6470
$A_{min} = 30.46$	0.5781	1.1560	0.5249	2.3620		2.5310	0.5624	1.2380	0.6194	1.6230	2.7560	1.5890	0.0595	1.0230	1.1960
					1.0000								0.6163		

$N = 7 \qquad R_{dB} = 0.28$ dB

	C1	C2	C3	C4	C5	C6	C7	C8	C9	C10	C11	C12	C13	C14	C15
$\Omega_s = 1.3902$	0.3493	0.6986	4.8560	21.85		7.0290	1.5620	0.2247	0.1123		1.0870	1.0370	0.2882	0.5386	2.7890
$A_{min} = 70.19$	0.4277	0.8555	1.0060	4.5310		3.9490	0.8777	0.7458	0.3729	2.7620	1.8440	1.4370	0.1530	0.8746	1.6680
	0.5047	1.0090	0.6544	2.9450	1.0000	2.9290	0.6509	1.0040	0.5021		2.4350	1.6080	0.0446	1.0130	1.4150
													0.3621		
$\Omega_s = 1.3250$	0.3525	0.7049	4.0970	18.44		6.8190	1.5150	0.2607	0.1304		1.1060	1.0460	0.2904	0.5524	2.6060
$A_{min} = 65.79$	0.4407	0.8815	0.9119	4.1030		3.8030	0.8452	0.8170	0.4085	2.6840	1.9130	1.4460	0.1478	0.8825	1.5760
	0.5221	1.0440	0.6146	2.7650	1.0000	2.8340	0.6297	1.0700	0.5350		2.5250	1.6120	0.0418	1.0120	1.3470
													0.3726		
$\Omega_s = 1.2521$	0.3575	0.7150	3.2890	14.80		6.5180	1.4480	0.3149	0.1574		1.1360	1.0590	0.2932	0.5731	2.3910
$A_{min} = 60.18$	0.4595	0.9189	0.8095	3.6430		3.6080	0.8018	0.9102	0.4551	2.5710	2.0130	1.4610	0.1399	0.8939	1.4710
	0.5449	1.0890	0.5723	2.5750	1.0000	2.7180	0.6040	1.1500	0.5751		2.6430	1.6170	0.0377	1.0120	1.2710
													0.3889		
$\Omega_s = 1.2062$	0.3621	0.7242	2.7980	12.59		6.2720	1.3940	0.3608	0.1804		1.1630	1.0700	0.2952	0.5908	2.2480
$A_{min} = 56.12$	0.4745	0.9490	0.7470	3.3620		3.4630	0.7696	0.9777	0.4888	2.4810	2.0950	1.4720	0.1333	0.9030	1.4030
	0.5615	1.1230	0.5468	2.4610	1.0000	2.6400	0.5866	1.2040	0.6024		2.7310	1.6210	0.0346	1.0110	1.2230
													0.4030		
$\Omega_s = 1.1547$	0.3696	0.7393	2.2630	10.18		5.9190	1.3150	0.4297	0.2148		1.2060	1.0880	0.2974	0.6177	2.0780
$A_{min} = 50.86$	0.4958	0.9916	0.6782	3.0520		3.2740	0.7276	1.0630	0.5319	2.3520	2.2140	1.4890	0.1234	0.9157	1.3230
	0.5825	1.1650	0.5195	2.3380	1.0000	2.5480	0.5661	1.2700	0.6349		2.8430	1.6270	0.0302	1.0100	1.1680
													0.4252		

REFERENCES

Huelsman, L. P.: *Theory and Design of Active RC Circuits,* McGraw-Hill, New York, 1968.

Thomas, L. C.: "The Biquad: Part I—Some Practical Design Considerations," *IEEE Transactions on Circuit Theory*, CT-18, May 1971, pp. 350–357.

Williams, A. B.: *Active Filter Design,* Artech House, Dedham, Mass., 1975.

————: *Electronic Filter Design Handbook,* McGraw-Hill, New York, 1981.

————: "Design Active Elliptic Filters Easily from Tables," *Electronic Design*, 19, No. 21, October 14, 1971, pp. 76–79.

Zverev, A. I.: *Handbook of Filter Synthesis,* Wiley, New York, 1967.

Chapter 4

TELECOMMUNICATIONS CIRCUITS

Robert C. Jones
Randall J. Hipp
Darin L. Kincaid
Glen M. Masker
Michael R. Sims
Jerri L. Smith
Mostek Corp.
Carrollton, Tex.

4–1 INTRODUCTION

4–1a Chapter Organization

Manufacturers of integrated circuits have provided the telecommunications industry with a wide selection of products. The products discussed here are representative of a majority of these devices. This chapter is organized such that the reader is first introduced to the device and its location within the overall system. A separate section is then devoted to the theory and application of each device type. A selection guide is included to provide the user with a source for information on particular products. This guide also contains the locations of many semiconductor firms that manufacture telecommunication devices. The reader is encouraged to contact the manufacturers on products of interest. In addition, the user should be aware that new products are continually becoming available.

4–1b Introduction to the Telephone System

The most familiar instrument to the user of a telephone system is the subscriber set or basic telephone. A subscriber set performs the dialing, ringing, transmitting, and receiving functions required at each user location. A typical subscriber set consists of seven main components:

1. Receiver
2. Transmitter
3. Speech network
4. Hook switch
5. Ringer
6. Dialer
7. Bridge rectifier (primarily used with electronic dialers)

The block diagram in Fig. 4-1 illustrates the interconnection of the seven main components within a subscriber set.

The transmitter and receiver are normally located in the handset section of a subscriber set. The transmitter converts user voice signals into electrical signals that are transmitted to the local switching center. The receiver converts electrical signals into sound. The signal at the receiver consists of the voiceband signals from the switching center and attenuated feedback from the transmitter. The feedback or ''sidetone'' function is performed by the speech network. The speech network also provides for separation of the transmit and receive signals at the subscriber set. Thus all signals between the switching center and subscriber set may be carried over a single wire pair. The speech network is discussed further in a separate section of this chapter.

The hook switch may be in either of two positions, on-hook or off-hook. These conditions correspond to idle and busy circuits, respectively, with the off-hook condition normally activated by lifting the handset. When the handset is lifted, a current-sensing device at the switching center detects the off-hook state. The switching center's logic circuitry will then turn off any ring signal and prepare to send and receive voice communication. If the subscriber is placing the call, the switching center will prepare to accept dial signals. The hook switch connects the telephone line to the ringer in an on-hook position and to the speech network in an off-hook position. In the off-hook position, the subscriber-set circuitry receives a DC bias from the power supply at the switching center. In the on-hook position, a ring signal may be initiated by a caller. An electrical signal of about 80 V_{rms} and 20–30 Hz is typically generated at the switching center to activate the ringer at a subscriber set.

The two methods commonly used to transmit dialing information to the switching

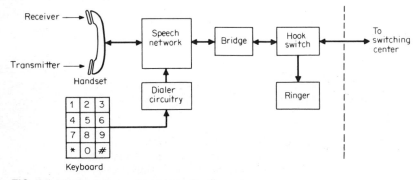

FIG. 4-1   Block diagram of a typical subscriber set.

center are pulse generation and tone generation. Rotary-type dialers generate pulses on the line, and these pulses are sensed and counted by the switching center. Electronic pulse dialers simulate the mechanical action of a rotary dialer. Tone dialers generate tone combinations of various frequencies. When electronic dialers are used in a subscriber set, a bridge rectifier is used to prevent damage to the dialer due to line reversal. The bridge provides the dialer with the proper polarity of the DC line bias. Electronic dialers are discussed further in separate sections of this chapter. Decoders used at the switching center to decode tone dialer signals are also discussed in this chapter.

The local switching center for most residential subscriber sets is a central office. A central office connects parties within its territory or routes a call to the appropriate switching center. A hierarchy of switching centers exists to provide the interconnection of all central offices. This hierarchy provides multiple paths between subscribers with calls normally routed through the lowest possible order of toll trunks and switching centers. Remote concentrators and private branch exchanges (PBXs) may exist between a subscriber set and the central office. Concentrators eliminate the need for a dedicated connection between every subscriber and the central office through the use of multiplexing and trunk sharing schemes. Private branch exchanges serve as switching centers for subgroups of subscribers, such as the employees of a business. A PBX will normally have access to the outside world through analog or digital trunks leased from the nearest central office.

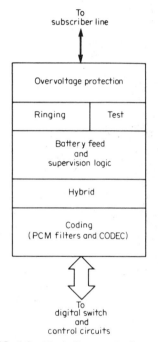

FIG. 4-2   Block diagram of a line card.

Central office architecture currently exists at many levels of technology. Although both older analog and newer digital switching schemes are in use today, the discussion here will primarily cover digital switching centers. The digital central office may be divided into two main sections, the line card and the digital switch. The line card implements the BORSCHT functions. BORSCHT is a mnemonic that represents the standard functions required of a line card in a digital switching system. These functions are defined below and a block diagram of the line card is shown in Fig. 4-2.

B—*Battery Feed:* Supplies the DC bias or loop current to the subscriber set.
O—*Overvoltage Protection:* Protects the line card from damage by high-voltage transients, such as those induced by lightning strikes near the transmission line.
R—*Ringing:* Controls the ringing signal induced on the subscriber line.
S—*Supervision:* Monitors the line to detect various subscriber set conditions.
C—*Coding:* Codes the subscriber's voice signal into digital data. This function is implemented by the PCM filter and CODEC discussed in separate sections of this chapter.

H—*Hybrid:* Performs the two-to-four wire conversion required between the two-wire subscriber line and the four-wire coding section.

T—*Test:* Performs tests on the line card and subscriber line to determine fault conditions.

The digital switch consists of random access memory and a network of computer and logic devices. The digital switch is responsible for observing the status of all subscriber lines and initiating any required interconnection between subscribers. The memory is used to store status information and provide a buffer for the voice data. The digital switch must also interface to toll trunks or tandem lines for calls between different cities or central offices.

4–2 PULSE DIALERS

4–2a Introductory Theory

The simplest type of dialer in use today is the pulse dialer, which uses a series of pulses to transmit dial signals to the central office. The original form of the pulse dialer is the mechanical rotary dialer, which contains a cam-and-gear shaft that rotates as the dial is turned. After the dial is released, a spring-powered motor returns the dial to its original position, with a governor regulating the return speed. Pulses are generated during the return by a pair of contacts that open and close at a rate of approximately 10 pulses per second (20 pps in Europe).

To meet U.S. telephone system requirements, pulse dialers must have the following characteristics:

1. The dial pulse signaling should consist of a sequence of momentary breaks in the telephone loop current corresponding to the numerical value of each digit, except digit "0," which should be represented by 10 break intervals.
2. The pulsing of the telephone loop current should operate at repetition rates between 8 and 11 pps, with 10 pps being nominal.
3. For an automatic dialer, the interdigital interval should be between 600 ms and 3 s.
4. During the break intervals of dial pulse signaling, the steady-state resistance from tip to ring with tip or ring grounded should be at least 50 kΩ.

These specifications can be found in such publications as the Bell Technical References PUB47001 and the EIA Standard RS-470.

Modern electronic pulse dialers perform the same functions as rotary dialers, but they offer several advantages over mechanical dialers. Electronic pulse dialers use push-button keyboards that can outpulse as they accept inputs, so they permit much higher entry rates than do rotary dialers. Most electronic pulse dialers have a first-in–first-out (FIFO) memory buffer that stores digits as they are entered and outpulses them in turn. Each memory buffer can store a limited number of digits, typically 17 to 20. Many pulse dialers also have a "wraparound" memory feature, permitting number sequences of any length to be dialed as long as the number of digits remaining to be outpulsed does not exceed the storage capability of the memory.

An additional feature available on most electronic pulse dialers is the redial function. The "last number dialed" is automatically stored in the memory buffer when the telephone enters the on-hook mode and may be redialed via a special keyboard input, often the * or # key. Many electronic pulse dialers offer additional features, such as programmable PBX pauses, pacifier tones to supply audible verification of a key input, pin-selectable dialing rates and make-break ratios, and options with regard to the type of keyboard and frequency reference used.

The electronic pulse dialer interface circuitry can be placed either in series or in parallel with the speech network. During outpulsing with a series pulse interface, the receiver is disconnected and the pulsing is accomplished by connecting and disconnecting the network, usually through a transistor. A series pulse interface is generally simple to implement and to obtain satisfactory muting, and it usually requires fewer components than a parallel interface. In a parallel pulse interface, the speech network is disabled and the pulsing occurs through a load resistor that is in parallel with, and of equivalent impedance to, the network. Because the network is isolated during pulsing, parallel pulse interfacing is useful for applications where the network cannot withstand pulsing transients. The main disadvantage of a parallel interface is that a loud pop may be heard in the receiver when the network is disabled. This pop can be objectionable to the user and careful design is required to eliminate it. Two pulse dialer application circuits, one using each of these interface methods, will be discussed in the following section.

4–2b   Application Solutions and Examples

In this section, two typical pulse dialer applications will be discussed. The first example will present a series pulse interface in which pulsing is performed through the speech network. The second example will demonstrate a parallel pulse interface, with pulsing occurring through a parallel load resistor rather than through the network.

Example 4–1   Design of a Series Pulse Interface Circuit

Design a circuit with series pulse interfacing.

Solution

One example of a series interface pulse dialer circuit is shown in Fig. 4-3. The pulse dialer used is the Mostek MK50992.

A current source is desired to present a high break impedance to the telephone line while guaranteeing sufficient current (≥ 150 µA) to power the MK50992. The current source for this application is composed of diodes CR_1 and CR_2, resistors R_4 and R_5, and transistor Q_2. The emitter-to-base voltage of Q_2 is approximately equal to one diode drop ($V_{CR} \approx 0.4$ to 0.7 V). Therefore, the voltage across resistor R_5 is also approximately equal to one diode drop. The current through R_5 to the pulse dialer is determined by the value of R_5 according to the equation

$$I_{typ} \approx \frac{V_{CR1}}{R_5} \tag{4-1}$$

When the telephone is in the on-hook mode, hook switches S_1 and S_2 are open. This disables the current source and eliminates any excess current flow through the base-emitter junction of Q_2. A large-value resistor (R_{11}) allows a small amount of current to maintain the memory on the MK50992.

When going off-hook, switches S_1 and S_2 close, tying the On-Hook/Test input (pin 17) to V^- (pin 6). The $\overline{Pulse}$ (pin 18) and $\overline{Mute}$ (pin 12) outputs drive external transistors to perform the muting and pulsing functions. The receiver is connected to the speech network through transistor Q_3. $\overline{Mute}$ causes this transistor to be held on until the outpulsing begins.

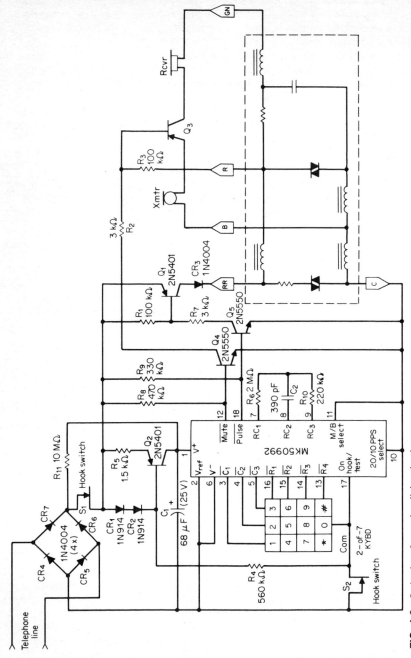

FIG. 4-3 Series interface pulse dialer circuit.

When $\overline{\text{Mute}}$ switches low, the receiver is disconnected from the speech network and the pops caused by breaking the line are isolated from the receiver. The $\overline{\text{Pulse}}$ output drives transistors Q_1 and Q_5 to make and break the line until the digit is completely outpulsed. $\overline{\text{Mute}}$ then switches high, returning the receiver to the speech network.

Example 4–2   Design of a Parallel Pulse Interface Circuit

Design a circuit with a parallel pulse interface.

Solution

The schematic diagram in Fig. 4-4 shows one method that can be used to perform a parallel pulse interface. In this application, the pulse dialer shown is the Mostek MK50981.

As in the previous example, a constant current source is used to supply the required operating current to the pulse dialer while maintaining a high break impedance. The current can be specified by properly selecting the value of R_1, according to the equation

$$I_{\text{typ}} \approx \frac{V_{CR_1}}{R_1} \tag{4-2}$$

In the on-hook mode, hook switches S_1 and S_2 are open, disabling the current source. Resistor R_3 supplies memory-retention current to the pulse dialer so that the redial function may operate properly.

To return off-hook, switches S_1 and S_2 are closed, tying On-Hook/Test (pin 15) to V^- (pin 6). The speech network is connected to the telephone line through transistor Q_5. $\overline{\text{Mute}}$ (pin 10) holds this transistor on until the outpulsing begins. The first break occurs when $\overline{\text{Mute}}$ switches low and the speech network is removed from the line. The pops caused by breaking the line are then isolated from the receiver. The $\overline{\text{Pulse}}$ output (pin 16) drives the Darlington pair Q_3 and Q_4 to make and break the line until the digit is completely outpulsed. $\overline{\text{Mute}}$ then switches high, returning the speech network to the line.

4–2c Design Considerations

In any type of subscriber-set application circuit some of the most important design considerations include:

1. Meeting the electrical specifications required by the telephone system to which the circuit will be interfaced.
2. Minimizing signal losses between the telephone line and the receiver and transmitter.
3. Obtaining optimal circuit performance with minimal parts count.

For pulse dialer application circuits, the telephone system specifications cover make-break ratio, pulse rate, and break impedance (see Sec. 4-2a). The make-break ratio is basically a function of the pulse dialer and is independent of the external circuitry. The pulse dialer used should therefore be selected to offer the make-break ratio required by the application. However, pulse rate and break impedance are dependent upon the external circuitry used, so the application circuit should be designed with these specifications in mind.

The pulse rate of a pulse dialer is directly proportional to the frequency of the

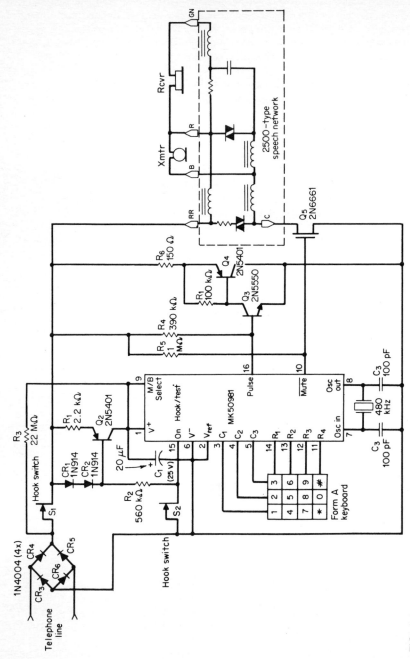

FIG. 4-4 Parallel interface pulse dialer circuit.

oscillator that is used. Most pulse dialers use either a ceramic resonator or an *RC* oscillator as a frequency reference. Ceramic resonators are very stable and are used in circuits requiring an accurate reference. *RC* oscillators, on the other hand, offer more versatility. The frequency of an *RC* oscillator (and therefore the pulse rate of the dialer) may be varied by changing the resistors and/or capacitors that make up the oscillator.

"Break impedance" is defined as the resistance presented to the telephone line by the application circuit during break. The break impedance is determined by the combination of all of the resistances in the circuit, including those of the pulse dialer and the speech network. Pulse dialer circuits frequently use a current source consisting of transistors, diodes, and resistors to supply current to the pulse dialer during make, while presenting a high impedance during break.

Another important design consideration for a pulse dialer circuit is the method of muting. During outpulsing, the receiver is muted so that the pops caused by breaking the line are isolated from the receiver. In most pulse dialer applications, the receiver is connected to the speech network through a muting transistor. This transistor turns on and off during pulsing to mute the receiver. The muting transistor can be either a bipolar transistor or an FET, depending upon the design goals of the circuit. Bipolar transistors are generally less expensive than FETs, but may cause greater signal loss than a comparable FET.

A frequent problem in applications using integrated-circuit dialers is device failure due to high voltage surges or transients on the line. There are a variety of transient protection devices on the market. The best one for any application will depend upon the specifics of the circuit.

To obtain optimal circuit performance with a minimum number of external components, it is important to carefully select the pulse dialer. There are a number of electronic pulse dialers on the market today with a wide variety of features. Some of these features are listed below. For a more complete listing of the pulse dialers offered by various manufacturers, see the Product Selection Guide (Sec. 4-9).

- Redial capability
- Pin-selectable pulse rates
- Pin-selectable make-break ratio
- Wraparound memory
- On-chip test capability for rapid outpulsing
- Power-up-clear
- Storable access pause
- Pacifier tone
- Pin-selectable interdigital pause

4-3   DUAL-TONE MULTIFREQUENCY ENCODERS

4-3a   Introductory Theory

Dual-tone multifrequency (DTMF) address signaling is used by the telephone industry to signal over the voice transmission path of a telephone system. DTMF signaling has various advantages over pulse signaling, such as faster dialing speeds and the ability to signal over any voice-grade transmission path. This method of signaling uses 16 distinct voiceband frequency signals, each consisting of two sinusoidal signals, one from a "low group" and one from a "high group" of frequencies. The characters that represent these DTMF signals are shown in Table 4-1.

The A, B, C, and D buttons are used in special applications and are not part of the common telephone keyboard.

In order to meet U.S. telephone system requirements, DTMF encoders must generate signals that will have the following characteristics when measured at a 600 Ω termination:

1. *Signal Levels*—DTMF signals should have a nominal level of −6 to −4 dBm per frequency. The minimum level for the low-group frequencies is −10 dBm, while it is −8 dBm for the high-group frequencies. The frequency pair should not have a signal level in excess of +2 dBm. The level of the high-group frequency component should equal or exceed that of the low-group frequency. This is a characteristic commonly called "preemphasis" or "twist." However, this difference in levels between the two frequencies should not be greater than 4 dB.

2. *Frequency Deviation*—Each of the 16 DTMF frequencies should be within ±1.2% of their nominal frequency values, or at worst case ±1.5%.

3. *Rise Time*—It should take no more than 5 ms (preferably 3 ms for automatic dialers) for each of the frequencies of the DTMF signal to go from minimum to 90% of the final magnitude of the two-frequency signal.

4. *Tone Distortion*—Tone distortion within the voiceband above 500 Hz should not exceed 10%. In this case, distortion is measured in terms of the total power of all extraneous frequencies accompanying the DTMF signal, relative to the sum of the power of the two fundamental frequencies.

These specifications are found in the Bell Technical Reference PUB47001 and in the EIA standard RS-470.

There are various ways a DTMF encoder may be designed to meet these DTMF signal characteristics. The most common type of encoder is the original *LC*-type encoder, which is still in wide use by many telephone companies. A common implementation of an *LC*-type encoder in a 500-type telephone set is shown in Fig. 4-5.

In the circuit shown, the DTMF encoder is within the dotted line. This DTMF encoder circuit consists of two independent tuned circuits, each consisting of a three-winding coil (A, A′, A″, and B, B′, B″) and a tuning capacitor (C_A and C_B). Windings A and B have four taps each. When a keyboard button is pushed, the corresponding S_A and S_B switches are actuated to select the desired DTMF frequencies at which the circuit will

TABLE 4–1 DTMF Frequencies

Nominal low-group frequencies (Hz)	Nominal high-group frequencies (Hz)			
	1209	1336	1477	1633
697	1	2	3	A
770	4	5	6	B
852	7	8	9	C
941	*	0	#	D

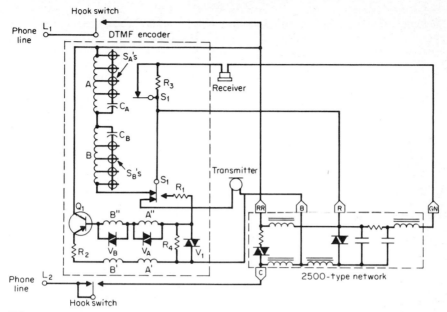

FIG. 4-5   *LC* tone encoder in a 2500-type application.

oscillate. Switch S_1 is common to, and actuated by, any keyboard button. Switch S_1, shown in the nontoning mode, produces four basic results when any button is pressed. First, it disconnects the receiver, leaving only R_3 to provide the user with a small amount of DTMF signal at the earpiece. Second, it disconnects the transmitter so that there is no interference of the tones caused by speech at the mouthpiece. Third, it interrupts the direct current through main windings A and B, thus shock-exciting the two tuned circuits. Fourth, it causes a bias voltage to be available to the transistor. Once the oscillator has been started, the two-frequency oscillation is sustained by feedback due to the transistor and transformer action between the secondary and tertiary windings of each coil. The transistor is operated linearly, and the amplitudes of the frequencies present at windings A'' and B'' are limited by varistors V_A and V_B. Thus, simultaneous oscillation at two frequencies is possible.

Due to innovations in the field of electronic integrated circuits, DTMF encoder circuits are now available in monolithic integrated-circuit form. Although there are many manufacturers of integrated DTMF encoders, the technique of DTMF tone generation used by the Mostek Tone II dialers (primarily the MK5087) will be the main point of discussion.

The MK5087 is a monolithic integrated circuit fabricated using the complementary-symmetry MOS (CMOS) process. A member of the Tone II family of integrated tone dialers, the MK5087 uses an inexpensive crystal reference to provide eight different audio sinusoidal frequencies, which are mixed to provide tones suitable for dual-tone multifrequency telephone dialing.

The MK5087 was designed specifically for integrated tone dialer applications that require wide supply operation with regulated output, auxiliary switching functions, single contact keyboard inputs, and a single tone inhibit option.

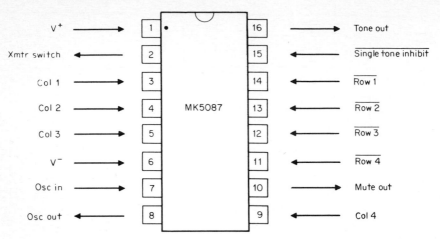

FIG. 4-6   MK5087 pin connections.

Keyboard entries to the Tone II family of integrated tone dialers cause the selection of the proper divide ratio to obtain the required two audio frequencies from the 3.58-MHz reference oscillator. Digital-to-analog (D/A) conversion is accomplished on-chip by a conventional R-2R ladder network. The tone output is a stair-step approximation to a sine wave and requires little or no filtering for low-distortion applications. The same op amp that accomplishes the current-to-voltage transformation necessary for the D/A converter also mixes the low- and high-group signals. Frequency stability of this type of

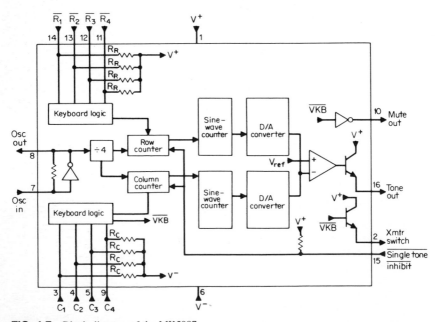

FIG. 4-7   Block diagram of the MK5087.

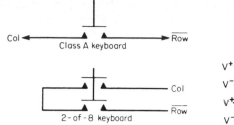

FIG. 4-8 Keyboard configurations for use with the MK5087.

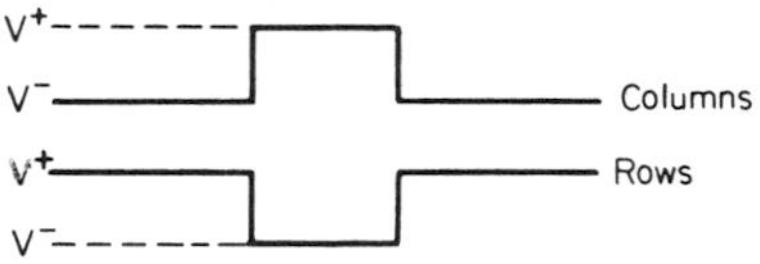

FIG. 4-9 Electronic input signals to the MK5087.

tone generator is such that no frequency adjustment is needed to meet standard DTMF specifications.

Pin connections for the MK5087 are shown in Fig. 4-6, and a block diagram is shown in Fig. 4-7.

The Xmtr Switch output (pin 2) is connected to the emitter of an on-chip bipolar transistor, whose collector is connected to V^+. With no keyboard input, this transistor is turned on and pulls pin 2 up to within V_{BE} of the V^+ supply. When a keyboard entry is sensed, this output goes open circuit (high impedance). The Xmtr Switch output switches regardless of the state of the Single-Tone Inhibit (pin 15) input.

The MK5087 features inputs compatible with the standard 2-of-8 keyboard, the inexpensive single-contact (Form A) keyboard, and electronic inputs. Fig. 4-8 shows how to connect to the two keyboard types, and Fig. 4-9 shows waveforms for electronic inputs. The inputs are static, i.e., there is no noise generation as occurs with scanned or dynamic inputs.

The internal structure of the MK5087 inputs is shown in Fig. 4-10. R_R and R_C pull in opposite directions and hold their associated input-sensing circuit turned off. When one or more row or column inputs are tied together, the input-sensing circuits sense the "$\frac{1}{2} V^+$ Level" and deliver a logic signal to the internal circuitry of the MK5087, causing the proper tone or tones to be generated.

When operating with a keyboard, normal operation is for dual-tone generation when any single button is pushed. Single tones are generated when one or more buttons in the same row or column are pushed, if the Single-Tone Inhibit input (pin 15) is pulled to V^+ (pin 1) or left floating. Activation of diagonal buttons will result in no tones being generated.

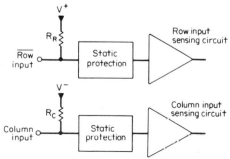

FIG. 4-10 Row and column inputs of the MK5087.

When the inputs to the MK5087 are electronically activated, as shown in Fig. 4-9, inputs to a single row and column will result in generation of that dual-tone digit. Input to a single column will result in that column tone being generated. Input to multiple columns will result in no tone being generated.

Activation of a single row is not sensed by the internal circuitry of the MK5087. If a single-row tone is desired, two columns must be activated along with the desired row.

The MK5087 contains an on-board inverter with sufficient loop gain to provide oscillation when working with a low-cost TV color-burst crystal. The inverter's input is Osc In (pin 7) and the output is Osc Out (pin 8). The circuit is designed to work with a crystal cut to 3.58 MHz to give the frequencies in Table 4-1. The oscillator is disabled whenever a keyboard input is not sensed. Any crystal frequency deviation from 3.58 MHz will be reflected in the tone output frequency period. This is shown in Table 4-2; however, most crystals do not vary more than $\pm 0.02\%$.

The Mute output is a conventional CMOS gate that pulls to V^- with no keyboard input, and pulls to the V^+ supply when a keyboard entry is sensed. This output is used to control auxiliary switching functions that are required to actuate upon keyboard input. The Mute output switches regardless of the state of the Single-Tone Inhibit input.

The Single-Tone Inhibit input (pin 15) is used to inhibit the generation of other than dual tones. It has a pull-up to the V^+ supply and, when left floating or tied to V^+, single or dual tones may be generated as described in the paragraph about row-column inputs. When forced to the V^- supply, any input situation that would normally result in a single tone now results in no tone, with all other chip functions operating normally.

The Tone output (pin 16) is connected internally in the MK5087 to the emitter of an *npn* transistor whose collector is tied to V^+. The input to this transistor is the on-chip op amp that mixes the row and column tones together. The level of a dual-tone output is the sum of the levels of a single-row and single-column output. This level is controlled by an on-chip reference that is not sensitive to variations in the supply voltage. The row and column output waveforms are shown in Figs. 4-11 and 4-12. These waveforms are digitally synthesized using on-chip D/A converters. Distortion measurement of these unfiltered waveforms will show a typical distortion of 9% or less. Spectral analysis

TABLE 4–2 MK5087 Output Frequency Deviation

	Standard DTMF (Hz)	Tone output frequency using 3.579645-MHz crystal	% Deviation from standard
Row			Low group
f_1	697	701.3	+0.62
f_2	770	771.4	+0.19
f_3	852	857.2	+0.61
f_4	941	935.1	-0.63
Col			High group
f_5	1209	1215.9	+0.57
f_6	1336	1331.7	-0.32
f_7	1477	1471.9	-0.35
f_8	1633	1645.0	+0.73

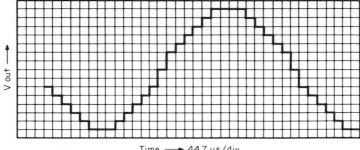

Time ⟶ 44.7 μs/div.

FIG. 4-11   Row 2 tone output of the MK5087.

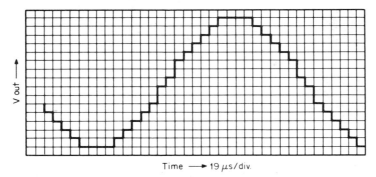

Time ⟶ 19 μs/div.

FIG. 4-12   Column 4 tone output of the MK5087.

of these waveforms will show that typically all harmonic and intermodulation distortion components will be −30 dB when referenced to the strongest fundamental (column tone).

A common method of dual-tone distortion measurement is the comparison of total power in the unwanted components (i.e., intermodulation and harmonic components) with the total power in the two fundamentals. For the MK5087 dual-tone waveform, THD is −20 dB maximum.

As noted earlier, in the discussion on *LC*-type DTMF encoders, it is necessary to mute the transmitter and receiver when generating DTMF signals in a telephone set. The transmitter needs to be switched out during toning so that unwanted speech will not interfere with DTMF signals being provided to the telephone line. The receiver must also be switched out or muted during tone generation so that only a comfortable amount of signal will be present at the user's ear. These switching functions can be accomplished in both the *LC*-type and the integrated-circuit encoder, using a keyboard with "common function switching." In such a scheme, any of the buttons on a keyboard will activate a set of mechanical switches to perform the necessary switching functions. However, in integrated DTMF encoders, such as the MK5087, auxiliary switching function outputs are provided. These outputs may be used to drive external transistors to perform the necessary switching functions. Thus, the need for an expensive keyboard with common function switching is eliminated.

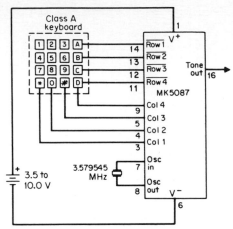

FIG. 4-13   A basic DTMF encoder circuit.

4–3b   Application Solutions and Examples

Example 4–3   Basic Implementation of a DTMF Encoder

Show the minimum number of components necessary for the basic implementation of a DTMF encoder.

Solution

Basic implementation of integrated DTMF encoders is generally quite simple. Most integrated DTMF encoders require only a frequency reference, a
power supply, and a keyboard for operation. For example, Fig. 4-13 shows
the operation of a Mostek MK5087 encoder, using only a keyboard, a 3.58-
MHz crystal (TV color-burst crystal), and a 3.5- to 10.0-V supply.

Example 4–4   Design of a Tone Dialing Telephone Using Common Function Switching

Design a DTMF encoder application for use in a tone dialing telephone
using a 2500-type network. Design the circuit to utilize mechanical switching
to mute the receiver and disconnect the transmitter.

Solution

The most common use for DTMF encoders is in tone dialing telephone sets.
DTMF tones are produced on the telephone line by modulating the telephone
loop current through a resistive load. When tone dialing, two functions must
typically occur in addition to the DTMF tone generation. The receiver should
be muted so that only a low level of the DTMF signals is provided to the
user. The transmitter should also be disconnected to increase the DC resistance of the speech network (to allow for the tone generation) and to
eliminate distortion of the DTMF tones during toning. These two functions
can be accomplished by using the standard telephone keyboard's mechanical
common function switches, as shown in Fig. 4-14. In this figure, keyboard
common function switch K_1 is used to mute the receiver by placing resistor
R_1 in series with it; K_2 is used to disconnect the transmitter. Switches S_1 and

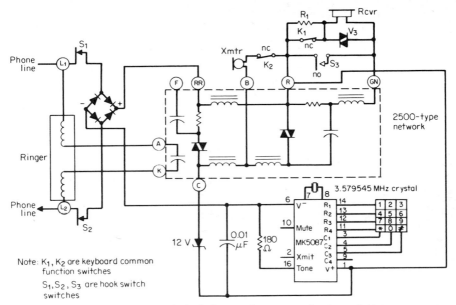

FIG. 4-14 DTMF encoder in a typical telephone, using common function switching.

S_2 are hook switches used to disconnect the telephone from the line upon hanging up. Switch S_3 is a hook switch that, upon hanging up, closes prior to S_1 and S_2 opening. Thus, the receiver is shunted so that no transients (caused by disconnecting the circuit) will be heard by the user.

Example 4–5   Design of a Tone Dialing Telephone Using Electronic Switching

Design a tone dialing telephone application for a DTMF encoder. The circuit should utilize electronic switching to mute the receiver and disconnect the transmitter.

Solution

The design of this circuit is similar to the design described in Example 4-4, except that electronic switching is used to perform the functions of muting the receiver and disconnecting the transmitter. The Mute output of the integrated DTMF encoder is used to drive transistors to perform the switching functions, as shown in Fig. 4-15. In this figure, Q_2 is used to disconnect the transmitter and Q_3 is used to mute the receiver.

Example 4–6   Design of an Acoustical Dialer

Design a battery-powered acoustical DTMF dialer.

Solution

(a) Many systems in use today require DTMF tone input from a telephone line to control an event or enter data. For instance, a business executive might need to use DTMF dialing from a remote telephone location to "talk" to a computer for remote order entry or status verification. Since many areas of the country do not have tone dialing, there is no convenient method to

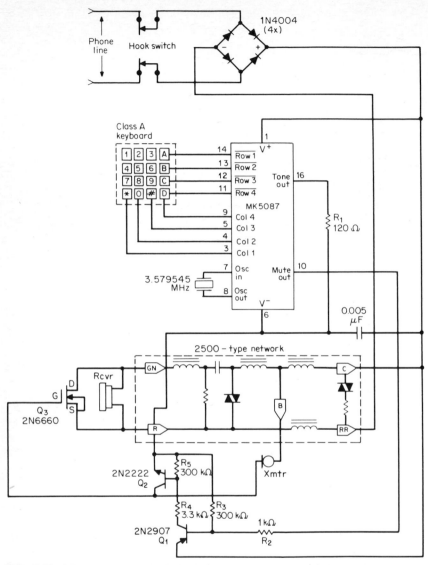

FIG. 4-15 DTMF encoder in a typical telephone, using electronic switching.

generate the required signals. The simplest solution to this problem is an acoustically coupled battery-powered DTMF tone generator. Fig. 4-16 shows a simple circuit that may be used for this application. The speaker may be any convenient type that will fit in the required space. Switch S_1 should be turned off whenever tone generation is not needed.

(b) Typical operation with the acoustical dialer will be to first dial the desired number via the rotary dial on the telephone. After the remote computer or

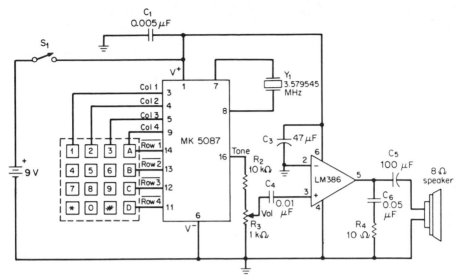

FIG. 4-16   Acoustic DTMF dialer.

data entry point answers the telephone, the acoustical dialer is held up to the transmitter of the telephone and the data is entered as desired.

4–3c   Design Considerations

When designing a circuit using a DTMF encoder there are several factors to be considered. One of the first factors to be considered when designing a DTMF encoder in a system is the power-supply requirements of the DTMF encoder. If the DTMF encoder is to be used in a telephone dialer powered by the telephone line, it should have a wide operating voltage range (i.e., 3.5 to 10.0 V) with low power consumption. The wide operating voltage range is necessary so that the DTMF dialer will operate properly over the wide range of telephone loop currents (typically 20 to 80 mA).

Another important factor to take into account is whether or not the output levels of the DTMF encoder are regulated or vary directly with power-supply variations. In an application where a fixed supply is not readily available, such as in a standard telephone dialer application, it is important for the DTMF encoder to have a regulated tone output in order to meet U.S. telephone system specifications. The range of values of load resistance that the tone output of the encoder is capable of driving should be considered along with the range of tone output levels. If the encoder is used in a telephone dialer, it is desirable that it be able to drive the telephone line through a small resistive load (i.e., 100 to 200 Ω) to produce DTMF tone levels within specifications. The distortion of the DTMF tones should be 10% or less to meet specifications (DTMF tone distortion is defined in Sec. 4-3a).

Keyboard input schemes should be considered when choosing a DTMF encoder for use in a particular application. DTMF encoders typically use either static or scanned keyboard inputs. Most encoders with static keyboard inputs require the use of a 2-of-8 keyboard with the common contact connected to either V^+ or V^-. This scheme typically makes the encoder less susceptible to RF interference than those using a scanned keyboard scheme. The advantage of scanned inputs is that an inexpensive Class A, or matrix, keyboard can typically be used. However, there are some encoders, such as Mostek's MK5087, that have static keyboard inputs that require only a Class A keyboard (but are more susceptible to RF interference). This is accomplished by having the row and column inputs pulled to opposite polarities of the supply, and having the MK5087 recognize a "$\frac{1}{2} V^+$ Level" as a valid input. Most encoders can be controlled electronically by presenting the desired row and column inputs with the necessary high or low logic levels. Some DTMF encoders also have inputs that require binary coded decimal (BCD) inputs to cause the generation of the desired tones.

The frequency reference and oscillator operation of the DTMF encoder should also be considered when designing a circuit. Most DTMF encoders use either a crystal or a ceramic resonator as a frequency reference. One of the most commonly used is the 3.58-MHz TV color-burst crystal. The 3.58-MHz crystal is generally more readily available and less expensive than crystals cut for other frequencies, such as 1.0 MHz. Ceramic resonators are typically quite inexpensive, but are not as accurate as a crystal reference. It is important to note that the oscillator of most DTMF encoders runs only when a key is depressed in order to lower standby current. This can cause problems when trying to "daisy-chain" the oscillator of several encoders, or to use the oscillator output signal to provide a reference frequency for other circuitry in the system. Many encoders stop their oscillator circuitry by turning on an internal transistor that pulls the oscillator input to one side of the supply. This can cause problems of loading down the input signal when trying to externally drive the oscillator circuitry.

The following is a listing of features useful in varied applications. These features are offered on DTMF encoders by the various manufacturers, as listed in the Product Selection Guide (Sec. 4-9).

- Wide operating voltage
- Class A or 2-of-8 keyboard inputs
- BCD inputs for selection of DTMF tone to be generated
- On-chip diode bridge
- 3.58-MHz frequency reference
- Inexpensive ceramic resonator for frequency reference
- Auxiliary switching output
- Separate Xmtr and Rcvr mute switching outputs
- High-band preemphasis on tone output
- Adjustable preemphasis on tone output
- Darlington tone output transistor
- Regulated tone output levels
- Separate high- and low-group tone outputs
- Separate on-chip bipolar transistor to use in construction of filter to meet European specifications
- Single-tone inhibit input

- Tone disable input

- Chip disable input

4-4 DTMF DECODERS

4-4a Introductory Theory

A DTMF decoder is necessary in any system using DTMF signaling. The purpose of a decoder is to decode a valid pair of signaling tones and provide output data corresponding to the DTMF signal received. To meet U.S. telephone system requirements, DTMF decoders must have the following characteristics (all references to units of dBm are with respect to 600 Ω).

1. The decoder should decode DTMF signals within ±1.5% of their nominal frequencies, and should not decode signals with either frequency deviating more than ±3.5% from the standard.
2. The decoder should decode DTMF tone bursts as short as 40 ms and recognize interdigital intervals as short as 40 ms. It should not recognize tone bursts or interdigital intervals shorter than 20 ms.
3. The decoder should decode DTMF signals in the presence of a dial tone that has each of its frequencies at a level of − 16 dBm ±3 dB.
4. The decoder should decode DTMF signals that have a power per frequency of − 25 to 0 dBm, with the high-frequency tone + 4 to − 8 dB relative to the low-frequency tone.
5. The decoder, in the presence of message circuit noise, should miss decoding less than 1 in 10,000 valid DTMF tone bursts.

These specifications are mainly found in the Bell Technical Reference PUB48002.

There are various ways to achieve the decoding of DTMF signals. One of the most obvious approaches is to use several precise, high Q, bandpass filters to separate each of the frequencies used in DTMF signaling. The drawback to this approach is that the design of the filters is complex, and the system is generally bulky and expensive if designed using discrete components. Even if integrated-circuit op amps are used in the construction of the filters, the circuit is still quite bulky and complex.

Another approach to designing a DTMF decoder is through the use of phase-locked-loop tone decoders. Seven or eight PLL tone decoders, each tuned to one of the DTMF frequencies (along with 12 or 16 NOR gates) can be used to form a relatively inexpensive DTMF decoder. The major drawback to this type of DTMF decoder is that it is fairly susceptible to "false hits" and is likely to drift in detection frequency due to the use of resistors and capacitors in its tuning. In other words, it is more likely to decode an invalid signal than are decoders using a different design.

High-performance DTMF decoders are currently available in monolithic integrated-circuit form. The Mostek MK5102 is one such DTMF decoder. The MK5102 requires a single + 5-V supply and a 3.58-MHz color-burst crystal as a frequency reference. To operate the MK5102 properly, an external band-separation filter is required to split the DTMF signal into its high-group and low-group components. However, band separation is not a very stringent requirement for the MK5102. As shown in Fig. 4-17, the MK5102 requires a band separation of only 33 dB. This allows for a signal-to-noise (S/N) ratio

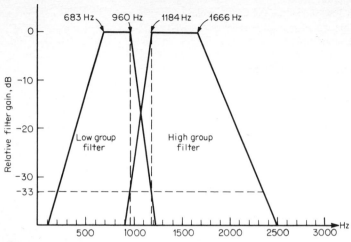

FIG. 4-17   DTMF band-separation filter requirements.

of 18 dB, a maximum twist of 6 dB, and a detection bandwidth of at least ±2%. A reduction of twist margin or S/N requirements will result in a correspondingly lower requirement for band separation. For example, with no twist margin, the band separation can be reduced to 27 dB. The plot shown in Fig. 4-17 depicts corner frequencies of 683, 960, 1184, and 1666 Hz, which represent a 2% deviation from the nominal DTMF frequencies of 697, 941, 1209, and 1633 Hz, respectively. This deviation is necessary to meet the DTMF decoder requirements stated earlier in this section.

The detection approach used in the MK5102 utilizes zero-crossing detection and digital period-counting. To increase the rejection of random noise and the residue from out-of-band components, an averaging scheme is used. Fig. 4-18a shows nine cycles of a symmetrical sine wave. If zero-crossings were the only detection criteria, and if the average period count obtained over nine periods were acceptable, then the signal in Fig. 4-18a would represent a valid tone. The jitter of the zero-crossings is integrated out by

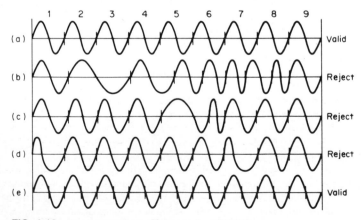

FIG. 4-18   Possible input waveforms to the MK5102.

the nine-period average. However, based on the simple nine-period average, the signal shown in Fig. 4-18*b* would be accepted as a valid tone. To improve rejection of this speech-type waveform, the nine-period detection time can be broken into three period-averaged subgroups, as indicated in Fig. 4-18*b*. By combining the nine-period average and the subgroup average criteria, 200 false hits are obtained on 30 min of standard speech tape. Fig. 4-18*c* represents a type of waveform that would produce a hit based on the nine-period and subgroup average algorithm. To improve rejection of this waveform, requirements must be placed on every single period in addition to the nine-period average and the subgroup average. However, the waveform of Fig. 4-18*d* will be detected using only these three criteria. Therefore, an additional requirement must be placed on each half-period of the waveform. Fig. 4-18*e* shows the only type of signal that will be accepted by a detection algorithm, which requires the following:

1. Valid nine-period average
2. Three valid subgroup averages
3. Valid single period
4. Valid half-period

Using these four criteria, the number of hits on a standard speech tape can be reduced to less than six.

The MK5102 has five output pins; the Strobe output (pin 4), which goes high whenever a valid DTMF signal is detected and is low otherwise, and four latched data outputs. These data outputs may have a 4-bit binary format, a dual 2-bit row/column format, or go to a high impedance state depending upon the state of the Format Control input. Using a decoder such as the MK5102, a high-performance DTMF receiver system can be obtained.

4–4b   Application Solutions and Examples

Example 4–7   Design of a PLL DTMF Receiver System

Design a DTMF receiver using 567 PLL tone decoders and NOR gates.

Solution

A DTMF receiver system can be designed using seven or eight 567 PLL tone decoders and 12 or 16 NOR gates, as discussed earlier in Sec. 4-4a. A circuit using this scheme is shown in Fig. 4-19. In this circuit each 567 decoder is tuned to one of the seven most common DTMF frequency components by resistor R_1 and capacitor C_1. Resistor R_2 is used to reduce the typical bandwidth of each of the decoders, and C_4 is used to decouple each of them. The 567 decoder outputs are normally high and go low when the frequency to which they are tuned is present. These outputs are connected to the NOR gates so that when any two tones comprising a valid DTMF signal are present, the corresponding NOR gate output will go high.

Example 4–8   Design of a DTMF Receiver System Using an Integrated DTMF Decoder

Design a high-performance DTMF receiver system using a DTMF decoder IC.

Solution

(a) Construction of a high-performance DTMF receiver system using a DTMF decoder IC, such as the MK5102, requires more than just the decoder

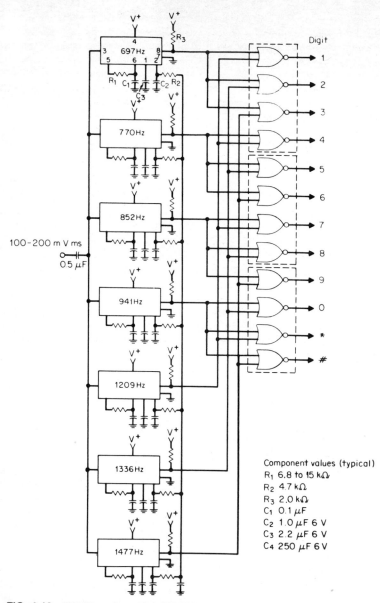

FIG. 4-19 DTMF receiver using 567 PLL decoders.

IC. Most integrated DTMF decoders require band-split filters (with the characteristics described in Sec. 4-4a), input squaring circuits, and, in some cases, an envelope decay detector to reduce the effects of ringing in hybrid filters. As described in Sec. 4-4a, to minimize the number of false hits, a detection algorithm must place stringent requirements on each half-period of the input waveform (high group or low group). To successfully meet these requirements, the duty cycle of the input waveform for the MK5102 must be between 49 and 51%. The input squaring circuit must therefore provide an output that accurately tracks the input without adversely affecting the duty cycle. Such a circuit, an inverting comparator with hysteresis, is illustrated in Fig. 4-20.

(b) In the squaring circuit shown in Fig. 4-20, C_1 is used to AC couple the filter output to the squaring circuit so that the DC bias present at the filter output will not affect the performance of the squaring circuit. R_3, R_4, and R_6 establish a bias level at about 2.5 V, and R_5 is used to provide the same bias level at the inverting input of the comparator used in the squaring circuit. The maximum input bias level at the inverting input is effectively the same as the voltage at the wiper of R_6. R_6 must be adjusted so that, for an input signal level of -28 dBm, the output duty cycle will be 50%. This adjustment compensates for the input offset voltage of the LM2901. R_L is the pull-up resistor for the open-collector output of the comparator. R_1 and R_2 set the hysteresis level.

(c) The output of the squaring circuit may be tied directly to the MK5102 if it meets the following requirements:

$$\text{Logic } \quad 1 \geq 4 \text{ V}$$
$$\text{Logic } \quad 0 \leq 1 \text{ V}$$

A squaring circuit with an output that does not meet these requirements must be capacitively coupled to the MK5102 with a 0.05-µF capacitor. The value of the coupling capacitor is critical because of the impedance of the bias circuit at the high-group or low-group input. The sudden appearance of a tone burst causes the DC bias point to shift upward. Until the DC bias returns to its normal level, the input comparator will not switch and the input signal will be ignored, causing an increase in the dual-tone detection time. Using a 0.05-µF capacitor will minimize the effect of this DC level shift. The peak-to-peak value of the coupled signal must be greater than 0.9 V, but less than the value of V^+.

(d) Since many commercially available hybrid filters exhibit a ringing char-

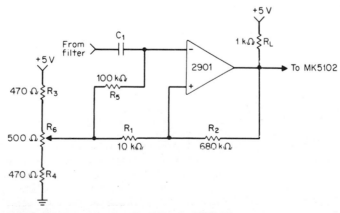

FIG. 4-20   Input squaring circuit for the MK5102.

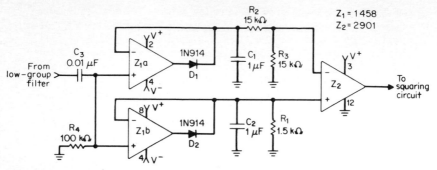

FIG. 4-21   Envelope decay detector for the MK5102.

acteristic at their output, additional circuitry is required to detect the beginning of ringing and squelch the output of the squaring circuit. The required
circuitry, an envelope decay detector, is shown in Fig. 4-21. The detector
consists of two precision rectifiers, two sample-and-hold circuits, and a
comparator. C_3 is used to couple the low-group filter output to the envelope
decay detector. Z_{1a}, D_1, C_1, R_2, and R_3 then rectify the incoming signal and
store a peak value. The $R_2/R_3/C_1$ time constant is set for 20 ms so that the
voltage at the inverting input of Z_2 will represent one-half the peak value of
the incoming signal. Z_{1b}, D_2, R_1, and C_2 also rectify the incoming signal and
store a peak value, but the time constant is set for 1.4 ms so that the voltage
at the noninverting input of Z_2 will represent the instantaneous peak value
of the incoming waveform. As long as the instantaneous value is greater
than one-half of the peak value, the comparator's output will be high.
However, as soon as the instantaneous value decreases to less than one-half
the peak value (this will occur as ringing begins), the comparator's output
will go low and inhibit the output of the squaring circuit. It is necessary to
provide only one envelope decay detector, since the MK5102 will treat the
absence of a low-group tone as an invalid dual tone. It is therefore interpreted
as an interdigital time.

(e) By putting together the filters, the squaring circuits, the envelope decay
detector, and the DTMF decoder, a high-performance DTMF receiver system
(as shown in Fig. 4-22) is obtained.

Example 4–9   Design of a Simple DTMF Receiver System

Design an inexpensive DTMF receiver system using as few components as
possible. Limit the number of ICs to two.

Solution

(a) Due to recent innovations in IC design using switched capacitor techniques, monolithic integrated-circuit DTMF band-split filters are now available. One such filter is the AMI S3525A. This device is a monolithic, CMOS,
switched-capacitor filter that uses a 3.58-MHz crystal as a time base and
has a buffered 3.58-MHz clock output that can drive the oscillator of a
DTMF decoder such as the MK5102. The S3525A also has on-chip comparators that can be used to construct adjustable squaring circuits.

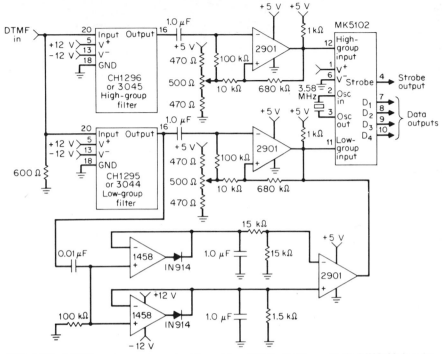

FIG. 4-22 DTMF receiver system using Cermetek or ITT North filters with the MK5102 decoder.

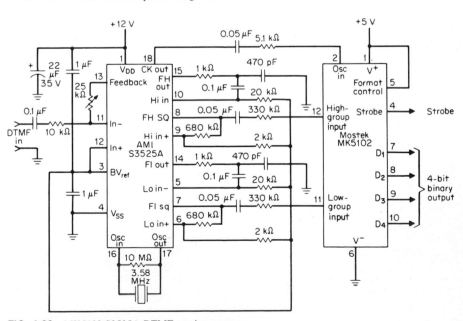

FIG. 4-23 MK5102/S3525A DTMF receiver system.

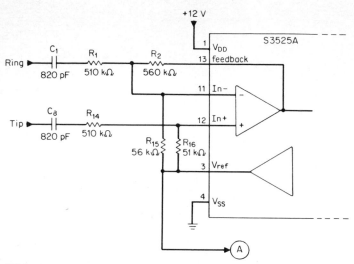

FIG. 4-24   High-impedance balanced connection to the S3525A.

(b) An entire DTMF receiver system may be constructed using only the S3525A DTMF band-split filter, the MK5102 DTMF decoder, and a few passive components, as shown in Fig. 4-23. Using this circuit configuration, the duty cycle of the signals provided to the MK5102 should be within the $50 \pm 1\%$ range required for reliable operation. For normal operation, the 25 kΩ potentiometer should be adjusted so that the filter has unity gain. The design of the squaring circuits is basically the same as that mentioned in Example 4-8(b) and shown in Fig. 4-20, except that the comparators are on-chip (on the S3525A) and the bias level to the inputs of the comparators is provided by the buffered output of an internal reference voltage.

(c) If direct connection of the filter to a telephone line, without the use of a transformer, is desired, the input circuitry shown in Fig. 4-24 provides a high-impedance, balanced connection with common mode rejection of 60 Hz and other noise on the line.

Example 4–10   Design of a Tone-to-Pulse Converter

Design a tone-to-pulse converter that will convert DTMF input signals into the output signals required for pulsing a telephone line as a rotary dial telephone would.

Solution

(a) In many parts of the country there are still central telephone offices that cannot accept DTMF signals. A tone-to-pulse converter would be very useful in situations such as these. The MK5102 has a 2-bit row/column code output option that is very useful when interfacing to a pulse dialer, as shown in Fig. 4-25. A pulse dialer is an electronic dialer that takes keyboard inputs and provides output signals to pulse the telephone line with the number of

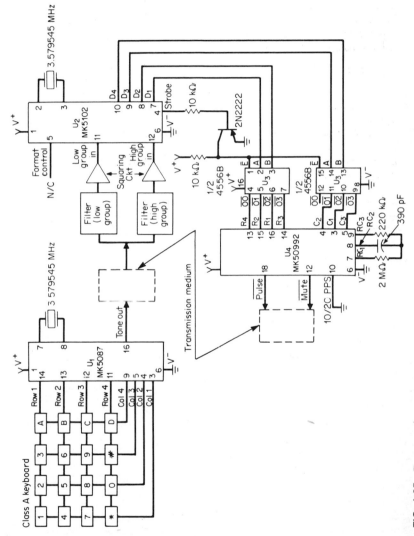

FIG. 4-25 Tone-to-pulse converter.

pulses corresponding to the key input. Pulse dialers are discussed in greater detail in Sec. 4-2.

(b) In this circuit, the MK5102, 4556B, and MK50992 combine to form a tone-to-pulse converter that allows the use of DTMF telephones in rotary exchanges. The DTMF tones are detected by the MK5102, which then generates the corresponding row/column code. Each 4556B then uses this 2-bit code to select one-of-four active low outputs. The MK50992 next interprets these signals as a valid key closure and generates a corresponding series of pulses.

Example 4-11   Design of a Remote Control System Using DTMF Signaling

Design a simple 16-channel remote control system using DTMF signaling.
Solution

(a) The 4-bit binary output of the MK5102 DTMF decoder can be used to drive a 4-to-16 line decoder to provide an output corresponding to each of the 16 possible DTMF signals. Thus each DTMF signal can control a different function.

(b) The circuit shown in Fig. 4-26 accomplishes a 16-channel remote control. In this circuit, a DTMF encoder with a 16-key keyboard is used to transmit 1-of-16 DTMF frequencies over a desired transmission medium to a DTMF receiver system using the MK5102 decoder. When a valid DTMF signal is detected, the Strobe output of the MK5102 goes high and the corresponding 4-bit binary code is latched into the data outputs. The 4514/4515 will then provide an output on one of the 16 outputs, corresponding to the 4-bit binary input it has received. As the Strobe and Inhibit connections show in Fig. 4-26, a 1-of-16 output will be present on the 4514/4515 only when a DTMF signal is present, causing the MK5102 Strobe output (pin 4) to be high. If a continuous output is desired, then the Inhibit input (pin 23) of the 4514/4515 should be pulled low.

(c) Using the circuit, a DTMF transmitter and a 16-key keyboard can be used to control any one of 16 different functions.

4-4c   Design Considerations

This section will discuss some of the factors to be considered when designing a circuit using a DTMF decoder. One of the first things to be considered in a DTMF receiver system design is noise. System noise will affect the operation of decoders, such as the MK5102, by causing the detection bandwidth to shrink. The instantaneous value of the low-group or high-group waveform is represented by the approximate relationship, $a = a_T$ $\sin \omega_T t + a_N \sin \omega_N t$, where a is the instantaneous amplitude of the overall waveform, a_T is the amplitude of the high-group or low-group component, and a_N is the amplitude of the noise. If the highly simplified noise term ($a_N \sin \omega_N t$) were removed, then the remaining term would represent a pure sine wave and the zero crossings of the waveform would be repeatable from cycle to cycle. All detection criteria would be present and the DTMF tone would be detected within a ± 2.0 to $\pm 2.9\%$ bandwidth. However, adding the noise term introduces instantaneous amplitude variations that will effectively alter the duty cycle of the sine wave by causing the zero-crossing points to jitter. If 0.5% jitter is caused by system noise, detection bandwidth will be decreased by 0.5%. Therefore, as the system noise level increases, the detection bandwidth will decrease.

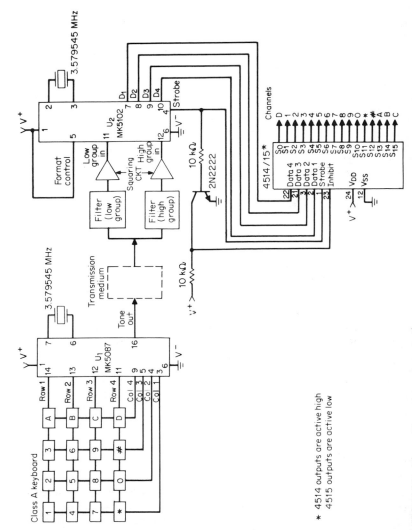

FIG. 4-26 Sixteen-channel remote control.

4-31

Just as system noise introduces jitter in the signal presented to the decoder, poor squaring circuits will also introduce jitter. Thus, poor squaring-circuit designs will cause a decrease in detection bandwidth. If the jitter is so bad that it exceeds the $50 \pm 1\%$ duty cycle input requirements of the MK5102 decoder, the circuit will not decode reliably.

An important factor to consider when designing the circuitry to be connected to DTMF decoder outputs is the format of the output data. Most DTMF decoders have a 4-bit binary output format that provides a standard BCD-type code for digit 1 through 9 inputs. However, digit 0 corresponds to 10 in BCD, and the output for digit D corresponds to 0 in BCD. This should be kept in mind when interfacing the decoder outputs with a device such as a BCD-to-decimal decoder.

The following is a listing of features useful in varied DTMF decoder applications. These features are offered on DTMF decoders by various manufacturers, as listed in the Product Selection Guide (Sec. 4-9).

- Decode all 16 standard DTMF digits
- Single power supply required
- Digital-detection schemes
- Pin-selectable output format
- Three-state outputs
- Latched outputs
- 4-bit binary output code
- Dual 2-bit row/column code
- 2-of-8 output code
- Early steering output
- Adjustable acquisition and release times
- Option to inhibit detection of * and # digits
- DTMF filters and decoder in one package

4–5   REPERTORY DIALERS

4–5a   Introductory Theory

The purpose of repertory dialers is to allow the user the possibility of automatically dialing one of many previously programmed numbers with a minimal amount of key entry commands. A repertory dialer may automatically dial, using pulse signaling as used in a rotary dial telephone, or it may use DTMF signaling. The automatic dialing may be accomplished with the use of mechanical or electronic equipment.

To meet U.S. telephone system requirements, repertory dialers must have the following characteristics:

1. Repertory dialers using dial pulse signaling.
    a. The dial pulse signaling should consist of a sequence of momentary breaks in the telephone loop current corresponding to the numerical value of each digit, except digit ''0,'' which should be represented by 10 break intervals.

b. The pulsing of the telephone loop current should operate at repetition rates between 8 and 11 pulses per second (pps), with 10 pps being nominal.

c. For an automatic dialer the interdigital interval should be between 600 ms and 3 s.

d. During the break intervals of dial pulse signaling, the steady-state resistance from tip to ring with tip or ring grounded should be at least 50 kΩ.

2. Repertory dialers using DTMF signaling.

a. The repertory dialer should have all of the DTMF signaling characteristics mentioned in Sec. 4-3a.

b. When automatically dialing, the DTMF tone bursts should have a minimum duration of 50 ms and minimum cycle time of 100 ms.

c. When automatically dialing, the interdigital interval should be between 45 ms and 3 s. A recommended dialing rate for automatic dialers is 7 to 10 tone bursts per second.

These specifications are typically found in such publications as EIA standard RS-470 and Bell Technical References PUB47001 and PUB47002. Various mechanical means have been used to create repertory dialers, but in this section we will only discuss electronic approaches. Most electronic repertory dialer systems are designed around a specially programmed microprocessor or similar control circuit. This control circuit will typically interface to external memory for storage of the repertory of telephone numbers. Some of these microprocessor repertory dialers have been programmed to generate dial pulse signals or DTMF signals on-chip, while others interface to pulse dialer ICs or DTMF dialer ICs to perform the actual telephone line signaling.

One example of a microprocessor-type repertory dialer is the Mostek MK5170. The MK5170 will interface to external static memory to provide a repertory of 10, 24, 50, or 100 20-digit telephone numbers. It has many options that make it very useful in the design of a full-feature repertory dialer. These features include the capability to interface to a 22-digit display, a digital real-time 12- or 24-hour clock, and a 100-hour timer.

Due to recent innovations in semiconductor design, there are now single-chip repertory dialers that require no external memory and operate directly off the telephone line, requiring only a small battery for memory retention. One such repertory dialer is the Mostek MK5175. The MK5175 will operate from the telephone line with supply voltages from 2.0 to 10.0 V. It will store ten 16-digit telephone numbers, including the last number dialed and will dial the numbers as a pulse dialer. The MK5175 will interface to Mostek DTMF dialers for DTMF repertory dialing. All of the repertory dialing features may be accessed through an inexpensive Form A keyboard or a standard 2-of-7 keyboard.

4–5b Application Solutions and Examples

Example 4–12   Design of a Simple Ten-Number Pulse Repertory Dialer

Design a simple repertory dialer that will have a ten-number repertory and will dial the numbers using pulse signaling.

Solution

(a) The simplest approach to this problem is to design a circuit using a repertory dialer IC, which was designed for this type of application. The Mostek MK5175 is one such device. The schematic diagram in Fig. 4-27 shows one method that can be used to interface the MK5175 with the telephone line. In the approach shown, the MK5175 is in the pulse dialer mode and the pulsing circuitry is in series with the speech network.

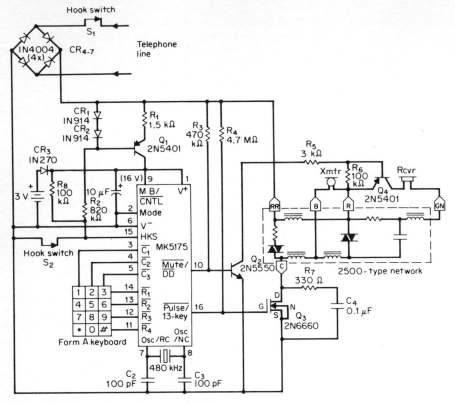

FIG. 4-27 Ten-number pulse repertory dialer.

(b) A current source is desired to present a high impedance to the telephone line while guaranteeing sufficient current to power the MK5175 while off-hook and dialing. The current source shown is constructed using diodes CR_1 and CR_2, resistors R_1 and R_2, and transistor Q_1. Other implementations, such as a constant current diode, may be considered. A 3-V battery has been included in the circuit to provide current to the MK5175 to retain the numbers stored in memory and to provide the power necessary for the on-hook entry and storage of numbers.

A diode bridge is used to ensure the proper voltage polarity for the MK5175, and hook switch S_1 is used to connect the circuit to the telephone line. Hook switch S_2 is used to provide the logic level necessary at HKS (pin 15) to set the MK5175 in its off-hook mode. Pin 2 (Mode) is connected to V^- to set the MK5175 in the pulse dialer mode. In this mode, pins 7 and 8 are defined as the oscillator pins and pins 9, 10, and 16 are defined as make-break (M/B) ratio select, $\overline{\text{Mute}}$, and $\overline{\text{Pulse}}$, respectively.

(c) The $\overline{\text{Pulse}}$ and $\overline{\text{Mute}}$ outputs drive external transistors to perform the outpulsing function. Resistor R_7 and capacitor C_4 are connected across transistor Q_3 for suppression of noise-producing sharp voltage rises generated during outpulsing. The receiver is connected to the speech network through transistor Q_4. $\overline{\text{Mute}}$ causes the transistor to be held on until outpulsing begins. When $\overline{\text{Mute}}$ switches low, the receiver is removed from the network. The transients caused by breaking the line are then isolated from the receiver. The $\overline{\text{Pulse}}$ output drives transistor Q_3 to make and break the line until the digit has been completely outpulsed. $\overline{\text{Mute}}$ then switches high, returning the receiver to the speech network.

(d) The operation of the MK5175 in this circuit is explained as follows:

Operation—During normal dialing, each digit is stored in the last number dialed (LND) buffer, location 0. The telephone number dialed can be left in this temporary LND buffer for later use or it can be copied into any of the other nine permanent memory locations.

Storage—Telephone numbers to be automatically dialed by the MK5175 may be entered into the LND buffer while either on-hook or off-hook. However, the MK5175 must be in the on-hook mode for a number to be copied into a permanent memory location. A number may be copied and stored by entering the key sequence **, followed by the address (1 to 9) of the memory location in which the number is to be stored. This operation requires 400 ms before going off-hook or reinitiating the store function. Information present in the LND buffer is erased when new data is entered and cannot be recalled.

Automatic dialing—The automatic dialing function is implemented by going off-hook and entering an *, followed by the address (1 to 9) of the desired telephone number. Dialing will begin with the depression of the address key. The LND buffer will contain the information last entered. A key sequence of * 0 will cause the last number entered to be redialed. More than one number sequence may be automatically dialed from memory without returning on-hook.

Pause/continue command—The MK5175 has a feature that allows an indefinite pause to be programmed into the first 15 digits of a number sequence by entering a # at the point in the sequence where a pause is desired. As the number is automatically dialed, the circuit will stop dialing when the pause is encountered. Any key entry after the interdigital pause (except the * key) will cause the MK5175 to continue dialing the remainder of the number. If more than one pause was originally programmed into the number, a corresponding number of continue commands must be entered to cause the entire number to be dialed.

(e) Some dialing examples for this circuit using the MK5175 would be as follows:

1. On-Hook, enter 323-6000
    Then enter * * 5
    323-6000 is stored in location 5
    ⋮
    Come off-hook
    Enter * 5
    323-6000 is automatically dialed

2. Off-Hook, dial 42 (PBX access code)
   While waiting for dial tone, enter #
   Dial 1-214-323-6000
   Busy/Hang up
   Enter * * 3
   (Number is stored in location 3)
   ⋮
   Come off-hook
   Enter * 3
   42 is dialed
   Wait for dial tone
   Enter 3 (continue command)
   1-214-323-6000 is dialed

Example 4–13   Design of a Simple Ten-Number Tone Repertory Dialer

Design a simple repertory dialer that has a ten-number repertory and will dial using DTMF signaling. Use a minimum number of components.

Solution

(a) An approach requiring only two 16-pin ICs could be developed by interfacing an MK5175 repertory dialer to a Mostek MK5380 DTMF dialer. The circuit shown in Fig. 4-28 illustrates one method that can be used to accomplish this.

(b) In the application shown in Fig. 4-28, tone mode operation of the MK5175 is selected by connecting Mode (pin 2) to V^+ (pin 1). In the tone mode only, a resistor (R_1) and a capacitor (C_1) are needed to provide the frequency reference for the MK5175. The oscillator frequency is given by the following equation:

$$f_{osc} = \frac{1}{R_1 C_1} \qquad (4\text{-}3)$$

A nominal frequency of 8 kHz will provide a tone rate of 100 ms on and 100 ms off, when automatically dialing a number. This tone rate is directly proportional to the oscillator frequency.

(c) In the tone mode, the MK5175 provides the user with the option of selecting whether the * and # keys or a 13th key (a control key) is used to initiate control functions. In this application, $\overline{\text{Pulse}}$/13-Key (pin 16) is connected to V^- (pin 6), thus selecting the 12-key mode. Therefore, the * and # keys are used in control functions and must be depressed twice consecutively in order for their corresponding DTMF tones to be produced by the MK5380.

(d) In the circuit shown in Fig. 4-28, the MK5175 and MK5380 are powered directly from the telephone line. A 3-V battery provides power for on-hook operation and memory retention. (Without a battery, the circuit will function as a nonrepertory tone dialer.)

The MK5175, in the tone mode, features a bidirectional keyboard scheme that passively monitors key inputs and simulates key closures so that a tone dialer will perform the repertory tone dialing function. Thus, the MK5175 and MK5380 keyboard inputs are connected together and share the same keyboard. The repertory dialer disables the tone dialer ($\overline{\text{Mute}}$/DD,

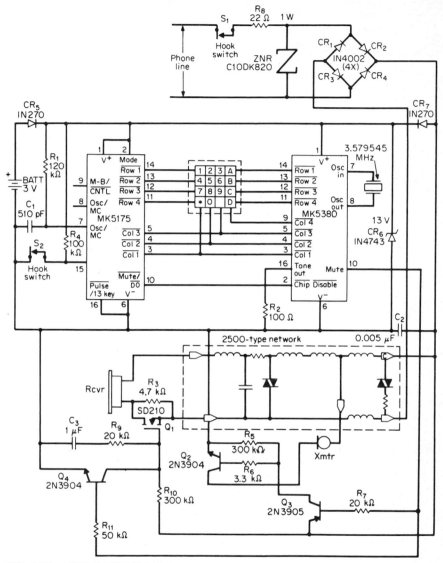

FIG. 4-28 MK5175/MK5380 DTMF repertory dialer.

pin 10) of the MK5175, pulls Chip Disable (pin 2) of the MK5380 low, and scans the keyboard whenever a command entry is detected. To prevent the keyboard input circuitry of the MK5380 from interfering with the scanning process of the MK5175, the chip disable feature of the MK5380 was designed so that the MK5380 keyboard inputs go to a high-impedance state when Chip Disable (pin 2) is pulled low.

During keyboard entries to the MK5380 (from the keyboard or the MK5175), the MK5380 Mute output switches high, activating the switching circuitry (composed of transistors Q_1–Q_4, resistors R_3–R_7 and R_9–R_{11}, and capacitor C_3) to mute the receiver and disconnect the transmitter. The receiver is muted so that only a low-level DTMF signal is provided to the user. The transmitter is disconnected to increase the DC resistance of the speech network to allow for the tone generation and to eliminate distortion of the DTMF tones due to noise present at the transmitter. The MK5380 produces tones on the telephone line by modulating the loop current through resistor R_2, which is connected from Tone Out (pin 16), to V^- (pin 6).

A diode bridge, composed of CR_1–CR_4, is used to protect the circuit from telephone line polarity reversals. A ZNR (Z_1), a resistor (R_8), a zener diode (CR_6), and a standard 2500-type speech network provide overvoltage and line-surge protection.

(e) The storage and automatic dialing operations are implemented in the same manner described in parts (d) and (e) of Example 4-12.

Example 4–14   Design of a Full-Feature Repertory Dialer System

Design a full-feature repertory dialer system that has the capability of storing as many as one hundred 20-digit numbers. This telephone should have the ability to display the location address and the number stored in that location. It also should have the following options:

1.   Dialing a stored number by the depression of a single key
2.   A real time 12- or 24-hour clock
3.   A 100-hour timer

Solution

(a) A block diagram for a full-feature repertory dialer system meeting these requirements is shown in Fig. 4-29. This diagram illustrates a repertory dialer system based on Mostek's MK5170 microprocessor-type repertory dialer IC.

(b) One of the first items to be considered in the design of a full-feature repertory system is the keyboard interface to the circuit. A 2-of-8 keyboard, using the interface shown in Fig. 4-30, provides the system control functions of number entry, Enter/Dial, Store, Infinite Pause, and Clear. A quad comparator (an LM2901 or equivalent) is used as a buffer between the keyboard and the MK5170. The noninverting input of each comparator is biased on $\frac{1}{2}$ 5 M (5 M is the 5-V memory supply, which has battery backup), and the appropriate noninverting inputs are tied to their assigned row or column. A key closure will pull two of the noninverting inputs within two diode drops of ground, thus causing the associated comparator outputs to go low. The MK5170 then senses these two low levels and, after identifying the key, drives the $\overline{\text{Dialing}}$ (pin 23) output low. This pulls the inverting inputs within one diode drop of ground and causes all of the comparator output transistors to turn off, thus isolating the keyboard from the MK5170 so that the MK5170 can apply row and column information to the dialer without interference from the keyboard.

(c) The next important factor to consider in the system design is the interface of the MK5170 to the pulse dialer, or DTMF dialer, and of the dialers to the telephone line. The dialer interfaces shown in Figs. 4-31 and 4-32 include level conversion circuitry as well as an inhibit dial feature. The active low

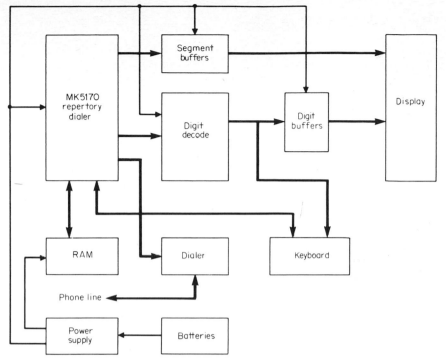

FIG. 4-29 Block diagram for a full-feature repertory dialer system.

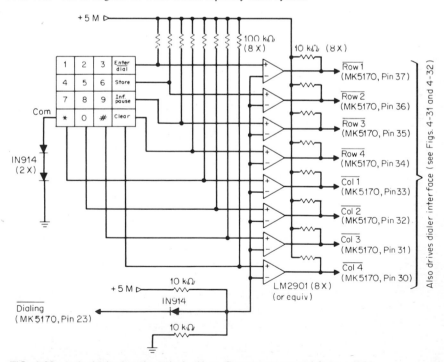

FIG. 4-30 A 2-of-8 keyboard terminal. (*Note:* Comparator powered from +5M.)

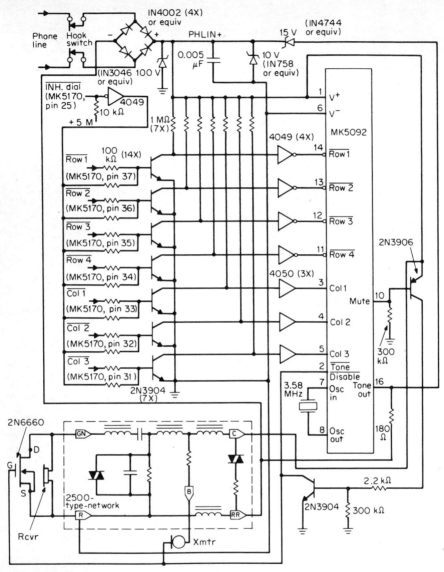

FIG. 4-31 DTMF dialer interface. (*Note:* 4049 and 4050 buffers are powered from PHLIN⁺.)

row and column signals ($\overline{\text{Row 1}}$/$\overline{\text{Col 3}}$) are applied to the bases of *npn* transistors. Because the power supply to the tone dialer can vary between 3.5 and 10.5 V and the power supply to the pulse dialer can vary between 2.5 and 6.0 V, level conversion is required between the 0- to 5-V row and column signals and the corresponding inputs to the MK5092 or MK50982. This required level conversion is provided by the seven *npn* transistors. When the MK5170 is required to isolate the dialer from the 2-of-8 keyboard

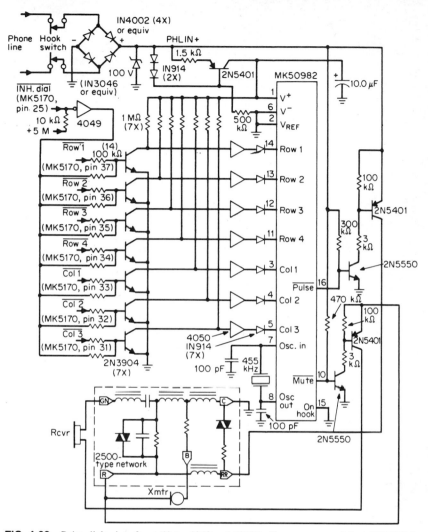

FIG. 4-32 Pulse dialer interface. (*Note:* 4049 and 4050 buffers are powered from PHLIN⁺.)

(i.e., during data entry), it does so by driving the $\overline{\text{Inh Dial}}$ (pin 25) line low. The $\overline{\text{Inh Dial}}$ signal is inverted and used to turn all of the *npn* transistors on so that the signals applied to the row and column inputs of the MK5092 and MK50982 will be at their inactive level. When the MK5170 starts a dialing cycle, the $\overline{\text{Inh Dial}}$ line will go high and the required sequence of row and column signals will be applied to the bases of the *npn* level converters. The 4050 buffers and 4049 inverting buffers shown in the dialer interface schematics are powered from the phone line and provide the buffering and logic inversion necessary to meet the input requirements of the MK5092 and MK50982, as well as permitting the use of 1-MΩ pull-up resistors at the collectors of the *npn* level converters. Using large-value pull-up resistors

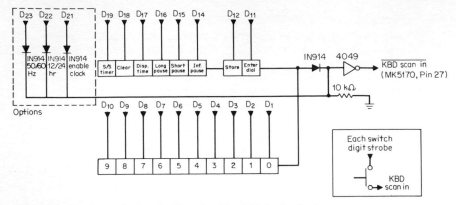

FIG. 4-33 Full-feature keyboard. (*Note:* See Fig. 4-35 for D_1–D_{23}.)

reduces the amount of current that the buffer circuitry draws from the phone line.

(d) The full-feature repertory dialer system of Fig. 4-29 implements all the features provided by the MK5170 by placing all of the keys and option diodes in a 3 × 24 matrix (shown in Figs. 4-33 and 4-34). $\overline{\text{Kbd Scan In}}$, $\overline{\#0-\#15}$, and $\overline{\#16-\#31}$ are scanned by the MK5170. When an active low level is detected on any of the three scan lines, the MK5170 determines which key has been pressed and takes appropriate action. Digit strobes for

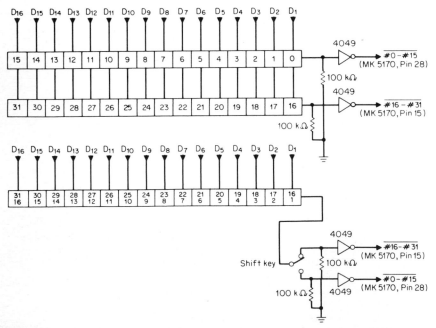

FIG 4-34 Two configurations for single key dialing. (*Note:* See Fig. 4-35 for D_1–D_{16}.)

scanning the input matrix are generated by the circuit shown in Fig. 4-35, which encodes the 5-bit digit select code (DS_0–DS_4) into 1-of-23 digit strobes (D_1–D_{23}). These positive pulses are then used to drive the bipolar digit drivers for the display as well as for scanning the input matrix. The Blank Display (pin 22) input is provided to inhibit the generation of digit strobes during memory access, since the segment and digit encoding lines are shared by memory addresses and read/write control (see Fig. 4-35). The suggested LED drive circuitry is shown in Fig. 4-36.

(e) The power-supply and time-base reference circuit, used for the real-time clock shown in Fig. 4-37, provides two independent 5-V power supplies:

1. +5 L, which powers the MK5170 and the display circuitry
2. +5 M, which powers the memory, the memory protect logic, and the keyboard buffers and is provided with a battery backup that consists of 6 NiCd cells and a charging circuit

The time-base reference consists of a comparator with hysteresis so that power line noise will not reach the MK5170.

(f) The memory protect logic shown in Fig. 4-38 provides two functions:

1. The MK5170 will be reset if its power supply dips to 4.75 V.
2. On power-up and during a power-supply dip to 4.75 V, $\overline{CE_1}$ (pin 13) and $\overline{CE_2}$ (pin 14) will be forced high so that no memory chip selects

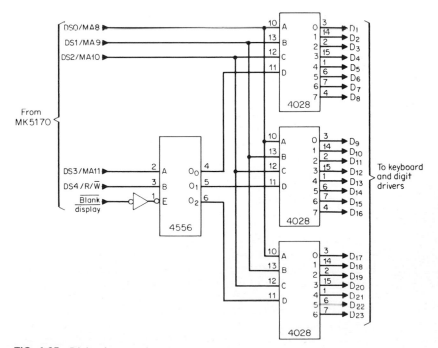

FIG. 4-35   Digit select encoder.

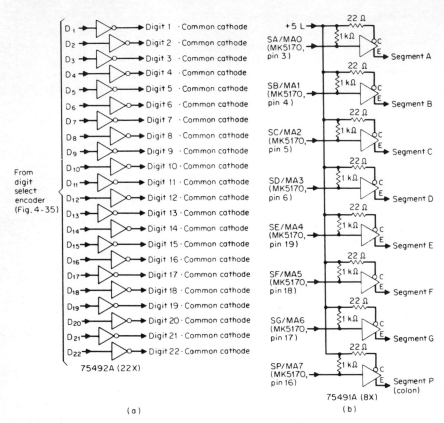

FIG. 4-36 LED drive circuitry. (*a*) Digit drive; (*b*) segment drive.

can be generated. This ensures that data stored in the MK4104 will not be destroyed. If a 2102 memory is used, the memory protect latch is not required. Note, however, that CE_1 and CE_2 must still be inverted to enable the memory. $\overline{CE_1}$ and $\overline{CE_2}$ will be enabled after $\overline{POC}$ (pin 39) is removed from the MK5170 and CE_1 goes high. Repertory size is determined automatically by the amount of memory installed so all that is required to alter the size of the repertory is to install memory devices, as shown in Fig. 4-39.

(g) A special on-hook circuit is shown for use with PBX systems (Fig. 4-40). This circuit prevents the MK5170 from seeing the momentary line disconnect exhibited by a PBX system as it switches to out-of-plant lines.

(h) Four types of functions are used to control the dialer:

1. Enter/Dial allows entry and dialing of telephone numbers
2. Store enables storage of entered number
3. Clear clears erroneous entries or begins a data entry sequence

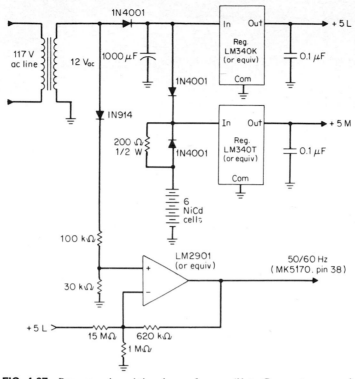

FIG. 4-37 Power-supply and time-base reference. (*Note:* Comparator powered from +5 M.)

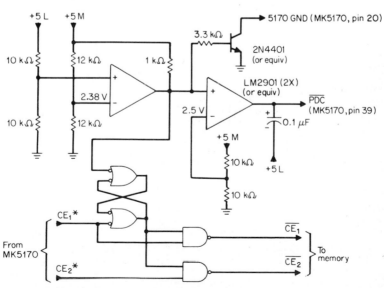

*CE₁ and CE₂ must be inverted to enable memory.

FIG. 4-38 Memory protect logic. (*Note:* All active devices powered from +5 M.)

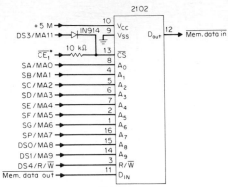

*See Fig. 4-38 for origin of $\overline{CE_1}$. All other signals interface directly with MK5170.

(a)

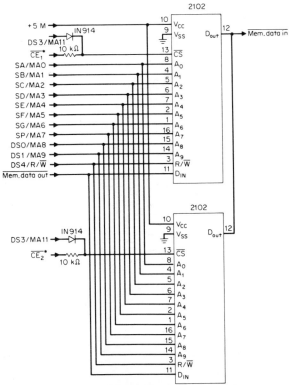

*See Fig. 4-38 for origin of $\overline{CE_1}$ and $\overline{CE_2}$. All other signals interface directly with MK5170.

(b)

FIG. 4-39 Memory connections to the MK5170.

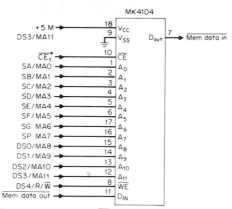

MK4104

+5 M → 18 V_{CC}
DS3/MA11 → 9 V_{SS} D_{out} 7 → $\overline{\text{Mem data in}}$

$\overline{CE_1}^*$ → 10 $\overline{CE}$
SA/MA0 → 1 A_0
SB/MA1 → 2 A_1
SC/MA2 → 3 A_2
SD/MA3 → 4 A_3
SE/MA4 → 5 A_4
SF/MA5 → 6 A_5
SG MA6 → 17 A_6
SP MA7 → 16 A_7
DS0/MA8 → 15 A_8
DS1/MA9 → 14 A_9
DS2/MA10 → 13 A_{10}
DS3/MA11 → 12 A_{11}
DS4/R/$\overline{W}$ → 8 $\overline{WE}$
$\overline{\text{Mem data out}}$ → 11 D_{IN}

*See Fig. 4-38 for origin of $\overline{CE_1}$. All other signals interface directly with MK5170.

(c)

MK4104

+5 M → 18 V_{CC} D_{out} 7 → $\overline{\text{Mem. data in}}$
9 V_{SS}

$\overline{CE_1}^*$ → 10 $\overline{CE}$
SA/MA0 → 1 A_0
SB/MA1 → 2 A_1
SC/MA2 → 3 A_2
SD/MA3 → 4 A_3
SE/MA4 → 5 A_4
SF/MA5 → 6 A_5
SG/MA6 → 17 A_6
SP/MA7 → 16 A_7
DS0/MA8 → 15 A_8
DS1/MA9 → 14 A_9
DS2/MA10 → 13 A_{10}
DS3/MA11 → 12 A_{11}
DS4/R/$\overline{W}$ → 8 $\overline{WE}$
Mem. data out → 11 D_{IN}

MK4104

18 V_{CC} D_{out} 7
9 V_{SS}

$\overline{CE_2}^*$ → 10 $\overline{CE}$
1 A_0
2 A_1
3 A_2
4 A_3
5 A_4
6 A_5
17 A_6
16 A_7
15 A_8
14 A_9
13 A_{10}
12 A_{11}
8 $\overline{WE}$
11 D_{IN}

*See Fig. 4-38 for origin of $\overline{CE_1}$ and $\overline{CF_2}$. All other signals interface directly with MK5170.

(d)

FIG. 4-39 (*Continued*)

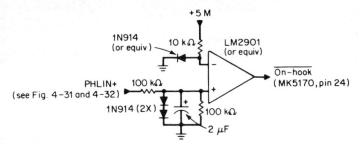

FIG. 4-40 On-hook circuitry for use with PBX systems. (Comparator powered from +5 M.)

4. Pause provides for intergroup pauses while dialing. Note that one of three lengths of pause may be selected:

    a. Infinite pause, which is terminated either by pressing the Enter/Dial button or by pressing one of the #0 through #31 keys

    b. Short pause (1.5 s), which is terminated either by a 1.5-s time-out or by the means described in *a*

    c. Long pauses (5 s), which are terminated either by a 5-s time-out or by the means described in *a*

Two additional buttons are used to control the clock and the timer. The Display Time button causes the MK5170 to display the current time. If this button is pressed and held for more than 3 s, then the MK5170 will enter the time-set mode. During the 3-s time-out, the colons will stop flashing. At the conclusion of the time-out, the colons will resume flashing and the time may be set via the number keys. S/S Timer provides for displaying the timer as well as starting and stopping it. If S/S Timer is pressed and the timer is not being displayed, the display will change to show the timer count. If the timer is being displayed when S/S Timer is pressed, the timer will start if it is not running and will stop if it is running.

An additional feature of the MK5170 is that once a minute the current time is stored in external RAM. The storage occurs on the minute transition. For example, as the time changes from 1:02 to 1:03, 1:02 will be stored. If AC power is then lost between 1:03:00 and 1:03:59, the clock will power-up at 1:02:00 when AC power is restored.

4–5c Design Considerations

When designing a circuit using a repertory dialer, several important factors should be considered to avoid problems. One of the first factors to be taken into account when designing a system using a repertory dialer is the power-supply requirements of the repertory dialer. If the repertory dialer requires an interface to other circuits, such as memory or dialer circuits, their power-supply requirements should also be considered. If a repertory dialer is to be powered by the telephone line, the dialer and its associated circuitry should have a wide operating voltage range (at least 3.5 to 10.0 V) with low power consumption. If the repertory dialer or its support circuitry requires a more rigid power supply, then a separate power supply powered by the AC line should be considered. In addition, if the power consumption is low, a battery may be considered. A battery backup system should definitely be considered to avoid the possibility of loss of data stored in memory due to a momentary loss in telephone line power or AC line power.

Another important factor to consider is whether or not the repertory dialer being used can provide DTMF signaling or pulse dialing options. It is important that the repertory dialer either provide the signals or the capability to interface to a dialer providing the dialing signals required for the system being designed.

Display possibilities should be considered when choosing a repertory dialer for a system that will have much more than a ten-number repertory. A display becomes very important when a large repertory is involved. The number of digits/telephone number that can be stored and displayed should also be considered.

Keyboard requirements should also be considered in the selection of a repertory dialer. Most repertory dialers require a 2-of-7 or 2-of-8 keyboard with the option of individual one-key dialing of stored numbers. With some repertory dialers, however, the one-key dialing is not an option, but a requirement. Therefore, separate keys would be required for each number in the repertory.

The following is a listing of features useful in varied applications. These features are offered on repertory dialers by various manufacturers, as listed in the Product Selection Guide (Sec. 4-9).

- Single +5-V supply
- Wide operating voltage range (2.0 to 10.0 V)
- Low power consumption
- Class A or 2-of-7 keyboard inputs
- Memory access with 2-of-8 keyboard or single key
- Ceramic resonator used as frequency reference
- *RC* oscillator used as frequency reference
- 3.58-MHz crystal used as frequency reference
- On-chip RAM for ten-number storage
- External RAM required for storage of up to 100 numbers
- Storable access pause
- Storage of 8, 16, 20, or 22-digit numbers
- Display interface capability
- Real-time clock option
- 100-hour timer option
- Stand-alone pulse dialer
- Pin-selectable dialing rate and make-break ratios
- Interfaces to DMTF dialer
- Stand-alone DTMF dialer

4–6   SPEECH NETWORKS

4–6a   Introductory Theory

A speech network is an essential part of the subscriber set. The speech network separates the receive and transmit signals and interfaces the transmitter and receiver to the line. Speech networks perform four basic functions. These can be defined as:

1.   2-to-4 wire conversion
2.   Loop-length compensation
3.   Sidetone balancing
4.   Surge protection

The "2-to-4 wire conversion" is the process by which the bidirectional signals of the telephone line (tip and ring, "2 wires") are separated into the transmit and receive signals (two wires each, totalling "4 wires"). Incoming signals from tip and ring are transferred to the two output lines (R and GN) that connect the receiver to the speech network. Signals produced by the transmitter travel through lines B and R and the network to tip and ring. There are actually only three transmission lines joining the receiver and the transmitter to the speech network because the R line is used by both the receiver and transmitter.

"Loop-length compensation" helps minimize the volume losses that occur in long-loop signaling. When the subscriber set is located near the central office (short-loop signaling), the loop-length compensation circuitry decreases the impedance of the speech network. This increases the load on the telephone line and causes the subscriber-set efficiency to drop. With long-loop signaling, the speech network impedance increases, reducing the load on the telephone line and increasing the subscriber-set efficiency. The overall effect of loop-length compensation is to minimize the difference in signal quality between long- and short-loop signaling.

The "sidetone balancing" circuitry serves to control the amount of the speaker's signal that is fed back to the receiver. This returned signal is called sidetone and the sidetone balancing circuitry provides a natural amount of feedback to the speaker's ear in order to maintain a normal conversation.

"Surge protection" helps prevent damage to the subscriber set from high-voltage transients, such as those induced by lightning near the lines.

The most familiar type of speech network in use today is the 500 or hybrid network, which is composed of discrete components. A schematic of this network and its connections to other parts of the subscriber set is shown in Fig. 4-41. One particular version of the 500 network that has become the industry standard for speech networks is the 2500-type network. The 2500-type network is made by Western Electric and Stromberg-Carlson and is identical to the 500 network, except that it does not contain resistor R_3 and capacitor C_4.

Every component of the 500 network, except capacitors C_1 and C_2, and resistor R_3, is involved in performing one or more of the four basic functions. Capacitor C_1, in conjunction with resistor R_1, helps to suppress the noise-producing sharp voltage rises that are generated during outpulsing when a rotary or pulse dialer is used. Capacitor C_2 prevents the loop current from flowing through the ringer, while resistor R_3 limits the current that flows through the transmitter.

The 2-to-4 wire conversion is performed by the inductors, capacitors C_3 and C_4, and resistor R_2. In the transmit stage, the telephone is off-hook and current flows into the

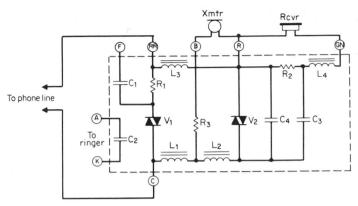

FIG. 4-41   500-type speech network connections.

transmitter. Capacitors C_3 and C_4 prevent direct current from reaching the receiver. Inductors L_1, L_2, and L_3 divide the voltage developed by the transmitter so that the voltages induced by them in L_4 are of the opposite polarity and effectively cancel each other. Voltages induced across resistor R_2 by inductor L_2 are arranged to oppose the voltages induced in L_4. As a result, the current flow through the receiver is small and the sidetone levels in the receiver during the transmit cycle are low.

In the receive cycle, the loop current passes through inductors L_1, L_2, and L_3 to produce additive voltages in L_4. The voltages in L_4 are approximately equal in magnitude and 180° out of phase with the voltages induced in L_2. The resultant voltage drop across R_2 is very small and maximum receiving signals are obtained with minimal power loss in R_2.

Loop-length compensation is performed by varistors V_1 and V_2. Varistors are discrete components that have the property of increasing in impedance when the currents to them are decreased. On long-loop signaling, where the loop currents are small, the varistor impedances increase, resulting in maximum set efficiency. With short loops and therefore large loop currents, the varistor impedances decrease, loading the telephone line more. The subscriber-set efficiency is decreased, thus minimizing the difference in quality between long- and short-loop signaling.

Sidetone balancing in the 500 network is provided by resistor R_2 and capacitors C_3 and C_4. These capacitors prevent the direct current that flows from the transmitter from reaching the receiver. Resistor R_2 allows a small amount of the transmitted signal to reach the receiver so that the speaker can hear his own voice in the earpiece. The sidetone level may be varied by changing the values of R_2, C_3, and C_4.

Finally, varistors V_1 (in conjunction with R_1) and V_2 also help provide surge protection for the network. When a high-voltage transient appears on the telephone line, the varistor impedances decrease, approaching zero, and effectively shorting out the high voltage.

The 500 network or the 2500-type network are used in the majority of telephone sets being built today. Two typical telephone applications using the 2500-type network will be discussed later in this section.

An alternate type of speech network, the monolithic integrated circuit or active speech network, is currently under development by several companies, including Mostek. Integrated-circuit speech networks have several potential advantages over hybrid networks, one of the most obvious being size. A single-chip monolithic speech network can completely replace the hybrid network, which means that the entire subscriber set can be significantly reduced in size.

Integrated-circuit speech networks can also be cost-saving devices. Some will be capable of interfacing to electret microphones, which are less expensive and more reliable than the standard carbon microphones. Active speech networks can also perform on-chip muting, eliminating the need for external muting transistors or the mechanical common function switching required by the hybrid network. Finally, the active speech network can be designed to directly interface with electronic tone and pulse dialers, thus greatly reducing the amount of external circuitry required.

As a result of the smaller size, decreased number of external components required, and ultimately, lower assembly costs, the integrated-circuit speech network can be significantly less expensive to use than the hybrid network. This decreased cost and the high reliability of integrated-circuit technology make the active speech network a promising product in modern telephone technology.

4–6b   Application Solutions and Examples

In this section, two typical telephone applications using the 2500-type speech network will be presented. The first example is a pulse dialer circuit using the MOSTEK MK50982 Pulse Dialer. The second example shows a typical tone dialer application circuit using the Mostek MK5380 Tone Dialer.

Example 4–15   Design of a Pulse Dialer Telephone Circuit

Design a typical pulse dialer telephone using the 2500-type speech network.

Solution

A typical pulse dialer telephone using the 2500-type speech network is shown in Fig. 4-42. In this application, the pulse dialer circuitry is in series with the speech network. A current source of some kind is desired to present a high impedance to the telephone line while guaranteeing sufficient current (≥ 150 μA) to power the MK50982. The current source in this circuit is composed of transistor Q_2 and resistor R_1. The amount of current to the pulse dialer is determined by the value of R_1 and the characteristics of transistor Q_2. As current through R_1 increases, the voltage across it (which corresponds to V_{GS} of the FET) also increases, thus regulating the amount of current that flows through Q_2. Diode CR_1 is used to block reverse current flow through the current source.

In the on-hook mode, hook switches S_1 and S_2 are open. An on-chip resistor pulls the On-Hook/Test input (pin 15) to V^+ (pin 1) and the current source is disabled. A large-value resistor (R_3) allows a small amount of current to maintain the memory on the MK50982.

In the off-hook mode, S_1 and S_2 are closed, tying pin 15 to V^- (pin 6). This sets the MK50982 in the normal mode, ready to accept key inputs. The $\overline{\text{Pulse}}$ (pin 16) and $\overline{\text{Mute}}$ (pin 10) outputs drive external transistors to perform the outpulsing function. The receiver is connected to the speech network through transistor Q_6. $\overline{\text{Mute}}$ causes this transistor to be held on until the outpulsing begins. When $\overline{\text{Mute}}$ switches low, the receiver is disconnected from the speech network and the pops caused by making and breaking the line are isolated from the receiver. The $\overline{\text{Pulse}}$ output drives transistors Q_3 and Q_5 to make and break the line until the digit is completely outpulsed. $\overline{\text{Mute}}$ then switches high, returning the receiver to the speech network.

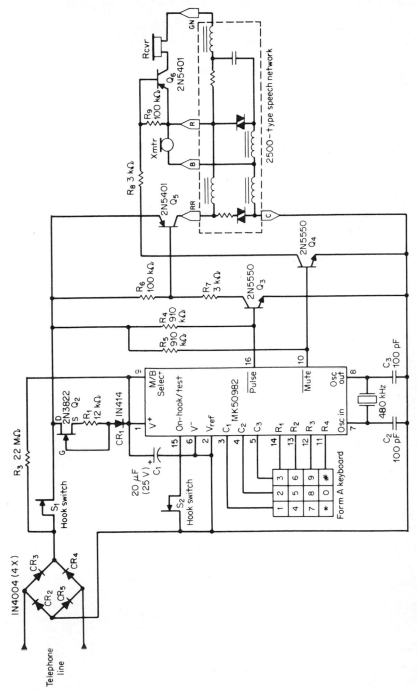

FIG. 4-42 Pulse dialer telephone circuit using a 2500-type speech network.

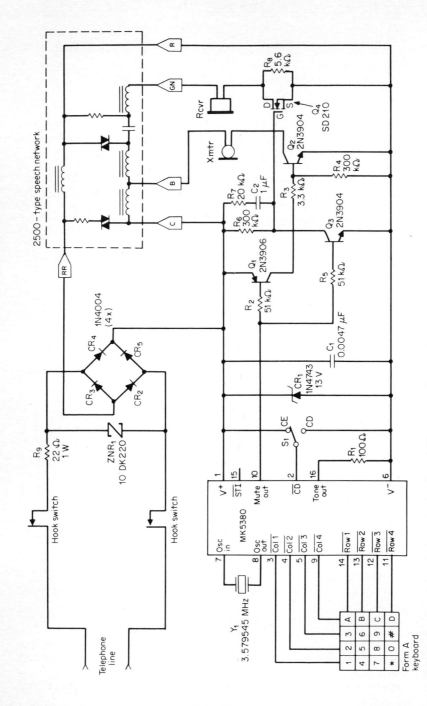

FIG. 4-43 Tone dialer telephone circuit using a 2500-type speech network.

Example 4–16   Design of a Tone Dialer Telephone Circuit

Design a typical tone dialer telephone using the 2500-type speech network.

Solution

Fig. 4-43 shows a typical tone dialer application using the 2500-type speech network and the Mostek MK5380 tone dialer. The MK5380 uses a 3.58-MHz color-burst crystal frequency reference to provide tones suitable for DTMF dialing.

The MK5380 is powered directly from the telephone line. A diode bridge composed of diodes CR_2–CR_5 is used to protect the MK5380 from telephone line polarity reversals, while zener diode CR_1 helps provide over-voltage and line-surge protection.

The MK5380 operates in the normal mode when the $\overline{\text{Chip Disable}}$ ($\overline{\text{CD}}$) input (pin 2) is either tied high (switch S_1 is in the CE position) or left floating. When $\overline{\text{CD}}$ is tied low (S_1 is in the CD position), the keyboard inputs go to a high-impedance state, tone generation is inhibited, and the amplifiers and oscillator are powered down.

In the normal mode, any keyboard entry activates the muting circuitry. The Mute output (pin 10) goes high, turning off transistors Q_1 and Q_2. This disconnects the transmitter, thereby increasing the DC resistance of the speech network and eliminating distortion of the DTMF signals during toning. In addition, when Mute Out goes high, Q_3 is turned on, Q_4 is turned off, and the receiver is muted so that only a small portion of the generated DTMF tones is heard by the user.

When the keyboard button is released, Mute Out goes low, reconnecting the transmitter and returning the receiver to its normal state. Resistor R_6 and capacitor C_2 provide a time delay to allow the transmitter to be reconnected while the receiver is still muted. This eliminates any objectionable pops that might otherwise be heard in the receiver when the transmitter is reconnected to the speech network.

In this application, the Single-Tone Inhibit input (pin 15) is left floating, allowing the MK5380 to produce single tones when two buttons in the same row or column are pressed. Tying pin 15 to V^+ (pin 1) will also allow single-tone generation. To prevent the generation of anything other than dual tones, pin 15 should be tied to V^- (pin 6). In this mode, any time two or more rows or columns are pressed, no tone will result.

4–6c   Design Considerations

In any subscriber-set application, two important design considerations for the speech network are

1. Minimizing signal losses across the receiver and transmitter
2. Minimizing the number of external parts while optimizing the circuit performance

Signal losses across the receiver and transmitter are often directly related to the type of muting circuitry used. This problem can be minimized by careful design.

In hybrid speech network applications, one or more muting transistors are often used to interface the receiver to the speech network. Bipolar transistors are most commonly used for these muting transistors because they are readily available and relatively inexpensive. However, to minimize the signal losses that occur in the receiver, Field Effect

Transistors (FETs) may replace the bipolar transistors. FETs are generally more expensive than bipolar transistors, but may improve the signal quality at the receiver.

An alternate method of minimizing the signal losses caused by external components is to select a speech network that performs its own muting. Some of the active or integrated-circuit speech networks currently under design will have this capability. Speech networks that perform on-board muting have the additional benefit of reducing the external parts count, and thus, the cost of the circuit.

Modern speech networks offer a variety of features that can optimize circuit performance while minimizing the external parts count. Some of these are listed below. For a more complete listing of both hybrid and active speech networks and their features, see the Product Selection Guide (Sec. 4-9).

- Voltage-regulated outputs to provide regulated supplies for dialers
- User-controllable sidetone level
- Frequency shaping and high-frequency emphasis
- Capability of interfacing to a variety of transducers, including dynamic receivers and transmitters

4–7 CODECs

4–7a Introductory Theory

The tremendous advance of digital technology over the years has led to the development and growth of digital switching and transmission systems within the telephone industry. Digital systems have provided economical multiplexing for transmission, increased noise immunity, and the advantages of solid-state switching. The integrated-circuit CODEC was developed to provide the transition from the inherently analog world of voice signals to one of a digital nature. Thus the general characteristics of the CODEC (enCOder/DECoder) can be anticipated by considering the requirements of an economical yet effective conversion between analog voice and digital signals. Two major administrations have published standards that outline requirements for telephone switching and transmission systems. AT&T or Bell System Standards are used throughout much of North America, while most of Europe and other areas follow CCITT standards. Thus two distinct types of CODECs may be found in the industry. The major differences between these types are discussed as the general CODEC characteristics are considered.

The basic function of the CODEC is to encode analog voice signals into digital words and decode digital words back to analog. Such conversions between continuous and discrete signals involve sampling rates and quantization. The sampling rate has been standardized to 8 kHz throughout most of the telephone industry. Sampling theorems state that a band-limited signal is uniquely represented by a set of samples taken at a rate of at least twice the frequency of the highest frequency component of the band-limited signal. For an 8-kHz sampling rate, incoming voice signals will need to be band-limited to 4 kHz (see PCM filters, Sec. 4-8). Although voice signals contain frequencies beyond 4 kHz, adequate representation of the human voice can be obtained from the voice frequencies in the range of 200 to 3400 Hz. Thus a sampling rate of 8 kHz maintains acceptable quality of voice signals. Both the AT&T and CCITT administrations have adopted an 8-kHz sampling rate.

When quantizing a sampled voice signal, improved reproduction over a wider range

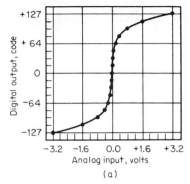
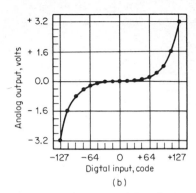

FIG. 4-44 Typical companding curves. (*a*) Encoder transfer characteristics; (*b*) decoder transfer characteristics.

is attained by compressing the signal when encoding and expanding the signal during decode. This technique of maintaining near constant signal-to-distortion ratios over wide dynamic ranges is called "companding" (COMpress/exPAND). Typical companding curves have a logarithmic nature and are approximated by line segments. Thus a CODEC's digital output code will consist of a sign-bit, chord-bits, and step-bits. The chord-bits will determine a particular chord or segment, while the step-bits determine location within a chord. Fig. 4-44 depicts both the encode and decode companding curves for a typical CODEC device. The companding curves specified by AT&T and CCITT differ slightly in the initial step size and the placement of the origin. AT&T defines a μ-law curve, while CCITT recommends an A-law curve. The "μ" and "A" are variables of logarithmic equations defining a family of companding curves. CODECs approximate the $\mu = 255$ and $A = 87.6$ companding curves.

The digital codes that have become standards are the 8-bit "sign-magnitude" and the "even-order bit-inversion" codes. These codes have a reduced probability of zero bits and zero strings as compared to the conventional sign-magnitude code. The increased ones density then aids in receiver clock recovery and timing on digital repeater lines. AT&T specifies the sign-magnitude code, with the least significant bit also serving as a signaling bit. CCITT recommends the inversion of even-order bits from the conventional sign-magnitude code. Differences in the methods of signaling, framing, and data rates of the mentioned administrations contribute to the variations of features and characteristics found in CODECs. The reader may obtain more information on AT&T and CCITT system specifications by consulting the reference section of this chapter.

Fig. 4-45 shows a typical block diagram for a single-channel CODEC. An alternate method of implementing the CODEC function is on a shared or multichannel basis. The shared-CODEC approach to system design typically uses faster, high-power components to encode/decode groups of channels through the use of an analog multiplexer. Although single-channel CODECs vary in design, the internal operation of the MK5116 diagramed in Fig. 4-45 will be discussed.

Using the block diagram, the path of signal flow can be traced. The analog input signal arrives at the sample-and-hold and a sample is stored on a capacitor array. The capacitor array is the equivalent of a 13-bit binary weighted capacitor ladder. This ladder also serves as a D/A converter by switching the capacitors between a reference voltage and ground. The difference between the analog sample and the first approximation by the successive approximation register (SAR) is fed to a comparator. The comparator's

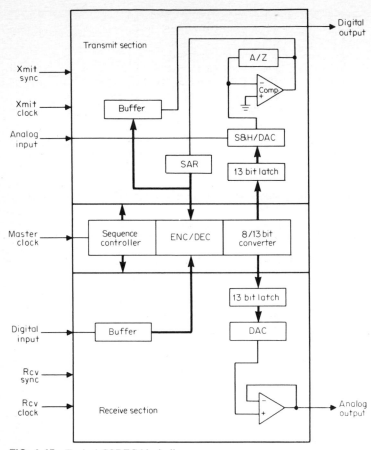

FIG. 4-45   Typical CODEC block diagram.

output represents the states of "too high" or "too low" and allows the SAR to make the next best approximation. The SAR outputs an 8-bit companded code which is converted to a 13-bit linear code. The DAC again performs the conversion from the 13-bit code to analog for comparison with the sampled signal. This loop continues until the correct 8-bit code has been generated and latched into the transmit buffer. The transmit data are shifted out of the buffer under control of the Xmit Sync and Xmit Clock.

Data is shifted into the receive buffer under control of the Rcv Sync and Rcv Clock. Although the receive section has a separate DAC, the 8–13-bit converter is shared with the transmit section. The 8-bit companded code is passed to the shared circuitry and the conversion to 13 bits is performed. The receive DAC converts the linear 13-bit code to analog and the resulting signal is buffered prior to output.

Most of the timing signals required for internal sequence control are generated from the master clock. The Xmit Sync controls the sample rate as an encode cycle is initiated on its low-to-high transition. The receive sync controls the reconstruct rate as a decode cycle is initiated on its high-to-low transition. Data are shifted from the device at the clock rate present on Xmit clock when Xmit Sync is in the high state. Rcv Clock and

Rcv Sync operate in a similiar manner when reading data into the device. Thus the serial data rate is variable and allows for time-division-multiplexing of several channels onto a common signal path.

The basic functions of sampling, quantization, and coding are inherent in the design of all CODECs. In addition, many auxiliary functions have been made available by various manufacturers. Included among the myriad of designs are such features as pin-programmable power-down and companding-law select, on-chip voltage references and time-slot computation, variable sample rate, and single-supply operation. A trend now being pursued by most manufacturers is to combine both CODEC and PCM filters onto a common substrate. It appears that this single-package concept, on-chip references, and the power-down function are features that will be most prevalent in future families of CODEC devices.

4–7b   Application Solutions and Examples

Example 4–17   Signal Level Calculations in a Telephone System

Determine the maximum signal power level at the analog input of a CODEC that has an acceptable input signal range of ± 2.5 V. Relate this level to the system reference test point.

Solution

(a) A telephone system contains elements of attenuation and gain such that the signal power level varies from point to point. One point, called the zero transmission level point (0 TLP), is defined as the reference test point of the system. 0 dBm0 implies that a level of 0 dBm (1 mW) exists at the 0 TLP. N dBm0 implies that a level of N dBm exists at the 0 TLP. Other TLPs are identified by the gain or loss at that TLP relative to the 0 TLP. Thus the signal level at a 6 dB TLP is 6 dB above the reference level (0 TLP). The maximum signal power level of a telephone system is typically $+3$ dBm0.

(b) The CODEC of this example has an acceptable input signal range of ± 2.5 V. The power of a 2.5-V peak signal is approximately $+7$ dBm at 600 Ω. This corresponds to the maximum system level of $+3$ dBm0. Relating the two we get:

$$+ 3 \text{ dBm0} \cong +7 \text{ dBm (at CODEC input)}$$
$$0 \text{ dBm0} \cong +4 \text{ dBm (at CODEC input)}$$

Therefore, the CODEC input is 4 dB above the reference level. Thus the CODEC input is a $+4$ dB TLP or $+4$ TLP.

Example 4–18   Data Rate Calculations for a Digital Transmission System

Determine the minimum data rate required to multiplex (TDM) the voice information from 24 CODECs onto a single twisted-pair.

Solution

(a) If we assume the standard sample rate of 8 kHz and 8 bits per sample, we get:

$$24 \text{ channels} \times \left(\frac{8000 \text{ samples}}{\text{channel-second}}\right) \times \left(\frac{8 \text{ bits}}{\text{sample}}\right) = 1.536 \text{ Mbits/s}$$

(b) The 1.536-Mbits/s calculation is for serial data transmitted in one direction. A second twisted-pair would be required for voice transmission in both

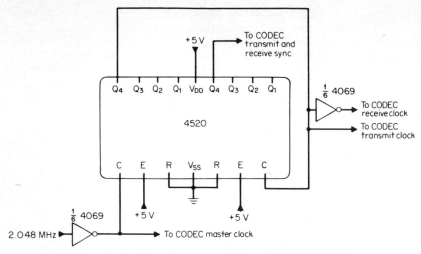

FIG. 4-46 Timing circuitry for performance evaluation of the MK5116 CODEC.

directions. The data rate specified by AT&T for the $T1$ links between central offices is 1.544 Mbits/s. This rate results from the insertion of a framing bit with every 24 channels of voice data. Preserving the 8-kHz sample rate results in a $T1$ data rate of 1.544 Mbits/s.

Example 4–19   Design of Timing Circuitry for Performance Evaluation of a CODEC

Design timing circuitry to supply a set of clocks and syncs to a MK5116 CODEC. The signals supplied should allow the device's performance to be evaluated by tying the digital output to the digital input.

Solution

The syncs should have a repetition rate of 8 kHz. By using an input clock rate of 2.048 MHz, the master clock requirement (1.5 to 2.1 MHz) is satisfied and the syncs can be generated by a counter (2.048 MHz ÷ 256 = 8 kHz). A 4520 dual binary counter may be connected as shown in Fig. 4-46 to provide the 8-kHz sync. Since the device transmits and receives during a logic 1 of syncs, the data clocks need to output/input 8 bits during the logic 1 state of the 8-kHz square-wave sync. Thus the data clocks can use the 128-kHz signal available at the divide-by-16 output of the 4520. The receive clock is inverted to allow the transmit data one-half period of settling time. Timing signals produced by the circuitry are shown in Fig. 4-47.

Example 4–20   Design of an End-to-End CODEC/Filter Demonstration Unit

Design a complete circuit to demonstrate the operation of a CODEC with a PCM filter. Use the MK5116 CODEC with a MK5912 PCM filter (PCM filters are discussed in Sec. 4-8).

Solution

(a) The timing circuit from Example 4-19 will be used to provide the timing signals to the CODEC. An oscillator circuit must be added to provide the

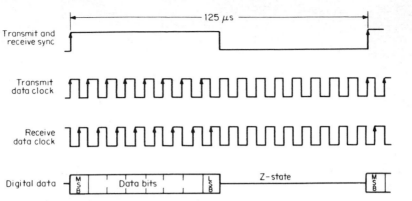

FIG. 4-47   Timing signals of Example 4-19.

2.048-MHz clock input. This clock also provides the clock required for the PCM filter. The complete demonstration circuit is shown in Fig. 4-48.

(b) The MK5116 requires voltage references for use in D/A conversions. A MC1403 is used to provide the positive reference and an op amp generates the negative reference, as shown in Fig. 4-48.

(c) The manufacturer of the MK5116 recommends a 3 kΩ resistor (R_4) to ground at the analog input of the device. Capacitors C_1 and C_7 remove any DC offset from the analog signals. Resistor R_3 reduces the input impedance to the MK5912 filter. All RC combinations have been chosen to provide negligible effect on voiceband signals.

(d) Capacitors C_2 thru C_6 and C_8 are power-supply decoupling capacitors and should be placed near the devices. Separate analog and digital grounds are used to reduce noise that might otherwise be coupled into the analog signals. These grounds should be tied together at the power supply.

(e) The analog input signal to the demonstration circuit may have a peak swing of ±1.77 V. The input signal is filtered and amplified by the transmit section of the PCM filter. The signal then passes to the analog input of the CODEC (pin 1, MK5116) and will have a peak swing of approximately ±2.5 V. The encoded data may be observed at the digital output (pin 8, MK5116) by using an oscilloscope with the external trigger connected to Xmit Sync (pin 6, MK5116). The reconstructed staircase waveform may be observed at the analog output of the CODEC (pin 13, MK5116). The signal then passes through the receive section of the PCM filter to the analog output of the demonstration circuit. Jumper J_1 can be used to connect the receive filter output to the input of the power amplifiers. The receive signal may then be observed at the differential output provided by pin 6 and pin 7 of the MK5912. Jumper J_2 deactivates the power amplifiers by connecting the input to V_{BB}.

Example 4–21   Block Diagram Layout for a Voice-Delay Unit

Outline the design of a voice-delay unit through the use of a block diagram. A CODEC and RAM should be used with calculations made to determine the amount of memory required for 250 ms of voice delay.

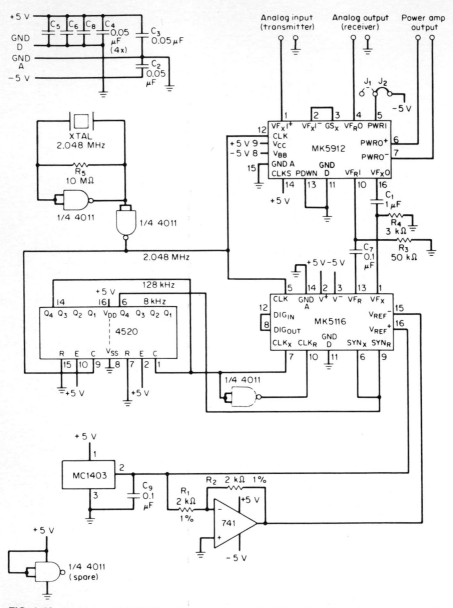

FIG. 4-48 End-to-end CODEC filter demonstration circuit. (*Note: C_1 and C_7 must be nonpolarized.*)

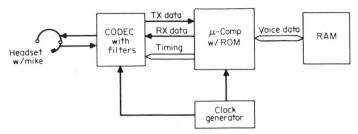

FIG. 4-49   Block diagram for a voice-delay unit.

Solution

(a) CODECs can be useful in many communications systems for which voice companding is desirable. The digital signals provide good noise immunity during data transmission and allow for data manipulation through the use of software. The audio-delay system of this example might be useful for speech therapy. It is known that patients with speech hesitation difficulties are aided by the sound of their speech delayed in time.

(b) The block diagram shown in Fig. 4-49 illustrates one possible architecture of a voice-delay unit. The microcomputer generates the data clocks and syncs. Voice data are stored into memory and retrieved after sufficient delay. The headset should allow the user to hear external sounds as well as the delayed voice.

(*c*) To determine the amount of memory required for 250 ms of voice delay, let us assume the standard 8-kHz sample rate. We get:

$$8000 \text{ samples per second} \times 250 \text{ ms} = 2000 \text{ samples}$$

Since each sample contains 8-bits, the memory requirement will be 2000 × 8 of RAM to store 250 ms of voice data.

4–7c   Design Considerations

A variety of CODECs are currently available to meet the requirements of various system architectures, with the desired system structure influencing the section of a particular CODEC. Single-channel, direct-mode devices seem to have gained popularity over shared CODECs and CODECs with time-slot computation features. Although the latter two have advantages in certain applications, single-channel, direct-mode devices are extremely flexible in architecture requirements and can be adapted to fit a number of alternate system designs.

A second factor in selection of a device is its ability to meet the requirements of the overall system. For CODECs, the system specifications of signal-to-distortion, gain-tracking, and idle-channel noise are of interest to the designer and will normally be specified by the manufacturer. Other important parameters are frequency response, cross-talk, and power-supply rejection. Although CODECs are often compared to system specifications as published by the AT&T or CCITT administrations, tighter specifications are normally imposed on all system fragments. Industry standard requirements for the CODEC do not currently exist. However, Table 4-3 reflects typical specifications required for CODECs used in private branch exchanges and central offices.

Printed circuit board layout should be done with care and proper consideration given to decoupling and ground planes. Many devices are supplied with separate digital and

TABLE 4–3 Typical CODEC Specifications

Parameter of interest	Typical system requirement	Typical CODEC requirement	Typical CODEC performance
Signal to distortion (end-to-end with message weighting)			
0 to −30 dBm0	33 dB	35 dB	39 dB
−40 dBm0	27 dB	29 dB	34 dB
−45 dBm0	22 dB	24 dB	29 dB
Gain tracking (end-to-end, sinusoidal)			
+3 to −40 dBm0	±0.5 dB	±0.4 dB	±0.1 dB
−40 to −50 dBm0	±1.0 dB	±0.8 dB	±0.1 dB
−50 to −55 dBm0	±3.0 dB	±2.5 dB	±0.2 dB
Idle channel noise (end-to-end)	23 dBrnC0 −65 dBm0p	18 dBrnC0 −68 dBm0p	12 dBrnC0 −75 dBm0p
Crosstalk loss	60 dB	75 dB	80–100 dB

analog grounds to reduce system noise problems. Coupling through parallel digital and analog runs should be avoided. Trace length into high-impedance inputs should also be minimized. Little concern for proper layout can often result in board performance well below that of the individual components.

For a listing of popular CODEC devices and their features, see the Product Selection Guide (Sec. 4-9).

4–8   PCM LINE FILTERS

4–8a   Introductory Theory

In a PCM communications system, PCM line filters perform the sensitive filter operations necessary for high-quality communications. PCM filters are very closely associated with the CODECs used in the conversion of voice signals between digital and analog representations. As described in Sec. 4-7, the analog input to the CODEC is sampled at a rate of 8 kHz. In any sampled data system of this type, one must be concerned with aliasing effects that arise when the frequency of the sampled signal is more than half the sampling frequency of the system. One function of a PCM line filter is to eliminate this problem by band-limiting the analog voice signal before it reaches the CODEC input.

The analog output of the CODEC is an approximation to the original analog signal through an 8-kHz D/A conversion process. The 8-kHz conversion results in a "staircase" waveform at the output of the CODEC. In theory, the reconstructed waveform exactly represents the original analog signal with the addition of high-frequency components due to the staircasing effect. These high-frequency components show up as a high degree of distortion that can be eliminated by postfiltering. Another effect caused by the staircase reconstruction is the attenuation of higher frequencies within the passband. A PCM line filter must compensate for these two effects.

In addition to the above features, a PCM line filter will normally provide 50–60-Hz rejection on the encode side of the system, since it is filtering signals coming from

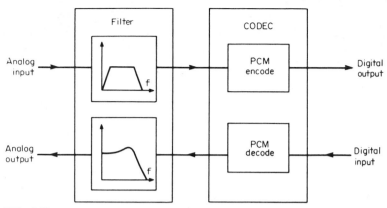

FIG. 4-50 CODEC/filter relationship.

the outside world. Note that this is not needed on the decode side due to the close proximity of the filter to the CODEC and the low susceptibility of the connecting wires to 50–60 cycle noise. Fig. 4-50 shows the physical relationship between the CODEC and the filters.

Many PCM filters available today perform in accordance with a strict set of guidelines set forth by AT&T and CCITT. A typical system requirement is shown in Fig. 4-51. These guidelines have become industry standards, with AT&T representing telephone communications in the United States and CCITT representing telephone communications in Europe. Until the technology became available to fabricate PCM filters on a chip, the transition to digital voice communications was slowed by the complicated circuitry required to implement these tight filter specifications. Excessive power consumption was also a drawback. Today single chips are available that perform all the PCM line filter functions with high reliability and extremely low power consumption.

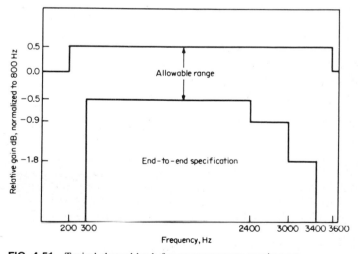

FIG. 4-51 Typical channel bank frequency response requirement.

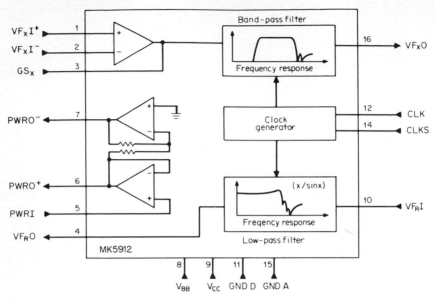

FIG. 4-52   Block diagram of a typical PCM line filter.

As an aid to understanding the need for PCM filters and how they function in a system, we will discuss the internal operation of a typical PCM line filter. Mostek's MK5912 PCM filter is one of the currently available switched-capacitor filters on the market. Fig. 4-52 shows a block diagram of the MK5912. The device contains both the transmit and receive filters, comes in a standard 16-pin dual-in-line package, and has a power dissipation of about 25 mW. In addition to the transmit and receive filters, the MK5912 contains a gain-setting op amp on the transmit side (upper) and unity-gain power amplifiers on the receive side (lower). The gain-setting op amp is useful for raising the telephone line signal levels (roughly ± 1.55-V peak) to the appropriate transmission level point of the CODEC (roughly ± 2.5-V peak for Mostek's family of CODECs). By setting this gain appropriately, maximum dynamic range and signal-to-noise ratio can be achieved. This op amp also aids in the 2-to-4 wire conversion process required at this interface. With the MK5912, the designer has the option of using an electronic 2-to-4 wire converter to interface tip and ring to the filter or using a hybrid transformer to provide this function. When using a transformer, the power amplifiers are needed to directly drive the 2-to-4 wire converter. If an electronic hybrid is used, the power amplifiers are not needed and can be disabled by applying V_{BB} to the PWRI input (pin 5).

The transmit filter section consists of a series combination of four individual filters. After the signal passes through the gain-setting amplifier, it is processed by an active *RC*, continuous-time, low-pass filter. This filter is necessitated by the fact that the switched-capacitor (SC) filters have a sampling nature and can exhibit aliasing effects. The first SC filter encountered is a high-pass filter. This section provides the 50–60-Hz rejection required by the *D*3 and CCITT specifications. The next SC filter establishes the 3-kHz roll-off and passband characteristics required by specifications. The output of an SC filter inherently has a certain amount of clock noise in it and must be followed by a smoothing, continuous-time, active *RC* filter. The sampling action of the CODEC could otherwise alias this noise into the signal passband. After this final filtering section, the signal typically passes from the filter to a CODEC for conversion into PCM data.

The receive filter consists of a single SC low-pass filter. This filter has been designed to be used with a CODEC having a zero-order-hold analog output. Signals reconstructed by this type of CODEC naturally exhibit a sin x/x frequency-response distortion accompanied by high-frequency noise components. The MK5912 receive filter serves the dual purpose of eliminating the high-frequency noise and introducing an $x/\sin x$ factor into the receive side of the system. The result is a flat passband from DC to 3 kHz.

The switched-capacitor filters in the transmit and receive sections are driven by a 256-kHz clock that is derived from the CLK input by the clock divider circuitry. The inputs to the clock divider circuitry consist of CLK (pin 12) and CLKS (pin 14). The master clock input (CLK) has the PCM system clock applied to it. Depending upon the frequency of this clock, the clock select pin (CLKS) is set to a logic level that sets up the clock divider circuitry to divide the master clock down to the 256-kHz internal clock.

The output of the receive filter can drive high-impedance electronic hybrids directly. For transformer hybrid applications, power amplifiers have been supplied. The output of the receive filter (VF_RO) can be tied into the input of the power amps ($PWRI$). The power amplifiers provide a low-impedance differential output and in doing so boost the receive signal by 6 dB. To set the receive circuitry gain, a voltage divider network is normally placed between VF_RO and $PWRI$. If the power amplifiers are not used, power consumption can be reduced by tying $PWRI$ to V_{BB}.

4–8b   Application Solutions and Examples

Example 4–22   Design of a 3-kHz Low-Pass Filter

Design a circuit using the MK5912 that attenuates frequencies above 3 kHz. Assume the signal to have a maximum amplitude of 0 dBm when referenced to 600 Ω.

Solution

(a) The circuit of Fig. 4-53 may be used. The MK5912 is designed for this type of application and provides an easy solution. Note that since the design only calls for attenuation above 3 kHz, the sin x/x characteristic of the receive filter has been ignored.

(b) The transmit section of the MK5912 is a bandpass filter and will not be used. As a good design practice, the gain-setting amplifier input should be connected as shown.

(c) The CMOS oscillator shown provides the clock frequency required by the clock generator. The standard frequency recommended by the manufacturer is 2.048 MHz, and the CLKS pin is biased accordingly. The power supplies are ±5 V ±5%.

The input to the receive filter is AC coupled to ensure maximum allowable input voltage swing. $PWRI$ is tied to V_{BB} to disable the power amps and minimize power consumption. $GRDD$ and $GRDA$ should be connected as close as possible to the power-supply ground to minimize clock noise.

Example 4–23   Design of a Telephone Line Interface to a PCM Filter

Design interface circuitry to allow connection of a MK5912 PCM filter to a standard telephone. Assume that a 1:1 voice-quality transformer is available.

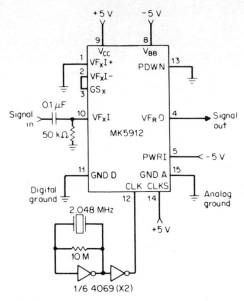

FIG. 4-53 Simple voiceband filter.

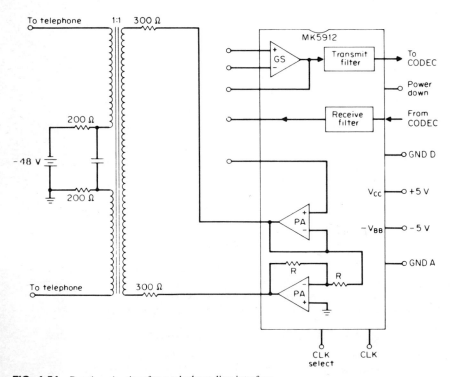

FIG. 4-54 Receive circuitry for a telephone line interface.

Solution

(a) As discussed in Sec. 4-7, the MK5116 has a TLP of 4 dB. This means that in addition to the 2-to-4 wire conversion necessary at the line interface, certain gain considerations must be accounted for between this interface and the CODEC. In other words, the maximum signal level on the telephone line is not equal to the maximum signal level that the CODEC will be expecting.

(b) The simplest portion of the circuitry is the receive side, which is shown in Fig. 4-54. Since the standard telephone line requires a 600 Ω termination and the output impedance of the power amps is extremely low, we must load the transformer accordingly. Two 300 Ω resistors are used instead of one 600 Ω resistor to maintain good longitudinal balance. Signals are transmitted from the power amps to the telephone, but signals coming from the telephone have no effect on the receive side of the filter due to the low output impedance of the power amps. This accomplishes a portion of the 2-to-4 wire conversion.

(c) The transmit circuitry is included in Fig. 4-55. This circuitry is somewhat trickier, since the input of the transmit filter must "listen" to the telephone but not to the output of the power amplifiers. R_5 accomplishes this by balancing the input to the gain-setting amplifier with respect to the midpoint of the power amplifier's differential output. Note that this input is not balanced with respect to the transformer. Keeping in mind that the trans-

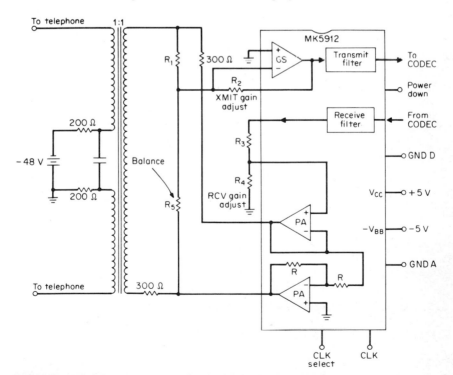

FIG. 4-55 Transmit and receive circuitry for a telephone line interface.

former appears to be a 600 Ω load to the filter and that one 300 Ω resistor is included on one side and not on the other, it can be shown that R_5 should be roughly equal to twice R_1. Both R_1 and R_5 should be chosen to be much greater than 600 Ω. This accomplishes the rest of the 2-to-4 wire conversion.
(d) Gain on the transmit side is now adjustable by varying R_2, and the gain on the receive side is adjusted by varying the R_3 and R_4 combination.

Example 4–24   Gain Considerations for a CODEC/Filter Line Interface

Select R_2, R_3, and R_4 in Example 4-23 to set the TLP of the CODEC at the recommended level.

Solution
(a) Many things must be considered when selecting the gain on the transmit side of the system (see Fig. 4-56). The combination of the 300 Ω resistors and the 600 Ω transformer causes us to lose half the signal immediately. Next we pick up some gain through the gain-setting amplifier, which is equal to R_2/R_1. Finally, the transmit filter provides 3 dB of gain (voltage gain = K_v = 1.414).
(b) On the other side, the receive filter has 0 dB of gain (K_v = 1). Then, the voltage divider (R_3 and R_4) gives a gain of $R_4/(R_3 + R_4)$. Next, when the signal gets converted to a balanced differential output, it picks up 6 dB (K_v = 2). And finally, we once again lose half the signal to the 300 Ω load resistors (see Fig. 4-57).
(c) In summary, for the transmit side, the total voltage gain is given by:

$$K_{vt} = (0.5)(1.414)(R_2/R_1) \tag{4-4}$$

And for the receive side:

$$K_{vr} = (2)(0.5)[R_4/(R_3 + R_4)] \tag{4-5}$$

Remember that when R_1 is chosen, this will set R_5 at $2R_1$.
(d) Designing for 0-dBm signal levels, we first note that 0 dBm is equal to

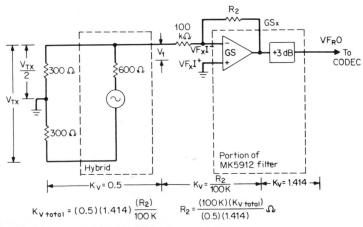

FIG. 4-56 Diagram for transmit gain considerations.

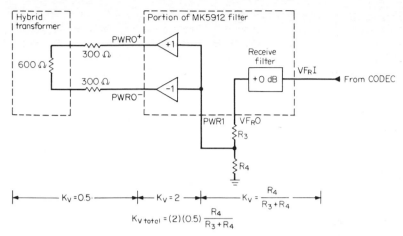

FIG. 4-57 Diagram for receive gain considerations.

0.77 V (rms). The MK5116 was designed so that 0 dBm0 on the transmit side is equal to 1.22 V (rms) and 0 dBm0 on the receive side is equal to 1.19 V (rms).

These conditions tell us that:

$$K_{vt} = (0.5)(1.414)(R_2/R_1) = 1.22 \text{ V}/0.77 \text{ V} = 1.58 \qquad (4\text{-}4)$$

$$\text{and } K_{vr} = (2)(0.5)[R_4/(R_3 + R_4)] = 0.77 \text{ V}/1.19 \text{ V} = 0.65 \qquad (4\text{-}5)$$

These equations now allow R_1, R_2, R_3, and R_4 to be chosen and the design is complete.

4–8c   Design Considerations

In the past, PCM line filter manufacturers have followed different lines in developing CODEC/filter combinations. Since each section of the CODEC needs its own filter, some CODEC/filter manufacturers have produced chip sets with the transmit filter on the same chip as the coder and grouped the receive pair on the other chip. Part of the reasoning behind this approach was to minimize crosstalk between the transmit and receive sections. On the other hand, several manufacturers have manufactured both filters on a single chip, while putting the entire CODEC on another chip. The telecommunications industry currently seems to favor the latter configuration. In the near future this point will not be so much of a factor, since telecommunications manufacturers are beginning to produce all four sections of the system on a single chip. Even then, the coder/filter–decoder/filter arrangement may continue to see use in systems where only a coder or decoder is needed.

Many designs call for an interface to the telephone line using a hybrid transformer. In this case, the designer may favor a filter that has power amplifiers available on the chip. This approach keeps the part count low and many filters are available with this feature. As electronic hybrids begin to be introduced into the industry, the power amplifiers will be less desirable due to the extra power consumption associated with them.

As an aid to the designer, the Product Selector Guide in Sec. 4-9 includes information and manufacturing sources of many currently available PCM line filters.

4–9 PRODUCT SELECTION GUIDE

This selection guide contains information on a variety of devices available for each product type. Each section presents two types of information. Presented first is Table 4-4, which is a list of the manufacturers supplying each product type. Following the list of manufacturers, popular related devices and features may be found in tabular form.

The list of manufacturers contains all of the information necessary to obtain further product data. The company name, address, and phone number are furnished. The source from which device data was obtained is also supplied.

The section of popular related devices includes part numbers, features, and manufacturer. Although the features listed vary due to the disparity of manufacturers, every effort has been made to standardize the terminology.

TABLE 4–4 Manufacturer Product Guide

Manufacturer	Pulse dialers	DTMF encoders	DTMF decoders	Repertory dialers	Speech networks	CODECs	PCM line filters
Advanced Micro Devices							
Advanced Micro Devices 901 Thompson Place Sunnyvale, CA 94086 (408) 732-2400						X	
Ref: AMD Preliminary Information Data Sheets							
AMI							
American Microsystems Inc. 3800 Homestead Road Santa Clara, CA 95051	X	X	X	X		X	
Ref: AMI 1980 MOS Products Catalog							
Cermetek							
Cermetek Microelectronics 660 National Avenue Mountain View, CA 94043 (415) 969-9433			X				
Ref: Cermetek Data Sheets							
CTS							
CTS Microelectronics 1201 Cumberland Avenue West Lafayette, IN 47906 (317) 463-2565			X				
Ref: CTS Data Sheets							

TABLE 4-4 (*Continued*)

Manufacturer	Pulse dialers	DTMF encoders	DTMF decoders	Repertory dialers	Speech networks	CODECs	PCM line filters
Fairchild Fairchild Camera and Instrument Corporation 464 Ellis Street Mountain View, CA 94042 (415) 962-5011 Ref: Fairchild Data Sheets				X		X	
Fujitsu Fujitsu America, Inc. Component Sales Division 910 Sherwood Drive Lake Bluff, IL 60044 (312) 295-2610 Ref: Fujitsu 1980 Preliminary Information Data Sheets						X	
General Instruments Microelectronics Division/General Instruments Corporation 600 West John Street Hicksville, NY 11802 (516) 733-3107 Ref: 1980 General Instruments Microelectronics Product Catalog	X	X	X	X			X
INTEL Intel Corporation 3065 Bowers Avenue Santa Clara, CA 95051 (408) 987-8080 Ref: Intel 1979 and 1980 Preliminary Information Data Sheets						X	X

ITT North

ITT Semiconductors
500 Broadway
Lawrence, MA 01841
(617) 688-1881

Ref: ITT Data Sheets

Mitel

Mitel Semiconductor
1745 Jefferson Davis Highway
Suite 603
Arlington, VA 22202
(703) 243-1600

Ref: Mitel Data Sheets and 1980 Preliminary Information Data Sheets

Mostek

Mostek Corporation
1215 W. Crosby Road
Carrollton, TX 75006
(214) 323-6000

Ref: Mostek 1931 Telecommunications Data Book

Motorola

Motorola Semiconductor Products Inc.
3501 Ed Bluestein Boulevard
Austin, TX 78721
(512) 928-6000

Ref: Motorola 1978 CMOS Data Book

Company							
ITT		X	X				
Mitel	X		X				
Mostek	X	X	X	X	X		X
Motorola	X	X				X	X

TABLE 4–4 (Continued)

Manufacturer	Pulse dialers	DTMF encoders	DTMF decoders	Repertory dialers	Speech networks	CODECs	PCM line filters

National
National Semiconductor Corporation
2900 Semiconductor Drive
Santa Clara, CA 95051
(408) 737-5000
Ref: National Semiconductor
  MOS/LSI Data Book
  and Data Sheets
	X	X	X			X	X

Nitron Nitron, Inc. 10420 Bubb Road Cupertino, CA 95014 (408) 255-7550 Ref: Nitron 1980 Data Sheets	X						
PMI Precision Monolithics Inc. 1500 Space Park Drive Santa Clara, CA 95050 (408) 985-6616 Ref: PMI 1980 Telecommunications Components Data Books						X	
Rockwell Rockwell International Corporation Microelectronic Devices P.O. Box 3669 Anaheim, CA 92803 (714) 632-3729 Ref: Rockwell-Collins 1977 Data Sheets	X		X				

Siemens

Siemens Corporation
186 Wood Avenue S.
Iselin, NJ 08830
(201) 494-1000

Ref: Siemens Data Sheets

Siliconix

Siliconix Inc.
2201 Laurelwood Road
Santa Clara, CA 95054
(408) 988-8000

Ref: Siliconix 1979 Telecommunications Data Book

Stromberg-Carlson

Stromberg-Carlson
P.O. Box 7266
Charlottesville, VA 22906
(804) 973-2200

SGS-ATES

SGS-ATES Semiconductor Corporation
240 Bear Hill Road
Waltham, MA 02154
(617) 890-6688

Ref: SGS-ATES 1979 and 1980 Preliminary Information Data Sheets

Teltone

Teltone Corporation
P.O. Box 657
10801-120th Avenue
Kirkland, WA 98033
(206) 827-9626

Ref: Teltone Data Sheets

TABLE 4-4 (*Continued*)

Manufacturer	Pulse dialers	DTMF encoders	DTMF decoders	Repertory dialers	Speech networks	CODECs	PCM line filters
Western Electric Western Electric 6200 East Broad Street Columbus, OH 43213 (614) 868-2258					X		

TABLE 4-5 Pulse Dialers

S2560A   18-pin   AMI

- CMOS for direct line operation
- *RC* oscillator used as frequency reference
- Dialing rate selectable
- Make-break ratio selectable (33/67 or 40/60)
- 20-digit memory storage
- Access pause
- Uses 2-of-7 or SPST matrix keyboard

AY-5-9100   18-pin   General Instruments

- 20-digit memory storage
- Dialing rate selectable
- Make-break ratio selectable
- Interdigital pause selectable
- Last number redial
- Access pause facility
- Companion rep dialer

AY-5-9151A   22-pin   General Instruments

- Supply operating range 2.5 to 5.0 V
- *RC* oscillator used as frequency reference
- Make-break and interdigital pause selectable
- Uses Form A keyboard
- Last number redial
- 22-digit and pause memory storage

AY-5-9152   22-pin   General Instruments

- Same as AY-5-9151A except:
- Make-break fixed at 40/60
- Two antiphase outputs for driving bistable relay

AY-5-9153A   28-pin   General Instruments

- Same as AY-5-9151A plus:
- Pin selectable 1-of-12 keyboard, 2-of-7 keyboard with common, or 4-bit binary with common inputs
- Repertory dialer when used with AY-5-9200
- 8-bit output for displaying number in storage

AY-5-9154A   28-pin   General Instruments

- Same as AY-5-9153 except:
- Make-break fixed at 40/60
- Two antiphase outputs for driving bistable relay

AY-9-9600   16-pin   General Instruments

- Operating power and voltage is 1 mW at 1.5 V
- Direct-line switching

MT4320   18-pin   Mitel

- Supply operating range 1.5 to 7.0 V
- Power dissipation: 100 μW at 1.5 V
- 20-digit memory storage
- Make-break ratio selectable
- Dialing rate selectable (10, 16, 20, and 932 Hz)

TABLE 4–5 (*Continued*)

MT4325/4326	18-pin	Mitel

- Last number redial
- Any key down causes exit from access pause
- Pacifier tone (300 Hz)
- Supply operating range 2.0 to 7.0 V
- 20-digit and pause memory storage
- Memory retention down to 1.0 V at 1 μA
- Make-break ratio selectable
- Dialing rate selectable (10 Hz, 14.9 kHz)

MK50981/50982/50991/50992	Mostek

- CMOS for direct line operation
- Uses 2-of-7 matrix or Form A type keyboard
- Inputs debounced
- Make-break ratio pin selectable
- Last number redial with an * of # input
- Provision for rapid testing
- On-chip voltage regulator
- Power-up-clear circuitry

MK50981	16-pin	Mostek

- Ceramic resonator used as frequency reference
- Pulse output
- Keyboard inputs active high
- Mute output

MK50982	16-pin	Mostek

- Ceramic resonator used as frequency reference
- Pulse output
- Keyboard inputs active high
- Mute output

MK50991	18-pin	Mostek

- *RC* oscillator used as frequency reference
- Pulse output
- Keyboard inputs active low
- Mute output

MK50992	18-pin	Mostek

- *RC* oscillator used as frequency reference
- Pulse output
- Keyboard inputs active low
- Mute output

MC14408/14409	16-pin	Motorola

- On-chip oscillator used as frequency reference
- 16-digit and pause memory storage
- Last number redial
- Dialing rate selectable (10, 20 pps)
- Selectable make-break ratio
- Requires MC14419

4-80

TABLE 4–5 (*Continued*)

MM5393/5394/53143/53144   National

- Direct-line operation
- Supply operating range to 2.0 V
- *RC* oscillator used as frequency reference
- Uses Form A keyboard
- 21-digit memory storage
- Interdigital pause selectable
- Last number redial

MM5393   18-pin   National

- Pacifier tone (600 Hz)
- Make-break ratio 38.5/61.5

MM5394   16-pin   National

- Make-break ratio 38.5/61.5

MM53143   18-pin   National

- Pacifier tone (600 Hz)
- Make-break ratio 33.3/66.7

MM53144   16-pin   National

- Make-break ratio 33.3/66.7

MM53190   20-pin   National

- Direct-line operation
- Uses 2-of-7 keyboard
- Ceramic resonator used as frequency reference
- Dialing rate selectable (10, 20 pps)
- Make-break ratio selectable
- Interdigital pause selectable
- Pacifier tone (632 Hz)
- Last number redial

NC2320/2321/2322   18/28/18-pin   Nitron

- Supply operating range, 2.5 to 5.5 V
- 3.58-MHz crystal oscillator used as frequency reference
- 20-digit memory storage
- Last number redial
- Make-break ratio selectable (33.3/66.7 or 40/60)
- Dialing rate selectable (10, 16, 20, and 932 pps)
- Interdigital pause selectable

The Loop Disconnect Dialer is available in three pin-out options. The NC2320 provides the functions most commonly required in the push-button telephone application. This option (M_1) remains at logic "1" throughout the dialing sequence. The NC2322 is identical to the NC2320, except that the (M_2) option is a logic "1" only during outpulsing, allowing the telephone line to be monitored during the IDP. The NC2321 offers both options, together with FD, IDP, and SYS CLK.

CRC8000/8001   16-pin   Rockwell

- 16-digit memory storage
- Dialing rate selectable (10, 20 pps)

TABLE 4–5 (*Continued*)

CRC8000/8001	16-pin	Rockwell

- Make-break ratio 40/60
- TTL compatible (CRC 8000)
- MOS compatible (CRC 8001)
- Provision for rapid testing

DF320/321/322	18/18/28-pin	Siliconix

- Supply operating range, 2.5 to 5.5 V
- Power consumption @ 3 V: active, 540 μW, standby, 3 μW
- Power-up-clear circuitry
- Inputs debounced
- 20-digit memory storage
- Last-number redial
- Make-break ratio selectable 33.3/66.7 or 40/60
- Dialing rate selectable (10, 16, 20 and 932 Hz)
- Interdigital pause selectable

The Loop Disconnect Dialer is available in three pin-out options. The DF320 provides the functions most commonly required in the push-button telephone application. M_1 is the masking option, which remains at logic ''1'' throughout the dialing sequence. The DF322 is identical to the DF320, except that M_2 is offered instead of M_1. The M_2 masking option is a logic ''1'' only during outpulsing, allowing the telephone line to be monitored during the IDP. The DF321 is a multioption version that offers both M_1 and M_2 together with FD, IDP, and SYS CLK.

TABLE 4–6 DTMF Encoders

S2559 A/B/C/D	16-pin	AMI

- Supply operating range 3.5 to 13.0 V (A/B)
- Supply operating range 2.7 to 10.0 V (C/D)
- Direct line-powered operation or small battery (9 V)
- 3.58-MHz crystal used as frequency reference
- On-chip mute drivers
- Uses 2-of-8 or Form A keyboard
- Dual-tone as well as single-tone operation

S2859	16-pin	AMI

- Supply operating range 3.0 to 10.0 V
- CMOS for direct line-powered operation
- 3.58-MHz crystal used as frequency reference
- Timing sequence for Xmit, Rec, Mute outputs
- Uses 2-of-8 or Form A keyboard
- Dual-tone and single-tone capabilities
- Darlington tone output

S2860	16-pin	AMI

- Optimized for constant supply (typically 3.5 V)
- CMOS for direct line-powered operation
- Designed for electronic telephone applications
- Uses 2-of-8 or Form A keyboard
- Dual-tone and single-tone capabilities
- 3.58-MHz crystal used as frequency reference

TABLE 4–6 (*Continued*)

S2861 A/B   16-pin   AMI

- Replaces S2559 with reduced distortion
- Supply operating range 2.5 to 10.0 V
- CMOS for direct line-powered operation
- 3.58-MHz crystal used as frequency reference
- Oscillator bias resistor on chip
- Uses 2-of-8 or Form A keyboard
- Dual-tone and single-tone capabilities
- Mode select (S2861A)
- Chip disable (S2861B)

AY-3-9400/9401/9410   14/16/16-pin   General Instruments

- Uses ceramic resonator as frequency reference
- High-group preemphasis selectable (AY-3-9410 only)
- Multikey lockout
- NMOS for direct line-powered operation
- Uses 2-of-8 keyboard
- High-group preemphasis
      3.52   dB   (AY-3-9400)
      2.00   dB   (AY-3-9401)
      3 or 6 dB   (AY-3-9410)

SBA5089   16-pin   ITT

- CMOS for fixed supply operation
- 3.58-MHz crystal used as frequency reference
- Single-tone inhibit selectable
- Tone disable selectable
- Uses 2-of-8 keyboard with V^- common

SBA5091   18-pin   ITT

- CEPT compatible
- 3.58-MHz crystal used as frequency reference
- Chip disable input
- Uses Form A keyboard
- On-chip transistor for use in constructing filter to meet CEPT distortion specs

---

MK5087   16-pin   Mostek

- CMOS for direct line-powered operation
- Mute output
- 3.58-MHz crystal used as frequency reference
- On-chip tone regulation
- Uses Form A or 2-of-8 keyboard
- Single-Tone Inhibit input

MK5089   16-pin   Mostek

- 3.58-MHz crystal used as frequency reference
- Designed for microprocessor and European applications
- Tone disable input
- Single-Tone Inhibit input
- Any Key Down output

MK5091   18-pin   Mostek

- CEPT compatible

TABLE 4–6 (*Continued*)

MK5091	18-pin	Mostek

- Bipolar transistor on-chip for use in constructing filter to meet CEPT specs
- 3.58-HMHz crystal used as frequency reference
- Single-Tone Inhibit input
- Tone Disable input
- Uses Form A or 2-of-8 keyboard
- CMOS for direct line-powered operation

MK5092	16-pin	Mostek

- Internal tone regulation
- Internal loop compensation and preemphasis
- 3.58-MHz crystal used as frequency reference
- Tone Disable input
- Single-Tone Inhibit input
- Uses Form A or 2-of-8 keyboard
- Mute output

MK5094	16-pin	Mostek

- Internal tone regulation
- Internal loop compensation and preemphasis
- 3.58-MHz crystal used as frequency reference
- Tone Disable input
- Single-Tone Inhibit input
- Uses Form A or 2-of-8 keyboard
- Mute output

MK5380	16-pin	Mostek

- Internal loop compensation and preemphasis
- 3.58-MHz crystal used as frequency reference
- Supply operating range 2.5 to 10.0 V
- Uses Form A or 2-of-8 keyboard
- Chip Disable input
- Single-Tone Inhibit input
- Mute output

MC14403/14405	16/18-pin	Motorola

- Supply operating range 3.0 to 12.0 V
- 3.58-MHz crystal used as frequency reference
- Oscillator is inhibited when no key is depressed
- Output level selectable
- Internal bridge rectifier (MC14405 only)
- Uses 2-of-8 keyboard
- Dual- or single-tone selectable

MC14410	16-pin	Motorola

- Supply operating range 4.4 to 6.0 V
- 3.58-MHz crystal used as frequency reference
- Uses 2-of-8 keyboard
- Multikey lockout
- Dual- or single-tone selectable

MM5395/53125	18-pin	National

- Supply operating range 3.0 to 5.0 V
- 3.58-MHz crystal used as frequency reference

TABLE 4–6 (*Continued*)

MM5395/53125	18-pin	National

- Uses 2-of-8 keyboard (MM5395)
- Uses Form A keyboard (MM53125)
- Multikey lockout with single-tone capability
- High-group preemphasis
- Open emitter output
- Mute output
- Direct line-powered operation

MM53130/53131/53132	18/18/16-pin	National

- Supply operating range 3.0 to 8.0 V
- 3.58-MHz crystal used as frequency reference
- Uses 2-of-8 or Form A keyboard
- BCD coded inputs (MM53130 only)
- Multikey lockout with or without single tone (MM53130 & MM53132 only)
- High-group preemphasis
- Open emitter output
- Separate receiver and transmitter Mute outputs
- Direct line-powered operation

TABLE 4–7 DTMF Decoders

AY-5-9800 Series	General Instruments

- Inherent discrimination $\geq 0.1\%$
- Digitally defined bandwidths
- Acquisition time typically 25 ms
- Interface directly with CP1600 microprocessor
- Programmable center frequencies and accuracies

AY-5-9800/9801/9804/9824	28-pin	General Instruments

- On-chip op amps
- 4-bit output code (AY-5-9801/9821)
- Binary output code (AY-5-9804/9824)

AY-5-9802/9822/9803/9823	40-pin	General Instruments

- On-chip op amps
- 1-of-16 output code (AY-5-9802/9822)
- 2-of-8 output code (AY-5-9803/9823)

AY-5-9805/9825/9807/9827/9808/9828	24-pin	General Instruments

- 4-bit output code (AY-5-9805/9825)
- 2-of-8 output code (AY-5-9807/9827)
- Binary output code (AY-5-9808/9828)

AY-5-9806/9826	40-pin	General Instruments

- 1-of-16 output code

MK5102-5/5103-5	Mostek

- CMOS circuitry for low power
- Detects all 16 standard DTMF digits
- 3.58-MHz crystal used as frequency reference

TABLE 4-7 (*Continued*)

MK5102-5/5103-5   Mostek

- Digital counter detection with period averaging
- Single +5-V supply
- 4-bit binary or 2-bit row/column output code
- Latched outputs

MK5102-5   16-pin   Mostek

- 18-dB S/N ratio (min)
- Tone coincidence duration 33 ms (min)

MK5103-5   16-pin   Mostek

- 14-dB S/N ratio (min)
- Tone coincidence duration 30 ms (min)

MT8820   24-pin   Mitel

- CMOS circuitry
- Detects all 16 standard DTMF digits
- 3.58-MHz crystal used as frequency reference
- Single +5-V or 12- to 15-V supply
- Detect times down to 20 ms
- 4-bit hex, 2-of-8 or 4-bit binary output code
- Built-in power-on reset
- Adjustable acquisition and release times

MT8860/8862/8863   Mitel

- CMOS circuitry for low power
- Detects all 16 standard DTMF digits
- 3.58-MHz crystal used as frequency reference
- Detect times down to 20 ms
- Latched outputs

MT8860   18-pin   Mitel

- Single +5-V or 8- to 15-V supply
- 4-bit binary output

MT8862/8863   24-pin   Mitel

- Single +5-V or 8- to 13-V supply
- Hex or 2-of-8 output codes

CRC8030   28-pin   Rockwell

- Detects all 16 standard DTMF digits
- 3.58-MHz crystal used as frequency reference
- Automatic internal reset when no tones are present
- Binary or 2-of-8 output code

M-927   40-pin   Teltone

- Detects all 16 standard DTMF digits
- Meets CCITT specifications
- Time-guarded rotary dial pulse counting
- 3.58-MHz crystal used as frequency reference
- Binary, 2-of-8, 1-of-12, or blank output code

TABLE 4–7 (*Continued*)

S3525A/B   18-pin   AMI

- Band-split filter and limiter
- CMOS circuitry
- 3.58-MHz crystal used as frequency reference
- Supply voltage range 10 to 13.5 V
- 52-dB dial-tone rejection
- 3.58-MHz buffered oscillator output (S3525A)
- 894.89-kHz square wave output (S2535B)

CH1295/1296   24-pin   Cermetek

- Low-group bandpass filter (CH1295)
- High-group bandpass filter (CH1296)
- ±12-V supply
- 50-dB dial-tone rejection
- 40-dB attenuation between high and low group

CTS600623/600624   20-pin   CTS

- Low-group bandpass filter (600623)
- High-group bandpass filter (600624)
- ±18-V supply
- 50-dB dial-tone rejection
- 40-dB attenuation between high and low group

ITT3040A/3041A   24-pin   ITT North

- Bandpass filters and hard limiters
- Low-group bandpass filter (3040A)
- High-group bandpass filter (3041A)
- 55-dB dial-tone rejection
- 40-dB attenuation between high and low group

ITT3044/3045   15-pin   ITT North

- Low-group bandpass filter (3044)
- High-group bandpass filter (3045)
- 55-dB dial-tone rejection
- 40-dB attenuation between high and low group

MT8865   16-pin   Mitel

- Band-split filter and hard limiter
- CMOS circuitry
- 3.58-MHz crystal used as frequency reference
- Supply voltage range 5 to 12 V
- 30-dB dial-tone rejection
- 38-dB attenuation between high and low group

AF121/122   24-pin   National

- Low-group bandpass filter (AF121)
- High-group bandpass filter (AF122)
- ±5- to ±18-V supply
- 40-dB dial-tone rejection
- 40-dB attenuation between high and low group

ITT3201   22-pin   ITT North

- Decoder, band-split filter, and limiter
- CMOS circuitry

4-87

TABLE 4–7 (*Continued*)

ITT3201	22-pin	ITT North

- Detects 12 or 16 standard DTMF digits
- 3.58-MHz crystal used as frequency reference
- Single + 12-V supply
- 4-bit hex or 2-of-8 output code

ITT3210	24-pin	ITT North

- Decoder, band-split filter, limiter, and dial-tone rejection filter
- 3.58-MHz crystal used as frequency reference

MH88210	50-pin	Mitel

- Decoder and band-split filter
- 14-dB S/N ratio
- Acquisition time adjustable down to 10 ms

TABLE 4–8 Repertory Dialers

S2562	40-pin	AMI

- CMOS circuitry
- 8- or 16-digit redial capability
- Stand-alone pulse dialer
- Interfaces with tone generator for DTMF dialing
- Two pin-selectable pulse rates
- Two pin-selectable interdigital pauses
- Stores sixteen 16-digit or thirty-two 8-digit numbers with external RAM
- Storable access pause
- *RC* oscillator used as frequency reference
- Power-fail detection
- BCD output for numerical displays

CET200	40-pin	Fairchild

- 12-digit redial capability
- Stand-alone pulse dialer
- Interfaces with tone dialers
- Two pin-selectable pulse rates
- Stores twenty 12-digit numbers with external RAM
- Access pause capability
- Overflow indicator
- Single + 5-V supply
- Two 12-digit scratchpad memories on-chip

AY-5-9200	16-pin	General Instruments

- Stores ten 22-digit numbers
- Interfaces with pulse dialers for pulse applications
- Interfaces with tone generators for DTMF dialing
- Storable access pause

TZ2001/2002/2003		General Instruments

- Microcomputer-based dialer
- On-board keyboard debounce circuitry

TABLE 4–8 (*Continued*)

TZ2001/2002/2003   General Instruments

- 12-digit last number redial
- *RC* oscillator used as frequency reference
- 32.768-kHz crystal used for real-time reference
- Requires external RAM for number storage

TZ2001   40-pin   General Instruments

- Stand-alone pulse dialer
- Stores sixteen 12-digit numbers with external RAM
- Real-time clock with display
- Elapsed-time timer/stopwatch
- Uses 8 × 4 matrix keyboard

TZ2002   40-pin   General Instruments

- Interfaces with tone generators for DTMF dialing
- Stores sixteen 12-digit numbers with external RAM
- Real-time clock with display
- Uses 8 × 4 matrix keyboard

TZ2003   40-pin   General Instruments

- Stand-alone pulse or tone dialer
- Stores thirty-two 16-digit numbers with external RAM
- Pin-selectable tone and pulse rates
- LED function indicator drivers
- Uses 4 × 4 matrix keyboard

MK5170   40-pin   Mostek

- Microcomputer-based dialer
- 20-digit redial capability
- Interfaces to pulse dialers for pulse applications
- Interfaces to tone generators for DTMF dialing
- Stores one hundred 20-digit numbers with external RAM
- 3.58-MHz crystal used as frequency reference
- 12/24 hour real-time clock with display
- LED display of number dialed and its address code
- 100-hour timer
- Single +5-V supply

MK5175   16-pin   Mostek

- CMOS circuitry for low power
- Supply voltage range 2 to 10 V
- Stores ten 16-digit numbers
- Line operation off-hook, battery operation on-hook
- Stand-alone pulse dialer
- Interfaces with tone generators for DTMF dialing
- Storable access pause
- Pin-selectable make/break ratio
- Uses 2-of-7 or Form A keyboard
- Tone mode may use SPST switch control key for 13-key mode
- Ceramic resonator used for frequency reference in pulse mode
- *RC* oscillator used as time base in tone mode
- Pulse output active low

TABLE 4–9 Speech Networks

Hybrid Speech Networks

2500 Network   Western Electric

- 2-to-4 wire conversion
- Sidetone balancing
- Loop-length compensation
- Surge protection

2500-Type Network   Stromberg-Carlson

- Same as 2500 network

Active Speech Networks

MK5242   16-pin   Mostek

- Direct telephone-line operation
- On-chip, 2-to-4 wire conversion
- Sidetone balancing with adjustable sidetone level
- Automatic loop-length compensation
- Direct interface to low cost, audio transducers
- Provides voltage-regulated output for dialers requiring fixed supply
- Direct interface to Mostek dialers
- Bipolar circuitry
- On-chip muting capability

LS156   16-pin   SGS-ATES

- On-chip, 2-to-4 wire conversion
- Meets CCITT specifications
- Automatic loop-length compensation
- Provides stabilized voltage to interface with tone generators
- Interfaces to dynamic receivers and transmitters

LS285A   14-pin   SGS-ATES

- On-chip, 2-to-4 wire conversion
- Sidetone balancing
- Automatic loop-length compensation
- Interfaces to dynamic receivers and transmitters

TABLE 4–10 CODECs

AM6070   18-pin   Advanced Micro Devices

- μ-255 law companding 12-bit D/A converter
- Output dynamic range of 72 dB
- 12-bit accuracy and resolution around zero
- Sign plus 12-bit range with sign plus 7-bit coding
- Microprocessor-controlled operations

AM6071   18-pin   Advanced Micro Devices

- A-law companding 8-bit D/A converter
- Output dynamic range of 62 dB
- Microprocessor-controlled operations

4-90

TABLE 4–10 (*Continued*)

AM6072	18-pin	Advanced Micro Devices

- μ-255 law companding 8-bit D/A converter
- Output dynamic range of 72 dB
- Tested to *D*3 compandor tracking specifications
- Microprocessor-controlled operations

5116/5116-1	16-pin	Fairchild

- μ-255 law companding CODEC
- CMOS circuitry
- ±5-V supplies
- Low-power dissipation: 30 mW
- Synchronous or asynchronous operation
- On-chip sample and hold
- On-chip offset null circuit
- Serial data output rate of 64 kbits/s to 2.1 Mbits/s
- Separate analog and digital grounds
- Sign plus magnitude coding (5116-1)
- Exceeds *D*3 channel bank transmission specifications
- Zero code suppression

5151	24-pin	Fairchild

- μ-255 law companding CODEC
- CMOS circuitry
- ±5-V supplies
- Low power dissipation: 30 mW
- Synchronous or asynchronous operation
- On-chip sample and hold
- On-chip offset null circuit
- Serial data output rate of 64 kbits/s to 2.1 Mbits/s
- Separate analog and digital grounds
- Exceeds *D*3 channel bank transmission specifications
- Zero code suppression

5156	16-pin	Fairchild

- A-law companding CODEC
- CMOS circuitry
- ±5 V supplies
- Low power dissipation: 30 mW
- Synchronous or asynchronous operation
- On-chip sample and hold
- On-chip offset null circuit
- Serial data output rate of 64 kbits/s to 2.1 Mbits/s
- Separate analog and digital grounds
- Exceeds CCITT specifications
- Even-order bit inversion data format

2910A	24-pin	Intel

- μ-255 law, companding, per-channel CODEC (8-bit)
- ±5- and +12-V supplies
- Low power dissipation: 33 mW
- On-chip sample and hold
- Output dynamic range of 78 dB
- CCITT G711 and G712 compatible, ATT T1 compatible with 8th-bit signaling
- Precision on-chip voltage reference
- Microcomputer interface with on-chip time-slot computation

TABLE 4–10 (*Continued*)

2911A	22-pin	Intel

- A-law, companding, per-channel CODEC (8 bit)
- ±5- and +12-V supplies
- Low power dissipation: 33 mW
- On-chip sample and hold
- Output dynamic range of 66 dB
- CCITT G711 and G732 compatible
- Precision on-chip voltage reference
- Microcomputer interface with on-chip time-slot computation
- Even-order bit inversion

MK5116	16-pin	Mostek

- μ-255 law, companding, per-channel CODEC (8 bit)
- CMOS circuitry
- ±5-V supplies
- Low power dissipation: 30 mW
- Synchronous or asynchronous operation
- On-chip sample and hold
- On-chip offset null circuit
- Serial data output rate of 64 kbits/s to 2.1 Mbits/s
- Separate analog and digital grounds

MK5151	24-pin	Mostek

- μ-255 law, companding, per-channel CODEC (8 bit)
- CMOS circuitry
- ±5-V supplies
- Low power dissipation: 30 mW _____
- Zero code suppression and sign-magnitude data format
- On-chip sample and hold
- On-chip offset null circuit
- Serial data output rate of 64 kbits/s to 2.1 Mbits/s
- Separate analog and digital grounds

MK5156	16-pin	Mostek

- A-law, companding, per-channel CODEC (8 bit)
- CMOS circuitry
- ±5-V supplies
- Low power dissipation: 30 mW
- Even-order bit inversion
- Synchronous or asynchronous operation
- On-chip sample and hold
- On-chip offset null circuit
- Serial data output rate of 64 kbits/s to 2.1 Mbits/s
- Separate analog and digital grounds

MC14404/14406/14407	24/28/24-pin	Motorola

- μ-255 law, companding, per-channel, 8-bit CODEC (MC14406)
- μ-255 and A-law, companding, per-channel, 8-bit CODEC (MC14404/14407)
- CMOS circuitry
- Supply voltage range: 10 to 16 V
- Low power dissipation: 80 mW
- Zero code suppression (MC14406/14407)
- Transmit and receive signaling (MC14406)
- On-chip auto zero

TABLE 4–10 (*Continued*)

MC14404/14406/14407	24/28/24-pin	Motorola

- μ-255 law digital format (MC14406/14407)
- A-law CCITT digital format (MC14404)

TP3001/3002	48/42-pin	National

- Two chip, companding shared CODEC system (8 bit)
- μ-255 law (TP3001)
- A-law (TP3002)
- Uses LF3700 and MM58100 (TP3001)
- Uses LF3700 and MM58150 (TP3002)
- Input and output sample and hold
- Auto zero circuitry
- Meets CCITT specifications
- Meets $D3$ channel bank specifications
- Serial data output rate of 64 kbits/s to 2.1 Mbits/s
- Provision for signaling bits (TP3001)

DAC-86	18-pin	PMI

- μ-255 law companding D/A converter
- Meets $D3$ tracking specifications

DAC-87	18-pin	PMI

- A-law companding D/A converter
- CCITT compatibility
- Sign plus 11-bit range with sign plus 7-bit coding
- 11-bit accuracy and resolution around zero
- Sign plus 66-dB dynamic range
- Outputs multiplexed

DAC-88	18-pin	PMI

- μ-255 companding D/A converter
- Meets $D3$ tracking specifications

SM61A/S291	24-pin	Siemens

- Two-channel, two-device, A-law CODEC system
- CCITT G711 compatible
- All digital interfaces TTL compatible
- 8-kHz sampling for one or two channels, or 16-kHz sampling for one channel
- Single-stage configuration capability

SM61C	28-pin	Siemens

- μ-255 law or A-law companding two-channel CODEC
- ±5- and +12-V supplies
- Mode of operation pin-selectable
- CCITT G711 and G732 compatible
- 8-kHz sampling for one or two channels, or 16-kHz sampling for one channel
- Digital interfaces TTL compatible
- Auto zero circuitry
- Low power consumption: 20 mW

DF331A/332A/334A	14-pin	Siliconix

- μ-255 law A/D converter (DF331A)
- μ-255 law D/A converter (DF332A/334A)
- 8-bit CODEC system with DF331A and DF332A or DF331A and DF334A

TABLE 4–10 (*Continued*)

DF331A/332A/334A	14-pin	Siliconix

- Low power dissipation: 11 mW
- 8-kHz sampling rate

DF341/342	14-pin	Siliconix

- A-law A/D converter (DF341)
- A-law D/A converter (DF342)
- Form 8-bit CODEC system
- On-board sample and hold (DF342)
- Low power dissipation: 11 mW
- Separate analog and digital grounds
- 8-kHz sampling rate

CODEC/Filter sets

MB6001/6002	18/16-pin	Fujitsu

- A/D converter and filter (MB6001)
- D/A converter and filter (MB6002)
- Pin-selectable μ-255 law or A-law CODEC system
- CMOS circuitry
- ± 2.5- and ± 5-V supplies
- Low power dissipation: 30 mW
- $D3$ and CCITT compatible
- 18-dB gain (MB6001)
- Built-in auto zero circuitry (MB6001)
- All digital inputs TTL compatible

S3501/3502	18-pin	AMI

- A/D converter and filter (S3501)
- D/A converter and filter (S3502)
- CMOS circuitry
- Forms μ-255 law CODEC system
- $D3$ and CCITT G711 or G733 compatible
- Auto zero circuitry (S3501)
- Serial data output rate of 56 kbits/s to 3.1 Mbits/s
- 8-kHz sampling rate

S3501A/3502A	18/16-pin	AMI

- Same as S3501/3502 system plus:
- CCIS compatible A/B signaling option

S3503/3504	18/16-pin	AMI

- A/D converter and filter (S3503)
- D/A converter and filter (S3504)
- CMOS circuitry
- Forms A-law CODEC system
- CCITT G711, G712, and G733 compatible
- Auto zero circuitry (S3503)
- Serial data output rate of 56 kbits/s to 3.1 Mbits/s
- 8-kHz sampling rate

TABLE 4-11 PCM Line Filters

ACF7270C/7271C	8-pin	General Instruments

- Transmit low-pass filter (ACF7270C)
- Receive low-pass filter (ACF7271C)
- Supply voltage range: ±12 to ±15 V
- Passband ripple: ±0.125 dB from 300 to 3 kHz
- Stopband attenuation: 32 dB from 4.2 to 100 kHz

ACF7272C/7273C	14-pin	General Instruments

- Transmit low-pass filter (ACF7272C)
- Receive low-pass filter (ACF7273C)
- Supply voltage range: ±12 to ±15 V
- Passband ripple: ±0.125 dB from 300 to 3 kHz
- Stopband attenuation: 32 dB from 4.2 to 100 kHz
- 60-Hz attenuation: 25 dB

2912	16-pin	Intel

- Transmit bandpass and receive low-pass filter
- ±5-V supply
- Low power consumption: 210 mW without power amplifiers
- AT&T $D3/D4$ compatible and CCITT G712 compatible
- 50–60-Hz rejection in the transmit filter
- Adjustable gain in both directions

2912A	16-pin	Intel

- Transmit bandpass and receive low-pass filter
- Lower power consumption: 50 mW without power amplifiers
- AT&T $D3/D4$ compatible and CCITT G712 compatible
- 60-Hz attenuation: 23 dB
- Adjustable gain in both directions

MK5912-1	16-pin	Mostek

- Transmit bandpass and receive low-pass filter
- ±5-V supply
- Low power consumption: 30 mW without power amplifiers
- 50–60-Hz rejection in the transmit filter
- Sin x/x compensation in the receive filter
- Adjustable gain in both directions

MK5912-3	16-pin	Mostek

- Transmit bandpass and receive low-pass filter
- ±5-V supply
- Low power consumption: 35 mW without power amplifiers
- 50–60 Hz rejection in the transmit filter
- Sin x/x compensation in the receive filter
- Adjustable gain in both directions

MC14413/14414	16-pin	Motorola

- Transmit bandpass and receive low-pass filter (MC14413)
- Transmit and receive low-pass filter (MC14414)
- Supply voltage range: ±5 to ±8 V
- Low power consumption: 30 mW
- Sin x/x compensation in the receive filter

TABLE 4–11 (*Continued*)

MC14413-1/14413-2/14414-1/14414-2	16-pin	Motorola

- Transmit bandpass and receive low-pass filter (14413-1/14413-2)
- Transmit and receive low-pass filter (MC14414-1/14414-2)
- Supply voltage range: ±5 to ±8 V
- Low power consumption: 30 mW
- Sin x/x compensation in the receive filter
- AT&T $D3/D4$ compatible (MC14413-2/14414-2)
- CCITT G712 compatible (MC14413-1/14414-1)

AF132	24-pin	National

- Transmit and receive low-pass filter
- Supply voltage range: ±9 to ±15 V

AF133/134	24-pin	National

- Transmit low-pass filter (AF133)
- Receive low-pass filter (AF134)
- Supply voltage range: ±12 to ±15 V
- AT&T $D3$ compatible

AF137	24-pin	National

- Transmit and receive low-pass filter
- Supply voltage range: ±5 to ±15 V

TP3040	16-pin	National

- Transmit and receive low-pass filter
- CMOS circuitry
- ±5-V supply
- Low power consumption: 30 mW without power amplifiers
- 50–60-Hz rejection in the transmit filter
- Sin x/x compensation in the receive filter
- AT&T $D3/D4$ compatible and CCITT G712 compatible

REFERENCES

AMI Application Note: "S2559 Digital Tone Generator," 79T01.

Bell System Technical Reference: "Electrical Characteristics of Bell System Network Facilities at the Interface with Voiceband Ancillary and Data Equipment," PUB 47001, August 1976.

Bell System Technical Reference: "Functional Product Class Criteria, PBX," PUB 48002, September 1978.

Bell System Technical Reference: "Digital Channel Bank Requirements and Objectives," PUB 43801, December 1978.

CCITT Orange Book: *Line Transmission*, Vol. III-2, 1977.

EIA Standard: "Telephone Instruments with Loop Signaling for Voiceband Applications," RS-470, January 1981.

MOSTEK Application Brief: "CODEC/Filter Demo Board," TCM-AB-05-80-01, January 1981.

MK5087 Data Sheet: "Integrated Tone Dialer," October 1980.

MK50981 Data Sheet: "Integrated Pulse Dialer with Redial," October 1980.

MK50982 Data Sheet: "Integrated Pulse Dialer with Redial," October 1980.

MK50992 Data Sheet: "Integrated Pulse Dialer with Redial," October 1980.

MK5102-5 Application Note: "Design Considerations for a DTMF Receiver System," August 1978.

MK5102-5 Preliminary Data Sheet: "Integrated Tone Receiver," July 1978.

MK5102/S3525A Application Brief: "DTMF Receiver System," TCM-AB-03-81-01, June 1981.

MK5116 Data Sheet: "μ-255 Law Companding CODEC," January 1981.

MK5170 Preliminary Data Sheet: "Repertory Dialer," August 1979.

MK5175 Target Specification: "Ten-Number Repertory Dialer," August 1981.

MK5380 Target Specification: "Integrated Tone Dialer," February 1981.

MK5912-3 Preliminary Data Sheet: "PCM Transmit/Receive Filters," June 1981.

PMI Telecommunications Components Data Book, Sec. 4: "Telephone Network: General Theory," 1979.

Schenker, L.: "Pushbutton Calling with Two-Group Voice-Frequency Code, *Bell System Technical Journal,* January 1960.

Signetics Analog Manual, Sec. IX: "Phase Locked Loops," 1976.

Chapter 5

PHASE-LOCKED LOOPS

Sid Ghosh
TRW Vidar Corp.
Mountainview, Calif.

5–1 PRINCIPLES OF PHASE-LOCKED LOOPS

5–1a Basic Transfer Functions

Fig. 5-1 shows the block diagram of a basic phase-locked loop (PLL). Most practical loops are not strictly linear in that the output voltage of the phase detector is not exactly proportional to the phase error, and the output frequency of the voltage-controlled os-

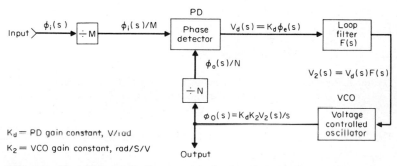

K_d = PD gain constant, V/rad

K_2 = VCO gain constant, rad/S/V

FIG. 5-1 Block diagram of a linear model of a phase-locked loop.

cillator (VCO) is not linearly related to the control voltage. Also, the loop behavior is extremely nonlinear when out of lock. We shall assume, however, that the loop is in lock and is linear. The assumption of linearity enables us to develop very useful mathematical characterizations of the loop that lead to important design tools, and a proper understanding of the operation of the loop.

A list of symbols used in the equations throughout this chapter is printed here for easy reference.

Nomenclature

B	bandwidth
$F(0)$	DC gain of filter
$F(s)$	loop filter transfer function (Laplace notation)
$H(s)$	loop transfer function (Laplace notation)
K	loop gain
K_d	phase detector gain constant
K_v	DC loop gain of PLL
K_2	VCO gain constant
Q_e	effective Q
T_p	pull-in time
$V_d(s)$	phase detector output voltage (Laplace notation)
$V_e(t)$	instantaneous error voltage
$V_i(t)$	instantaneous input voltage
$V_o(t)$	instantaneous output voltage
$V_2(s)$	filter output voltage (Laplace notation)
Δf_L	lock-in range
Δf_p	pull-in range
$\Delta \omega$	change in VCO output frequency
$\Delta \dot{\omega}$	frequency ramp
$\phi_e(t)$	instantaneous phase error
$\phi_o(t)$	instantaneous phase of output frequency
$\phi_i(t)$	instantaneous phase of input frequency
$\phi_e(s)$	phase error (Laplace notation)
$\phi_i(s)$	phase of input frequency (Laplace notation)
$\phi_o(s)$	phase of output frequency (Laplace notation)
ϕ_e	phase error
ϕ_i	phase of input frequency
ϕ_o	phase of output frequency
ω_c	carrier frequency
ω_m	modulating frequency
ω_n	natural frequency of loop
ζ	loop damping factor

The input frequency has an instantaneous phase $\phi_i(t)$, and the instantaneous phase of the VCO output frequency is $\phi_o(t)$. The input and output frequencies are divided by M and N, respectively, and after division their phases are compared in a phase detector (PD).

The phase error $\phi_e(t)$ can be expressed in Laplace notation as

$$\phi_e(s) = \frac{\phi_i(s)}{M} - \frac{\phi_o(s)}{N} \tag{5-1}$$

$$V_d(s) = K_d \phi_e(s) \tag{5-2}$$

where K_d (volts per radian) is the phase-detector gain constant.

The phase-detector output voltage is filtered by the loop filter $F(s)$, which rejects out-of-band noise and high-frequency signal components at the output of the PD. The filter output voltage is given by

$$V_2(s) = V_d(s)F(s) \tag{5-3}$$
$$= K_d F(s)\phi_e(s)$$

The corresponding change in VCO output frequency is

$$\Delta\omega = K_2 V_2(s)$$

where K_2 (radians per second per volt) is the VCO gain constant. Since $\Delta\omega = d\phi_o(s)/dt = s\phi_o(s)$

$$\phi_o(s) = \frac{K_d K_2 F(s)}{s} \tag{5-4}$$

Combining these equations, we get the basic loop transfer functions

$$\frac{\phi_o(s)}{\phi_i(s)} = H(s) = \frac{N}{M} \cdot \frac{K_d K_2 F(s)/N}{s + K_2 K_d F(s)/N}$$

$$= \frac{N}{M} \cdot \frac{KF(s)}{s + KF(s)} \tag{5-5}$$

where $K = K_2 K_d/N$. Also

$$\phi_e(s) = \frac{\phi_i(s)}{M} - \frac{\phi_o(s)}{N}$$

$$\frac{\phi_e(s)}{\phi_i(s)} = \frac{1}{M} \cdot \frac{s}{s + KF(s)} \tag{5-6}$$

We define here an important loop parameter K_v, called the DC loop gain of the PLL.

$$K_v = KF(0) \tag{5-7}$$

where $F(0)$ is the DC gain of the filter.

To investigate the performance of the loop further, we must have knowledge of the loop filter. The loop behaves like any closed-loop servo or feedback system. For proper operation of the loop there are three parameters that may have to be chosen independently, depending upon the application. These are the natural frequency of the loop ω_n, the damping factor ζ, and the DC loop gain K_v. We have defined K_v, and shall shortly discuss the relationship of the ω_n and ζ to the physical loop constants.

However, before proceeding to the next stage, we briefly mention the effects of the two dividers M and N at the inputs to the PD on the performance of the loop. It often becomes necessary to include these dividers. For example, the PLL may be used as a frequency synthesizer, where the output frequency is a multiple of the input frequency, $f_o = Nf_i$. In this case, N is a programmable divider. In a data communication system a PLL may be used to derive a stable and jitter-free clock from a very jittery input. If the jitter amplitude is very large (more than one time slot), it is necessary to divide the input clock (and also the VCO output) to reduce the jitter amplitude to less than one time slot of the divided clock so that the loop may not lose lock.

Unfortunately a large division ratio N in the feedback loop may have some undesirable effects.

1. The loop gain is reduced by N (Eq. 5-5), which slows down the transient response of the loop to any change at the input (see Fig. 5-6).
2. For a given jitter in the input phase, in the passband of the loop, the output jitter is N/M times the input jitter. If $N = M$ and the noise comes from the input, this does not result in any jitter magnification. But for noise originating at the PD input (not divided by M), the output magnification factor is N.
3. Since $\phi_o = N/M\phi_i$ ($\phi_e \cong 0$ in the steady state), there is a steady difference between the input and output phases (unless $N = M$).

It therefore follows that large division ratios in the feedback loop path should be avoided if possible. We refer to this later when discussing a prescaler in a frequency synthesizer. In the following discussions we assume that $M = N = 1$ for the sake of simplicity. For other values of M and N necessary loop equations may be derived directly from Eqs. 5-5 and 5-6.

Order of the Loop Transfer Function   The order of the loop is the highest power of s in the denominator of the loop transfer function and is determined by the loop filter.

First-Order Loop

If the loop filter is omitted altogether, $F(s) = 1$ and the basic loop transfer function becomes

$$H(s) = \frac{K}{s + K} \qquad (5\text{-}8)$$

This is a first-order loop. The only variable in the loop is $K_v = K$, and therefore the usefulness of this type of loop is very limited.

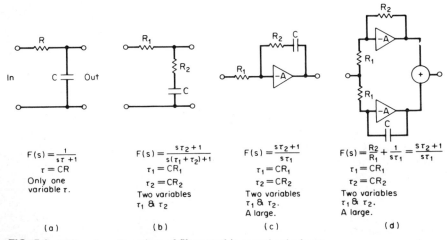

FIG. 5-2 Different configurations of filters used in second-order loops.

Second-Order Loop

Fig. 5-2 shows different types of filters that may be used to derive a second-order loop.

For the filter configuration in Fig. 5-2a

$$H(s) = \frac{\omega_n^2}{s^2 + 2\zeta\omega_n s + \omega_n^2} \tag{5-9}$$

where $\omega_n^2 = \dfrac{K}{\tau}$

$$\zeta = \frac{1}{2}\sqrt{\frac{\tau}{K}}$$

Here there are only two independent variables K and ζ, and therefore this configuration is not very suitable either.

For the configuration in Fig. 5-2b

$$H(s) = \frac{s\omega_n(2\zeta - \omega_n/K)}{s^2 + 2\zeta\omega_n s + \omega_n^2} \tag{5-10}$$

where $\omega_n^2 = \dfrac{K}{\tau_1 + \tau_2}$

$$\zeta = \frac{\omega_n}{2}\left(\tau_2 + \frac{1}{K}\right)$$

For the configuration in Fig. 5-2c, which is an active filter using an op amp,

$$H(s) = \frac{2\zeta\omega_n s}{s^2 + 2\zeta\omega_n s + \omega_n^2} \tag{5-11}$$

where $\omega_n^2 = \dfrac{K}{\tau_1}$

$$\zeta = \frac{\tau_2}{2}\omega_n$$

Both these latter configurations permit independent choice of ω_n, ζ, and K_v and are widely used in practical PLL designs. The active filter has the added advantage that the presence of the amplifier makes the DC loop gain K_v very high compared to that obtainable with the passive configuration. The active filter second-order loop is, therefore, the most attractive choice for most applications.

The active filter transfer function is given by

$$F(s) = \frac{s\tau_2 + 1}{s\tau_1} = \frac{\tau_2}{\tau_1} + \frac{1}{s\tau_1} = \frac{R_2}{R_1} + \frac{1}{sCR_1} \tag{5-12}$$

This is often called a proportional plus integral control, and the function may be synthesized by two active filters operating in parallel, as shown in Fig. 5-2d. As we see later, this configuration sometimes allows greater flexibility in the design of the loop.

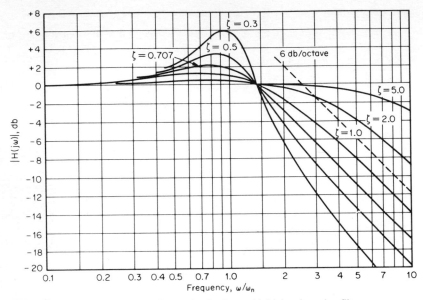

FIG. 5-3   Frequency response of second-order loop with high-gain active filter.

Fig. 5-3 shows the magnitude of frequency response of a second-order loop with a high-gain active filter. The loop acts as a low-pass filter to any variation in the input phase, due to noise or modulation. A compromise value for ζ is 0.7 for good frequency and transient response.

Third-Order Loop

Third-order loops, though not very common, are useful in certain applications. It is shown later that if the input to the loop is a frequency ramp (input frequency changing at a constant rate), the second-order loop produces a steady phase error, or dynamic lag. A third-order loop with a loop filter that is composed of two active integrators in tandem (Fig. 5-4a) produces zero dynamic lag, which is a definite advantage.

The transfer function of this loop is given by

$$H(s) = \frac{K(s\tau_2 + 1)(s\tau_4 + 1)}{s^3\tau_1\tau_3 + s^2K\tau_2\tau_4 + sK(\tau_2 + \tau_4) + K} \tag{5-13}$$

Fig. 5-4b shows yet another filter used in third-order loops. Without C_1, this filter is the same as in Fig. 5-2c. C_1 is sometimes added to provide additional loop attenuation at high frequencies.

The transfer function is given by

$$H(s) = \frac{K(s\tau_2 + 1)}{s^3\tau_1\tau_2 + s^2\tau_1 + Ks\tau_2 + K} \tag{5-14}$$

Compare the high-frequency attenuation of this transfer function ($\alpha 1/s^2$) with that of a second-order loop ($\alpha 1/s$).

Unlike the second-order loops, however, which are unconditionally stable, stability of the loop is an important design consideration in third-order loops.

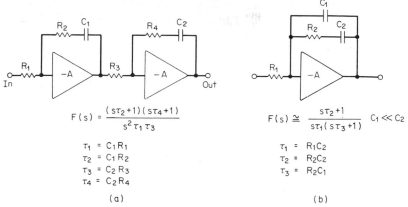

$$F(s) = \frac{(s\tau_2+1)(s\tau_4+1)}{s^2\tau_1\tau_3}$$

$$\tau_1 = C_1 R_1$$
$$\tau_2 = C_1 R_2$$
$$\tau_3 = C_2 R_3$$
$$\tau_4 = C_2 R_4$$

(a)

$$F(s) \cong \frac{s\tau_2+1}{s\tau_1(s\tau_3+1)} \quad C_1 \ll C_2$$

$$\tau_1 = R_1 C_2$$
$$\tau_2 = R_2 C_2$$
$$\tau_3 = R_2 C_1$$

(b)

FIG. 5-4   Useful filter configurations for third-order loop.

5–1b   Noise Performance

Noise Bandwidth   A loop characteristic of fundamental importance is its noise bandwidth. If the input phase to the phase detector is modulated by white noise of uniform spectrum Φ, then noise at the VCO output is given by

$$P_{no} = \frac{\Phi^2}{2\pi} \int_0^\infty |H(j\omega)|^2 \, d\,\omega$$

The integral in the above equation represents the bandwidth of the output noise, and is defined as noise bandwidth of the loop as follows:

$$B = \frac{1}{2\pi} \int_0^\infty |H(j\omega)|^2 \, d\,\omega$$

$$= \frac{\omega_n}{2}\left(\zeta + \frac{1}{4\zeta}\right) \text{ in hertz for a high-gain second-order loop} \qquad (5\text{-}15)$$

For a given ω_n, B is minimum when ζ is in the range 0.4–0.8.
The effective Q of the loop is given by

$$Q_e = \frac{f_0}{2B} \qquad (5\text{-}16)$$

where f_0 is the frequency of oscillation of the VCO. The susceptibility of the VCO output to any noise modulating the input depends on the effective Q, rather than the Q of the tank circuit in the VCO. For example, in a crystal oscillator the crystal Q may be very high ($>10^4$), but the effective Q of the loop may be only about 10^3. Therefore, noise inside the band $\pm f/B$ will appear at the output.

The design criterion for deciding on the noise bandwidth for a PLL depends on the application it is intended for. The PLLs are generally of two types, carrier tracking and modulation tracking. Carrier-tracking loops are designed to recover clean carrier or clock from the incoming signal, which may have frequency or phase modulation or considerable noise, and should therefore have as narrow a bandwidth as practicable. Modulation-tracking PLLs, on the other hand, are designed to work as discriminators, where the filter

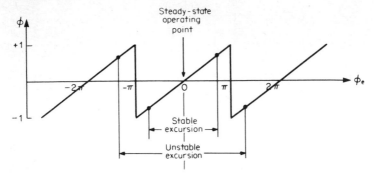

FIG. 5-5   Noise threshold in a PD with sawtooth characteristics.

output should reproduce the baseband spectrum, phase or frequency modulating the carrier. In these applications the loop bandwidth should be wider than the highest modulating frequency.

Noise Threshold   We have assumed so far that, in spite of the noise at the input, the loop is in lock, and the low-pass characteristics of the loop derived from linear transfer functions (Eq. 5-15) reject out of band noise. However, if the input noise is very high and exceeds a threshold (the input signal-to-noise ratio (SNR) is lower than a critical value), the loop starts slipping cycles, and if the SNR becomes even less, the loop may eventually lose lock. The operation of the loop under these conditions becomes highly nonlinear, and the noise reduction indicated by the concept of the noise bandwidth is no longer valid.

The noise threshold depends on the structure of the loop, especially the characteristics of the PD and the filter. We shall try to estimate the noise threshold of a loop on the assumption that the PD has sawtooth output characteristics, as shown in Fig. 5-5. We discuss different types of phase detectors and their output characteristics, including those with sawtooth outputs, in Sec. 5-3. Let us assume that under steady-state conditions ϕ_e is 0, but due to the input noise, ϕ_e fluctuates about its quiescent point.

If the excursion of ϕ_e is less than $\pm\pi$, the operating point will return to its quiescent value. If, however, ϕ_e, due to noise peaks, exceeds $\pm\pi$, the loop will slip a cycle and settle to a new quiescent operating point at 2π or $-\pi$. Under high noise conditions there may be several slips at a time and eventually the loop may lose lock altogether.

The performance of the loop in the presence of input noise depends not only on the magnitude of the noise, but also on its frequency spectrum. From Eq. 5-6, the relationship of the phase error ϕ_e to the input phase ϕ_i is of the nature of a high-pass function. If the frequency is well within the passband of the loop, the output phase will follow the input and ϕ_e will be 0. In this case, a relatively large amount of input noise will not affect the loop. Outside the passband, however, the output phase cannot follow the input, ϕ_e is nearly equal to ϕ_i, and the input noise threshold will be related to phase-detector characteristics, as discussed earlier.

5–1c   Tracking and Steady-State Errors

The tracking properties of a PLL indicate the phase error ϕ_e due to a change in the input phase ϕ_i.

From Eq. 5-6 we note

$$\phi_e(s) = \frac{s}{s + KF(s)} \cdot \phi_i(s)$$

For a change of input phase $\phi_i(s)$, the steady-state error voltage is given by (providing the loop remains in lock)

$$\phi_e(s) = \lim_{t \to \infty} s \lim_{s \to 0} \left[\frac{s}{s + KF(s)} \cdot \phi_i(s) \right] \qquad (5\text{-}17)$$

The input phase may change due to a step change in input phase

$$\phi_i(s) = \frac{\phi_i}{s}$$

a step change in input frequency $\Delta\omega$:

$$\phi_i(s) = \frac{\Delta\omega}{s^2}$$

frequency ramp at the input $\Delta\dot\omega$:

$$\phi_i(s) = \frac{\Delta\dot\omega}{s^3}$$

The response of the loop to these changes depends on the order of the loop.

A step change in the input phase will produce 0 phase error for all three orders of loop.

For a step change in the input frequency ω, the phase error for all three orders of loop is given by

$$\phi_e = \frac{\omega}{KF(0)} = \frac{\Delta\omega}{K_v} \qquad (5\text{-}18)$$

where K_v = DC loop gain.

For active filters, the DC loop gain is generally very high, and therefore the steady-state phase error due to a step change in the input frequency is negligibly small. The VCO frequency changes to the new input frequency, without producing a significant steady-state phase error.

For a ramp frequency change $\Delta\omega$ at the input, the steady-state phase error is given by

$$\phi_e(t) = \lim_{t \to \infty} \lim_{s \to 0} \frac{\Delta\dot\omega}{s[s + KF(s)]} \qquad (5\text{-}19)$$

In this case, the phase error will be nonzero for both first- and second-order loops, but will be zero for a third-order loop with $F(s)$ of the form

$$F(s) = \frac{G(s)}{s^2}$$

(see Fig. 5-4a). This is an important advantage of this type of third-order loop.

The response of the three types of loop to input changes has been summarized in Table 5-1.

TABLE 5–1 Summary of Steady-State Phase Errors in PLL

Input change	First order	Output change Second-order high gain	Third-order $F(s) = \dfrac{G(s)}{s^2}$
Step change in phase	0	0	0
Step change in frequency	$\dfrac{\Delta\omega}{K}$	0	0
Input frequency ramp	$\dfrac{\Delta\dot\omega t}{K}$	$\dfrac{\Delta\dot\omega}{\omega_n^{\,2}}$	0

The transient behavior of the loop when there is a sudden change in the input is also an important consideration in the design of the PLL. The transient response may be derived directly from Eq. 5-6. Figs. 5-6, 5-7, and 5-8 show the response of a high-gain second-order loop to the different input changes discussed above. One important conclusion to be drawn from these curves is that the smaller the natural frequency ω_n, the longer it takes the loop to stabilize to the steady-state value (assuming damping is chosen at its optimum of ≈ 0.7).

5–1d Acquisition

As we have mentioned before, most of the material developed so far is based on the assumption that the loop is in lock. "Acquisition" is the process of bringing the frequency of oscillation of the VCO in lock with the applied input frequency. As we have also

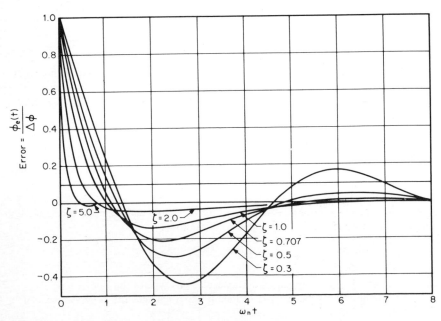

FIG. 5-6 Transient phase error $\phi_e(t)$ due to a step change in input phase $\Delta\phi$.

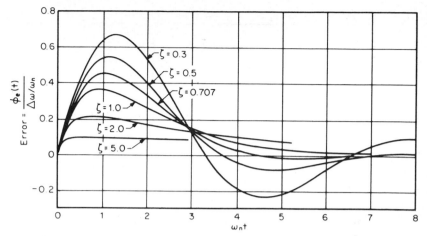

FIG. 5-7 Transient phase error $\phi_e(t)$ due to a step change in input frequency $\Delta\omega$.

noted earlier, the behavior of the loop under the out-of-lock condition is nonlinear and highly complex, and the mathematical analysis of the acquisition process is therefore very involved. However, an examination of physical operation of the loop when out-of-lock will help us understand the problem.

Under the out-of-lock condition the output of the phase detector will be at the beat frequency $\Delta f = f_1 - f_0$, where f_1 is the input frequency and f_0 is the VCO frequency, as shown in Fig. 5-9. The VCO frequency is itself frequency modulated by the beat frequency, and this explains the highly asymmetrical nature of the output waveform.

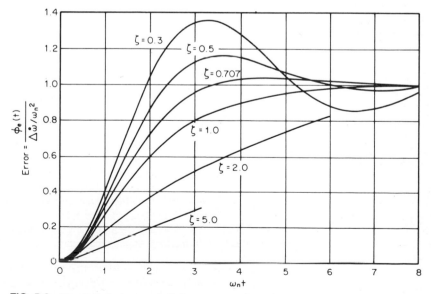

FIG. 5-8 Transient phase error $\phi_e(t)$ due to a ramp input frequency $\Delta\dot{\omega}$.

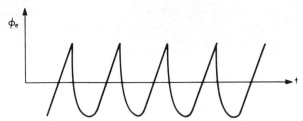

FIG. 5-9 Typical phase-detector output under out-of-lock condition at beat frequency between the input and the VCO frequency.

Let us first consider the case when Δf is much less than the filter bandwidth. The filter will pass the PD output unattenuated, and the peak-to-peak filter output will swing through the maximum range of the control voltage for which the loop is designed. As the output voltage reaches a value that makes the VCO frequency equal to the input and the slope of the swing is correct (either the positive or negative slope is correct for the servo operation of the loop), the VCO will lock onto the input frequency without slipping any cycles. The maximum difference between the input and VCO center frequency for which the loop will acquire lock without cycle slipping is called the lock-in range of the loop, Δf_L. It can be shown that

$$\Delta f_L = \frac{1}{\pi} \zeta \omega_n \qquad (5\text{-}20)$$

$$\Delta f_L < B$$

where B = loop bandwidth. Therefore, loop bandwidth should be wide for a large lock-in range.

If the beat frequency is much higher than the loop bandwidth, the filter output will be very small, there will be no frequency modulation of the VCO output, and the loop will never lock. However, at some intermediate input frequencies outside the lock-in range, there will be a finite filter output, even though it may not swing through the maximum control voltage. An examination of Fig. 5-9 shows that under these conditions the PD output will have a finite DC component. If an active filter is used, this DC component can slowly charge the integrator capacitor to an increasing bias voltage. If the sense of this voltage is correct and the input frequency is within the control range of the VCO, the loop may slowly walk into the lock state. The maximum difference between the input and VCO frequencies for which the loop will eventually pull into lock is called the pull-in range Δf_p.

The pull-in process is generally slow, and will only occur if the sense of the DC voltage is correct. A properly designed PLL should therefore not depend on the pull-in phenomenon, but should have a lock-in range that should accommodate the maximum anticipated difference between input and VCO center frequencies. This poses the most difficult problem in the design of a PLL. Most applications require that the loop bandwidth should be small to obtain a clean VCO output in the presence of large input noise. On the other hand, a large lock-in range requires a wide loop bandwidth. In many applications, it becomes necessary to use external aids to help the loop acquire acquisition.

Aided Acquisition In many applications, especially in RF carrier recovery, the VCO center frequency may differ from the input frequency by a relatively large amount

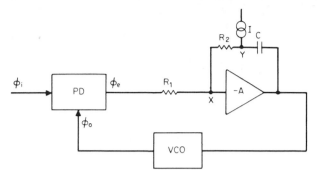

FIG. 5-10   Sweep method of acquisition.

due to the sensitivity of the oscillator circuits to component tolerances. The center frequency of the VCO, generally at the IF stage, may be 30 or 70 MHz, and it may be offset from the input frequency by as much as 100 kHz. The bandwidth of any practical PLL is generally much less, and may be of the order of 1 kHz. It therefore becomes necessary to use external aids to achieve acquisition. Yet another example is the clock recovery circuit in a data-transmission system, which often requires a PLL with extremely narrow bandwidth to produce jitter-free clock. Here external aids have also been found necessary to aid acquisition.

There are three methods that are generally used as aids to acquisition.

Sweep Method

Fig. 5-10 shows the principle of the sweep method of acquisition. A fixed current injected into the capacitor of an active second-order filter causes a voltage ramp to appear at the output. The ramp causes the VCO to sweep, and if the sweep rate is properly chosen, the loop will lock up when the VCO reaches the input frequency. The slewing current should be shut off after lock has been achieved. If it is left on, it is balanced by a compensating phase error. Note that the proper point of injection of the current is at point Y to produce a ramp. Sometimes it may be more convenient to inject the current at the summing point X, which is at fixed potential. A step current when switched on or off at point X will produce a step change in output voltage due to the damping resistor R_2. A sudden change in output voltage when the current is switched off after lock has been acquired may cause the loop to lose lock again. Therefore, if the current is injected at point X, it should be slowly turned on or off.

Variable Filter Bandwidth

Fig. 5-11 shows another method to aid acquisition. The loop bandwidth, and therefore the lock-in range, is large when the loop is out-of-lock. Once lock is achieved, the bandwidth is reduced very considerably. The bandwidth may be reduced by switching in R', or by switching in C'. The maximum allowable value of $R_1 + R'$ is limited by the bias current required for the op amp. The signal current through R_1 should be much larger than the bias current. If the bandwidth is controlled by switching in C', the switching in should be done through a variable resistor that should be gradually reduced to zero. This will eliminate the possibility of a sudden change in output voltage that would occur if the capacitor is simply switched in or out.

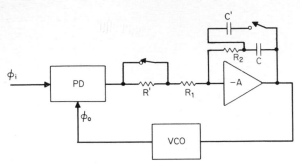

FIG. 5-11 Variable bandwidth method for aiding acquisition.

Frequency Detector Method

Probably the most elegant method for aided acquisition is to use a frequency detector (FD) in parallel with a phase detector, as in Fig. 5-12. The outputs of both phase and frequency detectors are summed, filtered, and then applied to the VCO. When the loop is out-of-lock, there is an output from the FD (preferably with a DC component) that applies a control voltage to the VCO, which helps to acquire acquisition. Once lock is achieved, the FD output becomes zero, and the PD controls the loop as in normal PLL operation.

Fig. 5-12 shows both the PD and FD sharing the same filter. If the filter is an active second-order one, this may be a disadvantage since the damping resistor tends to slow down acquisition by the FD. This is the case where the integral plus proportional configuration (Fig. 5-2d) may be very useful. Fig. 5-13 shows a phase–frequency-locked

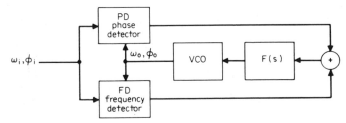

FIG. 5-12 A phase–frequency-locked loop.

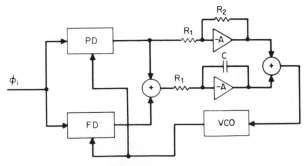

FIG. 5-13 Improved phase–frequency-locked loop.

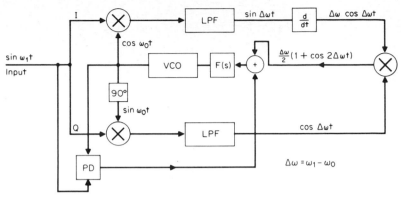

FIG. 5-14 A quadricorrelator used as a frequency detector operating in parallel with a phase detector.

loop, where the output from the FD is passed through the integral part of the filter only, but the PD uses both proportional and integral filtering.

A discriminator-type circuit (viz., a ratio detector) may be used as the frequency detector. However, this type of circuit is not very convenient for use in a loop, as described above. A frequency detector known as a "quadricorrelator" that has been used in many practical designs is shown in Fig. 5-14.

The input signal (sin $\omega_1 t$) is divided into two parallel arms, I (in-phase) and Q (quadriphase). In the I arm the input is multiplied directly with the VCO output (cos $\omega_0 t$), and in the Q arm it is multiplied by the VCO output after a 90° phase shift (sin $\omega_0 t$). The multiplier outputs will contain both sum and difference frequencies, and the low-pass filters in each arm are designed only to pass the difference components (sin $\Delta\omega t$ and cos $\Delta\omega t$, respectively, $\Delta\omega = \omega_1 - \omega_0$). The I arm output from the LPF is then differentiated, and the differentiated output is multiplied with the Q arm LPF output. This multiplier output has a DC output proportional to the frequency difference $\Delta\omega$, which is used to aid acquisition.

Even though the input signal to the quadricorrelator has been shown as a sine wave, the circuit has been used with *RZ* (return to zero) random input data. These are rectangular data pulses, less than one time slot wide, occurring randomly (not present in every time slot). We discuss the nature of these types of data further in Sec. 5-4. This represents a distinct advantage compared to conventional discriminators, where the input must be continuous (either sine wave or regular clock). The two multipliers at the input in Fig. 5-14 will then produce outputs only when data is a one at the input, and none when the data is zero. Therefore, if the spectrum of the random data contains a line at the clock frequency ω_1, the VCO will lock onto this frequency.

Disadvantages of the quadricorrelator are that it requires analog multipliers, and perhaps too many components. In Sec. 5-3 we discuss other frequency detectors implemented with digital circuits only. Nevertheless, the reader is invited to consider the very interesting structure of the quadricorrelator. This configuration, where the input signal is multiplied by in-phase and quadriphase VCO outputs, has been used in various forms for different purposes in phase-locked-loop designs. The Costas loop, used extensively in carrier recovery in RF communication, described in Sec. 5-2, is another example of this configuration. Yet another example is the quadrature phase detector, which is briefly described next.

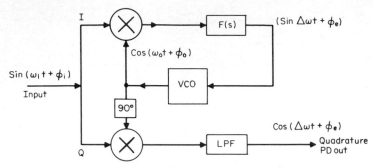

FIG. 5-15 Quadrature phase-detector circuit.

5–1e Quadrature Phase Detector

Fig. 5-15 illustrates the concept of a quadrature phase detector. As before, the input is multiplied by the in-phase and quadriphase VCO outputs in the I and Q arms, respectively. The output of the I arm is used as a phase detector to lock the VCO as shown. When the loop is out-of-lock, the filtered output of both arms will change at the beat frequency and will have no DC component. Under lock condition, the output of the I arm is proportional to sin ϕ_e ($\cong\phi_e$, if ϕ_e is small). This explains why the I arm can be used as a PD for the PLL. Under the same conditions, the Q arm output is proportional to cos ϕ_e ($\cong1$ for small ϕ_e). Therefore, if the bandwidth of the LPF in the Q arm is made very narrow, the Q arm output will be nearly zero when out-of-lock, but will have a steady DC voltage when in lock.

This output is sometimes used to indicate the in-lock condition of a PLL. The quadrature PD has also been used in an important class of practical circuits, called tone decoders, which are used to detect whether the input contains one or more of a set of tones or frequencies. We discuss tone decoders later in this chapter.

5–2 OTHER TYPES OF PHASE-LOCKED LOOPS

The conventional phase-locked loops we have investigated so far are the ones used in most applications requiring PLLs. In this section we discuss some special types of phase-locked circuits that are used in some special applications. For example, the Costas loop is widely used for carrier recovery in synchronous communication systems when the input spectrum may not contain any energy at the carrier frequency. A conventional PLL will fail to lock to the carrier frequency unless the input spectrum has a relatively strong line at that frequency. Similarly, in a very low-speed data-transmission system the use of the conventional PLL to recover the clock may be considered uneconomic, and a much simpler digital phase synchronizer may be adequate for the purpose.

5–2a Costas Loop for Carrier Recovery

In synchronous communication systems, information is transmitted by modulating a carrier with a baseband signal (the amplitude, frequency, or the phase of the carrier may be modulated). It is well known that at the receiver the baseband signal may be recovered by multiplying (or mixing) the modulated signal with the recovered carrier. (In RF transmission the demodulation generally takes place at an IF frequency rather than at the high RF carrier frequency.)

As an example, let us consider a suppressed carrier amplitude modulated signal. The modulating baseband signal is generally a band of frequencies, but for simplicity we consider it to be a single tone.

If ω_c is the carrier frequency and ω_m is the modulating frequency, the modulated carrier may be represented as

$$m(t) = \sin \omega_m t \sin \omega_c t \qquad (5-21)$$
$$= \tfrac{1}{2}[\cos(\omega_c - \omega_m)t - \cos(\omega_c + \omega_m)t]$$

It is clear, therefore, that the modulated carrier has no component at the carrier frequency ω_c, and a conventional PLL cannot lock to this frequency.

To demodulate the signal, let us assume we have recovered the carrier at the correct phase, and multiply the modulated signal with the recovered carrier.

$$m(t) = \sin \omega_m t \sin \omega_c t \cdot \sin \omega_c t$$
$$= \sin \omega_m t \frac{1 + \cos 2\omega_c t}{2} \qquad (5-22)$$
$$= \tfrac{1}{2} \sin \omega_m t + \tfrac{1}{2} \sin \omega_m t \cos 2\omega_c t$$

If this demodulated signal is now passed through a low-pass filter that will reject the high-frequency components around $2\omega_c$, the original modulating signal ω_m will be available at the filter output.

Carrier recovery at the receiver is generally done by either of two techniques, i.e., squaring the modulated signal or using a Costas phase-locked loop. The reader should appreciate that these techniques are useful mainly in suppressed carrier communication systems, such as the one we are discussing, single sideband modulation, digital frequency, or phase modulation. If the transmission does contain a carrier component, it can be recovered by simple narrow bandpass filtering.

If $m(t)$ in Eq. 5-21 is squared, we obtain

$$m(t)^2 = \sin^2 \omega_m t \sin^2 \omega_c t \qquad (5-23)$$
$$= \tfrac{1}{4}(1 + \cos 2\omega_m t)(1 + \cos 2\omega_c t)$$

The squared output therefore contains a double frequency component $\cos 2\omega_c t$. This component can be filtered out and divided by two to recover the carrier frequency. Note that the division process introduces a $\pm\pi$ phase ambiguity in the recovered carrier.

The Costas loop (Fig. 5-16) represents a very attractive alternative method for carrier recovery. Note the striking similarity between the Costas loop and the quadricorrelator circuit discussed earlier for frequency detection.

The input signal is divided into parallel I and Q arms, where it is multiplied by an in-phase, and quadriphase VCO output, which is assumed to be $\cos(\omega_c t + \phi_0)$. After low-pass filtering, the output of the Q arm is $\sin \omega_m t \cos \phi_e$ and that of the I arm is $\sin \omega_m t \sin \phi_e$. These two terms are multiplied again to obtain the phase-detector output.

$$\text{PD output} = \tfrac{1}{2} \sin^2 \omega_m t \sin 2\phi_e \qquad (5-24)$$
$$= \tfrac{1}{4}(1 + \cos 2\omega_m t)\sin 2\phi_e$$

The loop filter rejects the high-frequency terms, and its output is proportional to $\sin 2\phi_e$ and, if ϕ_e is small, the filter output is proportional to simply $2\phi_e$. This output is very similar to the filter output in a conventional loop. The servo action of the loop will tend to make ϕ_e zero (except for any steady offset).

Note that the Q arm output is the demodulated signal and that the I and Q arms

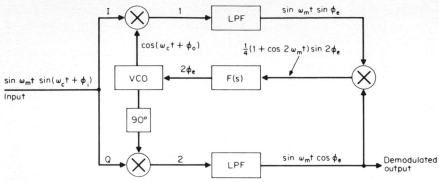

FIG. 5-16   Costas loop for carrier recovery.

operate in parallel to act as the phase detector. The Costas loop therefore combines the functions of carrier recovery and demodulation. With the squaring technique, the demodulation has to be done separately. Since the phase-detector output is proportional to $\sin 2\phi_e$, it cannot distinguish between a phase error of $\pm\phi_e$ or $\pm(\pi \pm \phi_e)$. Hence there is phase ambiguity of $\pm\pi$ in the recovered carrier, as in the multiplying circuit. In many applications, such as voice transmission, this phase reversal is of no consequence. However, with data transmission such reversals will result in a one being interpreted as a zero, and a zero as one. It is therefore usual practice in data transmission to precode the data before transmission, so that the received data can be interpreted correctly in spite of the ambiguity in the phase of the recovered carrier.

For reasons we have mentioned earlier, a carrier recovery circuit using a Costas loop will generally require external aid during initial acquisition. A sweep method is commonly used. However, it is relatively simple to add a frequency detector to the Costas loop. The outputs from the two LP filters in the I and Q arms may be further processed to form a quadricorrelator. However, the differentiator in the I arm of the quadricorrelator has the frequency response of a high-pass filter, and is likely to degrade loop performance in the presence of high input noise. An alternative arrangement is shown in Fig. 5-17.

The circuit shares the I and Q arms of the Costas loop up to the LP filter inputs. For the sake of simplicity, we shall assume that signal input has no modulation and is

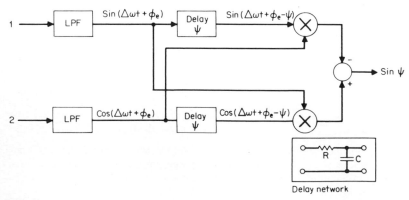

FIG. 5-17   Frequency detector for use with Costas loop.

given by

$$\sin(\omega_c t + \phi_i) = \sin[(\omega_0 + \Delta\omega)t + \phi_i]$$

where ω_0 is the VCO frequency and $\Delta\omega = \omega_c - \omega_0$.

The Q arm output of the LPF is $\cos(\Delta\omega t + \phi_e)$ and the I arm output is $\sin(\Delta\omega t + \phi_e)$. Output from each arm is delayed by a small angle ψ, and the delayed output of each arm is multiplied by the undelayed output from the other arm, so the difference between the two multiplier outputs is the frequency-detector output.

$$\text{FD output} = \sin(\Delta\omega t + \phi_e)\cos(\Delta\omega t + \phi_e - \psi)$$
$$- \cos(\Delta\omega t + \phi_e)\sin(\Delta\omega t + \phi_e - \psi) \qquad (5\text{-}25)$$
$$\cong \sin\psi$$

If the delay circuit is implemented with a single pole RC filter shown in Fig. 5-17, then the FD output is given by

$$\text{FD output} = \frac{\Delta\omega RC}{\sqrt{1 + \omega^2 R^2 C^2}} \qquad (5\text{-}26)$$
$$\cong \Delta\omega RC$$

if $\omega RC \ll 1$. Therefore, the output is proportional to the difference between input and VCO frequencies. This output may be used to aid acquisition by summation with the PD output in Fig. 5-16.

We have explained the operation of the Costas loop for carrier recovery in suppressed carrier amplitude modulated systems. This type of loop can be used for carrier recovery in other types of modulation as well, viz., single sideband, digital phase, or frequency modulation. The loop can be suitably modified to recover carrier from 4-phase or 8-phase modulated systems as well.

5–2b Digital Phase-Locked Loops

Digital phase-locked loops are discrete time versions of analog PLLs, and many different forms have been described in the literature. An interesting and useful configuration is shown in Fig. 5-18.

The phase detector operates in the same manner as in analog PLLs. The phase error is sampled and quantized in an A/D circuit. The quantized phase error is then processed in a digital discrete time filter. The discrete output of the filter is converted into analog samples by a D/A, and then held in a zero-order hold circuit the output of which controls the VCO.

This circuit may appear too complicated for most applications. However, the use of the digital filter makes this type of circuit extremely useful for some special applications. For example, in a digital network, an extremely stable reference clock (derived from an

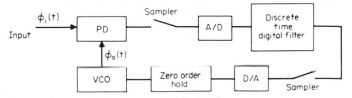

FIG. 5-18 A useful version of a digital phase-locked loop.

atomic source) is transmitted to many interconnected stations where local clocks are slaved to this reference. It is desirable that if for some reason there is a temporary interruption in the transmission of the reference clock, the slave clock frequency should stay very close to the reference frequency for an extended period of time (several days). If this can be achieved, then when the reference clock reappears the local PLL will lock on again very rapidly to the reference, and no data bits will be lost. In order that the local clock will not drift in the absence of an input, the filter output should not change, i.e., the integrator in the filter should act as a memory, with a very long time constant. In an analog integrator, the magnitude of the time constant is limited by the leakage current of the integrating capacitor and the bias current of the op amp. The digital integrator does not possess these imperfections and can act as a very suitable memory for the control voltage for the VCO.

This PLL configuration is therefore very attractive in applications where very large integrating time constants are desirable.

5–2c Digital Phase Synchronizers

There is another type of PLL, also referred to as a digital PLL, that is much simpler in structure than that of Fig. 5-18. To avoid any confusion in terminology, we refer to the simpler digital PLL as the digital phase synchronizer.

The principle of operation of this type of PLL can be explained by referring to Fig. 5-19. Let the input frequency be f_i, and the local oscillator frequency be f_o, which is nominally equal to Nf_i (where N is an integer). The oscillator output is divided in a variable divider to produce the output clock. The output clock samples the input and if the input phase advances relative to the output, the division ratio is reduced to $N - 1$. If, however, the input phase lags relative to the output, the division ratio is increased to $N + 1$.

If $x =$ fraction of time, division ratio is $N - 1$ and if $y =$ fraction of time, division ratio is $N + 1$, then

$$f_i = \frac{f_o}{x(N - 1) + y(N + 1)} \tag{5-27}$$

$$= \frac{f_o}{N - x + y} \quad (\text{since } x + y = 1)$$

Therefore if,

$$f_i > \frac{f_o}{N} \quad x > y$$

$$f_i < \frac{f_o}{N} \quad x < y$$

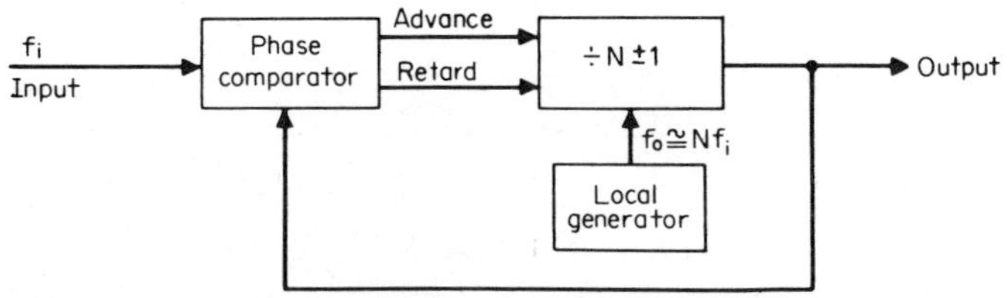

FIG. 5-19 Basic concept of digital phase synchronizer.

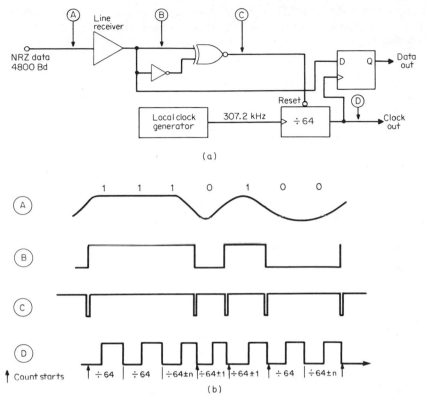

FIG. 5-20   Practical implementation of a digital phase synchronizer. (*a*) Circuit; (*b*) timing diagram.

The loop works as a first-order PLL, and there is an output jitter of one oscillator clock period ($1/f_o$). For small output jitter, f_o and N should be large ($N = 64$ is often used).

Fig. 5-20 illustrates a practical implementation of a variation of the concept of a digital phase synchronizer. The figure shows a circuit to generate a local clock that is synchronized to the received nonreturn to zero (NRZ) data at a baud rate of 4800. The received data (point A) is rounded and jittery due to imperfections and noise in the receive channel. The data is hard limited by the line receiver at point B, where it is still jittery. Both the positive and negative transitions in the data are used to produce very narrow negative-going pulses at point C. Note that this process is equivalent to differentiating the data transitions and reversing the polarity of all the positive-going pulses (i.e., differentiating and squaring). The spectrum of the NRZ data does not possess any line at the baud rate, but a strong line appears at the spectrum of the unidirectional differentiated pulses.

The local oscillator frequency is nominally 307.2 kHz (2400×64), but it is not synchronized with the input data. The output clock at point D is derived by dividing the oscillator output by a variable counter. The negative-going pulses at point C reset the counter every time to state zero. In the absence of the reset pulses, the counter will always divide by 64, but the reset pulses terminate the counter at more or less than 64 counts. The following characteristics of the output clock at point D are worth noting.

1. The *average* number of positive or negative transitions per second will be equal to the baud rate. These transitions may be used to further process the data, like retiming it or loading it into a buffer. The circuit shown will in effect center sample the data.
2. If there is one reset pulse for every data period at the input (the input pattern is 1 0 1 0 . . .), the division ratio will be close to 64 ± 1, 2 . . . , N, depending on the difference frequency (64 × baud rate − local clock rate) and input jitter. However, if there is a string of 1's or 0's at the input, there will be only one reset pulse at the end of the string. For data periods without reset pulses, the division ratio will be exactly 64, but the reset pulse at the end of the string will terminate the count at 64 ± n, where n can be large, depending on the number of 1's and 0's in the string. The circuit will eventually fail to synchronize if the strings are very long (say 25 1's or 0's in a row); therefore, it is essential for proper operation of the circuit to restrict the number of 1's or 0's in a string to a maximum number, say 8. But this problem is common to all circuits for recovering clock from random input data. Unrestricted NRZ data format is therefore considered unsuitable for data transmission. Phase-encoded data, which has at least one transition per data period, is much more attractive for purposes of both transmission and clock recovery.

5–2d   Injection-Locked Oscillators

Finally, we briefly mention the technique of injection locking an oscillator. It is well known that if a small external signal is injected at a suitable point in the feedback path of an oscillator, and the frequency of the external signal is very close to that of the oscillator, the oscillator locks itself to the injected frequency. The explanation of this phenomenon lies in the fact that the active device in the oscillator operates near the nonlinear region of its characteristics. The external signal shifts the phase of the feedback voltage, but not its magnitude. This changes the oscillator frequency, which in turn changes the phase of the output voltage. This process continues until the resultant phase of the external signal and output voltage is such that the oscillator frequency becomes equal to that of the impressed signal.

Except in some consumer circuits, injection-locked sinusoidal oscillators are seldom used in modern designs. However, injection locking is a very convenient and popular technique for synchronizing an astable multivibrator.

Fig. 5-21 shows a simple astable multivibrator that is designed to be synchronized to an external frequency. The multivibrator is implemented with CMOS NOR gates (CD 4001). The free-running frequency is determined by the timing components C and R, the supply voltage V_{DD}, and the transfer voltage of the gates $V_T(V_{DD}/2 \pm 40\%)$. The resistor $10R$ isolates the voltage waveform at point B from being affected by the internal protection circuitry at the input to gate 1.

The free-running waveforms at different points, assuming that the waveform at point B is completely isolated from gate 1 input, are shown in Fig. 5-21.

The free-running frequency is given by,

$$f_o = \frac{1}{t_1 + t_2} \tag{5-28}$$

where $t_1 = RC \ln \dfrac{V_{DD} + V_T}{V_T}$

$t_2 = RC \ln \dfrac{2V_{DD} - V_T}{V_{DD} - V_T}$

The waveforms become modified when the multivibrator is locked to the external signal f_1, as shown in the figure. For proper operation, the external signal frequency f_1 must be greater than free-running frequency f_0. But there is also an upper limit of f_1, about $2f_0$, which should not be exceeded for a reliable design. If the input frequency is in the permissible range, it will automatically terminate the self-timing of the multivibrator at the correct operating point.

This type of multivibrator is useful in designs that require a clock source that may be synchronized to an external signal, but it should free run in the absence of the clock

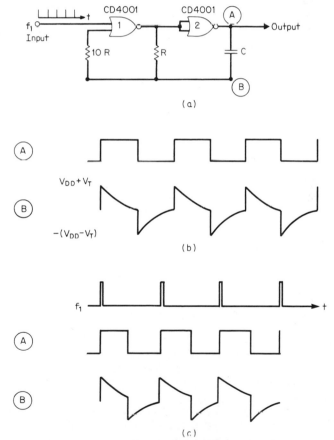

FIG. 5-21   Injection locking an astable multivibrator. (*a*) Schematic; (*b*) free-running timing diagram; (*c*) timing diagram when locked to f_1.

source. A DC-to-AC inverter is an example where such a clock source may be very useful.

5–3   PRACTICAL DESIGN OF PHASE-LOCKED LOOPS

Before describing some practical designs of phase-locked loops, it is worthwhile to examine some commonly used loop components, viz., the phase detectors, the voltage-controlled oscillators, and the factors determining the choice of a particular component for a given application. We have already discussed the different types of loop filters and indicated that the second-order-loop active filters are generally the usual choice in most practical designs for their superior performance.

5–3a   Phase Detectors

Various types of phase detectors have been described in literature, viz:

>   Sample-and hold phase detectors
>   Discriminator-type phase detectors
>   Multiplying-type phase detectors
>   Digital phase detectors

The characteristics of these phase detectors vary widely, and there is no single detector that is suitable for all applications.

Considerations that influence the choice of phase detectors for a particular application are:

- Nature of input signal. A cosinusoidal input requires phase detectors that may not be suitable for digital signals. Also, for digital signals, phase detectors used with regular clock inputs are often not suitable when the input is random data.

- Linearity of the output characteristics of the phase detector with the input phase error.

- Range of input phase error over which the output is linear. The wider the range over which the output is linear, the more useful the phase detector is as a control element in the loop, and the more immune it is to noise interference.

Of the various types mentioned above, the multiplying-type and the digital phase detectors are almost always used in practical designs, partly because they are widely available as integrated circuits; consequently, we describe only these two types here.

Multiplying Phase Detectors   The principle of operation of multiplying phase detectors may be explained by reference to Fig. 5-22. If the inputs to the PD are $V_i(t)$ and $V_o(t)$, then the output $V_e(t)$ (error voltage) is given by

$$V_e(t) = V_i(t) \cdot V_o(t) \tag{5-29}$$

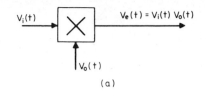

(a)

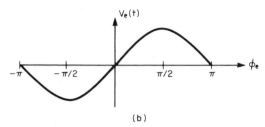

(b)

FIG. 5-22 Analog multiplier used as a phase detector.
(a) Configuration; (b) sinusoidal characteristics.

Let us assume that both the inputs are of the same frequency but differ in phase.

$$V_i(t) = \sin(\omega t + \phi_i)$$
$$V_o(t) = \cos(\omega t + \phi_o)$$

Then

$$V_e(t) = \sin(\omega t + \phi_i) \cdot \cos(\omega t + \phi_o) \qquad (5\text{-}30)$$
$$= \sin \phi_e + \text{high-frequency components}$$

where $\phi_e = \phi_i - \phi_o$.

If this output is now passed through a low-pass filter, the output of the filter will be the DC component $\sin \phi_e$. These characteristics of the multiplying phase detector are shown in Fig. 5-22b. The output is proportional to the sine of the input phase error. The input/output relationship is only linear for small values of ϕ_e ($\sin \phi_e \cong \phi_e$, for small ϕ_e), and the useful range is seen to be $\pm \pi/2$.

Monolithic analog four-quadrant multipliers are available that may be used as phase detectors; for example, Analog Devices AD 532, Intersil 8013, RCA 3091, etc. However, these are linear devices and for comparable performance they tend to be more expensive than the switching-type multipliers described below.

A switching-type modulator that is very commonly used in communication circuits is the ring modulator shown in Fig. 5-23. In this case, the amplitude of $V_o(t)$ is large compared with that of $V_i(t)$. The switching of the diodes D_1–D_4 is controlled by $V_o(t)$. The waveforms are as shown in the figure, and it can be seen $V_o(t)$ operates as a reversing switch on $V_i(t)$. The multiplier output is given by

$$V_e(t) = V_i(t) \cdot V_o(t)$$

and $V_o(t)$ may be expressed as a Fourier series of a zero mean square wave

$$V_o(t) = \frac{4}{\pi} [\cos(\omega t + \phi_o) - \tfrac{1}{3} \cos 3(\omega t + \phi_o) + \cdots] \qquad (5\text{-}31)$$

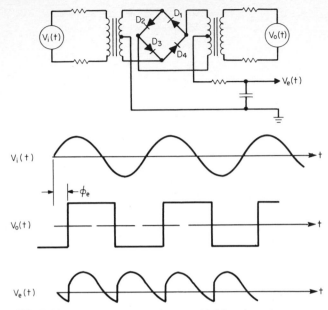

FIG. 5-23 Ring modulator used as a multiplying phase detector.

Note that $V_o(t)$ may be square or sinusoidal, but provided it is much larger than $V_i(t)$, it switches the diodes as if it were a square wave.

As before, the output has a DC term sin ϕ_e, though many more AC terms than in Eq. 5-30, and these must be filtered out. This type of detector has been used extensively at higher frequencies (IF stages, 30–70 MHz) because of availability of low-cost matched-diode quads (RCA 3019, National 3039) and their operability at very high frequencies.

The reader should recognize the circuit shown in Fig. 5-23 as a doubled balanced modulator. If $V_o(t)$ is a carrier frequency and $V_i(t)$ a modulating signal, the output is a suppressed-carrier, suppressed-baseband, double-sideband modulated signal.

It is possible to replace the transformers and the diodes in Fig. 5-23 by active elements (transistors or FETS). Fig. 5-24 shows a double balanced modulator implemented entirely with transistors. Basically the principle of operation of this circuit is exactly the same as the ring modulator, in that $V_o(t)$ switches the polarity of $V_i(t)$. This circuit is available in IC form; viz., MC 1596 (Motorola, Signetics). The same circuit has been used as phase detectors in many monolithic phase-locked loops available today (Exar XR 200 series, Signetics 560 series). It should be noted, however, that the operating frequency range of the monolithic phase detectors of this type (about 10 MHz) is much less than that obtainable with diode detectors.

Digital Phase Detectors   The multiplying phase detectors are useful if the inputs $V_i(t)$ and $V_o(t)$ are cosinusoidal. While they may be used when the inputs are in digital (or binary) form, the digital phase detectors are much more attractive when the inputs are digital, or may be easily converted to digital form.

The simplest digital phase detector may be implemented with an exclusive-OR gate (7486), as shown in Fig. 5-25.

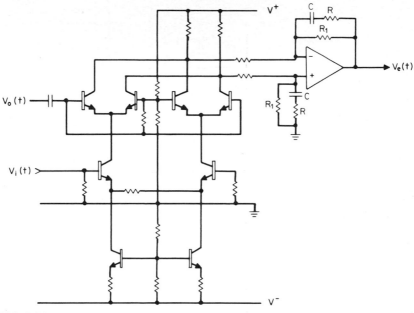

FIG. 5-24 Multiplying phase detector implemented with active devices only.

The following are the main characteristics of this type of phase detector:

- The input/output relationship is linear over the range 0 to π, and is triangular.

- The steady-state phase error should be $\pi/2$ for the loop to operate at the center of its linear region.

- The output frequency is double the input frequency.

- If the input is at logic 0, the output is at the center of the operating range.

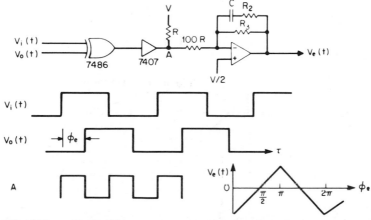

FIG. 5-25 An EXCLUSIVE-OR gate as a digital phase detector.

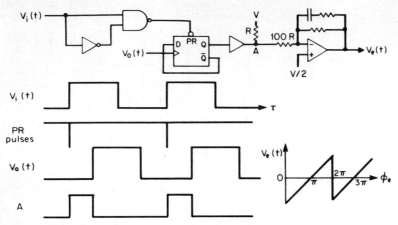

FIG. 5-26 A digital phase detector with increased operating range.

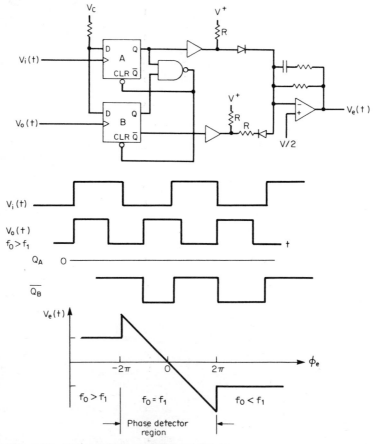

FIG. 5-27 Digital θ–F (phase–frequency) detector.

Note that the phase-detector characteristics shown in Fig. 5-25 are valid at frequencies well within the passband of the filter, implemented with the op amp. Also, the feedback resistor R_3 has been included for illustration only. In a properly designed PLL, the op amp is used as an integrator, not as a filter. Phase detectors of this type have been used in some monolithic phase-locked loops; viz., RCA CD 4046A.

Fig. 5-26 shows an improved design of a digital phase detector. Here the positive-going edges of the input $V_i(t)$ sets a D flip-flop and the positive-going edge of $V_o(t)$ toggles the flip-flop.

The important characteristics of this phase detector are:

- The input/output relationship is sawtooth in shape, and is linear over the region $0-2\pi$.

- The steady-state phase error should be π for the phase detector to operate at the center of its operating range.

- In the absence of input $V_i(t)$, the output is zero, i.e., at the center of its operating region.

- The output frequency is identical to the input frequency.

As before, the op amp has been shown as a low-pass filter rather than as an integrator.

An extremely useful and very popular digital phase-detector circuit is shown in Fig. 5-27. Here both $V_i(t)$ and $V_o(t)$ clock "1" into two separate flip-flops. However, a NAND gate senses when both flip-flops are high and then resets both of them. The output of the filter is the difference between the outputs of the two flip-flops, as shown in the figure.

An examination of the timing diagrams in Fig. 5-27 reveal the following remarkable characteristics of this phase detector.

1. It has a steady-state DC output when the frequency f_i of $V_i(t)$ is different from frequency f_o of $V_o(t)$, and the polarity of this DC output depends on whether f_i is greater or less than f_o. Thus the circuit acts as a digital frequency detector, and the frequency detector output is used in aiding acquisition in PLL. The reader should compare the simplicity of this frequency-detector circuit with that of a quadricorrelator shown in Fig. 5-14.
2. When $f_o = f_i$, the circuit acts as a phase detector and its input/output shows a linear relationship over the range -2π to $+2\pi$.
3. The steady-state phase error is zero for operation at the center of the operating region.
4. Since the useful range is $\pm 2\pi$, the peak input jitter should be less than a ± 1 time slot. However, if the jitter is zero mean (symmetric), much larger noise peaks will not be harmful.

A more detailed description of this phase–frequency (ϕ–F) detector is to be found in the Motorola Handbook (see references). This circuit is available as an IC package; viz., Motorola MC 4344/4044 in TTL version (operating frequency up to 15 MHz) and MC 12040 in an ECL version (operating frequency up to 80 MHz). This type of ϕ–F detector is also included in many monolithic PLLs; viz., RCA CD 4046A, Plessy SP 8921.

It should be pointed out that Fig. 5-27 illlustrates the basic concept of the ϕ–F detector, and the actual implementation has been done somewhat differently in the ICs mentioned above. Because of their extremely useful properties, these detectors are used very extensively in practical PLL designs.

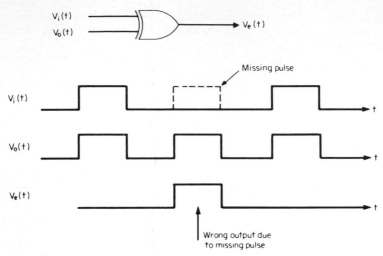

FIG. 5-28   Output of an EXCLUSIVE-OR phase detector when the input contains a missing pulse.

Digital Phase Detectors for Random Input Data

For operation of the phase detectors described so far, the input $V_i(t)$ must be a continuous signal, a cosinusoid or regular clock [$V_o(t)$ is always continuous]. If there is any missing transition at the input, the phase-detector output will be incompatible with the servo action of the loop. This is illustrated in Fig. 5-28 for an EXCLUSIVE-OR phase detector.

However, in many digital communication systems it becomes necessary to derive a clock with a PLL from an input that is random data; i.e., data that is not regular with respect to the clock but in which 1's and 0's occur randomly with some defined probability.

It is shown later that the random data must be the return-to-zero (RZ) type to possess a line in their spectrum at the bit rate, and only this type of random data may be used as an input to a PLL. If the random data is not of the RZ type, it must be processed to produce RZ pulses. We have already discussed two types of circuits that will operate satisfactorily with random input, viz., the quadricorrelator (Fig. 5-14) and the digital phase synchronizer (Fig. 5-20). A very simple phase detector that will operate with a random input is shown in Fig. 5-29. Note that the phase detector produces an output only when the input is a one, but none when the input is a zero.

- The input/output relationship is triangular and is linear over the range 0 to π (if the width of the input data is $T/2$, where $T = 1$/bit rate).

- The steady-state phase error is $\pi/2$.

- For proper control of the loop, there must be a sufficient number of ones in a given time, viz., long strings of zeros should be avoided.

This type of phase detector is included in one of the ϕ–F detector IC mentioned earlier; viz., Motorola MC 4344/MC 4044.

As we have already mentioned, the ϕ–F detector circuit of Fig. 5-27 will not operate with random input data. We have just described a phase detector suitable for such inputs.

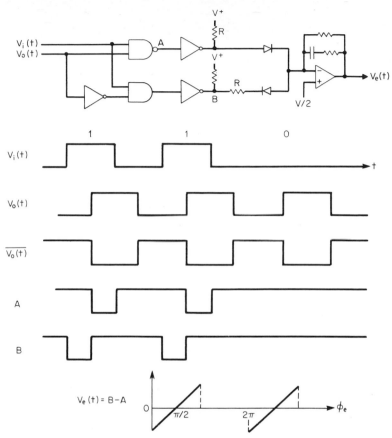

FIG. 5-29 A digital phase detector for use with random input data (RZ, pulse-width $= T/2$).

The question arises whether or not it is also possible to realize a digital frequency detector that will operate with random input data. Such circuits have been recently described in the literature (both Hogge and Mayo).

Fig. 5-30 illustrates the principle of operation of a digital frequency detector suitable for random inputs. The frequency detector is designed to operate in parallel with the phase detector of Fig. 5-29. We shall assume that the width of the RZ data is $T/2$, as before. In addition to the VCO output (CLK) that controls the phase detector, two other clocks are generated CLK^+ and CLK^-, respectively, which are advanced and delayed by ΔT ($\cong T/8$) relative to CLK.

Let us consider the effect of an isolated data bit on the operation of the circuit. If the loop is in lock, the bit will be in the phase-detector region and the output of the frequency detector will be zero. Now consider the case when the loop is out-of-lock and the input frequency f_i is greater than the oscillator frequency f_o. The phase of the data bit will advance relative to the clock, and a "1" will be clocked into FFB by CLK^-, which will set latch B and disable latch A. Latch B will remain set and latch A reset as the phase of the data bit advances, and a "1" is clocked in FFB, then both in FFB and

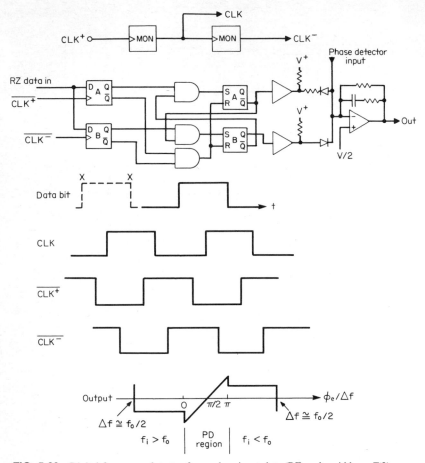

FIG. 5-30   Digital frequency detector for random input data (RZ, pulsewidth $= T/2$).

FFA, and finally only in FFA. The latches will be reset only when the bit enters the next phase-detector region (XX in the figure). Thus, for $f_i > f_o$, the frequency detector will have a net negative output. Similarly, for $f_i < f_o$, the output will be positive. These outputs may be used to help the loop acquire lock.

It is important to note some important differences between the characteristics of the phase- and frequency-detector circuits of Figs. 5-29 and 5-30, and the ϕ–F detector circuit of Fig. 5-27. The ϕ–F detector of Fig. 5-27 will produce a DC output so long as the input and output frequencies are different ($\pm \Delta f = f_i - f_o$). However, for the frequency detector of Fig. 5-30, the useful range is restricted approximately to $\pm \Delta f = f_o/2$. The reader can easily verify this from Fig. 5-29 by considering two successive data bits in relation to CLK (f_o).

Also, the phase detector of Fig. 5-29 has an operating region of $\pm\pi/2$ about its quiescent point, compared with $\pm 2\pi$ for the ϕ–F detector of Fig. 5-27. This type of detector is therefore more susceptible to input noise or jitter. For a PLL operating with the phase detector of Fig. 5-29 and the frequency detector of Fig. 5-30, the out-of-band

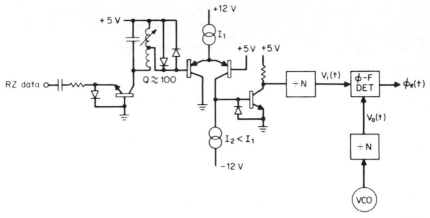

FIG. 5-31   Use of a θ–F detector when the data has a large amount of jitter.

input noise should be less than $\pm\pi/2$. However, there are practical transmission systems where the input data may have jitter that may extend over several time slots. From our previous discussion of noise threshold, the magnitude of input noise that the phase detector can tolerate will depend both on the magnitude and the spectrum of the noise. A technique that permits the use of a digital φ–F detector when the input data has a large amount of out-of-band jitter is shown in Fig. 5-31. A clock is derived from the RZ input data with an LC ringing circuit with a stable Q ($\cong 100$). The recovered clock, which is also jittery, is then divided by N so that the amount of jitter at the divider output is less than about $\pi/4$. This clock may then be used as an input to the φ–F detector. We examine this technique in further detail in Sec. 5-4.

5–3b   Voltage-Controlled Oscillators

The most commonly encountered voltage-controlled oscillators are listed below in order of decreasing stability:

> Crystal oscillators (VCXOs)
> LC oscillators
> RC multivibrators

The choice of an oscillator for a given application is influenced by two major considerations.

1. *Phase stability:*—Even though the frequency of the VCO is locked to that of the incoming signal, the relation of the output phase of the oscillator to that of the input signal depends on the free-running frequency of the oscillator. As the free-running frequency changes with temperature, or aging, the output phase will change, and in the event of very large changes, lock may be lost. For the same reason, the susceptibility of the VCO frequency to internal noise in the system will cause jitter or phase noise at the output of the VCO. Therefore, for good stability, the frequency of the VCO should not be affected appreciably by change of temperature, aging, or noise.
2. *Large control range:* It is desirable to be able to lock the VCO over a large frequency range. The larger the control range, the easier it is for the loop to acquire and maintain lock.

These two requirements are clearly conflicting. For a large control range, an oscillator must produce a large change in frequency for a given change in control voltage. Unfortunately, this also means that the oscillator is more susceptible to temperature change, aging, or noise. Crystal oscillators are the most stable of the oscillators mentioned above (control range $\cong 0.1\%$) and are almost always used for such critical applications as frequency synthesizers or clock synchronizers. *RC* multivibrators, on the other hand, have the largest control range (about $1:10^4$) and are used in such applications as FM demodulators and tone decoders.

Crystal Oscillators (VCXO)   There are many different designs of crystal oscillators, with varying degrees of sophistication, dictated by the degrees of stability required. For example, a simple crystal oscillator may be implemented with TTL or CMOS logic gates (short-term stability of about 10^{-3}). On the other extreme, there are oscillators of fairly complex designs with the crystal in a temperature-controlled oven to maintain stability of the order of 10^{-9}. The interested reader should consult Freking for the state-of-the-art design techniques for stable crystal oscillators.

A crystal oscillator designed for good stability should possess the following features:

- The crystal should be high Q, AT cut, and operate in fundamental series-resonant mode in the 3–20-MHz range (the stablest oscillators use 5-MHz temperature-controlled crystals with unloaded Q's of about 2×10^6).

- The crystal should work into a linear network, and if in the series-resonant mode, should have low source and load impedances.

- The crystal drive (or dissipation) should be fairly small (should not exceed 1 mW).

There are ICs that may be used as a crystal oscillator (e.g., Motorola MC 4024, Exar XR 215). However, most stable crystal oscillator designs are based on the use of discrete active devices.

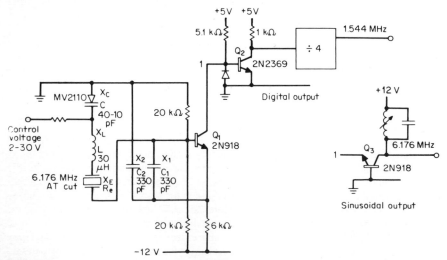

FIG. 5-32   A Colpitts oscillator used as a VCXO.

A very simple 6-MHz crystal oscillator circuit, known as a Colpitts oscillator, with a stability of the order of 10^{-5}, is shown in Fig. 5-32. This type of VCXO has been used widely in T carrier systems (first-level digital transmission systems in North America, with a bit rate of 1.544 MHz).

As with most oscillators, the operation of the circuit in Fig. 5-32 is nonlinear in that transistor Q_1 switches off during part of the cycle. However, assuming for simplicity, small-signal operation, the conditions of oscillation are

$$|X_1| \ll R_{in}$$

$$\left| \frac{X_1 X_2}{R_{in}} \cdot \beta \right| \geqq R_e \qquad (5\text{-}32)$$

where R_{in} = common emitter input impedance of Q_1
$\qquad = r_b + \beta \, r_e$
$\quad r_b$ = ohmic base resistance
$\quad r_e$ = emitter resistance (depends on
$\qquad\qquad$ emitter current)
$\quad \beta$ = common-collector current gain
$\quad R_e$ = effective series resistance of
$\qquad\qquad$ the crystal (ESR)

Also

$$X_1 + X_2 + X_C + X_L + X_E = 0 \qquad (5\text{-}33)$$

If $C_1 = C_2 = 330$ pF ($X_1 = X_2$) and $(R_{in}/\beta) \cong 30 \; \Omega$ for 1 mA of emitter current, then

$$\left| \frac{X_1 X_2}{R_{in}} \cdot \beta \right| = 200 \; \Omega$$

Therefore, with proper design margin, the maximum crystal ESR should be about 100 Ω. From Eq. 5-33 the total reactance in the crystal loop is zero. The inductor L is included to make the total reactance external to the crystal zero also ($X_L + X_1 + X_2 + X_C = 0$). This results in $X_E = 0$, i.e., the crystal operates at its series resonance. If L is not included, the crystal should be specified so that it resonates with about 20 pF in series (effective capacitance of C_1, C_2, and nominal C in series).

Some features of the oscillator circuit are worth mentioning. Even though the operation of the transistor Q_1 is highly nonlinear, the crystal itself works into a low impedance linear network consisting of C_1, C_2, C, and L. Also, the oscillator is effectively isolated from the output circuitry and its parasitic effects. The crystal dissipation may be estimated as follows. If the peak-to-peak voltage swing across the base emitter junction of Q_1 is e, and assuming this is is sinusoidal (not strictly correct), the rms crystal current is given by

$$I_c = \frac{e}{2\sqrt{2}X_1} \qquad (5\text{-}34)$$

Then the crystal dissipation is

$$P_c = \frac{e^2}{8 \, X_1{}^2} R_e \qquad (5\text{-}35)$$

If $e = 0.5$ V for 1 mA of emitter current and $R_e = 100$ Ω

$$P_c = 500 \ \mu W$$

Both crystal dissipation and the loop gain of the oscillator (Eq. 5-32) may be controlled with the emitter current. The oscillator of Fig. 5-32 may be pulled a minimum ± 1000 Hz from nominal as the control voltage varies from 2 to 30 V.

LC Oscillators   As previously mentioned, when cost and large tuning range are more important than stability, LC and RC oscillators are used. While RC multivibrators are suitable at relatively low frequencies (up to about 30 MHz), LC oscillators can operate at much higher frequencies (up to about 200 MHz). The circuits used for crystal oscillators may also be used for an LC oscillator by replacing the crystal with an LC combination and adjusting the component values. However, it may be more convenient to implement an LC oscillator using an IC.

Fig. 5-33 shows two LC oscillators, one using an op amp (RCA 3048) and the other an ECL IC (Motorola MC 1648). In each case, the frequency of oscillation is given by

$$f_0 = \frac{1}{2\pi\sqrt{LC_T}} \tag{5-36}$$

where $C_T = C_1 + C$ (capacitance of the varicap). Capacitors C_2, C_3, and C_4 are for decoupling.

The maximum frequency of oscillation of the first circuit is about 1 MHz and that of the second about 200 MHz.

RC Multivibrators   A wide selection of ICs are available for voltage-controlled RC multivibrators (Signetics 566, Exar XR 2209, National LM 566, Motorola 1658). Precision multivibrators are generally emitter coupled types, and Fig. 5-34 shows a simplified design of an emitter-coupled multivibrator and associated timing waveforms.

The frequency of oscillation is determined by the timing components R and C, and the control voltage V_C. The zener diodes are included to make the output independent of the control current I. In order that the frequency is determined by V_C

$$\frac{V_C}{R} \cdot R_1 > V_z$$

With this condition satisfied, the frequency of oscillation is given by

$$f_0 = \frac{V_C}{V_z} \cdot \frac{1}{4RC} \tag{5-37}$$

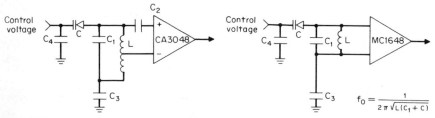

FIG. 5-33 LC oscillators using integrated circuits.

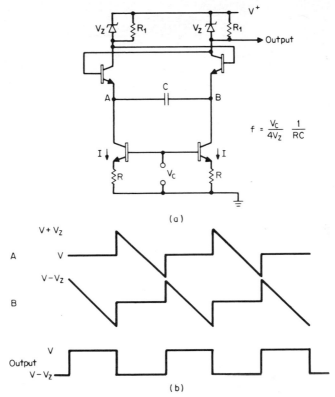

$$f = \frac{V_C}{4V_Z} \frac{1}{RC}$$

(a)

(b)

FIG. 5-34 Emitter-coupled multivibrator as a voltage-controlled oscillator. (*a*) Simplified schematic; (*b*) timing waveforms.

5–3c Design of Practical Phase-Locked Loops

So far we have discussed the components used in the design of a phase-locked loop. We now discuss two practical designs of such loops, one using a monolithic device (National LM 565) and the other using a crystal-controlled oscillator.

The LM 565 belongs to the very popular 560 series of monolithic phase-locked loop ICs. Even though the design described relates to the detection of FSK signals, the design equations may very easily be modified for other applications.

Fig. 5-35 shows the functional block diagram of the LM565. The relevant data for this device is as follows.

Phase Detector:

Input impedance	5 kΩ
Output impedance	3.6 kΩ
Sensitivity K_D	0.68 V/rad

Voltage-Controlled Oscillator: Free-running frequency

$$f_0 \cong \frac{1}{3.7\, R_0 C_0} \tag{5-38}$$

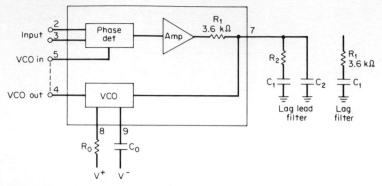

FIG. 5-35   Block diagram of the LM 565.

Closed-Loop Gain:

$$K = \frac{33.6 \, f_0}{V_T} \tag{5-39}$$

where V_T is the total supply voltage. (Note that the gain depends both on supply voltage and oscillator frequency.)

Demodulation Sensitivity: Output for $\pm 10\%$ deviation, 300 mV.

Loop Filters: Two types of second-order filters are generally used with this device (see Fig. 5-35). For the simple lag filter from Eq. 5-9,

$$\omega_n = \sqrt{\frac{K}{R_1 C_1}} \tag{5-40}$$

$$\zeta = \tfrac{1}{2}\sqrt{\frac{R_1 C_1}{K}} \tag{5-41}$$

For the lag-lead filter from Eq. 5-10,

$$\omega_n = \sqrt{\frac{K}{C_1(R_1 + R_2)}} \tag{5-42}$$

$$\zeta \cong \frac{\omega_n}{2} R_2 C_1 \tag{5-43}$$

$$C_2 \leqq 0.1 C_1 \tag{5-44}$$

(A 0.001-μF capacitor is required between pins 7 and 8 to prevent parasitic oscillation.)

Example 5–1   FSK Detection using a PLL

Teletype data is generally transmitted over the telephone lines as FSK signals (mark and space are transmitted as two different frequencies). A standard modem used for this purpose may transmit:

space frequency   2025 Hz
mark frequency   2225 Hz

This can clearly be considered as square wave on-off data, frequency modulating a carrier of 2125 Hz by ± 100 Hz (a total frequency deviation of $\cong 10\%$).

A typical data rate would be 110 baud (bits per second) so the baseband bandwidth would be 110 Hz. A detector is desired to recover the frequency-modulating data. The VCO control voltage of a PLL will perform this function when the VCO tracks the incoming frequency modulation.

Solution
We use an LM 565 to discuss the design of a PLL for detection of FSK signals. Interested readers should consult National Application Note AN 46 ("The Phase Lock Loop IC—A Communication Building Block") for further details.

The PLL must be a modulation-tracking type and should have a band-width large enough to accommodate the modulating signal without excessive rolloff.

From Carson's rule, the two-sided carrier bandwidth is

$$B \cong 2(\Delta f + f_m) \qquad (5\text{-}45)$$

where Δf = peak frequency deviation from center frequency
f_m = highest modulating frequency

so the loop bandwidth should be at least 210 Hz. In Section 5-4 we say that the loop bandwidth should be larger than this, but the postdetection filter bandwidth should be limited to the baseband. We make the loop bandwidth 300 Hz, and the postdetection filter bandwidth 100 Hz.

We assume $V_T = 12$ V(± 6 V). We also use a lag-lead filter for better performance. The design proceeds as follows:

VCO Timing Components

$$f_0 = 2125 \text{ Hz}$$

$$R_0 C_0 = \frac{1}{3.7 \times 2125} \qquad (5\text{-}38)$$
$$= 1.27186 \times 10^{-4}$$

If
$$C_0 = 0.0047 \text{ } \mu\text{F}$$
$$R_0 \cong 27 \text{ k}\Omega$$

Loop Filter

If we assume a loop bandwidth B of 300 Hz and let $\zeta = 0.7$, we can than compute

$$\omega_n = \frac{2B}{\zeta + \dfrac{1}{4\zeta}} = 568 \qquad (5\text{-}15)$$

$$\tau_2 = R_2 C_1 \cong \frac{2\zeta}{\omega_n} = 2.4648 \times 10^{-3} \qquad (5\text{-}10)$$

$$R_1 C_1 + R_2 C_1 = \frac{K}{\omega_n^2} = 1.8443 \times 10^{-2} \qquad (5\text{-}42)$$

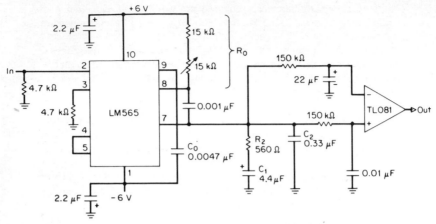

FIG. 5-36 A modulation-tracking phase-locked loop using an LM 565.

where $K = 5950$ from Eq. 5-39; therefore

$$R_1C_1 = 1.8443 \times 10^{-2} - 2.4648 \times 10^{-3}$$
$$= 1.5978 \times 10^{-2}$$

since $R_1 = 3.6 \text{ k}\Omega$ (internal)

then $C_1 = 4.4 \text{ }\mu\text{F}$

         $R_2 = 560 \text{ }\Omega$ (5-10)

let $C_2 = 0.33 \text{ }\mu\text{F}$ (5-44)

The complete design is shown in Fig. 5-36. The demodulated output across C_2 is 300 mV peak-to-peak (for 10% deviation). This signal is further *RC* filtered and applied to a TL081 amplifier, which restores the square wave data.

The LM 565 (and most other monolithic PLLs) uses a multiplying-type detector, hence it is important to ensure that the step changes in input frequency do not cause the phase-detector output to change by more than 90°. Since

$$\Delta\omega = 2\pi \times 200$$
$$\Delta\omega/\omega_n = 2.2$$

From Fig. 5-7, for a step change in input frequency,

$$\phi_e(t)\text{max} = \frac{\Delta\omega}{\omega_n} \times 0.45$$

$$= 0.99 \text{ rad}$$

$$\cong 57°$$

This should allow adequate performance margin for the PD.

Example 5–2 **Phase-Locked-Loop Design Using a Voltage-Controlled Crystal Oscillator (VCXO)**

As a second example we shall discuss the design of a phase-locked loop using a VCXO. These types of loops are widely used in digital networks to synchronize clocks at nodes located in different geographical areas to a reference clock. The nodal clocks should have very good frequency stability so that the free-running frequency stays close to the reference even if the reference fails. The loop should be carrier tracking, in that it should reject large amounts of jitter that may contaminate the reference, and therefore should have extremely narrow bandwidth.

Solution

A phase-locked loop used to generate a 1.544-MHz clock for the first level of digital hierarchy in North America (referred to as DS-1 level) is shown in Fig. 5-37. The crystal oscillator is as shown in Fig. 5-32, and the phase frequency detector is as shown in Fig. 5-27. The phase comparison usually takes place at 8 KHz. The large division ratio is partly so that a very large input jitter can be rejected and partly for administrative reasons (division to 64 KHz is also satisfactory).

We estimate the following loop constants:

$$K_D = 3.7/4\pi \text{ V/rad}$$

(Output changes $5 - 2V_d$ as phase changes $\pm 2\pi$.)

$$K_2 = 2\pi \times 2000/30 \text{ rad/(s} \cdot \text{V)}$$

(VCO frequency changes by ± 1000 Hz as the control voltage changes by 30 V.)

$$\text{Loop gain } K = K_D K_2/N = 0.159$$

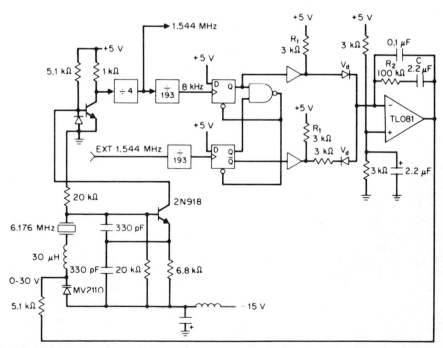

FIG. 5-37 A carrier-tracking phase-locked loop using a voltage-controlled crystal oscillator.

For the component values shown,

$$\omega_n = \sqrt{\frac{K}{R_1 C}} = 4.91 \qquad (5\text{-}9)$$

$$f_n = 0.78 \text{ Hz}$$

$$\zeta = \frac{\omega_n}{2} \cdot R_2 C = 0.53 \qquad (5\text{-}10)$$

The damping resistance is rather large and it is therefore important to include the 0.1-μF capacitor to attenuate the high-frequency components inside the loop.

The loop bandwidth is, therefore, about 2.5 Hz.

We now briefly discuss the effects of the very narrow bandwidth and small damping on the loop performance. From Figs. 5-7 and 5-8, a step change in input frequency or a frequency ramp will produce a rather large phase error in the loop. However, we have seen that this type of detector remains linear over a range of $\pm 2\pi$, and will stay in lock even for larger phase changes, so this is not a very critical consideration. But one undesirable effect of a very narrow band loop is that it takes a long time to pull in. An approximate estimate for pull-in time is given by,

$$T_p = \frac{(\Delta \omega)^2}{2 \zeta \omega_n^3} \qquad (5\text{-}46)$$

If the input frequency is offset by, say, 200 Hz from the free-running frequency,

$$\Delta \omega = \frac{2\pi \times 200}{193} = 6.511$$

The pull-in time for the design shown is therefore 0.34 s, which should be acceptable.

5–4 PHASE-LOCKED LOOP APPLICATIONS

The phase-locked loop is one of the most widely used building blocks in electronic systems. Indeed, its applications are so numerous that it is impossible to even attempt to mention them all here. Instead, we try to describe some of the more important applications in which phase-locked loops, especially IC phase-locked loops, have been used extensively.

5–4a Frequency Synthesizers

Frequency division multiplex (FDM) communication systems require carrier frequencies in discrete intervals for fixed channel spacings. For example, the 40-channel Citizens Band Broadcast requires carrier frequencies in the range 26.965 to 27.405 MHz in 10-kHz steps. Another example is the UHF television band (channels 14–83), which extends from 470 to 890 MHz in 6-MHz increments.

The simplest form of a frequency synthesizer using a PLL is shown in Fig. 5-38. The output frequency is controlled by the programmable counter setting, and is given by

$$f_o = N f_r \qquad (5\text{-}47)$$

where f_r = reference frequency

N = programmable counter setting

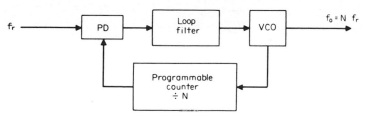

FIG. 5-38 Direct frequency synthesis using a PLL.

While this simple approach is often adequate for low-frequency applications, it has a big drawback in that the programmable counter has to divide the VCO output directly.

Due to the delays involved in the feedback path, the range of operation of programmable counters is much more restricted than straight counters. For example, the upper limit of ECL programmable counters is about 100 MHz, and that of CMOS counters is about 5 MHz.

One way to deal with high VCO frequencies is to use a fixed prescaler (divider) and divide both the reference and VCO frequencies by M (Fig. 5-39), so that f_o/M is within the counting range of the programmable counter. The disadvantage of this approach is that the large division ratio in the feedback loop reduces the PLL loop gain and increases its response time to a change in the counter setting (see Sec. 5-1).

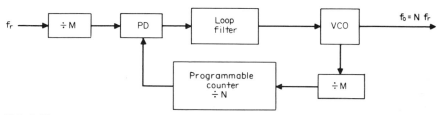

FIG. 5-39 Frequency synthesis with prescaling of reference and output frequencies.

There are two solutions to this problem. First, in Fig. 5-40 the VCO frequency is mixed with a stable offset frequency f_m to obtain the desired output frequency.

$$f_o = f_m + Nf_r \qquad (5\text{-}48)$$

The other solution, which is entirely digital, makes use of a two modulus prescaler. The principle underlying this technique may be explained by referring to Fig. 5-41. The prescaler, which is a high-frequency counter, divides by P or $P + 1$, depending on the

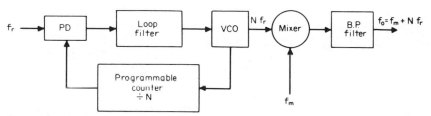

FIG. 5-40 Phase-locked-loop frequency synthesis by mixing up.

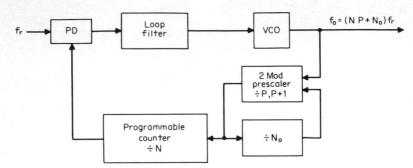

FIG. 5-41   Phase-locked-loop frequency synthesis using a two modulus prescaler.

state of the counter N_0. N and N_0 are relatively low-frequency down-counters, since their maximum operating frequency is f_o/P. Initially the counters are preset to N and N_0, respectively, and the prescaler divides by $P + 1$. After N_0 has counted down to zero, and N has counted down to $N - N_0$, the prescaler starts dividing by P. By the time N underflows, the total countdown is given by

$$N_T = (N - N_0)P + N_0(P + 1)$$
$$= NP + N_0 \qquad (N > N_0)$$

In practice, P is generally 10 (prescaler divides by 10 or 11) and N is a cascaded decade counter.

If $N = 10 N_2 + N_1$, then $N_T = 100 N_2 + 10 N_1 + N_0$. Therefore, the division ratio may be incremented by unit decimal numbers.

Monolithic PLL frequency synthesizers with all the necessary components in a single chip for citizens band radio transmitters and receivers are available (National MM 55104, Motorola MC 145112, and Plessy SP 8921, SP 8923). Fig. 5-42 illustrates the design of a frequency synthesizer to produce an output frequency in the 70–100-MHz range in steps of 0.1 MHz, using the 10/11 prescaler discussed earlier. Note the use of ECL devices in the prescaler and fast family of TTL devices for programmable counters.

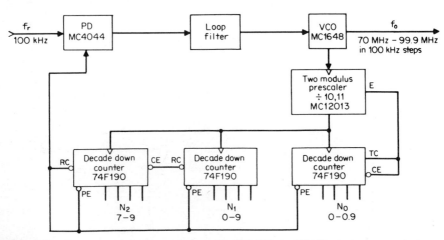

FIG. 5-42   Frequency synthesizer to generate 70–99.9 MHz in 100 kHz steps.

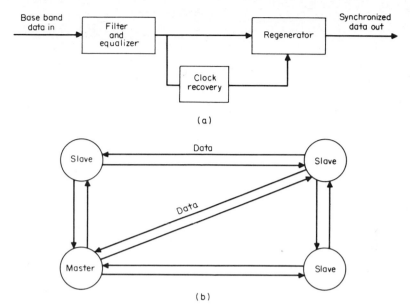

FIG. 5-43   Use of bit synchronization in a digital network. (*a*) Use of recovered clock to regenerate data; (*b*) clocks in several slave stations are derived from data originating in a master station.

5–4b   Bit Synchronizers

We have already discussed the use of phase-locked loops in carrier and clock recovery circuits in synchronous communication systems. Now we briefly discuss some of the problems associated with bit synchronization not mentioned before. Fig. 5-43 illustrates the need for clock recovery circuits in digital transmission systems. In Fig. 5-43*a* the recovered clock is used to regenerate the received data. Fig. 5-43*b* shows a master-slave digital network where the clocks in a number of geographically separated nodes are slaved to the clock in a master station.

We have shown in Fig. 5-37 that a very narrow-band crystal phase-locked loop may be used to derive clock from incoming data. The input to the loop has been assumed to be a continuous clock signal. However, if a phase detector of the type shown in Fig. 5-29 is used, and the incoming data is RZ type, the data may be used directly as the input to the loop.

However, most efficient data formats are not of the RZ type, and their spectra do not possess a line at the clock frequency. It therefore becomes necessary to apply some nonlinear process to convert the data to RZ format before applying to the loop. Fig. 5-44 shows one technique of converting NRZ data to RZ format and their spectra. Note the discrete line at the clock frequency in the RZ spectrum. The reader should recognize that the conversion process involves differentiating, squaring, and limiting.

As mentioned above, the RZ data may now be used directly to lock the loop, or may be used to excite a ringing (*LC*-tuned) circuit to recover the clock. The stable *Q* obtainable with an *LC* circuit is generally an order of magnitude lower than that with a properly designed PLL. Therefore, the clock recovered with a PLL is generally much cleaner than with a passive circuit, especially in a jittery environment. Indeed, in some

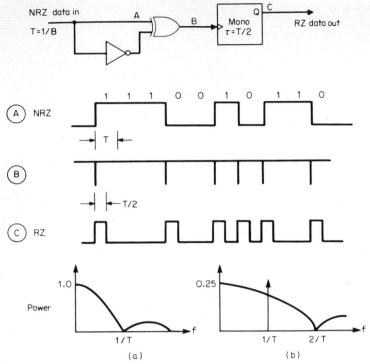

FIG. 5-44 Derivation of RZ data from NRZ data input. (*a*) Spectrum of NRZ data; (*b*) spectrum of RZ data (width $= T/2$).

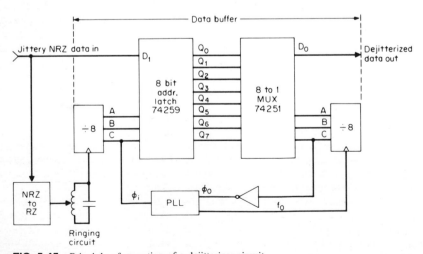

FIG. 5-45 Principle of operation of a dejitterizer circuit.

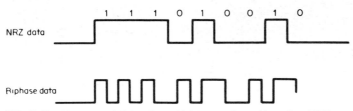

FIG. 5-46   Phase-encoded data has a higher rate of transitions than NRZ data, thus facilitating clock recovery.

applications, the clock is recovered with a relatively low-Q ringing circuit, and then this clock is used as an input to a PLL to derive a highly stable clock. Fig. 5-45 illustrates the use of this technique to dejitterize an extremely jittery data input. The input data is written into an 8-bit addressable latch with a clock that follows the input jitter (derived with a low-Q circuit). Each bit is stored in the latch for eight clock periods. A data bit is read out with a jitter-free clock nominally four clock periods after it has been written. This arrangement can therefore absorb jitter magnitude of approximately ±4 time slots. Clearly the size of the buffer will determine the amount of jitter that can be absorbed by this method.

It is obvious that for the ringing circuit or the PLL to operate satisfactorily the NRZ data must have a sufficient number of transitions during a given time. Unfortunately, with unrestricted NRZ data, there is no control over the number of ones or zeros in a string, and they may pose a serious problem for the clock recovery circuits. It is therefore customary to scramble the NRZ data before transmission to increase the probability of transitions, or use other encoding schemes aimed at aiding clock extraction at the receiver. For example, a phase-encoding scheme used for low-speed data transmission or storage (known as biphase encoding) is shown in Fig. 5-46. The encoded data has two transitions ($1 \rightarrow 0$ or $0 \rightarrow 1$) for every input 1, one transition for input 0. The high frequency of data transitions makes clock recovery relatively simple. However, the energy in the spectrum of the phase-encoded data is shifted toward a higher frequency band; therefore, this or a similar type of encoding is used only when bandwidth is not a limiting consideration.

5–4c   Frequency and Phase Demodulators

We have shown in Sec. 5-3 (Fig. 5-36) that if the input to the PLL is a frequency-modulated carrier, the demodulated signal may be recovered at the output of the loop filter. Here we discuss the principles underlying the use of a PLL to demodulate an angle-modulated carrier.

Note that for the PLL to operate in the linear mode, the input must have a strong component at the carrier frequency. For suppressed carrier systems, some nonlinear loops (i.e., Costas loop) have to be used.

Consider an angle-modulated input to a PLL,

$$V_i(t) = \sin[\omega t + \phi_i(t)] \qquad (5\text{-}49)$$

where ω = angular carrier frequency
$\phi_i(t)$ = instantaneous carrier phase modulated with a baseband signal

For frequency modulation, the instantaneous frequency deviation is given by

$$m(t) = \frac{d}{dt}\phi_i(t) \qquad (5\text{-}50)$$

In Laplace notation,

$$m(s) = s\phi_i(s) \tag{5-51}$$

For phase modulation, the instantaneous phase deviation is given by

$$m(t) = \phi_i(t) \tag{5-52}$$

or $$m(s) = \phi_i(s)$$

From Eqs. 5-3 and 5-6 we find that the filter output of the loop is given by

$$V_2(s) = K_d F(s)\phi_e(s)$$

$$= s\phi_i(s) \cdot \frac{K_d F(s)}{s + KF(s)}$$

$$= s\phi_i(s) \cdot \frac{H(s)}{K_2} \tag{5-53}$$

Comparing Eqs. 5-51 and 5-53, the filter output is directly proportional to the modulating signal modified by the loop bandwidth. If the carrier is phase modulated, the filter ouput has to be integrated (operation $1/s$) to recover the modulating signal. Fig. 5-47 shows a PLL configured to demodulate a frequency- or phase-modulated carrier. The loop should clearly be a modulation tracking loop. The bandwidth of the FM modulation may be estimated by applying Carson's rule on the bandwidth of a frequency-modulated carrier:

$$B \cong 2(\Delta f + f_m) \tag{5-45}$$

where Δf = peak frequency deviation
f_m = highest modulation frequency

For a phase-modulated carrier

$$B \cong 2f_m \tag{5-54}$$

For best performance (low threshold level) the loop bandwidth should be wider than the one-sided bandwidth indicated above and the S/N improvement should be achieved with a predetection filter (bandpass) and postdetection filter (low pass), as shown in Fig. 5-47. (Bandpass bandwidth $\cong B$, postdetection filter bandwidth f_m, and loop bandwidth $\cong B/2$.)

FSK is a special case of frequency modulation in that on-off (binary) data shifts the carrier frequency to two distinct values f_1 and f_2, corresponding to input one or zero. The demodulated output should also be binary, i.e., on-off.

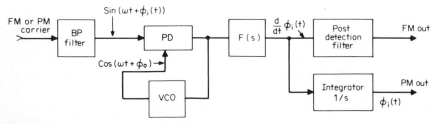

FIG. 5-47   Use of a PLL as a discriminator to demodulate an FM or PM carrier.

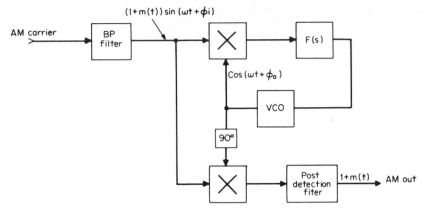

FIG. 5-48 Use of a PLL as a synchronous AM demodulator.

5–4d Coherent Amplitude Modulators

Let us consider an amplitude-modulated carrier given by

$$[1 + m(t)]\sin(\omega t + \phi_i) \tag{5-55}$$

Fig. 5-48 shows how the modulating signal may be recovered using a PLL. The de-modulated signal is available at the output of a quadrature phase detector (Fig. 5-15).

Assuming that the VCO output is $\cos(\omega t + \phi_o)$, and the phase detectors are multiplying types, the in-phase detector output is given by

$$V_e(t)_I = K_d[1 + m(t)]\sin \phi_e \tag{5-56}$$

ignoring high-frequency terms

$$V_e(t)_I \cong 0 \qquad \text{for } \phi_e \cong 0$$

This underlines a property of a PLL that to a first approximation the phase-detector output is unaffected by amplitude modulation of the carrier. However, the quadrature phase-detector output is given by, again ignoring the high-frequency terms,

$$V_e(t)_Q = K_d[1 + m(t)]\cos \phi_e \tag{5-57}$$
$$\cong K_d[1 + m(t)] \qquad \text{for } \phi_e \cong 0$$

which represents the modulating signal superimposed on a DC term (the DC term indicates that the loop is in lock). This type of amplitude demodulation is coherent since it is realized by multiplying the input with a recovered carrier in phase with the transmitted carrier, and should be distinguished from demodulation by envelope detection.

It should be noticed in Fig. 5-48 that the AM demodulator is outside the loop. The loop should therefore be carrier tracking; i.e., of very narrow bandwidth, so as not to introduce noise into the quadrature arm.

5–4e Tone Decoders

We have shown in Fig. 5-15 that a DC output of the quadrature phase detector indicates that the VCO is in lock with the input frequency. If the loop bandwidth is small, then there will be a lock indication only if the input frequency is very close to the VCO center frequency. The operation of PLL tone decoders is based on this principle.

Combination of tones are often used to transmit address (dialing) and supervision signals through a telephone network. For example, each digit dialed with a Touch-Tone telephone handset results in transmission of two frequencies to the switching office, as shown below:

Low group (Hz)	Digits		
697	1	2	3
770	4	5	6
852	7	8	9
941	*	0	#
High group (Hz)⟶	1209	1336	1477

There is a multiplicity of hardware in a telephone switching office to decode the digits from the received tone combinations. Fig. 5-49 shows one implementation using PLLs. Seven PLLs are required, and the center frequency of each VCO is adjusted to correspond to one of the seven transmitted tones. Note the separation between the frequencies in each group is about 10%, so that the loop bandwidth should be about 5%. The choice of the loop bandwidth is a compromise between the desired safeguard against

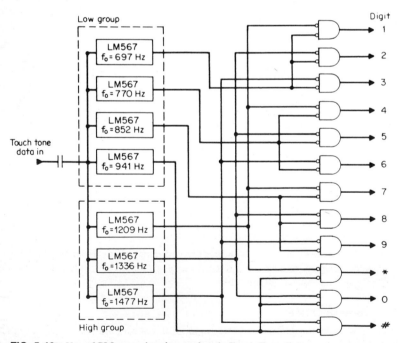

FIG. 5-49 Use of PLL tone decoders to decode Touch-Tone digits.

false lock to an adjacent frequency and the drift of the VCO center frequency with temperature and time, etc.

5–4f Choosing the Right PLL

It should be clear to the reader by now that the design parameters for PLLs for applications described here, or other possible applications, may be very different and should be carefully selected. One parameter that is common is the stability of the VCO center frequency. If this frequency drifts very much, the loop will eventually lose lock. For the loop to remain in lock, the difference between the VCO frequency and the input frequency should be within the pulling range of the loop.

Other parameters of the loop are generally dictated by the application the loop is intended for. For example, for FM demodulation the loop bandwidth should be relatively large, and the output should be linear. For FSK demodulation, on the other hand, the bandwidth should be large, but the output should be digital. Tone decoders and AM demodulators should include a 90° phase-shift network for the carrier and a quadrature phase detector. However, the AM detector output should be linear, while that of a tone decoder should be digital.

Many PLL ICs are available today that are tailor-made for different applications. Table 5-2 lists the more popular phase-detector and VCO devices as well as complete PLL ICs. Table 5-3 classifies PLLs according to application.

TABLE 5–2 Available ICs for PLL Applications

	IC	Manufacturer	Comments
Phase			
detectors			Maximum frequency
Multiplying	MC 1496	Motorola and	10 MHz
type	MC 1596	Signetics	10 MHz
Digital	MC 4344	Motorola	15 MHz
	MC 4044		15 MHz
	MC 12040		80 MHz
	11C 44	Fairchild	15 MHz
Voltage-			
controlled			
oscillators			Maximum oscillator frequency
LC	MC 1648	Motorola	200 MHz
RC	566	Signetics	1 MHz
	LM 566	National	1 MHz
	XR 205	Exar	4 MHz
	XR 2206		1 MHz
	XR 2207		1 MHz
	MC 1658	Motorola	80 MHz
			Maximum frequency
Monolithic	560, 561,and 562	Signetics	15 MHz
phase-locked	565		500 KHz
loops (all use	LM 565	National	500 KHz
RC multivibrators	XR 215	Exar	20 MHz
as VCO)	XR 2212	Exar	100 KHz
	CD 4046A	RCA	1 MHz (CMOS)

TABLE 5–3 IC Phase-Locked Loops Classified According to Applications

Application	IC	Manufacturer	Comments
Frequency synthesis and FM demodulation	565 LM 565 XR 215 XR 210 CD 4046A	Signetics National Exar Exar RCA	General purpose
FSK demodulator	XR 210 XR 2211	Exar Exar	Digital output
Tone decoders	XR 567 XR 2211 567 LM 567	Exar Exar Signetics National	Quadrature VCO, Quadrature PD, Digital output
AM demodulator	561	Signetics	Quadrature VCO, Quadrature PD, Linear output

Conclusion

Summarizing this chapter, we have mainly dealt with linear models of phase-locked loops, although we briefly mentioned some nonlinear models of loops as well; viz., Costas loop and digital loops. We described phase-locked-loop components and some practical designs of loops. Finally, we described some important applications of phase-locked loops.

The subject of phase-locked loops is very broad, both in theory and applications. The reader is encouraged to refer to the very extensive literature available on the subject (only a few examples of which are mentioned in the following References) for further study.

REFERENCES

Alfonso, J. A., A. J. Quitero, and D. S. Arantes: "A Phase-Locked Loop with Digital Frequency Comparator for Timing Signal Recovery," *National Telecommunication Conference Records,* 1979.

Bellisio, J. A.: "New Phase Locked Timing Recovery Method for Digital Regenerators," *ICC Conference Record,* June 1976.

Bennett, W. R., and J. R. Davey: *Data Transmission,* McGraw-Hill, New York.

Cahn, C. R.: "Improving Frequency Acquistion of a Costas Loop," *IEEE Transactions on Communications,* December 1977.

Costas, J. P.: "Synchronous Communications," *Proceedings IRE,* December 1956.

Freking, M. E.: *Crystal Oscillator Design and Temperature Compensation,* Van Nostrand-Reinholt, New York.

Gardner, F. M.: *Phaselock Techniques,* 2d ed., Wiley, New York.

Ghosh, S.: "Bandwidth of Phase Lock Loops," *Electronic Design,* March 15, 1978.

Ghosh, S., and C. Foster: "Phase Frequency Locked Loops Handle Random Input," EDN, February 1980.

Gupta, S. C.: "On Optimum Digital Phase Locked Loops," *IEEE Transactions on Communications*, April 1968.

Hogge, C. R.: "Carrier and Clock Recovery for 8 PSK Synchronous Demodulation," *IEEE Transactions on Communications*, May 1978.

Mayo, J. S.: "Experimental 224MB/S PCM Terminal," *Bell System Technical Journal*, November 1965.

Messerschmitt, D. G.: "Frequency Detectors for PLL Acquisition in Timing and Carrier Recovery," *IEEE Transactions on Communications*, September 1979.

Motorola Semiconductor, "Phase Locked Loop Data Library," Motorola Semiconductor Products, Inc., 1973.

Chapter 6

TIMING CIRCUITS

H. Ilhan Refioglu
Exar Integrated Systems Inc.
Sunnyvale, Calif.

6–1 INTRODUCTION

The introduction of the 555 timer made circuit designs for most timing applications possible, by using a few external components and, hence, brought a large degree of simplicity into timer design. The 555, with its virtues, such as simplicity, versatility, and economy, soon gained a widely accepted popularity somewhat comparable to that of the operational amplifier. The advent of the 555 brought up new timer applications and new circuit ideas based around the 555, using it as the basic building block of the system.

Soon the dual versions of the world's most popular IC timer became available in the market and were followed by other IC timers that offer more sophisticated features. Today, there is a wide selection of timer ICs available in the market, which offer improved accuracy, more versatility, programmability, and lower power consumption. These monolithic timing circuits form a timer product area that comprises a considerable family of devices that find a wide variety of applications in both linear and digital systems. In a large number of industrial control and test sequencing applications, these circuits provide direct, economical, and a much more reliable replacement for mechanical or electromechanical timing devices. They are also used for clock generation, as well as for pulse position and pulsewidth modulation. Some of the timer applications and market areas for timers are outlined in Fig. 6-1. A family tree of the commercially available monolithic

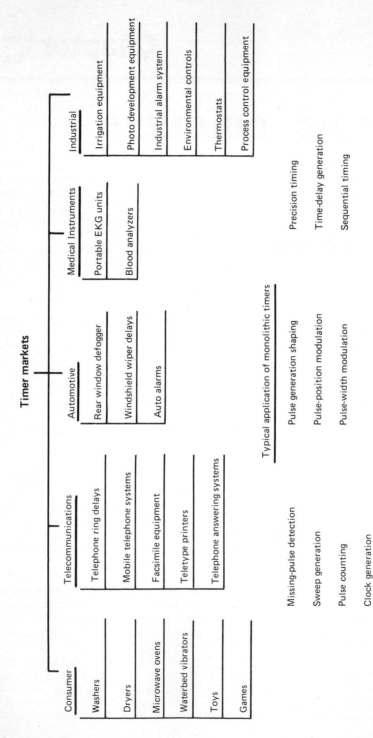

Timer markets

Consumer
- Washers
- Dryers
- Microwave ovens
- Waterbed vibrators
- Toys
- Games

Telecommunications
- Telephone ring delays
- Mobile telephone systems
- Facsimile equipment
- Teletype printers
- Telephone answering systems

Automotive
- Rear window defogger
- Windshield wiper delays
- Auto alarms

Medical Instruments
- Portable EKG units
- Blood analyzers

Industrial
- Irrigation equipment
- Photo development equipment
- Industrial alarm system
- Environmental controls
- Thermostats
- Process control equipment

Typical application of monolithic timers

- Missing-pulse detection
- Sweep generation
- Pulse counting
- Clock generation
- Pulse generation shaping
- Pulse-position modulation
- Pulse-width modulation
- Precision timing
- Time-delay generation
- Sequential timing

FIG. 6-1 Timer IC markets and typical applications of monolithic timers.

timers and their manufacturers is given in Fig. 6-2. There are three basic families of IC timer products.

1. *General-purpose timers:* These are basically single-cycle timers that operate by a single charging cycle of the external timing capacitor and are simple and low-cost devices that can be used in all but the most sophisticated of applications. They have excellent accuracy and very good stability. These units are capable of operating in both the monostable and astable modes over wide ranges.

    The dual and quad timers offer two and four timer circuits in one chip, respectively, without sacrificing the performance characteristics of each. They reduce the cost of system applications considerably. They can be used either independently or in conjunction with each other. Since they share a common silicon substrate, they offer excellent performance in matching and temperature tracking.

2. *Special-purpose timers:* This family of timer products is also of the single-cycle type and has evolved to meet specific applications. The high noise immunity timers are designed such that their operation does not get affected by the large transients and high electrical noise levels of industrial control environments.

    The low-power timer circuits offer considerable savings in power consumption over their conventional counterpart, the 555, and can operate down to a few volts without sacrificing key performance features such as timing accuracy and frequency stability, and are well suited for battery-operated applications, similar to those used in portable equipment and instrumentation.

3. *Timer/counters:* Timer/counters, or multiple-cycle timing circuits, have the additional capability of providing ultralong time delays. They employ the combination of a time-base oscillator (similar to the type used in most of the single-cycle timers) and a counter to generate the desired time delay. Timer/counters provide a more economical and practical solution for timing applications requiring long time delays in excess of several minutes.

6–2 FUNDAMENTALS OF IC TIMERS

Monolithic timers generate precise timing pulses, or time delays, whose length or repetition rate is determined by an external timing resistor R and a timing capacitor C. The timing interval is proportional to the external (RC) product and can be varied by the choice of the external R and C.

Based on their principle of operation, integrated-circuit timers can be classified into two categories; one-shot or single-cycle timers and multiple-cycle or timer/counters. Single-cycle timers operate by charging a timing current source, whereas the timer/counters combine a time-base generator with a counter stage to get long time delays.

6–2a Exponential-Ramp Timing Circuit

The simplest form of the one-shot-type timer is the exponential-ramp timing circuit. It can operate in either a monostable or astable (free-running) mode.

1. *Monostable mode of operation:* A block diagram of an exponential-ramp timing circuit operating in a monostable mode is shown in Fig. 6-3. Normally, all the components except the R and C shown in the figure are internal to the IC.

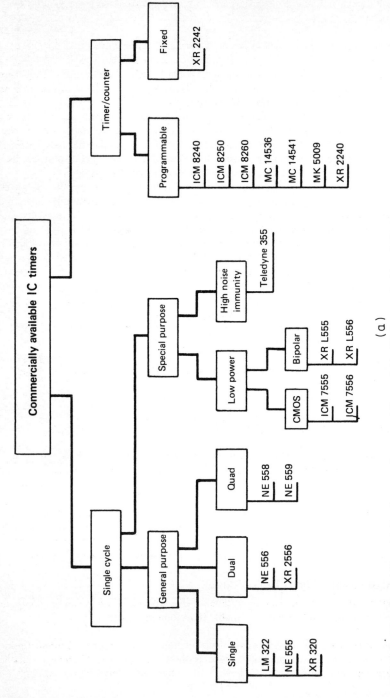

FIG. 6-2 Timer availability. (*a*) Family tree of commercially available IC timers. (*b*) List of IC timer manufacturers.

(*a*)

Timer type	Package	Manufacturer	Part no.
555	8-pin plastic MINIDIP	Signetics	NE555V
		Advanced Micro Devices	NE555V
		Exar	XR-555CP
		Fairchild	µA555TC
		Intersil	NE555V
		Lithic Systems	LS555
		Motorola	MC1455P1
		National	LM555CN
		Raytheon	RC555DN
		RCA	CA555CE
		Silicon General	SG555M
		Teledyne Semiconductor	555P
		Texas Instruments	SN72555P
556	14-pin plastic DIP	Signetics	NE556A
		Advanced Micro Devices	NE556A
		Exar	XR 556CP
		Fairchild	µA556PC
		Intersil	NE556A
		Lithic Systems	L556
		Motorola	MC3556P
		National	LN556N
		Raytheon	RC556DB
		Silicon General	SG556N
		Teledyne Semiconductor	556J
322	14-pin plastic DIP	National	LM322N
3905	8-pin plastic MINIDIP	National	LM3905N
2240	16-pin plastic DIP	Exar	XR-2240CP
		Fairchild	µA2240PC
		Intersil	ICL8240CPE
2250	16-pin Plastic DIP	Exar	XR-2250CP
		Intersil	ICL8250CPE
8260	16-pin plastic DIP	Intersil	ICL8260CPE
L555	8-pin plastic	Exar	XR-L555
		Signetics	
L556	14-pin plastic DIP	Exar	XR-L556
355	8-pin plastic DIP	Teradyne	Teradyne-355
7555	8-pin plastic DIP	Intersil	ICM-7555
7556	14-pin plastic DIP	Intersil	ICM-7556
2242	8-pin plastic DIP	Exar	XR-2242

(b)

FIG. 6-2 (*Continued*)

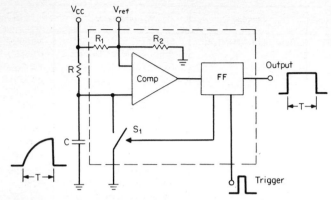

FIG. 6-3 Monostable mode of operation for exponential-ramp-type timing circuit.

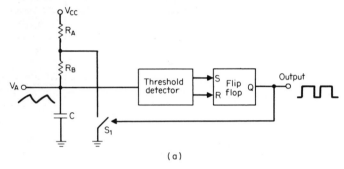

(a)

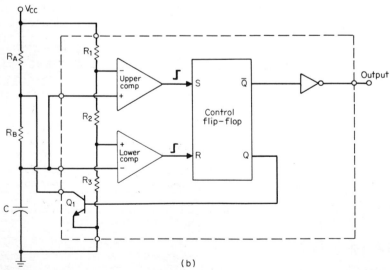

(b)

FIG. 6-4 Astable mode of operation for exponential-ramp-type timer. (a) Functional block diagram; (b) circuit schematic.

The operation of the circuit can be explained as follows. In its quiescent, or reset condition, the switch S_1, which is normally a grounded emitter *npn* transistor internal to the chip, is closed, clamping the capacitor to ground. The timer output in this state is low or near zero.

The timing cycle is initiated upon the arrival of a trigger pulse at the input. The trigger pulse sets the flip-flop, which in turn opens the switch S_1 across the capacitor C, and also causes the output to go to a high level. The output is now in its high state and the timer is in its unstable condition. Since the switch S_1 is open, the capacitor C will start to charge through the resistor R. The voltage across the capacitor rises exponentially toward the supply voltage V_{CC} with a time constant RC. When this voltage reaches the threshold voltage V_{ref}, the voltage comparator changes state, which resets the flip-flop, which in turn closes the switch S_1. The output reverts to its low state, the circuit returns to its standby condition, and the timing cycle of the monostable ends. The circuit stays at this stable condition until it is retriggered.

2. *Astable mode of operation:* The same basic timer circuit of Fig. 6-3 can be modified, as shown in Fig. 6-4, to give the astable mode of operation. Note that this block diagram now has two timing resistors and a slightly different threshold-detector arrangement, which involves two comparators to sense the two threshold levels associated with this type of operation.

    Operation of the circuit is as follows. Initially, assume the switch S_1 to be open and the output to be at its high state. Since R_B is connected in series with C and R_A, the capacitor charges toward V_{CC} through the series combination of R_A and R_B until the upper threshold is reached. During this charging period, which is determined by $(R_A + R_B)C$, the output is in its high state. When the exponential timing ramp across the capacitor reaches the upper threshold level, the flip-flop changes state. The discharge transistor (switch S_1) is turned on and the output goes low, and stays low, during the discharge period of $R_B C$. The capacitor C begins to discharge exponentially toward ground through R_B, until the voltage across the capacitor reaches the lower threshold level and retriggers the timer. The circuit once again reverts to its high output state, with S_1 opening and C charging toward V_{CC}. The timer now begins a new cycle. The circuit will continue to oscillate between the two threshold levels, with the output changing state with each threshold crossing. In this manner, a digital waveform is generated at the output of the timer. The waveform across the capacitor is an exponential ramp, as shown in the figure.

6–2b   Linear Ramp Timing Circuit

An alternate approach to the design of the simple exponential ramp timing circuit is the linear-ramp-type timer circuit shown in Fig. 6-5. This circuit, which is also a single-cycle timing circuit, operates on a similar principle. However, now the timing capacitor C is charged linearly with a constant current I and generates a linear-ramp waveform with a constant slope of (I/C). These types of circuits are also capable of operating in both astable and monostable modes.

6–2c   Performance Limitations of Single-Cycle Timers

The accurate timing intervals that can be obtained from commercially available one-shot-type timer ICs are limited to the range of from several microseconds to several minutes. For generating very short timing pulses (in the few microsecond range or lower) the internal time delays associated with the switching speeds of the comparators—especially the recovery delays of the lower comparator which usually consists of *pnp* transistors—

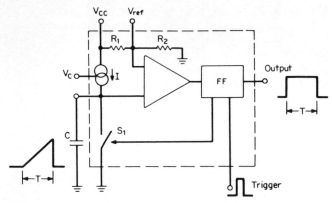

FIG. 6-5   Block diagram of the linear-ramp-type timer.

the flip-flop and discharge transistor (represented by the switch S_1) may contribute to additional timing errors and avoid reliable operation. For applications requiring time delays in the order of 1 μs or less, retrigerable one-shots of the TTL family, and CMOS versions of the basic monostable multivibrator, are better, since they offer faster response times. (These circuits, discussed elsewhere in this book, will not be covered in this chapter.)

For generating long time delays (in the several minute range) that require large values of R and C, the input bias currents of the comparators and the leakage currents associated with the timing capacitor or the internal discharge transistor may limit the timing accuracy of the circuit. In general, for timing applications requiring time delays in excess of several minutes, the multiple-cycle or timer/counter-type timer circuits provide a more economical and practical solution than the one-shot-type timer ICs.

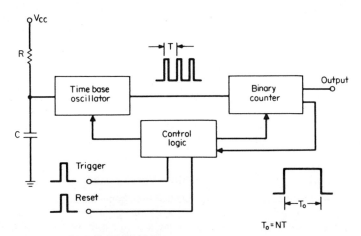

FIG. 6-6   Simplified block diagram of a timer/counter.

6–2d   Timer/Counter ICs

The timer/counter, or multiple-cycle timing circuits are the timers that have the additional capability of providing ultralong time delays. They use the combination of a time-base oscillator, which is basically very similar to the simple exponential-type timer circuit described before, and a binary counter to generate the desired time delay. Fig. 6-6 shows a simplified block diagram of a timer/counter IC, which is made up of three basic blocks: (1) a time-base oscillator, (2) a binary counter, and (3) a control flip-flop.

The operation of a timer/counter can be described as follows. When the circuit is at rest, or reset condition, the time-base oscillator is disabled and the counter is reset to zero. Once the circuit is triggered, the time-base oscillator is activated and produces a series of timing pulses whose repetition rate is proportional to the external timing resistor R and the capacitor C. These timing pulses are then counted by the binary counter, and when a preprogrammed count is reached, the binary counter resets the control flip-flops, stops the time-base oscillator, and ends the timing cycle. The total timing interval T_0 is then proportional to N times the (RC) product, where N is the preprogrammed count.

The combination of a stable time-base oscillator and a programmable binary counter on the same IC chip offers some unique application and performance features common to the timer/counter family. Some of these are outlined below.

- *Generating long delays with small capacitors:* For a given time delay setting, the timer/counter would require a timing capacitor C that is N times smaller than that needed for the one-shot-type timer, where N is the count programmed into the binary counter. Since large-value, low-leakage capacitors are quite expensive, this technique may provide substantial cost savings for generating long time delays in excess of several minutes.

- *Generating ultralong delays by cascading:* When cascading two timer/counters, one cascades the counter stages of both timers. Since the second timer/counter further divides down the counter output of the first timer, the total available count is increased *geometrically*, rather than arithmetically. For example, if one timer/counter gives a time delay of NRC, two such timer/counters, cascaded, will produce a time delay of N^2RC, where N is the count setting of the binary counter. Thus, a cascade of two timer/counter ICs, each with an 8-bit binary counter, can produce a time delay in excess of $32,000\ RC$.

- *Generating multiple delays from the same RC setting:* By using a programmable binary counter whose total count can be programmed between a minimum count of 1 to a maximum count of N, one can obtain N different time intervals from the same external RC setting.

- *Easy to set or calibrate:* Although timer/counters are normally used for generating long time delays or intervals, their accuracy characteristics are only determined by the characteristics of the time-base oscillator. The counter section does not affect the overall timing accuracy. Thus, time setting, or calibration for long interval timing, can be done quickly, without waiting for the entire timing cycle, by setting the accuracy of the time-base oscillator.

6–3   THE 555 TIMER IC

The 555, first introduced by Signetics in 1972, is a single-cycle timer that is very similar in structure and operating principles to the exponential-ramp timing circuit of Fig. 6-3. It is a general-purpose timer, capable of both the monostable and astable modes of operation over wide ranges.

6–3a   Internal Structure and Analysis

The functional block diagram of the 555 is given in Fig. 6-7. The operation of the circuit is as follows. The voltage-divider resistor chain, comprised of three identical resistors, sets up the threshold voltages of the upper and lower comparators as $\frac{2}{3} V_{CC}$ and $\frac{1}{3} V_{CC}$, respectively. The circuit's timing cycle starts by bringing the trigger input below the lower threshold value of $\frac{1}{3} V_{CC}$. The lower comparator changes state and the control flip-flop switches the output to its high level (which is $2V_{be}$ and an IR drop below V_{CC}) and turns off the discharge transistor Q_1. This allows the external capacitance C to charge toward V_{CC} through the external resistor R_A. When the voltage on the external capacitor reaches the upper comparator threshold voltage $\frac{2}{3} V_{CC}$, the upper comparator changes state, causing the control flip-flop to reset. The output returns low and Q_1 turns on to rapidly

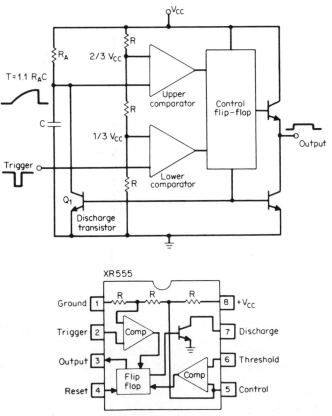

FIG. 6-7   The functional block diagram of the 555 and its package outline.

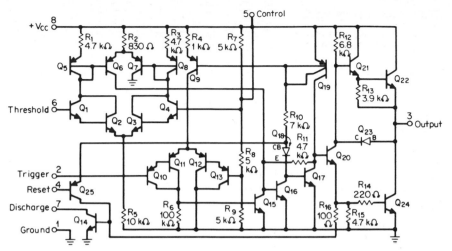

FIG. 6-8   The circuit schematic for 555.

discharge the external capacitor. The output is high only during the capacitor charging cycle, which is determined by the R_A and C values.

The circuit schematic for the 555 is given in Fig. 6-8. The resistor divider chain that determines the threshold voltages is comprised of 5 kΩ resistors R_7, R_8, and R_9, and biasing transistors Q_4 and Q_{13} of the upper and lower comparators, respectively. Transistors Q_1–Q_8 make up the upper comparator. Transistors Q_5–Q_8 are used as active loads to increase the gain of the comparator. Darlington differential input stages used in the comparator design provide high input impedance and low input currents, which allow a wide range of external timing resistor values to be used in applications.

The two comparator outputs are taken from the collectors of transistors Q_6, Q_{10}, and Q_{11}, and are fed to the control flip-flop that consists of Q_{16} and Q_{17}. Transistors Q_9 and Q_{19} are the biasing transistors for the lower comparator and the flip-flop.

The output stage of the 555 is the versatile totem-pole design, consisting of transistors Q_{20}–Q_{24}. It has the capability of sourcing or sinking 200 mA when operated from a 15-V power supply, and it can readily drive TTL inputs with a chip supply of 5 V.

The operation of the circuit can be explained as follows. A trigger input lower than $\frac{1}{3} V_{CC}$ applied to the base of Q_{10} (pin 2) turns Q_{10} and Q_{11} on and causes a positive-going output on the comparator load resistor R_6. This causes Q_{15} to turn on, causing its collector to go low, this sets the latch by turning Q_{16} off and turning Q_{17} on, which then causes the collector of Q_{17} and hence the output of the flip-flop to go low. For this state of the flip-flop, Q_{20} and the discharge transistor Q_{14} are off and the output (pin 3) is high. This set condition of the latch will remain until the circuit is reset.

When pin 6 reaches the upper threshold voltage level of $\frac{2}{3} V_{CC}$, the output of Q_6 goes high and turns on Q_{16}. Q_{16} removes the base drive of Q_{17}, turns it off, and resets the latch. An alternate way of resetting the latch is by pulling the reset input low (pin 4), which is the base of the reset transistor Q_{25}. This turns Q_{25} on, which then turns off Q_{17} by stealing its base drive, and reverse biases the collector base diode Q_{18}. Regardless of the method used, the reset state turns Q_{17} off and turns Q_{20} on. At this high state of the flip-flop, the discharge transistor Q_{14} and the output sink transistor Q_{24} are on and the output is at its low state, which is determined by the $V_{CE}(\text{sat})$ of Q_{24}.

6–3b   Monostable Mode

The basic modes of operation of the 555 are the monostable (one-shot) and the astable (free-running) modes. The circuit connection for the monostable mode of operation is given in Fig. 6-9.

In the monostable mode the circuit basically requires a timing resistor, a capacitor, and a bypass capacitor connected to the control terminals as external components. When a trigger pulse is applied to the trigger input (pin 2) that is less than $\frac{1}{3} V_{CC}$, the timer is triggered and starts its timing cycle. The output rises to a high level of approximately $V_{CC} - 1.6$ V; at the same time, C begins to charge toward V_{CC}. When the voltage across the capacitor crosses $\frac{2}{3} V_{CC}$, the timing period ends with the output reverting to its state of approximately zero volts. The timer is now ready for another input trigger. The timing diagram is also illustrated in the figure.

The timing expression for the monostable mode of operation can be derived quite simply. The general expression for voltage across a capacitor, exponentially rising from approximately zero volts to the supply voltage, is

$$v_c = V_{CC}(1 - e^{-t/\tau}) \tag{6-1}$$

where $\tau = RC$ is the time constant. Since the timing period ends when the voltage across the capacitor reaches $\frac{2}{3} V_{CC}$, the equation for this condition becomes

$$v_c = \tfrac{2}{3} V_{CC} = V_{CC}(1 - e^{-T/RC}) \tag{6-2}$$

Solving this equation for T, the expression for the timing period can be determined as

$$T = RC \ln 3 = 1.0986RC \tag{6-3}$$

or simply

$$T = 1.1RC \tag{6-4}$$

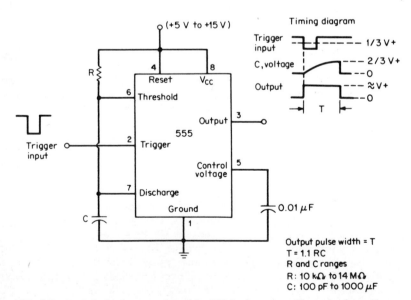

FIG. 6-9   Typical circuit connection of the 555 for the monostable mode of operation and its timing diagram.

This is the basic equation for the timing period of the 555, operating as a monostable timer.

6-3c Astable Mode

The circuit connection for the astable mode of operation of the 555 is given in Fig. 6-10. The basic difference from the monostable connection is that the threshold (6) and trigger (2) pins are connected together so that both upper and lower comparators can sense the voltage on the capacitor, and also another timing resistor R_B which determines the discharge period of the capacitor, is included.

Operation of the circuit is as follows: Initially, the voltage across the capacitor C will be low, causing the timer to be triggered by pin 2. This forces the output to go high and turns off the discharge transistor, causing the capacitor to charge toward V_{CC} through the series combination of R_A and R_B. The capacitor charges exponentially toward V_{CC} until the voltage across it reaches $\frac{2}{3} V_{CC}$. At this time, the upper threshold is reached causing the output to go low and the discharge transistor to turn on. The capacitor then starts to discharge exponentially to ground through R_B and the low impedance of the discharge transistor, until it reaches the lower threshold point of $\frac{1}{3} V_{CC}$. This triggers the timer once again, starting a new cycle. The timer then continues to oscillate, generating a digital waveform at the output and forming an exponential (triangular) timing ramp across the capacitor, as shown in Fig. 6-10.

The timing expression for the astable mode of operation can also be derived in a similar fashion. The expression for the voltage across the capacitor, rising exponentially from $\frac{1}{3} V_{CC}$ (the lower threshold) to V_{CC} is given as

$$v_c = \tfrac{1}{3} V_{CC} + \tfrac{2}{3} V_{CC}(1 - e^{-t/\tau}) \qquad (6\text{-}5)$$

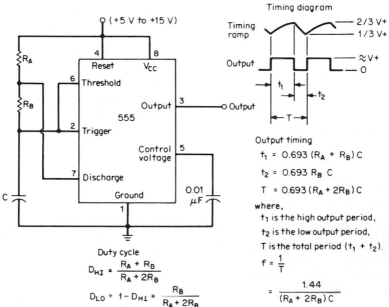

FIG. 6-10 Typical circuit connection of the 555 for the astable mode of operation and its waveforms.

TABLE 6–1 Comparison of One-Shot IC Timers

	LM 322	NE 555	NE 556	NE 558	NE 559	Teledyne 355	XR 320	XR 2556
Supply range	4.5–40 V	4.5–16 V	4.5–16 V	4.5–16 V	4.5–16 V	11–16 V	4.5–20 V	4.5–16 V
Supply current @ 5 V	2.5 mA	3 mA	6 mA	27 mA @ 15 V	12 mA @ 12 V	15 mA @ 16 V	2 mA	6 mA
Timing accuracy	1%	1%	1%	2%	2%	1%	1%	1%
Temperature stability	30 ppm/°C	50 ppm/°C	50 ppm/°C	150 ppm/°C	150 ppm/°C	300 ppm/°C	100 ppm/°C	50 ppm/°C
Supply stability	Not specified	0.05%/V	0.05%/V	0.1%/V	0.1%/V	0.4%/V	0.1%/V	0.01%/V
Output current (source)	50 mA	200 mA	150 mA			100 mA	100 mA	200 mA
Output current (sink)	50 mA	200 mA	150 mA	100 mA	100 mA	100 mA	100 mA	200 mA
Package	14 pin	8 pin	14 pin	16 pin	16 pin	8 pin	14 pin	14 pin
Positive trigger	Yes	No	No	No	No	No	Yes	No
Negative trigger	No	Yes	Yes	Yes	Yes	Yes	Yes	Yes
Modulation	Yes	Yes	Yes	Yes	No	Yes	Yes	No
Multiple logic output	Yes	No	No	No	No	No	Yes	No
Astable operation	No	Yes	Yes	Yes	Yes	Yes	Yes	Yes
Monostable operation	Yes	Yes	Yes	Yes	Yes	Yes	Yes	Yes
Linear-Ramp output	No	No	No	No	No	No	Yes	No
Technology	Bipolar	Bipolar	Bipolar	Bipolar	Bipolar	Bipolar	Bipolar	Bipolar
Original manufacturer	National	Signetics	Signetics	Signetics	Signetics	Teledyne	Exar	Exar
Other sources		AMD Exar Fairchild Intersil Motorola National RCA Texas Inst.	AMD Exar Fairchild Intersil Motorola National Raytheon	Exar	Exar			

Note: IC types are identified using the product designation of their original manufacturer and whenever applicable, alternate suppliers or second sources are also indicated.

where $\tau = (R_A + R_B)C$. The charging time t_1 of the total timing period T can be found by substituting $v_c = \frac{2}{3} V_{CC}$ in Eq. 6-5.
Then,

$$v_c = \frac{2}{3} V_{CC} = \frac{1}{3} V_{CC} + \frac{2}{3} V_{CC}(1 - e^{-t_1/(R_A + R_B)C}) \tag{6-6}$$

from which the charge time t_1 can be derived as

$$t_1 = \tau \ln 2 = 0.693 \,(R_A + R_B)C \tag{6-7}$$

Similarly, the expression for the voltage across the capacitor, decaying exponentially from $\frac{2}{3} V_{CC}$ to ground (zero), is given as:

$$v_c = \frac{2}{3} V_{CC} \, e^{-t/\tau} \tag{6-8}$$

where $\tau = R_B C$. The discharge time t_2 of the total timing period T can be found by setting $v_c = \frac{1}{3} V_{CC}$ in Eq. 6-8. Then,

$$v_c = \frac{1}{3} V_{CC} = \frac{2}{3} V_{CC} \, e^{-t_2/\tau} \tag{6-9}$$

Thus, t_2 can be derived as

$$t_2 = \tau \ln 2 = 0.693 \, R_B C \tag{6-10}$$

Therefore, the total timing period

$$T = t_1 + t_2 = 0.693 \,(R_A + 2R_B)C \tag{6-11}$$

From this timing period the frequency of oscillation of the timer, in its astable mode of operation, can be derived as

$$f = \frac{1}{T} = \frac{1.44}{(R_A + 2R_B)C} \tag{6-12}$$

6–4   OTHER SINGLE-CYCLE TIMERS

In this section we will discuss some of the commercially available single-cycle type of timers other than the 555. Since the 555 was discussed in detail in regard to its internal structure, operating principles, and modes of operation, and since there are enough similarities between the timers of this section and the 555, the discussions are very brief, and comparisons and references to the 555 are made whenever applicable.

In addition, to give an overview of the single-cycle (one-shot) type timers, the comparison of electrical specifications, performance characteristics, basic features, package types, and original and other manufacturers of the commercially available timer ICs are presented in Table 6-1.

6–4a   The 320 Linear-Ramp-Type Timer

The XR-320 is a monolithic timing circuit that operates on the linear-ramp generation principle. The functional block diagram and the package outline of the circuit are given in Fig. 6-11. Its control flip-flop offers more flexibility than the 555. It can be triggered (set) directly by either a positive-going or a negative-going pulse on two separate input lines (set terminals) of the flip-flop. The discharge transistor Q_0, internal to the chip, is normally in the "on" (saturated) state and the voltage across the timing capacitor is clamped to ground. When a trigger input is applied, Q_0 is turned "off" and C is charged linearly rather than exponentially by a temperature-compensated composite *pnp* constant

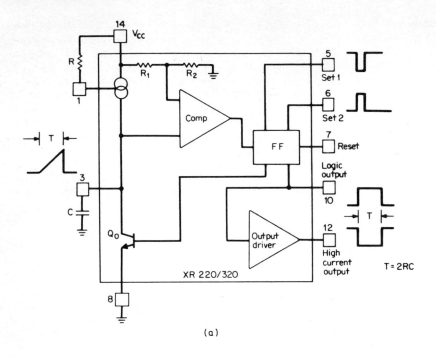

(a)

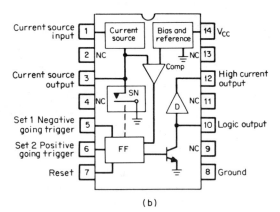

(b)

FIG. 6-11   The XR-320 timer. (*a*) Functional block diagram; (*b*) package outline.

current source. The constant current I is set by an external resistor R connected to V_{CC} and generates a linear-ramp waveform across the capacitor with a slope of (I/C). The constant current I can be controlled by an external control voltage V_C applied to the current source.

After a time duration T of precisely $2RC$ determined by the current source and comparator thresholds, the voltage across C reaches the threshold level, upon which the comparator changes state, the flip-flop is reset, Q_0 is turned on, and the output reverts to its original (stable) state. The capacitor is discharged through Q_0 and the timing cycle is completed.

The circuit provides two independent complementary outputs: one of them is the open collector type, which is a medium current output (up to 10 mA), that requires a pull-up resistor, and the other is a high current (up to 100 mA) totem-pole design similar to the 555. With no trigger pulse applied, the output at pin 10 is low (near ground potential) and the output at pin 12 is high (near V_{CC}). Upon triggering, the outputs change state and pin 10 becomes high and pin 12 becomes low. The outputs remain in this switched (unstable) condition during the timing cycle and revert to their original (stable) state after the termination of the timing cycle. Once the circuit is triggered, it becomes immune to subsequent triggering until the entire timing cycle is completed. The timing cycle can be reset at any time simply by grounding pin 7.

Typical circuit connections of the XR-320, in a monostable mode of operation for both positive and negative triggering conditions and their respective input and output waveforms, are shown in Fig. 6-12.

By shorting pins 3 and 5, the XR-320 will operate in a free-running or astable (self-triggering) mode. In this mode of operation, the circuit functions as a stable clock pulse generator with an output repetition rate of approximately $1/2RC$. Typical circuit connections for free-running operations and the corresponding waveforms for the self-triggered operation are shown in Fig. 6-13. As can be seen in the figure, one cycle is not precisely equal to $2RC$ because of the additional retrace period, which is the very small but finite discharge time of the capacitor through the low impedance of the discharge transistor Q_0 before the timer gets ready for the next linear-ramp charging interval.

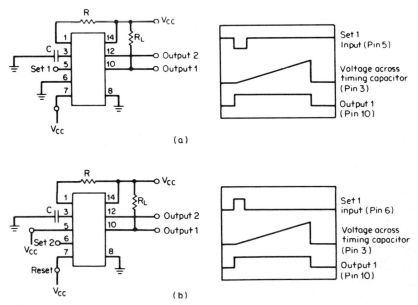

FIG. 6-12 Circuit connections of XR-320 for monostable operation. (*a*) Negative trigger; (*b*) positive trigger.

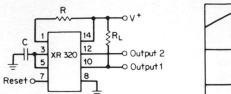 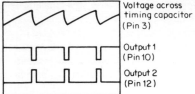

FIG. 6-13   Typical circuit connection of the XR-320 for astable mode of operation and the corresponding waveforms.

6–4b   The Versatile 322/3905 Precision Monostable Timers

These timer circuits, manufactured by National Semiconductor, feature significant improvements over the 555. A functional block diagram of the LM322 timer is given in Fig. 6-14. As shown in the figure, an internal voltage regulator provides a constant 3.15-V supply for the external timing resistor and the comparator threshold. Since comparator thresholds are proportional to the supply voltage in the 555 and the XR-320, transients and supply variations during, and especially near, the end of the timing interval may cause timing errors. The 322 design eliminates supply-voltage fluctuations as a source of error by virtue of the internal voltage regulator, which provides a buffer against power-supply variations.

Another virtue of the internal voltage regulator is the lower power dissipation of the circuit at higher supply voltages. The 555 and the 320 are essentially resistive and hence their power consumption increases with supply voltage. The internal circuitry of the LM322, however, draws a constant supply current of less than 4.5 mA at 3.15 V and its regulator maintains a nearly constant current load for any supply voltage between 4.5 and 40 V.

The threshold voltage in the 322/3905 is 2.0 V and is derived from the internal voltage regulator by two voltage divider resistors of 4 and 6.9 kΩ. In the 322, this point (pin 7) is made externally available to trim the timing period, if desired. In the 8-pin 3905, this feature is not available.

The expression for the voltage across the capacitor in the LM322 can be given by the equation

$$v_c = V_{ref} (1 - e^{-t/\tau}) \tag{6-13}$$

where $V_{ref} = 3.15$ V is the reference voltage output of the voltage regulator and $\tau = RC$ is the time constant. Since the timing period ends when the voltage across the capacitor reaches the comparator threshold voltage, $V_{th} = 0.632 V_{ref} = 2.0$ V, the timing expression can be found by solving the equation

$$v_c = 0.632 V_{ref} = 2.0 = 3.15 (1 - e^{-T/\tau}) \tag{6-14}$$

Thus, for the LM322/LM3905, the timing expression is

$$T = RC \tag{6-15}$$

The range of delay times can be extended, in the case of the LM322, by increasing the comparator's normally low current levels. This is achieved by connecting the boost pin (pin 11) to V_{CC}, which increases the comparator current levels by orders of magnitude. This option allows accurate microsecond timing, but it is not available on LM3905.

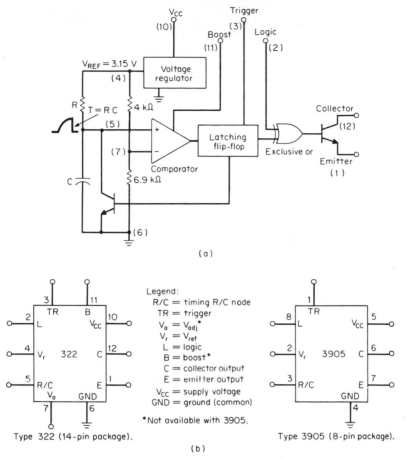

FIG. 6-14 The LM322 timer. (*a*) Functional block diagram; (*b*) pinouts and terminal designations.

The operation of the circuit is very similar to the 555. Triggering in the 322 and 3905 is accomplished by a positive pulse. This pulse sets the latch, turns off the discharge transistor, and initiates the timing cycle. The capacitor starts to charge through the timing resistor R toward the voltage reference. When this exponentially rising voltage across the capacitor reaches the threshold voltage of the comparator, the comparator changes state, resets the latch, and terminates the timing cycle.

The signal from the latch, however, does not simply yield a high output, as in the case of the 555-type timer, due to an exclusive or gate arrangement used between the latch and the output. One of the two inputs of the gate is internally connected to the output of the latch, whereas the other input is connected as the logic input pin. The function of this gate is to control the state of the timer output by its logic input pin. When the logic pin is high (1), the output is "off" during the timing cycle and "on" otherwise. When the logic pin is low (0), the output transistor is "on" during the timing cycle and "off" otherwise.

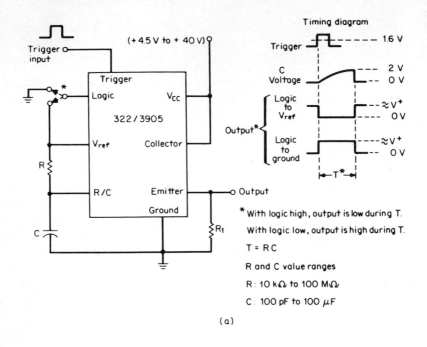

(a)

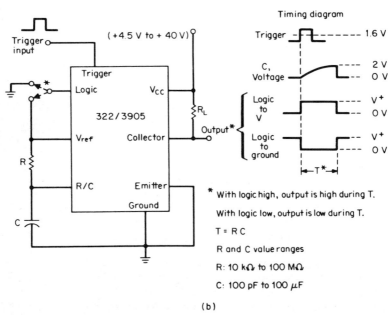

(b)

FIG. 6-15 The 322/3905 monostable mode of operation. (*a*) Output taken from emitter. (*b*) Output taken from collector.

This family of timers from National Semiconductor also offers a great degree of flexibility in their output circuitry. A current driver, floating, *npn* power transistor with built-in current limiting is used in the output stage. Both collector and emitter pins are brought outside the package for external connection. Thus, the output can function as either an open collector stage or as an emitter follower for driving up to 40 V and 50 mA, into loads referred to ground or the supply. With the versatility of this output circuitry, a wide variety of load driving requirements can be met. Fig. 6-15 shows the typical circuit connection of the chip for a monostable mode of operation, using both the open-collector output and the emitter-follower output options. Waveforms for both options are also illustrated in the figure.

The biggest shortcomings of the versatile LM322 and LM3905 chip are that they basically only operate in a monostable mode and they do not offer the astable mode of operation.

6–4c   The 556 Dual Timer

The NE556 timer, first introduced by Signetics, contains two independent 555-type timing circuits on a single monolithic chip. Supplied in a 14-pin dual-in-line package with common ground and power-supply pins, the two timer sections can be used either independently or in conjunction with each other. The matching and temperature tracking characteristics between each timer section of the chip are superior to those available from two separate timer packages.

Each timer section has independent set, reset, and modulation controls, and each independent output can source or sink 200 mA of current. The performance characteristics of each half of the 556 match the performance of the 555, and the pin functions of each half are the same as their 555 counterpart. Each half of the circuit can also operate in a free-running (astable) mode by connecting the circuit for a self-triggering mode, as in the case of the 555.

The package outline of the chip and its typical circuit connection for monostable and astable modes of operation are illustrated in Fig. 6-16.

6–4d   Quad Timing Circuits (NE558/NE559)

The quad timing circuits contain four independent timer sections on a single monolithic chip. Each of the timer sections on the chip is entirely independent, and each one can produce a time delay from microseconds to minutes, as set by an external *RC* network. They provide a cost effective alternative to single timer ICs in applications requiring a multiplicity of timing or sequencing functions. Each timer has its separate timing, trigger, and output terminals, but all four timers in the IC package share a common supply voltage, ground, reset, as well as modulation control terminals. All four timing sections can be used simultaneously, or each timer section can be cascaded or connected in tandem with other timer sections for sequential timing applications without requiring coupling capacitors, since the quad timer circuits are edge triggered devices.

The equivalent circuit schematic for the 558/559 timers, with their package outline, are given in Fig. 6-17. As can be seen from the figure, the major difference between the two circuits is their output stage. The 558 has open-collector output stages, whereas the 559 has emitter-follower outputs in a Darlington configuration. Thus, in a normal mode of operation, the 558 family of quad timers requires a pull-up resistor to V_{CC}, whereas the 559 family requires a pull-down resistor from each output to ground. Although each output can handle a current of up to 100 mA individually with more than one output active, the total output current capability is limited by the power dissipation rating of the

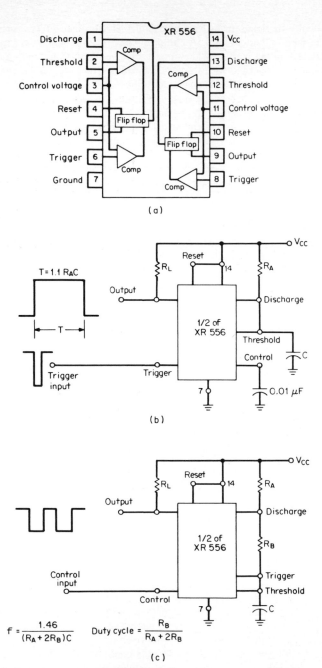

FIG. 6-16 The 566 timer. (*a*) Package outline; (*b*) circuit connection for monostable operation; (*c*) circuit connection for astable operation.

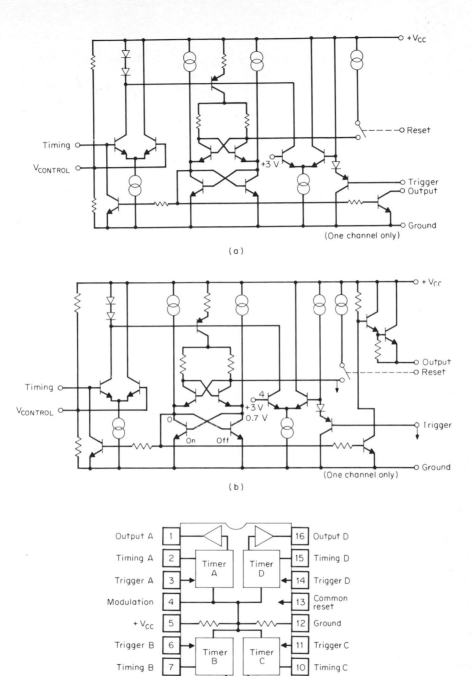

FIG. 6-17 The 558/559 timers. (*a*) Equivalent circuit schematic for the 558 timer. (*b*) Equivalent circuit schematic for the 559 timer. (*c*) Package outline for 558/559 timers.

IC package. The outputs are normally at low state and go to high state during the timing interval.

In the monostable, or one-shot mode of operation, it is necessary to supply an external timing resistor and a capacitor for each section of the timer IC. The timing terminals of those timer sections not being used can be left open circuited.

Each timer section of the quad timer IC has its own trigger input. The trigger level is nominally set at about 1.5 V (three volts minus $2V_{be}$), and the trigger input is edge triggered on the falling edge of an input trigger pulse. In other words, for proper triggering, the triggering signal must first go high and then go low.

The reset control pin (pin 13) is common to all four timer sections and resets all of the timer sections simultaneously. The reset voltage must be brought below 0.8 V to ensure a reset condition. When the reset is activated, all the outputs go to the low state. While the reset is active, the trigger inputs are inhibited, but after reset is finished, the trigger voltage must be taken high and then low to implement triggering.

6–4e The 355 Industrial Timer

The Teledyne Semiconductor 355 timer is designed to be used as an accurate time-delay device, or as an astable oscillator, in industrial control environments. It triggers on the negative edge of the low-going trigger pulse. It is pin compatible with the 555 timer. The functional block diagram and the package outline of the circuit are given in Fig. 6-18.

The operation of the circuit is very similar to the 555. The output logic level is normally in a low state and goes high during the timing cycle. It has a current sink or

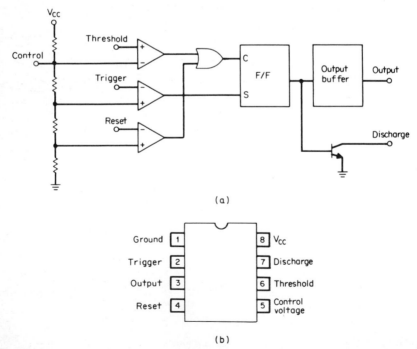

(a)

(b)

FIG. 6-18 The 355 industrial timer. (a) Functional block diagram; (b) package outline.

source capability of 100 mA. The timing cycle is initiated by lowering the DC level at the trigger terminal below $0.45\ V_{CC}$. Once triggered, the circuit is immune to additional triggering until the timing cycle is completed, which occurs when the voltage level at the threshold input reaches $\frac{2}{3}\ V_{CC}$. At that point, the threshold comparator changes state, resets the internal flip-flop, and initiates the discharge cycle. The timing cycle, or the frequency of oscillation, can be controlled or modulated by using the control voltage terminal. This terminal is internally biased at $\frac{2}{3}\ V_{CC}$ and the control signal for frequency modulation, or pulsewidth modulation, is applied to this terminal.

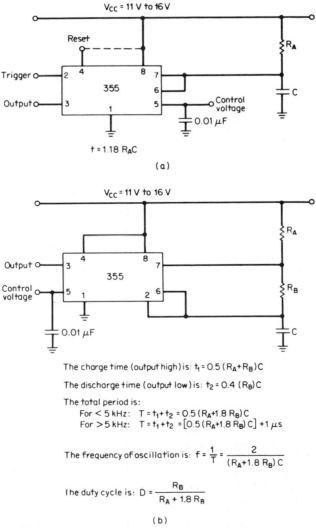

FIG. 6-19 Typical circuit connection of the 355 timer. (a) Monostable operation; (b) astable operation.

The discharge terminal corresponds to the collector of the discharge transistor. During the charging cycle, the discharge transistor is off and this terminal behaves as an open circuit; during discharge, the transistor is on and it becomes a low-impedance path to ground through which the capacitor discharges.

The timing cycle can be interrupted by grounding the reset terminal. When the reset signal is applied, the output goes low and remains in that state while the reset voltage is applied. When the reset signal is removed, the output remains low until retriggered.

Typical circuit connection diagrams for astable and monostable modes of operation are given together with the timing and frequency expressions in Fig. 6-19.

6–5   MICROPOWER TIMER ICs

These are the low-power timer circuits that can perform the general-purpose timing functions, with a minimum amount of power dissipation, since they offer more than an order of magnitude reduction in standby current drain. The power-consumption savings of these types of micropower timers, over their conventional counterpart, the 555, is displayed in Fig. 6-20. As shown in the figure, low-power timers also offer a wide operating supply-voltage range. They can operate down to a few volts, whereas the minimum operating voltage for the regular 555-type timer is 4.5 V. This feature enables micropower timers to operate safely and reliably with two 1.5-V NiCd batteries.

Micropower timer ICs provide a direct pin-for-pin replacement for their standard counterparts. They offer low threshold, trigger, and reset currents, since they operate at much lower current levels. They do not exhibit supply-current transients (spikes) during output transitions, which is commonly observed in the standard 555-type timers, due to the large crowbar currents produced in the output drivers during output switching. Improvement on the supply-current transients is shown in Fig. 6-21, by comparing the supply-current characteristics for the conventional 555 family with a low-power timer during an output transition. As shown in the figure, the conventional 555 timer can produce 300 to 400 mA of supply-current spikes during switching, whereas low-power timers are virtually transient free.

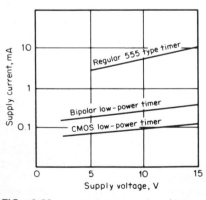

FIG. 6-20 Approximate supply-voltage vs. supply-current characteristics of regular 555-type timers and low-power timers.

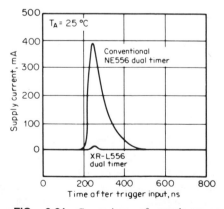

FIG. 6-21 Comparison of supply-current transient of conventional 555-type timer with micropower timers.

TABLE 6-2 Comparison of Low-Power One-Shot IC Timers

	ICM 7555	ICM 7556	XR L555	XR L556
Supply range	2–18 V	2–18 V	2.7–15 V	2.7–15 V
Supply current @ 5 V	80 μA	160 μA	190 μA	380 μA
Timing accuracy	2%	2%	1%	1%
Temperature stability	50 ppm/°C	50 ppm/°C	50 ppm/°C	50 ppm/°C
Supply stability	1.0%/V	1.0%/V	0.5%/V	0.5%/V
Output current (source)	Not specified	Not specified	50 mA	50 mA
Output current (sink)	Not specified	Not specified	2 mA	2 mA
Package	8 pin	14 pin	8 pin	14 pin
Positive trigger	No	No	No	No
Negative trigger	Yes	Yes	Yes	Yes
Modulation	Yes	Yes	Yes	Yes
Multiple logic output	No	No	No	No
Astable operation	Yes	Yes	Yes	Yes
Monostable operation	Yes	Yes	Yes	Yes
Linear-ramp output	No	No	No	No
Technology	CMOS	CMOS	Bipolar	Bipolar
Original manufacturer	Intersil	Intersil	Exar	Exar
Other sources			Signetics	Signetics

Note: IC types are identified using the product designation of their original manufacturer and whenever applicable, alternate suppliers or second sources are also indicated.

In Table 6-2 an overview of the commercially available low-power single-cycle timer ICs is presented. In the table, basic electrical specifications, performance characteristics, features, and the manufacturers of some of the commercially available micropower timers are summarized.

6–5a The 7555/7556 CMOS Timers

The ICM7555/6 are CMOS *RC* timers capable of producing an accurate time delay or frequency. The ICM7556 is a dual version of the ICM 7555, with the two timer sections operating independently of one another, sharing only V_{CC} and ground. The 7555 and 7556 devices have the identical pin configurations as the regular 555 and 556 devices, respectively. In most applications, they are direct replacements for their standard counterparts. They operate in both astable and monostable modes, just as do the 555 and 556, but they can be used with higher impedance timing elements for longer *RC* time constants, due to the high input impedances of the circuit, achieved by the MOS technology. (The trigger, threshold, and reset currents are typically 20 pA.) The 7555/7556 timers offer a wide operating supply voltage range of 2 to 18 V with a very low supply current of typically 80 μA for 7555 and 160 μA for 7556.

The block diagram and the equivalent circuit of the chips are given in Fig. 6-22.

6–5b The L555/L556 Low-Power Timers

The XR-L555 is the micropower version of the popular 555-type timer especially designed for applications requiring very low power dissipation. It is directly pin compatible with the basic 555 timer. However, it exhibits one-fifteenth the power dissipation and can operate down to 2.7 V with a typical supply current of 150 μA without sacrificing such key features as timing accuracy and frequency stability.

The equivalent schematic diagram and the functional block diagram of the L555 are given in Fig. 6-23.

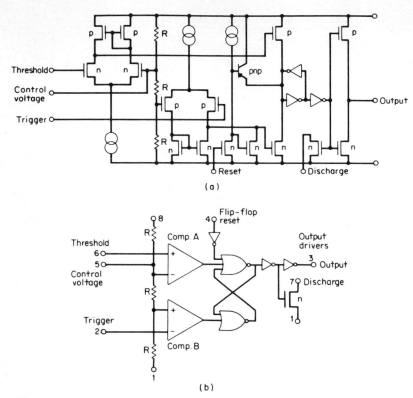

FIG. 6-22   The ICM 7555/7556 CMOS timer. (*a*) Equivalent circuit diagram; (*b*) block diagram.

To achieve low-power operation, certain design techniques were used. The voltage divider chain that determines the threshold voltage uses very high-value resistors, as well as diodes to save real estate (chip area). Current mirrors and current sources are used instead of resistors, whenever possible, to reduce current without sacrificing performance and to achieve low supply-voltage operation. Inclusion of Q_{37} between the flip-flop output and the output source transistor Q_{36} avoids supply-current spikes during switching. The output circuit can source up to 50 mA and can sink 2 mA or drive TTL circuits.

The XR-L556 is the dual of the L555 and is the micropower version of the 556-type dual timer. It is directly pin compatible with the basic 556-type dual timer circuit. It is especially designed for applications requiring multiple timing functions with very low power dissipation.

6–6   COUNTER/TIMER ICs

The timer circuits discussed in the previous sections are of the single-cycle type and produce time delays determined by the single charging of the external timing capacitor. In timing applications requiring delays in excess of several minutes, these circuits require excessively large and impractical values of timing capacitance and resistance.

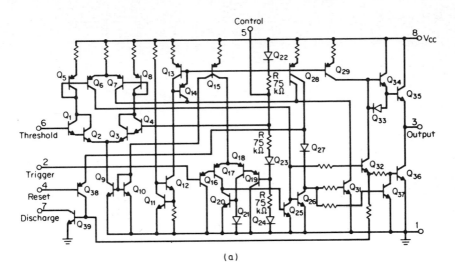

(a)

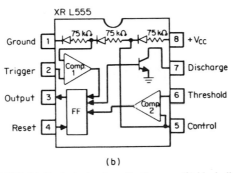

(b)

FIG. 6-23   The XR-L555. (*a*) Equivalent schematic diagram; (*b*) block diagram.

For example, using the maximum values of $R = 10$ MΩ and $C = 100$ μF for timing components, the maximum output period that can be achieved by the 555-type timer is

$$T = (1.1)(10^7)(10^{-4}) = 1100 \text{ s} \cong 18 \text{ min}$$

However, at these maximum limits, precision components are difficult to obtain and when available they tend to be expensive. In the above 1-μF range, capacitors with better than 5% tolerance are difficult to get and even the tantalum types have a tendency to leak, adversely affecting the precision of the timing cycle.

In addition to the problems associated with the external components, the leakage currents of the internal discharge transistor and the base currents of the comparators become comparable to the charge and discharge currents of the capacitor at these very high timing component values. These problems may get even more serious when the device is operated at low supply voltages and extremes of temperature.

A better alternative to overcome the problem of achieving long time delays is to use timer/counter-type products, which provide a more economical, practical, and accurate solution than the one-shot-type IC timers. Timer/counters contain a precision time-base

TABLE 6-3 Comparison of Timer/Counters

	ICL-8240	ICL-8250	ICL-8260	MC-14536	MC-14541	MK-5009	XR-2240	XR-2242
Supply voltage	4–18 V	4.5–18 V	4.5–18 V	3.0–18 V	3.0–18 V	4.5–5.5 V -9.6 to -14.4 V	4–15 V	4–15 V
Supply current	4 mA	4 mA	4 mA	Depends on operating conditions	Depends on operating conditions	6 mA	4 mA	4 mA
Timing accuracy	0.5%	0.5%	0.5%	2%	2%	Not specified	0.5%	0.5%
Temperature stability	200 ppm/°C	200 ppm/°C	200 ppm/°C	Not specified	Not specified	2000 ppm/°C	200 ppm/°C	200 ppm/°C
Supply stability	0.08%/V	0.08%/V	0.08%/V	2%/V	Not specified	0.3%/V	0.08%/V	0.08%/V
Package	16 pin	16 pin	16 pin	16 pin	14 pin	16 pin	16 pin	8 pin
Programming capability	1-255	1-99	1-60	1-255	$(2^8, 2^{10}, 2^{13}, 2^{16})$	1-15	1-255	None (128)
Count code	Binary	BCD	Binary	Binary	Binary	Binary	Binary	Binary
Trigger polarity	Positive	Positive	Positive	Positive	Positive	Positive	Positive	Positive
Modulation	Yes	Yes	Yes	Yes	Yes	Yes	Yes	Yes
Technology	Bipolar	Bipolar	Bipolar	CMOS	CMOS	MOS (P-channel)	Bipolar	Bipolar
Manufacturer	Intersil	Intersil	Intersil	Motorola	Motorola	MOSTEK	Exar	Exar

Note: IC types are identified using the product designation of their original manufacturer and whenever applicable, alternate suppliers or second sources are also indicated.

oscillator that drives a built-in counter stage. There are basically two categories of timer/counter ICs: programmable and fixed timer/counters.

The programmable timer/counter ICs are comprised of a timing section made up of a 555-type oscillator, followed by a counter section. The timing period of the counter is externally programmable by the user. In principle, the timer portion of the chip generates a basic timing pulse of period T. This is then multiplied (or counted) by the counter to effectively increase the timing period by a desired multiplication factor that is externally adjustable or variable. Hence, the total timing interval can be programmed as desired.

The fixed timer/counter ICs also operate on the same principle and have a very similar circuit (chip) structure. The basic difference is in the counter section of the chip, which does not have the programming feature.

In Table 6-3 a comparison of commercially available timer/counter ICs is presented, summarizing their important characteristics and features.

6–6a   The 2240 Binary Programmable Timer/Counter IC

The XR-2240 programmable timer/counter is a monolithic controller capable of producing ultralong time delays without sacrificing accuracy for time delays from microseconds up to five days. Two timing circuits can be cascaded to generate time delays up to three years.

The time delay is set by an external RC network and can be programmed to any value from 1 RC to 255 RC. In astable operation, the circuit can generate 256 separate frequencies, or pulse patterns, from a single RC setting and can be synchronized with external clock signals. Both the control inputs (pins 10 and 11) and the outputs (pins 1 to 8) are compatible with TTL and DTL logic levels. The package outline and the simplified circuit schematic of the chip are given in Fig. 6-24.

The circuit is comprised of three basic sections: (1) a time-base generator that generates a sequence of precisely spaced timing pulses, (2) a programmable 8-bit binary counter, and (3) control logic. The time-base section is made up of the 555-type clock oscillator. The buffered output of the time base is provided at pin 14, and is also internally connected to the binary counter section.

The counter consists of eight binary stages, with a buffered output available from each. Each output is low for the multiple of the time-base period shown. The outputs of the binary counter sections are buffered open-collector-type stages that can be shorted together and pulled up to the power supply by a resistor in a "wired-or" configuration to facilitate external programming. If a number of counter outputs are externally shorted together, the combined output would be high only when all of the connected outputs are high. In this manner, any time delay between 1 RC and 255 RC can be programmed by the proper interconnection of the binary counter outputs.

The third subsection is the control logic (essentially a flip-flop), which is the "brain" of the timer. It consists of a latch that is set and reset by pins 11 (trigger) and 10 (reset), respectively. This circuit controls the timer/counter and the time base oscillator. Upon receiving a positive-going trigger pulse at pin 11, it actuates the time-base oscillator and enables the counter section, setting all the counter outputs to the low state. When a positive-going reset pulse is applied to pin 10 of the control flip-flop, it disables the time-base oscillator, resets all counter stages, and ends the timing cycle.

The operation of the circuit is as follows. When the timer is at standby, or reset condition, the time-base oscillator (which operates similarly to an astable 555 timer) is inhibited, or keyed-off, and all the counter outputs are in the high state. When the circuit

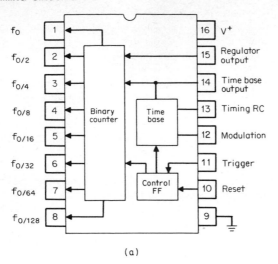

(a)

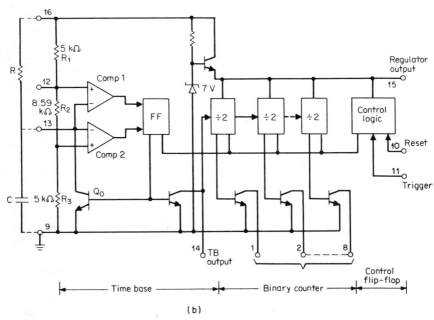

(b)

FIG. 6-24 The XR-2240 programmable timer/counter. (*a*) The block diagram and package outline of the XR-2240. (*b*) The simplified circuit schematic of the XR-2240.

is enabled by a positive-going pulse at pin 11, the time-base section is activated, all of the counter outputs are set to low, and the timing cycle is started. The timing capacitor C starts to charge exponentially through the external resistor R. When the upper threshold level V_A is reached, the capacitor is rapidly discharged back to the lower threshold level V_B through the discharge transistor Q_0. When the lower threshold is reached, the flip-flop changes state, turning off the discharge transistor and the time-base oscillator will then

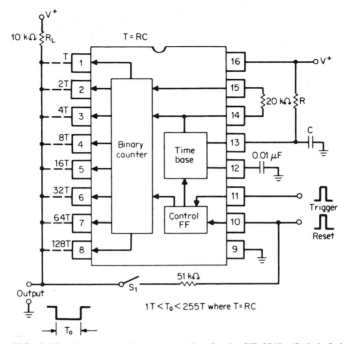

FIG. 6-25   Generalized circuit connection for the XR-2240. (Switch S_1 is closed for monostable operation and is left open for astable mode of operation.)

continue to cycle, producing a series of timing pulses. In this manner, the time-base circuit generates a series of clock pulses with a period T_o set by the external R and C. The internal bias resistors R_1, R_2, and R_3, which determine the threshold levels, are chosen such that the period of the time-base output pulse train is equal to the RC product, independent of supply-voltage changes.

In most timing applications, one or more of the counter outputs is connected back to the reset terminal, as shown in Fig. 6-25, with S_1 closed. In this manner, the timing pulses of the time base are counted by the binary counter until a preset count is reached or completed. At this point, the control logic, or control flip-flop, is reset automatically by the counter, which in turn disables the time base and completes the timing cycle, returning the circuit to its rest state.

If none of the counter outputs are connected back to the reset terminal (switch S_1 open), the circuit would operate in its astable or free-running mode, subsequent to a trigger input.

It should be noted that the circuit is immune to additional trigger inputs until the completion of the timing cycle. It can be reset to the standby condition by applying a positive-going reset pulse to pin 10. Both the trigger and the reset thresholds are compatible with TTL logic levels.

The circuit can be programmed by shorting the open collector-type outputs (pins 1 through 8) of the binary counter to a common pull-up resistor, to form a wired-or connection. The combined output will be low as long as any one of the outputs is low. In this manner, the time delays associated with each counter output can be summed by simply shorting them together to a common output bus, as shown in Fig. 6-25. For example, if only pin 6 is connected to the output and the rest left open, the total duration of the timing cycle T_o would be $32T$. Similarly, if pins 1, 5, and 6 were shorted to the output bus, the total time delay would be

$$T_o = (1 + 16 + 32)T = 49T$$

In this manner, by proper choice of counter terminals connected to the output bus, one can program the timing cycle to be

$$1T \leq T_o \leq 255T$$

where $T = RC$.

Fig. 6-26 shows the output waveforms at some of the circuit terminals, subsequent to the application of a trigger input. The duration of the timing period is determined by the choice of the binary counter output connected to the reset terminal. For example, to obtain a time delay of $8T$, pin 4 would be connected to the reset terminal (pin 10); thus, at the end of a count of 2^3 or 8, the counter at pin 4 would change back to the high state and complete the timing cycle.

6–6b   Other Programmable Timer/Counter ICs

The ICL 8240/8250/8260 is a family of monolithic programmable timers that evolved from the XR-2240 counter/timer. While the 8240 is a straight second source of the XR-2240, the 8250 and 8260 have a different programming range or output count selection.

As in the 2240 timer/counter, each device consists of an accurate, low-drift oscillator, a counter section of master-slave flip-flops, and appropriate logic and control circuitry, all on one monolithic chip. There are no functional differences in the basic operating modes and principles between the 8250, 8260, and 2240 (8240), with the exception of the counting. These units are also very useful for generating ultralong time delays with relatively inexpensive RC components.

The 8250 is specifically designed for decimal counting and delays. The counter section of the chip is an 8-bit design arranged to count in decimal fashion, over two decades, allowing selection of time delays from 1 RC to 99 RC. The carry-out gate allows expansion to 9999 RC or more, by cascading two or more 8250s.

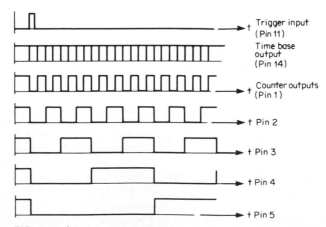

FIG. 6-26 Timing diagram of output waveforms for the XR-2240.

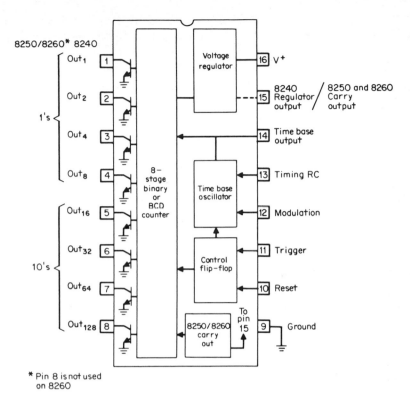

FIG. 6-27   The ICL 8240/8250/8260 programmable timer/counter.

The block diagram of this family of timers is given in Fig. 6-27. As illustrated in the figure, there are still eight stages in the counter section with a similar output configuration to that of the 2240. In the 8250, however, the counters are arranged in groups of four bits. The first four bits (1, 2, 4, and 8) count in BCD form from 0 to 9. This "units decade" drives the "tens decade" (the second four-bit set, 10, 20, 40, and 80), which counts from 0 to 90.

The carry-out terminal (pin 15) is used in applications where two or more 2250s are to be cascaded. This pin goes low at a count of 100 and is used to drive the counter input of a succeeding 2250, which will then count up to thousands.

The 8260 is optimized to time accurate delays in seconds, minutes, and hours. This device is also basically BCD programmable, but its maximum count is 59. Three 8260s can be cascaded, using the carry-out pin to count seconds, minutes, and hours. The major difference in this circuit is that in the tens decade (the second four-counter) the 80-output (pin 8) is deleted (not used), since only 10, 20, and 40 are needed for a count of 59. This is also illustrated in Fig. 6-27.

6–6c   The 2242 Fixed Timer/Counter

The XR-2242 long range timer is derived from the XR-2240 chip and is an 8-pin package version of the 2240. Intermediate counter output pins are omitted, sacrificing the programming capability, to reduce size and cost. The simplified schematic diagram and the functional block diagram of the circuit are given in Fig. 6-28.

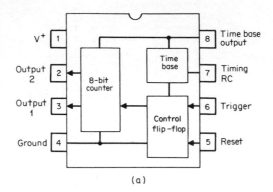

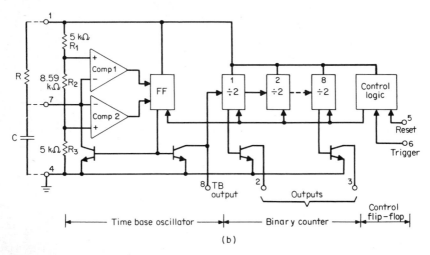

FIG. 6-28   The XR-2242 fixed timer/counter. (*a*) Simplified schematic diagram; (*b*) block diagram.

   The circuit is capable of producing ultralong time delays from microseconds to days. For a given external RC network connected to the timing terminal, the circuit produces an output timing pulse of $128\ RC$. If two circuits are cascaded, a total time delay of $32,768\ RC$ is obtained, which can be used to generate time delays up to a year.

   Generalized circuit connections for the XR-2242 and the corresponding timing diagram for output waveforms are given in Fig. 6-29. In monostable timer applications, the output terminal (pin 3) is connected back to the reset terminal (pin 5). In this manner, after 128 clock pulses, this output goes to high state, resets the circuit, and completes the timing cycle. Thus, subsequent to triggering, the output at pin 3 will produce a total timing pulse of $128\ RC$ before the circuit resets itself to complete the timing cycle. During the timing interval, the first-stage output, or the secondary output at pin 2, produces a square wave output with the period of $2\ RC$, as shown in Fig. 6-29.

   If the output at pin 3 is not connected back to the reset terminal, the circuit continues to operate in an astable mode subsequent to a trigger input, as in the case of the 2240 timer.

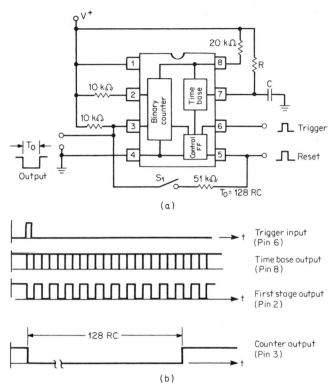

FIG. 6-29   Monostable operation of the XR-2242. (*a*) Generalized circuit diagram; (*b*) timing diagram.

6-7   TIMER APPLICATIONS

So far, various classes and categories of commercially available IC timers, their internal structure, principles of operation, basic features, and performance characteristics have been discussed. In this section we discuss the application of monolithic timer circuits, giving specific emphasis to the selection criteria in choosing the right timer chip for the job. Several design examples for various applications are also given in this section.

Because of its versatility, the monolithic IC timer offers a very wide variety and range of applications in circuit or system design. If the performance characteristics and limitations of the timer IC chosen for a certain application are not carefully considered, the total system performance may be degraded. Similarly, if the timing function is overspecified with an excessive amount of precision and high performance, particularly in regard to its stability and accuracy requirements, then the system cost will increase unnecessarily.

The key selection criterion in choosing the right timer for the job involves finding the monolithic IC that results in the lowest overall system cost for a given performance requirement. However, given the wide range of commercially available IC timers, this selection process is often neither easy nor obvious. By blindly choosing the lowest cost IC, a system designer might overlook other cost factors, such as the cost of external

TABLE 6–4 Major Applications of Timing Circuits (Selection Chart for Timer ICs)

Major Applications	XR-320	NE-555	XR-L555 ICM-7555	NE-556	XR-L556 ICM-7556	XR-2556	NE-558	NE-559	XR-2240	XR-2242	Teradyne 355
Interval timing											
Short interval (microseconds to seconds)	√	√	√	√	√	√	√	√			√
Long interval (seconds to days)			√	√	√	√	√	√			
Programmable time delays									√	√	
Delayed timing				√	√	√	√	√	√		
Pulse generation/shaping											
Pulse shaping		√	√	√	√	√	√	√			√
Pulse-position modulation		√	√	√	√	√					√
Pulsewidth modulation		√	√	√	√	√					√
Pulse counting				√	√	√	√	√			
Delayed pulse generation				√	√	√	√	√	√		
Oscillation/clock generation											
Clock generator		√	√	√		√	√				√
High-current oscillator		√	√	√		√				√	√
Low-voltage oscillator			√		√						
Voltage-controlled oscillator		√	√	√	√	√	√	√			√
Tone-burst generator				√	√	√	√	√	√		
Ultralow-frequency oscillator										√	
Programmable oscillator									√	√	
Dual oscillator				√	√	√	√	√	√		
Ramp generation											
Linear-ramp generator	√										
Staircase generator						√					

components (precision resistors and capacitors) necessary for the timing function. This fact holds particularly true in the case of long-interval timing, which requires high-value low-leakage capacitors.

An IC timer is an extremely versatile function block that suits hundreds of different types or classes of applications. However, for classification purposes a very large majority of applications for IC timers can be grouped into one of the following categories:

1. Interval or event timing
2. Pulse generation and shaping
3. Oscillation or clock generation
4. Ramp or sweep generation

A summary of the major timer applications and recommended timer ICs are given in Table 6-4, which provides a general guideline to the designers and users in the selection of a timer IC. Basic categories of applications are discussed in greater detail in the following sections.

6–7a Interval or Event Timing

In such an application one uses the IC timer either to control the time interval between events or the duration of an event. A typical example of such an application would be to control the opening or closing of an electromechanical relay or sequencing of lights.

Most timing requirements fall within a few microseconds to several minutes. For such applications, the basic 555-type timer is often the best choice, based on its low cost, versatility, and its availability from a multitude of IC manufacturers. The circuit diagram given in Fig. 6-30 can be used for such a general-purpose interval timing application.

Designs involving battery-operated or portable equipment require a timer that can implement general-purpose timing functions with a minimum amount of power dissipation. For such applications, the micropower versions of the basic 555, such as Exar's XR-L555 or Intersil's ICM-7555, are recommended since they offer more than an order of magnitude reduction in standby current drain. Low power applications requiring absolute minimum supply voltage and power dissipation are better served by the ICM-7555. However, the XR-L555 (second sourced by Signetics) offers higher accuracy, stability, and output current capability than the CMOS ICM-7555, making it the optimum choice for applications demanding high performance and load drive capability. Since both circuits are pin-to-pin equivalents of the 555 timer, the circuit of Fig. 6-30 would also be applicable for these applications.

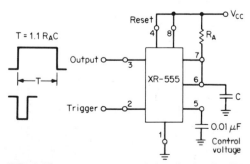

FIG. 6-30 Interval or event timing circuit using 555-type timer IC.

In some industrial control environments, particularly those involving electrome-chanical relays or rotating machinery, a timer IC might require a high degree of immunity to electrical noise spikes or transients. For such applications, the 355-type timer manu-factured by Teledyne is recommended. It is basically a modified version of the basic 555-type exponential timer and it features increased trigger and reset levels made com-patible with standard high-noise-immunity logic families.

Example 6–1 Design of Monostable Timer

Design a timer circuit that operates with a 15-V supply and turns on an LED for a duration of approximately 10 ms every time it receives a negative trigger pulse.

Solution

(a) The timer IC that is best suited for such an application is the 555 timer. The circuit of Fig. 6-30 will be used. Using the timing relation for the 555 timer (Eq. 6-4), the value of the timing components can be found as

$$R_A C = \frac{T}{1.1} = \frac{10 \text{ ms}}{1.1} = 9.1 \times 10^{-3}$$

Selecting a standard capacitor value of 0.22 µF, the timing resistor R_A can be found to be

$$R_A = \frac{9.1 \times 10^{-3}}{0.22 \times 10^{-6}} = 41.4 \text{ k}\Omega$$

Thus, the nearest standard value resistor of 39 kΩ is chosen for the circuit.
(b) The output of the 555, when triggered, is high and approximately 1.6 V $[2V_{BE} + V_{CE}(\text{sat})_{pnp}]$ below V_{CC}. Considering that most LEDs require about 20 mA of operating current and a forward voltage drop V_f of 1.4 V, the series resistor that limits the LED current can be found as follows:

$$R_S = \frac{V_{\text{out}} - V_f}{I_{\text{LED}}} = \frac{(15 - 1.6) - 1.4}{20} = 600 \ \Omega$$

Hence, a standard resistor value of 620 Ω is used in the circuit.
The resulting circuit is shown in Fig. 6-31. To improve the noise immunity of the device, a 0.01-µF capacitor is connected between pin 5 and ground. If a reset option is not required, then this pin (pin 4) should be connected to the supply.

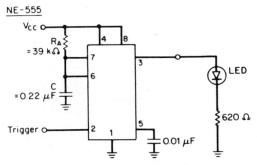

FIG. 6-31 Timing circuit of Example 6-1.

Example 6–2 Noise-Immune Monostable

What are the modifications recommended for the circuit of Example 6-1 to operate in a noisy environment?

Solution

(a) Instead of the 555 timer, the 355 industrial timer is recommended.

(b) Since the timing relation of 355 is slightly different than the 555, the timing elements may require slight modification. For this case,

$$R_A C = \frac{T}{1.18} = \frac{10 \text{ ms}}{1.18} = 8.48 \times 10^{-3}$$

Again, selecting the same capacitor value of 0.22 µF, the timing resistor R_A is found to be

$$R_A = \frac{8.48 \times 10^{-3}}{0.22 \times 10^{-6}} = 38.5 \text{ k}\Omega$$

Again, 39 kΩ is used for R_A.

(c) The output voltage level of the 355 is the same as the 555 when triggered (i.e., during the timing cycle), and hence the output circuitry does not require a modification.

The resulting circuit is shown in Fig. 6-32.

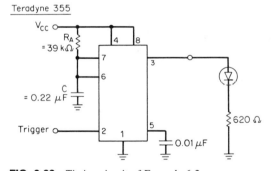

FIG. 6-32 Timing circuit of Example 6-2.

Example 6–3 Low-Power Monostable Timer

Design a low-power timer circuit using an XR-L555 that operates with a 5-V supply and turns on a solid-state relay for 100 ms when triggered. The minimum voltage required to turn on the relay is 3 V, and the control current range of the relay is from 5 to 13 mA. The coil resistance is 400 Ω.

Solution

(a) Since the L555 is a pin-to-pin equivalent of the 555 timer, the circuit of Fig. 6-30 is used. The timing elements can be determined as

$$R_A C = \frac{100 \text{ ms}}{1.1} = 9.1 \times 10^{-2}$$

Choosing a capacitor value of 4.7 µF, the timing resistor is found to be:

$$R_A = \frac{9.1 \times 10^{-2}}{4.7 \times 10^{-6}} = 19.4 \text{ k}\Omega$$

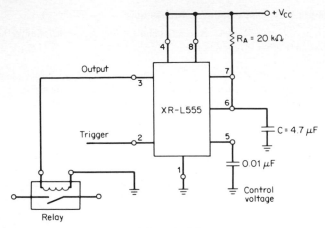

FIG. 6-33   Timing circuit of Example 6-3.

Hence, a standard value of 20 kΩ is selected for R_A.

(b) For 5-V operation, the output of the L555 during the timing cycle is about 3.4 V. Hence, the control current available to turn on the relay is

$$I_{on} = 3.4 \text{ V}/400 \text{ } \Omega = 8.5 \text{ mA}$$

which is within the specified range.

The resulting circuit is shown in Fig. 6-33.

To meet interval or event timing requirements in the minutes, hours, or days range, timer/counter ICs present the most economical approach because they can produce long time delays using a small capacitor and hence provide considerable savings to the system designer. For such uses, the low-cost XR-2242 long delay timer, which operates on the timer/counter principle, is the most cost effective circuit.

Ultra-long time delays can be generated by cascading two XR-2242 timers, as shown in Fig. 6-34. In this configuration, the counter section of Unit 2 is cascaded with the counter output of Unit 1 to provide a total count of 32,768 clock cycles before the output (pin 3) of Unit 2, changes state. In the circuit, the output (pin 3) of Unit 1 is directly connected to the time-base output (pin 8) of Unit 2 through a common pull-up resistor. In this manner, the counter section of Unit 2 is triggered every time the output of Unit 1

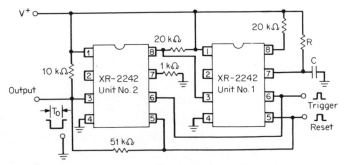

FIG. 6-34   Cascaded operation of two XR-2242 timer circuits.

makes a positive-going transition. The time-base section of Unit 2 is disabled by connecting pin 7 of Unit 2 to ground through a 1 kΩ resistor. The reset and trigger terminals of both units are connected together for common controls.

Example 6–4   Long Delay Timer

Using the cascaded configuration of the XR-2242 timer counter IC, determine the timing component values that will achieve a time delay of about 1 hr.

Solution

The timing relation for the two stages is

$$T = (128)(256)RC = 32,768\ RC$$

For a time delay of 1 hr = 3600 s, the value of the timing components are determined as

$$RC = \frac{3600}{32,768} = 0.110$$

Selecting a capacitor value of 10 μF,

$$R = \frac{0.110}{10 \times 10^{-6}} = 11\ k\Omega$$

Many interval and event timing applications require sequencing of timing functions, i.e., one timer completes its operation and initiates the next timer, and so on. Because these applications use a multiplicity of timer circuits, they are best served by dual timer ICs, such as the NE556, or quad devices such as the NE558 or NE559.

Fig. 6-35 shows a typical implementation of the quad timer in a sequential timing application. For illustration purposes, the 558 timer is used in this example. Note that, when triggered, the circuit produces four sequential time delays, where the duration of each output is independently controlled by its own RC time constant. Yet, all four outputs can be modulated over a 50:1 range and remain proportional over this entire range. Since each timer section is edge triggered, the sections can be cascaded by direct coupling of respective outputs and trigger inputs.

Certain timing applications require that the start of the timing pulse be delayed by a specific time from the occurrence of the trigger. Again, this is accomplished by using a dual timer where one section can be used to set the initial "delay" subsequent to the trigger, and the second section can be used to generate the actual timing pulse.

Fig. 6-36 shows such a sequential timing (delayed pulse generator) application of the 556 timer. In this application, the output of one timer section (Timer 1) is capacitively coupled to the trigger terminal of the second, as shown in the figure. When the timer is triggered at pin 6, its output at pin 5 goes high for a time duration of $T_1 = 1.1R_1C_1$. At the end of this timing cycle, pin 5 goes low and triggers Timer 2 through the capacitive coupling C_C between pins 5 and 8. Then the output at pin 9 goes high for a time duration $T_2 = 1.1R_2C_2$. In this manner, the unit behaves as a "delayed one-shot" where the output of Timer 2 is delayed from the initial trigger at pin 6 by a time delay of T_1.

Example 6–5   Sequential Timer Design

Using the circuit configuration of Fig. 6-36, determine the timing component values for both timer sections to generate a 100-μs pulse 10 ms after the trigger signal is received.

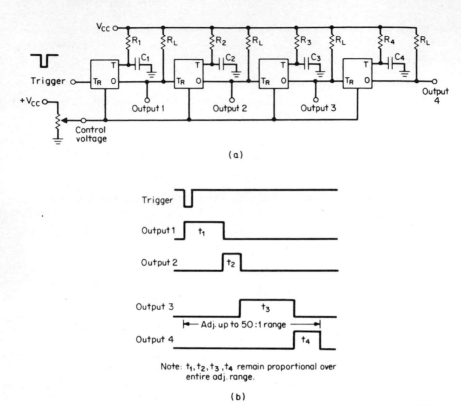

(a)

Note: t_1, t_2, t_3, t_4 remain proportional over entire adj. range.

(b)

FIG. 6-35 Using the 558 timer as a four-stage sequential timer with voltage control capability. (*a*) Circuit connection; (*b*) timing waveform.

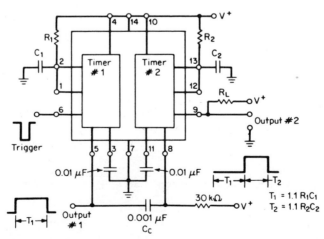

FIG. 6-36 Using the 556 timer as a delayed one-shot or a two-stage sequential timer.

Solution
(a) The delay time T_1 is 10 ms. Hence, the timing components for Timer 1 can be found from

$$R_1C_1 = \frac{10 \text{ ms}}{1.1} = 9.1 \times 10^{-3}$$

Select a capacitor value of 0.01 μF. Then

$$R_1 = \frac{9.1 \times 10^{-3}}{0.01 \times 10^{-6}} = 9.1 \times 10^5 \, \Omega = 910 \text{ k}\Omega$$

(b) The pulsewidth is 100 μs. Thus,

$$T_2 = 100 \text{ μs}$$

The timing components for Timer 2 are determined from

$$R_2C_2 = \frac{100 \text{ μs}}{1.1} = 9.1 \times 10^{-5}$$

Again, select a capacitor value of 0.01 μF. Then

$$R_2 = \frac{9.1 \times 10^{-5}}{0.01 \times 10^{-6}} = 9.1 \times 10^3 \, \Omega = 9.1 \text{ k}\Omega$$

Some timing applications require that the timing interval be digitally programmable, without switching additional precision resistors and/or capacitors into the circuit. Such a function can be easily achieved by using a programmable timer/counter circuit, such as the 2240, where output duration can be programmed from 1.0 RC to 255 RC (in 1 RC increments), where R and C are the external timing components.

Example 6–6   Programmable Timer

Program the XR-2240 timer/counter IC so that a time interval of 2 min is achieved for $T = RC = 1$ s.

Solution
The overall time interval is $T_o = 120$ s. Using the relation

$$T_o = NRC = NT$$

the programming coefficient (integer) N is found to be 120. Hence, $64T$, $32T$, $16T$, and $8T$ outputs of the XR-2240 chip must be used (i.e., connected to the supply with a pull-up resistor) and all the rest of the outputs should be left open. The resulting circuit diagram is given in Fig. 6-37.

6–7b   Pulse Generation and Shaping

Another popular class of applications for one-shot-type timers is pulse shaping or stretching.

As a pulse stretcher, an IC timer is operated in its monostable mode and is triggered by a series of input pulses whose repetition period is longer than the timing period of the IC. The output from the timer will then have the same repetition rate as the input pulse train, except that each output pulse will now have a uniform duration or length as set by the RC time constant of the timer. Any of the general-purpose one-shot timers can function in this application, depending on the required input trigger level and polarity. In most

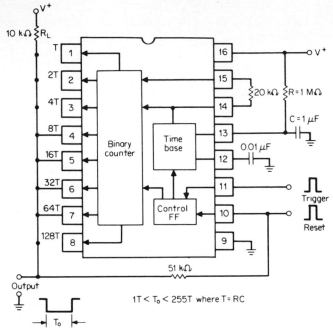

FIG. 6-37 Timer/counter IC configuration of Example 6-6.

applications, the basic 555-type timer is often the most economical solution because of its low unit cost and ready availability.

A variation on the basic pulse-stretcher design is the delayed pulse generator of the previous section (Example 6-5), which converts the input pulse train to a different pulse sequence, which has the same repetition rate but a different duration and a different phase. A dual timer circuit or a quad timer serves this application best. The first timer, which is triggered by the input signal, sets the phase difference or delay between the input and the output pulse sequence; and the second timer, which is triggered at the trailing edge of the first one, sets the output pulsewidth. For applications that demand low power, the ICM-7556 and XR-L556 would be the appropriate choice.

Pulse-blanking circuits selectively interrupt or blank out a pulse train. Such an application can be performed using a dual timer IC, such as the NE-556, where one section of the timer can be operated as a pulse stretcher triggered by the input pulse train; and the second timer section can be triggered by a separate timing signal and serve as an enable/disable control for the first timer, thus interrupting or blanking its output during its timing interval.

As an extension of the pulse-blanking application, the IC timers can also be used for frequency division. Fig. 6-38 shows a typical circuit connection for a dual timer IC, such as the NE-556, for its operation as a frequency divider and pulse shaper. In this case, one of the timer sections (Timer 1) is used as the frequency divider section and its output (at pin 5) is used to trigger the second timer section, which serves as a pulse stretcher and determines the duty cycle of the output. Frequency division is done by setting the timing period T_1 of the first timer to be longer than the period of the input

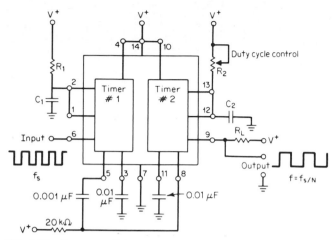

FIG. 6-38   Frequency divider and pulse shaper using a dual timer.

pulse T_P. Once triggered, the timer is insensitive to additional trigger inputs until the conclusion of the timing cycle. Thus, for example, if the timer period is set to 4.5 ms and the period of the input pulse train is 1.0 ms, only every fifth input pulse will trigger Timer 1 of Fig. 6-38 and produce an output. This will then result in an output whose frequency is one-fifth of the input.

Example 6–7   Timer Frequency Divider

The dual timer circuit of Fig. 6-38 is to be used to convert the 5-kHz square wave to a 1-kHz 100-μs pulse train. Determine the component values of the circuit and draw the resulting waveforms.

Solution

(a) The period of the incoming square wave is 0.2 ms or 200 μS. Since frequency division results from Timer 1's insensitivity to retriggering during its cycle time, to achieve times-five division, the timing period of Timer 1 should be less than five times but larger than four times the period of the input signal (i.e., timing period of the timer should be between 0.8 ms and 1 ms). Choosing 0.9 ms as the design target,

$$R_1 C_1 = \frac{0.9 \text{ ms}}{1.1} = 0.82 \times 10^{-3}$$

Choosing a capacitor value of 0.01 μF, R_1 can be found to be 82 kΩ.

(b) The pulsewidth of the frequency divided output waveform is determined by the timing period of Timer 2. To achieve 100 μs

$$R_2 C_2 = \frac{0.1 \text{ ms}}{1.1} = 91 \text{ μs} = 9.1 \times 10^{-5}$$

Choosing a 1000-pF capacitor, R_2 is found to be 91 kΩ. The resulting waveforms are shown in Fig. 6-39.

By using a potentiometer for R_2 the duty cycle of the output can be varied by adjusting the output pulse width.

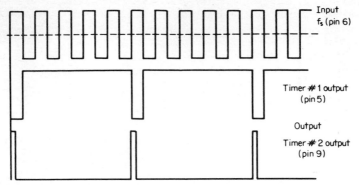

FIG. 6-39   Waveforms of Example 6-7.

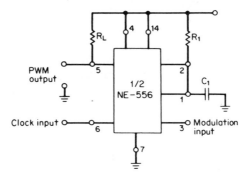

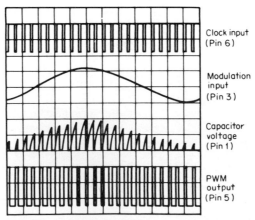

FIG. 6-40   A circuit for pulsewidth modulation and the resulting waveforms.

It has to be noted that if the pulse-shaping feature is not needed, the two timer sections of the dual timer can be used as a frequency divider, since they are electrically independent. Thus, if the trigger terminals of both timer sections are connected to a common input, two independent outputs at frequencies $f_1 = f_S/N_1$ and $f_2 = f_S/N_2$ can be produced, where N_1 and N_2 are the division factors for respective timer sections that are set by the external timing resistors and capacitors of each section.

Certain pulse-generation and pulse-shaping applications require modulating the pulsewidth of an output pulse sequence without affecting its repetition rate. Such a requirement can be met by a one-shot timer of the 555-type operating in its monostable mode and being triggered by a fixed-frequency input pulse train. Output pulses are generated at the same rate as the input pulse train, except the output pulsewidth is determined by the timing components R_1 and C_1. The width of the output pulses from the timer IC can be modified without affecting the repetition rate by applying a control voltage to the modulation terminal of the timer circuit, as shown in Fig. 6-40. The actual circuit waveforms generated in this manner are also shown in the figure. The control characteristics associated with the modulation terminals are depicted in Fig. 6-41.

When using a dual timer for pulse-width modulation, an external clock signal is not necessary, since one section can be operated in its astable mode and serve as the clock generator. Fig. 6-42 is used as the recommended connection for such an application. In this case, Timer 2 is used as the clock generator and Timer 1 is used as the pulsewidth modulator section.

As a final example of applications in the pulse-generation and pulse-shaping category, pulse position modulation is considered. It involves the generation of a pulse sequence whose pulsewidth is constant (and usually very narrow), but whose repetition

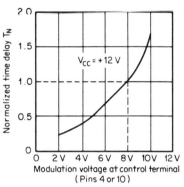

FIG. 6-41 555-type timer control characteristics—normalized time delay vs. control voltage.

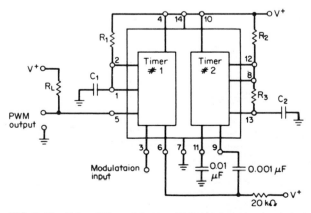

FIG. 6-42 Pulsewidth modulation with internal clock using a dual timer.

rate is modulated. A dual or quad timer IC easily implements such a function where the second timer generates the narrow output pulses when triggered by the output of the first timer. The first timer section is operated in its free-running (i.e., astable) mode, and its frequency is then externally modulated by applying a control voltage to its modulation terminal.

6–7c   Oscillation or Clock Generation

Timers can be operated in a free-running or "self-triggering" mode to generate periodic timing pulses. Since the output pulsewidth or the frequency can be controlled by the choice of external resistors and capacitors, these circuits make excellent low-cost clock oscillators for a number of digital systems.

In clock oscillator designs, an IC generates a fixed-frequency output waveform with a nearly 50% duty cycle. The NE-555 timer, whose output duty cycle can be controlled by the choice of two external resistors, is ideally suited for such an application for clock frequencies up to 100 kHz.

The 555-type timer, which can provide up to 200 mA of current drive, is also well suited for high-current oscillator applications that require the circuit output to be able to source or sink high load currents ($\geqq$100 mA) in order to drive electromechanical relays or capacitive loads.

Voltage-controlled-oscillator (VCO) circuits find a wide range of applications in phase-locked-loop systems. The 555-type timer can be used as a VCO by applying the proper control voltage to its modulation terminal (pin 5) and operating the IC in its self-triggering mode.

Battery-operated or remote-controlled instruments often require a low-power clock oscillator. Low-threshold CMOS logic circuits normally require stable clock oscillators that can operate with very little power consumption and with a low voltage (3 V) power

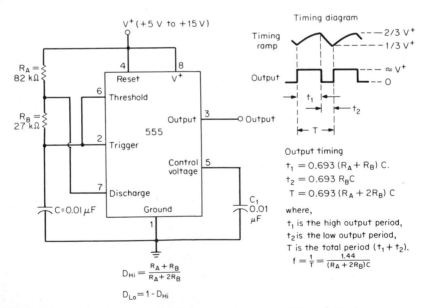

FIG. 6-43   The astable timer of Example 6-8 and its associated waveforms.

supply. The XR-L555 and ICM-7555 micropower timers, which operate with only a small fraction of the power drain of the conventional 555-type timer and can operate with supply voltages as low as 2 to 2.5 V, are very well suited for these types of applications.

Example 6–8   555-Type Oscillator

Design a general-purpose 1-kHz digital waveform oscillator with an 80% duty cycle (i.e., 80% of the time the output is at a high level).

Solution

The 555-type timer IC that is operated in its astable mode as shown in Fig. 6-43, is very well suited for such an application. The frequency expression, as given in Eq. 6-12, is

$$f = \frac{1}{T} = \frac{1.44}{(R_A + 2R_B)C}$$

Choosing a capacitor value of 0.01 µF and substituting $f = 1$ kHz, the frequency expression becomes

$$R_A + 2R_B = 1.44 \times 10^5 \qquad (6\text{-}16)$$

The expression for the high duty cycle is

$$D_{hi} = \frac{t_1}{T} \qquad (6\text{-}17)$$

Substituting Eqs. 6-7 and 6-12 results in

$$D_{hi} = \frac{R_A + R_B}{R_A + 2R_B} \qquad (6\text{-}18)$$

Similarly, the expression for low duty cycle is

$$D_{lo} = \frac{t_2}{T} \qquad (6\text{-}19)$$

and substituting Eqs. 6-7 and 6-12 gives us

$$D_{lo} = \frac{R_B}{R_A + 2R_B} \qquad (6\text{-}20)$$

For this example, $D_{hi} = 0.80$ and $D_{lo} = 0.20$. Either one of Eqs. 6-19 or 6-20 can be used to determine the resistor ratio R_A/R_B.

For the duty cycle value of this problem, these equations yield

$$R_A = 3R_B \qquad (6\text{-}21)$$

Substituting this back into Eq. 6-16 results in

$$5R_B = 1.44 \times 10^5$$
or
$$R_B = 28 \text{ k}\Omega$$

Thus, from Eq. 6-21 R_A can be found to be

$$R_A = 84 \text{ k}\Omega$$

Note that since 28 kΩ and 84 kΩ are not standard resistor values, 27 kΩ and 82 kΩ resistors should be used for R_B and R_A, respectively.

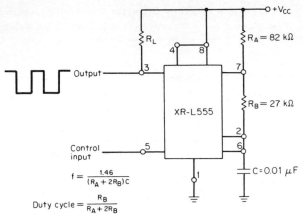

FIG. 6-44   The timer circuit of Example 6-9.

Example 6–9   Low-Power Oscillator Design

Design a 1-kHz clock oscillator for a CMOS logic circuit that operates with two NiCd batteries (approximately 2.7 V). The duty cycle should be 80%.

Solution

The 555-type timer cannot be used for such an application, since it cannot operate well below 5 V. Hence, the optimum choice would be either XR-L555 or ICM-7555 timers, which combine low voltage operation with low power consumption.

Since both chips are pin-to-pin equivalents of the 555 timer, the same timing element values of Example 6-8 can be used. The resulting circuit diagram for the case of the XR-L555 timer IC is given in Fig. 6-44.

Example 6–10   A 10-kHz Square Wave Oscillator

Using the regular 555 timer, design a square wave oscillator that operates at 10 kHz.

Solution

A similar circuit to that of Example 6-8 is used for this application. By choosing a capacitor value of 100 pF, the frequency expression (Eq. 6-12) for $f = 10$ kHz becomes

$$R_A + 2R_B = 1.44 \times 10^6 \qquad (6\text{-}22)$$

By looking at the duty cycle expressions of Example 6-8 it can be seen that the only way to achieve a 50% duty cycle that corresponds to a square wave (i.e., to make $t_1 = t_2$ in Fig. 6-43) is by making R_B large with respect to R_A. Thus, the requirement to achieve a square wave is

$$R_B \gg R_A$$

By selecting $R_B = 100R_A$ and substituting back in Eq. 6-22

$$201R_A = 1.44 \times 10^6$$

or, $$R_A = 7.2 \text{ k}\Omega$$

Then, $$R_B = 720 \text{ k}\Omega$$

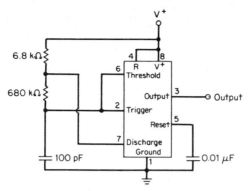

FIG. 6-45 The timer circuit of Example 6-10.

Hence, the following standard resistor values are chosen:

$$R_A = 6.8 \text{ k}\Omega \text{ and } R_B = 680 \text{ k}\Omega$$

If frequency accuracy is important, than a potentiometer should be used for R_B or a trimmer for C. The resulting circuit is shown in Fig. 6-45.

Certain applications require a stable ultralow-frequency clock oscillator whose frequency can be as low as one cycle per day. The programmable timers and the XR-2242-type timer provide ideal solutions for these applications with their long period output waveforms.

The programmable timer ICs are also very well suited for certain test instrumentation designs that require generation of a pseudo random binary data pattern, which would then repeat itself periodically. The XR-2240-type programmable timer/counter, which provides eight separate "open collector" outputs, can perform such a function by selective shorting of one or more of its outputs to a common pull-up resistor in a "wired-or" configuration.

In Fig. 6-46, the XR-2240 timer/counter IC is shown connected as a binary pattern generator. Some of the achievable bit sequences are also illustrated in the figure. The circuit, connected in a free-running, self-triggering mode, is designed for continuous operation. The circuit self-triggers automatically when the power supply is turned on and continues to operate in its free-running mode indefinitely.

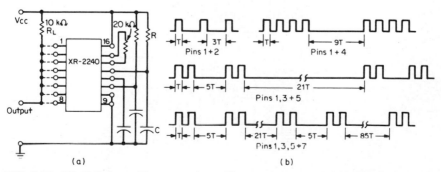

FIG. 6-46 XR-2240 timer chip connected as a binary pattern generator. (*a*) Circuit; (*b*) timing diagram.

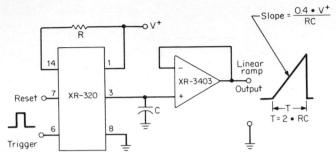

FIG. 6-47   Linear-ramp generation.

The pulse pattern repeats itself at a rate equal to the period of the highest counter bit connected to the common output bus. The minimum pulsewidth contained in the pulse train is determined by the lowest counter bit connected to the output.

6–7d   Ramp or Sweep Generation

In a number of timing applications it is necessary to generate an analog voltage that is proportional to the time elapsed during the timing cycle. This function is particularly useful for generating linear sweep voltages for oscilloscopes or *XY* recorder display applications and it can be accomplished either linearly or digitally.

A linear-ramp voltage can be obtained by charging a capacitor with a precision current source. Most IC timers, such as the 555-type timer, charge the timing capacitor through a resistor rather than through a current source. This results in an exponential- rather than a linear-ramp output across the capacitor. However, the XR-320 timer uses an internal constant current generator to charge the timing capacitor. Thus, upon triggering, it generates a positive-going ramp that rises from ground level to approximately 80% of the positive supply. Fig. 6-47 shows a simple circuit configuration using the XR-

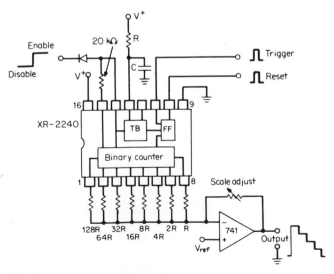

FIG. 6-48   Staircase generator application.

320 timer and a buffer operational amplifier that can generate such a linear-ramp waveform, with a duration of $2RC$, as set by the external time constant. The output is of sufficiently low impedance to drive an XY recorder or oscilloscope display.

In many applications a digitally generated ramp, or staircase voltage, is preferred over a linear- or continuous-ramp waveform. Such a signal can be easily obtained using a programmable timer/counter IC, such as the XR-2240, along with a current-summing op amp and a string of binarily weighted resistors, as shown in Fig. 6-48. In such an application, the open collector outputs of the timer/counter operate as grounded switches. When the circuit is triggered, all the outputs (pin 1 through pin 8) go to a low state and then binarily change state, with all outputs going to a high state after 256 clock cycles. This causes the output of the current-summing op amp to generate a negative-going staircase output of 256 equal steps. The time duration of each step is equal to $1.0RC$, which is set externally. The staircase can be stopped or "frozen" at any desired level by means of an external enable/disable command, as shown in the figure. This unique capability of counter/timers finds a wide variety of applications for digital sample/hold circuits or analog/digital conversion.

Chapter 7

IC POWER-MANAGEMENT CIRCUITS

Robert C. Frostholm Account Manager
Automotive Marketing
National Semiconductor Corp.
Santa Clara, Calif.

The author was with Signetics Corp. when most of the material was written.

7–1 INTRODUCTION TO POWER-MANAGEMENT CIRCUITS

Until recently the term voltage regulator was used to describe that part of a system that provided a constant output voltage independent of the input supply voltage, output load current, and temperature. Today, integrated circuits are being developed that do these functions and much more. To encompass this next level of circuit integration the term power-management circuits is being applied.

Power-management circuits, which include the conventional voltage regulators and references, now also include switched-mode power controllers, supervisory circuits, power monitors, voltage-sensing and protection circuits, current sense latches, and many others. These new functions have been incorporated into highly sophisticated power systems for improved operation, increased reliability, and better energy efficiency. The present rush by the semiconductor industry to penetrate this burgeoning new market is resulting in a wide selection of reasonably priced LSI circuits. The new LSI devices are helping to upgrade power-supply system performance and to reduce costs so much that they are now finding wide acceptance in standard, low-cost, open-frame supplies.

This chapter discusses the more popular industry standard voltage-regulator circuits in sufficient detail to provide system design guidelines. The major thrust, however, is to introduce the latest high-technology power-management devices, with primary concentration on the switched-mode power-control circuits.

The recent energy crisis has stimulated the demand for more efficiency in all aspects of our economy, not the least of which is electronic systems. Linear regulators, such as the series pass devices that are discussed here, are very energy inefficient and consequently are falling behind in the design-in race, with a projected annual growth rate of 4.5% through 1985, compared to a projected 29% growth rate for switchers.

7–1a   Evolution of IC Regulators

The first integrated-circuit regulator was produced about a decade ago. As crude as it was, it was the only device available for quite some time and consequently received much attention. Even today as a universal regulator building block the LM723 is probably one of the hottest selling analog ICs. Industry estimates indicate that all manufacturers of this circuit currently supply about 2 million units per month to the electronics marketplace.

The LM723 was virtually a stand-alone for almost five years until it was determined how to incorporate peripheral circuitry on the chip, including a high current pass transistor. Thus evolved the three-terminal, series pass voltage regulator family known as the 7800 series. The introduction of this family of fixed-voltage positive regulators was soon followed by the 7900 series of fixed-voltage negative regulators. The success of these two families was unmatched. Priced at well under one dollar, with good line and load regulation specifications, as well as current capability of as much as 1 A, these devices were particularly well suited for such devices as local card regulators. The tremendous advantage here is that the requirements for the main supply and its associated voltage regulators could be significantly reduced, affecting part count and cost.

A small extension of both three-terminal regulator families was the dual polarity tracking regulators. This evolution put a positive and negative regulator into a single package with an additional amplifier that caused one output to track another. The dual polarity tracking regulators are particularly useful in systems requiring both a positive and negative supply voltage and consistent tracking between them.

Also an evolution of the three-terminal family was the shunt regulators and precision reference sources. The popularity of shunt regulators has declined dramatically recently, primarily due to competition from more efficient designs. Precision reference sources, on the other hand, are seeing a rebirth as the state-of-the-art data-conversion circuits expand into more and more applications. The D/A converters require precision reference sources with extremely low temperature coefficients and accurate initial set-point accuracy.

7–1b   Types of Regulators

Most all of the above mentioned ICs are low power. Often where larger currents are required, series power pass elements are added. Some manufacturers have even been successful in incorporating these high-current transistors on the IC itself. The penalty that must be paid for this advantage is power dissipation. The high-current regulators (and actually most all of the three-terminal family) require careful mounting to ensure a minimal thermal resistance. The temperature at which an IC operates has a direct correlation to long-term reliability. The more heat that can be drawn out of the chip, the better the reliability.

Series and shunt regulators operate basically in the same manner. A DC output voltage is sensed and compared to an internal reference voltage. An error voltage is generated, which is then used to control a variable resistance (the active device) that keeps the DC output voltage constant (zero error voltage). In controlling the output voltage these devices dissipate a great deal of power across the active element. The efficiency of a linear series regulator may often be as low as 25 to 40%.

Switched-mode power supplies (SMPS) on the other hand, can achieve efficiencies of 75 to 80% with no difficulty at all. SMPS are a modern version of the old electro-mechanical vibrator (like the ones used to supply power to the old car radios). Instead of a mechanical switch, a powerful switching transistor chops the input voltage.

There are three basic configurations for SMPS: flyback, forward, and double forward (many times called push-pull). The proper choice depends upon the application and performance requirements. The flyback is the simplest and least expensive. It is a good choice for multiple output supplies, since only one diode and one capacitor are required per output. The push-pull supply is the most complex, requiring two complimentary signals to provide the switching function.

The semiconductor industry as a whole is emphasizing development of new SMPS control circuits and peripheral housekeeping and protection circuits. Those devices that have emerged as early industry standards are discussed in detail.

It is not possible to cover in detail or even briefly mention every IC used in the field of power management. The devices chosen for this chapter represent, in the author's opinion, the more important integrated circuits in this field and in most cases, the most popular from a user's standpoint.

7–1c Glossary

Before getting too far into the world of power-management circuits, it is worth reviewing the vocabulary of this field of integrated circuitry. Toward this end, a glossary of terms common to power-management circuits follows.

Ambient Temperature   The temperature of the air surrounding an operating power source.

Converter   A device that delivers DC power when energized from a DC source. AC is generated in intermediate steps in the DC/DC process. Rectifying the inverter output results in a DC/DC converter.

Crowbar   A way to short circuit a power supply's output electrically, and thus prevent overvoltage from damaging the load should the power supply fail.

Current Limiting   To restrict the maximum power-supply output current to a preset value during overload conditions and automatically restore the output once the overload is removed.

Derating   To reduce the rated power output as a function of elevated ambient temperatures.

Drop-out Voltage   The input/output voltage differential at which the device ceases to regulate against further reductions in the input.

EMI (RFI)   Electromagnetic interface (radio-frequency interference). Unwanted high-frequency energy caused primarily by the switching components in the power supply. EMI can be conducted through the input or output lines, or radiated through space.

ESR (Equivalent Series Resistance)   The amount of resistance in series with an ideal (lossless) capacitor, representing the equivalent circuit of a real capacitor. In general, the lower the ESR, the more effective the capacitor as a filter. ESR is a prime determinant of ripple in switching supplies.

Efficiency   The ratio of output power to the input power in percent. In a multiple-output switching power supply, efficiency is a function of total output power (the particular division of power among the lower voltage and the higher voltage outputs) and input power.

Feedforward   A technique for monitoring the input of a switching regulator and adjusting the switching oscillator to compensate for changes in the input line prior to their having a detrimental effect on the output. (Generally, the use of feedforward gives a 15- to 20-times improvement to line regulation.)

Flyback Converter   A SMPS in which the energy storage element (inductor) is connected in parallel with the load.

Foldback Current Limit   A technique to reduce the output current to its short-circuit limit as the load resistance decreases.

Forward Converter   A SMPS in which the energy storage element (inductor) is connected in series with the load.

Heat Sink   A cooling element for electronic devices that provides a low path of thermal resistance from the device.

Hiccup Mode   When a fault occurs, the start/stop function is initiated. If the fault remains, the system will continually shutdown and restart. This action is called the hiccup mode, and it effectively limits the energy during fault conditions.

Input Current   The current flowing into the device with a specified voltage applied to the input.

Input/Output Voltage Differential   The voltage between the unregulated input and the regulated output.

Input Voltage Range   The maximum and minimum input voltages for which a power source will operate within its specifications.

Inverter   A device that delivers AC power when energized from a source of DC power. It may be frequency, amplitude, or pulsewidth modulated to vary output voltage.

Line Regulation   A change in DC output voltage due to an input voltage variation, with all other factors held constant. Expressed as a percentage of the nominal DC output voltage, it is also called source-voltage effect.

Load Regulation   A change in DC output voltage produced by varying external load current, with all other factors held constant. Expressed as a percentage of the nominal DC output voltage, it is also called load effect.

Overcurrent Limiting   A protection mechanism that limits the output current of any given output without affecting output voltage.

Overshoot   A transient voltage change exceeding the normal regulation limits. It can occur when a power supply is turned on or off, or with a step change in line voltage or load.

Overvoltage Protection   Shutting down output when the output voltage exceeds a specified value. This feature is especially important for 5-V logic supplies.

Peak Transient Output Current   The maximum peak current that can be delivered to a load during transient conditions, such as electric motor starts.

Quiescent Current   The current consumed by the device that is not delivered to the load. ($I_Q = I_{in} - I_{out}$.)

Rated Output Current   The maximum current that can be drawn from the output of the supply for specified regulation or temperature change.

Regulation   The percent of output voltage change resulting from a specified change of input voltage, output load, temperature, or time. The resulting specifications are line regulation, load regulation, temperature coefficient, and stability.

Ripple   The periodic AC noise component at the power source DC output. Unless specified separately, this specification may include random voltage noise. It is best expressed as volts peak-to-peak.

Short-Circuit Protection   Any output power limiting system that helps a power source sustain no damage under short-circuit load conditions and automatically returns the output to normal when the short circuit is removed. Output voltage and/or output current can be precisely limited to preclude damage caused by overload and short circuit.

SMPS   Switched-mode power supply.

Soft Start   A technique used in SMPS to limit the duty cycle of the pulsewidth modulator during the first few cycles of operation to eliminate dangerous transients during turn-on.

Start/Stop   A protective feature of SMPS to shut the system down as soon as a fault occurs and keep the system down during the next several cycles of the pulsewidth modulator. The system should then restart via a soft-start mode.

Switching Frequency   The rate at which the source voltage is switched in a switching regulator or chopped in a DC/DC converter.

Switching Regulator   A high-efficiency DC/DC converter, consisting of inductors and capacitors, which store energy as well as a switch (or switches) that opens and closes to regulate a voltage across a load. The switch duty cycle is generally controlled by a feedback loop to stabilize the output voltage.

Temperature Coefficient (tempco)   Average change in output voltage for a change in ambient temperature, expressed as a percentage of nominal output voltage. Tempco is derived by measuring at the extremes of the temperature range, and does not represent the maximum possible rate of change.

Temperature Range, Operating   The range of environmental temperatures (usually in °C) over which a power supply can safely operate.

Thermal Resistance   The resistance of a material or materials to the conduction of heat measured in °C/W.

(a) $R_{\theta JC}$ = Thermal resistance junction to case.
(b) $R_{\theta JA}$ = Thermal resistance junction to ambient.
(c) $R_{\theta CA}$ = Thermal resistance case to ambient.
(d) $R_{\theta CS}$ = Thermal resistance case to heat sink.

Tracking, Dual Output   A feature of many dual output supplies, it causes a change of one output to shift the other output so that each maintains the same absolute voltage relationship (ignoring polarity) with respect to common.

Warm-up Time   The time required after turn-on for the output voltage to reach its equilibrium value within the specified output accuracy.

7–2   BASIC PRINCIPLES OF OPERATION OF POWER-MANAGEMENT CIRCUITS

7–2a   Industry Standard Universal Building Block: The LM723

Nearly a decade after its introduction, the LM723 remains one of the most popular integrated-circuit voltage regulators on the market. The popularity of the device can be attributed to its versatility. It can be used as a positive or negative voltage regulator, and with a few external components, many convenient safety and protective features can be added.

A block diagram of the LM723 is shown in Fig. 7-1. The five basic blocks are:

1. A bias supply
2. An internal voltage reference source
3. An error amplifier
4. An output stage
5. A current limit element

The bias supply for the LM723 is derived from an internal zener diode (actually a regulator within a regulator). This supplies the bias voltage for the LM723 circuitry. An internal 7.15-V reference source V_{ref} is also derived from a zener diode. Additional circuitry in the chip provides temperature compensation for this reference (typically $+2.4$ mV/°C temperature coefficient). The error amplifier is a differential pair of transistors.

Normally the reference voltage V_{ref} or some voltage derived from V_{ref} is applied to the noninverting input ($+$In) and a voltage proportional to the output is applied to the inverting input ($-$In). The feedback loop is then closed using the output stage. The output stage must be compensated with a capacitor from the *comp* pin to either ground or the inverting input. Current limiting is achieved by using a current sense resistor in the output. As the current reaches an undesirable level, the voltage developed across the sense resistor causes an internal transistor to begin to conduct and remove base drive from the output stage, which limits the output current of the chip.

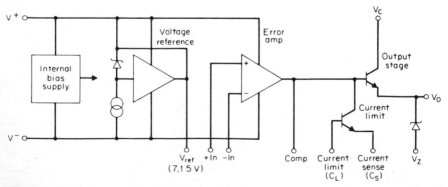

FIG. 7-1   Block diagram of the LM723 universal regulator.

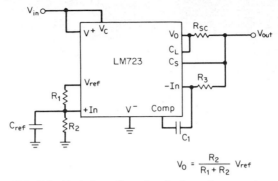

$$V_0 = \frac{R_2}{R_1 + R_2} V_{ref}$$

FIG. 7-2 Typical configuration for low-voltage regulator ($2 \text{ V} < V_{out} < 7 \text{ V}$).

Recommended unity-gain compensation is about 0.005 µF. If V_o is greater than V_{ref}, the closed-loop gain must increase. The compensation capacitor should be reduced proportionally to the increase in gain.

An internal zener diode is provided between V_z and the output V_o with its cathode on the V_o side. If the over current sense function is not going to be used, a second zener is available using the emitter-base junction between the C_S and C_L pins. The anode of this zener is C_L and may be connected directly to the output. This provides both positive and negative 6.2-V zener references with respect to V_{out}.

Minimum V_{in} is 9.5 V to ensure proper operation of the internal reference and zener supplies. To reduce power consumption V_c need only be 3 V greater than the regulated output voltage (providing $V_{in} = 9.5$ V).

Fig. 7-2 shows a typical low-voltage configuration. Resistors R_1 and R_2 divide V_{ref} and apply it to the noninverting input of the error amplifier. A capacitor C_{ref} may be added to improve ripple rejection and reduce the output noise voltage. For adjustable output, R_1 and R_2 can be replaced by a potentiometer network.

The design equations are

$$V_o = \frac{R_2}{R_1 + R_2} V_{ref} \qquad (7\text{-}1)$$

where $R_1 + R_2 > 1.5 \text{ k}\Omega$; and

$$R_3 = \frac{R_1 R_2}{R_1 + R_2} \qquad (7\text{-}2)$$

(Resistor R_3 is used to ensure minimum drift and can be eliminated to reduce component count by using a direct connection between C_s and $-\text{In}$.)

$$R_{sc} = \frac{V_{sense}}{I_{limit}} \qquad (7\text{-}3)$$

where V_{sense} can be obtained from Fig. 7-3 and I_{limit} is the desired short-circuit current. If R_{sc} is set to zero, an output current of 150 mA is possible. Compensation capacitor C_1 is typically 100 pF.

The following example demonstrates the design of a low-voltage regulator using the LM723.

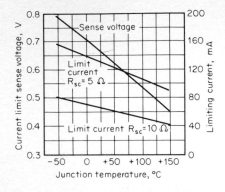

FIG. 7-3 Relationship between sense voltage, limit current, R_{sc} (*Fairchild Camera and Instrument Corp., Mountain View, Calif.*) and junction temperature.

Example 7–1   Design of 5-V Regulator Using LM723

Design a regulator using the LM723 to satisfy the following requirement:

$$V_{in} = 15 \text{ V}$$
$$V_{out} = 5 \text{ V}$$
$$I_{out} = 50 \text{ mA}$$
$$T_a = 25°C$$
$$I_{limit} = 75 \text{ mA}$$

Solution

Using the circuit of Fig. 7-2, let $R_2 = 3 \text{ k}\Omega$ and $V_o = 5$ V, then

$$R_1 = R_2 \frac{(V_{ref} - V_o)}{V_o} \tag{7-1}$$

$$= \frac{3 \times 10^3 \, (7.15 - 5)}{5} = 1.29 \text{ k}\Omega$$

also

$$R_{sc} = \frac{V_{sense}}{I_{limit}} = \frac{0.66}{0.075} = 8.8 \text{ }\Omega \tag{7-3}$$

where V_{sense} was obtained from Fig. 7-3.

The resulting circuit is shown in Fig. 7-4.

For output voltages in excess of 7 V, the circuit in Fig. 7-5 should be used. The output voltage is given by

$$V_o = \frac{R_1 + R_2}{R_2} V_{ref} \tag{7-4}$$

Resistors R_3 and R_{sc} are determined by Eqs. 7-2 and 7-3, respectively.

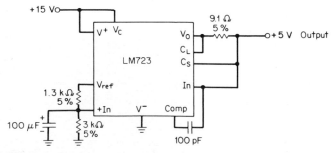

FIG. 7-4 The 5-V regulator of Example 7-1.

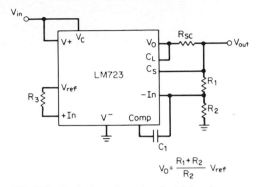

FIG. 7-5 Typical configuration for high-voltage regulator (7 V $< V_{out} <$ 37 V).

An external series pass transistor may be added to boost the output current capability. Two such circuits are shown in Fig. 7-6 using *pnp* and *npn* transistors.

The LM723 can also be used as a shunt regulator, as shown in Fig. 7-7. Special attention should be addressed to resistor R_4 to ensure that it can handle the large power dissipation required in this mode.

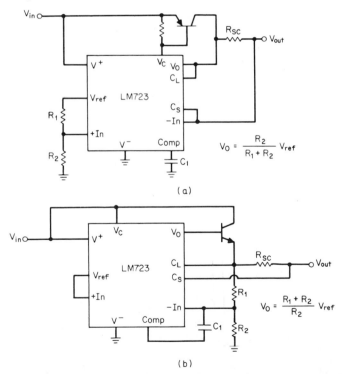

FIG. 7-6 Using external pass transistors for current boosting. (*a*) *pnp* configuration (2 V $< V_{out} <$ 7 V); (*b*) *npn* configuration (7 V $< V_{out} <$ 37 V).

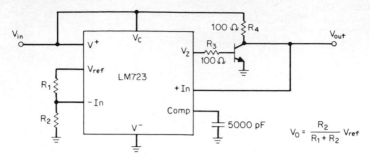

FIG. 7-7   Shunt regulator ($2\text{ V} < V_{\text{out}} < 7\text{ V}$).

Foldback current limiting can be accomplished in several different ways. Fig. 7-8 shows one of the simplest. The base of the current-limit transistor is driven by a voltage from a resistor divider from V_o to ground. The short-circuit current then becomes

$$I_{\text{sc}} = \frac{V_{\text{sense}}}{R_{\text{sc}}}\left[\frac{R_3 + R_4}{R_4}\right] \tag{7-5}$$

and the maximum current possible under normal operating conditions is

$$I_m = I_{\text{sc}} + \frac{V_o}{R_{\text{sc}}}\frac{R_3}{R_4} \tag{7-6}$$

where V_{sense} is obtained from Fig. 7-3 and is typically 0.65 V.

If a short-circuit current limit and maximum operating current are given, then the values of R_3 and R_4 must be determined. First compute

$$\alpha = \left[\frac{I_m}{I_{\text{sc}}} - 1\right]\frac{V_{\text{sense}}}{V_{\text{out}}} \tag{7-7}$$

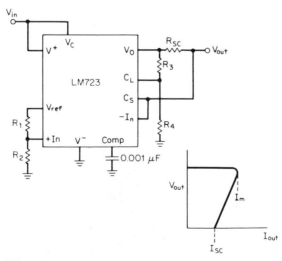

FIG. 7-8   Positive regulator with foldback current limiting ($2\text{ V} < V_{\text{out}} < 7\text{ V}$).

Choose a value for R_4. Then

$$R_3 = \frac{\alpha}{1 - \alpha} R_4 \qquad (7\text{-}8)$$

and
$$R_{sc} = \frac{V_{sense}}{I_{sc}} \left[\frac{1}{1 - \alpha} \right] \qquad (7\text{-}9)$$

The values of R_1 and R_2 are determined from Eq. 7-1.

7–2b   Series Pass Regulators

The voltage reference portion of a regulator actually sets the quality of the system. Any instability, noise, or temperature drift of the reference will show up directly at the output. The LM723 previously discussed used a zener reference source. The devices to be discussed in this chapter use zener and bandgap references, the bandgap having slightly better noise specifications.

Fixed Regulators   Positive three-terminal regulators come in three current ranges: 100-mA, 500-mA, and 1-A output capability. These families of devices all use bandgap references derived from a predictable temperature, current, and voltage relationship in a base-emitter junction.

The three-terminal voltage regulators incorporate three types of protective circuitry on the chip. The current-limit feature protects the IC against short-circuit conditions. Excessive input/output differential is protected by the safe-area limit feature, and thermal limit protects the chip against excessive junction temperature.

Fig. 7-9 shows how these protective elements combine in the chip to provide full protection.

As the current through the pass transistor Q_1 increases, so does the voltage drop across R_3. Eventually, this voltage will increase to the point where Q_2 will begin to

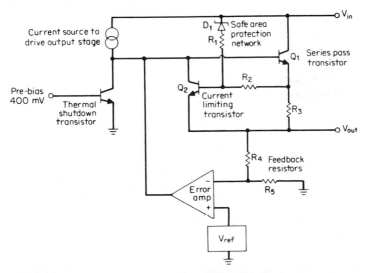

FIG. 7-9   Basic three-terminal regulator protection circuitry.

conduct and remove base drive from Q_1, consequently reducing the output current of the chip.

The safe-area protection circuitry guards against excessive collector-emitter voltage across the pass transistor (input/output differential). When the input/output voltage differential is such that the zener diode begins to conduct, current will flow through D_1, R_1, R_2, and R_3 in parallel with the Q_1, R_3 path. The voltage drop across R_3 is therefore a function of Q_1 current and the input/output voltage differential.

Thermal shutdown is accomplished by placing a temperature sensing transistor in close proximity to the series pass element. The sensing transistor is biased just below its turn-on point. As the junction temperature increases, this transistor will begin to conduct and pull base drive current away from Q_1.

Adjustable Regulators   The three-terminal regulators offer a variety of fixed output voltage options. From time to time, applications arise that require voltages not available as a standard output option. For these applications the four-terminal devices are particularly well suited. The IC is the same die as the three-terminal device, except that the two resistors on the chip that program the output voltage are not used and a fourth lead is made available where the user can apply two external resistors to program the output voltage. These are the 78G and 79G 1-A and 78MG and 79MG 1/2-A parts.

Since the four-terminal adjustable devices can be programmed for any voltage from 5 to 30 V, many users will stock these devices in lieu of the three-terminal regulators as a means of keeping inventories to a minimum and maintaining great flexibility in terms of output adjustment.

If nonstandard output voltages greater than the nominal value are required from a three-terminal device, a pair of resistors may be added as shown in Fig. 7-10. The output voltage is given by

$$V_{out} = V_{reg} \left[1 + \frac{R_2}{R_1} \right] + I_Q R_2 \qquad (7\text{-}10)$$

also

$$R_1 = \frac{V_{reg}}{I_{R_1}} \qquad (7\text{-}11)$$

and

$$R_2 = \frac{V_{out} - V_{reg}}{I_{R_1} + I_Q} \qquad (7\text{-}12)$$

It is desirable to have the current through R_1 much greater than I_Q (regulator quiescent current) to eliminate the effects of the variations that occur to I_Q with line and load

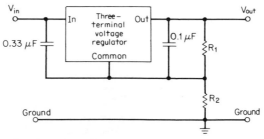

FIG. 7-10   Nonstandard boosted output for three-terminal regulator.

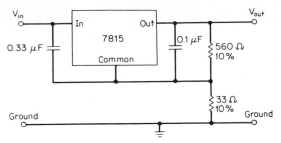

FIG. 7-11 Boosted three-terminal regulator of Example 7-2.

changes. Therefore if we set $I_{R_1} = 5I_Q$, then

$$R_2 = \frac{V_{out} - V_{reg}}{6I_Q} \qquad (7\text{-}13)$$

and

$$R_1 = \frac{V_{reg}}{5I_Q} \qquad (7\text{-}14)$$

Example 7–2   Boosting Three-Terminal Regulator Output

Obtain 16 V from a 15-V three-terminal regulator.

Solution

Assume from the data sheet $I_Q = 5.1$ mA, then

$$R_2 = \frac{V_{out} - V_{reg}}{6I_Q} = \frac{16 \text{ V} - 15 \text{ V}}{30.6 \text{ mA}} = 32.7 \ \Omega \qquad (7\text{-}13)$$

$$R_1 = \frac{V_{reg}}{5I_Q} = \frac{15 \text{ V}}{25.5 \text{ mA}} = 588 \ \Omega \qquad (7\text{-}14)$$

The resulting circuit is shown in Fig. 7-11.

The disadvantage of this approach is that the low values of resistors necessary to swamp out I_Q can cause excessive power dissipation.

A four-terminal device is shown in Fig. 7-12. To program the output an external resistor divider network is also used. The output is given by

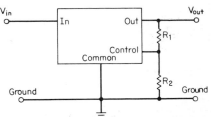

FIG. 7-12 Four-terminal adjustable regulator.

$$V_{out} = V_{control} \frac{R_1 + R_2}{R_2} \qquad (7\text{-}15)$$

where $V_{control} = 5$ V for positive four-terminal devices. The recommended current in R_2 is 1 mA. Therefore R_2 is normally 5 kΩ and

$$V_{out} - 5 \text{ V} + 0.001 \times R_1 \qquad (7\text{-}16)$$

By replacing R_1 with a potentiometer, a continuously variable output is possible with a maximum V_{out} of 30 V.

The positive device has an internal reference established by a string of bandgap references to create a nominal 5-V control voltage. The negative adjustable devices have nominal references of -2.23 V.

The output is determined by

$$V_{out} = (-2.23 \text{ V}) \frac{R_1 + R_2}{R_2} \qquad (7\text{-}17)$$

If R_2 is 2.2 kΩ, then

$$V_{out} = -(2.23 \text{ V} + 0.001R_1) \qquad (7\text{-}18)$$

The following two examples illustrate the design of positive and negative four-terminal regulators.

Example 7–3   Design of Positive Four-Terminal Regulator

Design a positive four-terminal regulator to provide a V_{out} of $+11.3$ V.
Solution
Using a 78G- or 78MG-type device with a $V_{control}$ of 5 V and a recommended current of 1 mA through R_2,

$$R_2 = 5 \text{ k}\Omega$$

and
$$R_1 = \frac{V_{out} - 5 \text{ V}}{0.001} = \frac{11.3 \text{ V} - 5 \text{ V}}{0.001} = 6.3 \text{ k}\Omega \qquad (7\text{-}16)$$

The resulting circuit is shown in Fig. 7-13.

Example 7–4   Design of Negative Four-Terminal Regulator

Design a negative four-terminal regulator to obtain a V_{out} of -17.6 V.
Solution
Using a 79G or 79MG device, the $V_{control}$ is given as -2.23 V and the recommended current through R_2 is 1 mA. Therefore

$$R_2 = 2.2 \text{ k}\Omega$$

and
$$R_1 = -\frac{V_{out} + 2.23 \text{ V}}{0.001} \qquad (7\text{-}18)$$

$$= -\frac{-17.6 \text{ V} + 2.23 \text{ V}}{0.001} = 15.4 \text{ k}\Omega$$

The resulting circuit is shown in Fig. 7-14.

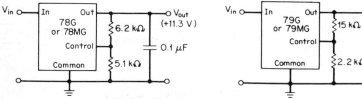

FIG. 7-13 Positive four-terminal regulator of Example 7-3.

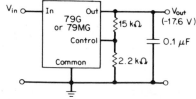

FIG. 7-14 Negative four-terminal regulator of Example 7-4.

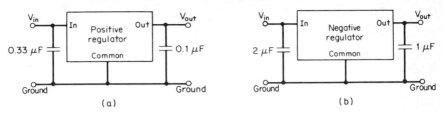

FIG. 7-15   Bypassing regulator input/output. (*a*) Positive three-terminal regulator. (*b*) Negative three-terminal regulator.

Input/Output Bypassing   Bypassing of the input and output ports of the three- and four-terminal devices is very important. Input bypass capacitors of 0.33 µF are required for the positive regulators if the device is located an appreciable distance from the main power supply. Output bypassing is not required, but a 0.1 µF will improve the transient response of the regulator, and its use is considered good design practice.

With the negative three- and four-terminal devices, the use of bypass capacitors is mandatory. An input bypass capacitor of 2 µF and an output bypass of 1 µF should be Mylar, ceramic, or tantalum and have good high-frequency characteristics. Be sure to observe proper polarity of these capacitors if electrolytics are used. Typical configurations are shown in Fig. 7-15.

Regulator output impedance is in the order of 0.1 Ω or less, but increases as a function of frequency above 10 kHz due to the gain rolloff of the error amplifier. A tantalum electrolyic bypass capacitor connected to the regulator output will maintain low impedance for frequencies up to 1 MHz. A ceramic capacitor should be placed in parallel with the tantalum capacitor for driving fast switching loads to compensate for the rising impedance of the electrolytic capacitor above 1 MHz. If switching loads are distributed over a large area, additional ceramic bypass capacitors should be located at the loads. Very large value output bypass capacitors should not be used unless adequate measures are taken to prevent the output from rising above the input, or to avoid discharging the bypass capacitor through the series pass transistor of the regulator if the input is accidentally grounded. A reverse-biased diode connected from input to output is normally sufficient to achieve this protection.

A very important precaution to be taken with the three- and four-terminal devices is the application of power. *Always* disconnect power when inserting or removing a regulator from a socket. If this is not possible, then the common terminal should be connected prior to or simultaneously with the input and disconnected simultaneously or after the output and the input. The regulator can be damaged if the input is disconnected and the charge on the output bypass capacitor is allowed to discharge through the chip via the common terminal.

Current-Source Configuration   A three-terminal regulator can be used as a constant current source with the addition of two external resistors, as shown in Fig. 7-16.

In the constant current regulator configuration, the current is determined by which three-terminal device is used and resistor R_1.

$$I_{out} = \frac{V_{out}}{R_1} + I_Q \qquad (7\text{-}19)$$

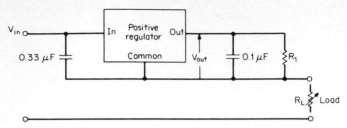

FIG. 7-16   Constant current regulator, using a three-terminal device.

Boosting Output Current   Many applications require currents in excess of the capability of the three-terminal regulators. External series pass transistors, such as those shown in Fig. 7-17a and 7-17b, can substantially boost the output current capability. Fig. 7-17a shows the basic pass configuration, while Fig. 7-17b has short-circuit protection for the pass transistor added.

The value of R_1 establishes the point at which the pass transistor will begin to conduct and is found from

$$R_1 = \frac{V_{BE \ (pass \ transistor)}}{(I_{reg} \ max) - \left(\dfrac{I_{out}}{\beta_{(pass \ transistor)}}\right)} \tag{7-20}$$

β must be >10

$$I_{out} \ max = \beta\left(I_{reg} \ max - \frac{V_{BE}}{R_1}\right)$$

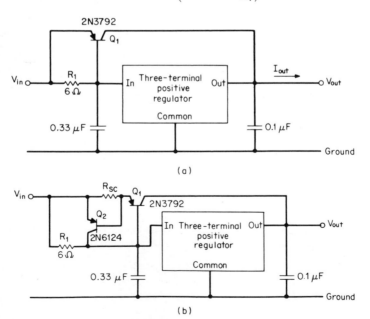

FIG. 7-17   External current boosting. (a) Series pass current boost; (b) current boost with short-circuit protection.

In Fig. 7-17*b* the short-circuit current is determined by R_{sc} and the base emitter voltage of Q_2.

Larger output currents can be achieved by putting several regulators in parallel, as shown in Fig. 7-18. When paralleling three-terminal regulators, the devices should be preselected for initial output accuracy. Ideally, matching should be within a few millivolts. If matching is not possible, the current will not be shared equally. The result will be that some regulators will operate at or near their current limit, while others are at or near their quiescent no-load levels.

The three-terminal regulators have maximum input voltage limits that must not be exceeded. If it is not possible to keep the input below the maximum rated limits, additional external circuitry may be added to protect the device. Fig. 7-19 shows a typical circuit to protect the regulator under high input voltage conditions. The zener breakdown voltage is determined by the minimum input voltage of the regulator and the V_{BE} of the transistor.

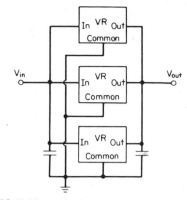

FIG. 7-18 Paralleling three-terminal regulators to increase current.

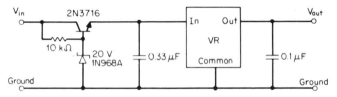

FIG. 7-19 High-voltage input protection circuit.

7–2c   Precision Reference Sources

The voltage regulators described previously all incorporated internal voltage references on the IC. The references were made as accurate and as noise free as possible. But the technology required to make precision references is different from that used in voltage regulators, and the need for truly precise references led to the development of the precision reference source.

Precision reference sources offer some unique advantages to the user. The LM336 gives a regulated output, with no external components, over the input voltage range of 3.5 to 30 V. Temperature coefficients of 10 ppm/°C make it particularly well suited as a reference source for all 8-, 10-, and 12-bit D/A converters. The output voltage of the LM336 is 2.5 V, while the AD581 provides a 10-V output over the input voltage range of 12 to 40 V. Both devices use bandgap references.

Specifying voltage references has been a point of contention for some time. The two most critical parameters are temperature coefficient (change in output voltage over temperature) and initial set-point accuracy (deviation from specified output voltage at 25°C).

Some manufacturers specify an error band over the operating temperature range of the device and guarantee the output will remain within that error band. Others establish a tolerance for the initial output at 25°C and a tolerance for that output over temperature. The latter seems to be the preferred method, since it takes a total error band and allocates part of it to initial accuracy and part to temperature drift. In some systems the set-point accuracy of the chip is not important because of board-level adjustments that are made to compensate for deviations from initial accuracy, and the system performance can be guaranteed to a tighter limit.

Many references have the capability to both sink and source current. It is important to be sure that the device chosen for a particular application has short-circuit protection to both the positive supply and the ground.

Fast turn-on is also an important feature in a precision reference. Temperature coefficient is a key parameter, therefore it is important that during turn-on (when the device is at room ambient temperature), the output settles quickly to its set point. A good rule of thumb is to look for a device that settles to ± 1 mV within 200 μs after turn-on.

Some reference sources, such as the National LM399, have an internal heater to maintain very tight temperature coefficients over a limited temperature range. Above the heater temperature the temperature coefficient of the reference gets much worse, since the heater no longer has the stabilizing effect.

A drawback to the heated reference is that the chip consumes high current, especially during turn-on. Turn-on current may be as high as 200 mA, and in normal operation quiescent current may be as much as 15 mA. The use of an on-chip heater also increases the turn-on time to as much as 2 s.

7–2d Dual Tracking Regulators

As the use of on-card regulators expands into more and more applications, the limitations of conventional three-terminal regulators become increasingly evident. Initially, these three-terminal regulators seemed to be the answer to on-card regulation. For each voltage required one simply added the proper value device. If a "nonstandard" voltage was required, then a four-terminal adjustable device was used.

The increasing complexity of analog systems has led to situations where many different functions could be included on one board. For example, it is not uncommon for a single board within a system to include many of the following functions: operational amplifiers, comparators, sense amplifiers, signal processors such as driver gates, multiplexers, switches, D/A and A/D converters, microprocessors, etc. This "card-level" LSI presents a heavy burden for the power source. Several different voltages will be required on the card, both positive and negative. A sensible approach to solving this problem is to take advantage of the various dual polarity regulators on the market today. There are several to choose from, including two from Signetics, three from National, two from Raytheon, one from Motorola. Although most offer fixed output voltages, many are easily adjustable to suit each individual application.

In many applications the performance of the system can be improved by selecting the proper dual device. For example, all of the regulators discussed here offer some degree of improvement over the common three-terminal devices (78MXX and 79MXX) in the areas of line and/or load regulation. Some dual devices, such as the LM325/6/7, go to extremes to offer 0.06% line and load specs. However, the technology of recent years has developed ICs with high tolerances to slight variations in the supply lines, as is evidenced by high power-supply rejection ratios. Therefore super line and load regulation specs may not be what is needed. Ignoring component insertion costs, a dual polarity regulator can result in substantial savings when compared with the three-terminal devices and their associated external circuitry.

TABLE 7-1 Selection Guide for Dual Regulators

Part type	Nominal output	Output adjustment range	External components required
Signetics NE5554N	±15 V	+0 to +20 V −5 to −20 V	One resistor per output
Signetics NE5553N	±12 V	+0 to +20 V −5 to −20 V	One resistor per output
National LM325N	±15 V	None	N/A
National LM326N	±12 V	None	N/A
National LM327N	+5 V, −12 V	None	N/A
Raytheon RC4194D	None	±50 mV to ±42 V	One external resistor
Raytheon RC4195NB	±15 V	None	N/A
Motorola MC1468G	±15 V	±14.5 to ±20 V	Two resistors per output

If the component count and board space required are of concern, substantial savings can be made here also. A dual polarity system using a positive and negative three-terminal regulator requires a total of six components vs. as few as three with many of the dual polarity devices described.

Programmability   Another important feature available in the dual polarity regulators is programmability. Aside from the RC4194, which requires external resistance to program any output, most of the devices discussed here are preset to some nominal pair of output voltages. Table 7-1 details the nominal outputs and the range over which they may be adjusted. Also, it is not necessary to maintain balance with some devices, such as the Signetics' family. This allows the flexibility of programming unusual voltage combinations, for example, +13.8 V and −5.2 V. Such flexible programmability also allows the user to save on inventory costs, as several combinations can now be achieved with only a single IC.

Programming the dual polarity devices is as easy as programming the gain of an op amp. Fig. 7-20 shows a typical block diagram of a dual polarity regulator. Internal resistors R_1 through R_4 provide the preset output voltages. External resistors R_A through R_D are used to shunt the internal resistors to modify the gain configurations of the operational amplifiers.

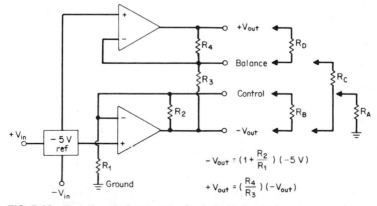

$$-V_{out} = (1 + \frac{R_2}{R_1})(-5 \text{ V})$$

$$+V_{out} = (\frac{R_4}{R_3})(-V_{out})$$

FIG. 7-20   Typical equivalent circuit of a dual polarity regulator.

TABLE 7–2 Preset Options and Minimum Shunt Values for NE5553 and NE5554 Dual Regulators

Part type	R_1	R_2	R_3	R_4	Preset outputs	R_A	R_B	R_C	R_D
NE5553	5 kΩ	7 kΩ	14 kΩ	14 kΩ	±12 V	4.36 kΩ	0	3.5 kΩ	0
NE5554	5 kΩ	10 kΩ	14 kΩ	14 kΩ	±15 V	10 kΩ	0	3.5 kΩ	0

Note: Values for R_1–R_4 are approximate internal values. Values for R_A, R_B, R_C, and R_D are minimum values for shunt resistors to prevent output from exceeding ±20 V.

The simplified block diagram of a dual polarity regulator is nothing more than a reference and a pair of op amps. To change the output voltages, shunt resistors R_A through R_D are placed across R_1 through R_4, as shown in Fig. 7-20. Table 7-2 shows values of R_1 through R_4 for the various preset options and suggests minimum shunt resistor values to prevent damage to the device by preventing the output from exceeding the recommended ±20 V.

To adjust the negative voltage, shunt R_2 to lower the negative output below its preset value and shunt R_1 to increase the negative output above its preset value.

To adjust the positive voltage, shunt R_4 to reduce the positive output below its preset value, and shunt R_3 to increase the positive output above its preset value.

Although fairly good output currents are available from the standard devices (±100 to ±300 mA), occasionally the need arises for increased current capabilities. Fig. 7-21 shows a typical approach to beefing up the current capabilities. Only one resistor and one power pass transistor is required per output.

It is also possible to increase the current capability on one side only if there happens to be a particularly heavy current imbalance required within the system. This might be expected if there is a lot of logic on the card being supplied. A heavy demand for +5 V would result.

7–2e   Low Dropout Regulators

In many applications where the typical 2 to 2.5 V input-output differentials of series-pass regulators could not be tolerated, alternate solutions were sought. Switching regulators were clearly an overkill for most of these applications. National Semiconductor recognized this need early and began with the development of the LM2930. The key element of this device is that it uses a high current *pnp* pass transistor. An input-output voltage differential of 600 mV is the result. Although only capable of 150 mA continuous and peak currents of typically 400 mA, this chip was the pioneer of low dropout devices.

Another important feature of an *pnp* pass device is the fact that the IC is automatically

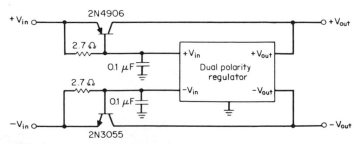

FIG. 7-21   High current dual polarity regulator.

internally protected against reverse transients. If the device is plugged in backwards into a powered system, no damage will result.

The LM2931, another *pnp* pass regulator, was designed primarily for battery powered and automotive applications. The device is available as a fixed 5 V, three terminal device and as a five terminal adjustable output with remote T^2L (2.0 V and 0.8 V thresholds) on/off control capability. Maximum input voltage is 65 V. The device boasts a reverse supply protection of -30 V.

7–2f Switched-Mode Power-Supply Controllers

Introduction to SMPS Controllers For years, analog and digital approaches have competed to solve complex electronic signal processing problems. Motor speed controls, tachometers, timers, and a host of other functions previously considered analog have now been invaded by digital solutions. So too must the linear power supply bow to a digital solution, the switched-mode power supply (SMPS). Better power efficiency, smaller size, and lighter weight have caused rethinking in the power-supply design field.

Series pass regulators have dominated the market for a long time. They are inexpensive, easy to understand, and consequently easy to use, and available from many sources. These linear regulators operate with a pass element (transistor) in a continuously conducting mode and this equates to continuous power dissipation. Another way of looking at it is to compare the total available output power to the input power. The difference between the two is the power dissipated as heat during normal operation of the IC, as shown in Fig. 7-22.

The dropout voltage of the IC establishes the minimum possible input/output voltage differential. This differential is usually about 2.5 V. More often than not, it is impossible to achieve a design where the regulator input is just above the dropout point. Operating at this point would provide maximum efficiency. Any increase above this point by the input voltage increases the power dissipation of the regulator and thus reduces the efficiency.

Switched-mode-type regulators, on the other hand, are very energy efficient. They are basically DC/DC converters where a DC voltage is applied to a chopper (generally operating above 20 kHz to minimize the size of the inductive elements). The chopped DC is then fed to a transformer where the voltage may be stepped up or down as required. (The transformer also provides isolation.) The secondary output voltage is then rectified and filtered to provide the desired DC output level. The chief function of the control circuitry is to sense the output and adjust the duty cycle of the chopper's switching transistors to keep the output constant regardless of changes in input voltage and load current.

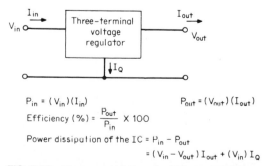

$$P_{in} = (V_{in})(I_{in}) \qquad\qquad P_{out} = (V_{out})(I_{out})$$

$$\text{Efficiency (\%)} = \frac{P_{out}}{P_{in}} \times 100$$

$$\text{Power dissipation of the IC} = P_{in} - P_{out}$$
$$= (V_{in} - V_{out})\,I_{out} + (V_{in})\,I_Q$$

FIG. 7-22 Power considerations for three-terminal devices.

Unlike their forerunners, the electromechanical vibrator, switchers incorporate no moving parts and thus have far longer lifetimes. The primary disadvantage of switchers has been cost, since they require considerably more components than do the linear or series pass systems. Consequently, the role of the switcher has been to satisfy high power requirements, while linear regulators have dominated the low power arena.

But recent advances in semiconductor technology have made switchers smaller, lighter, more efficient, and less expensive than ever before. They now can compete in some applications with linear regulators where previously they could not. Whether measured in terms of cost per watt or watt per cubic inch, switchers are winning more and more design-ins. One of the chief reasons is the advent of switched-mode power-supply control circuits. These ICs incorporate most of the control circuitry for switching regulators into a single low-cost integrated circuit.

Basic SMPS Types   There are two basic types of converters used in switched-mode power supplies: the forward converter and the flyback converter. In both types of converters an inductor is used as an energy storage element. In the forward converter the inductor is connected in series with the load. Thus energy is passed to the load and the coil during the on condition of the chopper. In the flyback converter, the coil is connected in parallel with the load. Energy is stored in the coil during the on period and transferred to the load during the off period. These are sometimes known as series or parallel converters, respectively.

Fig. 7-23 shows the basic configurations for both a forward converter system and a flyback converter system. Each approach has its advantages and disadvantages. In the forward converter, for example, the switching transistor conducts current to the load only during the on condition, and the peak value of V_{CE} that the device must withstand is only equal to the input DC voltage. Also the inductor can be smaller and the capacitor has a lower ripple current to deal with. Disadvantages include difficulty in achieving isolation and the full input DC being applied to the load in the event of a shorted switching transistor.

The advantages of the flyback converter are the opposite of the disadvantages of the forward converter. Input/output isolation is very easy to achieve by adding a secondary to the inductor. Also it is not necessary to protect the load against excess voltage in the event of a shorted switching transistor. However, there are disadvantages. The peak value

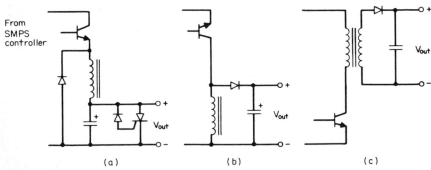

FIG. 7-23   Basic converter systems. (*a*) Forward converter with switching transistor short-circuit protection. (*b*) Flyback converter. (*c*) Flyback converter with isolation.

of V_{CE} (the switching transistor) must withstand the sum of the input DC voltage and
the output voltage ($V_{CE} = V_{i,max} + V_o$). Thus
both the inductor and diode have to pass
higher peak current and withstand higher
peak voltage. The choke is larger, and the
capacitor must pass higher ripple current.
And, of course, the higher switching voltages and currents generate increased
amounts of noise. A selection guide is shown
in Fig. 7-24.

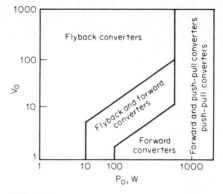

FIG. 7-24 Converter-type selector guide.

The NE5560 Control Circuit   The
role of the switched-mode power-supply
control circuit is to provide the drive
for the switching transistor in a DC/DC
converter. Ideally, the control circuit should do more than simply provide a frequency-
stable output signal with an adjustable duty cycle.

The Signetics NE5560 is a monolithic IC that incorporates all the control and housekeeping (protection) functions likely to be required in a switched-mode power supply. It
can be configured in either a forward or flyback configuration. The output stage is single
ended, with open collector and open emitter. Minimum guaranteed output current is
40 mA. The block diagram of the NE5560 is shown in Fig. 7-25.

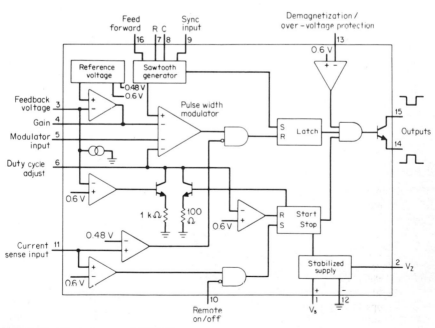

FIG. 7-25 Block diagram of NE5560.

Device Description

Sawtooth Generator

The sawtooth generator is an externally programmed oscillator using a single RC network. The frequency range is 50 Hz to 100 kHz, with a duty-cycle adjust range from 0 to 98%. The sawtooth waveform swings from a minimum of 1.1 V to a maximum of 5.6 V.

The sawtooth generator consists of a programmable current source (programmed by an external R), which determines a constant current to charge the external capacitor C. The voltage on the capacitor is sensed by a pair of comparators. A comparator with a 5.6-V threshold sets a latch and a comparator with a 1.1-V threshold resets the latch. The external synchronization feature is a TTL-compatible input, which prevents setting of the latch. The synchronization frequency must be lower than the free-running frequency of the sawtooth generator. Synchronization is desired when more than one controller is used in a system. Synchronized oscillators will minimize intermodulation RFI from the switching transistors.

Stabilized Supply

The stabilized power supply operates in either of two configurations, voltage fed or current fed. In the voltage-fed mode the maximum supply voltage is 18 V and the chip draws less than 10 mA. However, many applications for switching regulators involve very high voltages, in which case the NE5560 can be powered from a current source. For example, if the only supply available was 300 V DC, a series resistor to pin 1 could supply the circuit with 30 mA maximum. Minimum current required in the current-fed mode is 10 mA. When the current applied to pin 1 exceeds 10 mA, a string of internal zener diodes turns on, which limits the voltage on pin 1 to between 20 and 30 V. It is important not to exceed 30 mA at pin 1 in the current-fed mode.

Current Sensing

The NE5560 incorporates an overload protection feature. If too much current is drawn from a switched-mode power supply, the output transistor can be damaged. A built-in pair of comparators monitors the SMPS output current via an external sensing resistor. The threshold of the first comparator is 480 mV. When the voltage on pin 11 exceeds 480 mV, the output of the pulsewidth modulator becomes gated, shutting off the output and effectively reducing the duty cycle for one cycle. If the conditions persist, the procedure repeats itself during each subsequent cycle. However, as the duty cycle becomes smaller, the storage time of the power transistor becomes a factor and the current can again increase until it passes the threshold of the second comparator (600 mV). The second comparator immediately inhibits the output pulses to the power transistor, via the start/stop circuit.

Soft Start

The function of the start/stop circuit is to inhibit the output pulses when a fault occurs and to initiate a "soft start" action when the fault indication is removed.

When the 600-mV comparator in the overcurrent protection circuitry toggles, the external capacitor on pin 6 discharges through an internal 100 Ω resistor. When the voltage on pin 6 drops to 600 mV, the start/stop circuit resets, allowing the soft-start action to begin. The time constant determined by $T = C_{ext} \times 100\ \Omega$ determines the dead time. During soft start C_{ext} begins to charge up to V_z via a resistor divider from pin 2. As C_{ext}

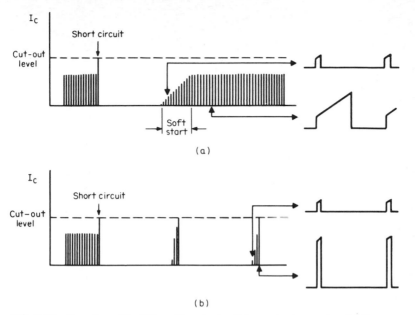

FIG. 7-26   Operation of the "hiccup" mode circuit for overload protection. (*a*) Temporary fault condition; (*b*) permanent fault.

begins charging, the voltage on pin 6 increases. This increase is detected by the pulsewidth modulator, which again begins providing pulses to the output, but with a greatly reduced duty cycle, which increases gradually as the voltage on pin 6 increases. If the fault condition still exists, the entire cycle of events repeats itself. This procedure is called the "hiccup" mode. Fig. 7-26 shows the waveform of the soft-start action for a temporary and a permanent fault condition.

Microprocessor-Compatible Remote Control

An external remote on/off pin is provided to duplicate the inhibit and soft start intentionally. The remote on/off function is a TTL-compatible input. When left floating or tied to a voltage greater than 2 V, the IC functions normally. When pulled below 0.8 V, the output is inhibited and will remain that way until the "low" is removed.

The remote on/off pin allows sequential power-up and power-down in systems where more than one SMPS is used. Also in applications where it is desired to power-down a particular portion of a system, the remote on/off function can be used, specifically in microprocessor-controlled systems.

The output stage consists of an output transistor (open emitter, open collector), a gate, and a latch. The latch is set by the flyback of the sawtooth generator and is reset by either the current-limit circuit or the pulsewidth modulator. This approach eliminates the possibility of double pulsing. Output current is 40 mA at less than 0.5 V $V_{CE(sat)}$.

A maximum duty-cycle adjust pin (pin 6) is provided to program a maximum duty factor to prevent a duty cycle in excess of 50% in forward converter configuration, which can cause saturation of the magnetics, and to limit the maximum standoff voltage rating of the switching transistor in flyback converters. A DC voltage on pin 6 sets the duty cycle. A resistor divider network from V_Z to pin 6 and pin 6 to ground sets the maximum

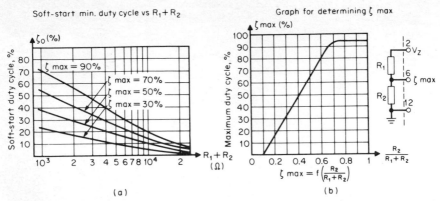

FIG. 7-27   Controlling duty cycle. (*a*) Fallback minimum duty cycle as a function of the original duty cycle and $R_1 + R_2$. (*b*) Determining the maximum duty cycle.

duty cycle and reduced duty cycle that the chip reverts to under fault conditions. Fig. 7-27 shows how pin 6 also sets the minimum duty cycle when a loop fault occurs. It is important to remember that the duty cycle must be large enough to ensure that at maximum load and minimum input voltage the resulting feedback voltage on pin 3 must exceed 0.6 V and the duty cycle must be small enough to limit the amount of energy to the output when a loop fault occurs.

The duty cycle can also be altered by the pulsewidth modulator modulation input (pin 5). This affects the duty cycle the same as pin 6, but has no influence on the start/stop circuit.

The feedforward pin on the 5560 has a provision to compensate for line variations in the input voltage. Pin 16 is used to sample the input voltage to the converter. When the voltage on pin 16 exceeds V_Z, it increases the charging current for the timing capacitor on pin 8. The higher the voltage, the larger the charging current and consequently the shorter the duty cycle. Conversely, if the voltage on the feedforward pin decreases, the duty cycle increases to compensate for this change. Ideally, the 5560 should be operated with the feedforward in its active area; i.e., between V_Z and V_{CC} such that it has plenty of headroom to compensate for variations in the line voltage, up or down. The frequency of the oscillator is not affected by increasing the voltage on pin 16, because the system also changes the upper trip level of the sawtooth waveform.

Switching power supplies also use feedback techniques to detect perturbations in the output and to internally modify the duty cycle to correct any deviations that may be detected. An error amplifier is provided on chip to sense the output voltage via a resistor divider network. The gain of the error amplifier is controlled by an external resistor between pins 3 and 4. A 0.003-μF cap is recommended from pin 4 to ground for loop compensation. Typical open-loop gain of this amplifier is 60 dB. A Bode plot is shown in Fig. 7-28.

Special protection features not found in any other SMPS controller circuits include a completely protected feedback loop. With other systems, if the loop opens, the gain increases, which will increase the duty cycle dramatically and possibly destroy the output stage. With the NE5560, if the loop opens, an internal current source pulls the pin 3 voltage up, giving the false impression that the output voltage is high. This information is delivered to the pulsewidth modulator and the duty cycle of the output is reduced to a safe level, preventing a runaway condition.

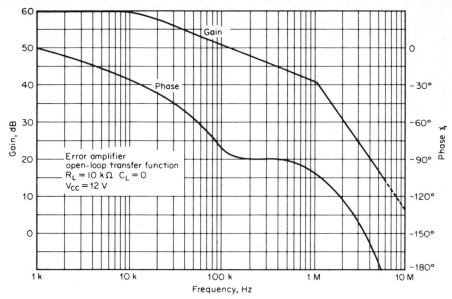

FIG. 7-28 Bode plot for NE5560 error amplifier.

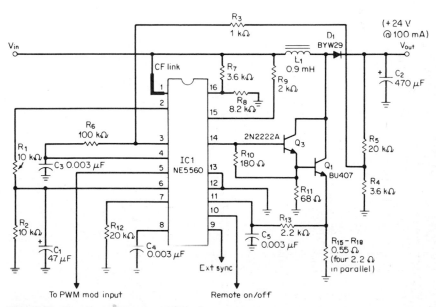

FIG. 7-29 Step-up configuration of flyback converter.

7-27

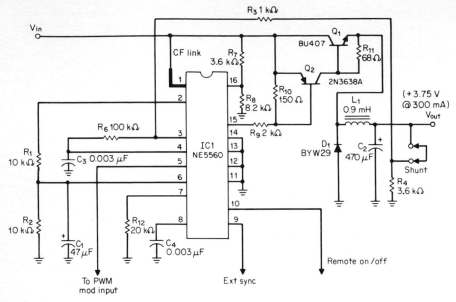

FIG. 7-30   Step-down configuration of forward converter.

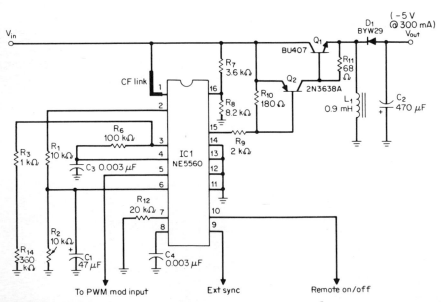

FIG. 7-31   Inverting configuration of flyback converter.

7-28

A second safety feature on the loop protects the system in the event that the feedback loop somehow gets shorted to ground. In this case, an internal comparator senses that the amplifier input (pin 6) is below 0.6 V. When this comparator is activated, the soft-start capacitor on pin 6 gets discharged through an internal 1 kΩ resistor. This short remains as long as the voltage on pin 3 remains below 0.6 V, which results in a greatly reduced duty cycle (a function of the forced voltage on pin 6), further protecting the switching supply.

Fig. 7-29 shows a typical flyback converter in the step-up configuration using external switching transistors. Output voltage V_o is 24 V and can be adjusted by changing the value of R_5. The output will change one volt for every 1-kΩ change in R_5. Switching frequency is 20 kHz, and the output current can range from 0 to 100 mA.

The circuit of Fig. 7-30 is a step-down configuration where the output voltage is approximately 3.75 V and can be increased by one volt for each additional 1-kΩ of shunt inserted. The switching frequency is also 20 kHz, and the output current capability is up to 300 mA. The circuit of Fig. 7-31 delivers -5 V at 300 mA from a positive input. The core used for L_1 in these configurations is a Ferroxcube 2213-PL00-3C8.

A variety of SMPS ICs is available. A device selection guide is provided in Table 7-7 of Sec. 7-4. An application note is available from Signetics (AN130 by Lester Hadley), which describes in great detail the design for a double-forward converter using a NE5560.

7–3 POWER DISSIPATION

7–3a Thermal Considerations

To fully utilize the various regulators available today, careful attention must be paid to ensure proper heat removal. Junction to case and junction to ambient thermal resistance and maximum operating junction temperature are key parameters to achieve efficient thermal management.

Electronic cooling devices (heat sinks, coolers, dissipators) are generally rated by a statement of their thermal resistance. That is, temperature rise per unit of heat transfer or power dissipated, expressed in units of degrees Celsius per watt. For a particular application it is necessary to determine the thermal resistance that the cooler must have in order to maintain a junction temperature that is not detrimental to the operation, performance, and reliability of the semiconductor device.

The basic relation for heat transfer or power dissipation may be stated as follows:

$$P_D = \frac{\Delta T}{\Sigma R_\theta} \tag{7-22}$$

where

P_D = the power dissipated by the semiconductor device in watts
ΔT = the temperature difference or driving potential that causes the flow of heat
ΣR_θ = the sum of the thermal resistances of the heat flow path across which ΔT exists

The above relationship may be stated in the following form:

$$P_D = \frac{T_J - T_A}{R_{\theta JC} + R_{\theta CS} + R_{\theta SA}} \tag{7-23}$$

$$P_D = \frac{T_C - T_A}{R_{\theta CS} + R_{\theta SA}} \tag{7-24}$$

$$P_D = \frac{T_S - T_A}{R_{\theta SA}} \tag{7-25}$$

where

T_J = the junction temperature, in °C (maximum is usually stated by manufacturer of semiconductor device)

T_C = case temperature of semiconductor device, in °C

T_S = temperature of mounting surface (cooler) in thermal contact with semiconductor device, in °C

T_A = ambient temperature, in °C

$R_{\theta CS}$ = thermal resistance through the interface between the semiconductor device and the surface on which it is mounted, in °C per watt

$R_{\theta SA}$ = thermal resistance from mounting surface to ambient or thermal resistance of cooler, in °C per watt

$R_{\theta JC}$ = thermal resistance from junction to case of semiconductor device, in °C per watt (usually stated by manufacturer of semiconductor device).

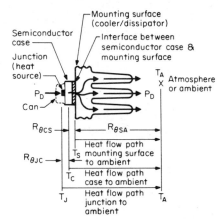

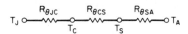

These equations are generally used to determine the required thermal resistance of the cooler ($R_{\theta SA}$) since the heat dissipation, maximum junction, and/or case temperature and ambient temperature are known or set. A typical situation is shown in Fig. 7-32. The common practice is to represent the above system with a network of series resistances as shown.

FIG. 7-32   Schematic diagram indicating location of the various heat flow paths, temperature, and thermal resistances.

7–3b   Heat-Sink Selection

Thermal data for two heat sinks are given in Fig. 7-33. The use of these curves in heat-sink selection is best demonstrated by the following examples:

Example 7–5   Heat Sink for TO-220 Case

TO-220 case style dissipating 5 W

$R_{\theta JC}$ = 3.0°C/W (from semiconductor manufacturer)
T_J max = 150°C
T_A max = 50°C

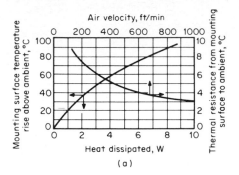

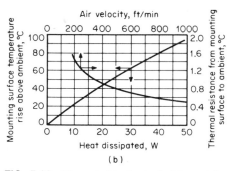

FIG. 7-33 Thermal data for typical heat sinks. (*Thermalloy, Inc., Dallas, Tex.*) (*a*) 6030B heat sink for TO-220 package. (*b*) 6500B heat sink for TO3 package.

Solution

To find the proper heat sink to keep the semiconductor junction from exceeding 150°C in a natural convection:

$$P_D = \frac{T_J - T_A}{R_{\theta JC} + R_{\theta CS} + R_{\theta SA}} \tag{7-23}$$

Solving for $R_{\theta SA}$

$$R_{\theta SA} = \frac{T_J - T_A}{P_D} - (R_{\theta JC} + R_{\theta CS})$$

Assume the device is mounted without an insulator and with Thermalcote. The thermal resistance from case to mounting surface ($R_{\theta CS}$) can be obtained from Fig. 7-34 for a TO-220 case style.

$$R_{\theta CS} = 1.0°C/W \text{ at 6-in-lb mounting-screw torque}$$

Therefore

$$R_{\theta SA} = \frac{150°C - 50°C}{5 \text{ W}} - (3.0°C/W + 1.0°C/W)$$

$$R_{\theta SA} = 16°C/W$$

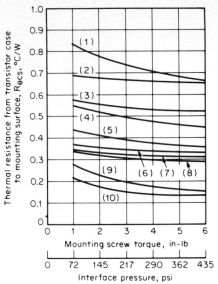

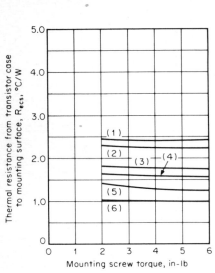

Legend:

(1) Thermalfilm II, 0.002 in thickness
(2) Thermalfilm I, 0.002 in thickness
(3) Mica, 0.003 in thickness
(4) Mica, 0.002 in thickness
(5) Hard anodized, 0.020 in thickness
(6) Bare joint—no finish

FIG. 7-34   Thermal resistance of TO-220 case (with Thermalcote compound).

Legend:

(1) Silicon rubber, 0.012 in thickness
(2) Thermalfilm II, 0.002 in thickness
(3) Thermalfilm I, 0.002 in thickness
(4) Silicon rubber, 0.006 in thickness
(5) Mica, 0.003 in thickness
(6) Mica, 0.002 in thickness
(7) Aluminum oxide, 0.062 in thickness
(8) Hard anodized, 0.020 in thickness
(9) Beryllium oxide, 0.062 in thickness
(10) Bare joint—no finish

FIG. 7-35   Thermal resistance of TO-3 case (with Thermalcote compound).

A Thermalloy P/N 6030B at 5-W power dissipation has a mounting surface temperature of 66°C above ambient. Therefore

$$R_{\theta SA} = \frac{66°C}{5\ W} = 13.2°C/W$$

which meets this requirement for natural convention.

Example 7–6   Heat Sink for TO-3 Package

TO-3 case style dissipating 30 W

$$R_{\theta JC} = 1.0°C/W \text{ (from semiconductor manufacturer)}$$
$$T_J \max = 150°C$$
$$T_A \max = 50°C$$

Solution
From the previous example

$$R_{\theta SA} = \frac{T_j - T_A}{P_D} - (R_{\theta JC} + R_{\theta CS}) \qquad (7\text{-}23)$$

Assuming the device is mounted with a hard anodized washer (0.020 in thick) and with Thermalcote, the thermal resistance from case to mounting surface ($R_{\theta CS}$) can be obtained from Fig. 7-35 for a TO-3 case style; therefore

$$R_{\theta SA} = \frac{150°C - 50°C}{30\ W} - (1.0°C/W + 0.28°C/W)$$
$$R_{\theta SA} = 2.05°C/W$$

A Thermalloy P/N 6500B at 30-W power dissipation has a mounting surface temperature of 60°C above ambient from Fig. 7-33; therefore

$$R_{\theta SA} = \frac{60°}{30\ W} = 2°C/W$$

which meets this requirement for natural convection.

Example 7–7   Extruded Heat Sink

Determine the required length of 6500B extrusion to keep junction temperature from exceeding 150°C with natural convection for 40-W dissipation in Example 7-6.

Solution

$$R_{\theta SA} = \frac{T_j - T_A}{P_D} - (R_{\theta JC} + R_{\theta CS}) \qquad (7\text{-}23)$$

$$R_{\theta SA} = \frac{150 - 50}{40} - (1.0°C/W + 0.28°C/W)$$

$$R_{\theta SA} = 1.22°C/W$$

Fig. 7-36 presents data for determining the approximate thermal resistance of an extruded heat sink based on the 3-in thermal performance. Determine the thermal resistance ratios for use with Fig. 7-36.

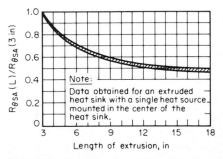

$$\frac{R_{\theta SA}(L)}{R_{\theta SA}(3\ in)} = \frac{1.22°C/W}{2.05°C/W}$$
$$= 0.60$$

FIG. 7-36   Thermal resistance for extruded heat sink.

where L (length) $\simeq 9.0$ in. Therefore, you should specify approximately 9.0 in of the 6500B extrusion.

Example 7–8   Forced Air Convection Ambient

TO-3 case style dissipating 50 W.

$$R_{\theta JC} = 1°\text{C/W (from semiconductor manufacturer)}$$
$$T_J \text{ max} = 150°\text{C}$$
$$T_A \text{ max} = 50°\text{C}$$

Determine the proper heat skin to keep the semiconductor junction from exceeding 150°C with forced convection ($\simeq$ 600 f/min).
Solution

$$R_{\theta SA} = \frac{R_J - T_A}{P_D} - (R_{\theta JC} + R_{\theta CS}) \tag{7-23}$$

Assume the device is mounted without an insulator and with Thermalcote, then from Fig. 7-35

$$R_{\theta SA} = 0.125°\text{C/W at 6-in-lb mounting-screw torque}$$

Therefore

$$R_{\theta SA} = \frac{150°\text{C} - 50°\text{C}}{50 \text{ W}} - (1°\text{C/W} + 0.125°\text{C/W})$$
$$R_{\theta SA} = 0.88°\text{C/W}$$

Using Fig. 7-33, Thermalloy P/N 6500B at 600-f/min air velocity has a $R_{\theta SA} = 0.68°\text{C/W}$, which meets this requirement.

7–4   VOLTAGE-REGULATOR SELECTION TABLES

The following tables aid the selection of the proper voltage regulator to meet the power-supply requirements. The devices are classified by part number in Table 7-3 and by output current in Table 7-4. Adjustable devices are compared in Table 7-5 and special purpose regulators are given in Table 7-6. SMPS devices are presented in Table 7-7.

TABLE 7-3 Voltage-Regulator Selection Guide By Device Number

Device	Function and polarity	Input voltage range (V)	Output voltage range (V)	Output current max (A)	Line regulation (%)	Load regulation (%)	Quiescent current (mA)	Ripple rejection min (dB)	Dropout voltage max (V)
LM104	Adjustable negative	−8 to −50	−0.015 to −40	0.012	0.1	0.1	5	60	2
LM105	Adjustable negative	8.5 to 50	4.5 to 30	0.012	0.06	0.1	2	60	3
LM109	Fixed positive	7 to 35	4.7 to 5.3	1	1	2	4.2		2
LM209	Fixed positive	7 to 35	4.7 to 5.3	1	2	2	4.2		2
LM304	Adjustable negative	−8 to −40	−0.035 to −30	0.02	0.1	0.2	5	60	3
LM305	Adjustable positive	8.5 to 40	4.5 to 30	0.012	0.06	0.1	2	60	3
LM305A	Adjustable negative	8.5 to 50	4.5 to 40	0.045	0.06	0.4	2		3
LM339	Fixed positive	7 to 35	4.8 to 5.2	1	2	2	5.2		2
LM376	Adjustable positive	9 to 40	5 to 37	0.025	0.05	0.5	2.5		3
LM723	Precision	9.5 to 40	2 to 37	0.125	0.5	0.2	4	58	3
NE550	Precision	8.5	2	0.150	0.3	0.2	3	75	3
7805	Fixed positive	7 to 35	4.8 to 5.2	1	1	1	8	62	2.5
7806	Fixed positive	8 to 35	5.75 to 6.25	1	1	1	8	59	2.5
7808	Fixed positive	10 to 35	7.7 to 8.3	1	1	1	8	56	2.5
7812	Fixed positive	14 to 35	11.5 to 12.5	1	1	1	8	55	2.5
7815	Fixed positive	17 to 35	14.4 to 15.6	1	1	1	8	54	2.5

TABLE 7-3 (Continued)

Device	Function and polarity	Input voltage range (V)	Output voltage range (V)	Output current max (A)	Line regulation (%)	Load regulation (%)	Quiescent current (mA)	Ripple rejection min (dB)	Dropout voltage max (V)
7818	Fixed positive	20 to 35	17.3 to 18.7	1	1	1	8	53	2.5
7824	Fixed positive	26 to 40	23 to 25	1	1	1	8	50	2.5
7885	Fixed positive	10.5 to 35	8.2 to 8.8	1	1	1	8	54	2.5
78G	Adjustable positive	7.5 to 40	5 to 30	1	0.75	1	5	62	3
78H05	Fixed positive	8.5 to 20	4.8 to 5.2	5	1	2	10	60	3.5
78H05A	Fixed positive	25	4.8 to 5.2	5	1	1	10	60	2.2
78H12	Fixed positive	15.5 to 20	11.5 to 12.5	5	1	2	10	60	3.5
78H15	Fixed positive		14.4 to 15.6	5	1	2	10	60	3.5
78HG	Adjustable positive	8.5 to 25	5 to 20	5	1	1	10	60	3.5
78L05	Fixed positive	7.2 to 30	4.8 to 5.2	0.1	2	1	3.8	49	2.2
78L09	Fixed positive	11.2 to 30	8.64 to 9.36	0.1	2	1	4	43	2.2
78L12	Fixed positive	14.2 to 30	11.5 to 12.5	0.1	2	1	6	42	2.2
78L15	Fixed positive	17.2 to 35	14.4 to 15.6	0.1	2	1	6	39	2.2
78L18	Fixed positive	20.2 to 35	17.3 to 18.9	0.1	2	1	3		2.2
78L24	Fixed positive	26.2 to 40	23.1 to 24.9	0.1	2	1	3		2.2

Device	Type		Column3	Column4	Column5	Column6	Column7	Column8	Column9
78L26	Fixed positive	4.3 to 30	2.5 to 2.7	0.1	2	1	5.5	43	2.2
78L62	Fixed positive	8.4 to 30	5.95 to 6.45	0.1	2	1	3.9	46	2.2
78L82	Fixed positive	10.4 to 30	7.9 to 8.5	0.1	2	1	4	44	2.2
78M05	Fixed positive	7.5 to 35	4.8 to 5.2	0.5	1	1	6	62	2.2
78M06	Fixed positive	8.5 to 35	5.75 to 6.25	0.5	1	1	6	59	2.3
78M08	Fixed positive	10.5 to 35	7.7 to 8.3	0.5	1	1	6	56	2.3
78M12	Fixed positive	14.5 to 35	11.5 to 12.5	0.5	1	1	6	55	2.3
78M15	Fixed positive	17.5 to 35	14.4 to 15.6	0.5	1	1	6	54	2.3
78M20	Fixed positive	22.5 to 40	19.0 to 21	0.5	1	1	6	53	2.3
78M24	Fixed positive	26.5 to 40	23 to 25	0.5	1	1	6	50	2.3
78MG	Adjustable positive	7.5 to 40	5 to 30	0.5	0.75	1	5	62	3
7905	Fixed negative	−7.3 to −35	−4.8 to −5.2	1	1	1	2	54	2.3
7906	Fixed negative	−8.3 to −35	−5.75 to −6.25	1	1	1	2	54	2.3
7908	Fixed negative	−10.3 to −35	−7.7 to −8.3	1	1	1	2	54	2.3
7912	Fixed negative	−14.5 to −35	−11.5 to −12.5	1	1	1	3	54	2.3
7915	Fixed negative	−17.6 to −35	−14.4 to −15.6	1	1	1	3	54	2.3
7918	Fixed negative	−20.7 to −35	−17.3 to −18.7	1	1	1	3	54	2.3
7924	Adjustable negative	−27 to −40	−23 to −25	1	1	1	3	54	2.3

TABLE 7-3 (*Continued*)

Device	Function and polarity	Input voltage range (V)	Output voltage range (V)	Output current max (A)	Line regulation (%)	Load regulation (%)	Quiescent current (mA)	Ripple rejection min (dB)	Dropout voltage max (V)
79G	Adjustable negative	−7 to −40	−2.23 to −30	1	1	1	2	50	2.3
79HG	Adjustable negative	−7 to −40	−2.25 to −24	5	1	1	5	50	2
79M05	Fixed negative	−7.5 to −35	−4.8 to −5.2	0.5	1	1	2	54	2.3
79M06	Fixed negative	−7.35 to −35	−5.75 to −6.25	0.5	1	1	2	54	2.3
79M08	Fixed negative	−9.4 to −35	−7.7 to −8.3	0.5	1	1	2	54	2.3
79M12	Fixed negative	−13.6 to −35	−11.5 to −12.5	0.5	1	1	3	54	2.3
79M15	Fixed negative	−16.7 to −35	−14.4 to −15.6	0.5	1	1	3	54	2.3
79M20	Fixed negative	−22.1 to −40	−19 to −20	0.5	1	1	3.5	54	2.3
79M24	Fixed negative	−26.1 to −40	−23 to −25	0.5	1	1	3.5	54	2.3
79MG	Adjustable negative	−7 to −30	−2.23 to −30	0.5	1	1	2.5	50	2.3
SH123	Fixed positive	20	4.8 to 5.2	3	2	2	10	60	2.2
SH223	Fixed positive	20	4.8 to 5.2	3	2	2	10	60	2.2
SH323	Fixed positive	20	4.8 to 5.2	3	2	2	10	60	2.2
SH1705	Fixed positive	25	4.8 to 5.2	5	1	1	10	60	2.2

TABLE 7-4 Fixed Voltage-Regulator Selection Guide by Output Current

Device	Output voltage (typ) (V)	Temperature*	Line regulation (max) (mV)	Load regulation (max) (mV)	Ripple rejection (min) (dB)	Quiescent current (mA)	Input voltage range (V)	Dropout voltage (typ) (V)
				Fixed positive 100 mA				
78L26	2.6	C	100	50	43	5.5	4.8 to 35	2.2
78L05	5.0	C	150	60	41	5.5	7.2 to 35	2.2
78L62	6.2	C	175	80	40	5.5	8.4 to 35	2.2
78L82	8.2	C	175	80	39	5.5	10.4 to 35	2.2
78L09	9.0	C	188	90	38	5.5	11.2 to 35	2.2
78L12	12	C	250	100	37	6.0	14.2 to 35	2.2
78L15	15	C	300	150	34	6.0	17.2 to 35	2.2
78L18	18	C	300	170	33	6.0	20.2 to 40	2.2
78L24	24	C	300	200	31	6.0	26.2 to 40	2.2
				Fixed positive 500 mA				
78M05	5.0	M	50	50	62	6.0	8.0 to 35	2.5
78M05	5.0	C	100	100	62	6.0	7.5 to 35	2.5
78M06	6.0	M	60	60	59	6.0	9.0 to 35	2.5
78M06	6.0	C	100	120	59	6.0	8.5 to 35	2.5
78M08	8.0	M	60	80	56	6.0	11 to 35	2.5
78M08	8.0	C	100	160	56	6.0	10.5 to 35	2.5
78M12	12	M	60	120	55	6.0	15 to 35	2.5
78M15	15	M	60	150	54	6.0	18 to 35	2.5
78M15	15	C	100	300	54	6.0	17.5 to 35	2.5
78M20	20	M	60	200	53	6.0	23 to 40	2.5
78M20	20	C	100	400	53	6.0	22.5 to 40	2.5
78M24	24	M	60	240	50	6.0	27 to 40	2.5
78M24	24	C	100	480	50	6.0	26.5 to 40	2.5
				Fixed negative 500 mA				
79M05	−5.0	M	50	100	54	2.0	−7.5 to −35	2.5
79M05	−5.0	C	50	100	54	2.0	−7.3 to −35	2.3
79M06	−6.0	M	60	120	54	2.0	−8.5 to −35	2.5

TABLE 7-4 (Continued)

Device	Output voltage (typ) (V)	Temperature*	Line regulation (max) (mV)	Load regulation (max) (mV)	Ripple rejection (min) (dB)	Quiescent current (mA)	Input voltage range (V)	Dropout voltage (typ) (V)
79M06	−6.0	C	60	120	54	2.0	−8.3 to −35	2.3
79M08	−8.0	M	80	160	54	2.0	−10.5 to −35	2.5
79M08	−8.0	C	80	160	54	2.0	−10.3 to −35	2.3
79M12	−12	M	80	240	54	3.0	−14.5 to −35	2.5
79M12	−12	C	80	240	54	3.0	−14.3 to −35	2.3
79M15	−15	M	80	240	54	3.0	−17.5 to −35	2.5
79M15	−15	C	80	240	54	3.0	−17.3 to −35	2.5
79M20	−20	M	80	300	54	3.5	−22.5 to −40	2.3
79M20	−20	C	80	300	54	3.5	−22.3 to −40	2.5
79M24	−24	M	80	300	54	3.5	−26.5 to −40	2.5
79M24	−24	C	80	300	54	3.5	−26.3 to −40	2.3
Fixed negative 3.0 A								
LM145	−5.0	M	15	75	66	3.0	−20	2.8
LM345	−5.0	C	25	100	66	3.0	−20	2.8
Fixed positive 1.0 A								
7805	5.0	M	50	50	68	6.0	8.0 to 35	3.0
7805	5.0	C	100	100	62	8.0	7.5 to 35	2.5
LM309	5.0	C	50	100	...	10	7.0 to 35	2.0
LM309	5.0	M	50	100	...	10	7.0 to 35	2.0
7806	6.0	M	60	60	65	6.0	9.0 to 35	3.0
7806	6.0	C	120	120	59	8.0	8.5 to 35	2.5
7808	8.0	M	80	80	62	6.0	11 to 35	3.0
7808	8.0	C	160	160	56	8.0	10.5 to 35	2.5
7885	8.5	M	85	85	60	6.0	11.5 to 35	3.0
7885	8.5	C	170	170	54	8.0	11 to 35	2.5
7812	12	M	120	120	61	6.0	15 to 35	3.0
7812	12	C	240	240	55	8.0	14.5 to 35	2.5
7815	15	M	150	150	60	6.0	18 to 35	3.0
7815	15	C	300	300	54	8.0	17.5 to 35	2.5

7818	18	M	180	180	59	6.0	21 to 35	3.0
7818	18	C	360	360	53	8.0	20.5 to 35	2.5
7824	24	M	240	240	56	6.0	27 to 40	3.0
7824	24	C	480	480	50	8.0	26.5 to 40	2.5

Fixed negative 1.0 A

7905	−5.0	M	50	50	54	2.0	−7.8 to −35	2.8
7905	−5.0	C	100	100	54	2.0	−7.3 to −35	2.3
7906	−6.0	M	60	60	54	2.0	−8.8 to −35	2.8
7906	−6.0	C	120	120	54	2.0	−8.3 to −35	2.3
7908	−8.0	M	80	80	54	2.0	−10.8 to −35	2.8
7908	−8.0	C	160	160	54	2.0	−10.2 to −35	2.3
7912	−12	M	120	120	54	3.0	−14.8 to −35	2.8
7912	−12	C	240	240	54	3.0	−14.3 to −35	2.3
7915	−15	M	150	150	54	3.0	−17.8 to −35	2.8
7915	−15	C	300	300	54	3.0	−17.3 to −35	2.3
7918	−18	M	180	180	54	3.0	−20.8 to −35	2.8
7918	−18	C	360	360	54	3.0	−20.3 to −35	2.3
7924	−24	M	240	240	54	3.0	−26.8 to −40	2.8
7924	−24	C	480	480	54	3.0	−26.3 to −40	2.3

Fixed positive 2.0 A

UA78CB	13.8	C	150	150	50	8.0	17 to 25	2.5

Fixed positive 3.0 A

LM123	5.0	M	25	100	...	20	7.5 to 20	2.5
LM223	5.0	M	25	100	...	20	7.5 to 20	2.5
LM323	5.0	C	25	100	...	20	7.5 to 20	2.5

Fixed positive 5.0 A

78H05	5.0	C,M	120	50	60	10	8.5 to 25	3.5
78H05A	5.0	C,M	25	50	60	10	7.8 to 25	2.3
78H12	12	C	...	120	60	10	15.5 to 25	3.5
78H15	15	C	30	30	60	10	18.5 to 25	...

TABLE 7-5 Adjustable Voltage-Regulator Selection Guide by Output Current

Device	Output current (mA)	Output voltage range (V)	Temperature*	Line regulation ($\%V_i$)	Load regulation ($\%V_o$)	Ripple rejection (dB)	Quiescent current (mA)	Input voltage range (V)	Dropout voltage (V)
				Positive adjustable					
LM105	12	4.5 to 30	M	0.06	0.1	1.0	2.0	8.5 to 50	3.0
LM305	12	4.5 to 30	C	0.06	0.1	1.0	2.0	8.5 to 40	3.0
LM376	25	5.0 to 37	C	0.1	0.5	1.0	2.5	9.0 to 40	3.0
LM305A	45	4.5 to 40	C	0.06	0.4	...	2.0	8.5 to 50	3.0
LM723	150	2.0 to 37	M	0.3	0.15	58	3.5	9.5 to 40	3.0
LM723	150	2.0 to 37	C	0.5	0.2	58	4.0	9.5 to 40	3.0
78MG	500	5.0 to 30	M	1.0	1.0	62	5.0	7.5 to 40	3.0
78MG	500	5.0 to 30	C	1.0	1.0	62	5.0	7.5 to 40	2.5
78G	1000	5.0 to 30	M	1.0	1.0	68	5.0	7.5 to 40	2.5
78G	1000	5.0 to 30	C	1.0	1.0	62	5.0	7.5 to 40	3.0
LM117	1500	1.2 to 37	M	0.01	0.1	66	10.0	3 to 40	1.5
LM317	1500	1.2 to 37	C	0.01	0.1	66	10.0	3 to 40	1.5
LM150	3000	1.2 to 33	M	0.01	0.3	66	5.0	35	2.2
LM350	3000	1.2 to 33	C	0.03	0.5	66	10.0	35	2.2
LM138	5000	1.2 to 32	M	0.01	0.3	60	5.0	35	2.6
LM338	5000	1.2 to 32	C	0.03	0.5	60	10.0	35	2.6
78HG	5000	5.0 to 24	C	1.0	1.0	60	10.0	8.5 to 25	3.5
LM196	10,000	1.25 to 15	M	0.01	1.0	60	10.0	20	3.5
LM396	10,000	1.25 to 15	C	0.02	1.0	60	10.0	20	2.75
				Negative adjustable					
LM104	25	−0.15 to −40	M	0.1	5 mV	1.0	5.0	−8.0 to −50	2.0
LM304	25	−0.035 to −30	C	0.1	5 mV	1.0	5.0	−8.0 to −40	2.0
79MG	500	−2.25 to −30	M	1.0	1.0	50	2.5	−7.0 to −30	2.5
79MG	500	−2.23 to −30	C	1.0	1.0	50	2.5	−7.0 to −30	2.3
79G	1000	−2.23 to −30	M	1.0	2.0	50	2.0	−7.0 to −40	2.8
79G	1000	−2.23 to −30	C	1.0	2.0	50	2.0	−7.0 to −40	2.3
LM137	1500	−1.2 to 37	M	0.02	1.0	66	3.0	−40	1.8
LM337	1500	−1.2 to 37	C	0.04	1.0	66	6.0	−40	1.8
79GH	5000	−2.25 to −24	C,M	1.0	1.0	50	5.0	−7.0 to −40	2.0

*Operating junction temperature range: C = commercial temperature range, 0°C to +125°C. M = extended military, −55°C to +150°C.

TABLE 7-6 Special Purpose Regulators

Device	Function	Input voltage range (V)	Output voltage range (V)	Output current max (A)	Line regulation (%)	Load regulation (%)	Quiescent current (mA)	Ripple rejection (dB)	Dropout voltage (V)
LM325	Dual polarity tracking	±30 V	±15 V	100 mA	0.06	0.06	8	66	2.0
LM326	Dual polarity tracking	±30 V	±12 V	100 mA	0.06	0.06	8	66	2.0
LM2930	Low dropout regulator	26 V	5 V	150 mA	0.2	1	1	66	0.6
LM2931	Low dropout regulator	26 V	3 to 24 V	150 mA	0.2	1	1	66	0.6

TABLE 7-7 SMPS IC Selection Chart

Manufacturer	Device number	Input voltage range	Output configuration	Output current	OSC frequency range	Duty-cycle adjustment range	Remote on/off	Soft start	Current-fed operation capability	Over-current sensing	Cycle-by-cycle current limit	Over-voltage sensing	Bulk line sense	Internal error amp protection
Fairchild	78540	2.5 to 40 V	Single ended	1-A peak	Not applicable	Not applicable	No	No	No	Yes	No	No	No	No
Ferranti	ZN1066	+5 V	Push-pull single	120 mA	0.5 Hz to 500 kHz	0 to 100%	Yes	Yes	No	Yes	Yes	Yes	No	No
Motorola	MC3420	10 to 30 V	Push-pull	50 mA	500 Hz to 100 kHz	0 to 100%	Yes	Accomp Ext‡	No	Accomp Ext‡	No	Accomp Ext‡	No	No
Philips	TDA1060	10 to 30 V*	Single	40 mA	50 Hz to 100 kHz	0 to 98%	Yes	Yes	Yes	Yes	Yes	Yes	No	Yes
Signetics	NE5560	10 to 30 V*	Single	40 mA	50 Hz to 100 kHz	0 to 98%	Yes	Yes	Yes	Yes	Yes	Yes	No	Yes
	NE5561	10 to 30 V*	Single	40 mA	50 Hz to 100 kHz	0 to 98%	No	Accomp Ext‡	Yes	No	No	No	No	Yes
	NE5562	10 to 30 V*	Single push-pull	40 mA 100mA	50 Hz to 500 kHz	0 to 98%	Yes	Yes	Yes	Yes†	Yes	Yes	Yes	Yes
	NE5563	10 to 30 V*	Single push-pull	40 mA 100 mA	50 Hz to 500 kHz	0 to 98%	Yes	Yes	Yes	Yes†	Yes	Yes	Yes	Yes
Silicon General	SG3524	8 to 40 V	Push-pull	100 mA	300 kHz	0 to 45%	Yes	No	No	No	No	No	No	No
	SG3525	8 to 35 V	Push-pull	200 mA	100 Hz to 400 kHz	0 to 45%	Yes	Yes	No	No	No	No	No	No
	SG3526	8 to 35 V	Push-pull single ended	100 mA	1 Hz to 400 kHz	3 to 50%	Yes	Yes	No	Yes	No	No	Yes	No
Texas Instruments	TL494	7 to 40 V	Push-pull single ended	125 mA	Not applicable	Not applicable	No	No	No	Yes	No	Yes	No	No
	TL497	7 to 40 V	Single	500 mA	Not applicable	Not applicable	Yes	Yes	No	Yes	No	Yes	No	No

*For V_{CC} greater than 18 V, device must be operated in current-fed mode.
†NE5562 and NE5563 have capability to accumulate number of overcurrent instances per unit time.
‡Accomplish externally.

REFERENCES

Fairchild: *Voltage Regulator Handbook*, Fairchild Semiconductor Corp.

Frostholm, R. C.: "Versatile Switching Regulator," *Radio Electronics Magazine*, Gernback Publishers, February 1980.

Ferroxcube: "Linear Ferrite Materials and Components," Ferroxcube Corp.

Ferroxcube: *Linear Ferrite Magnetic Design Manual*, Ferroxcube Corp.

Hayt, W. H.: *Engineering Electronics*, 3rd ed., McGraw-Hill, New York.

Millman, J., and H. Taub: *Pulse and Digital Circuits*, McGraw-Hill, New York.

Pressman, A. I.: *Switching and Linear Power Supply Power Converter Design*, Hayden, Rochelle Park, N.J., 1977.

Signetics: "Switched Mode Power Supply Control Circuits," Signetics Corp., Sunnyvale, Calif.

Thermalloy: "Semiconductor Accessories," Thermalloy Inc., Dallas, Tex.

Chapter 8

A/D AND D/A CONVERSION

Peter D. Bradshaw Director of Advanced Applications
Array Technology Inc.
San Jose, Calif.

Author was with Intersil Inc. when this chapter was written.

8–1 INTRODUCTION

"A/D conversion" is the conversion of analog signals to a digital form, and "D/A conversion" is the generation of an analog signal from digital data. The analog signals will be in the form of either a voltage or a current, while the digital signals will generally be binary and encoded either as straight binary or as binary-coded decimal (BCD) digits. For display-oriented applications, particularly for A/D conversion, the digital signals are frequently coded in a format suitable for directly driving the display, such as a seven-segment pattern or a bar-graph arrangement. Clearly, other formats, such as dot matrix, are possible, though not currently available. The relationship between the analog and digital values may be linear, but in some cases is set up to be a specific nonlinear relationship.

Such conversion steps are frequently included in more complex systems for measurement and control. A wide range of complexities are encompassed by these systems. Perhaps among the simplest would be a digital multimeter (DMM), where a few resistors and switches for measurement conditioning, and a display for output, together with a power supply or battery, are added to a suitable A/D converter, as shown in Fig. 8-1. A much larger system would be represented by a process-control system, as used in oil refineries, paper mills, power stations, etc., as shown in Fig. 8-2.

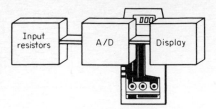

FIG. 8-1 Digital multimeter.

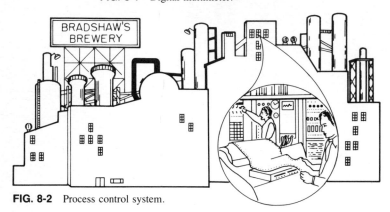

FIG. 8-2 Process control system.

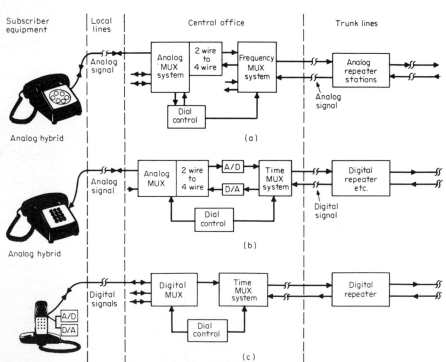

FIG. 8-3 Evolution of the telephone system. (*a*) Traditional system (analog throughout); (*b*) modern system (digital between central offices); (*c*) future system (digital between subscribers).

The remarkable capabilities and low cost of digital transmission of data are leading to changes in traditionally analog systems. Thus the "hi-fi" audio field is about to be revolutionized by the introduction of digital recording and playback equipment for consumer use, at a price that is a small fraction of that of the professional and studio systems introduced just a few years ago. Digital signal processing is used extensively in TV studio operations on video signals. The same impetus is causing extensive changes to the future form of the telephone system along the lines shown in Fig. 8-3.

In such extensive systems, the A/D and D/A converters, though key parts of the system, are imbedded in a number of other elements, such as computers, transmission and switching networks, temporary and long-term storage elements, and many more. The population and size of these other elements can greatly exceed that of the A/D/A portion, and sometimes even dwarf it. Nevertheless, usually the information in the rest of the system, especially if the main part is digital, is only as good or as usable as the precision and capabilities of this key part will allow.

Although recognizing that this handbook is not likely to be read in serial or detective novel style, the discussion of the various parts of these systems is organized into as logical a flow as possible, so that "key" portions of a subsystem are described earlier in the chapter than the subsystem itself, and those other portions that are less key (easier to design and/or less critical to the overall performance parameters) appear after the key portions. Where this "plan" does not give guidance, the order of description is based on the order in which the elements are most commonly found in practical system signal flow patterns.

8–2 PRINCIPLES OF D/A CONVERSION

The conversion of a digital signal to the corresponding analog output (voltage or current) can be achieved by several methods. We may loosely categorize these as "static" and "time division." In static methods, the digital signal closes a series of switches on a fixed basis (for a fixed digital input) to control the currents or voltages. For time-division switching, on the other hand, a switch is closed on a dynamic basis such that the average of a voltage or current over a substantial time corresponds to the desired value. Each of these techniques has certain advantages and disadvantages, and will be discussed separately. Further categorization is based on whether the constant in the switching process is a current or a voltage, whether the output signal is a voltage or a current, and also whether the reference level of the conversion is built in ("complete DAC") or an external signal that must be applied ("multiplying DAC").

8–2a Current-Switching DAC

This type of DAC is characterized by a set of current sources that can be switched into an output leg. Each current source corresponds to one bit of the digital input. A simple 4-bit DAC of this type is shown in Fig. 8-4. Several techniques are possible for setting up the relative weighting of the current sources corresponding to the various bits forming the output. The one shown in Fig. 8-4 uses a series of binary-weighted individual resistors to establish binary-weighted currents in the collectors of transistors Q_7-Q_{10}. These currents are then switched either to the output or to a supply line via the input logic control and Q_1-Q_4. The fifth current source Q_6 is used to set up an appropriate reference level via the op amp A_1, the resistor R_s, and the external reference at V_{ref}. The op amp drives the

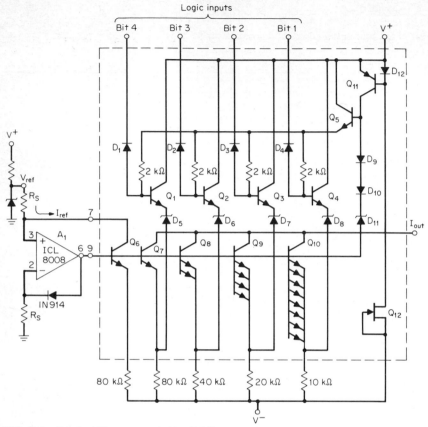

FIG. 8-4 Simple 4-bit current-switching DAC.

base bias line to achieve the necessary balance at its input, when (allowing for the V_{os} of A_1)

$$I_{Q_6} = \frac{V_{ref} + V_{osA_1}}{R_s} = 125 \ \mu A \tag{8-1}$$

The current in Q_7 will match this, since the voltages, resistors, etc., in the emitter circuits are all equal. As indicated, Q_8 has twice the emitter area of Q_7 or Q_6, while its emitter resistance is one-half of the value for the others. We may consider Q_8 and its resistor split in half when each part would carry the same current as either Q_7 or Q_6, so that the total would be twice as much. This total will not be affected by merging the two parts back together, so the collector current of Q_8 will be exactly two times that given by Eq. 8-1, or

$$I_{Q_8} = 2 \times I_{Q_7} = 2 \times I_{Q_6} = 2 \left(\frac{V_{ref} + V_{osA_1}}{R_s} \right) \tag{8-2}$$

This matching of V_{be} drops in transistors carrying ratioed currents by area ratioing is extremely important in precision analog circuitry, such as a DAC like this. Similarly, Q_9

has four times the emitter area of Q_7 and one-quarter the emitter resistor, giving precisely four times the collector current for I_{Q_9}. The binary weighting continues with Q_{10} having eight times the area and one-eighth the emitter resistor, so $I_{Q_{10}}$ is eight times I_{Q_7}. These calculations assume, of course, that the digital inputs are all low, so that Q_1–Q_4 are cut off. If any digital input is high, the corresponding transistor in the Q_1–Q_4 set diverts the resistor current from the current source elements through Q_5 to V^+. Thus the output current at I_{out} will be given by the sum of selected currents as

$$I_{out} = D_1 I_{Q_{10}} + D_2 I_{Q_9} + D_3 I_{Q_8} + D_4 I_{Q_7} \tag{8-3}$$

where D_n represents the digital input. In terms of the reference current I_{Q_6}, we may write the above equation as follows:

$$I_{out} = (D_1 2^3 + D_2 2^2 + D_3 2^1 + D_4 2^0) I_{Q_6} \tag{8-4}$$

Further substituting from Eq. 8-1 for I_{Q_6}, we find the result

$$I_{out} = (D_1 2^3 + D_2 2^2 + D_3 2^1 + D_4 2^0) \frac{V_{ref}}{R_s} \tag{8-5}$$

where we have ignored A_1's offset voltage error V_{osA_1}. Note that the maximum output current is just $(2^4 - 1) I_{Q_6}$ when all the digital inputs are low.

An alternative method of setting up the binary weighting of the currents is to use an "R-$2R$" ladder network. The principle of this ladder is shown in Fig. 8-5. Note that as shown all the ladder legs must terminate at the same voltage. If this relationship is maintained, the current will be split exactly in half at each ladder node, since inspection will show that the rest of the ladder looks like $2R$ at each node, the same value as the ladder element. The ladder can be of any length, with one node for each bit. A terminating resistor is needed at the end corresponding to the least significant bit. The current in this leg, added to the $(2^n - 1)I_{LSB}$ total of all the active legs brings the whole current drawn by the ladder to $2^n I_{LSB}$ as expected, where n is the number of bits, and I_{LSB} is the current corresponding to the least significant bit. A major advantage of this ladder system over the binary-weighted resistor arrangement of Fig. 8-4 is the small range of resistor values required (only 2:1 as compared to 2^n:1), assisting in temperature coefficient matching and also in monolithic construction.

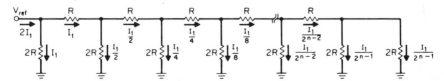

FIG. 8-5 R-$2R$ ladder network.

The requirement for equal terminating voltages on all legs, whether switched to the output or not, leads to some extra complexity in the current-source setting and switching portions of the circuit. Fig. 8-6 shows the basic circuit of a popular 8-bit DAC using this kind of R-$2R$ ladder in the current sources for the more significant bits. The least significant bits, where accuracy is less important, are set up by ratiocd transistor areas. Note that a complementary current output can be obtained very easily in this type of arrangement.

A third technique for achieving binary weighting in a current-switching DAC is shown in Fig. 8-7. Here a string of equal currents are switched into the nodes of an

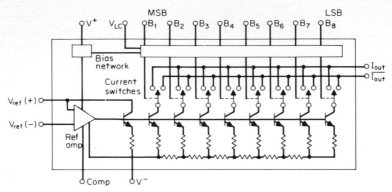

FIG. 8-6 An 8-bit DAC using an *R*-2*R* ladder.

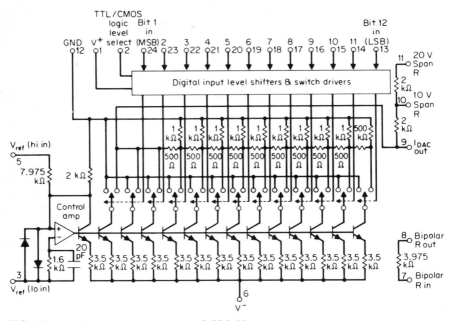

FIG. 8-7 Equal currents switched into an *R*-2*R* ladder.

R-2*R* ladder network. The output current or voltage is weighted according to the distance down the ladder to the corresponding bit current. The currents actually being switched are quite high for all bit values, ensuring a low switching time, so that the resistor values are again within a narrow range. The main disadvantages are the loss of the high output compliance (independence of the output current on the output voltage level) of the circuits of Figs. 8-4 and 8-6 and the need of a second ladder network if complementary output currents are desired. A further advantage, not very clear from the schematic, is that all the current-setting transistors are equal in size.

8-2b Voltage-Switching DAC

So far all the DAC systems we have discussed have operated by switching a current into one node or another. The next group covered basically operate by switching one node between the voltages on two other nodes. A common arrangement for this kind of DAC is shown in Fig. 8-8. This configuration can be built very efficiently with MOSFET

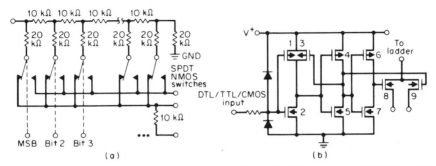

FIG. 8-8   Common voltage-switching DAC. (*a*) Ladder and switches; (*b*) typical CMOS switch and driver.

switches and CMOS logic for the drivers. This construction, combined with thin-film resistors, is eminently suited to monolithic fabrication. Note again the use of an *R*-2*R* ladder network. The particular arrangement of the elements is, in general, much more versatile than the others previously given here. In addition to the straight voltage-switched operation, shown in Fig. 8-9, it can be (and most frequently is) used to simulate the

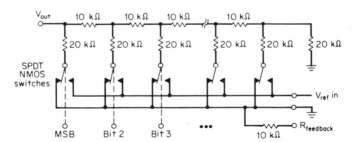

FIG. 8-9   Voltage-switched operation of the DAC of Fig. 8-8.

current-switching type of operation, as in Fig. 8-10. In fact, switch-resistance-induced nonlinearities in the connection of Fig. 8-9 severely limit its use at high accuracies. The resistance of a MOS switch is a function of the voltage between the gate and the channel, and the gate voltage will be set by the logic supply, whereas the channel voltage is that of the two nodes.

The connection of Fig. 8-10 does not suffer from this problem at all, and is that normally used in high-accuracy circuits. However, this configuration lacks output compliance, since any voltage between the output node legs will cause serious errors in the ladder currents. This will generally require the use of a fairly precise operational amplifier,

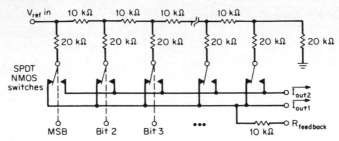

FIG. 8-10 Simulated current-switched operation of Fig. 8-8.

or some other load acting as a "virtual ground." As we shall see, this is a fairly common situation in DAC applications, in any case, so it is not a major disadvantage in the use of this type of DAC.

8–2c   Other Types of DACs

Several other types of DAC can be constructed. One of the simplest is the time-switching DAC. This could be considered a 1-bit DAC of any of the types discussed above, but where the input data are manipulated in such a way as to generate an average output of the desired level. A simple form of this DAC is shown in Fig. 8-11. The counter and register drive the set and reset of an R/S flip-flop, whose output will therefore be high for the number of clock pulses corresponding to the contents of the register, and low for the remainder of the counter's full-count cycle. The accuracy is limited only by the errors associated with the output switch and clock jitter (apart from the reference input, of course; see Sec. 8-2e). However, the filter requires either a long time constant or many poles to get the required level of ripple content to less than 1 LSB. More sophisticated designs can break the output waveform up into smaller pieces, so that the total number of clock periods for which the output is high per full count remains the same, but the low-frequency components in the output are very small, and a simple filter with a time

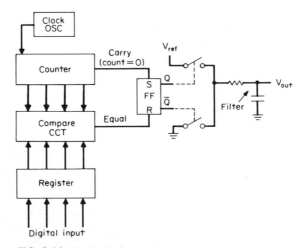

FIG. 8-11 A simple time-switching DAC.

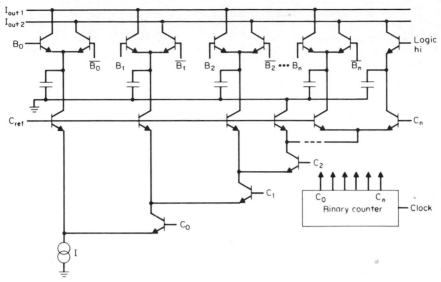

FIG. 8-12 Binary division by current switching and averaging.

constant only a little longer than the full-count time will maintain a low ripple content. Clearly, the conversion time can never fall below this full-count time.

Another time-switching technique can be used to derive a set of binary-weighted currents by time division of one constant current flow. Thus the current is fed to the MSB for one-half the time, the second bit for one-quarter the time, the third bit for one-eighth, etc., the final LSB amount remaining after the LSB itself is dumped. The current flows in each leg are then time averaged, as shown in Fig. 8-12, before being switched into the output in the conventional manner. An alternative scheme splits the current into two approximate halves, whose destinations are continuously traded (averaging out the error) at each binary stage. Both these schemes suffer from requiring a number of carefully designed averaging circuits, but clearly benefit from the output conversion time not depending at all on the timing of the switching system.

Another kind of DAC is based on the switched-capacitor techniques that are becoming widely used in filter systems. The basis of these circuits is shown in Fig. 8-13, which shows a typical stage of a switched-capacitor system. The input voltages are summed

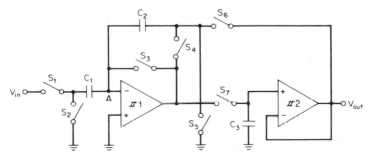

FIG. 8-13 Switched-capacitor cell.

into the output at a level depending essentially on the capacitance ratios, which can be a relatively well-controlled parameter in an IC. By setting up a suitable configuration, a group of these circuits can successively generate binary-weighted divisions of an incoming reference voltage and sum the appropriate combinations to generate the desired output. Such circuits are currently being used mainly in low-accuracy digital telephone system CODEC and filter combinations, but one can soon expect moderate-speed low-cost devices of somewhat higher accuracy to become available.

A very simple DAC can be constructed from a string of equal resistors, as shown in Fig. 8-14. Because of the large number of components needed, this type is suitable only at or below about 8 bits of resolution, and is mainly used in successive approximation converters, which are discussed below.

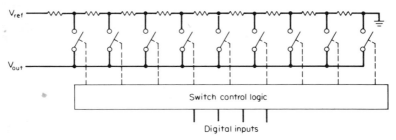

FIG. 8-14 Resistor string DAC.

8–2d Current Output vs. Voltage Output DACs

The next important division between DACs derives from whether the output signal is a voltage or a current. Clearly, the current-switching DAC of Fig. 8-4 basically has a current output, while that of Fig. 8-9 has a voltage output. The whole picture, however, is more complicated than that. If the output node of the current-switching DAC of Fig. 8-7 is left open, it will be a voltage output DAC, whereas if it is tied to a virtual ground, it will be a current output DAC. Similarly, the voltage-switching DAC in Fig. 8-8 provides a current output in Fig. 8-10. In general, a true current output DAC with good voltage compliance (e.g., Fig. 8-4 or Fig. 8-6) can become a voltage output DAC by the use of a simple load resistor, as shown in Fig. 8-15, and even one with poor output compliance (Fig. 8-10) can be converted via an op amp and feedback resistor with the circuit of Fig. 8-16.

The virtual ground at the inverting input node of the op amp ensures that the accuracy of the DAC is maintained. Precise match between the voltage output and the reference voltage input to the resistor network can be assured by including the feedback resistor in the network assembly. This is done in virtually all available parts. For high-accuracy systems, it is important to ensure that the inherent errors in the op amp are small enough (see Chap. 1). In particular, offset voltage, offset voltage drift with time and temperature, input bias current in the (potentially variable) DAC output impedance, and errors due to finite gain should all be considered.

8–2e Multiplying vs. Complete DACs

The DAC circuits shown so far all rely on some externally supplied V_{ref}, and the output voltage or current will depend proportionally on this V_{ref}. In principle, DACs could be designed to depend just as well on an I_{ref}, but normally where an actual current is the

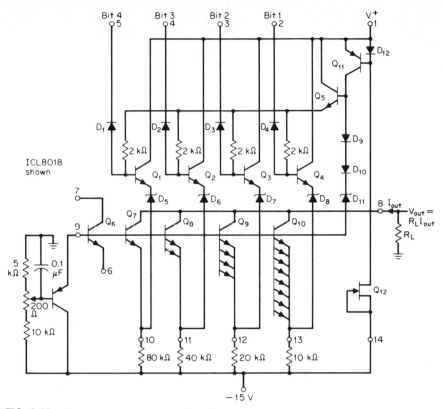

FIG. 8-15 Generating a voltage output DAC from a current output device.

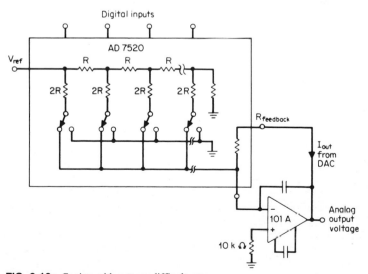

FIG. 8-16 Coping with a more difficult case.

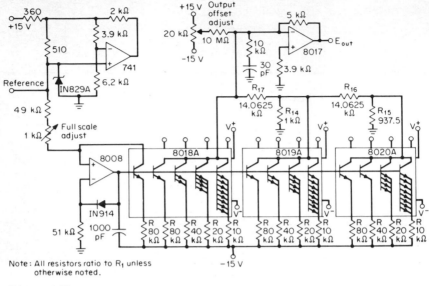

Note: All resistors ratio to R_1 unless otherwise noted.

Tolerance table

R_1	10 kΩ	0.1% ABS	R_6	20 kΩ	0.195%	R_{12}	40 kΩ	Ratio to R_{10} 1%
R_2	20 kΩ	0.0122%	R_7	40 kΩ	0.391%	R_{13}	80 kΩ	Ratio to R_{10} 1%
R_3	40 kΩ	0.0244%	R_8	80 kΩ	0.781%	R_{14}	1 kΩ	1% ABS
R_4	80 kΩ	0.0488%	R_9	80 kΩ	0.1%	R_{15}	937.5 Ω	1% ABS
R_5	10 kΩ	0.096%	R_{10}	10 kΩ	0.5% ABS	R_{16}	14.0625 kΩ	Ratio to R_{15} 1%
			R_{11}	20 kΩ	Ratio to R_{10} 1%	R_{17}	14.0625 kΩ	Ratio to R_{14} 0.1%

FIG. 8-17 A 12-bit DAC built from quad switches.

fundamental reference, a resistor, frequently combined with an op amp, is provided to derive it from a voltage, as shown in Fig. 8-17. Converters that include a reference voltage are frequently called "complete," to distinguish them from those not including a reference. The latter are normally called "multiplying" DACs, since their function can be considered that of multiplying an externally applied signal (the V_{ref} input) by a digital fraction. Although at first sight this might appear to be an outbreak of advertising jargon, there are many applications of DACs in which an external signal must inherently be used as "reference intput," so that a fixed built-in reference would be a serious inconvenience. In fact, almost all "complete" DACs bring the internal reference out, so it must be jumpered back in to a separate reference intput pin to allow use in these applications.

Most monolithic circuits with accuracy specifications of better than 10-bits do not offer a built-in reference source. The basic reason for this is that the technology requirements for good DAC performance and for good reference voltage generation are relatively incompatible. At the time of writing, no CMOS-based reference circuit comes close to offering a temperature coefficient low enough for 12-bit accuracy over any reasonable temperature range, even assuming that the now-common laser trimming of an initial (room temperature) value was done. The bipolar reference circuits currently available with good temperature coefficients are either screened by extensive (and expensive) temperature testing (e.g., the ICL8069, AD580, etc.) or use constant temperature heated substrates (e.g., the LM199, ICL8075-9, etc.). Neither of these techniques is suitable for use as part of a larger and more complex IC. Discarding a large percentage of complete

DACs (in both senses of the word!) for inadequate temperature coefficient would greatly increase the price of the remaining good parts, while the power dissipation and reliability penalties of a constant high temperature DAC are certainly unfavorable. In both cases, a separate reference circuit will continue to be the better solution until a more accurate and reproducible reference concept in IC form is realized.

8–2f Some Practical Examples of DACs

Several of the basic schematics shown earlier are those of specific practical devices. Thus Fig. 8-6 is a simplified schematic of the DAC-08 8-bit DAC, while Fig. 8-7 shows the HA572 12-bit device. Similarly, Fig. 8-8 covers the AD7520/21/30/31 family, as well as the laser-trimmed AD7541 circuit. However, many practical devices use a combination of the techniques discussed above, and others add some special features to the basic principles.

The classical module and hybrid 10- to 16-bit DAC has been constructed for many years with a circuit similar to that shown in Fig. 8-17, and this still makes the fastest devices at 12-bit or higher accuracy. Appropriately enough, perhaps, the operation is somewhat of a cross between Figs. 8-4 and 8-7. The individual quad switches each handle binary-weighted currents, but the outputs from each quad are summed in a ladder network. Minor changes in the ladder component values can generate a decimal weighting function, giving a BCD-coded DAC. The ICs required can generally be bought in matched sets to achieve specific levels of performance, and pretrimmed resistor networks are available from several vendors to match the various IC families of this form.

Two developments in DACs in recent years have led to significant increases in the accuracy available in monolithic IC form. The first of these is laser trimming of the component values. In this process, resistor values are adjusted by the metamorphosis of a portion of the thin-film resistor material by the intense heat of a focused laser beam. Usually this is done in wafer form (though occasionally in partially assembled form) and combined with the testing process to achieve a successive trim and test algorithm. In some devices, leads are broken instead, using the laser beam, or diode links are shorted by energetic pulses, etc. The AD7541 is one of the best known examples of a device built using this process, though several others are also available. The basic schematic of this part is the same as that of the AD7520/1, shown in Fig. 8-8, but 12-bit linearity is readily available, as compared with only 10-bits for the untrimmed device.

The other technique, typified by the recently announced ICL7134, is the use of a PROM in the device to control a correction system, so that the errors of the individual part can be calibrated out after assembly. The block diagram of this device is given in Fig. 8-18. Note that in addition to the basic DAC of the (CMOS) standard Fig. 8-8 type, there are two similar small DACs, one to program out the gain errors and the other to correct the nonlinearity of the main DAC. By storing these corrections for each possible value of the most significant few bits, superposition errors caused by internal resistances to the common summing points, as well as small nonlinearities in the resistor, can also be corrected. In this manner, 14-bit linearity is achieved in monolithic form. In principle, by using an EPROM (or EEPROM), recalibration to correct for long-term drift of the characteristics is possible. Another unusual feature of this DAC is the separation of the reference voltage feed to the MSB from that to the remaining portions of the ladder. This allows the generation of a bipolar output by inverting the voltage fed to the MSB with an op amp, if desired.

Several recent devices have used novel modifications of the basic DAC schematics to achieve a monotonic characteristic, though not necessarily providing nonlinearity at

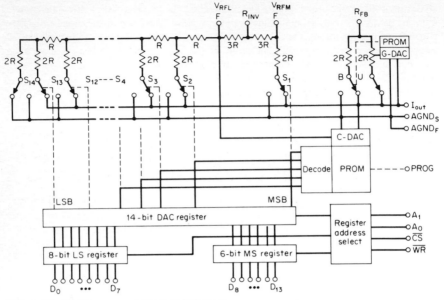

FIG. 8-18   PROM-corrected 14-bit CMOS DAC.

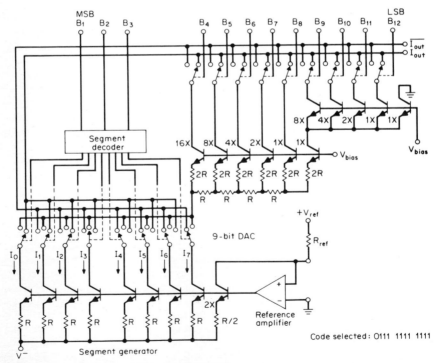

Code selected: 0111 1111 1111

FIG. 8-19   A 12-bit monotonic untrimmed DAC.

the same level of accuracy. The AM6012 achieves this by modifying the circuit of Fig. 8-5, as shown in Fig. 8-19. The most significant three bits' current sources are replaced by eight nominally equal current sources. The most significant three bits of the input are decoded to direct between 0 and 7 of these, in order, directly to the output, and the next, in order, to the ladder network for the remaining bits. Thus the remaining bits interpolate between the successive summed values of these eight current sources, thereby ensuring 12-bit monotonicity with an untrimmed process that need be no more than 9-bit accurate.

A similar device, based on the arrangement of Fig. 8-8, is shown in Fig. 8-20. Here, a voltage-switched DAC has its two inputs switched between two points on a simple resistor string, so that once again the lower bits interpolate between values set by the more significant bits (in this case, 4 bits). This circuit, the AD7546, offers 16-bit resolution with monotonicity, although the linearity can be much lower.

One class of DACs not covered so far is the "companding DAC," used in digital

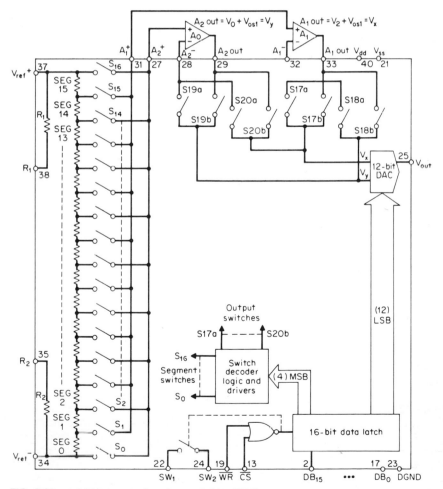

FIG. 8-20   A 16-bit monotonic voltage-switching DAC.

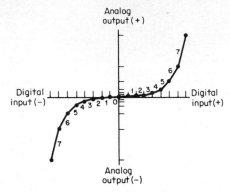

FIG. 8-21 Companding DAC input/output characteristic.

telephone systems. These DACs have an output function, as shown in Fig. 8-21, that offers much more effective resolution at low values than at high values, in return for fewer input bits. This has been found to give acceptable speech quality with a significantly lower digital data rate than would be required for a corresponding linear system. The schematic of such a circuit is shown in Fig. 8-22.

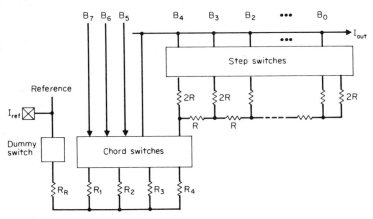

FIG. 8-22 A companding DAC circuit.

8-3 IMPORTANT DAC SPECIFICATIONS

The three key specifications for a DAC are the resolution, the linearity, and the settling time. The "resolution" indicates the number of bits of the digital input, and thus the number of distinct outputs available. The smallest increment in the analog output (on average) is then the reference voltage divided by this number, i.e., 2^n. The "linearity" specifies the deviation of the output from the ideal value, and is usually given in LSB-related terms. Note that this "ideal" is commonly expressed either as a "best straight line" or as a line between the endpoints (the output with inputs of all zeros and the output

TABLE 8–1 Device Selection Chart

D/A converters

	8 bit	10 bit	12 bit	14 bit
High speed	DAC-08	. . . .	HA562	. . . .
Low cost	AD7520/30	AD7521/31	AD7541	ICL7134

A/D converters

	8 bit	10 bit	12 bit	14 bit	16 bit
Flash (ultra high speed)	TDC1007J				
Successive approximation:					
High speed		AD573	AD572 AD574	ICL7115	
Low cost	ADC0801/4, AD570				
Integrating (binary)		. . . .	ICL7109	ICL7104-14	ICL7104-16
(decimal)		ICL7136/7 ($3\frac{1}{2}$ Dig)	ICL7135 ($4\frac{1}{2}$ Dig)		

Multiplexers

	4 channel	8 channel	16 channel
Single		IH6108/DG508	IH6116/DG506
Overload protected		IH5108	
Differential	IH6208/DG509	IH6116/DG507	
Overload protected	IH5208		

Analog switches

General purpose	DG180/191 family
High performance	IH5140/5 family
Low cost	IH5009/38 families

with inputs of all ones). The latter specification is tougher to meet, and is generally more desirable, especially since most DACs do very well at the zero output endpoint in absolute terms (see Fig. 8-23).

Another specification often confused with linearity is monotonicity, which means that the output will always increase for an increase in the digital input (not always true for nonlinear DACs). A nonlinearity of $<\frac{1}{2}$LSB guarantees monotonicity, but not vice

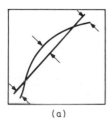

(a)

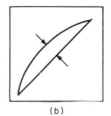

(b)

FIG. 8-23 ''Endpoint'' vs. ''best straight line'' linearity. (a) Best straight line; (b) endpoint.

versa. In theory, it would be possible to build a monotonic DAC in which all steps except one were vanishingly small, and whose nonlinearity would then be close to full scale! For some control systems and similar applications, monotonicity with reasonable nonlinearity is adequate, but in most cases a DAC's linearity should be at least close to, and preferrably better than, its resolution.

"Settling time" is often specified both for small steps and for large steps in the input data. In both cases, the settling specified should be to 1 LSB or less. Most DACs will feed some digital switching signals from the controlling elements directly to the output line through the internal capacitances, causing output "glitches" even for small-value transitions. The external digital signals themselves can also be fed directly to the output through plain capacitive coupling in the package, etc. These glitch and feedthrough problems can be overcome by using a sample-and-hold circuit on the output, at the expense of some timing and sequencing requirements.

Characteristics, such as voltage or current output and complete vs. multiplying, are important in terms of convenience and often economy, but a device that is of the "wrong" type for any given application can be fairly readily converted to the other type, as we have shown. The same is true of various other specifications that can fill up a data sheet, such as supply current, logic thresholds, and so on, as well as such interface conveniences as built-in data latches, which are very useful if microprocessor system compatibility is important. However, these are fairly obvious specifications, and no more need be said about them here.

8–3a Glossary of Terms

Bipolar Output This is a device or configuration in which the output can swing either positive or negative, depending on the digital input. Usually the input is coded in offset binary, but sometimes in two's complement form.

Digital Feedthrough Error caused by direct capacitive coupling of digital input (or output) signals to the analog output.

Feedthrough Error Error caused by capacitance coupling from V_{ref} to the output in a multiplying DAC with a zero digital input. Not the same as digital feedthrough.

Gain Ratio of a multiplying DAC's output voltage to the V_{ref} input. This corresponds to an effective scale factor error. For a complete DAC, this will often be absorbed into the reference value specification.

Monotonicity The property of always increasing the output for an increase in the digital input. This is guaranteed by $<\frac{1}{2}$ LSB linearity, but not vice versa. Can be a useful property in its own right, but do not confuse it with linearity.

Nonlinearity Error contributed by deviation of the DAC transfer function from a straight line. This straight line may be specified as "best" or "between endpoints." For a multiplying DAC this should hold true over a full V_{ref} range.

Resolution Value of the LSB. Thus a DAC with n bits of resolution has an LSB value of $V_{ref}/2^n$. Resolution does not imply linearity, and the two terms should not be confused.

Settling Time Time required for the output function of the DAC to settle, preferrably to $<\frac{1}{2}$ LSB for a given digital input change, e.g., zero to full scale.

8–4 PRINCIPLES OF A/D CONVERSION

Once again, a division can be made between A/Ds that directly operate on the input voltage and those that use time-division techniques to perform the conversion. Nearly all A/D converters are voltage input types, and the exceptions can be treated with the techniques shown above, so the distinctions made between DACs in this regard are not useful here. It is more common to split up the ADCs according to their fundamental methods of conversion, as we do here.

The major methods of A/D conversion are called "parallel" or "flash," "successive approximation," and "integrating," and the converters using them are normally so labeled. We describe each of these in turn, before discussing briefly some hybrid and also some other techniques. Before doing this, it may be useful to review some typical characteristics of these various A/D conversion techniques. Fig. 8-24 shows a plot of speed (in samples per second), accuracy (in bits), and price in three dimensions. If accuracy is looked at as a percentage, then all three scales can be considered logarithmic. It is interesting to note that none of the "zones" occupied by the three main types of converter overlap, and that the price goes up very steeply at the highest speeds. Since the capabilities do not overlap, it is not surprising that the areas of application do not either, and indeed the low-speed market is dominated by integrating converters, while the video processing, TV, and radar systems use parallel converters almost exclusively. The middle ground is occupied by the successive approximation types.

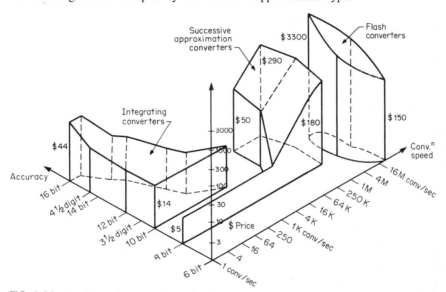

FIG. 8-24 A price/performance chart of A/D converters.

8–4a Parallel (Flash) Conversion

This technique may be thought of as a "brute force" method of A/D conversion. One comparator is provided for each possible level in the result, and the output is decoded into the appropriate binary form (Fig. 8-25). An ordinary analog comparator can be considered as a 1-bit parallel converter, and if it is a latched device, it even has a latched output! Usually these types of converter use a "pipelined" internal architecture, so that

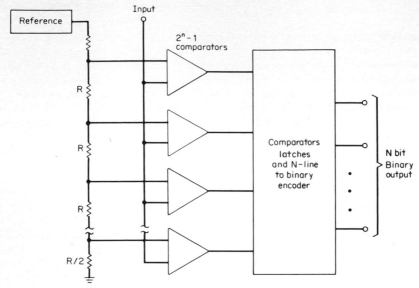

FIG. 8-25   A parallel (Flash) A/D converter.

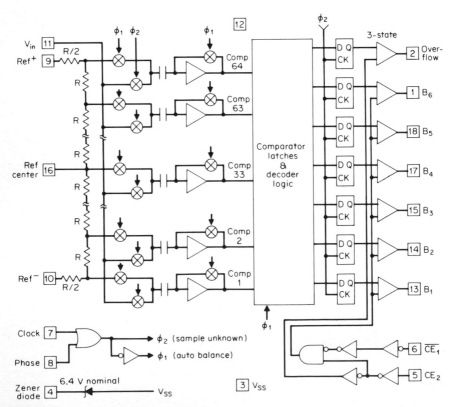

FIG. 8-26   An auto-zeroed parallel converter.

the digital processing on one result can be done in parallel with acquisition of a new input. The technique is very fast, with a new result available on every clock pulse. On the other hand, the large number of comparators needed (255 or 256 for an 8-bit converter) leads to a relatively expensive device. Historically, this type of converter has only been available as a large rack-mounted piece of equipment, but in the last few years a number of ICs have become available offering between 4 and 9-bits of resolution.

In addition to the sheer complexity of a multibit parallel converter, other limitations on its accuracy are the offset voltages in the comparators. The difference between adjacent levels can be only a few millivolts, and if the "sum-of-offsets" of a pair of adjacent comparators exceeds this, a logically inconsistent signal is fed to the logic decode tree. Even if the logic is arranged so as to accommodate this, an error must result. This problem is compounded in IC form by the necessity of keeping the comparator die area small, both to maintain speed and to restrain die size and yield. Recently, a new IC converter, using CMOS technology, that solves this problem by auto-zeroing each comparator during part of the conversion cycle has become available. As can be seen in Fig. 8-26, a capacitor is tied from the relevant reference point to the input of each comparator, whose output is tied back to its input. Thus the capacitor is charged to the sum of the reference point and the comparator offset. During the other part of the cycle, the capacitor is tied to the input signal, while the comparator loop is broken, allowing the difference between the input and the reference point to drive the comparator output. The large switching input currents in the capacitors tend to balance out to some degree, but the low effective input impedance is not usually a problem in the kind of high-speed system where these devices are used.

This type of converter is widely used in radar and TV signal processing, typically combined with a first-in–first-out (FIFO) buffer for TV frame synchronization or a fast Fourier transform arithmetic system for radar cross-section analysis, etc.

8–4b Successive Approximation Converters

The successive approximation converter is based on the use of a DAC with a logic system that drives the DAC until its output matches the input. The logic input to the DAC then corresponds to the desired digital output value. A simple block diagram of the system is shown in Fig. 8-27, where the "successive approximation register" is the logic that

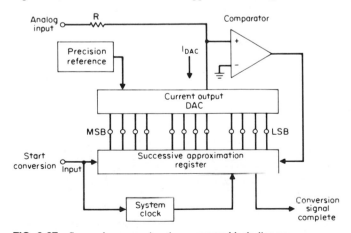

FIG. 8-27 Successive approximation converter block diagram.

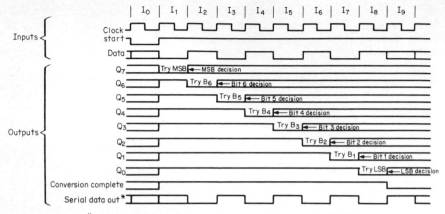

FIG. 8-28   Timing diagram of operation.

performs the required algorithm. The comparator compares the input signal and the DAC output, feeding the logical result back to the register, whose output at the end of the conversion is the desired result.

A timing-diagram presentation of the operation of the converter is shown in Fig. 8-28. As can be seen, the successive approximation register (SAR) starts by setting the MSB and clearing all the other bits. This value is one-half of full scale, and after one clock period the comparator will indicate to the SAR whether the input is above or below this value. The SAR will then keep the MSB, if above, and clear it, if below, and also set the second bit high. This process will be repeated until the LSB has been set and tested. Fig. 8-29 shows the progress of these successive approximations in analog form, and the way in which the trial value converges on the correct result.

For completeness, in addition to the "logic analyzer" and "oscilloscope" presentations, we include a "flow diagram" presentation in Fig. 8-30. If this is implemented in software, with the appropriate hardware connections, a computer can be used in place of the SAR to perform successive approximation conversions. It is fairly easy, in fact, to set up such a system to behave either as a DAC or a successive approximation ADC under software control.

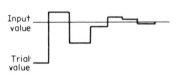

FIG. 8-29   The successive approximations.

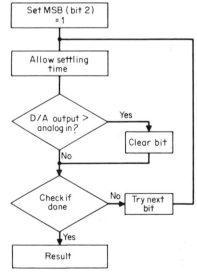

FIG. 8-30   Successive approximation flow diagram.

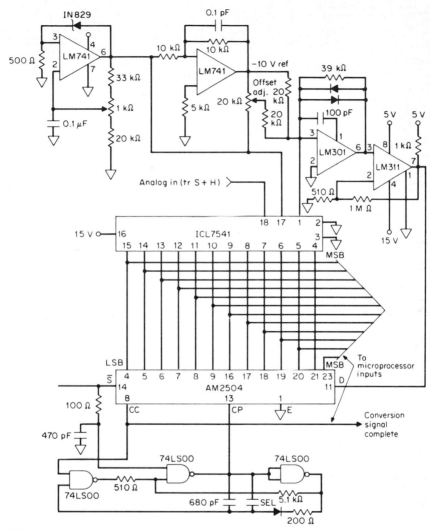

FIG. 8-31 Practical successive approximation converter.

A practical successive approximation ADC is shown in Fig. 8-31. This uses the DAC of Fig. 8-8 together with an AM2504 SAR driven by a standard type of comparator. The "feedback" resistor, used as described in Sec. 8-2d to provide voltage output, is here used as the input resistor. This ensures the same kind of accuracy as offered in this mode for the DAC itself. The only additional sources of error are in the comparator input terms, which should be specified with care. An alternative arrangement with better speed and accuracy is given in Sec. 8-4e.

8–4c The Integrating Converter

This type of converter uses a "time-ratio" system to convert the voltage ratio of the input and reference. There are several forms of integrating converter, but all of them relay on timing ramps on the output of an analog integrator driven by the respective signals. The most popular integrating converter is the "dual slope" type, for which a very basic block diagram is shown in Fig. 8-32. The input of the integrator is switched to ground, to the input signal, or to the reference, and the output of the integrator is fed through a comparator into the logic and timing system. This system also controls the input switching and the output latching, etc.

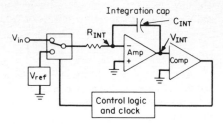

FIG. 8-32   A basic integrating converter.

The conversion takes place in three phases, as shown in Fig. 8-33. The first is a "zero phase." During this phase, the output of the integrator is zeroed, and usually the offset voltages in the system are also zeroed out at the same time, so that this phase becomes an "auto-zero phase." (The circuitry to do this is not shown in Fig. 8-32.) The second phase is the "input integrate phase" or just "integrate phase." During this phase, the input of the integrator is switched to the input signal. Thus the variable input signal is applied to the integrator for a fixed time controlled by the clock and timing system. During the third phase, the "reference integrate" or "deintegrate phase," the integrator input is fed from the fixed reference voltage for a variable time, namely the time to return the integrator output to its initial value. The time required to do this is latched in the logic as the result.

The equation governing this operation is as follows:

$$V_{int} = \frac{V_{in} \cdot N_{int}}{R_{int} \cdot C_{int}} = \frac{-V_{ref} \cdot N_{de}}{R_{int} \cdot C_{int}} \tag{8-6}$$

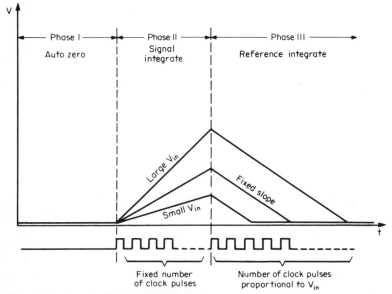

FIG. 8-33   The three phases of the conversion.

where N_{xx} refers to the number of counts in the corresponding phase of the conversion, and V_{int} is the voltage on the integrator output. This can be rearranged to give

$$N_{de} = N_{int} (V_{in}/V_{ref}) \tag{8-7}$$

Note that N_{de} will be the result.

The only sources of error in a well-designed dual slope converter are clock jitter and the reference voltage. A typical practical example of such a device is shown in Fig. 8-34, which shows the analog section of a popular $3\frac{1}{2}$-digit A/D converter. The integrator and comparator section is clear, and the switch section, although more complex, is also easy to see. An input buffer is provided, ensuring an extremely high input impedance rather than the resistor value in Fig. 8-32. The auto-zero system feeds the output of the comparator back to the integrator negative return, zeroing the input follower-buffer, the integrator, and the comparator also.

The digital section of this converter is given in Fig. 8-35. The oscillator and divider control the switch timing, with additional inputs from the polarity latch and zero-crossing detector. The latched value is decoded into 7-segment form for direct driving of a display. Other variations of this basic circuit provide binary outputs for microprocessor interfaces at up to 12-bit accuracy in one chip, or 16 bits in two-chip systems, and multiplexed BCD at up to $4\frac{1}{2}$ digits.

Several variations of this basic technique are also available in monolithic and two-chip form. The "charge-balancing" converter uses a very similar block diagram (Fig. 8-36), but the integrate and deintegrate cycles are combined in overlapped pieces. The auto-zero operation is done with a 50% duty cycle reference input applied, while the convert cycle alternates cycles with the reference applied for most of a group of counts and cycles with only a small reference time. A typical pattern would use an auto-zero cycle with four counts of "ref" followed by four counts of "no-ref," while the "convert" cycles are either one "ref" with seven "no-ref" or seven "ref" with one "no-ref" counts. Thus eight total counts and two transitions are included in each cycle. The "convert" time uses these two in such a way as to keep the integrator output close to "zero." After the main conversion is done, the accumulated result will be in units of six counts, so a short "fine" or "vernier" cycle of single "ref" or "no-ref" counts without input ensues to accommodate the residual on the integrator output and give a one-count resolution. The main advantage of this technique is that the effective integrator

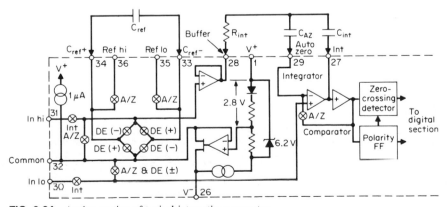

FIG. 8-34 Analog section of typical integrating converter.

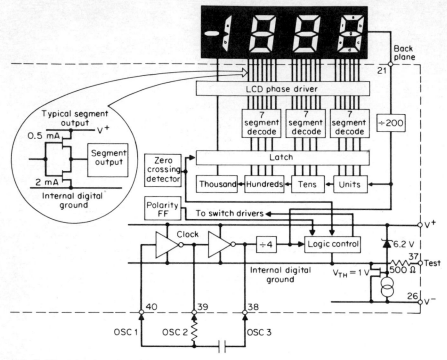

FIG. 8-35 Digital section of integrating converter.

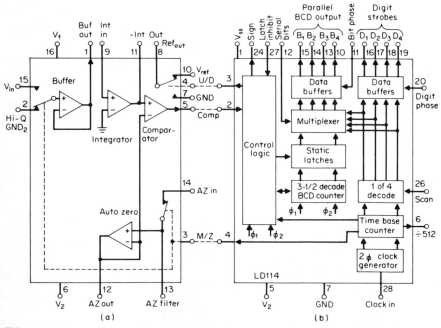

FIG. 8-36 The charge-balancing converter. (a) LD111 analog processor; (b) LD114 digital processor.

swing, as seen by the comparator, is many times greater than in a dual slope system, thus considerably easing the comparator design problems.

Analog sections for both the dual slope and the charge-balancing converter systems are available, and can be interfaced with a microprocessor to do the counting and control functions. Care is needed to ensure that the processor keeps even track of the timing, and if instruction loops are used for this, interrupts usually must be disabled during critical times. Nevertheless, the flexibility of digital processing makes this an attractive option in cases where special treatment of the data is required, and lots of otherwise idle processing time is available.

8–4d   Other Types of A/D Converters

Several other types of A/D converters are also used in certain applications. Some of them are essentially combinations of other converters, and the most important of these is probably the two-step converter shown in Fig. 8-37. This circuit is basically a successive approximation type, but using a flash converter as the comparator. The multibit result from the first conversion is subtracted from the input, using an accurate DAC, and the residual is multiplied up and fed to the second conversion. The result is a digital sum of the results of the two conversions. The accuracy is close to twice the number of bits of the flash converter alone (some overlap is necessary), while the speed is a little under half that of the flash converter, but much faster than more conventional successive approximation devices of the same accuracy. The second stage may be performed in the same flash device as the first, or a second device may be used. Converters of this type are available in hybrid and modular form, and, in principle, monolithic construction is possible.

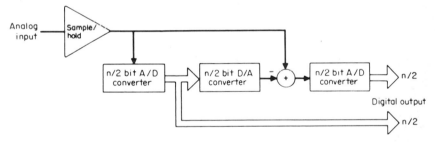

FIG. 8-37   A two-stage high-speed converter.

Another interesting A/D conversion technique is the so-called "cyclic" converter, where a single stage performs a 1-bit conversion (a comparator), subtracts the bit value, and doubles the residual, so that the next stage can repeat the process. The succession of identical stages is attractive, and by suitable configuration of each stage the result can be obtained in "Gray code" form, a very advantageous form for this type of converter, since the transfer characteristic has no discontinuities and only one bit changes between successive digital values. The required transfer characteristic is shown in Fig. 8-38, together with a simple implementation.

Another type of A/D converter once popular, although now rarely used, is the tracking converter, shown in Fig. 8-39. This is a close cousin of the successive approximation converter, but with an up-down counter instead of the successive approximation register. The counter is clocked up or down on each clock pulse, depending on the comparator output, thus causing the DAC output to follow the analog input; hence the name "tracking."

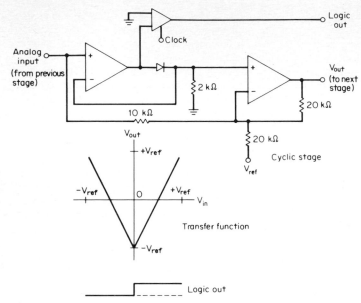

FIG. 8-38 A cyclic converter and its transfer characteristics.

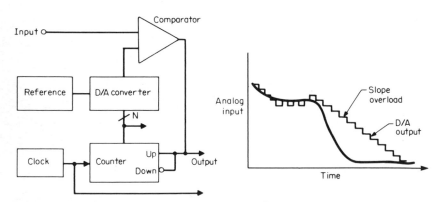

FIG. 8-39 The tracking A/D converter.

Obviously, the digital value can only follow the input at the rate of 1 LSB per clock, but if this condition is met, the value tracks the input with less than one clock delay. The software version of the successive approximation ADC can be reprogrammed to perform tracking conversions, if the input changes slowly, and successive approximation conversions on a quickly changing input, to maximize the rapidity of reading.

8—4e Some Practical A/D Converters

Once again, the circuits shown in the preceding discussion have mostly been of actual devices. Figure 8-26 is of an RCA CA3300 converter, while Fig. 8-34 shows the analog sections of the Intersil ICL7106, ICL7107, and ICL7126 converters, and a very similar arrangement is used for the ICL7109, ICL7116, ICL7117, and ICL7135 devices. The corresponding digital sections differ from that of Fig. 8-35 mainly in the details of the counting bases (binary or decimal), the control and statusing information, and the actual output formats for the other respective devices. Similarly, Fig. 8-36 shows an LD111/114 pair. Other devices use the arrangement of Fig. 8-27 with DACs discussed above, and need little further discussion. For example, the traditional hybrid or module ADC has been constructed for many years (and still is) using the DAC of Fig. 8-17 in Fig. 8-27.

Using the same order of description as in the previous sections, we should look at other practical flash converters. The devices currently available include a family from TRW offering accuracy up to 9 bits and conversion speeds up to 30 MHz! The Seimens

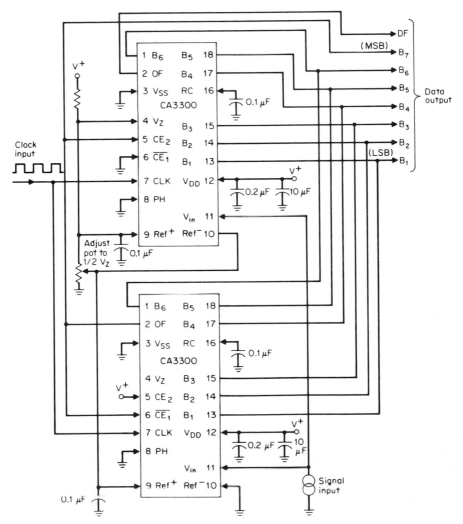

FIG. 8-40   Stacking flash converters for more resolution.

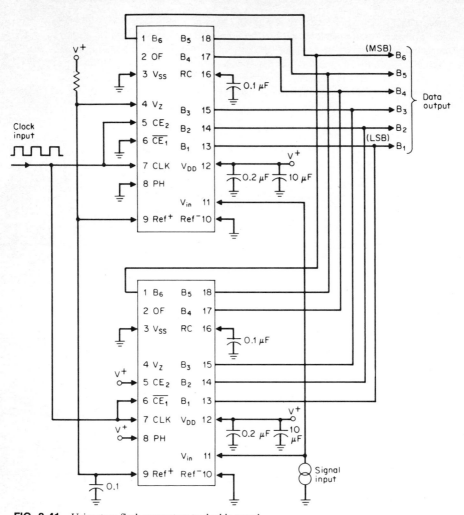

FIG. 8-41 Using two flash converters to double speed.

SDA5010 gives nominally 100-MHz conversion at 6 bits. All of these conform to the block diagram of Fig. 8-25, with such features as pipelined architecture (where digital processing is done on data latched on a previous clock pulse, possibly in several steps), and overflow outputs to allow vertical stacking of converters for more resolution. An example of the latter is shown in Fig. 8-40. An increased conversion rate is possible by running two devices with opposite phases of clock, so that two results are obtained in each full clock period, as suggested in Fig. 8-41.

Successive approximation converters exhibit a number of other practical variations that need some exploration. Already mentioned in the discussion of DACs is the system shown in Fig. 8-42, frequently used in 8-bit ADCs and often combined on a die with an input multiplexer (as in the ADC0808) or a small microcomputer (e.g., the I8022). These

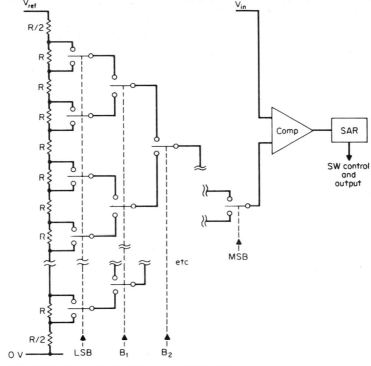

FIG. 8-42 Simple 8-bit ADC using a resistor ladder.

use a string of equal resistors similar to that used in a flash converter, but use an SAR to find the tap point that matches the input value. Although avoiding "missing codes" (see below), the linearity is usually barely adequate for a full 8-bit accuracy.

Such a long ladder makes extension to more accuracy difficult. A more versatile scheme is shown in Fig. 8-43, where a resistor ladder is combined with a set of ratioed capacitors to achieve the same result more efficiently. The input value is compared with a value derived from two points on the ladder, weighted by the capacitor ratio, in an auto-zeroed comparator, under control of the SAR. This arrangement is used in the ADC0801–4 family.

Among higher accuracy converters, the same laser trimming advances that improved DAC performance are also useful. Although the PROM system shown in Fig. 8-18 can be combined with an SAR to achieve comparable results, a novel modification of the normal operation of a successive approximation ADC has some significant advantages; the new Intersil ICL7115 is an example. A block diagram of this device is shown in Fig. 8-44. The most radical departure from the "normal" is in the DAC, which has a radix of about 1.8, rather than the usual 2.0 binary weight. This value ensures that if a marginally incorrect decision is made by the comparator, the remaining terms left for comparison can correct the error. To utilize this, each comparison value has added to it a temporary increment, which is removed after the trial. In addition, the result is built up in an adder, and is based on the actual analog value of each leg (stored in the PROM) being added

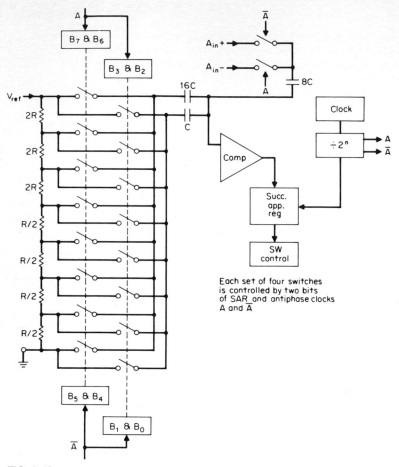

FIG. 8-43 ADC0801: A mixed capacitive-resistive successive approximation ADC.

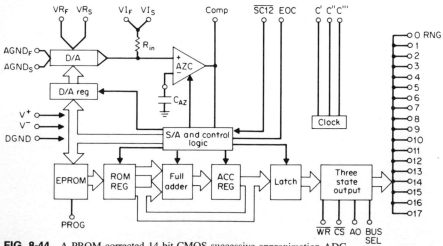

FIG. 8-44 A PROM-corrected 14-bit CMOS successive approximation ADC.

in as it is retained. The net result is a digitally calibrated ADC requiring more steps (17 possible legs, trials, and cycles are needed to ensure 14-bit accuracy), but more than compensating by allowing faster comparator speed. This device is constructed with CMOS technology and offers low power consumption and three-state outputs, together with convenient microprocessor interface, a feature of several modern ADCs, including that of Fig. 8-43.

One area that needs care in designing successive approximation ADCs is the arranging of the phase relationships for a bipolar device when using a bipolar DAC, such as the ICL7134. The MSB needs to be treated carefully, since it is opposite in effect on the output to all the other bits. Fig. 8-45 shows the correct connection, using a pair of AM25(L)03s as the SAR. These parts include an inverted MSB, which is very useful both in this case and if "two's complement" binary coding is required. Note that the

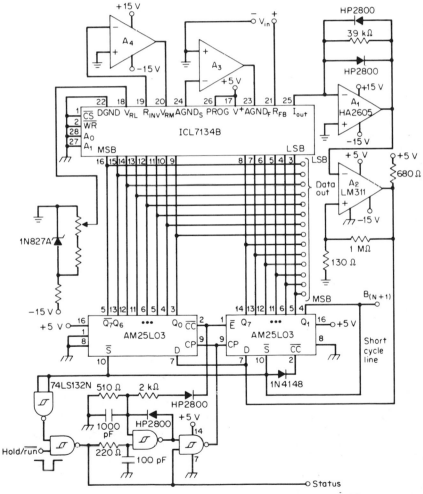

FIG. 8-45 Phasing a 14-bit successive approximation ADC using a bipolar DAC.

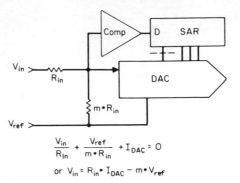

$$\frac{V_{in}}{R_{in}} + \frac{V_{ref}}{m \cdot R_{in}} + I_{DAC} = 0$$

$$\text{or } V_{in} = R_{in} \cdot I_{DAC} - m \cdot V_{ref}$$

FIG. 8-46   Offsetting a bipolar successive approximation ADC.

oscillator frequency is changed between the most significant and least significant parts of the cycle, to optimize the conversion time for the settling time of the comparator. Also a two-stage comparator is provided to generate a virtual ground at the output of the DAC, which reduces the settling time at the input of the comparator. This is a significant benefit if the total capacitance at that node is appreciable, as it typically is with CMOS DACs. The same care in phasing is needed when using a DAC with a polarity-switched output, such as the DAC-100.

A more common technique for achieving a biopolar successive approximation ADC is to offset the input with a suitable resistor that is tied to V_{ref}. The circuit of Fig. 8-46 shows this done on a converter based on a current-switching DAC. This resistor needs to match the reference and input resistors, and is usually provided in the resistor networks common in this configuration. The standard output code in this case would be "offset binary," but by inverting the MSB (the inverted value is available on most SARs) a "two's complement" output is obtained.

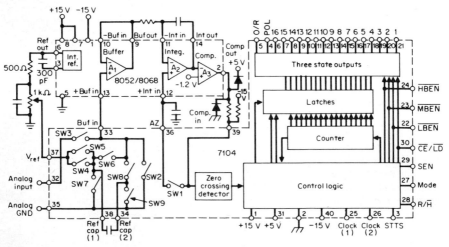

FIG. 8-47   A 16-bit integrating converter with binary output.

In integrating converters, the main variations from the devices shown in the previous section relate to the output formats, which range from 7-segment display drives through multiplexed BCD and microprocessor bus-compatible binary outputs to handshaking byte- and even bit-serial outputs. A 16-bit two-chip binary output device offering either a standard or handshake-type bus-oriented interface is shown in Fig. 8-47. This device typifies the opposite end of the integrating converter spectrum that is shown in Figs. 8-34 and 8-35. Several of these devices are now second sourced, and they dominate the digital panel- and multimeter field.

8–5 IMPORTANT A/D CONVERTER SPECIFICATIONS

The key specifications of an ADC are similar to those of a DAC, though there are differences. The resolution again indicates the number of bits (or its equivalent in digits) of the digital output, and thus the number of distinct inputs that can be distinguished. The smallest distinguishable input is then the "full-scale input" divided by this number, i.e., 2^n for a binary system or 10^n for a decimal system. Note that the usual designation of a "$3\frac{1}{2}$" or "$4\frac{1}{2}$" digit converter should strictly be "$3\frac{1}{3}$" or "$4\frac{1}{3}$," since the normal upper limit is 2000 or 20,000 counts. Also note that a polarity output is often given separately, which effectively doubles the maximum number of resolvable points, and adds the equivalent of 1 bit.

The linearity specifies the deviation of the input/output relationship from the ideal. This can be on a "best straight line" basis, and for integrating ADCs, is often based on separate "best" lines for positive and negative inputs, with a small discontinuity at zero. Differential linearity specifies how much larger or smaller than ideal the analog step is between adjacent digital values. A somewhat related specification, usually found only on successive approximation converters, relates to missing codes. If a successive approximation ADC is constructed with a nonmonotonic DAC (see Sec. 8-3), it will turn out that there will be certain output codes that will not occur. This arises if the differential nonlinearity exceeds 1 LSB.

The speed of an ADC is usually specified as the conversion speed, the maximum rate at which new results can be obtained. There are sometimes limitations on how independent these results can be. Thus, in the case of many parallel or flash converters, a separate bandwidth specification is given, being the maximum frequency that the digital output will follow, with a certain loss over the response at a much lower frequency. This loss arises if the comparator inputs cannot follow the input as fast as new conversions are obtained, so each conversion depends to some degree on the previous input. A similar effect occurs with some integrating converters, where a small residual error can be retained on the auto-zero system from the previous conversion, especially if it resulted in an overrange result. Generally successive approximation ADCs are not subject to this kind of problem, although it can arise in the sample-and-hold circuit frequently used with them (see below).

Another specification frequently misused is that of being "ratiometric." It is trivially obvious that the digital output will depend on the reference value, but a ratiometric converter is one in which the result inherently depends on the ratio of the input and the reference and is not dependent on the matching of any components. Thus the dual slope converters of Figs. 8-34 and 8-46, the flash converters of Figs. 8-25 and 8-26, and the successive approximation converter in Fig. 8-42, are all ratiometric, while the successive approximation ADCs of Figs. 8-43, 8-44, and 8-45 are strictly not, since the ratio of some capacitor or resistor pair can change the scale factor if it drifts with time, temperature, or state of the tide, etc. Similarly, the charge-balancing converter of Fig. 8-36 depends on a pair of resistors for its scale factor, so it is not ratiometric in the strict sense of the word.

Other characteristic specifications can generally be readily understood, and do not need separate discussion here. If a microprocessor bus is to be driven, it is worth checking the output drive capability of the device, since, in common with many LSI peripheral devices, the output current is often not enough to drive a large system bus, and a bus driver circuit may be called for. Convenience features such as a differential input are frequently provided on higher accuracy parts, and a differential reference input is also often available, simplifying many applications. The dynamic range of reference values that can be used is limited on the low end by the noise and offset at the input, and should be watched carefully. Input impedance and supply current are obvious, and although a remarkable range of power dissipations are covered by A/D converters, certainly 2 W to maybe 500 μW, no special mystery lies here!

8–5a Glossary of Terms

Bandwidth The maximum input frequency that can be converted with the specified reduction in digital output level (referred back to an analog level). Note that "3 dB down" is only 2-bit accuracy! This specification is usually found only on parallel (flash) converters, or other devices intended for "video" or TV processing.

Conversion Rate The maximum rate at which conversions can be performed correctly. The specification may assume that the input value is not changing radically between conversions, depending on the type and details of the converter. See Bandwidth.

Differential Nonlinearity The difference between the input ranges corresponding to adjacent outputs and the ideal value (1 LSB).

Full-Scale Input The maximum input value that will give a valid reading. For parallel and successive approximation converters, this is usually the same as the reference, while most dual slope converters have a full-scale input of twice the reference voltage.

Missing Codes The existance of theoretical output values that will not occur in the device. Usually specified as the opposite, i.e., "no missing codes," and found in specifications for successive approximation A/Ds, although some other types of converters can suffer from this problem also. Related to differential linearity, since if this is under $\frac{1}{2}$ LSB, there should be no missing codes.

Nonlinearity Error contributed by deviation of the transfer characteristic (strictly the center of each output value's input range) from the ideal straight line.

Ratiometric Strictly, a converter in which the scale factor between the input and reference values does not depend on any component values or ratios, so that it is not subject to manufacturing variation, time, or temperature drift, etc. Sometimes misuse to refer to a device in which the scale factor is (nominally) a cardinal number (e.g., 2:1, or 1:2). The attribute is useful when converting the results of certain kinds of transducers whose output is proportional to a driving voltage that can also be used as a reference for the A/D.

Resolution Value of the LSB. Thus an ADC with n bits of resolution has an LSB value of V_{fs} (the full-scale input) divided by 2^n; similarly for a decimal system. Resolution is not the same as linearity or accuracy, and they should not be confused.

Roll-Over Error   The difference between the magnitude of the readings for equal-value positive and negative inputs. This specification is common on integrating converter specifications.

8-6 OTHER CIRCUITS USED IN D/A AND A/D CONVERSION

There are a number of auxiliary circuits frequently associated with D/A converters and A/D converters, but that are not covered elsewhere in this book. The most notable of these are analog multiplexers and switches, and track-and-hold amplifiers. Other elements, such as transducer preamplifiers, programmable gain amplifiers, and high-power output amplifiers, can be constructed from devices such as op amps, chopper-stabilized amplifiers, and instrumentation amplifiers, combined with switches, and resistor networks, etc., based on standard techniques. The discussion here focuses on the uses and specific features that distinguish these elements in these applications from the standard.

8-6a Transducer Preamplifiers

The main difference here is in the increased accuracy and linearity usually demanded of digitally oriented systems. Achievement of these differences is assisted by the general improvement in op amp characteristics, notably the recent introduction of a low-cost chopper-stabilized device. A circuit using this device is shown in Fig. 8-48. This circuit also shows the use of a transducer driving voltage as a reference supply, also enhancing accuracy and stability with a ratiometric converter.

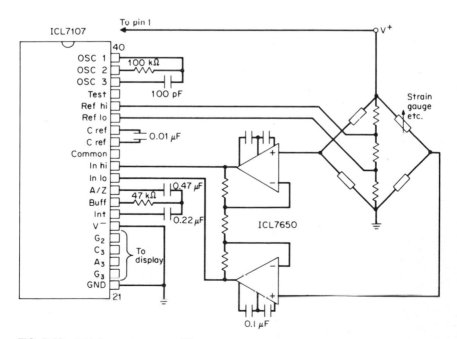

FIG. 8-48   A high-accuracy preamplifier.

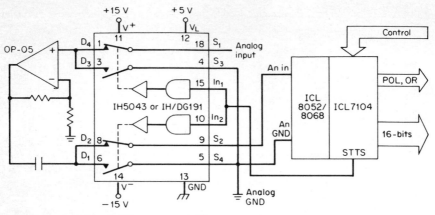

FIG. 8-49 Sample-and-difference preamplifier.

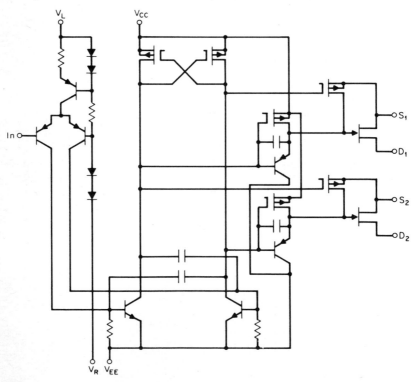

FIG. 8-50 JFET-based analog switch.

An alternative technique for removing the input offset and drift from a preamp is shown in Fig. 8-49, where the status output from an integrating converter is used to control a "sample-and-difference" amplifier. Use of a low-noise op amp ensures a good performance for the overall system, since the effective input noise of the A/D is reduced by the preamp gain. Use of this type of system in front of a multiplexer, though possible, adds some complexity to the switching, unless the system can wait for each new channel to be "sampled."

8–6b Analog Multiplexers and Switches

These devices allow circuit configurations to be altered under logic control. Two switch technologies dominate the field, the JFET (usually with bipolar driver) type, of which a typical example is shown in Fig. 8-50, and the CMOS type, an example of which is shown in Fig. 8-51. The former is normally constructed as a hybrid, leading to higher cost than the latter monolithic type, which generally offers better specifications and is now gaining in popularity. Several different switch arrangements are possible in the standard devices, and many families are pin-compatible, allowing good interchangeability.

Most multiplexers are CMOS, since hybrid construction would be very difficult for such devices. A typical device is shown in Fig. 8-52. Both single-channel and differential multiplexers with up to 16 input channels are available, most of which feature "enable" inputs as well as the "address" inputs to facilitate expansion. Recent introductions have featured a variety of "fault protection" arrangements to guard against an excessive input

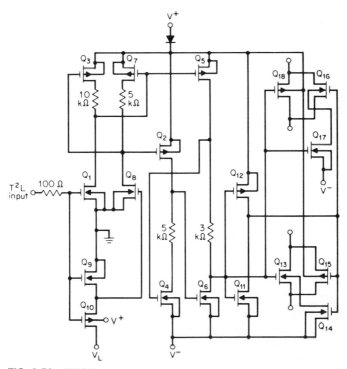

FIG. 8-51   CMOS-based analog switch.

on one channel from disturbing another input channel, and in some cases protecting the output also. One such system is shown in Fig. 8-53.

The majority of the switches and multiplexers will handle signals between +10 and −10 V, and many of them up to ±15 V. Typical ON resistances of switches are in the 30–75 Ω range, while for multiplexers these are usually 500–1000 Ω. The leakage currents are frequently well below 1 nA for each input or output, although they generally rise above at the higher temperature ranges. The logic input currents and voltages are normally tailored to interface easily with normal logic families, such as TTL and CMOS, and many families offer both high-speed and low power consumption. An important feature offered by many devices is "break-before make" operation, which ensures that two channels will not be inadvertently tied together during a transition.

Expansion of the number of channels is readily accomplished by the kind of technique shown in Fig. 8-54. However, the increase of

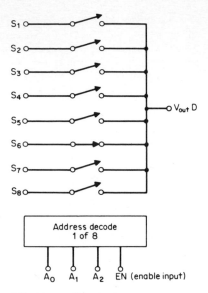

3 line binary address inputs
(1 0 1) and EN HI
Above example shows channel 6 turned on

FIG. 8-52   Analog multiplexer.

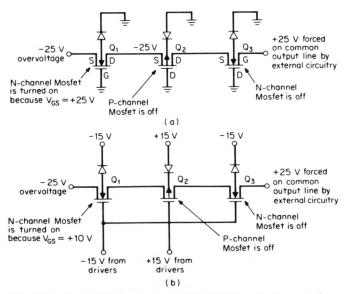

FIG. 8-53   Fault protection scheme of IH5108/5208 multiplexers. (*a*) Overvoltage with MUX power off. (*b*) Overvoltage with MUX power on.

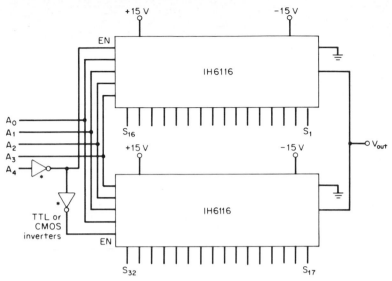

FIG. 8-54   Expanding a multiplexer.

output leakage and capacitance can make both the static and dynamic errors significant in large multichannel systems, so it is often preferrable to provide submultiplexing, as shown in Fig. 8-55. The ON resistance penalty of the analog switch can be more than compensated by the reduced leakage and capacitance on the output.

Apart from the obvious selection of input signals for processing, or the destination of output signals, analog switches and multiplexers are used to control the gain of amplifiers, construct sample-and-hold amplifiers, and for many other tasks as well. Some of these have already been shown, and others will appear in later sections of this chapter.

8–6c   Sample-and-Hold (Track-and-Hold) Amplifiers

Although strictly speaking one would expect a "Sample-and-Hold" amplifier to take a sample of an input at one instant and hold it thereafter until restrobed, as compared with a device that tracked the input until told to hold it, the former name is now so widely used to describe the latter function that rescue of rationality may be nigh impossible! Fortunately, the names describe the function accurately enough so that we can go ahead and look at the techniques for performing this task, whatever we call it. One circuit for doing this is given in Fig. 8-56, which shows a monolithic device that reflects a long-popular hybrid and module arrangement. The input amplifier drives a "hold capacitor" C_h during the "track" time so that the output amplifier tracks the input signal. Upon changing to the "hold" mode, the capacitor holds the appropriate value to sustain the correct "held" output. The input parameters are controlled by the input amplifier, while the output amplifier needs a very low input bias current to minimize the "droop rate" of the output. Its input offset voltage, however, is divided by the open-loop gain of the input amplifier, so it can be neglected. It is generally desirable to make some provision for closing a loop around the input amplifier in the hold mode to reduce the excursions required to recover from saturation on returning to the track (sample) mode. The main remaining source of error is then the charge injection into the hold capacitor on strobing

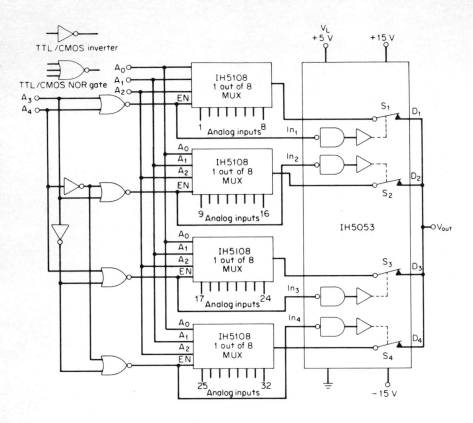

Decode truth table

A_4	A_3	A_2	A_1	A_0	On switch
0	0	0	0	0	S_1
0	0	0	0	1	S_2
0	0	0	1	0	S_3
0	0	0	1	1	S_4
0	0	1	0	0	S_5
0	0	1	0	1	S_6
0	0	1	1	0	S_7
0	0	1	1	1	S_8
0	1	0	0	0	S_9
0	1	0	0	1	S_{10}
0	1	0	1	0	S_{11}
0	1	0	1	1	S_{12}
0	1	1	0	0	S_{13}
0	1	1	0	1	S_{14}
0	1	1	1	0	S_{15}
0	1	1	1	1	S_{16}

Decode truth table

A_4	A_3	A_2	A_1	A_0	On switch
1	0	0	0	0	S_{17}
1	0	0	0	1	S_{18}
1	0	0	1	0	S_{19}
1	0	0	1	1	S_{20}
1	0	1	0	0	S_{21}
1	0	1	0	1	S_{22}
1	0	1	1	0	S_{23}
1	0	1	1	1	S_{24}
1	1	0	0	0	S_{25}
1	1	0	0	1	S_{26}
1	1	0	1	0	S_{27}
1	1	0	1	1	S_{28}
1	1	1	0	0	S_{29}
1	1	1	0	1	S_{30}
1	1	1	1	0	S_{31}
1	1	1	1	1	S_{32}

FIG. 8-55 Using a submultiplexer to reduce errors.

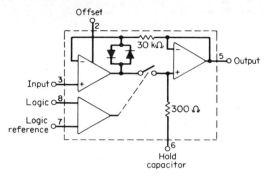

FIG. 8-56   LF198 sample (track)-and-hold amplifier.

the device, which can be reduced by careful design, especially with the use of "dummy" switches to cancel it out.

Two other configurations should be mentioned here. Fig. 8-57 shows an "inverting" track-and-hold amplifier that has the advantage of maintaining a virtual ground at the sensitive node where charge injection and leakage can cause problems, while Fig. 8-58 shows a device using the same amplifier as input and output device, which is switched between the two functions by the hold control.

One important specification of a sample (track) -and-hold amplifier is the aperture time. Naturally when the logic input demands a hold mode, the circuit takes some finite

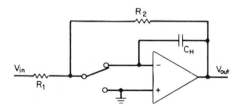

FIG. 8-57   Inverting track-and-hold amplifier.

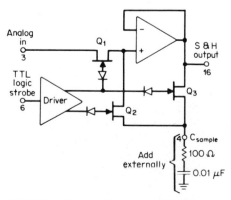

FIG. 8-58   One-amplifier track-and-hold device.

time to respond. Generally this is not too important in system operation, but when it is, it can be accommodated by sending the signal early or perhaps by delaying the analog input slightly. However, variation in this time is a real problem in some applications, and this "aperture jitter" needs to be carefully controlled in these cases. Particular attention should be paid to variation of the delay with signal level, as this can lead to a "skew" in the results.

The applications of these devices are many and varied, though most fit into two categories. The first is to be used in front of successive approximation A/D converters. The usefulness of a track-and-hold in front of a successive approximation ADC is illus-

trated in Fig. 8-59, which shows several possible input waveforms and the trial value, all of which will lead to the same digital value (see Fig. 8-29 and Sec. 8-4b). The digital result does correspond to the analog input value at some time during the conversion, but this time is not well defined, which can be a serious problem in many waveform analysis systems. By using a sample-and-hold (track-and-hold), the input waveform is held steady during the conversion, and the time is controlled by the beginning of the hold mode.

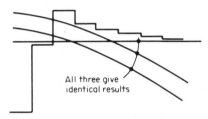

FIG. 8-59   The need for a sample (track)-and-hold with a successive approximation A/D converter.

Another common use is as an output device on D/A converters. Many DACs generate output glitches when the digital input changes, and strobing the hold mode during the

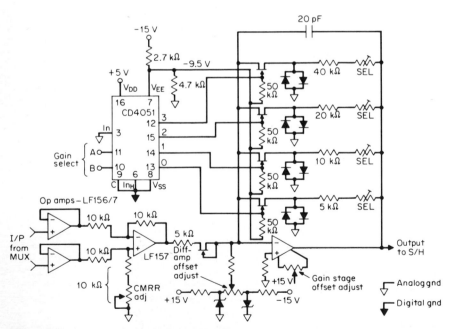

FIG. 8-60   Traditional programmable gain amplifier.

transitions will eliminate this. In systems with many analog outputs, one DAC with an analog multiplexer feeding a number of sample-and-hold (track-and-hold) devices can be more economical than individual DACs.

8–6d Programmable Gain Amplifiers

Programmable gain amplifiers are used to precondition the signal before feeding it to an ADC when the required gain is not precisely determined beforehand or a wider dynamic range than available from the ADC is desired. The traditional configuration, shown in Fig. 8-60, is just a standard inverting amplifier (often preceded by an instrumentation amplifier) whose gain is switched by means of an analog switch or multiplexer between various values set by a network of resistors.

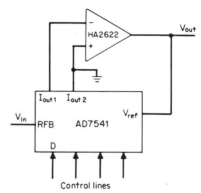

A more interesting configuration of a device to perform the same task is presented in Fig. 8 61. This one uses a CMOS DAC as a feedback element around an op amp. The gain is programmed by the digital input to the DAC, and the accuracy is controlled by the accuracy of the DAC at low gains. Note that at high gains the gain accuracy degrades, since 1-LSB error is a larger proportion of the fedback amount, if that amount is small, than at low gains, where it is large.

FIG. 8-61 DAC-based programmable gain amplifier.

8–7 COMPLETE DATA-ACQUISITION SYSTEMS

The growth of microcomputer systems has led to a surge in the construction of data-acquisition systems on a single board, which was specifically designed so as to fit into, and interface with, the more popular microcomputer backplanes. For example, boards of this type are available for, among others, the STD bus, the multibus, and the S100 bus, as well as the LSI-11 and PDP-11 minicomputer buses. These boards are generally constructed along the lines shown in Fig. 8-62 and incorporate both full A/D and D/A systems. Digital input and output lines are also often incorporated. The control of the multiplexer channel, the programmable gain, and the track-and-hold and A/D converter may be done entirely by software, although simpler systems may be less flexible. The board access addresses may occupy space in either the memory or the peripheral (input/output) address spaces of the computer. These two boards are called a "memory-mapped" and a "I/O-mapped" configuration, respectively, and the most versatile boards allow user setting of both the address and the mapping by jumpers or DIP-switches, etc.

The system described above operates best on a moderate number of inputs, each located fairly close to the computer system, and is well suited to the acquisition of a large amount of data on each channel in a short time. There are, however, many applications for data-acquisition systems where the sources of the analog data are well spaced out, often large in number, and yet the data rate from each is quite low. Such tasks as monitoring the temperature in a large building or an oil refinery, or the flammable gas concentration in a coal mine, for instance, meet these latter criteria much better than the former ones. To answer this need, a number of systems oriented about serial transmission

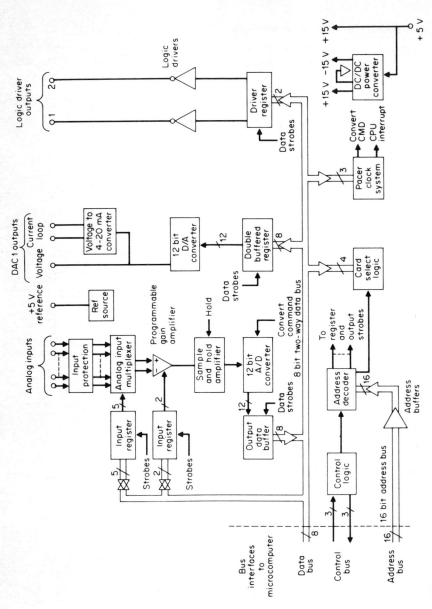

FIG. 8-62 Typical data-acquisition system. (Full system frequently has several DAC outputs, each with its own register.)

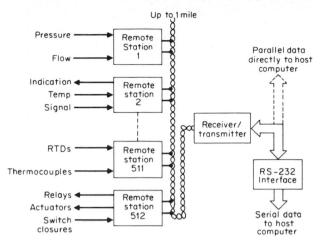

FIG. 8-63 A serially oriented data-acquisition system.

of digital data from a number of "remote" data-acquisition centers, each with its own preamplifiers, multiplexers, and an A/D converter, have been developed. A typical example of this type of system is shown in Fig. 8-63, and the block diagram of the so-called "remote" station is shown in Fig. 8-64. This station includes an input multiplexer, an A/D converter (dual slope, since the data rate is low), and a microcomputer to handle the protocol of serial data transmission along a single twisted pair. In this particular system, up to 256 identical remote stations may be placed on one twisted pair, at distances up to several miles, and the protocol allows two types of board to be connected so that altogether 512 remote stations can be accommodated. Depending on the details of the configuration, these remote stations can all be polled in about 6 s, which is perfectly adequate for the intended systems. The installation costs of such a system are very low, compared to those of a system similar to that shown in Fig. 8-62.

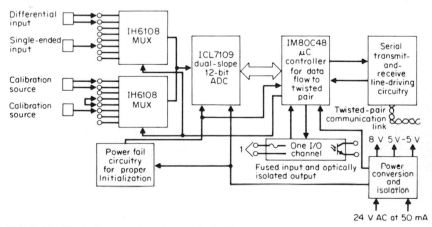

FIG. 8-64 The analog remote station of Fig. 8-63.

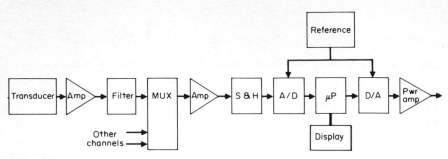

FIG. 8-65 The pattern of most data-acquisition systems.

In general, a useful way to think about data-acquisition systems is presented in Fig. 8-65. Most systems can be fitted to this pattern, although some of the elements are frequently missing or trivial (e.g., the transducer for a digital voltmeter might become merely a test lead!).

It is interesting to note that the recent developmental thrust of module and hybrid devices has been toward absorbing more of these elements into more complete ''bricks.'' In IC form, the same trend is occurring in lower accuracy (8-bit) systems, with multiplexers included on ADCs and even static RAM on a DAC. A harbinger of the future may be found in the so-called ''analog processor,'' which includes a DAC (configurable as a successive approximation ADC), input and output sample-and-hold (track-and-hold), and a programmable microcomputer all on one die. William Blake's line ''To see a world in a grain of sand . . .'' may not be so far out as we thought!

Chapter 9

SSI LOGIC CIRCUITS

Hamil Aldridge
Paradyne Corp.
Largo, Fla.

9–1 INTRODUCTION

The trend in the integrated-circuit industry has steadily been toward more complex, more dense integrated-circuit chips. This trend has been made possible through new circuit technologies and improved fabrication processes.

Small circuit integration (SSI) is being replaced by medium-scale integration (MSI), large-scale integration (LSI), and very large scale integration (VLSI). However, SSI is still needed in most systems requiring simple logical functions. It is the purpose of this chapter to help designers to select and apply SSI circuits for their design requirement.

The characteristics for each logic family are presented to help designers select the best logic family for a design application.

Typical design problems with step-by-step solutions are also included as design examples.

9–2 TYPES OF LOGIC FAMILIES

Although several IC technologies have been introduced over the years, three have reached and maintained prominence: TTL logic (transistor-transistor logic), CMOS logic (complementary metal-oxide semiconductor logic), and ECL logic (emitter-coupled logic).

9–2a   TTL Technology

TTL integrated circuits have gained wide acceptance for the last several years. Introduced by Texas Instruments in 1964, this family of integrated circuits quickly reached broad popularity because of its compromise between speed and power consumption.

The TTL family has added several members to its standard TTL line, including the low-power TTL, high-speed TTL, high-speed Schottky TTL, low-power Schottky TTL, and most recently advanced Schottky TTL, and advancement low-power Schottky. All TTL lines use the same basic circuit configuration and are compatible.

Each series represents a compromise between speed and power. Since the speed-power product is approximately constant, increased power dissipation must be traded off for increased speed, and vice versa. This is because to achieve higher speeds and lower propagation delays, the circuit resistor values have to be reduced. Thus, reducing resistor values means increased power consumption.

One means of moving to an improved speed-power curve is to improve the circuit design. Schottky clamp diodes do just that. By preventing the circuit transistors from entering saturation, transistor storage time is reduced. This yields improved speeds without increasing power; therefore, the Schottky family operates more efficiently.

The TTL family is available in two operating ranges. They are as shown in the following table:

	Series	Temperature range	Power-supply range
Military	54XX	-55 to $+125°C$	$+4.5$ V DC to $+5.5$ V DC
Industrial	74XX	0 to $+ 70°C$	$+4.75$ V DC to $+5.25$ V DC

9–2b   CMOS Technology

Complementary symmetry metal oxide semiconductors (CMOS) enjoy popular use because of their low power dissipation and their ability to operate over a wide supply-voltage range. A CMOS device is fabricated using two metal-oxide semiconductor (MOS) gates. One is an n-channel MOS gate and the other is a p-channel MOS gate. The way these two gates are connected gives the device its name (complementary-symmetry metal-oxide semiconductor).

The unique feature of the CMOS gate is that no current flows through it when it is at either a one or zero state. Therefore, power is only dissipated when the CMOS gate is in the process of switching states. Consequently, power dissipation is proportional to the frequency at which the gate is switched.

The tradeoff for the inherent low power dissipation of CMOS is speed. As with all other MOS integrated circuits, CMOS is suitable for medium-speed applications up to 7 MHz.

CMOS logic is generally available in two operating ranges. For the RCA CD4000A Series, the two ranges are given in the following table:

	Temperature range	Power-supply range
Ceramic	-55 to $+125°C$	$+3$ V DC to $+12$ V DC
Plastic	-40 to $+ 85°C$	$+3$ V DC to $+12$ V DC

9–2c   ECL Technology

Emitter-coupled logic (ECL) is best known for its high-speed operation. ECL is a non-saturating form of digital logic that eliminates transistor storage time as a speed-limiting characteristic, permitting very high-speed operation. However, as a tradeoff for the non-saturating mode, ECL is the least efficient of the three families and dissipates the most power.

ECL utilizes a pair of input transistors: one in a conductive state and the other non-conductive. Switching is accomplished by means of a signal appearing across a common emitter resistor, which is how the name emitter-coupled logic is derived.

The MECL I family was the first digital monolithic integrated-circuit line produced by Motorola. Introduced in 1962, MECL I was considerably beyond the state-of-the-art at that time. No other form of logic could approach the performance of MECL I. As a result, several high-performance systems used the MECL I logic family.

Now, two decades later, Schottky TTL technology has narrowed the performance gap. Motorola has since added MECL II, MECL III, and MECL 10,000 to their ECL family. Each provided improvements over the previous series in performance or ease of use. The final result yields typical propagation delays of 1 ns and 500-MHz flip-flop toggle rates for their MECL III series.

ECL is available in the three operating ranges listed in the following table:

Series	Temperature range	Power-supply range
1. MC10500 MC10600 MCM10500 MC1648M MC12500	-55 to $+125°C$	$V_{EE} = -5.2 \text{ V} \pm 0.010 \text{ V}$
2. MC10100 MC10200 MC1600 MC12000	-30 to $+85°C$	$V_{EE} = -5.2 \text{ V} \pm 0.010 \text{ V}$
3. MC10100 MC1697A MC12000	0 to $+75°C$	$V_{EE} = -5.2 \text{ V} \pm 0.010 \text{ V}$

9–3   CHARACTERISTICS OF LOGIC FAMILIES

The selection of the optimum logic family is a key part of any design. Some designs require high-speed operation, others low power consumption, and yet other designs may require low cost. This section provides the necessary information, charts, and curves to aid the designer in this selection process.

9–3a   Typical Gate Structure

Fig. 9-1 shows a typical two input gate for each logic family. Each gate schematic is representative of the input and output circuitry for each logic family. This information can be useful to the designer when interfacing to nonstandard circuits.

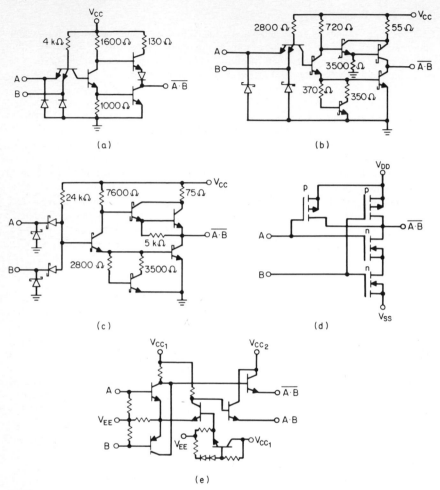

FIG. 9-1 Typical gate structure for each logic family. (*a*) 74; (*b*) 74S; (*c*) 74LS; (*d*) CMOS; (*e*) ECL.

9–3b Voltage Transfer Characteristics

Typical voltage transfer characteristics for each logic family are shown in Fig. 9-2. These curves contain several items of interest for the circuit designer, such as output on and off voltages as a function of input on and off voltages and DC noise margin. In addition these types of curves sometimes also show:

1. Variations in the transfer characteristics as a function of power supply.
2. Variations in the transfer characteristics as a function of temperature.
3. The switching power (if the supply current is plotted on the same figure).
4. Hysteresis characteristics, if applicable (Schmitt trigger).

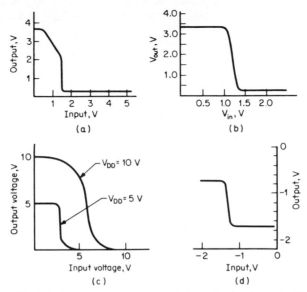

FIG. 9-2 Voltage transfer characteristic by family. (*a*) TTL; (*b*) Schottky TTL; (*c*) CMOS; (*d*) ECL.

Manufacturer's data sheets rarely present the voltage transfer characteristics in graphical form. Instead, they specify a recommended operating zone for voltage transfer characteristics (see Fig. 9-3). Point (a) specifies the minimum input voltage (VIH) required to produce a maximum low voltage (VOL) at the gate's output. Point (b) specifies the maximum input voltage (VIL) required to produce a minimum high voltage (VOH) at the gate's output. Typical and worst-case input voltages are usually presented where

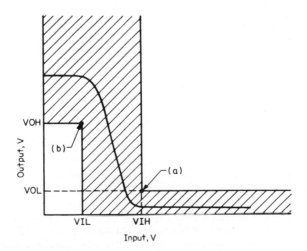

FIG. 9-3 Voltage transfer function for typical inverter gate.

applicable. This information is particularly important to the designer when interfacing different families.

Two devices are compatible if the following inequalities are satisfied:

$$\text{VOL (Driver)} \leq \text{VIL (Receiver)} \tag{9-1}$$
$$\text{VOH (Driver)} \geq \text{VIH (Receiver)} \tag{9-2}$$

9–3c   Speed or Propagation Delay

The speed at which a logic family can operate is usually an important factor when designing a system. Speed is normally specified as "propagation delay," which is defined as the time required for a signal to propagate through a device. For an inverter gate, it is the delay from a point on the input waveform to the same point on the output waveform (see Fig. 9-4). This point may typically be chosen half-way between a logical low level and a logical high level (called the 50% point).

Note that two delay times are specified. One, t_{plh}, is the propagation delay time when the output changes from a low state to a high state, while the other, t_{phl}, is the propagation delay time when the output changes from a high state to a low state.

The total propagation delay through a circuit may be found by summing the individual propagation delays for each device in that circuit. Therefore, it is important for the designer to determine the transition state for each device. Fig. 9-5 shows the range of propagation delay times for each family.

Manufacturers use a second method for specifying speed called the "toggle rate" or "toggle frequency." The maximum toggle frequency is the fastest that a device

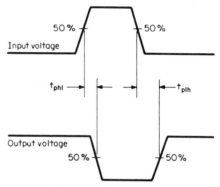

FIG. 9-4   Propagation delay for a typical inverting gate.

such as a flip-flop can be switched to an alternate state. Switching rates faster than the maximum toggle rate result in undetermined output states that are naturally undesirable.

9–3d   Power Dissipation

Power dissipation becomes particularly important when stringent power-supply currents must be met or when circuit thermal dissipation becomes a critical requirement.

Power dissipation is defined as the product of the supply voltage and the mean current supplied to the circuit. Power dissipation is commonly specified as power dissipation per gate. To estimate the total power dissipation, the power dissipation per gate must be multiplied by the number of equivalent gates in the system or circuit.

Power dissipation varies with operating speed across the logic families. Fig. 9-6 demonstrates the power dissipation vs. frequency for each logic family. Note that for TTL the power dissipation per gate is constant until the frequency reaches the 5-MHz range, and then increases sharply with frequency. The power dissipation per gate varies linearly with frequency for the CMOS family. Therefore, the designer's operating frequency should be used when comparing power dissipation among logic families.

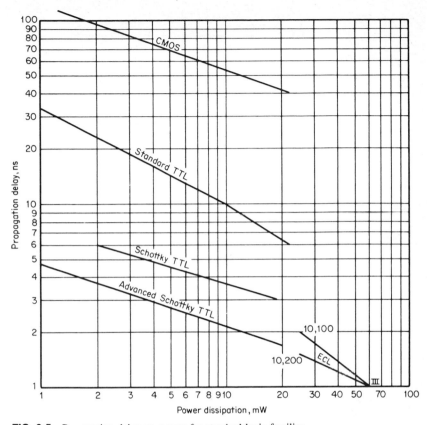

FIG. 9-5 Propagation delay vs. power for standard logic families.

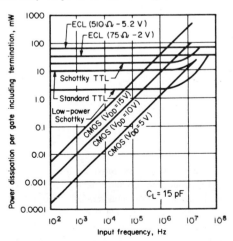

FIG. 9-6 Power dissipation per logic gate vs. frequency for each logic family.

TABLE 9–1 Logic Family vs. Noise Immunity

Logic Family	VNIL (V)	VNIH (V)
Standard TTL	0.4	0.4
S-TTL	0.3	0.7
CMOS	0.95	0.95
ECL	0.175	0.145

9–3e Noise Immunity

No logic system is absolutely perfect. Therefore, noise is a fact of life that the designer must contend with. Noise can create unwanted logic states and cause faulty system operation. The problem of eliminating unwanted noise problems may be attacked two ways. One approach leads to the reduction of the noise source. Transmission-line techniques, decoupling, and shielding are but a few of the methods employed to reduce the noise at its source. The second way involves making the noise receiver less susceptible to unwanted noise. A logic family's noise immunity relates to its ability to operate properly in a noisy environment. Generally, slower logic families are less susceptible to noise because they are slower to respond to noise spikes.

Two types of noise immunity are of interest. The first is known as DC noise immunity and relates to the static input voltage levels that a logic device must see for proper operation. Referring back to Eq. 9-1, the amount that VOL (Driver) is less than VIL (Receiver) is referred to as the input low noise margin (VNIL) and is expressed as

$$\text{VNIL} = |\text{ VIL MAX (Receiver)} - \text{VOL MAX (Driver) }| \qquad (9\text{-}3)$$

Likewise, from Eq. 9-2, the amount that VOH (Driver) exceeds VIH (Receiver) is referred to as the input high noise margin (VNIH) and is expressed as

$$\text{VNIH} = |\text{ VOH MIN (Driver)} - \text{VIH MIN (Receiver) }| \qquad (9\text{-}4)$$

Table 9-1 compares VNIL and VNIH for each logic family. The CMOS family is best followed by standard TTL, S-TTL, and finally ECL.

The second type of noise immunity is the AC noise immunity. Data sheets rarely specify AC noise immunity because of the many factors that affect it. Unlike DC immunity, AC noise immunity must deal with duration as well as amplitude. If unwanted noise changes a device's input long enough, the device will respond by changing its output state. Fig. 9-7 demonstrates the effect of pulsewidth on the CMOS and TTL family's noise immunity. Note that more amplitude is required as the width of the pulse decreases.

9–3f Loading

In any given design, several logic blocks must be interconnected to realize a logical function. Loading refers to the number of logic devices that a logic device may drive. It may best be explained in terms of fan-out and fan-in. "Fan-out" is a measure of a logic device's drive capability. "Fan-in" is a measure of the input load that a logic device presents. Let N represent the number of input devices to be driven, then starting with the basic requirement

$$\text{Output drive} \geq \text{total input load} \qquad (9\text{-}5)$$

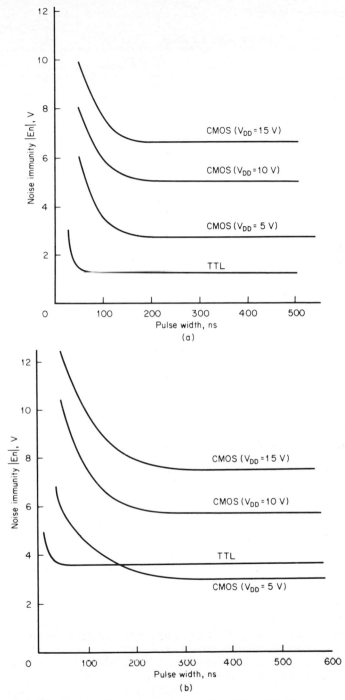

FIG. 9-7 Typical AC noise immunity vs. pulse width. (*a*) Positive pulse; (*b*) negative pulse.

TABLE 9–2 Loading Chart for Logic Families

Driver	Receiver						
	TTL	S-TTL	LS-TTL	AS-TTL	ALS-TTL	CMOS (5 V)	ECL
TTL	10	8	40	8	40	*>100	†
S-TTL	12	10	50	10	50	*>100	†
LS-TTL	5	4	20	4	20	*>100	†
AS-TTL	12	10	50	10	50	*>100	†
ALS-TTL	5	10	20	4	20	*>100	†
CMOS	0	0	1	0	1	>100	†
ECL	†	†	†	†	†	†	‡

*Assumes R pull-up is used.
†Not directly compatible because of logic-level differences—requires a level shifter circuit.
‡See manufacturer's design guidelines.

and expressing this in terms of N, fan-out, and fan-in,

$$\text{Fan-out} \geq N \cdot \text{fan-in} \tag{9-6}$$

Dividing both sides by fan-in and transposing terms, we have

$$N \leq \frac{\text{fan-out}}{\text{fan-in}} \tag{9-7}$$

This expression states that the number of driven devices (N) must be less than or equal to the integer number represented by dividing the driving device's output capability (fan-out) by the input devices load capability (fan-in). Table 9-2 compares the loading characteristics for each device member in the three families of logic.

9–3g   Cost/Availability

The cost of a logic family usually becomes important when other characteristics (speed, power) do not necessarily dictate the logic family type.

Price and availability are usually closely related. Lower prices indicate large quantities, wide usage, and good availability. High prices usually indicate less availability due to high demand and limited supply or the manufacturer's inability to produce the parts in volume.

9–4   DEFINITION OF GENERAL PARAMETERS

Digital IC data sheets are normally divided into three sections by the IC manufacturers. The first section presents a brief technical description of the function of the part and may include a truth table, simplified schematic, and logic diagram. The second section deals with absolute maximum ratings, and the third section presents a list of electrical characteristics. The first section is usually self-explanatory; however, sections two and three are not quite as straightforward. This section will help the designer to interpret the absolute maximum ratings and electrical characteristics.

9–4a   Absolute Maximum Ratings

Absolute maximum ratings define the limits to which a device may be stressed without causing permanent damage to the part. Typical parameters covered under this section are supply voltage, input voltage, operating free air temperature range, and storage temperature range. Other parameters may be added to this section for specialized parts.

It is important to note that these limits are not operating limits and that under no circumstances should the designer exceed these ratings.

9–4b   Electrical Characteristics

These parameters specify the manufacturer's recommended operating range. This section includes such information as input device requirements, output device requirements, supply current requirements, and switching characteristics. The designer should always note the conditions under which parameters are measured. Typical parameters are normally measured at nominal power-supply levels and at 25°C. Maximum or minimum parameters should be measured at the worst-case operating supply level and the worst-case operating temperature range.

A load circuit is presented in this section of the manufacturer's data sheet with a footnote. The designer must determine whether his or her application falls within the limits of the test circuit. If not, additional testing will be required to determine the proper parameters for the application. When designing within compatible families, loading and propagation delay times are of concern to the designer. However, when interfacing the logic family to a nonstandard device or circuit, each parameter must be carefully checked for proper operating limits.

9–5   GATES

9–5a   AND Gate

The AND gate provides an output that is the logical AND of the inputs. If all inputs are a Logical 1, then the output is a Logical 1. Table 9-3a shows the standard symbol, boolean expression, and truth table for a two-input AND gate. Note from the truth table that inputs A and B must be a Logical 1 in order for the output Y to be a Logical 1. The boolean expression is simply another way to express the truth table. The Y output is a Logical 1 when inputs A and B are a Logical 1. Although our example used a two-input AND gate, four- and eight-input AND gates are typically available.

9–5b   NAND Gate

The NAND gate can be thought of as a NOT-AND gate function. If an inverter or "not" function were attached to the output of the AND gate function above, the results would be a NAND gate. If all inputs are a Logical 1, then the output is a Logical zero. Table 9-3b shows the standard symbol, boolean expression, and truth table for a two-input NAND gate. Note the bubble on the gate's output, the bar ($-$) over the right side of the boolean expression, and the output Y of the truth table. These symbols indicate the "not" function in relation to its counterpart, the AND gate. NAND gates are typically available in two-, three-, four-, and eight-input configurations.

9–5c   OR Gate

The OR gate provides an output that is the logical OR of the inputs. If any one of the inputs is a Logical 1, then the output is a Logical 1. Table 9-3c shows the standard symbol, boolean expression, and truth table for a two-input OR gate. Note from the truth table that if input A or input B is a Logical 1, then the output is a Logical 1. Again, the

TABLE 9–3 Basic Logic Elements

Function	Standard symbol	Boolean expression	Truth table
(a) AND	A, B → Y	$Y = A \cdot B$	A B Y 0 0 0 0 1 0 1 0 0 1 1 1
(b) NAND	A, B → Y	$Y = \overline{A \cdot B}$	A B Y 0 0 1 0 1 1 1 0 1 1 1 0
(c) OR	A, B → Y	$Y = A + B$	A B Y 0 0 0 0 1 1 1 0 1 1 1 1
(d) NOR	A, B → Y	$Y = \overline{A + B}$	A B Y 0 0 1 0 1 0 1 0 0 1 1 0
(e) XOR	A, B → Y	$Y = A\overline{B} + \overline{A}B$ $Y = A \oplus B$	A B Y 0 0 0 0 1 1 1 0 1 1 1 0
(f) XNOR	A, B → Y	$Y = \overline{A} \cdot \overline{B} + AB$ $Y = \overline{A \oplus B}$	A B Y 0 0 1 0 1 0 1 0 0 1 1 1
(g) Inverter (NOT)	A → Y	$Y = \overline{A}$	A Y 0 1 1 0

boolean expression restates the truth table: output Y is a Logical 1 when input A or input B is a Logical 1. OR gates are available in two-, three-, four-, and eight-input configurations.

9–5d NOR Gate

The NOR gate can be thought of as a NOT-OR gate function. The NOR gate is equivalent in function to an OR gate with an inverter or "not" function placed at its output. If any one of the inputs is a Logical 1, then the output is a Logical 0. Table 9-3d shows the standard symbol, boolean expression, and truth table for a two-input NOR gate. Note the bubble on the gate output, the bar over the right side of the expression, and the Y output of the truth table. It is the exact complement of the OR gate's truth table. NOR gates are available in two-, three-, four-, and eight-input configurations.

9–5e EXCLUSIVE-OR Gate (XOR)

The EXCLUSIVE-OR gate provides an output that is the logical "exclusive" OR of the inputs. If any one of the inputs is a Logical 1, then the output is a Logical 1. Any other

TABLE 9–3   (Continued)

Function	Standard symbol	Boolean expression	Truth table

(*h*) AND/OR

$Y = (A \cdot B) + (C \cdot D)$

A	B	C	D	Y
0	0	0	0	0
0	0	0	1	0
0	0	1	0	0
0	0	1	1	1
0	1	0	0	0
0	1	0	1	0
0	1	1	0	0
0	1	1	1	1
1	0	0	0	0
1	0	0	1	0
1	0	1	0	0
1	0	1	1	1
1	1	0	0	1
1	1	0	1	1
1	1	1	0	1
1	1	1	1	1

(*i*) AND/OR/ INVERT

$Y = \overline{(A \cdot B) + (C \cdot D)}$

A	B	C	D	Y
0	0	0	0	1
0	0	0	1	1
0	0	1	0	1
0	0	1	1	0
0	1	0	0	1
0	1	0	1	1
0	1	1	0	1
0	1	1	1	0
1	0	0	0	1
1	0	0	1	1
1	0	1	0	1
1	0	1	1	0
1	1	0	0	0
1	1	0	1	0
1	1	1	0	0
1	1	1	1	0

set of input conditions results in a Logical 0 output. Table 9-3*e* shows the standard symbol, boolean expression, and truth table for a two-input EXCLUSIVE-OR gate. The "⊕" symbol in the boolean expression means EXCLUSIVE-OR.

An understanding of the relationship between the OR function and the EXCLUSIVE-OR function may help the reader to understand how the name EXCLUSIVE-OR was derived. The commonly named OR (INCLUSIVE-OR) function provides an output Logical 1 if any input (one or more) is a Logical 1. Any input (one or more) is inclusive because it includes every condition in which the input(s) is a Logical 1 (see truth table OR function). In contrast, the EXCLUSIVE-OR function generates an output Logical 1 if and only if one input is a Logical 1, and excludes the case when more than one input is a Logical 1 (see XOR truth table). Hence, the name EXCLUSIVE-OR is derived.

Two interesting functions are generated by the XOR function. From its truth table the reader may note that the *Y* output is a Logical 0 when the two inputs are the same. Thus, the XOR gate may be used as a logical compare function. Secondly, note that if the *B* input is a Logical 0, the *Y* output follows the *A* input. In contrast, note that if the

B input is a Logical 1, the Y output complements the A input. Thus, a programmable invert/noninvert function is available.

EXCLUSIVE-OR gates are available in a two-input configuration.

9–5f EXCLUSIVE-NOR Gate (XNOR)

The EXCLUSIVE-NOR can be thought of as a NOT-XOR function. The output is the complement or inverse of the XOR output. The output Y is a Logical 0 when and only when one of the inputs, A or B, is a Logical 1. All other cases result in a Logical 1 at the gates' output. Table 9-3f shows the standard logic symbol, boolean expression, and truth table for a two-input XNOR gate. XNOR gates are available in a two-input configuration.

9–5g Inverter (NOT) Gate

The output of the inverter gate complements or NOT's the input. If the input is a Logical 1, then the output is a Logical 0. Conversely, if the input is a Logical 0, then the output is a Logical 1. Table 9-3g shows the standard logic symbol, boolean expression, and truth table for the inverter gate. Note that the inverter function may be implemented with other NOT functions (NAND, NOR, XOR, XNOR), as well as the inverter gate. For each of the above cases, the reader should refer to the truth table in Table 9-3 to determine what to do with the unused inputs.

9–5h AND/OR Gate

The AND/OR function is slightly different from the previous cases. Two functions are realized with the AND/OR gate. The first level is the AND function, followed by the OR function. Table 9-3h shows the standard logic symbol, boolean expression, and truth table for a two-wide two-input AND/OR gate. The output Y is a Logical 1, if node E and/or F is a Logical 1. Node E is a Logical 1 if inputs A and B are Logical 1's. Node F is a Logical 1 if inputs C and D are Logical 1's. Therefore, output Y is a Logical 1 if inputs A and B are a Logical 1 *or* if inputs C and D are a Logical 1.

9–5i AND/OR/INVERT

Three logical functions are realized with the AND/OR/INVERT gate. Level one provides the AND function of the grouped inputs. Level two logically OR's each AND function output, and level three inverts or NOT's the result. Table 9-3i gives the standard logic symbol, boolean expression, and truth table for a dual two-input AND/OR/INVERT gate. Note from the truth table that inputs A and B or inputs C and D must be logical 1's to force the output Y to a Logical 0. The reader should note that the AND/OR/INVERT function is the complement of the AND/OR function covered above. The AND/OR/INVERT gate configuration is typically available in combinations of two, three, or four inputs and two-wide or four-wide arrays.

Example 9–1 Decoder Design

Design a decoder circuit with inputs I0, I1, I2 and outputs 00, 01, 02, 03, which satisfies the truth table shown in Fig. 9-8.

Solution

a) Map first output 00 onto a Mahoney map (refer to Marcus reference).

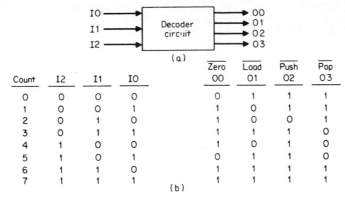

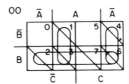

FIG. 9-8 Decoder of Example 9-1. (*a*) Block diagram; (*b*) truth table.

b) Couple all possible entries on the map as follows:

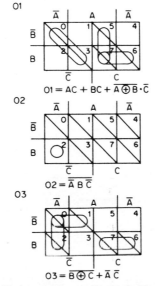

c) Interpret the results and write the minimum sum-of-products expression.

$$O0 = B + (\overline{A}C + A\overline{C})$$

d) $O0 = B + A \oplus C$

Repeat steps a, b, and c for outputs O1, O2, and O3.

$O1 = AC + BC + \overline{A \oplus B} \cdot \overline{C}$

$O2 = \overline{A} B \overline{C}$

$O3 = \overline{B \oplus C} + \overline{A} \, \overline{C}$

The decoder circuit will then appear as shown in Fig. 9-9.

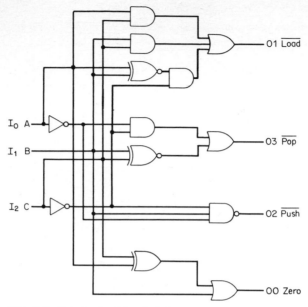

FIG. 9-9 Solution for Example 9-1.

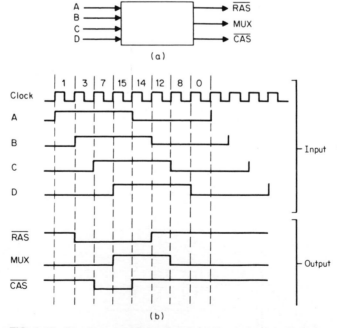

(a)

(b)

FIG. 9-10 Requirement of Example 9-2. (*a*) Block diagram; (*b*) timing diagram.

Example 9–2   Timing Waveform Generation

Given the Johnson counter output sequence shown in Fig. 9-10, generate the following timing signals $\overline{RAS}$, MUX, $\overline{CAS}$.

Solution

a) Generate the truth table from the timing chart.

Count	D	C	B	A	$\overline{RAS}$	MUX	$\overline{CAS}$
0	0	0	0	0	1	0	1
1	0	0	0	1	1	0	1
2	0	0	1	0	X	X	X
3	0	0	1	1	0	0	1
4	0	1	0	0	X	X	X
5	0	1	0	1	X	X	X
6	0	1	1	0	X	X	X
7	0	1	1	1	0	0	0
8	1	0	0	0	1	0	1
9	1	0	0	1	X	X	X
10	1	0	1	0	X	X	X
11	1	0	1	1	X	X	X
12	1	1	0	0	1	1	1
13	1	1	0	1	X	X	X
14	1	1	1	0	0	1	1
15	1	1	1	1	0	1	0

b) Map each output from the truth table.

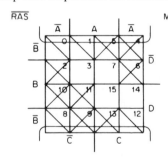

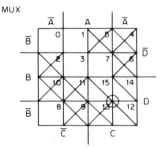

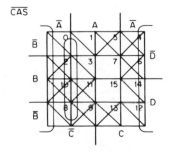

c) Write the boolean expression for each output.

$$\overline{RAS} = \overline{B}$$
$$MUX = C \cdot D$$
$$\overline{CAS} = \overline{A} + \overline{C}$$

or

$$\overline{CAS} = \overline{A \cdot C}$$

d) Realize the hardware equivalent from these boolean expressions. The resulting circuit is shown in Fig. 9-11.

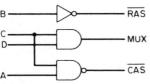

FIG. 9-11   Circuit solution of Example 9-2.

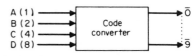

FIG. 9-12   Binary-to-decimal code converter of Example 9-3.

Example 9–3   Binary-to-Decimal Code Converter

Design a circuit that converts binary weighted (1248) code to decimal code, as illustrated in Fig. 9-12.

Solution
a) Define the truth table, covering all possible input and output combinations.

Count	0	1	2	3	4	5	6	7	8	9	10	11	12	13	14	15
Input
A	0	1	0	1	0	1	0	1	0	1	0	1	0	1	0	1
B	0	0	1	1	0	0	1	1	0	0	1	1	0	0	1	1
C	0	0	0	0	1	1	1	1	0	0	0	0	1	1	1	1
D	0	0	0	0	0	0	0	0	1	1	1	1	1	1	1	1
Output
$\overline{0}$	0	1	1	1	1	1	1	1	1	1	X	X	X	X	X	X
$\overline{1}$	1	0	1	1	1	1	1	1	1	1	X	X	X	X	X	X
$\overline{2}$	1	1	0	1	1	1	1	1	1	1	X	X	X	X	X	X
$\overline{3}$	1	1	1	0	1	1	1	1	1	1	X	X	X	X	X	X
$\overline{4}$	1	1	1	1	0	1	1	1	1	1	X	X	X	X	X	X
$\overline{5}$	1	1	1	1	1	0	1	1	1	1	X	X	X	X	X	X
$\overline{6}$	1	1	1	1	1	1	0	1	1	1	X	X	X	X	X	X
$\overline{7}$	1	1	1	1	1	1	1	0	1	1	X	X	X	X	X	X
$\overline{8}$	1	1	1	1	1	1	1	1	0	1	X	X	X	X	X	X
$\overline{9}$	1	1	1	1	1	1	1	1	1	0	X	X	X	X	X	X

b) Map each output $\overline{0}$ thru $\overline{9}$ onto a Mahoney map.

f1 = $\overline{0}$

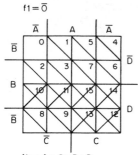

f1 = A + C + B + D
or
f1 = $\overline{A} \cdot \overline{B} \cdot \overline{C} \cdot \overline{D}$

f2 = $\overline{1}$

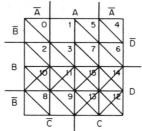

f2 = $\overline{A}$ + C + B + D
or
f2 = $\overline{A \cdot B \cdot \overline{C} \cdot \overline{D}}$

f3 = $\overline{2}$

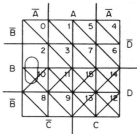

f3 = $\overline{B}$ + C + A
or
f3 = $\overline{\overline{A} \cdot B \cdot \overline{C}}$

f4 = $\overline{3}$

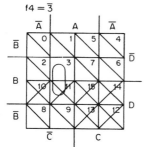

f4 = C + $\overline{B}$ + $\overline{A}$
or
f4 = $\overline{A \cdot B \cdot \overline{C}}$

f5 = $\overline{4}$

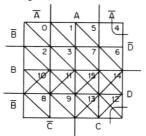

f5 = $\overline{C}$ + B + A
or
f5 = $\overline{\overline{A} \cdot \overline{B} \cdot C}$

f6 = $\overline{5}$

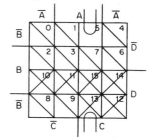

f6 = $\overline{C}$ + B + $\overline{A}$
or
f6 = $\overline{A \cdot \overline{B} \cdot C}$

f7 = $\overline{6}$

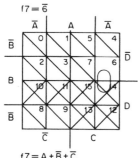

f7 = A + $\overline{B}$ + $\overline{C}$
or
f7 = $\overline{\overline{A} \cdot B \cdot C}$

f8 = $\overline{7}$

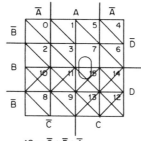

f8 = $\overline{C}$ + $\overline{B}$ + $\overline{A}$
or
f8 = $\overline{A \cdot B \cdot C}$

f9 = $\overline{8}$

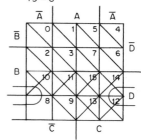

f9 = A + $\overline{D}$
or
f9 = $\overline{\overline{A} \cdot D}$

f10 = $\overline{9}$

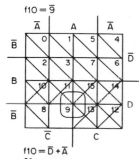

f10 = $\overline{D}$ + $\overline{A}$
or
f10 = $\overline{A \cdot D}$

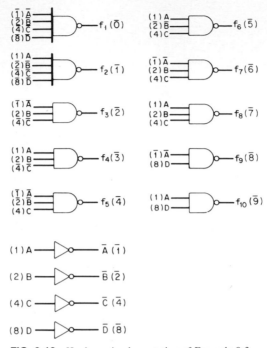

FIG. 9-13   Hardware implementation of Example 9-3.

c) From the boolean equation, synthesize each logical function. The solution is shown in Fig. 9-13.

9–6   LATCHES AND FLIP-FLOPS

Bistable devices have long been used to store singular events. The two basic bistable devices are the latch and the flip-flop.

Latches operate asyncronously, providing an output that immediately responds to the input. This type of device may be termed "DATA DRIVEN" because the output state is determined solely by the input data. RS-type latches are usually made from cross-coupled NAND gates; however, other gate types may be used, as shown in Fig. 9-14.

D-type latches (or "transparent" latches) operate slightly differently from the RS-type latch. One data (D) input is provided in contrast to two inputs (R and S) for the RS-type latch. An additional (G) input is used to enable the input data. A high on the enable input allows the output to follow the input. A low on the enable holds the output at its present value and is independent of data input changes during this time. Fig. 9-15 shows the logical symbol and truth table for the D-type latch.

The flip-flop operates syncronously, with the output following the input at a pre-scribed time as determined by the input clock. The clock can be thought of as sampling the data at a time defined by the transition of the clock. Therefore, flip-flops are classified as "clock driven" devices, a trait extremely important to syncronous operation. Flip-flops today use one of three possible clocking mechanisms. A short discussion of these three mechanisms follows.

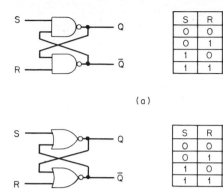

S	R	Q	Q̄
0	0	1	1
0	1	1	0
1	0	0	1
1	1	Q	Q

(a)

S	R	Q	Q̄
0	0	Q	Q
0	1	1	0
1	0	0	1
1	1	0	0

(b)

FIG. 9-14   Gate latches. (a) NAND gate implementation; (b) NOR gate implementation.

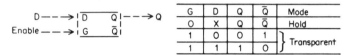

G	D	Q	Q̄	Mode
0	X	Q	Q̄	Hold
1	0	0	1	Transparent
1	1	1	0	Transparent

FIG. 9-15   D-type latch.

DC Coupled or Edge Triggered   An edge-triggered device samples the data input when the clock crosses the device's DC threshold. Devices are designed to operate on the positive edge or the negative edge of the clock, but not both.

Master-Slave   The master-slave clocking mechanism can best be explained using two elements. For example, assume data is clocked into the first element (or master) on the positive edge of the clock. The output of the first element is then clocked into the second element (or slave) and presented to the output.

AC Coupled   The clock is capacitively coupled to the flip-flop device. This provides DC isolation from the internal clock circuitry.

Example 9–4   Switch Debouncer

Design a switch debounce circuit for the switch shown in Fig. 9-16a. The output will go low when the switch is in position B.
Solution
One tried-and-proved solution requires cross strapping two gates into an RS-latch configuration, as shown in the Fig. 9-16b. Note that the switch has no drive capability; therefore, pull-up resistors are required. Note that a delay through two gates occurs once the switch has been closed.
   A second and more recent solution to the switch debounce problem uses two inverter gates as shown in Fig. 9-16c. A cursory glance at the inverter solution might cause some concern regarding the short-circuit condition. Nevertheless, most TTL gates possessing totem-pole output configurations are capable of sustaining a short-circuit current on the order of 100

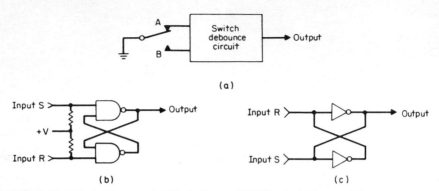

FIG. 9-16   Switch debouncer. (*a*) Block diagram; (*b*) RS latch; (*c*) inverter debouncer.

mA for short periods of time. This period is the thermal time constant of the device, and it typically lies between 2 and 10 s, depending upon the manufacturer. The alternate solution, however, reduces this time to the propagation delays of the gates. For most TTL gates, this time is on the order of 10 to 30 ns per gate, so even the worst-case devices are <60 ns.

9–6a   D Flip-Flop

The edge-triggered D flip-flop transfers the data input (D) to the Q and $\overline{Q}$ output on the transition of the clock. Examples of a positive edge-triggered flip-flop are the TTL 7474 and the CMOS CD4013 devices. Fig. 9-17 shows the logic symbol and truth table for a positive edge-triggered D flip-flop.

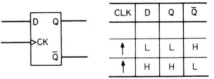

CLK	D	Q	$\overline{Q}$
↑	L	L	H
↑	H	H	L

FIG. 9-17   Positive edge-triggered D flip-flop.

9–6b   J-K Edge-Triggered Flip-Flop

The JK edge-triggered flip-flop operates much like the D edge-triggered flip-flop. The inputs J and K are transferred to the output on the transition of the clock. This flip-flop

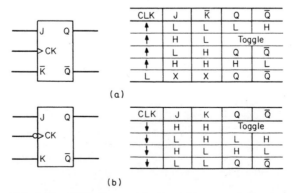

CLK	J	$\overline{K}$	Q	$\overline{Q}$
↑	L	L	L	H
↑	H	L	Toggle	
↑	L	H	Q	$\overline{Q}$
↑	H	H	H	L
L	X	X	Q	$\overline{Q}$

(a)

CLK	J	K	Q	$\overline{Q}$
↓	H	H	Toggle	
↓	L	H	L	H
↓	H	L	H	L
↓	L	L	Q	$\overline{Q}$

(b)

FIG. 9-18   JK flip-flops. (*a*) Positive edge triggered; (*b*) negative edge triggered.

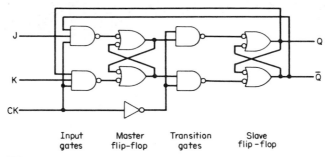

FIG. 9-19   Master-slave equivalent circuit.

is available in positive edge-triggered and negative edge-triggered devices. The 74S109 device is an example of a positive edge-triggered JK flip-flop. Fig. 9-18*a* shows the logic symbol and truth table for a flip-flop of this type.

The 74112 device is representative of the negative edge-triggered JK flip-flops. Fig. 9-18*b* reveals the logic symbol and truth table for the negative edge-triggered JK device.

A more complex type of JK flip-flop is the master-slave flip-flop. This device is actually made up of two flip-flops in series. Fig. 9-19 illustrates the equivalent circuit, and the timing is shown in Fig. 9-20.

A high on the clock line enables the input gates and allows the *J-K* inputs to be transferred to the master flip-flop. A subsequent low on the clock line enables the transition gates and passes the master flip-flop's output to the slave flip-flop. Note that the low transition on the clock line closes the input gates and freezes the information in the master flip-flop. Typical examples of the master-slave flip-flop are the 74107, CD 4027, and MC10135 devices.

The designer should be aware of restrictions when using the master-slave JK flip-flop. If the flip-flop is set ($Q = H, \overline{Q} = L$) and the clock is high, a high on the K input at any time will cause the master flip-flop to reset. Likewise, the master flip-flop will be set if the clock is high, the flip-flop has been previously reset, and a high occurs at the J input. Limiting the time that the clock is high is one method that the designer may use to minimize this problem.

As shown in the truth table for each flip-flop type, manufacturers may provide a clear input or clear and preset inputs. Irrespective of the input states, a low on the clear line forces the flip-flop into a reset condition ($\overline{Q} = 1, Q = 0$). Likewise, a low on the preset line forces the flip-flop into a set condition ($\overline{Q} = 0, Q = 1$). A low on both the clear and preset lines results in an indeterminate condition.

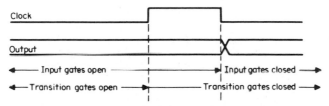

FIG. 9-20   Master-slave timing diagram.

Example 9–5   Parallel Counter

Design a parallel counter to produce the following state sequence: 0, 1, 3, 7, 15, 14, 12, 8, 0, 1 . . .

Solution

a) List the state sequence in binary.

	State sequence							
	0	1	3	7	15	14	12	8
Q_A	0	1	1	1	1	0	0	0
Q_B	0	0	1	1	1	1	0	0
Q_C	0	0	0	1	1	1	1	0
Q_D	0	0	0	0	1	1	1	1

b) Draw the general solution. The circuit is shown in Fig. 9-21.

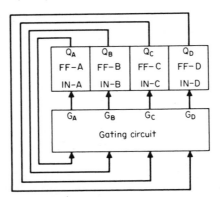

FIG. 9-21   General solution for Example 9-5.

c) Prepare Mahoney maps for G_A, G_B, G_C, and G_D.

d) Select flip-flop type and configuration. For this example, a D-type flip-flop will be used, which implies a steering configuration. The truth table for the D-type flip-flop is shown below.

Truth Table

D	Q
0	0
1	1

e) Using the table above, map each state for Q_A using the following guide-line: Mark a "1" in the appropriate square if the Q_A flip-flop is a "1" for the next count. Mark a "0" in that square if the Q_A flip-flop is a "0" for the next count.

f) Repeat the procedure in Step e for Q_B, Q_C, and Q_D.

g) Any count not in the state sequence is a constraint.

h) Express each map as the minimum sum of products.

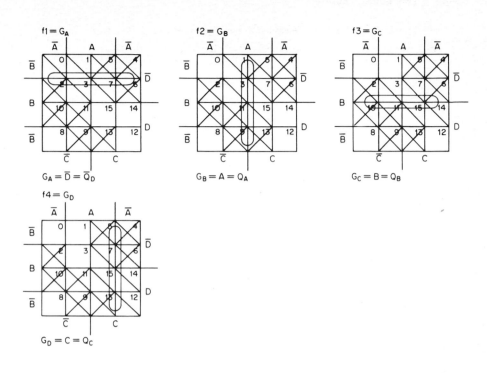

$f1 = G_A$

$G_A = \overline{D} = \overline{Q}_D$

$f2 = G_B$

$G_B = A = Q_A$

$f3 = G_C$

$G_C = B = Q_B$

$f4 = G_D$

$G_D = C = Q_C$

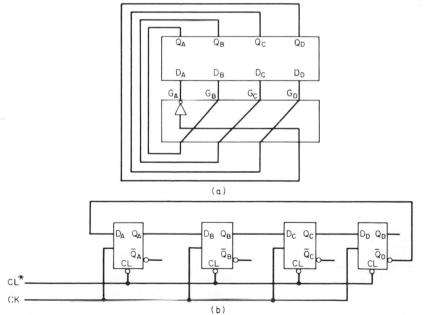

(a)

(b)

FIG. 9-22 Circuit of Example 9-5. (*a*) General solution; (*b*) Moebius (twisted ring) counter.
*Note: This counter design is not self correcting. Therefore, CL must be used to initialize the counter to 0 state.

i) Synthesize each boolean expression into hardware form. This results in Fig. 9-22a.

Redrawing this figure yields a solution for a common Moebius or twisted ring counter, as shown in Fig. 9-22b.

Alternate Solution

d) Referring back to Step d of the first solution, select a different flip-flop type and configuration as an alternate solution. A JK flip-flop used in a complementing configuration will be used. The truth table is shown below.

J/K	Q
0	No change
1	Toggle

e) Using the truth table above, map each state for Q_A using the following guideline: Mark a "1" in the appropriate square if the Q_A flip-flop changes state when advancing to the next count. Mark a "0" in that square if there is no change when advancing to the next count.

f) Repeat the procedure in Step e for Q_B, Q_C, and Q_D.

g) Any count not in the state sequence is a constraint.

h) Express each map as the minimum sum of products as follows:

f1 = G_A

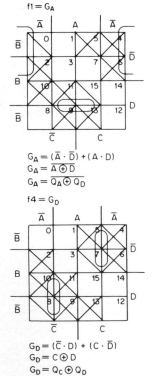

$$G_A = (\bar{A} \cdot \bar{D}) + (A \cdot D)$$
$$G_A = \overline{A \oplus D}$$
$$G_A = \overline{Q_A \oplus Q_D}$$

f2 = G_B

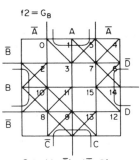

$$G_B = (A \cdot \bar{B}) + (\bar{A} \cdot B)$$
$$G_B = A \oplus B$$
$$G_B = Q_A \oplus Q_B$$

f3 = G_C

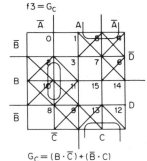

$$G_C = (B \cdot \bar{C}) + (\bar{B} \cdot C)$$
$$G_C = B \oplus C$$
$$G_C = Q_B \oplus Q_C$$

f4 = G_D

$$G_D = (\bar{C} \cdot D) + (C \cdot \bar{D})$$
$$G_D = C \oplus D$$
$$G_D = Q_C \oplus Q_D$$

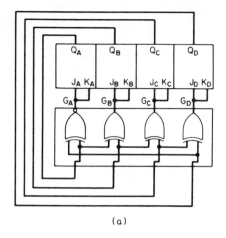

(a)

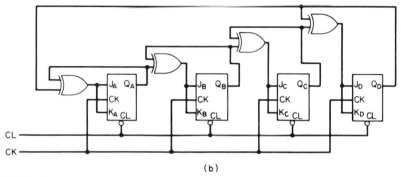

(b)

FIG. 9-23   Alternate circuit for Example 9-5. (a) General solution; (b) counter. *Note: This counter design is not self correcting. Therefore, CL must be used to initialize the counter to 0 state.

i) Synthesize each boolean expression into hardware form as shown in Fig. 9-23a. Redrawing this circuit results in the more accepted counter circuit shown in Fig. 9-23b.

Example 9-6   Sequential-Level Output

Design an event recording circuit that captures an external interrupt signal and holds the interrupt until acknowledged by the interrupted processor (see timing diagram of Fig. 9-24a).

Solution

a) *Select Device(s).* An examination of the timing diagram indicates the need for a level type output device. This device must have an output that changes from a low to high state based on an edge transition (low to high) on input χ_2. Furthermore, this output must return to its original low state as a result of a low-going pulse on input χ_1. An edge-triggered flip-flop meets the device requirements.

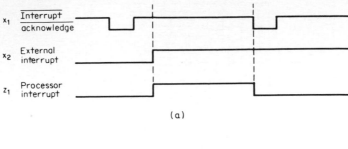

(a)

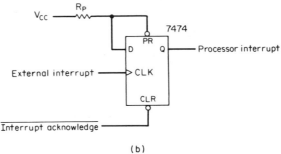

(b)

FIG. 9-24   Sequential-level output circuit of Example 9-6. (*a*) Timing diagram;
(*b*) circuit solution.

b) *Implement Device(s)*. Assign input χ_2 to the clock input of the flip-flop.
This implies that a positive edge-triggered flip-flop, such as a 7474, may
be used. The output Q must transition from a low state to a high state.
Therefore, the steering input (D) must be tied high. The χ_1 input must be
assigned to the clear input of the flip-flop to return the Q output to a low
state.
c) The circuit solution is shown in Fig. 9-24*b*. The reader should note that
this is just one of several possible solutions.

Example 9–7   Sequential-Pulse Output

Design a pulse blanker circuit that deletes the first clock output pulse after
the application of a reset signal.
Solution
a) *Select Device(s)*. An examination of the timing diagram of Fig. 9-25*a*
indicates that a pulse blanker signal must activate when the χ_1 input pulse
occurs and remain active until the next falling edge transition of the χ_2 input.
Select a negative edge-triggered flip-flop to perform this pulse blanker function.
b) *Implement Device(s)*. Assign the χ_1 input to the preset function of the
flip-flop. Assuming an AND gate will be used to generate the Z_1 output, a
low-level must be assigned to the active state of the pulse blanker output.
Therefore, the Q function will be assigned as the pulse blanker output.
Assigning the χ_2 input to the CLK function of the flip-flop provides the
necessary Q transition on the trailing edge of the χ_2 signal. The steering

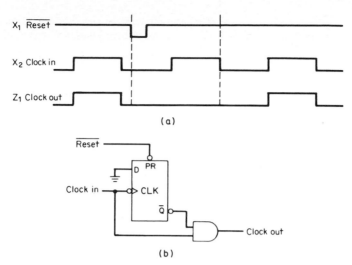

FIG. 9-25 Sequential pulse circuit. (*a*) Timing diagram; (*b*) hardware solution.

input must be low to produce the proper output upon a transition of the input CLK.

c) The circuit solution is illustrated in Fig. 9-25*b*.

9–7 ONE SHOTS

The one shot (monostable multivibrator) provides an output pulse as a result of a single input transition. The duration of the output pulse is a function of an *RC* time constant associated with the one-shot device. This unique device allows the designer to produce a pulsed output whereby the duration is independent of the timing constraints imposed by its surrounding circuitry. This is particularly useful when simple decoding techniques cannot readily produce the desired pulsewidth.

9–7a Nonretriggerable

One shots are classified by two types, a nonretriggerable type and a retriggerable type. Fig. 9-26 illustrates the operation of a nonretriggerable one shot. A transition from a low to a high at the input triggers or initiates the beginning of a one-shot cycle. The output immediately changes from a low to a high and remains there for the duration of the pulse as defined by the *RC* time constant of the one shot. During this ON time, additional trigger transitions are ignored. However, trigger transitions during the recovery time must be avoided. This recovery time is needed to return the programming capacitor (C_{ext}) to its

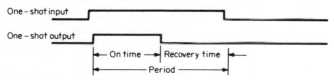

FIG. 9-26 Nonretriggerable one shot.

correct initial value. Therefore, the recovery time must be added to the ON time of the one shot to determine the period. This relationship between the ON time and the period is referred to as the duty cycle.

$$\% \text{ duty cycle} = \frac{\text{ON time}}{\text{ON time} + \text{recovery time}} \times 100\% \qquad (9\text{-}8)$$

or

$$\% \text{ duty cycle} = \frac{\text{ON time}}{\text{period}} \times 100\% \qquad (9\text{-}9)$$

Should the recommended duty cycle be exceeded, the output becomes indeterminable and jitter results.

Nonretriggerable one shots, such as the 74LS221 device, are capable of operating with duty cycles as high as 90% if R_{ext} equals R_{ext} (max).

9–7b   Retriggerable

Retriggerable one shots operate much like the nonretriggerable type, with one important difference. Unlike the nonretriggerable type, the retriggerable one shot accepts a retrigger transition from the input during the output ON time. The output remains on and a new cycle is started. Fig. 9-27 illustrates this mode of operation.

The one-shot output will remain on as long as the time between trigger transitions is less than the one-shot's ON time.

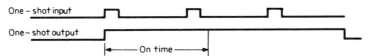

FIG. 9-27   Retriggerable one shot.

Example 9–8   Positive Edge-Triggered One Shot

Design a circuit that generates an output pulse each time a positive transition occurs on clock A, as shown in Fig. 9-28a.

Solution
a) Select a nonretriggerable one shot (74LS221) to perform the required function. Assign the B input to the clock A signal and the Q output to the output A signal.
b) Calculate the % duty cycle. From Fig. 9-28a

$$\text{One Shot Period} = 104.1 \ \mu\text{s}$$
$$\text{Pulse Duration} = 60 \ \text{ns}$$

$$\% \text{ duty cycle} = \frac{t_{ON}}{t_{period}} \times 100\%$$

$$\% \text{ duty cycle} = \frac{60 \times 10^{-9} \ \text{s}}{104.1 \times 10^{-6} \ \text{s}} \times 100\%$$

$$\% \text{ duty cycle} = 0.06\%$$

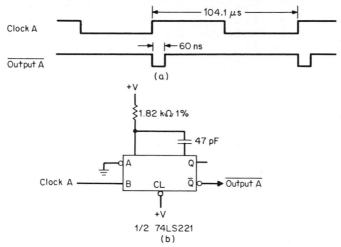

FIG. 9-28   Positive edge-triggered one shot. (*a*) Timing diagram; (*b*) circuit.

c) Select R_T and C_{ext}. The pulse duration is defined by the equation

$$t_w = \ln 2 \cdot R_T \cdot C_{ext}$$

$$t_w \cong 0.7 \cdot R_T \cdot C_{ext} \tag{9-10}$$

$$R_T C_{ext} \cong \frac{T_w}{0.7}$$

$$R_T C_{ext} \cong \frac{60 \times 10^{-9} \text{ s}}{0.7} = 85.71 \times 10^{-9} \text{ s}$$

Using 5% standard capacitor values for C_{ext}, calculate R_{ext} using Eq. 9-10. Let $C_{ext} = 47$ pF; then $R_T = 1.82$ kΩ. Fig. 9-28*b* shows the resulting circuit.

Example 9–9   Pulse Stretcher

Design a pulse-stretcher circuit that generates an output *Y* for each occurrence of the input *X*, as shown in Fig. 9-29*a*.

Solution

a) Select a nontriggerable one shot (74LS221) to perform the required function. Assign the *A* input to the input *X* signal and the *Q* output to the output *Y* signal.

b) Calculate the % duty cycle. From Fig. 9-29*a*

$$\text{One Shot Period} = t_{period} = 6.2 \text{ s}$$

$$\text{Pulse Duration} = t_{ON} = 400 \text{ ms}$$

$$\% \text{ duty cycle} = \frac{t_{ON}}{t_{period}} \times 100\%$$

$$= \frac{0.4}{6.2} \times 100\%$$

$$\% \text{ duty cycle} = 6.5\%$$

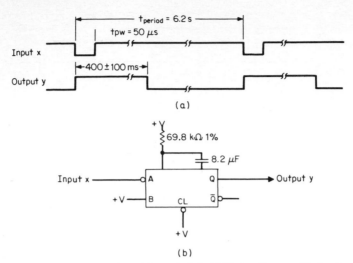

FIG. 9-29 Pulse stretcher of Example 9-9. (*a*) Timing diagram; (*b*) circuit.

c) Select R_T and C_{ext}. The pulse duration (t_w) is defined by the equation

$$t_w = \ln 2 \cdot (R_T \cdot C_{ext})$$

$$t_w \cong 0.7 \cdot (R_T \cdot C_{ext}) \tag{9-10}$$

$$R_T \cdot C_{ext} \cong \frac{t_w}{0.7}$$

$$= \frac{400 \times 10^{-3} \text{ s}}{0.7} = 571 \times 10^{-3} \text{ s}$$

letting $C_{ext} = 8.2\ \mu F$, then $R_T = 69.8\ k\Omega$. Fig. 9-29*b* shows the final circuit with the selected values.

9–8 SPECIAL-PURPOSE CIRCUITS

A special class of circuits has been provided by IC manufacturers that offers solutions to specific problems encountered by the designer. This special class of devices addresses such problems as fault-free operation in a noisy environment, high-performance information transmission techniques, and interfacing of noncompatible logic families. Each circuit type is discussed. The designer should note that this class of circuits covers a broad range of devices and is not limited to this discussion.

9–8a Schmitt Trigger

The Schmitt trigger is a special class of gate. Improved noise immunity and wave-shaping capability characterize the Schmitt trigger gate. The transfer curve for an inverter-type gate is shown in Fig. 9-30.

To better understand the transfer curve, its transition operation will be discussed.

Assume point *a* as a starting point. As the input voltage V_{in} increases, the output voltage V_{out} does not begin to switch until point *b*. As V_{in} continues to increase, V_{out}

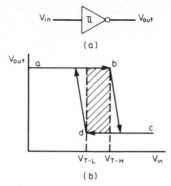

FIG. 9-30 Schmitt trigger inverter gate. (*a*) Logic symbol; (*b*) transfer curve.

switches to a low state, ending at point *c*. To turn the gate output off, V_{in} starts at point *c* and decreases toward *d*. When V_{in} reaches point *d*, the gate output (V_{out}) begins to switch to a high state. V_{in} continues to decrease, ending at point *a*. The shaded area on the transfer curve indicates the amount of input hysteresis for the Schmitt trigger circuit. Several common gate types are provided with the Schmitt trigger capability. The 7413 NAND gate and the 7414 inverter gate are two TTL Schmitt trigger examples.

The Hex Schmitt Trigger (MC14584B) and the Dual Schmitt Trigger (MC14583B) are two examples in the CMOS family.

9–8b Line Drivers

Line drivers are a special gate type that perform a specific function. Cables carrying digital signals must be treated as transmission lines when the component wavelengths of the digital signals become shorter than the electrical wavelength of the cable. This calls for cables with a low characteristic impedance ($Z_0 < 100$ Ω) and devices capable of interfacing with them. TTL-compatible gates are not designed to drive low-impedance lines in the 50 to 100 Ω range. Line drivers are available to handle this task. Fig. 9-31 shows a typical line driver configuration. Note that the receiving end of the transmission line is terminated in its characteristic impedance and will, therefore, exhibit no reflections. This "parallel termination" configuration allows the use of multiple receivers over a distributed bus. A configuration of this type would be used for cable lengths of 2 to 20 in and baud rates up to 10 Mbits/s.

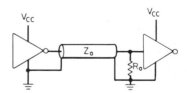

FIG. 9-31 Single-ended simplex operation.

9–8c Level Shifters

From time to time, the design engineer is faced with mixing logic families. One part of the system may require high-speed operation, while another part may allow slower—lower power devices. Proper interfacing between logic families requires that proper voltage levels, current levels, and noise margins be maintained. Fig. 9-32 shows each solution for interfacing the TTL, CMOS, and ECL families.

TTL to CMOS

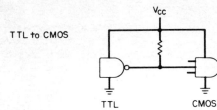

RCA provides a bidirectional CMOS/TTL interface level converter (CD40115) which provides this function without the use of external pullup resistors.

CMOS to TTL

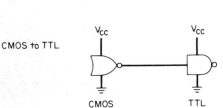

TTL to ECL

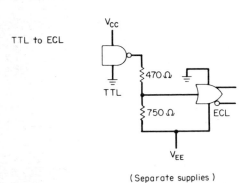

(Separate supplies)

For ECL gates having internal 2 kΩ pulldown resistors, change 750 Ω resistor to 1.2 kΩ

Motorola and Fairchild provide a Quad TTL to ECL translator (MC10124/F10124).

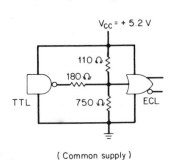

(Common supply)

FIG. 9-32 Level shifters.

ECL to TTL

Motorola and Fairchild provide a
Quad MECL to TTL translator
(MC10125/F10125)

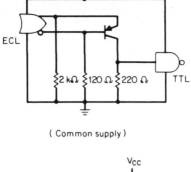

(Common supply)

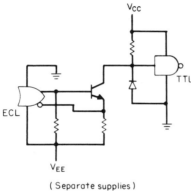

(Separate supplies)

CMOS to ECL

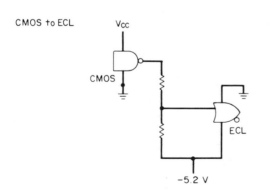

FIG. 9-32 *(Continued)*

9-9 DEVICE SELECTION CHARTS

TABLE 9-4 Preferred TTL SSI Circuits

Function	Description	Speed or propagation delay	Power	Part number
Gates				
Inverter	Hex	9.5 ns	2 mW	74LS04
		3 ns	19 mW	74S04
NAND	Quad 2-input	9.5 ns	2 mW	74LS00
		3 ns	19 mW	74S00
	Triple 3-input	9.5 ns	2 mW	74LS10
		3 ns	19 mW	74S10
	Dual 4-input	9.5 ns	2 mW	74LS20
		3 ns	19 mW	74S20
	8-input	17 ns	2.4 mW	74LS30
		3 ns	19 mW	74S30
NOR	Quad 2-input	10 ns	2.75 mW	74LS02
		3.5 ns	29 mW	74S02
	Triple 3-input	10 ns	4.5 mW	74LS27
	Dual 4-input	10.5 ns	23 mW	7425
AND	Quad 2-input	12 ns	4.25 mW	74LS08
		4.75 ns	32 mW	74S08
OR	Quad 2-input	12 ns	5 mW	74LS32
		4 ns	35 mW	74S32
AND/OR/INVERT	2-wide/2-input	12.5 ns	2.75 mW	74LS51
		3.5 ns	28 mW	74S51
Flip-flops				
JK negative edge triggered	Dual with preset and clear	45 MHz	10 mW	74LS112
		125 MHz	75 mW	74S112
	dual with clear	45 MHz	10 mW	74LS73
JK positive edge triggered	Dual with $\bar{K}$	33 MHz	10 mW	74LS109A
	Input, preset, and clear	33 MHz	45 mW	74109
JK pulse triggered	Dual with clear	20 MHz	50 mW	7473
		30 MHz	80 mW	74H73
	Dual with preset and clear	20 MHz	50 mW	7476
		30 MHz	80 mW	74H76
D positive edge triggered	Dual with preset and clear	3 MHz	4 mW	74L74
		25 MHz	43 mW	7474
		110 MHz	75 mW	74S74
Latch	Quad $\bar{S} - \bar{R}$	12 ns	19 mW	74LS279
One shots				
Nonretriggerable	Dual	20 ns to 70 s	23 mW	74LS221
Retriggerable	Dual	45 ns to ∞	60 mW	74LS123
Special purpose				
Schmitt trigger	Hex	15 ns	9 mW	74LS14
	Quad 2-input positive NAND	15 ns	9 mW	74LS132
		8 ns	45 mW	74S132
	Dual 4-input positive NAND	16.5 ns	9 mW	74LS13

TABLE 9–4 (*Continued*)

Function	Description	Speed or propagation delay	Power	Part number
Special purpose				
Buffer inverting	Octal	10 ns	130 mW	74LS240
		5 ns	450 mW	74S240
Buffer noninverting	Octal	10 ns	135 mW	74LS244
				74S244
50 Ω/75 Ω line drivers	Quad 2-input positive NOR	7 ns	28 mW	74128
	Dual 4-input positive NAND	4 ns	44 mW	74S140
Level shifters	TTL/ECL			MC10124/ F10124
	ECL/TTL			MC10125/ F10125
	TTL/CMOS			CD40115 14504B
	CMOS/TTL			CD40115

Note: Preferred parts in table meet 0 to 70°C operating temperature requirements. SN 54XXX-type parts are available with an operating temperature range of −55 to +125°C.

TABLE 9–5 Preferred CMOS SSI Circuits

Function	Description	Speed or pair propagation delay @ $V_{DD} = +5.0$ VDC (C-load = 50 pF)	Power per package @ 100 kHz	Part number
Gates				
Inverter	Hex-Buffer	110 ns	900 μW	MC14049UB CD4049B
	Hex	130 ns	900 μW	MC14069UB CD4069UB
NAND	Quad 2-input	250 ns	600 μW	MC14011B CD4011B
	Triple 3-input	320 ns	450 μW	MC14023B CD4023B
	Dual 4-input	320 ns	300 μW	MC14012B CD4012B
	8-input	400 ns	150 μW	MC14068B CD4068B
NOR	Quad 2-input	250 ns	600 μW	MC14001B CD4001B
	Triple 3-input	320 ns	450 μW	MC14025B CD4025B
	Dual 4-input	320 ns	300 μW	MC14002B CD4002B
	8-input	400 ns	150 μW	MC14078B CD4078B

TABLE 9–5 (Continued)

Function	Description	Speed or pair propagation delay @ $V_{DD} = +5.0$ VDC (C-load = 50 pF)	Power per package @ 100 kHz	Part number
Gates				
AND	Quad 2-input	320 ns	600 μW	MC14081B CD4081B
	Triple 3-input	320 ns	450 μW	MC14073B CD4073B
	Dual 4-input	320 ns	300 μW	MC14082B CD4082B
OR	Quad 2-input	320 ns	600 μW	MC14071B CD4071B
	Triple 3-input	320 ns	450 μW	MC14075B CD4075B
	Dual 4-input	320 ns	300 μW	MC14072B CD4072B
AND/OR/INVERT	Dual expandable	565 ns	300 μW	MC14506B
EXCLUSIVE-OR	Quad	350 ns	150 μW	MC4070B CD4070B
EXCLUSIVE-NOR		350 ns	150 μW	MC14077B CD4077B
Flip-flops				
JK Positive edge triggered	Dual with preset and clear	1.5 MHz	400 μW	MC14027B CD4027B
D positive edge triggered	Dual with preset and clear	4 MHz	375 μW	MC14013B CD4013B
Latch	Quad-positive or negative edge triggered	220 ns	500 μW	MC14042B CD4042B
One shots				
Retriggerable	Dual	$t_{pw} - 10$ μs to ∞	7 mW $R_L = 2$ kΩ $C_L = 1$ μF	MC14538B CD4538B
Special purpose				
Schmitt trigger inverter	Hex	250 ns	900 μW	MC14584B
Schmitt trigger NAND	Quad 2-input	250 ns	600 μW	MC14093B CD4093B
Level shifters	Dual CMOS to TTL	1300 ns 20 ns	665 μW 20 mW	MC14583B CD40115
	TTL to CMOS	60 ns 320 ns	20 mW 12.5 mW	CD40115 MC14504B
Buffer	Hex	127 ns	900 μW	MC14050B CD4050B
	Hex-tri state outputs	150 ns	1250 μW	MC14503B

Note: The preferred parts listed above are available for two operating temperature ranges: −40 to +85°C and −55 to +125°C.

TABLE 9–6 Preferred ECL SSI Circuits

Function	Description	Speed or propagation delay	Power per package	Part number
Gates				
Inverter	Hex	4 ns	255 mW	MC10195
	Hex with enable	2.9 ns	210 mW	MC10189
NOR	Quad 2-input	2.9 ns	135 mW	MC10102
				F10102
		1.6 ns	291 mW	MC1662
	Quad 2-input with Strobe	2.9 ns	135 mW	MC10100
				F10100
	Quad 3-input	2.0 ns	99 mW	F95004
	Triple 4-3-3-input	2.9 ns	109 mW	MC10106
				F10106
		2.0 ns	57 mW	F95106
	Dual 3-input-3-output	3.5 ns	198 mW	MC10111
				F10111
				MC10211
				F10211
		2.5 ns	146 mW	F95111
AND	Hex	4.0 ns	255 mW	MC10197
OR	Quad 2-input	2.9 ns	135 mW	MC10103
		1.6 ns	291 mW	MC1664
	Dual 3-input-3-output	3.5 ns	198 mW	MC10110
				F10110
		2.5 ns	146 mW	F95110
		2.5 ns	198 mW	MC10210
				F10210
OR/NOR	Quad	2.9 ns	135 mW	MC10101
				F10101
	Triple NOR + one OR/NOR	2.0 ns	99 mW	F95102
	Triple OR + one NOR		99 mW	F95103
	Triple 2-3-2-input	2.9 ns	109 mW	MC10105
				F10105
		2.0 ns	57 mW	F95105
	Dual 1-4-5-input	2.9 ns	73 mW	MC10109
				F10109
		2.0 ns	57 mW	F95109
	Dual 3-input-3-output	2.5 ns	198 mW	MC10212
	Triple 3-input 3-output	2.0 ns	57 mW	F95003
OR/AND	Dual two-wide-3-input	3.4 ns	135 mW	MC10118
				F10118
	Dual two-wide 2-3-input + invert	3.4 ns	135 mW	MC10117
				F10117
	Four-wide 4-3-3-3 input	3.4 ns	135 mW	MC10119
				F10119
	Four-wide + invert	3.4 ns	135 mW	MC10121
				F10121
OR/NOR	Dual	1.6 ns	146 mW	MC1660
		2.0 ns	52 mW	F95002
	Dual 4-5	1.3 ns	156 mW	MC1688

TABLE 9–6 (*Continued*)

Function	Description	Speed or propagation delay	Power per package	Part number
Gates				
EXCLUSIVE-OR	Quad	4.5 ns	218 mW	MC10113 F10113
	Triple 2-input	2.3 ns	286 mW	MC1672
EXCLUSIVE-OR/	Triple	3.7 ns	146 mW	MC10107 F10107
EXCLUSIVE-NOR		2.5 ns	109 mW	F95107
EXCLUSIVE-NOR	Triple 2-input	2.3 ns	286 mW	MC1674
Flip-flops				
JK master-slave	Dual	4.5 ns	354 mW	MC10135 F10135
D master-slave	Single	270 MHz	250 mW	MC1670
	Dual	4.5 ns	291 mW	MC10131 F10131
		3.3 ns	338 mW	MC10231 F10231
	Hex	4.5 ns	572 mW	MC10176 F10176
D-type latch	Dual	2.5 ns	286 mW	MC1668
	Quad	5.6 ns	390 mW	MC10168
	Quad-negative transition	5.4 ns	390 mW	MC10133 F10133
	Quad-positive transition	5.6 ns	390 mW	MC10153 F10153
	Quad-common clock	4.0 ns	182 mW	MC10130 F10130
	Dual	2.5 ns	286 mW	MC1666
Special purpose				
Buffer		2.0 ns	218 mW	MC10188
Level shifters	Quad TTL to ECL	6.0 ns	351 mW	MC10124 F10124
	Quad ECL to TTL	6.0 ns	468 mW	MC10125 F10125
	Triple ECL to CMOS	12.5 ns	499 mW	MC10177
One shot	Single retriggerable	2.8 ns	520 mW	MC10198

Note: The preferred parts listed above are available in three operating temperature ranges as defined by the table below:

Ambient temperature range	Device family type
0 to 70°C	MC 10100 SERIES F10 K SERIES F95 K SERIES
−30 to +85°C	MC10100 SERIES MC10200 SERIES MC1600 SERIES
−55 to +125°C	MC10500 SERIES F10 K SERIES

REFERENCES

Fairchild: *The TTL Application Handbook*, Fairchild Semiconductor, Mountain View, Calif., 1973.

Fleming, D.: Code Conversion-Application Bulletin, Fairchild Semiconductor, Mountain View, Calif., 1967.

Greenfield, J.D.: *Practical Digital Design Using ICs,* Wiley, New York, 1977.

Marcus, M.P.: *Switching Circuits for Engineers,* 2d ed., Prentice-Hall, Englewood Cliffs, N.J., 1967.

Meggerson, Jr., L.: *Switch Bounce Eliminator Does Double Duty,* EDN, November 1, 1970, p. 48.

Meiksin, Z.H.: *Electronic Design with Off-the-Shelf Integrated Circuits,* Parker Publishing Co., West Nyack, N.Y., 1980.

Motorola: "MECL Data Book," Series B, 3d Printing, Motorola, Inc., Phoenix, Ariz., 1982.

Norris, B.: *Digital Integrated Circuits and Operational-Amplifier and Optoelectronic Circuit Design*, McGraw-Hill, New York, 1976.

RCA: "COS/MOS Integrated Circuits," RCA Corporation, Somerville, N.J., 1980.

Stout, D.F.: *Handbook of Microcircuit Design and Application*, McGraw-Hill, New York, 1980.

Texas Instruments: *The TTL Data Book for Design Engineers,* 2d ed., Texas Instruments, Inc., Dallas, Tex., 1976.

Texas Instruments: "Advanced Schottky, Advanced Low-Power Schottky," Texas Instruments, Inc., Dallas, Tex., 1979.

Chapter 10

MSI LOGIC CIRCUITS

Peter Alfke Director
Applications Engineering
Advanced Micro Devices Inc.
Sunnyvale, Calif.

The author developed many of these ideas and circuits while at Fairchild
Camera & Instrument Corp., who kindly gave permission to use material
previously published in the *TTL Applications Handbook* and in the Fairchild
Journal of Semiconductor Progress.

10–1 INTRODUCTION TO MSI

10–1a History of MSI

The name MSI (medium-scale integration) was coined in the mid-sixties when IC technology had advanced to the point where more than a few gates or flip-flops could be
integrated on one monolithic IC. Circuits of 10 to 100 gate complexity were called MSI
to distinguish them from the older SSI (small-scale integration) and from the upcoming
LSI (large-scale integration), which have gate complexities above 100 gates.

10–1b MSI Technologies

MSI circuits became available in several technologies (TTL, Schottky TTL, low-power
Schottky TTL, ECL, and CMOS) and are even today very popular as the most versatile
form of logic. MSI offers the advantage of a fairly high level of integration, which means
low package count, small size, and low power consumption, while maintaining high
performance and full design flexibility.

By utilizing MSI circuits plus a few gates and flip-flops the designer can not only
implement exactly the systems solution that he or she thought of, but can also achieve
a 5-to-1 saving in component count and PC board area and at least a 2-to-1 saving in
power consumption, not to mention the substantial reduction in design and debug time,
compared to the traditional SSI design.

Now, 10 to 15 years after their original introduction, MSI circuits have lost some of their glamor. MSI designs are no longer evaluated against obsolete SSI designs, but rather against MOS microprocessors and against microprogrammable bit slices in bipolar technology.

In many cases, those two approaches offer a better solution, especially when a function is both complex and slow and/or must be easily modifiable.

MSI circuits are used today:

- In applications that require high performance (minicomputers, disk controllers), often together with microprogrammed bit slices

- In specialized applications that take advantage of the wide range of speed-power tradeoffs available in MSI devices using ECL, TTL, and CMOS technology

- In small applications or in one-of-a-kind designs where it is not worth while to use a microprocessor

- As support circuits (also called ''glue'') in microprocessor circuits, where MSI performs address decode, status decode, address-data multiplexing, or similar functions

TTL is the oldest and still the most popular form of MSI, if we include the low-power Schottky variation that appeared in the mid-seventies.

What started as a fractured approach and bitter fight between competing semiconductor houses (T.I. always had the largest number of devices, Fairchild had the more consistent and better thought out features, Signetics had some popular ''designer's choice'' circuits, National introduced the tristate output flavor, and AMD offered better electrical parameters) has now been united under the T.I.-originated 7400 nomenclature. All the others have sacrificed their pride and accepted the 7400 number series (5400 for military temperature range), and every TTL manufacturer has a complete line of 7400-type TTL MSI circuits. T.I., National, Fairchild, Signetics, AMD, and Motorola have highly overlapping TTL MSI product lines.

CMOS appeared much later. It lingered for many years as an RCA specialty used for military and aerospace applications, until Motorola, National, and Fairchild jumped on the bandwagon and CMOS became popular for industrial applications as well. The original RCA 4000 series circuits were not very systems oriented, featuring inconsistent polarities and strange functions combined with parametric and even functional differences between allegedly identical devices from different manufacturers. Belatedly, some of the popular TTL functions were added to the CMOS MSI family, but the competing manufacturers never achieved the same degree of commonality as in TTL MSI.

ECL is less popular. It offers much higher speed than TTL or even Schottky TTL, but at the penalty of reduced noise margins, more demanding and costly interconnections, and higher power consumption. Except for some prescaler and phase-locked-loop circuits that are used in all areas of radio communications, including TV tuners, ECL is used only in high-performance instrumentation and testing applications, and it dominates mainframe computers. Motorola was for a long time the principal supplier of ECL with its MECL I, II, and III families. Their MECL 10,000 family is the most successful; even though it's slower than MECL III, it's much easier to use.

Fairchild introduced the 100K line of subnanosecond SSI, MSI, and even LSI circuits, thus offering much higher speed.

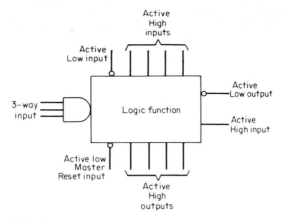

FIG. 10-1 Example of logic notation.

This chapter describes TTL-MSI applications, but the user can substitute the equivalent CMOS-MSI circuit in most cases and ECL-MSI in some cases.

10–1c Nomenclature and Notation

A word on nomenclature and logic notation. MSI circuits represent functional building blocks, though their name cannot possibly convey all the features of each device. It is therefore especially important to draw these blocks in a consistent, clear, and unambiguous way.

The logic used to represent the MSI devices follows MIL Std 806B for logic symbols. MSI elements are represented by rectangular blocks with appropriate external AND/OR gates when necessary, as shown in Fig. 10-1. A small circle at an input means that the specific input is active Low; i.e., it produces the desired function, in conjunction with other inputs, if its voltage is the lower of the two logic levels in the system. A circle at the output indicates that when the function designated is true, the output is Low. Generally, inputs are at the top and left, while outputs appear at the bottom and right of the logic symbol. An exception is the asynchronous Master Reset in some sequential circuits, which is always at the left-hand bottom corner.

Inputs and outputs are labeled with mnemonic letters, as illustrated in Table 10-1. Note than an active Low function labeled outside of the logic symbol contains a bar over the label, while the same function inside the symbol is labeled without the bar. When several inputs or outputs use the same letter, subscript numbers starting with zero are used in an order natural for device operation.

This nomenclature is used throughout this chapter and may differ from nomenclature used on data sheets (notably early 7400 devices), where outputs use alphabetic subscripts or use number sequences starting with one.

10–2 COMBINATORIAL FUNCTIONS

MSI circuits can be classified as either combinatorial or sequential. In combinatorial circuits, the output is only a function of the input conditions. There is no storage and no effect of previous history. Multiplexers, decoders, and arithmetic circuits are examples

TABLE 10–1 Logic Nomenclature

Label	Meaning	Example
I_X	General term for inputs to combinatorial circuits	E I_0 I_1 I_2 I_3 I_4 I_5 I_6 I_7 / S_0 / S_1 / S_2 / 74151 / 8-input / multiplexer / Z Z
J,K S,R D P	Inputs to JK, SR, and D flip-flops, latches, registers, and counters	0 1 / E $D_0 D_1 D_2 D_3$ / 1/2 9308 / dual 4-bit / latch / MR $Q_0 Q_1 Q_2 Q_3$ ‖ 0 1 / E $D_0 D_1 D_2 D_3$ / 1/2 9308 / dual 4-bit / latch / MR $Q_0 Q_1 Q_2 Q_3$
A_X, S_X	Address or Select inputs, used to select an input, output, data route, junction, or memory location	A_0 A_1 A_2 A_3 / 7442A / 1-of-10 / decoder / $Q_0 Q_1 Q_2 Q_3 Q_4 Q_5 Q_6 Q_7 Q_8 Q_9$
$\overline{E}$	Enable, active Low on all TTL/MSI	$\overline{E}$ — E / $A_0 A_1 A_2 A_3 A_4$ $B_0 B_1 B_2 B_3 B_4$ / 9324 / 5-bit / comparator / A>B A<B A=B
$\overline{PE}$	Parallel Enable, a control input used to synchronously load information in parallel into an otherwise autonomous circuit	$\overline{PE}$ — / PE $P_0 P_1 P_2 P_3$ / J 74195 / CP 4-bit / universal Q_3 / K shift register / MR $Q_0 Q_1 Q_2 Q_3$

of combinatorial functions. Sequential circuits contain data storage elements, such as latches or flip-flops, so their outputs are therefore determined not only by the input conditions, but also by the previous state, the history of the circuit. Latches, registers, memories, and counters are examples of sequential functions.

10–2a Multiplexers

Digital multiplexers are combinatorial (nonmemory) devices controlled by a selector address that routes one of many input signals to the output. They can be considered semiconductor equivalents to multiposition switches or stepping switches.

Multiplexers are used for data routing and time division multiplexing. They can also generate complex logic functions. A single multiplexer package can replace several gate

TABLE 10–1 (*Continued*)

Label	Meaning	Example
$\overline{MR}$	Master Reset, asynchronously resets all outputs to zero, overriding all other inputs	PE $P_0 P_1 P_2 P_3$ / CEP 74160 / CET BCD decode TC / CP counter / MR $Q_0 Q_1 Q_2 Q_3$ / $\overline{MR}$
$\overline{CL}$	Clear, resets outputs to zero but does not override all other inputs	E D / A_0 74259 / A_1 8-bit addressable latch / A_2 / CL $Q_0 Q_1 Q_2 Q_3 Q_4 Q_5 Q_6 Q_7$ / $\overline{CL}$
CP	Clock Pulse, generally a High-to-Low-to-High transition. An active High clock (no circle) means outputs change on Low-to-High clock transition	PE $P_0 P_1 P_2 P_3$ / CEP 74161 / CET 4-bit binary TC / CP counter / MR $Q_0 Q_1 Q_2 Q_3$
CE, CEP, CET	Count Enable inputs for counters	
Z_x, O_x, F_x	General terms for outputs of combinatorial circuits	$I_{0a} I_{1a} I_{2a} I_{3a} I_{0b} I_{1b} I_{2b} I_{3b}$ / S_0 9309 dual 4-input / S_1 multiplexer / Z_a Z_a Z_b Z_b
Q_x	General term for outputs of sequential circuits	PE $P_0 P_1 P_2 P_3$ / CEP 74160 / CET BCD decode TC / CP counter / MR $Q_0 Q_1 Q_2 Q_3$
TC	Terminal Count output (1111 for up binary counters, 1001 for up decimal counters, or 0000 for down counters)	

packages, saving printed circuit board area, interconnections, propagation delays, power dissipation, design effort, and component cost.

Table 10-5 contains a listing of the more popular multiplexers. These devices may be reconfigured by parallel connections of inputs. The 74153 dual 4-input multiplexer, for example, can be used as two independent 2-input multiplexers by connecting the inputs in parallel, as shown in Fig. 10-2.

Data Routing Multiplexers can be utilized to route digital data under control of data-selection inputs. The following example illustrates an application of this technique.

Example 10–1 Multiplexer Selection of BCD Counter Bank Contents

Design a circuit to display the contents of one of two multidigit BCD counter banks using multiplexers for data routing.

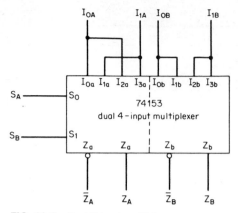

FIG. 10-2 Dual 2-input multiplexer.

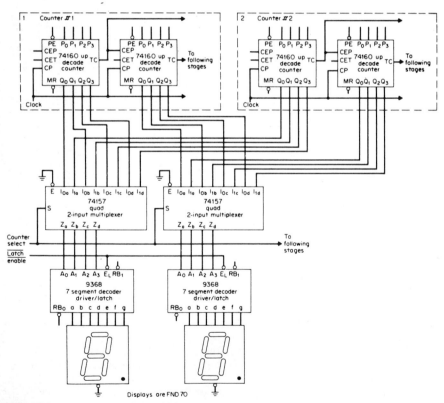

FIG. 10-3 Data routing using multiplexers.

Solution

One typical solution is shown in Fig. 10-3. This system displays the contents of one of the two multidigit BCD counter banks. The 74157 multiplexers select one of the two counters: when the counter Select line is Low, counter 1 is selected; when it is High, counter 2 is selected. The multiplexer outputs feed into the 9368 BCD to 7-segment decoder drivers with input latches.

The display follows the selected counter when the Latch Enable input is Low. When this line is High, the display is no longer affected by input changes, but retains the information that was applied prior to the Low-to-High transition of the Latch Enable. The 9368 interfaces directly with common cathode LED displays.

Multiple-Word Data Bussing   Five 9309 dual 4-bit multiplexers connected as shown in Fig. 10-4 can be used to switch two bits of data from one of 16 words to a 2-bit data bus. The address supplied to the S_0, S_1, S_2, S_3 inputs selects the word to be transferred. If 12-bit words are to be transferred to a 12-bit bus, the circuit is repeated six times. The complementary outputs of the 9309 are used at both levels in order to minimize the through delay. (The Z output is derived from the $\bar{Z}$ output through an additional inverter and is therefore delayed by one additional gate delay.) The two inversions of the two multiplexer levels cancel, so that data is not inverted.

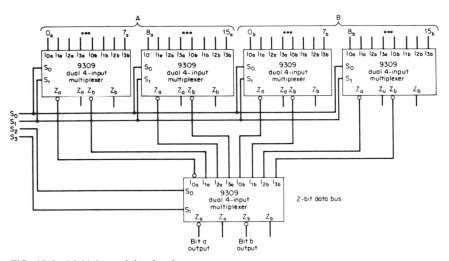

FIG. 10-4   Multiple-word data bussing.

Time Multiplexing   Alone, the 74153 and the 74151 permit time multiplexing of a maximum of four and eight data lines, respectively. By cascading these devices in two or more levels, the number of inputs can be increased. The circuit of Fig. 10-5 shows two levels of multiplexers cascaded to implement a 32-input multiplexer with a delay of about 50 ns. It can be expanded to the 64-input multiplexer without adding delay. In the 32-input multiplexer, the 74151 Enable can be used to gate the selected data out. Note that the negative outputs are used at both levels to improve through delay. Since the assertion output is generated by reinverting the negative output, it is therefore slower.

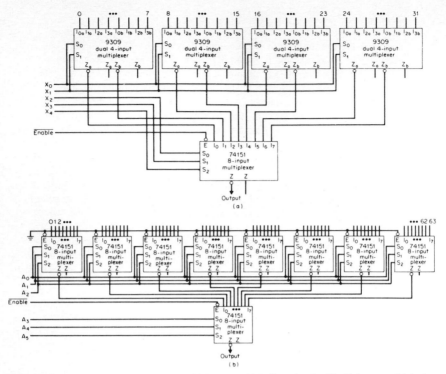

FIG. 10-5   Time multiplexing. (a) The 32-input multiplexing circuit; (b) 64-input multiplexing circuit.

Multiplexers as Function Generators   In most digital systems there are areas, usually in the control section, where a number of inputs generate an output in a highly irregular way. In other words, an unusual function must be generated that is apparently not available as an MSI building block.

In such cases, many designers tend to return to classical methods of logic design with NAND and NOR gates, using boolean algebra, Karnaugh maps, and Veitch diagrams for logic minimization. Surprisingly enough, multiplexers can simplify these designs.

For N input variables, a total of $2^{(2^N)}$ different functions are obtainable. Therefore:

- The 74157 quad 2-input multiplexer can generate any four of the 16 different functions of two variables

- The 74153 and 9309 dual 4-input multiplexers can generate any two of the 256 different functions of three variables

- The 74151 and 74152 8-input multiplexers can generate any one of the 65,536 different functions of four variables

- The 74150 16-input multiplexer can generate any one of the over 4 billion different functions of five variables

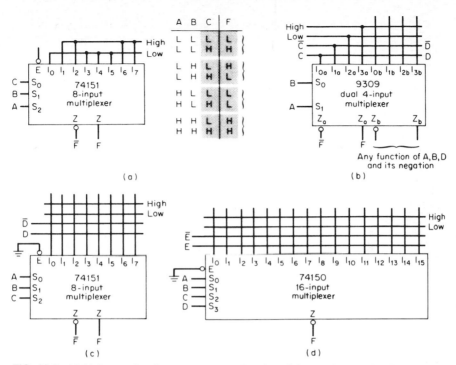

FIG. 10-6 Multipliers as function generators. (*a*) Function of three variables using the 74151. (*b*) Function of three variables using the 9309. (*c*) Function of four variables using the 74151. (*d*) Function of five variables using the 74150.

If a function has a certain regularity, adders or a few NAND, NOR, AND, OR, EX-CLUSIVE-OR, and inverter gates are possibly more economical. However, for a completely random function, the multiplexer approach is more economical. It is certainly more compact, flexible, and easier to design.

Function generation with multiplexers is best explained with examples. An 8-input multiplexer, such as the 74151 illustrated in Fig. 10-6a, can obviously generate any possible function of three variables. The desired function is written as a truth table. The variables A, B, and C are applied to the Select inputs S_0, S_1, S_2, and the eight inputs are connected to either a High or a Low level, according to the required truth table. This method is simple, but inefficient. The same function can also be generated by one-half of a dual 4-input multiplexer, such as the 74153 or 9309. For this purpose, the truth table is divided into four blocks, as shown. Within each block, inputs A and B are constant, but output F can exhibit any of four characteristics:

- Low for both input codes independent of C
- High for both input codes independent of C
- Identical to C
- Identical to $\overline{C}$

Therefore, the function can be implemented by a 4-input multiplexer, as shown in Fig. 10-6b, using the input variables A and B as Select inputs S_0 and S_1, and feeding the appropriate input with one of four signals: either a High, a Low, or the input variables C or $\overline{C}$. The other half of the multiplexer can be used to generate any other function of the variables A, B, and any third variable, though not necessarily C.

The same reasoning can be applied to a function of four variables, as indicated in Fig. 10-6c. An 8-input multiplexer, such as the 74151, can generate any of the 65,536 (2^{16}) possible functions of the four variables A, B, C, and D.

A 16-input multiplexer, such as the 74150 in Fig. 10-6d, can generate any of the more than 4 billion (2^{32}) possible functions of the five variables A, B, C, D, and E.

Switch-Setting Comparator   Sometimes it is necessary to compare a coded value with the actual setting of a multiposition switch for equality. For example, a 3-bit code could be used to represent one of eight possible switch positions. This code is compared with the setting of an eight-position switch, and an output signal is generated to indicate equality. One way to do this is to feed the switch outputs into a priority encoder to generate a 3-bit code corresponding to the switch setting. This code can then be compared with the input code, using an identity comparator.

In Fig. 10-7 the same result is achieved with a single package, using an 8-input multiplexer (74151). The input code is used for address inputs, and the switch outputs are the data inputs to the multiplexer. Because the common terminal of the switch is at ground, the corresponding multiplexer input signal for any given switch setting is grounded. The input code selects a particular input, and the multiplexer output indicates whether this selected input is at ground or open. Pull-up resistors on the switch outputs are recommended to improve noise immunity; however, they are omitted in the drawing for clarity. Although the Enable input of the multiplexer is shown grounded, in practice it could be used for gating the output.

X of Y Pattern Detector   The detection of a specific number (or a specific set) of ones among many inputs is a common design problem, particularly with error-correcting codes and when reading parallel data from multitrack digital tape decks and disks. A straightforward gate-minimized design is quite complex and usually inefficient. Multi-

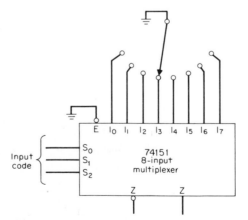

FIG. 10-7   Switch-setting comparator.

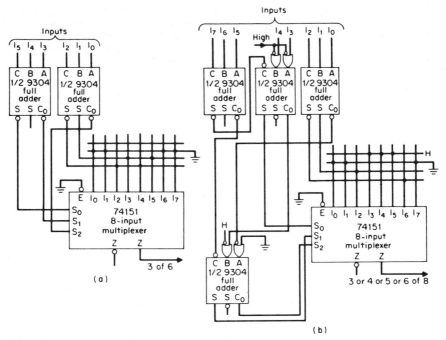

FIG. 10-8 *X* of *Y* pattern detector. (*a*) Three-of-six detector. (*b*) Three-, four-, five-, or six-of-eight detector.

plexers or adders can simplify such designs to some degree, but the most cost efficient design uses a combination of both.

The designs of Fig. 10-8 use full adders to reduce the number of inputs to four variables, and then use an 8-input multiplexer to generate any desired function of these four variables. The result is an output that is High for a specified number (or specified set) of High inputs.

In Fig. 10-8*a* two MSI packages (9304 and 74151) generate a High output when three (and only three) of the six inputs are High.

In Fig. 10-8*b* three MSI circuits generate a High output when three, four, five, or six of the eight inputs are High.

This combination of adders and multiplexers reduces package count to less than half of the equivalent conventional implementation. It also makes this circuit easily programmable for the detection of different patterns.

Seven-Segment-to-BCD Conversion   Multiplexers can be used to perform code conversion. The following example illustrates a unique solution to a conversion requirement.

Example 10–2   Simple, Fast, and Economical 7-Segment-to-BCD Converter

MOS calculator chips offer sophisticated logic and arithmetic capabilities, as many as 24 digits of BCD data storage, and provide outputs for multiplexed numeric display. Because of these features, and their extremely low cost, these chips are being increasingly used in areas beyond the intent of

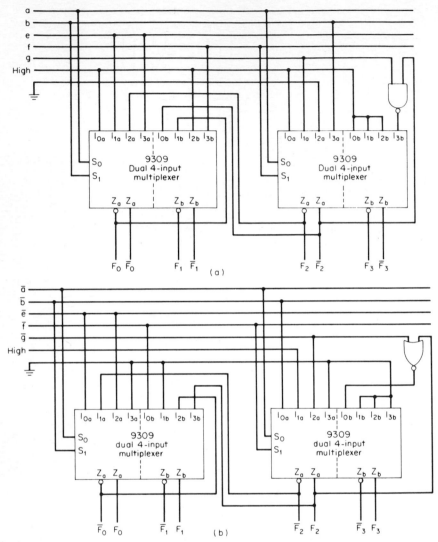

FIG. 10-9 A simple, fast, and economical 7-segment-to-BCD converter. (*a*) Active High segment inputs. (*b*) Active Low segment inputs.

their original design. They are appearing, for example, in digital control, data collection, and data conversion systems. In these applications, a 7-segment encoded output format is often not useful and must be converted back to BCD. A circuit is desired to perform this task.

Solution

Several approaches have been suggested for this conversion, but they have been unnecessarily complicated and expensive.

The circuit shown in Fig. 10-9 uses less than three integrated circuits to convert the 7-segment encoded input signal to BCD outputs. Both active

High and active Low outputs are available simultaneously; a zero is generated when the seven segments are blank. The simplicity of this approach stems from a careful analysis of input redundancies. Only five of the seven segments are actually required to define the character uniquely, even allowing for the different styles of numbers six and nine used in different calculators.

Two dual 4-input multiplexers (9309) with True and Complement outputs are used to reencode the 7-segment inputs, requiring only one additional gate. The circuit of Fig. 10-9a accepts active High (positive logic) inputs where V_{in} is greater than $+2.4$ V for an active segment. Voltage V_{in} is more negative than $+0.4$ V, and is capable of sinking the TTL current for an inactive segment. The circuit of Fig. 10-9b accepts active Low signals.

10–2b Decoders

There are two categories of decoders, logic decoders and display decoder/drivers. Logic decoders are MSI devices controlled by an address. They select and activate a particular output as specified by the address. Display decoders and display decoder/drivers generate numeric codes, such as 7-segment, and then provide the codes to a driver or drive the displays directly.

Logic decoders are the type discussed here. They are available in many configurations and are used extensively in the selective addressing structures of memory systems. They are also used for data or clock routing, demultiplexing, and can act as minterm generators in random and control logic.

Memory Addressing   The most obvious use of the 74139 dual 2-to-4 line decoder is in logic decoding and memory addressing. As shown in Fig. 10-10, the decoder supplies

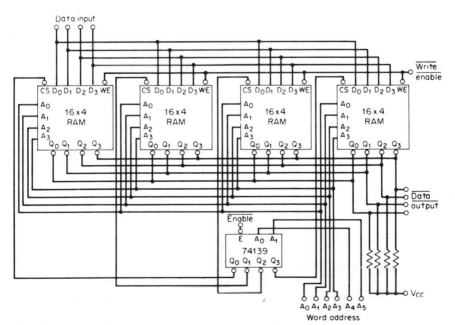

FIG. 10-10   Using the 74139 for memory addressing.

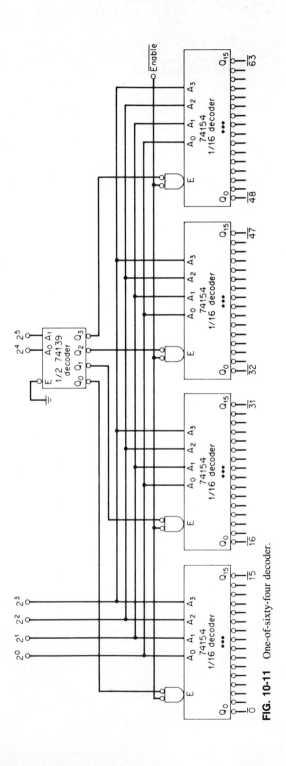

FIG. 10-11 One-of-sixty-four decoder.

the extra decoding necessary to address a word in a 64-word semiconductor memory. One 1-of-4 decoder is used to decode the two most significant bits of memory address and to enable the appropriate memory units. The four least significant bits are decoded on the memory (3101A, 93403, 74189, or 27S02). The high fan-out capability of the 74139 allows it to drive ten memory units with a word length of 40 bits without additional buffers.

One-of-Sixty-Four Decoder   The 74139 can be used to make a 1-of-64 decoder out of four 74154 1-of-16 decoders. Each of the four 74154s shown in Fig. 10-11 is selected by one of the 74139 decoder outputs. Thus, the two most significant bits are decoded by the 1-of-4 decoder and are used to select the appropriate 74154 decoder. The dual AND Enable of the 74154 permits the use of one Enable for selection and the other for strobing. It is preferable to frame decoder address changes at the last level for higher Enable switching speeds.

Four-Phase Clock Generator   Clock demultiplexing for clock distribution and generation is readily accomplished with the 74139. Fig. 10-12 shows a 4-phase clock generator producing nonoverlapping clock pulses for TTL circuitry or to drive MOS circuitry through interfaces. Note that the Enable is used as the clock input, eliminating glitches by framing address changes that occur when the flip-flops, registers, or counters change state on the rising clock edge.

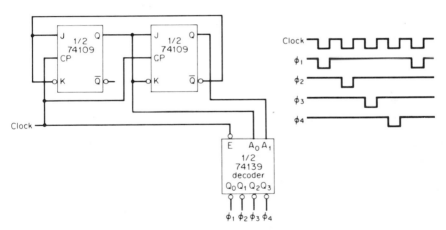

FIG. 10-12   Four-phase clock generator.

Function Generation   Each half of the 74139 generates all four minterms of two variables. These four minterms are useful in some applications, replacing logic functions and thereby reducing the number of packages required in a logic network. All the gate functions which the 74139 can replace are shown in Fig. 10-13, together with a nine's complement circuit utilizing these gate functions.

Switch Encoding   The following example illustrates how utilization of multiplexers can drastically reduce circuit complexity.

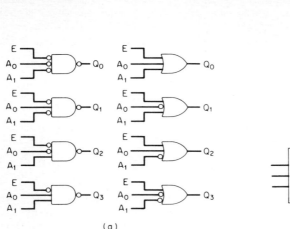

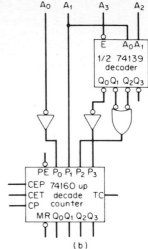

(a)   (b)

FIG. 10-13 Function generation.

Example 10-3   A Scanning Thumbwheel Switch Encoder

Thumbwheel switches are becoming increasingly popular for remote programming of counters, displays, industrial control systems, etc. To reduce the number of interconnections between the switches and the destination, it is desirable to use multiplexing techniques. Ten decades of BCD thumbwheel switches unmultiplexed would require more than 40 interconnections, while a multiplexed system requires less than 20 interconnections. A multiplexed solution is therefore desirable.

Solution

The conventional method of multiplexing uses BCD (or any 4-bit code) thumbwheel switches, each with a diode in series with the four outputs. These are connected to four parallel bus lines to the system output. The wiper arm of each switch is then selected from a decoder. Since the code is generated by the switch, such a conventional system requires different thumbwheel switches for different codes, some of which are considerably more expensive than others, e.g., nine's complement.

The system of Fig. 10-14 requires no diodes and uses standard, low cost, single-pole decade switches. The ten outputs are bussed to a simple encoder that generates the code required; the schematic shows BCD, but the nine's complement is equally simple.

The wiper arm of each switch is separately addressed by the active Low output of the 9302 open collector decoder. Nine pull-up resistors at the encoder inputs ensure proper noise immunity. Open collector decoder outputs are required, since two or more switches might be in the same position, thus interconnecting several decoder outputs. The address applied to the decoder determines which switch is addressed; its position appears at the outputs of the four NAND gates. This system uses fewer and simpler parts and fewer solder joints than a conventional system, providing improved reliability.

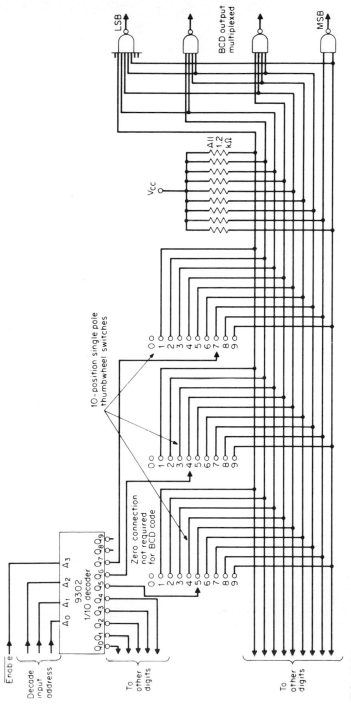

FIG. 10-14 A scanning thumbwheel switch encoder.

10-17

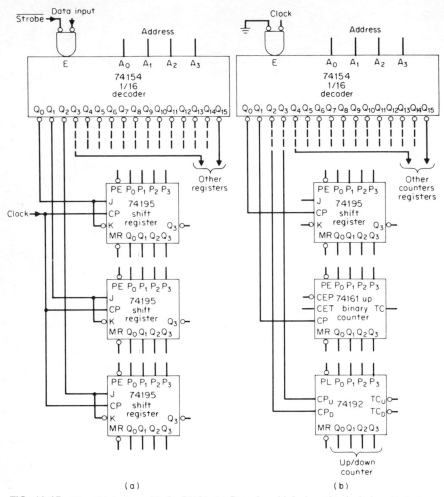

FIG. 10-15   Demultiplexing with the 74154. (*a*) Data demultiplexing; (*b*) clock demultiplexing.

Data Demultiplexing   The 74154 decoder shown in Fig. 10-15*a* can select a particular output under the control of an address, and the active Low Enable can be used as a data input that is routed to a specified output under control of the address input. If the address configuration selects output zero, then this output goes Low if the AND Enable is active, and goes High if it is inactive. Therefore, when the data is inserted into one input of the active Low AND Enable gate, it is switched to the output under control of a strobe present on the other AND input. Thus, the decoder performs a demultiplexing function. Note that all unselected outputs are High.

Clock Demultiplexing   Many applications of this demultiplexing principle are possible. Fig. 10-15*b* shows the 74154 decoder serving as a clock demultiplexer. Under control of the address, the clock is routed to the appropriate register or counter. If the address to the decoder changes after the Low-to-High clock transition, there are no glitches or spikes on unselected outputs.

10–2c   Encoders

Encoders are circuits with many inputs that generate the address of the active input. If a system design guarantees only one encoder input active, the encoder logic is very simple and can be implemented with gates (see Fig. 10-16).

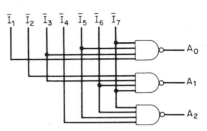

FIG. 10-16   Encoding with gates.

As several inputs can be active at one time, a simple encoder would generate the logic OR of their addresses, which is probably undesirable (i.e., inputs 2 and 4 active would generate address 6).

Priority encoders generate the address of the active input with the highest priority. The priority is preassigned according to the position at the inputs.

The 74148 8-input priority encoder of Fig. 10-17 is a multipurpose device useful in a wide variety of applications, such as priority encoding, priority control, decimal or binary encoding, code conversion, analog-to-digital, and digital-to-analog conversion. A priority encoder can improve computer systems by providing the computer with high-speed hardware priority-interrupt capabilities.

The 74148 provides 3 bits of binary-coded output representing the position of the highest order input, along with an output indicating the presence of any input. It is easily expanded through input and output enables to provide priority encoding over many bits.

The 74148 accepts eight active Low inputs and produces a 3-bit binary weighted output code representing the position of the highest order active input. Thus, when two or more inputs are simultaneously active, the input with the highest priority is encoded and the other inputs are ignored. In addition, all inputs are OR tied to provide a group signal indicating the presence of any Low input signal. This group signal is Low whenever any input is Low and the encoder is enabled.

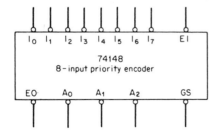

Leads	
$\bar{I}_0$	Priority (active low) input
$\bar{I}_1 - \bar{I}_7$	Priority (active low) inputs
$\overline{EI}$	Enable (active low) input
$\overline{EO}$	Enable (active low) output
$\overline{GS}$	Group signal (active low) output
$\bar{A}_0, \bar{A}_1, \bar{A}_2$	Address (active low) outputs

	$\overline{EI}$	$\bar{i}_0$	$\bar{i}_1$	$\bar{i}_2$	$\bar{i}_3$	$\bar{i}_4$	$\bar{i}_5$	$\bar{i}_6$	$\bar{i}_7$	$\overline{GS}$	$\bar{A}_0$	$\bar{A}_1$	$\bar{A}_2$	$\overline{EO}$
(Disabled)	H	x	x	x	x	x	x	x	x	H	H	H	H	H
(No active input)	L	H	H	H	H	H	H	H	H	H	H	H	H	L
	L	x	x	x	x	x	x	x	L	L	L	L	L	H
	L	x	x	x	x	x	x	L	H	L	H	L	L	H
	L	x	x	x	x	x	L	H	H	L	L	H	L	H
	L	x	x	x	x	L	H	H	H	L	H	H	L	H
	L	x	x	x	L	H	H	H	H	L	L	L	H	H
	L	x	x	L	H	H	H	H	H	L	H	L	H	H
	L	x	L	H	H	H	H	H	H	L	L	H	H	H
	L	L	H	H	H	H	H	H	H	L	H	H	H	H

FIG. 10-17   The 74148 priority encoder.

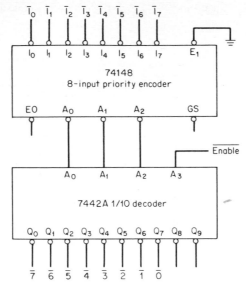

FIG. 10-18 Linear priority encoder.

The 74147 has nine inputs, but lacks the ''zero'' and Enable inputs and the Group Select and Enable outputs required for most applications.

Linear Priority Encoder The linear encoding network of Fig. 10-18 accepts eight active Low inputs and produces a single active Low output corresponding to the highest order input. The network consists of a 74148 to establish the address of the highest order input and a 7442A to decode this address and activate the appropriate output. This method offers a considerable package reduction over discrete linear priority networks and is easily expandable by adding more encoders and decoders. A 16-input encoding network requires only two 74148s, a 1-of-16 decoder (74154), and one gate package.

Digital-to-Analog Conversion Using Rate Multipliers Digital-to-analog conversion, although normally performed using a special IC designed for this purpose, can also be accomplished using rate multipliers. The following two examples illustrate the technique where a 74148 priority encoder is utilized.

Example 10-4 Digital-to-Analog Conversion with a Binary Rate Multiplier

The 74148 can be used for digital-to-analog conversions. In this conversion technique, a rate multiplier is formed and the output is integrated. This method is very economical for multiple D/A conversions because each additional channel of conversion requires only one multiplexer and one integrator.

Solution

In the converters of Fig. 10-19, the 8 bits of binary data are sampled (rate multiplied) and in the course of 256 clock periods, converted to a PDM signal that is fed to an integrator producing the analog output. Each 8-bit digital input is independently sampled by an 8-input multiplexer. The 74148 supplies a code sequence to each multiplexer such that the most significant

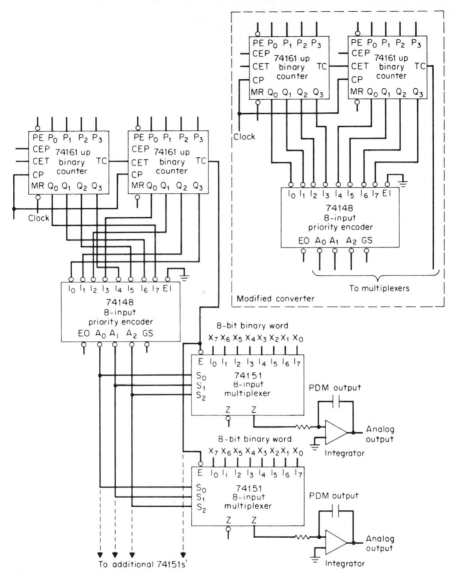

FIG. 10-19 Digital-to-analog conversion with a binary rate multiplier.

binary input is sampled for 50% of the count cycle, the next most significant input is sampled for 25% of the cycle, and so on. This sampling retains the weighting inherent in the binary code.

The converter in Fig. 10-19 generates a well-interlaced PDM signal with a narrow bandwidth that is easily integrated. The output can follow digital input data changes faster than in the other approach shown in the insert. The output of the modified converter is not well interlaced and generates a wide bandwidth PDM signal. It therefore requires a

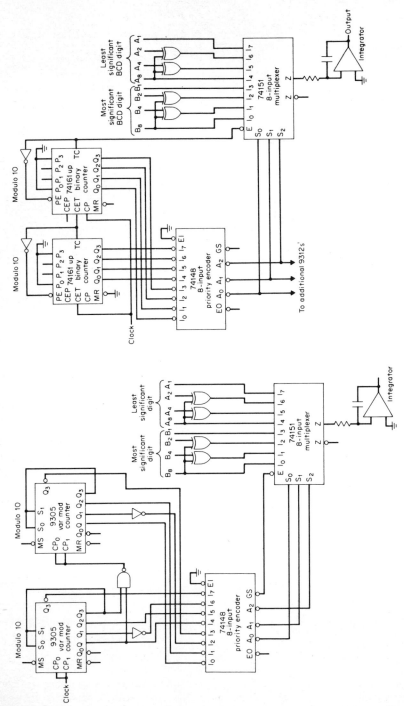

FIG. 10-20 Decimal D/A conversion using a BCD rate multiplier.

longer integrating time constant, but it only has a maximum of eight logic changes per conversion cycle (vs. 256 changes) and is much less sensitive to switching delays, rise and fall times, etc.

At high speeds, switching delays in the priority encoder and multiplexer introduce errors in the PDM output. A resynchronizing D flip-flop at each multiplexer output eliminates these cascaded delays. The maximum clock rate must allow enough time after the counter transition for the propagation delays in the priority encoder and multiplexer. The output of the multiplexer can be integrated or fed to integrating-type devices, such as panel meters, solenoids, or motors.

For each additional channel, a multiplexer and integrator are required. To expand the conversion to more bits, the counter, priority encoder, and multiplexer must be expanded. For example, a 16-bit converter requires a 16-bit counter, a 16-input priority encoder, and a 16-bit multiplexer. As before, each additional channel only requires the addition of a multiplexer and integrator.

Example 10–5   Decimal Digital-to-Analog Conversion Using a BCD Rate Multiplier

BCD 8421 code to analog signal conversion is similar to binary conversion. All the advantages of the binary D/A conversion are retained and only one additional gate package per channel is required. The extra package is needed to manipulate the BCD input data slightly so that correct sampling occurs.

Solution

In the 2-digit BCD D/A converters or rate multipliers shown in Fig. 10-20 a complete conversion occurs every 100 clock pulses. The most significant digit is sampled 90% of the time and the least significant digit is sampled 10% of the time. To obtain the correct weighting, the A_1, A_2, A_4 inputs are sampled, respectively, for one, two, and four sample times. The A_8 input is sampled two sample times alone and in addition is OR tied with the A_4 and A_2 inputs. Therefore, if the A_8 input is a one, the output is High for eight clock pulses. The PDM output is fed to an integrator to produce an analog output. The two decimal converters shown here differ in the same way as the binary converters shown in Fig. 10-19. The first converter produces a well-interlaced pattern, and the second has fewer transitions per conversion cycle.

Keyboard Encoders   Keyboards are becoming increasingly popular as input devices for digital systems, often as substitutes for banks of rotary switches or push buttons. Usually only one key is activated at a time. The address of the active key can then be encoded and transmitted to the digital system on fewer wires.

The design of an encoder for a full ASCII keyboard is fairly well known, since several semiconductor manufacturers sell complete MOS/LSI keyboard encoders that are tailored to the ASCII keyboard requirements. For smaller keyboards, however, these LSI chips represent an expensive overkill and may also be too inflexible. Therefore, it seems worthwhile to explore several cost effective designs for small keyboards with 10 to 64 keys that encode the key strokes in a binary code and provide the proper interface to a digital system.

The design of a keyboard encoder must cope with the following problems:

- Inherently asynchronous key depressions that occur at a very limited rate (less than 10 per second) but can change at any moment

- Mechanical contact bounce whenever contact is made
- Two-key rollover that results when the second entry key is depressed before the previous one has been released

Ten-Key Encoding Using Gates

The simplest, but not the best approach to keyboard encoding is to use TTL NAND gates that require contact closures to ground. This leads to the most straightforward design, as shown in Fig. 10-21, but is not recommended for the following reasons:

- It does not distinguish between "all keys up" and "key zero down"
- It generates erroneous output codes if more than one key is depressed
- It is difficult to debounce
- It requires many input pull-up resistors
- Its parts count becomes prohibitive for more than 16 contacts

The design in Fig. 10-22 eliminates the first of these disadvantages by generating an active Low output signal (address). Thus, the "all keys up" condition generates an "all Low" output equivalent to a binary 15, and therefore different from "zero." All the other disadvantages remain, so obviously this design is only meaningful for up to 15 keys.

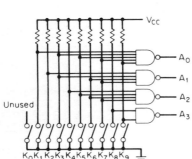

FIG. 10-21 Keyboard encoder with gates.

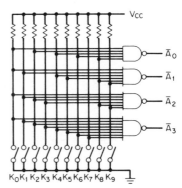

FIG. 10-22 Improved keyboard encoder.

Ten-Key Encoding Using 74148

Fig. 10-23 shows how 10 keys can be encoded with a 74148 priority encoder plus one gate package. A 2-input NAND gate disables the 74148 when contacts K_8 or K_9 are closed, and is used to produce the A_3 output code. When K_8 or K_9 are not closed, the encoder is enabled and encodes inputs $\bar{I}_0$–$\bar{I}_7$, normally. This decimal encoder has active High outputs representing the highest order input. However, just inserting the two inverters in the A_0 and A_3 lines instead of the A_1 and A_2 lines provides active Low outputs.

The Enable Output (EO) is Low if no key is activated. If more than one key is depressed, only the key with the highest number is encoded. This is not as desirable as 2-key rollover, but it prevents the generating of erroneous codes. Note, however, that wrong output codes can be generated for a few nanoseconds following any key depression.

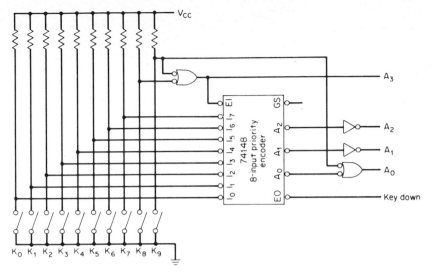

FIG. 10-23   Keyboard encoder using priority encoder.

This circuit is better than the gate configurations, but it is still difficult to debounce, requires pull-up resistors on all inputs, and becomes prohibitively expensive for more than 16 keys.

To solve these problems both economically and reliably, a better approach is to abandon these combinatorial ideas and use a sequential scanning method instead. This increases the response time from nanoseconds to milliseconds, which is generally acceptable.

A Simple 10-Key Scanning Encoder

The scanning encoder of Fig. 10-24 consists of a counter, decoder, contacts, and a controlled oscillator. When all keys are up, the oscillator free runs at about 1 kHz, causing the counter to count modulo 10 and activate the open-collector outputs, one after the other, of the 9302 1-of-10 decoder. The 10-key contacts are connected to the decoder outputs and their common terminal is pulled to V_{CC} through a 1-kΩ resistor.

Depressing any key will cause this common terminal to be pulled close to ground as soon as the counter state becomes identical to the number of the depressed key. This causes the oscillator to stop with a Low level on the clock input to the 7490 decade counter. The oscillator time constant provides some bounce protection; however, under unfavorable circumstances the counter might make one additional complete scan before settling. A Low output signal on the Valid Code line indicates that the counter output corresponds to the number of the depressed key.

Two-key rollover protection is inherent in this design. If a second key is depressed while the counter is still locked onto the first one, the second key is ignored until the first one is released and the counter searches for the other key depression. If two or more keys are depressed simultaneously within 10 ms, or if two additional keys are depressed while the first one is still down, the system cannot resolve the entry sequence but still produces valid codes.

This simple scanning circuit has only a few remaining drawbacks—lack of perfect bounce suppression and difficulty in distinguishing between key bounce and repetitive

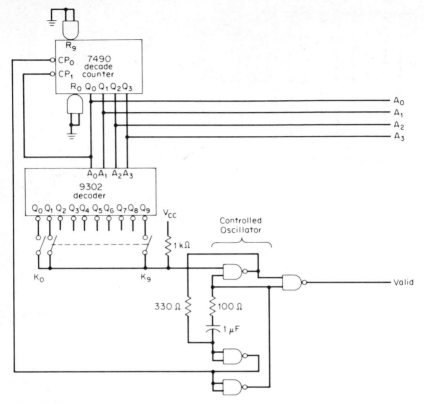

FIG. 10-24   A 10-key scanning encoder.

entry of the same key. To solve these problems, a retriggerable monostable can be added to the circuit so that it can distinguish between short and long times, as shown in Fig. 10-25.

As long as all keys are up, the monostable is continuously retriggered by the clock pulses. When a key is depressed and the oscillator is stopped, the monostable eventually times out. A High level on the $\overline{Q}$ output (Valid) indicates that the counter outputs correspond to the depressed key. When the key is released, the monostable is triggered on the first rising clock edge and terminates the Valid signal one-half clock period before the counter changes state. Thus there is no output ambiguity.

Some digital systems require a pulse, not an edge, to enter data; this can be generated by replacing the 9601 monostable in Fig. 10-25 with a 9602 dual monostable, as shown in Fig. 10-26.

Scanning Keyboard Encoders for 16 or More Keys

The simple 10-key scanning encoder design can be expanded for 16 and even more keys by adding more counter stages and decoders, using the 9302 1-of-10 decoder as a 1-of-8 decoder with the A_3 input as an active Low Enable input. This brute-force design is not recommended for more than 16 keys, since a scanning matrix encoder requires fewer

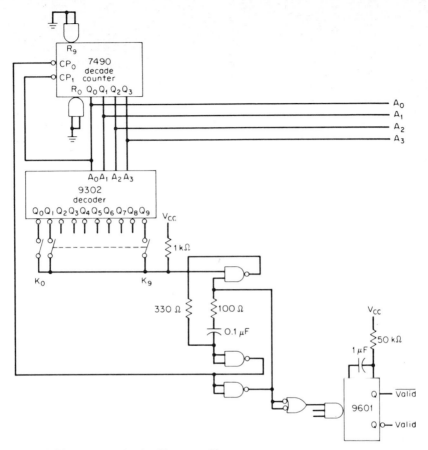

FIG. 10-25   Scanning circuit with monostable.

parts and significantly fewer wires. An 8×2 matrix encoder for 16 keys and an 8×8 matrix encoder for 64 keys are shown in Figs. 10-27 and 10-28, respectively. Note that these circuits require no diodes at the matrix intersections, since no more than one or two keys should be depressed simultaneously. If three or more keys are depressed, a wrong code could be generated; however, this is no real drawback, since the system cannot even resolve the sequence in which these keys were depressed.

Computer Handshake

Some digital systems (computers) require a more sophisticated interface between the keyboard encoder (peripheral) and the receiving logic (processor); Fig. 10-29 shows one possible design. When a key is detected and the bounce has settled, the monostable times out and sets the edge-triggered flip-flop. This generates a Ready signal to the computer and also prevents the scanner from advancing, even if the key is released, until the computer has acknowledged the received data with a strobe pulse that resets the Ready flip-flop.

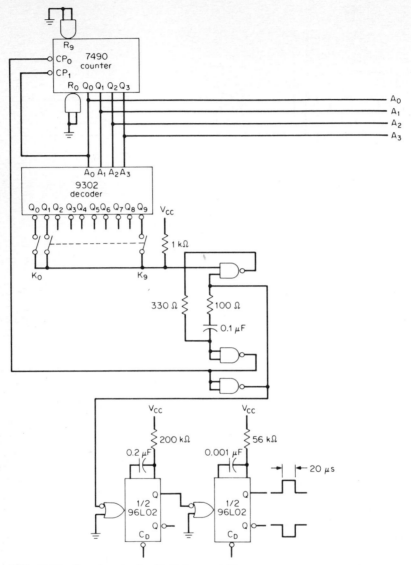

FIG. 10-26   Scanning circuit with dual monostable.

10–2d   Operators

The term "operators" describes a broad category of combinatorial (nonmemory) devices that perform logic operations, such as AND, OR, EXCLUSIVE-OR, invert, and arithmetic operations, such as add, subtract, multiply, divide, and compare the magnitude of two operands or generate/check parity.

Because operators tend to be used in the heart of digital systems, they strongly influence system design and architecture. It is important to investigate the large number

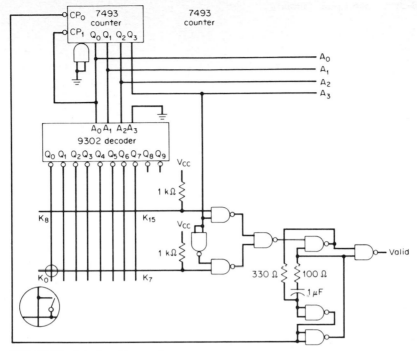

FIG. 10-27 8×2 matrix encoder.

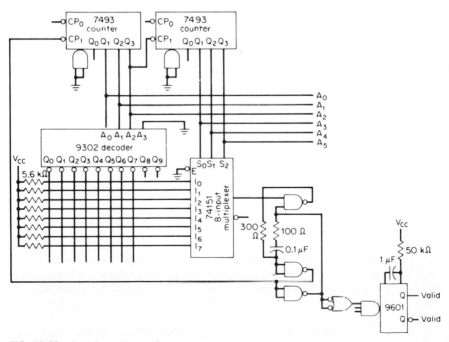

FIG. 10-28 8×8 matrix encoder.

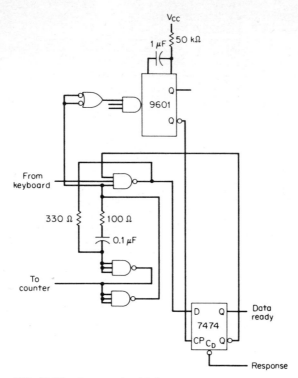

FIG. 10-29   Computer handshake.

of alternate devices before settling on a system design. These devices represent compromises of speed, cost, part count, and connection complexity. The following points are some major design considerations.

- *Speed.* Slower systems usually require fewer and less expensive components and are less sensitive to noise. The system designer should always attempt to use all available time; perhaps by changing to serial architecture, or to incrementing counters, etc.

- *Codes.* Binary arithmetic is simpler than decimal arithmetic. BCD and Excess-3 codes are preferred for decimal operation. Special codes (BCD and Excess-3 Gray) require extensive conversion before use in arithmetic operations.

- *Negative Numbers.* For addition and subtraction, negative numbers are best represented as complements, one's or two's complement in binary notation, nine's or ten's complement in decimal notation. The easiest to generate are one's and nine's complement; however, two's and ten's complements permit faster and simpler arithmetic. For multiplication and division, and for human interfacing (input/output), negative numbers are best represented in signed magnitude form.

- *Versatility.* When several different operations are to be performed, a well-designed arithmetic logic unit (ALU) may be able to execute them in sequence. For example, an ALU can count by incrementing or decrementing a register, or it may be used to control a display multiplexer, etc.

Terminal Confusion   The signals used in digital systems are described in several different and sometimes confusing terms. A logic signal can be either Active (= True) or Not Active (Not True =. False). Digital circuits, on the other hand, are defined for voltage levels that are either High (more positive) or Low (less positive or more negative). Either of these levels can be considered Active (True), the opposite level is then Not Active (False).

MIL Std 806 has established a clear symbology: The High level is considered Active, unless a small circle ("bubble") at the input or output describes the opposite assignment (Low = Active).

In nonarithmetic circuits, the symbols "0" and "1" are unnecessary and confusing because some people think that a one implies a High level, others think of it as an Active (True) signal, and some mistakenly think that it must mean both Active and High.

Therefore, this chapter generally does not use zero and one, but uses the terms Active and Not Active for systems descriptions, and the terms H and L for circuit descriptions and truth tables.

In arithmetic (binary and BCD) systems the terms zero and one cannot be avoided, since they have a mathematical significance. They have to be related to the logic terms in a consistent and unambiguous way.

Arithmetic 1 = Active = True
Arithmetic 0 = Not Active = Not True = False

The rules of Mil Std 806B are then used to describe whether a High level means a one (active High, no bubble) or whether a Low level means a one (active Low, with a bubble at the input or output of the logic symbol).

Functions of Adders   A full adder produces sum and carry outputs as a function of the three inputs A, B, and C. The center truth table of Fig. 10-30 describes the electrical function in terms of High and Low. The two logic truth tables and the two logic symbols describe this circuit in terms of either active High or active Low logic levels. Any logic network that performs binary addition or subtraction can be described in terms of active High as well as in terms of active Low inputs and outputs.

Such equivalence is a basic feature of adder structures and is true regardless of the number of bits and the method of carry propagation. It applies to a single full adder as well as to a complex ALU system.

Carry Signals in Parallel Binary Adders   High-speed digital systems perform addition and subtraction on parallel words of typically 8 to 64 bits. The result of an addition or subtraction at any bit position, however, depends not only on the two operand bits in that position but also on the less significant operand bits. More specifically, the result depends on the carry from the less significant bit positions.

Ripple Carry

In the simplest scheme, each position receives a potential carry input from the less significant position and passes a potential carry on to the more significant position. Thus

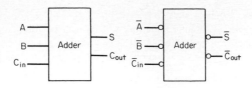

Active High logic function					Electrical function					Active Low logic function				
A	B	C_{in}	S	C_{out}	A	B	C_{in}	S	C_{out}	$\bar{A}$	$\bar{B}$	$\bar{C}_{in}$	$\bar{S}$	$\bar{C}_{out}$
0	0	0	0	0	L	L	L	L	L	1	1	1	1	1
1	0	0	1	0	H	L	L	H	L	0	1	1	0	1
0	1	0	1	0	L	H	L	H	L	1	0	1	0	1
1	1	0	0	1	H	H	L	L	H	0	0	1	1	0
0	0	1	1	0	L	L	H	H	L	1	1	0	0	1
1	0	1	0	1	H	L	H	L	H	0	1	0	1	0
0	1	1	0	1	L	H	H	L	H	1	0	0	1	0
1	1	1	1	1	H	H	H	H	H	0	0	0	0	0

FIG. 10-30   Electrical functions of adders.

the worst-case delay for the addition of two n-bit numbers is n-1 carry delays plus one sum delay. This technique is used with simple adders, such as the 9304 and the 7483 4-bit ripple-carry adder. It uses a minimum of hardware, but it is rather slow.

Carry Lookahead[1]   Addition and subtraction can be made much faster if more logic is used at each bit position to anticipate the carry into this position instead of waiting for a ripple carry to propagate through all the lower positions. An adder constructed with carry anticipation is called a "carry lookahead adder." The carry lookahead technique is explained as follows:

The carry into position 0 is C_0
The carry into position 1 is $C_1 = A_0 \cdot B_0 + C_0(A_0 + B_0)$
The carry into position 2 is $C_2 = A_1 \cdot B_1 + C_1(A_1 + B_1)$

If the two auxiliary functions & and V are defined

$$\&_i = A_iB_i$$
$$V_i = A_i + B_i$$

then the carry equations are:

$$C_1 = \&_0 + V_0C_0$$
$$C_2 = \&_1 + V_1(\&_0 + V_0C_0)$$
$$C_3 = \&_2 + V_2(\&_1 + V_1\&_0 + V_1V_0C_0)$$

or, in general terms:

$$C_{i+1} = \&_i + V_i\&_{i-1} + V_iV_{i-1}\&_{i-2} + V_iV_{i-1}V_{i-2}\&_{i-3} + \cdots$$

The anticipated carry into any position can thus be generated in two gate delays (counting AND/OR/INVERT as one gate delay), one gate delay to generate all the & and V functions, and a second gate delay to generate the anticipated carry. The sum/difference outputs are generated in one additional delay for a total of three gate delays, *independent of word length*.

[1]The material in reduced type is taken from the Fairchild *TTL Applications Handbook*.

The auxiliary functions & and V can be interpreted as

    $\&$ = Carry Generate–AB generates a carry, independent of any incoming carry

    V = Carry Propagate–$A + B$ pass on an incoming carry

This "brute force" carry lookahead scheme is conceptually simple, but, due to the large number of interconnections and the heavy loading of the & and V functions, becomes impractical as the word length increases beyond five or six bits.

    The same concept, however, can be applied on a higher level by dividing the word into practical blocks of 4-bit lengths, using carry lookahead within each block, generating new auxiliary functions G, Carry Generate and P, Carry Propagate which refer to the whole block. G is obviously the carry out of the most significant position of the block. P is defined as Carry Propagate through the block i.e., P is True if a carry into the block would result in a carry out of the block. For a block size of four bits (the 9340 and the 9341/74181)

$$G = \&_3 + V_3\&_2 + V_3V_2\&_1 + V_3V_2V_1\&_0$$
$$P = V_3V_2V_1V_0$$

Neither of these functions is affected by the incoming carry; they will therefore be stable within two gate delays and can be used to supply carry information to the more significant blocks. The carry into block n is:

$$C_n = G_{n-1} + P_{n-1}G_{n-2} + P_{n-1}P_{n-2}G_{n-3} + \cdots$$

This carry in signal is used in the internal carry lookahead structure:

$$C_0 = C_n$$
$$C_1 = \&_0 + V_0C_n$$
$$C_2 = \&_1 + V_1\&_0 + V_1V_0C_n$$
$$C_3 = \&_2 + V_2\&_1 + V_2V_1\&_0 + V_2V_1V_0C_n$$

    The TTL MSI carry lookahead arithmetic logic units, the 9340 and the 9341/74181 use this 2-level carry lookahead, but because of connection differences, they differ in partitioning. The 9340 incorporates the carry in logic in the adder device, but limits it to inputs from three less significant blocks. This gives full carry lookahead over 16 bits, using four 9340 packages.

    The 9341 has more logic flexibility, which requires three additional mode control inputs. It can not therefore, contain any carry in logic. It is contained in a separate device, the 9342/74182. Only one 9342 is needed to achieve full carry lookahead over 16 bits.

    Number Representation   All presently available TTL/MSI adders and ALUs work on binary numbers. Operation in other number systems, such as BCD, Excess 3, etc. is achieved by additional logic and/or additional cycles through the binary adder.

    There is only one way to represent positive binary numbers, but negative binary numbers can be represented in three ways.

- Sign Magnitude—The most significant bit indicates the sign (0 = positive, 1 = negative). The remaining bits indicate the magnitude, represented as a positive number.

        Sign   LSB

         0 1 1 0 1   = $+13$

         1 1 1 0 1   = -13

    This representation is convenient for multiplication and division, and may be desirable for human-oriented input and output, but, for addition and subtraction, it is inconvenient and rarely used.

- Ones Complement—Negative numbers are bit inversions of their positive equivalents. The most significant bit indicates the sign (0 = positive,

1 = negative). Thus $-A$ is actually represented as $2^n - A - 1$. The ones complement is very easy to form, but it has several drawbacks, notably a double representation for Zero (all Ones or all Zeros)

• Twos Complement—This is the most common representation. It is more difficult to generate than ones complement, but it simplifies addition and subtraction. The twos complement is generated by inverting each bit of the positive number and adding one to the LSB.

```
Sign          LSB
0   1   1   0   1   = +13
1   0   0   1   1   = -13
```

Thus an n-bit word can represent the range from $+(2^{n-1} - 1)$ to $-(2^{n-1})$

A 4-bit word can represent the range from 0111 = +7 to 1000 = -8.

Addition and Subtraction of Binary Numbers Addition of positive numbers is straightforward, but a carry into the sign bit must be prevented and interpreted as overflow. When two negative numbers or a negative and a positive number are added, the operation depends on the negative number representation. In twos complement methods, addition is straightforward, but it must include the sign bit. Any carry out of the sign position is simply ignored.

```
+14  01110        + 7  00111        -4  11100
- 7  11001        -14  10010        -3  11101
+ 7  00111        - 7  11001        -7  11001
```

If ones complement notation is used, the operation is similar, but the carry out of the sign bit must be used as a carry input to the least significant bit (LSB). This is commonly called "end-around carry".

```
+14  01110        + 7  00111        -4  11011
- 7  11000        -14  10001        -3  11100
     00110        - 7  11000            10111
+        1                          +        1
+ 7  00111                          -7  11000
```

In twos complement subtraction the arithmetic is performed by inverting; i.e., ones complement the subtrahend and adding, and by forcing a carry into the least significant bit (LSB).

```
 +14   01110       +  7   00111       -6   11010
-(+ 7) -00111     -(+14) -01110      -(+8) -01000
       01110              00111             11010
     +11000            +10001            +10111
+        1        +        1        +        1
 + 7   00111       -  7   11001       -14  10010
```

In ones complement methods, subtraction is performed by inverting; i.e., ones complement the subtrahend and adding, using the Carry Out of the sign position as carry input to the LSB (end-around carry).

```
 +14   01110       +  7   00111       -6   11001
-(+ 7) -00111     -(+14) -01110      -(+8) -01000
       01110              00111             11001
     +11000            +10001            +10111
       00110        -  7   11000            10000
+        1                          +        1
 + 7   00111                          -14  10001
```

It is interesting to note that the Carry Out of the sign position occurs when the result does *not* change sign; *no* carry occurs when the sign changes, implying a "borrow".

Serial Binary Addition   The most versatile full adder circuit is the 9304, two completely independent full adders. One of these adders has an additional set of opposite polarity inputs. The 9304 is used for serial addition and for addition of more than two variables.

Half a 9304 dual full adder and one-half of a dual flip-flop perform serial binary addition, as shown in Fig. 10-31. For active High operands the carry flip-flop must be set when the least significant bit is applied. For active Low operands, the flip-flop must be reset when the least significant bit is applied.

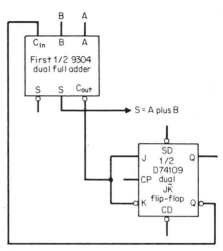

FIG. 10-31   Serial binary addition.

Serial Binary Addition/Subtraction   The most obvious design of a serial adder/subtractor inverts the *B* input for subtraction, using the other half of the 9304 as a conditional inverter, as illustrated in Fig. 10-32. This design requires either a second pass for end-around carry or it requires that the carry flip-flop starts out set for add, reset for subtract (with active High operands, opposite with active Low operands).

This second pass is avoided by using two EXCLUSIVE-OR gates in the data path, thereby effectively using the adder with active High operands in one mode and active

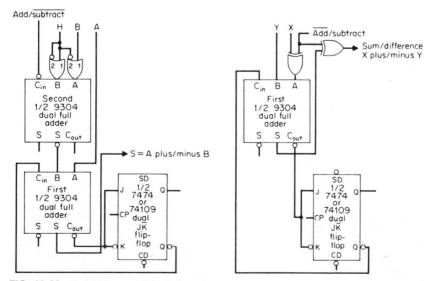

FIG. 10-32   Serial binary addition/subtraction.

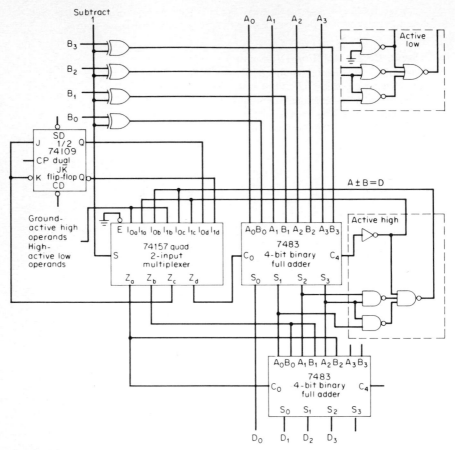

FIG. 10-33   BCD adder/subtractor.

Low operands in the other. For both addition and subtraction, the carry flip-flop must start out set for active High operands, reset for active Low operands.

The 7483 BCD Adder/Subtractor   The 7483 consists of four cascaded full adders. They add 4 bits of A with 4 bits of B plus a carry input, generating four sum bits and a carry output. There are no control inputs and the speed is limited by the internal ripple carry structure. However, this low-cost 4-bit adder in a 16-lead package is useful in medium-speed parallel binary systems and in character-serial BCD arithmetic.

The circuit of Fig. 10-33 performs BCD-corrected addition and subtraction on 4 bits (one digit) in parallel. For addition, the control input (Subtract) is Low, and the first 7483 4-bit ripple-carry adders adds the B_{0-3} inputs to the A_{0-3} inputs, generating the binary sum on outputs S_{0-3} and the binary carry on output C_4. Whenever the binary sum exceeds 9, i.e., when $S_3(S_2 + S_1) + C_4$, a decimal carry is generated by the gating structure shown, setting the carry flip-flop and forcing a binary 6 onto the B inputs of the second 7483 4-bit adder. The outputs $D_0 - D_3$ represent the BCD corrected sum $D = A$ plus B.

For subtraction, the control input (Subtract) is High, inverting B_{0-3} inputs to the first 7483 adder. The 74157 multiplexer feeds the $\overline{Q}$ output of the carry flip-flop into the Carry In of the first 7483, which performs $\overline{\text{Carry}}$ plus A plus $\overline{B}$, the well-known algorithm for binary subtraction. The Carry Out (C_4) signal is inverted before it is routed through the multiplexer into the $J \cdot \overline{K}$ input of the Carry Borrow flip-flop. Whenever this flip-flop is being set, the binary result at S_{0-3} requires correction by subtracting 6 or adding 10. This is performed in the second 7483 by routing the C_4 signal into the C_0 (weight 2) and the B_2 input (weight 8).

The outputs D_{0-3} represent the BCD corrected result $D = A$ minus B. Since BCD addition is an asymmetrical function, the circuit must be slightly modified for active Low operands (see Fig. 10-33).

The 74181 Arithmetic Logic Unit   The 74181 ALU, shown in Fig. 10-34, is a parallel 4-bit MSI device that can perform 16 arithmetic and all 16 possible logic operations on two 4-bit parallel words. The significant arithmetic operations are add, subtract, pass, increment, decrement, invert, and double. The operation is selected by four select lines S_0–S_3 and a mode control line M, which is Low for arithmetic operations and High for logic operations. The device has a Carry In, a Carry Out for ripple carry cascading of units, and two lookahead auxiliary carry functions, Carry Generate and Carry Propagate, for use with the carry lookahead 74182. An open collector $A = B$ output is also provided that can be AND tied to the $A = B$ outputs of other ALUs to detect an all High output condition for several units.

74181 Operation

In the logic of the 74181 ALU, four identical AND/OR networks gate the A and B input operands with the four select lines S_{0-3} to produce the required first-level auxiliary AND and OR functions. These are then used to generate the sum and carry functions. Internal carry lookahead gives high speed. The $A = B$ output is generated by sensing the all-one condition at the F outputs. When control M is in the High state, carries are inhibited from propagating and logic functions are generated at the outputs. The functions available with the device form a closed set such that inversion of the logic inputs produces a function that is still in the set. Therefore, the device performs the same logic and arithmetic functions in the active High representation as it does in the active Low representation, but with a different select code. If a mixed representation is employed, the majority of useful functions are still available. The operation tables for each mode are shown in Table 10-2.

The 74182 Carry Lookahead Generator   The 74181 ALU can be used in a variety of carry modes. The simplest of these is in a ripple-carry mode, where the Carry In C_{in} of an ALU is driven by the Carry Out signal C_4 from the previous ALU. This method of propagating the carry is slow for large word lengths, but has the advantage that additional carry circuits are not required. If several levels of lookahead are permitted and extra logic is used, the speed of the ALU can be improved. The 74181 gives the auxiliary carry functions Carry Generate and Carry Propagate, which can be used with the 74182 to give complete carry lookahead or ripple-block lookahead. In this latter mode, the ALU is split into 16-bit blocks, each with its own lookahead, with carries allowed to ripple between the clocks. The 74182 accepts up to four sets of Carry Generate and Carry Propagate functions and a Carry In, and provides the three Carry Out signals required by the ALUs and also the next level auxiliary functions. These auxiliary functions generated by the carry lookahead circuit allow further levels of lookahead. Unfortunately,

TABLE 10-2 Operating Modes of 74181 ALU

First (upper) diagram and table — active-LOW data

74181 4-bit arithmetic logic unit. Inputs: $\overline{A_0}\,\overline{B_0}\,A_0\,B_0$, $\overline{A_1}\,B_1\,A_1\,B_1$, $\overline{A_2}\,B_2\,A_2\,B_2$, $\overline{A_3}\,B_3\,A_3\,B_3$, C_0, M, S_0, S_1, S_2, S_3. Outputs: C_4, $A=B$, G, P, F_0, F_1, F_2, F_3, $\overline{F_0}$, $\overline{F_1}$, $\overline{F_2}$, $\overline{F_3}$.

S_0	S_1	S_2	S_3	Logic ($M=H$)	Arithmetic ($M=L$, C_0 = Inactive)	Arithmetic ($M=L$, C_0 = Active)
L	L	L	L	$\overline{A}$	A minus 1	A
H	L	L	L	$\overline{A \cdot B}$	$A \cdot B$ minus 1	$A \cdot B$
L	H	L	L	$\overline{A} + B$	$A \cdot \overline{B}$ minus 1	$A \cdot \overline{B}$
H	H	L	L	Logic '1'	minus 1 (2s comp.)	Zero
L	L	H	L	$\overline{A + B}$	A plus $(A + \overline{B})$	A plus $(A + \overline{B})$ plus 1
H	L	H	L	$\overline{B}$	$A \cdot B$ plus $(A + \overline{B})$	$A \cdot B$ plus $(A + \overline{B})$ plus 1
L	H	H	L	$\overline{A \oplus B}$	A minus B minus 1	A minus B
H	H	H	L	$A + \overline{B}$	$A + \overline{B}$	$A + \overline{B}$ plus 1
L	L	L	H	$\overline{A} \cdot B$	A plus $(A + B)$	A plus $(A + B)$ plus 1
H	L	L	H	$A \oplus B$	$A \cdot \overline{B}$ plus $(A + B)$	$A \cdot \overline{B}$ plus $(A + B)$ plus 1
L	H	L	H	B	A plus B	A plus B plus 1
H	H	L	H	$A \cdot B$	$A \cdot B$ plus $(A + B)$	$A \cdot B$ plus $(A + B)$ plus 1
L	L	H	H	Logic '0'	A plus A $(2 \times A)$	A plus A $(2 \times A)$ plus 1
H	L	H	H	$A \cdot \overline{B}$	A plus $A \cdot B$	A plus $A \cdot B$ plus 1
L	H	H	H	$A + B$	A plus $A \cdot \overline{B}$	A plus $A \cdot \overline{B}$ plus 1
H	H	H	H	A	A	A plus 1

Second (lower) diagram and table — active-HIGH data

74181 4-bit arithmetic logic unit. Inputs: $A_0\,B_0\,A_1\,B_1\,A_2\,B_2\,A_3\,B_3$, C_0, M, S_0, S_1, S_2, S_3. Outputs: C_4, $\overline{C_4}$, $A=B$, X, Y, F_0, F_1, F_2, F_3.

S_0	S_1	S_2	S_3	Logic ($M=H$)	Arithmetic ($M=L$, C_0 = Inactive)	Arithmetic ($M=L$, C_0 = Active)
L	L	L	L	$\overline{A}$	A	A plus 1
H	L	L	L	$\overline{A + B}$	$A + B$	$A + B$ plus 1
L	H	L	L	$\overline{A} \cdot B$	$A + \overline{B}$	$A + \overline{B}$ plus 1
H	H	L	L	Logic '0'	minus 1 (2s comp.)	Zero
L	L	H	L	$\overline{A \cdot B}$	A plus $A \cdot \overline{B}$	A plus $A \cdot \overline{B}$ plus 1
H	L	H	L	$\overline{B}$	$(A + B)$ plus $A \cdot \overline{B}$	$(A + B)$ plus $A \cdot \overline{B}$ plus 1
L	H	H	L	$A \oplus B$	A minus B minus 1	A minus B
H	H	H	L	$A \cdot \overline{B}$	$A \cdot \overline{B}$ minus 1	$A \cdot \overline{B}$
L	L	L	H	$\overline{A} + B$	A plus $A \cdot B$	A plus $A \cdot B$ plus 1
H	L	L	H	$\overline{A \oplus B}$	A plus B	(A plus B) plus 1
L	H	L	H	B	$(A + \overline{B})$ plus $A \cdot B$	$(A + \overline{B})$ plus $A \cdot B$ plus 1
H	H	L	H	$A \cdot B$	$A \cdot B$ minus 1	$A \cdot B$
L	L	H	H	Logic '1'	A plus A $(2 \times A)$	A plus A $(2 \times A)$ plus 1
H	L	H	H	$A + \overline{B}$	$(A + B)$ plus A	$(A + B)$ plus A plus 1
L	H	H	H	$A + B$	$(A + \overline{B})$ plus A	$(A + \overline{B})$ plus A plus 1
H	H	H	H	A	A minus 1	A

74181 — 4-bit arithmetic logic unit (active-LOW data)

Pins — inputs: $\bar A_0$ B_0 $\bar A_1$ B_1 $\bar A_2$ B_2 $\bar A_3$ B_3, C_0, M, S_0 S_1 S_2 S_3; outputs: C_4, $A=B$, $\bar F_0$ $\bar F_1$ $\bar F_2$ $\bar F_3$, F_0 F_1 F_2 F_3, G, P.

S_0	S_1	S_2	S_3	$M=H$ (Logic)	$M=L$ (Arithmetic)	$M=L$ (Arithmetic)
L	L	L	L	$\bar A$	A minus 1	A
H	L	L	L	$\bar A + B$	$A\cdot\bar B$ minus 1	$A\cdot\bar B$
L	H	L	L	$\bar A\cdot B$	$A\cdot B$ minus 1	$A\cdot B$
H	H	L	L	Logic '1'	minus 1 (2s comp.)	Zero
L	L	H	L	$\bar A\cdot B$	A plus $(A+B)$	A plus $(A+B)$ plus 1
H	L	H	L	B	$A\cdot\bar B$ plus $(A+B)$	$A\cdot\bar B$ plus $(A+B)$ plus 1
L	H	H	L	$A+B$	A plus B	A plus B plus 1
H	H	H	L	$A\oplus B$	A plus $(A+\bar B)$	A plus $(A+\bar B)$ plus 1
L	L	L	H	$\bar B$	A minus B minus 1	A minus B
H	L	L	H	$A+\bar B$	$A\cdot B$ plus $(A+\bar B)$	$A\cdot B$ plus $(A+\bar B)$ plus 1
L	H	L	H	B	$A+B$	$A+B$ plus 1
H	H	L	H	$A\oplus B$	A plus A $(2\times A)$	A plus A $(2\times A)$ plus 1
L	L	H	H	Logic '0'	A plus $A\cdot\bar B$	A plus $A\cdot\bar B$ plus 1
H	L	H	H	$A\cdot B$	A plus $A\cdot B$	A plus $A\cdot B$ plus 1
L	H	H	H	$A+\bar B$		
H	H	H	H	A	A minus 1	A

74181 — 4-bit arithmetic logic unit (active-HIGH data)

Pins — inputs: A_0 $\bar B_0$ A_1 $\bar B_1$ A_2 $\bar B_2$ A_3 $\bar B_3$, $\bar C_4$, $A=B$, C_0, M, S_0 S_1 S_2 S_3; outputs: $\bar C_4$, $A=B$, $\bar F_0$ $\bar F_1$ $\bar F_2$ $\bar F_3$, F_0 F_1 F_2 F_3, X, Y.

S_0	S_1	S_2	S_3	$M=H$ (Logic)	$M=L$ (Arithmetic)	$M=L$ (Arithmetic)
L	L	L	L	$\bar A$	A	A plus 1
H	L	L	L	$\overline{A+B}$	$A+B$	$A+B$ plus 1
L	H	L	L	$\bar A\cdot B$	$A+\bar B$	$A+\bar B$ plus 1
H	H	L	L	Logic '0'	minus 1 (2s comp.)	Zero
L	L	H	L	$\overline{A\cdot B}$	A plus $A\cdot\bar B$	A plus $A\cdot\bar B$ plus 1
H	L	H	L	$\bar B$	$(A+B)$ plus $A\cdot\bar B$	$(A+B)$ plus $A\cdot\bar B$ plus 1
L	H	H	L	$A\oplus B$	A minus B minus 1	A minus B
H	H	H	L	$A\cdot\bar B$	$A\cdot\bar B$ minus 1	$A\cdot\bar B$
L	L	L	H	$\bar A+B$	A plus $A\cdot B$	A plus $A\cdot B$ plus 1
H	L	L	H	$\overline{A\oplus B}$	A plus B	A plus B plus 1
L	H	L	H	B	$(A+\bar B)$ plus $A\cdot B$	$(A+\bar B)$ plus $A\cdot B$ plus 1
H	H	L	H	$A\cdot B$	$A\cdot B$ minus 1	$A\cdot B$
L	L	H	H	Logic '1'	A plus A $(2\times A)$	A plus A $(2\times A)$ plus 1
H	L	H	H	$A+\bar B$	$(A+B)$ plus A	$(A+B)$ plus A plus 1
L	H	H	H	$A+B$	$(A+\bar B)$ plus A	$(A+\bar B)$ plus A plus 1
H	H	H	H	A	A minus 1	A

10-39

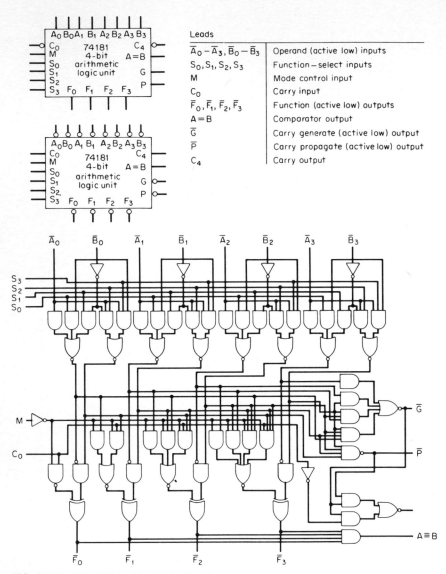

Leads	
$\overline{A}_0 - \overline{A}_3, \overline{B}_0 - \overline{B}_3$	Operand (active low) inputs
S_0, S_1, S_2, S_3	Function − select inputs
M	Mode control input
C_0	Carry input
$\overline{F}_0, \overline{F}_1, \overline{F}_2, \overline{F}_3$	Function (active low) outputs
A = B	Comparator output
$\overline{G}$	Carry generate (active low) output
$\overline{P}$	Carry propagate (active low) output
C_4	Carry output

FIG. 10-34   The 74181 arithmetic logic unit.

to satisfy signal polarities, a penalty of two gate delays is incurred for each level of lookahead, and the auxiliary functions are rarely used over more than two levels of lookahead. The logic symbols and logic diagram of the 74182 carry lookahead circuit are shown in Fig. 10-35. The auxiliary logic functions in the active High case are Not Carry Generate and Carry Propagate. They have been labeled X and Y, respectively. Of course, they are connected in the same manner as the active Low case. In this logic design, the auxiliary functions are used to generate the three Carry Out signals and the two auxiliary functions required for further levels of lookahead.

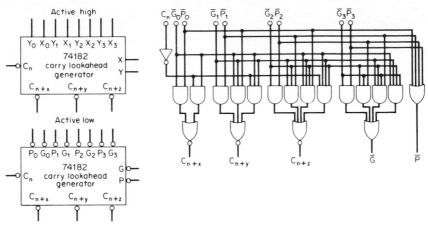

FIG. 10-35   The 74182 carry lookahead generator.

Carry Lookahead Circuit   The single carry lookahead circuit of Fig. 10-36 is used with four 74181 ALUs to perform arithmetic operations with complete carry lookahead over 16-bit words. For word lengths of 20 and 24 bits, the fastest speed is achieved by only using a single 74182 as above and letting the carry ripple through the additional one or two 74181s. For word lengths of 28 and 32 bits, the fastest speed is achieved by using two 74182s, constructing two blocks similar to the 16-bit block above, and letting the carry ripple from the first block to the second. Only when the word length exceeds 32 bits is there a speed advantage in using three levels of carry lookahead.

The 8 × 8 Bit Binary Multiplier   The circuit of Fig. 10-37 performs the conventional shift-and-add algorithm for binary multiplication. It accepts two 8-bit words (A_{0-7} and B_{0-7}) and generates the 16-bit product C_{0-15} after 10 clock pulses. The system is self-contained, requiring a continuously running clock, and generates a Ready signal

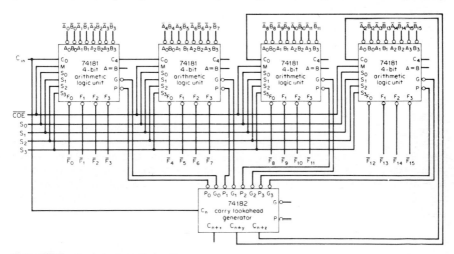

FIG. 10-36   Carry lookahead circuit.

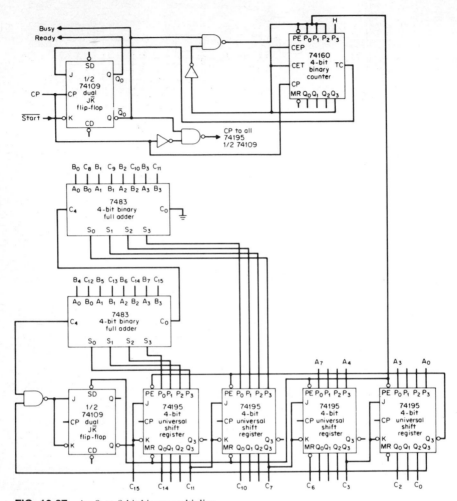

FIG. 10-37   An 8 × 8 bit binary multiplier.

that indicates when the product is available at the 16 outputs. In the idle mode, the 74160 control counter is stopped in position 0 and the $\overline{Q}_0$ Busy output is Low, inhibiting clock pulses to the input/output register. A High-to-Low transition on the $\overline{\text{Start}}$ input initiates a multiplication. The next Low-to-High clock transition resets Q_0 (Ready), generating $\overline{Q}_0$ = Busy and enabling the clock to the input/output registers. It also activates the Parallel Enable ($\overline{\text{PE}}$) inputs of both the 74160 control counter and the A register, as well as the Master Reset ($\overline{\text{MR}}$) inputs of the remaining output register. After the next clock pulse, the 74160 control counter is loaded with code 8, the A register is loaded with the 8 bits of factor A, and the remaining registers are cleared. The next eight clock pulses perform the actual multiplication. Each clock pulse does the following:

- Increments the 74160 control counter
- Right shifts the eight right-hand bits of the input/output register

- Right shifts the entire C register when the LSB of the A register is zero

- Adds factor $B(B_{0-7})$ to the contents of the eight leftmost positions of the C register (C_{8-15}) and inserts the sum one position further to the right when the LSB of the A register is one

When the control counter has reached TC (position 15), it sets Q_0, which generates Ready, and removes the Busy output, indicating that the product is available at the outputs C_{0-15}.

Combinatorial Multipliers   For very fast systems that cannot tolerate the through delay of conventional shift-and-add multiplication, a number of dedicated LSI circuits are available that perform direct, combinatorial multiplication as follows:

2 × 4 bits:	Am 25S05 by AMD
8 × 8 bits:	MM67558 by MMI
	Am25S558 by AMD
	MPY8HJ by TRW
12 × 12 bits:	MPY12HJ by TRW
16 × 16 bits:	MPY16HJ by TRW
	TDC1010 by TRW
	Am29516 by AMD

These circuits are used in digital filters, radar, and sonar systems, and in various instrumentation applications implementing the fast Fourier transform (FFT).

The detailed explanation of these multipliers is beyond the scope of this book. Applications information is available from the manufacturers:

Advanced Micro Devices, Sunnyvale, Calif.
Monolithic Memories, Inc., Sunnyvale, Calif.
TRW, Redondo Beach, Calif.

Comparators   Comparator systems fall into two classes:

- Identity comparators, which detect whether or not two words are identical.

- Magnitude comparators, which also detect which of the two words is larger. Magnitude comparators are more complex and tend to be slower.

All comparators are defined in binary terms, but they can obviously be used with BCD or any other monotonic code without change.

One EXCLUSIVE-OR and one flip-flop form the serial identity comparator of Fig. 10-38a. The flip-flop must start out reset. As long as the A and B inputs are identical, the output of the EXCLUSIVE-OR is Low, leaving the flip-flop in its reset state. When $A \neq B$ the flip-flop is set and stays set until a new cycle is initiated by asynchronously clearing the flip-flop. The state of Q after the last bit has been clocked indicates the result of the comparison:

$$Q: A \neq B \qquad \overline{Q}: A = B$$

Obviously the bit sequence does not affect the identity comparison.

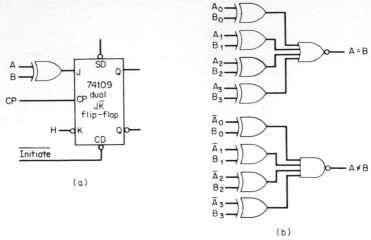

FIG. 10-38 Identity comparators. (*a*) Bit serial operation; (*b*) parallel operation.

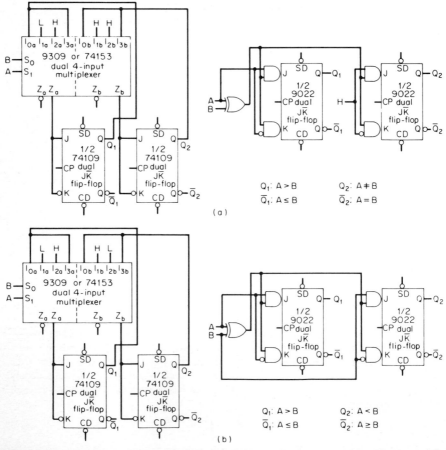

FIG. 10-39 Bit serial magnitude comparison—LSB first. (*a*) Basic circuit; (*b*) rearrangement.

10-44

Parallel identity comparison is most efficiently performed with quad EXCLUSIVE-OR gates with outputs NORed or NANDed. The NAND configuration is faster, but requires opposite polarities of the two operands. Both arrangements are illustrated in Fig. 10-38b.

Bit Serial Magnitude Comparison—LSB First

Magnitude comparison discriminates between three possible conditions: A is greater than B, A is less than B, and A equals B, usually encoded on two output signals.

A serial magnitude comparator for LSB first is most efficiently implemented by either a dual 4-input multiplexer and a dual flip-flop or by an EXCLUSIVE-OR gate and a dual flip-flop with Enable, as in Fig. 10-39a.

Assuming active High notation, Q_1 is set by $A \cdot \overline{B}$, reset by $\overline{A} \cdot B$, and unaffected by

$$A \cdot B \text{ or } \overline{A} \cdot \overline{B} \ (A = B)$$

Q_2 is set by $A \neq B$ and unaffected by $A = B$.

Thus, if both flip-flops start out reset, their state after clocking in the most significant bit indicates the result of the comparison. A slight rearrangement of the same basic circuit (Fig. 10-39b) generates a different set of outputs.

Bit Serial Magnitude Comparison—MSB First

Magnitude comparison is also possible when the serial words come in "backward," with their most significant bits first (Fig. 10-40a). In this case, the first bit where A differs from B determines the result. This circuit sets Q_1 when $A \cdot \overline{B} \cdot \overline{Q_2}$, i.e., if A is greater than B and all previous bits have been $A = B$, leaving Q_1 unaffected under all other conditions.

It sets Q_2 if $A \neq B$, but does not reset it until a new comparison is initiated by clearing both flip-flops.

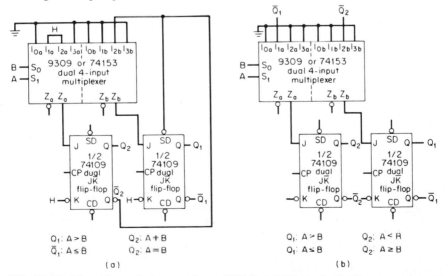

FIG. 10-40 Bit serial magnitude comparison—MSB first. (*a*) Basic circuit; (*b*) rearrangement.

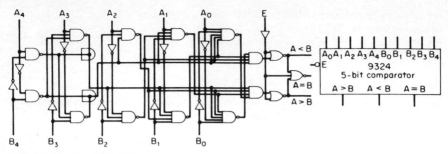

FIG. 10-41   The 9324 5-bit comparator.

A slight rearrangement of basically the same circuit in Fig. 10-40*b* generates a different set of outputs:

$$Q_1: A \text{ is greater than } B \qquad Q_2: A \text{ is less than } B$$

High-speed parallel systems require a direct magnitude comparison over many bits in parallel. In a computer this function is usually performed by the arithmetic logic unit. Subtracting B from A yields a negative result if A is less than B, a positive result if A is greater than B, and zero if $A = B$. If an isolated parallel comparison is needed, it is performed most economically by the 9324 5-bit magnitude comparator or the 7485 4-bit magnitude comparator.

The 9324 5-Bit Comparator

The 9324 of Fig. 10-41 is a 5-bit (or expandable 4-bit) magnitude comparator. It accepts two 5-bit numbers, A_{0-4} and B_{0-4}, and generates three mutually exclusive, active High outputs: A greater than B, A less than B, and A equal to B. When the active Low Enable input is High, all outputs are forced Low. The delay from the operand inputs to the "A is less than B" and "A is greater than B" outputs is a maximum of five gate delays, or approximately 40 ns. The "$A = B$" output is derived from the other two outputs and is therefore delayed by another gate. The 9324 might be ripple expanded as an expandable 4-bit comparator, but since it is a true 5-bit comparator, it can be expanded in parallel, resulting in much faster operation at no extra cost. Parallel comparator arrays are shown in Fig. 10-42 for up to 25 bits.

Error Detection/Correction   Whenever digital data is transferred from one location to another, there is a probability for error due either to device failure or noise. There are numerous ways to handle errors at the system level. Some systems detect errors and request retransmission of data. In other systems, retransmission may be impossible or prohibitively expensive. In such systems, the receiving equipment must not only be able to detect, but also correct the error.

Both error detection and error correction rely on the transmission of redundant information. This requires additional bits of data and lowers the overall efficiency of transmission. In parallel systems additional wires, transmitters, and receivers are required, whereas serial transmission systems use additional time to transmit the redundant information. All these methods cannot completely eliminate errors, but as the percentage of redundant data bits or the sophistication of the error-detection or correction algorithm increases, the probability of undetected or uncorrected errors decreases.

Parity Generation

The simplest and most common method of dealing with errors is the addition of a single extra bit, called a parity bit, chosen such that the total number of ones in the word (counting the parity bit) is odd (in an odd parity system) or even (in an even parity system). Odd parity is generally preferred, since it ensures at least one "1" in any word. At the receiving end, the parity of the word is examined. If any single bit in the word was changed, the detector indicates wrong parity. However, if an even number of errors occurs, this simple method cannot detect it. The parity bit provides only single error detection.

In the serial parity generator of Fig. 10-43, a flip-flop is toggled for every "1" in the data word, and the state of this flip-flop is inserted as a trailing parity bit. On the receiving side, the parity checker has an equivalent flip-flop. Its state is interrogated after the data has been received. Both circuits are easily adapted for odd or even parity systems.

For parallel systems it is necessary to generate the modulo 2 sum of many inputs simultaneously. This requires an array of cascaded EXCLUSIVE-OR circuits. The 74180, 74280, 9348, and 8262 circuits are specifically designed for this function. These are 8- to 12-input parity checkers or generators used in error detecting and correcting applications on parallel data.

Error Correction Using Hamming Codes

A parity bit can only detect single errors. It cannot reliably detect multiple errors and it cannot correct single errors either. A single redundant bit does not carry enough information to do so. It is possible, however, to add more redundant information to the data, formulated such that errors are not only detected but also corrected.

A data word containing an error-correcting field of redundant information is called a Hamming code. It uses several parity bits generated and arranged such that a unique set of parity errors results from an error in any given bit position. For example, three redundancy bits can have a total of eight different states. Since one of these states must indicate "no error," the other seven states can be used to locate an error in any one of seven transmitted bits. Three of the transmitted bits are the redundancy bits themselves, leaving four data bits in which an error can be uniquely detected and also corrected. The coding of the parity bits is done cleverly so that the pattern of parity errors is the binary address of the bit in error. In general, a Hamming code contains $2^m - 1$ bits, m of which are the Hamming or check bits, and $2^m - m - 1$ are the data bits. For example:

Total bits	Hamming bits	Data bits
7	3	4
15	4	11
31	5	26

Thus three additional parity (Hamming) bits can provide single-error correction for 4-bit data words. The seven bits are arranged in the following way:

$$P_0 P_1 D_0 P_2 D_1 D_2 D_3$$

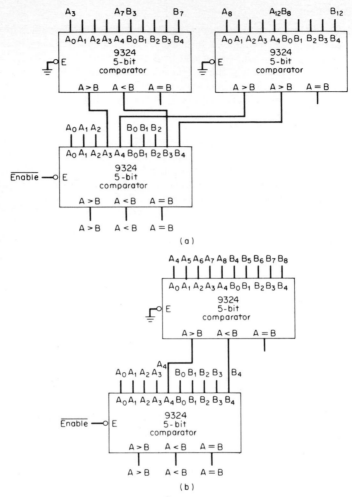

FIG. 10-42   Parallel comparator arrays. (*a*) Ten to thirteen bits; (*b*) 6 to 9 bits; (*c*) 14 to 17 bits; (*d*) 22 to 25 bits.

where D_0, D_1, D_2, D_3 are the four data bits

P_0 is odd parity over bits D_0, D_1, D_3
P_1 is odd parity over bits D_0, D_2, D_3
P_2 is odd parity over bits D_1, D_2, D_3

At the receiving end the three parity bits are again generated from the data bits using an identical scheme. Then these three parity bits are compared with the three transmitted parity bits. If they all match, there was no single error. If they differ, the pattern of mismatches is interpreted as a binary address of the bit in error.

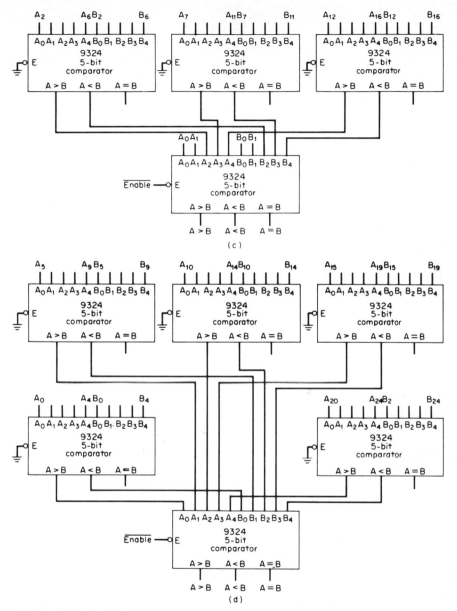

FIG. 10-42 (*Continued*)

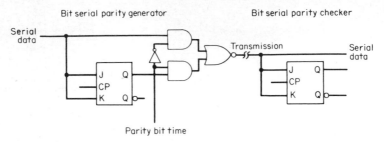

FIG. 10-43 Serial parity generation and detection.

A practical system would avoid the additional comparison and generate the error address (E_{0-2}) by including the received parity bits in the parity check:

E_0 is the odd parity over bits $P_0 D_0 D_1 D_3$
E_1 is the odd parity over bits $P_1 D_0 D_2 D_3$
E_2 is the odd parity over bits $P_2 D_1 D_2 D_3$

This Hamming code can detect and correct single errors, but it will fail on double errors. It would correct the wrong bit. If, however, one more overall parity bit is added, it is also possible to detect (but not correct) double errors. When the receiver finds the overall parity check correct and the error address is zero, there was no error. If the overall parity check is wrong and the error address is not zero, there was a single error that can be corrected. If, however, the overall parity check is correct, but the error address is not zero, then there was a noncorrectable double error.

There are three dedicated LSI devices specifically designed for Hamming error correction of single errors and detection of double errors.

The 64-pin Fujitsu MB 1412A operates on 8 bits and is expandable to up to 64 bits.

The 48-pin AMD 8160 operates on 16 data bits and is expandable to 32 and 64 bits. It supports byte write operations, and has syndrome outputs for error logging as well as diagnostic modes for memory testing.

The 28-pin TI 74630 also operates on 16 data bits. It does not support byte write operation and lacks the syndrome outputs and diagnostic features of the 8160.

For details see the manufacturers' literature:

Fujitsu America, Santa Clara, Calif.
Advanced Micro Devices, Sunnyvale, Calif.
Texas Instruments, Inc., Dallas, Tex.

Code Conversion Numbers can be represented in a large variety of codes. The binary code is the most natural, the simplest, and the one most commonly used in high-speed computer systems. For convenience this code is often grouped in 3-bit groups (octal code) or in 4-bit groups (hexadecimal code), but since these are just different ways of interpreting the binary code, all its features are retained.

Unfortunately, a different numbering system, based on the number 10, is in everyday use, and also mixed numbering systems are used for some special applications (time, angles, etc.). This has created a need for binary-to-BCD and BCD-to-binary converter circuits.

The number of bits and digits involved, the time available, and the amount of general-purpose (perhaps even microprogrammed) logic available in the system are important factors in selecting one of the many different methods available for code conversion.

Any arbitrary code can be converted into any other arbitrary code by using a read only memory (ROM) as a lookup table. This method is very fast with bipolar ROMs, but in most cases it is unnecessarily expensive, since most codes show some kind of regularity. Cheaper and fewer MSI circuits can take advantage of this regularity and provide a more economical solution.

Binary adders are used in high-speed parallel BCD-to-binary conversion. Every bit in a BCD number can be expressed as a binary number, and their sum is the binary equivalent of the whole BCD number.

Two-Digit BCD to 7-Bit Binary Converter Using Adders

Converting a 2-digit BCD number into a 7-bit binary number is accomplished simply and economically with two 4-bit adders. The necessary interconnections are determined by first expressing each of the weighted BCD bits in terms of numbers that are powers of 2.

$$80 = 64 + 16 = 2^6 + 2^4$$
$$40 = 32 + 8 = 2^5 + 2^3 \quad \text{etc.}$$

Arranging the BCD and binary numbers in an orderly array, as shown in Table 10-3, makes it easy to see which of the BCD inputs must be summed into the various

TABLE 10-3 BCD-to-Binary Conversion using Adders

BCD	2^0 (1)	2^1 (2)	2^2 (4)	2^3 (8)	2^4 (16)	2^5 (32)	2^6 (64)	2^7 (128)	2^8 (256)	2^9 (512)	2^{10} (1024)	2^{11} (2048)	2^{12} (4096)	2^{13} (8192)
1	X													
2		X												
4			X											
8				X										
10		X		X										
20			X		X									
40				X		X								
80					X		X							
100			X			X	X							
200				X			X	X						
400					X			X	X					
800						X			X	X				
1000				X		X	X	X	X	X				
2000					X		X	X	X	X	X			
4000						X		X	X	X	X	X		
8000							X		X	X	X	X	X	

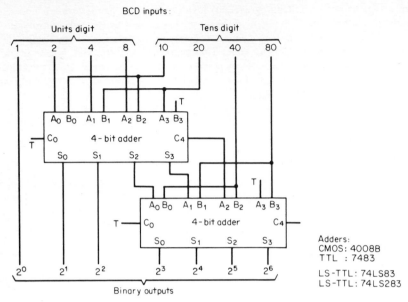

FIG. 10-44   Two-digit BCD-to-binary conversion.

binary outputs. For example, the 2^0 output is just the least significant bit of the unit's BCD digit, while inputs 2 and 10 must be summed to product the 2^1 output. Notice that the 2^3 sum has more than two inputs (8, 10, and 40) and therefore cannot be formed in a single adder stage. Thus, for the 2^3 output, the sum is partially formed in the first adder package and completed in the second, as shown in Fig. 10-44. Inputs marked with a T must be terminated Low for active-High inputs and terminated High for active Low inputs.

Three-Digit BCD to 10-Bit Binary Converter

The parallel BCD-to-binary converter of Fig. 10-45 uses four 7483 4-bit ripple-carry adders to sum all the binary equivalents of the 12 bits in a 3-digit BCD number and generates a 10-bit binary number.

  As indicated in Table 10-3, there are four inputs to the binary eight. This would normally require a considerably more complex adder structure, but since the BCD bits of weight four and eight are mutually exclusive, they can be ORed outside of the adder array and the eight can be split into two fours. Carry lookahead adders can be used for faster operation. This method is practical for three to four digits (four digits require 10 adders). Beyond this, the complexity of the adder structure is prohibitive.

Example 10-6   Eight-Bit Binary to 3-Digit Decimal Display Decoder

The popularity of 8-bit microprocessors has created a demand for 8-bit binary-to-decimal display converters, since a 3-digit number is not only easier to read, interpret, and remember than an 8-bit binary word, but also requires less panel space for the readout. A low-complexity circuit is desired to perform this conversion.

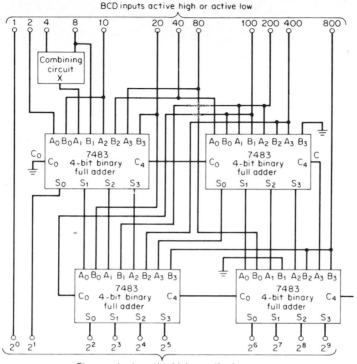

FIG. 10-45 Parallel BCD-to-binary converter.

Solution

ROMs and PROMs are particularly well suited for such code conversion, but a brute-force textbook design would require a 256 × 10 ROM plus three 7-segment decoder/drivers. The circuit of Fig. 10-46 achieves the same result with only a 256 × 4 PROM, three 7-segment decoder/drivers with input latches (9368 or 9374), and two gate packages.

The total number of PROM bits is reduced by excluding the least significant bit from the code conversion (LSB$_{in}$ = LSB$_{out}$), and by combining the I_7 input with one PROM output to generate the three possible values of "hundreds" information according to the small truth table. This reduces the PROM requirement to 128 × (3 + 4 + 1) bits.

Such a PROM is not commercially available, but a 256 × 4 PROM can be used in a time-multiplexed arrangement by utilizing the latches in the 9368s or 9374s to demultiplex the PROM output information. The schematic shows this design in detail.

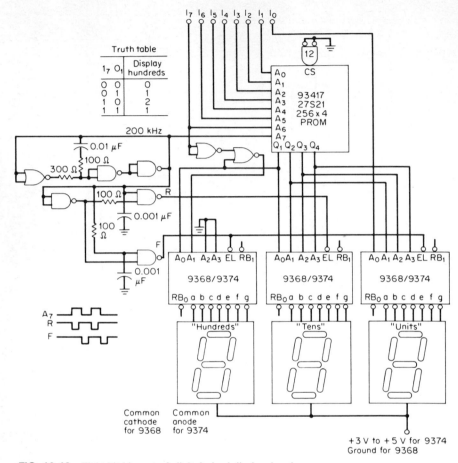

FIG. 10-46   Eight-bit binary to 3-digit decimal display decoder.

Serial-In–Serial-Out BCD-to-Binary Converter

A well-known algorithm generates the binary equivalent of a BCD number by repeatedly dividing it by two. The series of least significant bits generated is the binary output, least significant first. This algorithm can be implemented with 74195 shift registers and some gates or adders, as shown in Fig. 10-47.

When a BCD number is stored in the 74195 shift register, with its LSB in the Q_3 stage, a right shift effectively divides it by two. A problem arises if the LSB of the more significant digit is a one, implying a value of 10 with respect to the first digit. Shifting this one into the Q_0 position changes the ten to an eight, instead of dividing it by two. To correct for this, a three must be subtracted from the new contents of the 74195 register. The circuit shown provides a gate-minimized implementation of this algorithm using the parallel inputs of the 74195 for the correction. It converts a 4-digit (less than 10,000) BCD number into its 14-bit binary equivalent. Operation is started by bit-serially shifting in the three least significant BCD digits (LSB of the LSD first) while the Convert input is Low. The actual conversion starts when the three digits have been shifted in and the

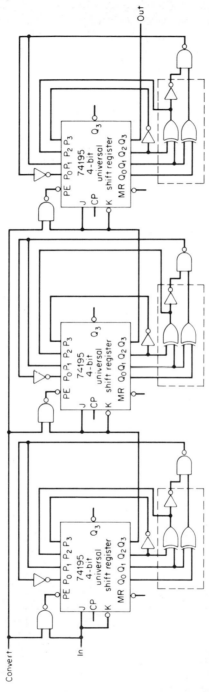

FIG. 10-47 Serial-in–serial-out BCD-to-binary converter.

LSB of the most significant digit is being applied to the serial input. At this point, the Convert input is made High, activating the three correction networks whenever there is a one to be shifted into any of the registers. The next 14 clock pulses shift out the binary result, LSB first. This circuit can be used for any number of digits. It requires only one 4-bit shift register with a conversion network for each decimal digit except the MSD.

Bit-Serial Binary-to-BCD Converter

The reverse of the BCD-to-binary algorithm is used for binary-to-BCD conversion. The binary word is shifted, most significant bit first, into a shift register consisting of several series-connected 74195s. Each shift doubles the contents of the registers in terms of BCD notation. Therefore, a correction is required whenever any of the 4-bit registers contains a number greater than four, which when shifted generates a non-BCD code. This correction is performed by adding three to the contents of the register and inserting the sum one bit downstream into the parallel data inputs. By adding 11 and then ignoring the most significant bit, the same 4-bit adder also detected whether or not the correction is necessary. A binary number is completely converted when its LSB has been shifted in, but the shift register must be long enough to hold the BCD result, always longer than the binary number. This circuit can be used for any number of bits and digits. It requires only one 74195 4-bit shift register, one 7483 4-bit adder, and one inverter for each resulting BCD digit. This arrangement is shown in Fig. 10-48.

Gray Code Conversions

Binary codes are not particularly suited for electrical or electrooptical encoder systems (angular position shaft encoders, etc.) because a movement from one state to the next often results in more than one bit change (from seven to eight, the binary code changes from 0111 to 1000). Such bit changes can never really be simultaneous, so the encoder generates erroneous transient codes when switching between certain positions. This prob-

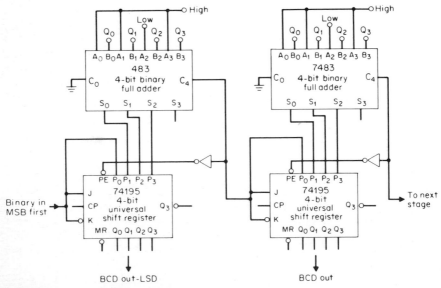

FIG. 10-48 Bit-serial binary-to-BCD converter.

TABLE 10-4 Code Comparison

Decimal	Binary	Gray	Excess 3 Binary	Excess 3 Gray
0	0000	0000	0011	0010
1	0001	0001	0100	0110
2	0010	0011	0101	0111
3	0011	0010	0110	0101
4	0100	0110	0111	0100
5	0101	0111	1000	1100
6	0110	0101	1001	1101
7	0111	0100	1010	1111
8	1000	1100	1011	1110
9	1001	1101	1100	1010
10	1010	1111		
11	1011	1110		
12	1100	1010		
13	1101	1011		
14	1110	1001		
15	1111	1000		

lem is avoided with a Gray code where only one bit changes between adjacent states. The Gray code is a nonweighted code and awkward for other applications. It must be converted to binary or BCD before any arithmetic can be performed. The Gray code is compared to the binary code in Table 10-4.

In Gray-to-binary serial conversion, a flip-flop that toggles for every one performs the conversion. The most significant bit, however, must come in first. Gray-to-binary parallel conversion is performed by a series of EXCLUSIVE-OR gates. These circuits are shown in Fig. 10-49.

In binary-to-Gray serial conversion, a flip-flop acts as a 1-bit delay element and an EXCLUSIVE-OR gate is used between the present and the previous binary bit. Note that, in this case as well as in Gray-to-binary serial conversion, the most significant bit must

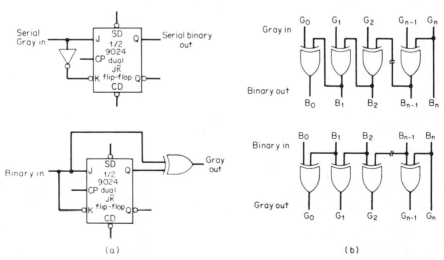

FIG. 10-49 Gray code conversions. (*a*) Serial MSB first; (*b*) parallel.

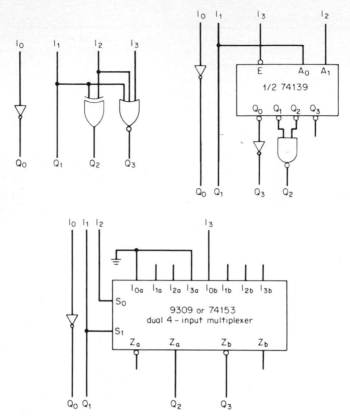

FIG. 10-50 Generating nine's complements.

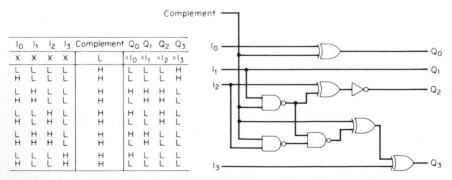

I₀	I₁	I₂	I₃	Complement	Q₀	Q₁	Q₂	Q₃
X	X	X	X	L	=I₀	=I₁	=I₂	=I₃
L	L	L	L	H	L	L	L	H
H	L	L	L	H	L	L	L	H
L	H	L	L	H	H	H	H	L
H	H	L	L	H	L	H	H	L
L	L	H	L	H	H	L	H	L
H	L	H	L	H	L	L	H	L
L	H	H	L	H	H	H	L	L
H	H	H	L	H	L	H	L	L
L	L	L	H	H	H	L	L	L
H	L	L	H	H	L	L	L	L

FIG. 10-51 Controlled nine's complement circuit using two gate packages.

come in first. Binary-to-Gray parallel conversion is performed by a series of EXCLUSIVE-OR gates.

Decimal systems use Excess 3 Gray Code because this code has the feature of changing only one bit at a time, even on a nine-to-zero transition. Excess 3 Gray Code is detected or generated in the same manner as Gray codes, but a three is added to the binary value for binary-to-Excess-3 conversion and a three is subtracted (i.e., adding binary 13) from the binary value for Excess-3-to-binary conversion.

Generating Nine's Complement

The one's complement of a binary number is easily generated by inverting each bit. The equivalent in a decimal (BCD) system, nine's complement, is not that easy. The three circuits of Fig. 10-50 convert a 1-digit BCD input into its nine's complement. They use about one equivalent gate or MSI package per digit (decade). The controlled nine's complement circuit of Fig. 10-51 uses two gate packages and either generates the nine's complement or passes the BCD inputs through unchanged.

10–3   SEQUENTIAL CIRCUITS

10–3a   Latches

Latches are the simplest data storage devices. The basic latch circuit consists of two cross-coupled gates, usually NAND gates. Three forms of latches are shown in Fig. 10-52.

A Low on the $\overline{S}$ input of the basic latch shown in Fig. 10-52a sets the latch (Q High, $\overline{Q}$ Low), a Low on the $\overline{R}$ input resets it. When both inputs are High, the latch stays in its previous state. Using two more gates, as shown in Fig. 10-52b, the latch can be strobed or enabled. When the Enable input is High, the S or R input affects the latch. When E is Low, the latch is not affected by the inputs.

By generating $R = \overline{S}$ (using an additional inverter) the latch is changed to the D-type of Fig. 10-52c. The Q output follows the D input as long as E is High, but the latch stays locked when E goes Low.

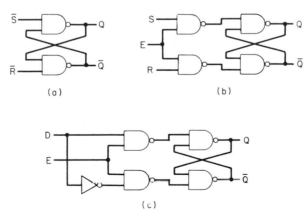

(a)   (b)

(c)

FIG. 10-52   Latches. (a) Basic latch; (b) strobed latch; (c) D-type latch.

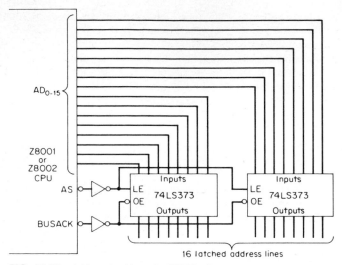

FIG. 10-53 Address latching of a Z8000 microprocessor.

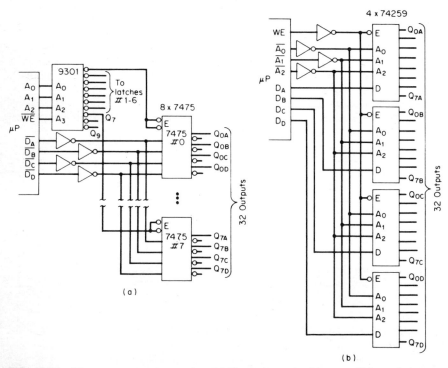

FIG. 10-54 Microprocessor port expansion. (*a*) Classical remedy; (*b*) more efficient solution.

Latches are transparent, i.e., when enabled the output changes as the inputs change. Latches should not, therefore, be used in applications where the latch output feeds back into the input, since this could create a race condition (oscillation). Registers should be used in such cases.

Latches are used to retrieve stable Address information from the time multiplexed Address/Data busses of many popular microprocessors (8085, 8086, Z8000).

The Z8000, shown in Fig. 10-53, uses a 16-bit time-shared Address/Data bus that must be demultiplexed; that is, latches for use with standard memories. $\overline{AS}$ is the obvious control signal for address latching and two 74LS373 Octal Transparent Latches are the best choice for this function. Since addresses are not yet guaranteed valid when $\overline{AS}$ goes Low, it is not possible to use the falling edge of $\overline{AS}$ to clock the addresses into edge-triggered registers. The rising edge of $\overline{AS}$ might be used as a clock, but this delays address availability by almost 100 ns. Transparent latches are the better choice.

Expanding Microprocessor Output Capability

The 74259 (9334) is an 8-bit latch with individual outputs from each latch, but a single, 3-bit addressed data input. This device offers more efficient storage when parallel input is not required and can be used for microprocessor output expansion.

Small microprocessor systems are often limited by their output capabilities, both in the number of output lines and in their drive (sink) current. The classical remedy is a number of quad latches, e.g., eight 7475s driven by four buffered data outputs and selected by a decoder, as illustrated in Fig. 10-54a. This expands one microprocessor output port (eight lines) to 32 TTL outputs at the expense of 10 TTL packages. Obviously, only one set of four TTL outputs can be changed at a time.

A cheaper and more compact solution that achieves the same results with only five TTL packages is shown in Fig. 10-54b. This circuit uses four 74259 8-bit addressable latches and one hex inverter. Note that the four TTL outputs that can be changed simultaneously are now on different packages. The 74LS259 low-power Schottky and the 4724 CMOS addressable latches are equivalent devices. The 4724 eliminates the need for the hex inverter, but offers less output drive.

10–3b   Registers

Registers are data storage devices that are more sophisticated than latches. They use edge-triggered flip-flops and are therefore nontransparent; i.e., the outputs change as a result of a clock edge according to input signals that were present before this clock edge. Outputs can therefore be fed back to inputs without incurring any race conditions. (The asynchronous data inputs of the 7494 and 7496 registers do not follow this rule and must be used more carefully.)

The 74195 is the most versatile 4-bit register. Serial and parallel operations are totally synchronous, and additional flexibility is provided by separate J and $\overline{K}$ serial inputs that form a D input when tied together. The fourth bit has both output polarities brought out.

Quad D Flip-Flop or Dual 2-Bit Register

When the 74195 is operated in the parallel mode it appears as four common-clocked D flip flops, as in Fig. 10-55a. These four flip-flops can be externally interconnected to form other combinations, as in the dual 2-bit configuration of Fig. 10-55b.

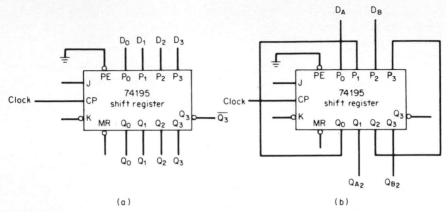

(a)   (b)

FIG. 10-55 The 74195 4-bit universal shift register. (*a*) Quad D flip-flop; (*b*) dual 2-bit register.

Left/Right Shift Register The synchronous parallel inputs of the 74195 can be used to produce a register that shifts left or right on each clock. As shown in Fig. 10-56, the Q_1, Q_2, and Q_3 outputs are connected to the P_0, P_1, P_2 inputs so that each element now shifts right when the Parallel Enable is High and left when it is Low. For left shifting, Q_0 is the serial data output and P_3 is the serial data input.

Counting with Shift Registers The 74195 4-bit universal shift register can be used for a wide variety of counting applications. Twisted ring counters offer glitch-free decoding of any individual state with one inverter and one 2-input NAND gate. Decoding any group of adjacent states (2, 3, 4, 5, 6, or 7) is equally simple. The unused states of these counters are nonpersistent; i.e., the counter reverts into its operating loop if accidentally set to an unused state. This technique is shown in Fig. 10-57.

Twisted Ring (Johnson or Moebius) Reversible Counters Twisted ring reversible counters are possible with shift registers and multiplexers. Individual or adjacent states are easily decoded without glitches with 2-input NAND gates and inverters. Again, all unused states are nonpersistent. Counters for modulo 8 and modulo 6 are given in Fig. 10-58.

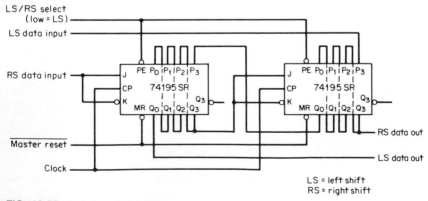

FIG. 10-56 Left/right shift register.

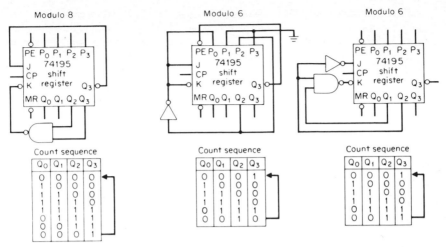

FIG. 10-57 Counting with shift registers.

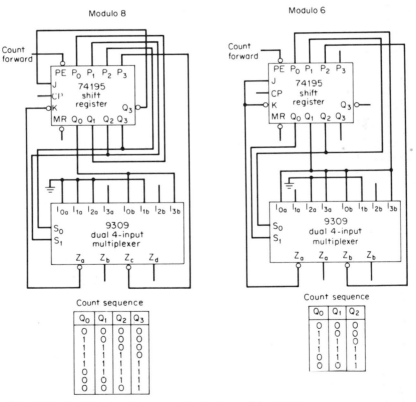

FIG. 10-58 Twisted ring (Johnson or Moebius) reversible counters.

Fast Direction Detector Two digital signals, A and B, are necessary and sufficient to detect and indicate the direction of a moving object. These two inputs could come from the voltages applied to the coils of a stepping motor, or they could be signals from two photocells detecting sprocket holes, etc.

The direction of movement can be detected very simply by applying one signal as the clock, the other as the Data Input to an edge-triggered type-D flip-flop, but such a detector has an inherent delay of up to one full period of B, and therefore does not follow direction changes very well. This can be improved by the circuit of Fig. 10-59, which detects and indicates the direction on every transition of each of the two input signals as early and as accurately as theoretically possible. This circuit uses a high-speed oscillator driving a dual 2-bit shift register fed from the input signals. There are four synchronous outputs: A_1 and B_1, the synchronized equivalents of the input signals; A_2 and B_2, their counterparts, delayed by one clock cycle, or a few 100 ns.

These four outputs are used to determine the direction as follows:

A_1	A_2	B_1	B_2	
			Forward	
H	L	L	L	A goes High while B is Low
L	H	H	H	A goes Low while B is High
H	H	H	L	B goes High while A is High
L	L	L	H	B goes Low while A is Low
			Reverse	
H	L	H	H	A goes High while B is High
L	H	L	L	A goes Low while B is Low
L	L	H	L	B goes High while A is Low
H	H	L	H	B goes Low while A is High

This seemingly complicated logic can be implemented very efficiently with two EXCLUSIVE-OR gates and two NAND gates.

$$\text{Forward} \qquad \overline{(A_1 \oplus B_2)(A_2 \oplus B_1)}$$
$$\text{Reverse} \qquad \overline{(A_1 \oplus B_2)}\,(A_2 \oplus B_1)$$

The remaining two NAND gates are used as a cross-coupled latch to store the direction information.

Asynchronous Data Trap with Independent Data Transfer Many digital systems, particularly computer peripherals, require a storage device that can accept new input data while maintaining the previous output, and transfer the trapped data to the outputs at a later time. A long word can thus be assembled with multiple sequential memory accesses, but the outputs all change simultaneously.

A relatively unknown feature of 74160 through 74163 synchronous counters allows them to be used as a 4-bit data trap. The mode control inputs (CET, CEP, and $\overline{PE}$) are not edge triggered. This approach is illustrated in Fig. 10-60.

When the CET or CEP inputs are permanently disabled (Low) and the clock input is Low, the four master latches receive information from the respective data inputs (P)

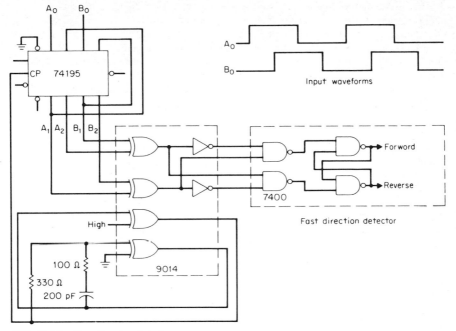

FIG. 10-59 Fast direction detector.

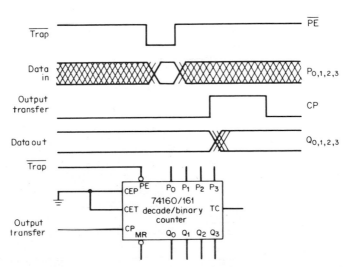

FIG. 10-60 Asynchronous data trap with independent output transfer.

as long as the Parallel Enable ($\overline{PE}$) is Low. When $\overline{PE}$ goes High, the four data bits are trapped in the four master latches, but the four outputs remain in their previous states until the clock input is brought High.

Between the rising edge of $\overline{PE}$ and the rising edge of the clock input, each of the four Master/Slave flip-flops stores the old and the new data statically for any desired length of time. The set-up time of the data inputs with respect to the $\overline{PE}$ input going High is less than 30 ns. The output delay (from clock High to output changing) is less than 23 ns, and the clock High time must be 17 ns or more.

The 9310 and 9316—the original synchronous counters from which the 74160 series was copied—have the same feature, but all Schottky and low-power Schottky counters in the 74160 and 9310/16 family are fully edge triggered and cannot be used in this special application.

Simple Pseudorandom Sequence Generator A simple pseudorandom sequence generator is illustrated in Fig. 10-61. The circuit uses a 9328 and 9300 (74195) shift register and recycles every 50 ms with a 20-MHz clock frequency. The required feedback connection can be expressed as

$$\overline{Q_2 \oplus Q_{19}} = \overline{Q_2}\overline{Q_{19}} + Q_2 Q_{19}$$

To provide this logic without additional gates, Q_2 is fed into the Parallel Enable of the 9300 shift register connected to shift even when parallel loading takes place. When Q_2 is Low, the input to the shift register is $\overline{Q_{19}}$, when Q_2 is High, the input is Q_{19} via the normal $J\overline{K}$ inputs.

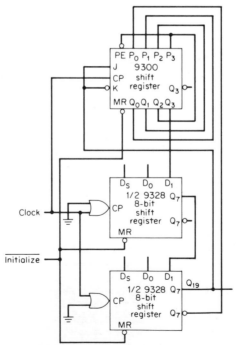

FIG. 10-61 Simple pseudorandom sequence generator.

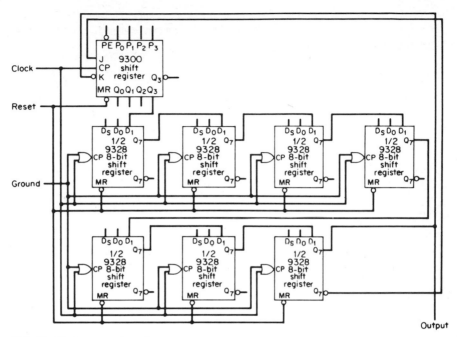

FIG. 10-62   Long pseudorandom sequence generator.

Long Pseudorandom Sequence Generator   The 9328 can be used as part of a long shift counter to provide a pseudorandom sequence, as shown in Fig. 10-62. This counter passes through 2^{60}-1 states, so many that even at a clock frequency of 20 MHz the counter will not repeat until more than 18 centuries have elapsed.

10–3c   Counters

MSI counters are usually 4-bits long and, when enabled, increment or decrement their content by one. Counters can be classified in many ways:

> *Synchronous vs. asynchronous.* In synchronous counters all changes occur as a result of one clock edge, minimizing through delay and output skew. In the simpler asynchronous counters, one flip-flop toggles the other, resulting in substantial through-delay and output skew.
>
> *Presettable vs. nonpresettable.* Presettable (or loadable) counters can be loaded with any value. In the better synchronous counters, this loading is also synchronous, affected by the same clock edge that is used for counting.
>
> *Up/down vs. up only.* Up/down counters offer more versatility but sacrifice some other feature when they are confined to a 16-pin package.
>
> *Binary vs. decimal.* Binary counters are simpler, but decimal counters are more practical in human readable applications.

The 74160 through 74163 synchronous counters (originally introduced as the 9310 and 9316) are especially well suited for synchronous counting.

These counters are fully synchronous, i.e., every change occurs as a result of the rising clock edge. Even the parallel loading is synchronous, enabled by a Low on $\overline{PE}$. Loading overrides counting.

The maximum value (i.e., 9 for the 74160 and binary 15 or F for 74161) is decoded and activates the TC (Terminal Count) output. There are two Count Enable inputs.

The counter increments only when both CEP (Count Enable Parallel) and CET (Count Enable Trickle) are High. The difference between CEP and CET is that a Low on CET forces TC Low, whereas CEP does not effect TC. The 74160 and 74161 have an asynchronous clear input (MR), whereas the clear input on the 74162 and 74163 is synchronous.

Multistage Synchronous Counters   For multistage counting, all less significant stages must be at their terminal count before the next more significant counter is enabled. The 74160 and 74161 internally decode the terminal count condition, which is ANDed with the CET input to generate the TC output. This arrangement, shown in Fig. 10-63, allows series enabling by connecting the TC output (Enable signal) to the CET input of the following stage. This setup requires very few interconnections but has a drawback: the counter chain is fully synchronous, but since it takes time for the Enable to ripple through the counter stages, maximum counting speed is reduced. This drawback can be overcome by proper use of the CEP and CET inputs. The CEP input of the 74160 and 74161 is internally ANDed with the CET input and connected to the R and S inputs of the individual flip-flops within the counter. This feature makes it possible to build a multistage counter that can operate as fast as a single counter stage. The advantage of the ''enable while counting'' method is best seen by assuming that all stages except the second and last are in their terminal condition. As the second stage advances to its terminal count, an Enable is allowed to trickle down to the last counter stage, and it has the full

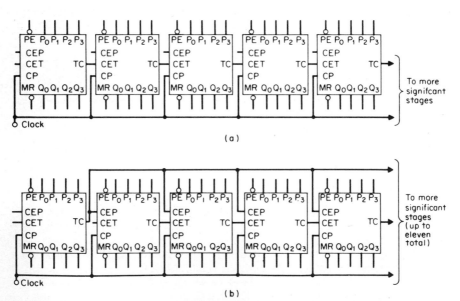

FIG. 10-63   Multistage counting. (a) Slow multistage counting scheme for 74160 through 74163 counters. (b) High-speed multistage counting scheme for 74160 through 74163 counters.

cycle time of the first counter to reach it. When the TC of the first stage goes active (High), all CEP inputs are activated, allowing all stages to count on the next clock.

Multistage Programmable Counters   In the multistage programmable decimal and binary counters, shown in Fig. 10-64, the state prior to Terminal Count (TC-1) is decoded and activates the $\overline{PE}$ input. Therefore, the next clock pulse does not increment the counter to Terminal Count (all nines for decimal, all ones for binary), but rather loads the program value into the counter. The counters are programmed with the nine's or one's complement of the count modulos, instead of the more complicated ten's or two's complement used in the conventional approach. The maximum count frequency is limited by the delay in TC decoding and the set-up time of the $\overline{PE}$ input. This can be improved with an additional flip-flop, as shown next.

The maximum count frequency of a programmable counter can be improved by decoding the TC-2 state of the counter and synchronizing this state in a fast flip-flop, such as the 74S109. This method is illustrated in Fig. 10-65.

The clock pulse that increments the counter to TC-1 also resets this flip-flop, thus activating the $\overline{PE}$ input. The next clock pulse loads the counter with the program value. Guaranteed count frequency can be as high as 25 MHz, limited only by the sum of the t_{pd} of the flip-flop plus the set-up time of the $\overline{PE}$ inputs.

The programmable counters, shown in Figs. 10-64 and 10-65, suffer from a decrease in maximum counting speed when they are programmed with certain unfavorable numbers that do not allow enough time for the delay of the TC ripple chain.

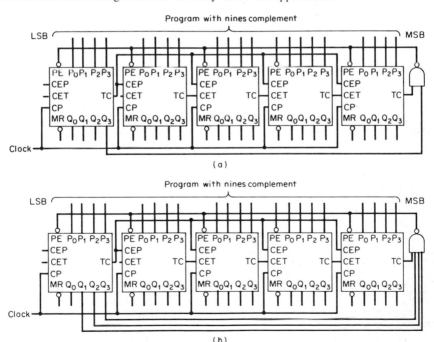

FIG. 10-64   Multistage programmable counters. (*a*) Decimal (74160, 74162); (*b*) binary (74161, 74163).

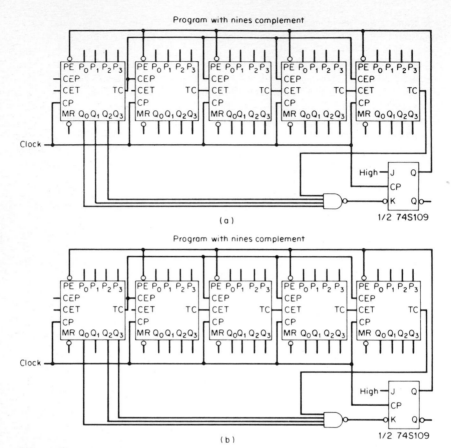

FIG. 10-65 Multistage programmable counters. (*a*) Decimal (using 74160, 74162); (*b*) binary (using 74161, 74163).

For example, assume that a BCD counter is programmed for modulo 90. The counting sequence is

MSD ⌐   ⌐ LSD
99996
99997
99998   activates $\overline{PE}$
99909   nine's complement of 90 is loaded
99910
etc.

The ripple TC output from the MSD must disappear during one clock period (when 99909 has been loaded).

If the clock period is shorter than this ripple delay, the next clock pulse reloads and the counter divides by the wrong number. This problem is overcome by a second flip-flop, as shown in Fig. 10-66.

The dual flip-flop provides additional time for the TC outputs to ripple Low, since

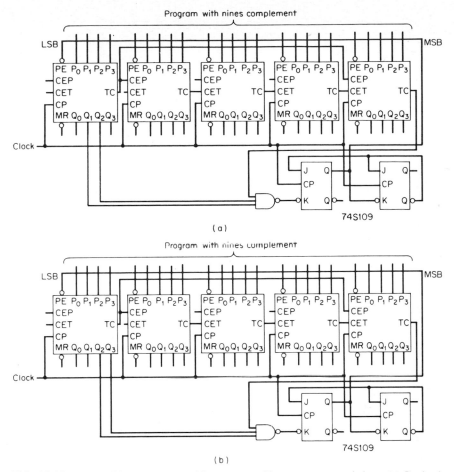

FIG. 10-66 Fast multistage programmable counters without program restrictions. (*a*) Decimal (using 74160, 74162); (*b*) binary (using 74161, 74163).

it activates the $\overline{PE}$ signal for two clock pulses instead of one. The two flip-flops form a modulo 3 counter and are normally set. TC-3 is decoded and activates the reset ($\overline{K}$) input of the first flip-flop. The next clock pulse increments the counter to TC-2 and resets the first flip-flop. This activates the $\overline{PE}$ inputs and the reset ($\overline{K}$) input of the second flip-flop. The next clock pulse loads the program value into the counter and resets the second flip-flop. The following clock pulse loads the counter again and it sets both flip-flops. The next clock pulse increments the counter.

Counters with 50% Duty-Cycle Output   Four circuits that divide by 6, 10, 12, and 14 are shown in Fig. 10-67. The Q_3 output provides a 50% duty-cycle output. No additional gates are required, except in the divide-by-14 circuit. In addition, all the count sequences start on 0000 and end on 1111, which means the Master Reset ($\overline{MR}$) input and the Terminal Count (TC) output still function properly.

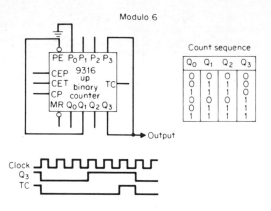

Modulo 6

Count sequence

Q₀	Q₁	Q₂	Q₃
0	0	0	0
0	1	1	0
1	1	1	0
0	0	0	1
0	1	1	1
1	1	1	1

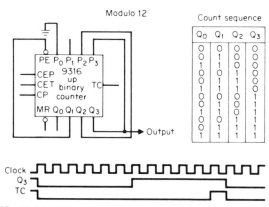

Modulo 12

Count sequence

Q₀	Q₁	Q₂	Q₃
0	0	0	0
0	1	0	0
1	1	0	0
0	0	1	0
0	1	1	0
1	1	1	0
0	0	0	1
0	1	0	1
1	1	0	1
0	0	1	1
0	1	1	1
1	1	1	1

FIG. 10-67   Some 50% duty-cycle output counters.

Synchronously parallel loading the 74161 forces the counter to skip some of the states it would otherwise count through. In each circuit, either the Q_1 or Q_2 output is connected to the active Low Parallel Enable ($\overline{PE}$) input. Whenever this output is Low, the counter loads instead of counting on the next clock pulse.

74192/74193 Up/Down Counters   The 74192 is an up/down decade counter and the 74193 is an up/down 4-bit binary counter. Both devices are synchronous dual-clock up/down counters with asynchronous Parallel Load, asynchronous overriding Master Reset, and internal Terminal Count logic that allows the counters to be easily cascaded without additional logic. The 74192 and 74193 can be used in many up/down counting applications, particularly when the initial count value must be loaded into the counter and multistage counting is required.

Counting is synchronous, with the outputs changing state after the Low-to-High transition of either the count up clock (CP_U) or count down clock (CP_D). The direction of the count is determined by the clock input, which is pulsed while the other clock input is High.

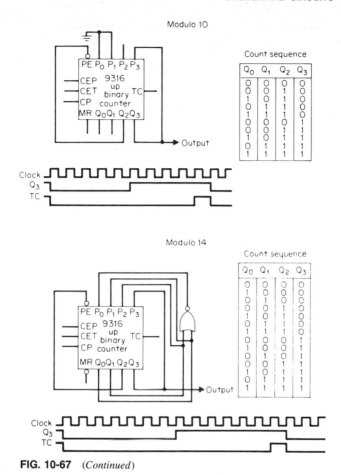

FIG. 10-67   (*Continued*)

The 74192 and 74193 have an asynchronous parallel load capability, permitting the counter to be preset. When the Parallel Load ($\overline{PL}$) and the Master Reset (MR) inputs are Low, information present on the parallel data inputs (P_0, P_1, P_2, and P_3) is loaded into the counter and appears on the outputs regardless of the conditions of the clock inputs. When the Parallel Load input goes High, this information is stored in the counter, and when the counter is clocked it changes to the next appropriate state in the count sequence. The parallel inputs are inhibited when the Parallel Load is High and have no effect on the counter. A High on the asynchronous Master Reset (MR) input overrides both clocks and Parallel Load and clears the counter. Obviously, for predictable operation, the Parallel Load and Master Reset must not be deactivated simultaneously.

The 74192 and 74193 have Terminal Count Up ($\overline{TC}_U$) and Terminal Count Down ($\overline{TC}_D$) outputs that allow multistage ripple binary and decade counter operations without additional logic. The Terminal Count Up output is Low while the up-clock input is Low and the counter is in its highest state (9 for 74192, 15 for the 74193). Similarly, the Terminal Count Down output is Low while the down-clock input is Low and the counter is in state zero.

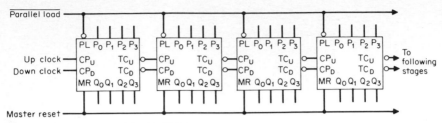

FIG. 10-68   Cascading of 74192, 74193 counters.

The counters are cascaded by feeding the Terminal Count Up output to the up-clock input and the Terminal Count Down output to the down-clock input of the following (more significant) counter, as shown in Fig. 10-68. Therefore, when a 74193 counter is in state 15 and counting up or in state 0 and counting down, a clock pulse will change the counter's state on the rising edge and simultaneously clock the following counter through the appropriate active Low terminal output. The operation of the 74192 is the same, except when counting up, clocking occurs on state nine. The delay between the clock input and the Terminal Count output of each counter is two gate delays (typically 18 ns). Obviously, these delays are cumulative when cascading counters. When a counter is reset, the Terminal Count Down output goes Low if the down clock is Low and, conversely, if a counter is preset to its terminal count value, the Terminal Count Up output goes Low while the up clock is Low.

Example 10–7   Light-Controlled Up/Down Counting

Many industrial or scientific applications require a count of objects traveling past a sensor. A circuit is therefore required to count moving objects as they move between a light source and phototransistors.

**Solution**
The circuit shown in Fig. 10-69 permits a count of objects passing in either direction and allows for reversals in movement or nonuniform movement. Each object passing from bottom to top increments the counter. Any object passing between the light source and the two phototransistors is counted as long as the object is large enough to cover both transistors simultaneously. This circuit can cope with any erratic movement, even reversal of direction. Hex inverters serve as a clock generator and as phototransistor amplifiers. The dual flip-flop and 3-input NAND gates are used to route the phototransistors' signals to the up/down counters.

When an object moves from bottom to top it covers phototransistor two first, bringing line B Low. This stores a zero in the 2-bit shift register. As the object continues, phototransistor one is then covered and brings line A High. As the object moves even further, it uncovers phototransistor two, bringing line B High again. The next clock pulse loads a 1 into the first bit of the shift register. This one–zero combination in the shift register and High level on line A are decoded and gated with the clock to increment the counter. For an object moving from top to bottom, the sequence is reversed and the counter decremented.

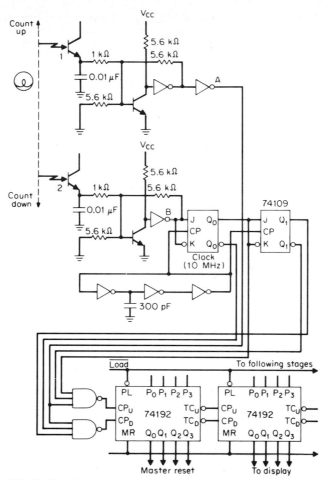

FIG. 10-69 Light-controlled up/down counting.

10–3d Design of a Simple Programmed Logic Controller

There is a growing trend to build electronic control equipment not with specialized circuitry, but with a computerlike architecture that uses regular circuitry and stores its program in memory (ROMs or RAMs). Complex equipment (numeric control, etc.) uses commercial computers or minicomputers, while smaller and slower applications use the emerging microcomputers that are now being offered by several semiconductor manufacturers. Even the cheapest calculators implement logic in a programmed, ROM-oriented way, all on a single chip.

This section describes some easy-to-understand TTL/MSI-oriented designs for a small dedicated controller. This controller is applicable where a minicomputer would be too expensive and a microcomputer would be too slow, too cumbersome to program, or too complicated to understand. This concept uses one or two dozen inexpensive TTL/MSI circuits plus one or two read only memories, field or mask-programmed, and can implement practically any control function with up to 16 inputs and up to 50 outputs.

Example 10–8 Washing Machine Controller

A simple, open-loop controller, as found in every washing machine, is a good beginning. Here a synchronous motor drives a reduction gear, which in turn drives a drum with programming pins or cams that activate the output switches (Fig. 10-70a). An electronic implementation is desired.

Solution

The electronic equivalent of this pin-drum controller is shown in Fig. 10-70b, where an oscillator (motor) drives a $\div 256$ counter (gearbox) addressing a ROM (drum) with eight outputs. If the objective were to generate eight arbitrarily changing, completely random outputs, the design would stop here. Fortunately the real world does not usually require outputs that change in a completely random fashion. Rather, the requirement is to be able to activate and hold certain outputs (solenoids, valves, lights, etc.), starting at a certain position in the program, and deactivate them later at a different position. For this purpose the ROM represents an overdesign. It is simple to reduce the number of ROM outputs and/or increase the number of system outputs by using additional inexpensive MSI components.

The ROM outputs can be interpreted as addresses and instructions. As shown in the example of Fig. 10-70b, the first four outputs are an address activating, through a 74154 1-of-16 decoder, any one of up to 16 MSI circuits. The remaining four ROM outputs are used as instructions to the selected MSI circuit. Address 15 activates the first 4-bit register, changing its four outputs to the associated 4-bit instruction code coming out of the ROM. Address 14 selects another 4-bit register, while address 13 selects a 74259 8-bit addressable latch. The 4-bit instruction determines which output is to be changed and to what level it is to be changed. For an insignificant increase in cost, the number of outputs has been increased from 8 to over 64, with the constraint that only one group can be changed simultaneously.

This is still a very unsophisticated open-loop controller. It can be improved by adding a controlled speed reduction, consisting of a presettable counter, as shown in Fig. 10-71. One instruction can change the instruction rate to any one of 16 values, maintaining it there until it is changed again. The real power of this design is shown, however, when feedback, or—in programming terms—a conditional jump capability is included, as in the circuit of Fig. 10-72. One of the 16 addresses is used to interrogate the status of eight input lines, and the associated instruction defines which input is to be interrogated and which level is the desired one. The subsequent ROM output is then not interpreted as an address/instruction pair, but rather as a program jump address. If the input under test has the expected level (High or Low), this jump address is loaded into the program counter and the program continues from this new address. If the input under test does not have the expected level, the jump address is ignored and the program continues without executing the jump.

Obviously this design can be made even more sophisticated by adding arithmetic capabilities, data memory, address stacks, etc., but carrying this

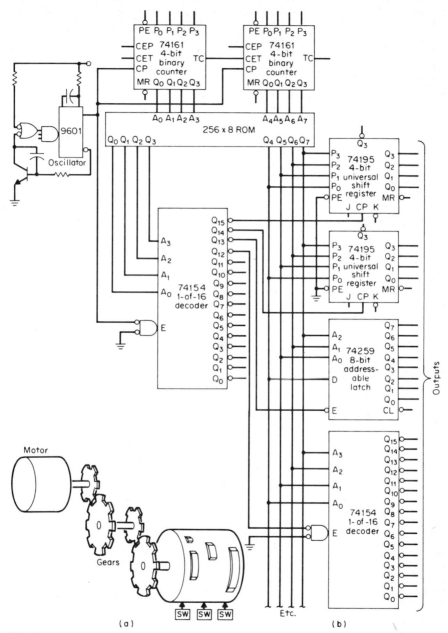

FIG. 10-70 Washing machine controller. (a) Simple open-loop controller. (b) Programmed logic controller, open loop.

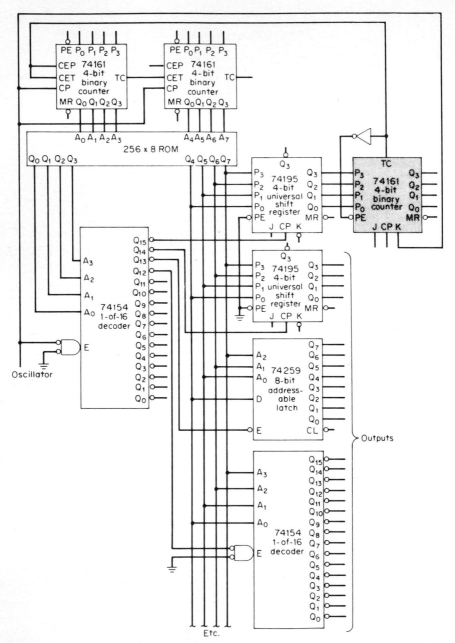

FIG. 10-71 Programmed logic controller, open loop, variable speed.

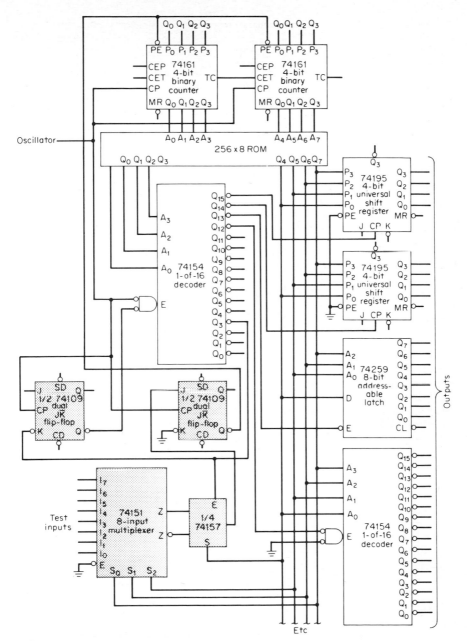

FIG. 10-72 Programmed logic controller, conditional jump.

too far would defeat the basic value of this design, its simplicity and economy. The advantage of this approach over conventional logic implementation lies in the flexibility that it gives to the circuit designer.

The design of a small control system usually starts with a clear knowledge of the number of outputs and inputs required and their electrical characteristics. But the exact definition of *how* the control inputs affect the outputs (under all normal and abnormal circumstances) takes most of the time and leads to most of the usual errors.

The classical logic design can only start when the system design is finished, and will require extensive changes if the system design is changed due to mistakes or new requirements.

The programmed controller, however, can be designed, constructed, and tested as soon as the required inputs and outputs are defined, essentially simultaneously with the detailed systems design. System design, programming, and circuit design can be done in parallel, significantly reducing turnaround time. System changes can be implemented by changing the (P)ROM, and can be tested and verified in hours instead of weeks.

10–4 DESIGNING WITH MSI

10–4a The Impact of MSI on Logic Design

In the days of vacuum tubes, transistors, diodes, and even SSI, the art of logic design was clearly defined and measured. The logic designer attempted to implement a design with a minimum number of components, using such established techniques as Karnaugh maps, Veitch diagrams, and boolean algebra. System design, logic design, and component selection were independent, requiring little interaction on the part of the designers. Medium-scale integration, standard circuits with 20- to 100-gate complexity, has radically changed this relationship and made system design, logic design, and component selection heavily interdependent, each influencing and influenced by the others. It is no longer sufficient—or even important—to minimize the number of gates and flip-flops. It is far more important to select the proper complex integrated circuit that can perform the desired function most economically. It may even be appropriate to redefine subsystems to accommodate more sophisticated and more cost effective components. In addition, the higher levels of integration also offer reduced power consumption and improved system reliability. Logic design, no longer an isolated art, has left its ivory tower and is more demanding, but at the same time far more stimulating and rewarding.

The logic designer today must be involved in system design, must know about the complex components available, and must be aware of the economical impacts of semiconductors, printed circuit boards, connections, and power supplies. This knowledge, and the tradeoffs represented, are necessary to achieve this goal . . . the lowest system cost for specified performance.

10–4b General System Design Rules

- *Adapt system architecture to performance required and components used.* Use parallel architecture and fast components for highest speed. Use serial architecture and slow components for slow systems, which reduces cost and power consumption. Use parallel architecture with slow components or serial architecture with fast components for intermediate-speed requirements.

- *Avoid asynchronous systems; convert them to synchronous.* Synchronous systems are easier to design, debug, and service. They are more reliable than asynchronous systems. A simple, inexpensive clock generator using less than one gate package may be sufficient to solve an inherently asynchronous problem in a synchronous manner.

- *Use extra care with all clock signals to counters and registers and with trigger inputs to monostables.* Avoid clock gating as much as possible; use the synchronous Enable inputs instead. Beware of the glitches on the outputs of decoders and similar combinatorial logic. Avoid slow rise times (less than 50 ns) and watch out for double pulses (overtones) from crystal oscillators. Most problems with inherently slow systems can be traced to double triggering of registers and monostables due to poor clock and trigger signals. The designer of slow systems must be constantly aware of the fact that modern components are capable of 10- to 50-MHz operation and that they react to trigger spikes invisible on an oscilloscope used for displaying slow events.

- *Minimize the use of monostables and avoid RC elements in any signal path.* Monostables are often used as a ''quick and dirty'' corrective for an improperly designed system. Monostables are inherently linear circuits with limited noise immunity, which is a major disadvantage in noisy digital environments. A carefully designed synchronous system using edge-triggered devices rarely needs a monostable.

- *MSI designs should be based directly on system block diagrams.* A gate-minimized logic design hides the basic system structure, and a direct conversion to MSI is bound to be inefficient. It is always better to discard gate-minimized logic design and design with MSI directly from the original system block diagrams.

- *Imaginatively explore MSI functional capabilities.* The name applied to most MSI circuits merely describes the primary function of that device. A well-defined MSI device is far more versatile than the obvious function indicated by its name. A synchronous presettable counter can be used as a shift register, a decoder can be a data demultiplexer, and a multiplexer can be an efficient function generator. MSI devices are surprisingly versatile and this versatility should be used to advantage.

10-4c MSI Selection Tables

TABLE 10-5 Multiplexers

Quad 2 input	Dual 4 input	8 input	16 input
		TTL	
74157	74153	74151	74150
74158	74253	74251	
74257	74352	74152	
74258	74353	9312	
74298	9309	9313	
9322		25LS2535	

TABLE 10–5 (*Continued*)

Quad 2 input	Dual 4 input	8 input	16 input
	CMOS		
4019	4539	4512	
4519			
	ECL		
10159	10174	10164	100164
10158			
	Triple 3-input	Dual 8-input	
	10071	100163	

TABLE 10–6 Decoders

Dual-1-of-4	1-of-8	1-of-10	1-of-16
	TTL		
74139	74259	7442	74154
74155	7445	7445	9311
74156	7442	9301	
9321	74137	9302	
25LS2539	74138	25LS2537	
	74145		
	9301		
	9302		
	25LS2538		
	CMOS		
4052	4051	4028	4514
4555			4515
4556			
	ECL		
10171	10161		
10571	10561		
10172	10162		
10572	10562		

TABLE 10–7 Operators

	TTL
Dual full adder	9304
Quad serial adder/subtractor	74LS385 (25LS15)
4-bit adder	7483
4-bit ALU	74181
Carry lookahead	74182
4×2 Two's complement multiplier	25S05
8×8 multiplier	MM67558
	Am25S558
	MPY8HJ
12×12 multiplier	MPY12HJ

10-82

TABLE 10–7 (*Continued*)

TTL	
16 × 16 multiplier	MPY16HJ
	TDC1010
	Am29516
4-bit magnitude comparator	7485
5-bit magnitude comparator	9324
8-input parity	74180
	74280
9-input parity	8262
12-input parity	9348

CMOS	
4-bit adder	4008
BCD adder	4560
4-bit ALU	4581
13-input parity	4531
8-bit input parity	4532

ECL	
Dual adder/subtractor	10180
4-bit ALU	10181
	100181
Carry lookahead	10179
	100179
5-bit magnitude comparator	10166
9-bit magnitude comparator	100166
9 + 9-input parity	100166
11-input parity	10170
12-input parity	10160

TABLE 10–8 Latches

4 bit	4 + 4 bit	8 bit		
TTL				
7475	74116	74LS373		
7477	74256	74LS573		
74196	9308	74LS259		
74197		74LS533		
74279		9334		
74375				
9314				
CMOS				
4042	4723	4724		
4043				
4044				

2 bit	3 bit	4 bit	5 bit	6 bit
ECL				
10130	100130	101331	10175	100150
		10153		
		10168		

TABLE 10–9 Registers

4 bit	6 bit	8 bit	16 bit
		TTL	
74173	74174	74164	9328
74175	74378	74165	
74178		74166	
74194		74198	
74195		74199	
74295		74273	
74298		74299	
74379		74323	
74395		74322	
74398		74374	
74399		74377	
9300		74574	
25LS2519		25LS2520	
		CMOS	
4035		4014	
40194		4015	
40195		4021	
		4034	
		ECL	
10000		100141	
10141			

TABLE 10–10 Counters

	TTL asynchronous	
Decade	4-bit binary	Divide by 12
7490	7493	7492
74176	74177	
74196	74197	
74290	74293	
74390	74393	
74490		

	TTL synchronous	
74160	74161	
74162	74163	
9310	9316	
74168 (up/down)	74169 (up/down)	
74190 (up/down)	74191 (up/down)	
74192 (up/down)	74193 (up/down)	
74568 (up/down)	74569 (up/down)	

CMOS asynchronous	
Multistage binary	BCD
4020 (14 bit)	4553 (3 digit)
4024 (7 bit)	4534 (5 digit)
4040 (12 bit)	
4045 (21 bit)	

TABLE 10–10 (*Continued*)

CMOS asynchronous	
Multistage binary	BCD
4727 (7 bit)	
4521 (24 bit)	

CMOS synchronous		
Decade	4-bit binary	Miscellaneous
4017	4029 (up/down)	4526
40160	40161	(programmable)
40162	40163	4022 (octal
40192 (up/down)	40193	counter timer)
4518	4516 (up/down)	
4510 (up/down)	4520	

Chapter 11

MICROPROCESSORS

Dr. William R. Warner
Dave Kohlmeier
Don Birkley
Tektronix Corp.
Beaverton, Oreg.

11–1 INTRODUCTION

This chapter will outline the process of selecting the appropriate microprocessor for a given application (see Fig. 11-1). It will also present some criteria for determining whether or not a microprocessor should be used in the application at all.

The range of costs and capabilities of microprocessors is very wide and growing rapidly. From the 4-bit single-chip microcomputers, such as the TMS-1000 series, serving low-cost high-volume applications, such as electronic games and microwave ovens, to the new generation 16-bit CPUs such as the Motorola 68000 and Zilog Z-8000, which have the computing power of a not very old mainframe computer, a wide range of applications can be addressed.

For low-volume simple tasks the microprocessor may be a less cost effective solution than conventional logic, or if speed is a big consideration, the microprocessor may be too slow. However, with microcomputers costing under $10 and single-board computers costing under $200, the computer solution is attractive for many applications. To help you decide whether a microprocessor is appropriate for your application, use the guidelines and comparisons in this chapter.

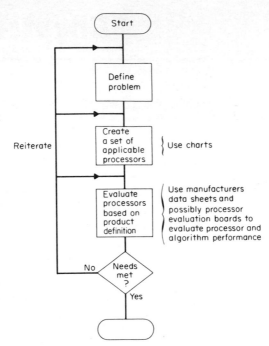

FIG. 11-1   The selection process.

11–1a Glossary of Terms

Address Space   The address space of a processor is the amount of memory that the processor can address without additional hardware. Some of the new processors also use the term ''address space'' to designate one of several completely separate regions of memory that are only accessible to programs running in a particular mode.

Algorithm   Exact step-by-step statement of the solution of a problem.

Asynchronous   Not synchrononized by a master clock. A typical example would be a TTY or RS-232-type terminal, where the characters are transmitted at nonsynchronized intervals, and the individual bits are transmitted at an agreed-upon rate. The transmitter sends data at a rate dependent on its own internal clock, and the receiver samples the line based on its own internal clock, once it detects that data is being sent.

ACIA   (asynchronous communications interface adapter) A device similar to a UART that interfaces the microprocessor bus to an asynchronous device such as a terminal. (Note: A terminal would not attach to the ACIA directly, but through a buffer.)

A/D   Analog to digital.

ASCI   American Standard Interchange Code. A 7-bit code used to represent a set of printable characters and certain control information. Commonly used with TTY-like data terminals.

Assembler   A program that converts a mnemonic form of machine language into the binary object format for execution by the computer.

Baud   Commonly used to mean bits per second. Strictly, it is signal elements per second.

BCD   (binary coded decimal) A code where 4 bits are used to encode only the digits 0 to 9.

Boot   (bootstrap, or boot ROM) The initializing program that starts computer operations. Typically, this program only starts the loading of a more complete operating system or program.

Byte   A group of 8 bits.

CAS   Column address strobe (see RAS).

Checksum   A data field added after a block of data that is used to check for errors in the data. Typically, the checksum would be the truncated sum of all the bytes in the data field, although a number of different algorithms are used.

CPU   (central processing unit) In a microcomputer, the microprocessor. In computers, in general, the portion of the computer dedicated to general arithmetic and control functions.

CRC   (cyclic redundancy check) A method of calculating checksums that is based on binary polynomials. It provides much better error detection than a standard "checksum."

Cross-program   A program, typically a compiler or assembler, that resides on one computer, but generates programs or data for a different computer. Compilers that run on large mainframe computers are available. They generate code for small microcomputers that do not have the resources to run a compiler.

Cycle-stealing   A method of transferring data to or from memory without the CPU running the bus cycle. Typically done by a peripheral that holds the processor off the bus for a cycle, while it transfers its data (a form of DMA).

D/A   Digital to analog.

DAC   Digital to analog converter.

DMA   (direct memory access) The process of a peripheral device accessing main memory directly instead of through the CPU.

DOS   Disk operating system.

Emulator   An emulator is a device that plugs into the prototype circuit under test and replaces the microprocessor chip. The designer is then able to control the prototype through the emulator.

EPU   (extended processing unit) A device that, closely coupled with a central processing unit, expands the CPU's capabilities (e.g., a floating-point processor for a microprocessor).

Fetch   The first memory operation of an instruction cycle, where the instruction is brought from memory to the CPU.

FFT   (fast Fourier transform) A numerical analysis algorithm used frequently in signal processing and related fields.

FIFO   First-in–first-out.

GPIB   (general-purpose interface bus) The IEEE 488 interface bus.

Half-duplex   Communication technique where data can travel only one direction at a time.

I/O   Input/output.

K   When referring to an amount of memory, 1K = 1024. The closest binary unit to 1000 decimal.

Kernel   The group of software routines that perform the basic operating systems functions. Generally would include provisions for starting other programs, loading programs, allocating resources, and similar functions. It would not include such programs as assemblers, compilers, file systems, etc.

Object-Code   The machine-level representation of a program, the output of a compiler or assembler.

RAS   (row address strobe) The signal used to indicate a portion of an address is presented to the multiplexed address pins of a memory.

RS-232   An industry standard interface for data transfer. The standard is published by EIA (Electronic Industries Association).

SDLC   (synchronous data-link control) A communications protocol for synchronous data links.

SBC   (single-board computer) A number of manufacturers supply single circuit boards with a microprocessor, RAM, ROM, I/O ports, and extra memory sockets. These boards can be used to evaluate a microprocessor or can be used directly in low-volume applications as a part of the finished product.

Source code   The human readable form of a computer program. The input to an assembler or compiler.

Synchronous   Synchronized by a master clock.

TTY   Common abbreviation for Teletype, or similar device.

UART   (universal asynchronous receiver transmitter) A device used to convert parallel data (e.g., from a microprocessor) to or from serial data (e.g., a terminal I/O port).

11–2   THE SELECTION PROCESS

It should be stated clearly and forcefully that the selection process is twofold:

> The problem definition
> The selection of the processor

First let us address some of the considerations in defining the problem.

The selecting of a microprocessor from a myriad of products on the market cannot be anything more than a guess, unless the application in question is completely defined. In defining the application, both the present and future must be considered. More often than not the initial specification does not take into consideration the changes that may take place during the development cycle or after the product is in the marketplace. This section gives the reader a few things to consider when deciding whether the application is fully described.

11–2a Communications

Communications overhead computation is not an easy task, so it is often overlooked, even though many texts are available on the subject. It is important that the designer realize the potential degradation of processor performance that occurs when the processor is asked to handle a communication task. Some areas for concern are enumerated in the following sections.

Protocols Will the product be required to communicate via a hardware link that has a specific protocol? There can be a substantial software overhead if the processor is required to format a buffer with specific control information, insert data-link escape characters, compute checksums or CRC sums, or other computational intensive tasks. If this type of problem is being considered, then be sure to investigate the special protocol chips that are available.

Mass Storage Devices Even the smartest disk controllers do nothing in terms of file management, so the addition of a floppy disk will require complex support routines and again processor time.

Diagnostics General-purpose computers to the smallest controllers should include diagnostics in the initial design. How will the results of the tests be communicated? Again, this creates software overhead.

Co- or Multiprocessors This may dictate processor selection in itself. To design in the "hooks" for multiple processor communication requires, in most cases, specific information on the processors involved, especially in the coprocessor format. In a multiboard configuration, the bus that is designed from the start will determine the flexibility of future system upgrades. It cannot be overemphasized that the interconnect specification of most systems in the marketplace today is the limiting factor in upgrading to higher levels of performance.

Quantity In some cases, the quantity of data that needs to be moved by the processor might suggest different solutions. Some might be: the addition of a communication controller, the selection of a processor with the ability to do block moves of data, or a direct memory access (DMA) device.

The definition of the communication requirements of a processor must be thorough or the selection of a processor will be based on an incomplete set of data.

11–2b Control: I/O

If the computer is going to be controlling external equipment, then it is very important to define the interface between that equipment and the computer. That interface consists of two major components: the hardware interface and the software interface.

First, in designing the hardware interface consider how control and data are transferred to the peripheral. The device may be memory mapped, or connected via any one of a number of general interface chips, such as ACIAs or PIAs, or it may interface through a more specialized device, such as a disk controller. If you are interfacing to a nonstandard device, be sure to give consideration to protecting the computer system from noise or malfunction of the peripheral equipment.

The hardware/software interface is just as important as the hardware interface. Great consideration should be given to the software interface since the overall system performance is often limited by programming considerations.

The software often has to consider two levels of control: the control of the I/O ports that talk to the external world, and what is said to the external devices that make them work. In designing your system be sure to consider the hardware/software tradeoffs. If you place a large control burden on the software, can the processor you are considering handle the loading? Would it be cheaper to use a more powerful microprocessor and/or more memory, or would the functions be better handled by peripheral hardware?

11–2c Computation

Consider the complexity of computations required for your application.

1. In what form is your data acquired and/or stored, binary or BCD? If the data is pure binary, be sure the processor can easily handle the precision of your data items. It is not all that difficult and time-consuming to write multibyte arithmetic routines, but (depending on the size of the project) the increased reliability, decreased amount of code and memory, and easier development time will probably justify the higher cost of a processor with a larger word size.

2. Is your data primarily numeric or character in nature? If the data is primarily character, then the 8-bit processors can handle the job nicely.

3. Is speed an important design factor? Such applications as signal processing frequently require complex calculations, such as fast Fourier transforms (FFTs), to be performed in some critical time. If you have a tight speed requirement, evaluate the algorithm you will be using with the candidate processors to see if their processing speed is adequate.

4. Is your data floating point? There are a couple of important considerations here. First, there is little point in reinventing floating-point software. Check to see whether a high-level language with floating-point arithmetic is available. If not, or if you are constrained for other reasons to using assembly language, check for the availability of floating-point subroutines, either in source-code form or position-independent ROMs.

   Second, is the processor speed adequate to allow software floating-point calculations? If not, then consider the possibility of using external floating-point hardware to do the calculations.

11–2d Real Time?

There are applications where speed of response to an external request is crucial.

The number and frequency of these external requests will determine the way in which they are handled. Polling of I/O devices is the cheaper hardware solution, but it is slow when compared to vectored interrupts. Is there a need for a real time clock? A time stamp on user files in a general-purpose machine is a nice feature. Does the processor have the time to handle a clock interrupt and the routines to support it? What is the maximum number of devices that will be allowed to interrupt the processor? The maximum load on the processor needs to be computed, that is, the amount of processing that can be requested before data is lost. A computer with eight terminals transmitting at 2400 bytes per second while at the same time maintaining a real-time clock and communicating with a hard disk presents a specific need in the search for a processor.

11–2e Environment

Each application will vary in its operational environment. Temperature ranges of an industrial controller will be much broader than those of a general-purpose computer. Power consumption of a solar-powered application will require a certain type of processor. Whether a processor needs to be radiation hardened will also narrow the range. Physical size might need to be limited. The examples listed above are in no way complete, but are there to acquaint the reader with the process of complete problem definition. Once the problem is defined, the selection process becomes one of evaluating tradeoffs.

11–3 CONSIDERATIONS

11–3a Register Architecture

Number of registers   One of the major considerations in assessing the suitability of a microprocessor to execute the type of software intended for it is the register architecture. The first and perhaps most obvious attribute to look at is the number of registers. A common tendency here might be to say ''the more registers, the better,'' but there are more important considerations. Consider two extreme examples that follow.

1. The Mostek 3870 has a register bank of 64 general-purpose registers; however, most arithmetic operations must be performed in the one ''D'' register. Sixteen of the general-purpose registers are directly addressable, and of those eight are more or less reserved for specific purposes. The remaining 48 registers must be accessed indirectly through another address register. Fig. 11-2 illustrates the 3870 architecture.
2. The Motorola 68000 has only eight general-purpose data registers. There is no distinction, however, in how any of these eight registers are accessed, and virtually any arithmetic operation can be performed directly in any one of these registers (see Fig. 11-3).

Is the 3870 ''better'' because it has more registers, or is the 68000 ''better'' because it addresses the registers it has more easily? There are obviously more considerations than just the number of registers, yet the 3870 with 64 registers can be used in small control applications with no additional RAM, while the 68000's generality gives it much greater computing power.

11–3b Precision of Registers

One of the most important questions to consider about the register architecture of the microprocessor is the precision of the registers. If a large amount of computation is going to be done, then it is very desirable that the registers be sufficiently large to allow single register arithmetic operations. The tradeoff of the smaller, less expensive processor is that for multiple precision arithmetic, more code space and more execution time is required. Also, unless a high-level language that handles the multibyte precision automatically is used, more time will be required to write and debug the software.

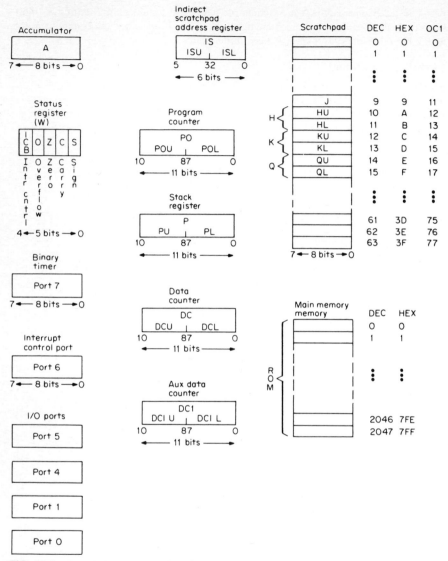

FIG. 11-2 The 3870 programmable registers, ports, and memory map.

The maximum unsigned integer vs. register size is:

4 bits	15
8 bits	255
16 bits	65535
32 bits	4294967295

Some of the 8-bit processors have provisions for doing 16-bit arithmetic in a pair of 8-bit registers or in special 16-bit registers.

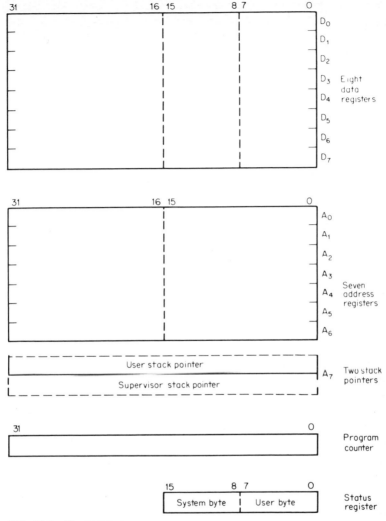

FIG. 11-3   The 68000.

11–3c   Symmetry

Processor symmetry refers to the interchangibility of instructions, addressing modes, and registers.

> If $R0$ can be moved to $R4$, can $R4$ be moved to $R0$?
>
> If $R0$ can be loaded via any addressing mode, can any other register be similarly loaded?
>
> Can arithmetic operations be performed in any of the general-purpose registers?
>
> Can data be moved from any register to any register, any register to memory, or memory to memory?

"Yes" answers to these questions indicate that the processor does indeed have a high degree of "symmetry" in its instruction set and register architecture. Symmetry is a desirable feature in a processor, as it generally makes coding easier and faster.

11–3d Address Space

The "address space" of a processor is the amount of memory that it can address without the aid of external bank select hardware. The address space of a number of common processors is shown in Table 11-1.

As you can see, there is a wide range of capabilities available in terms of address space alone. The amount of memory you will need depends primarily on your intended application. A simple electronic game program might fit in 1 or 2 kbytes of code, while a moderately complex multi- or general-purpose program may easily exceed 64 kbytes.

The new 16-bit processors, such as the 68000 and Z-8000, have another feature, that is, multiple address spaces. This is an old feature of mainframe and minicomputers that, like many other "big computer" concepts, has filtered into the microprocessor arena. Let's consider the Motorola 68000 for an example.

There are three "function code" lines pined out of the 68000. These lines encode (among other things) whether the 68000 is in user mode or supervisor mode, and whether the 68000 is doing a program access or a data access. This information can be used to decode the memory so that there are actually four separate and distinct memory spaces, each with a space of 16 Mbytes (supervisor-program, supervisor-data, user-program, and user-data). This division can give a total address space of 64 Mbytes.

Address space partitioning can be a very valuable feature, especially in general-purpose and multiuser computer systems. This is so because it not only expands the memory space, but also provides a mechanism for protecting the operating system code from access or alteration by a user program, because a program running in user mode cannot generate a reference to supervisor address space.

11–3e Stack

A very important feature of a processor is its stack architecture. Older machines, such as the 8008 and the 2650, have internal fixed-length stacks that limit the depth of subroutine

TABLE 11–1 Address Space of Various Processors

Company	Processor	Address Space
Intel	8048	1K
Mostek	3870	2K
Mos Technology	6503	4K
Intersil	6100	4K
Signetics	2650	32K
National	SC/MP	64K
Intel	8080	64K
Intel	8085	64K
Zilog	Z-80	64K
Motorola	6800	64K
Mos Technology	6502	64K
RCA	1802	64K
TI	9900	64K
Intel	8086	1M
Zilog	Z-8001	8M
Motorola	68000	16M

nesting, while the newer machines have stack pointer registers that access external memory for the stack, allowing nesting and argument passing limited only by the memory available.

11-4 INSTRUCTION SETS

11-4a Classes of Instructions

Math Most processors are fairly similar in terms of the instructions that are available. The more powerful processors have multiply and divide instructions. If your application is going to use BCD arithmetic, check to see that BCD is supported on the chip you choose.

Logic AND, OR, XOR, and NOT are fairly standard among all of the processors. Also convenient, but not always available, are a variety of bit test and set operations. For operating system and real-time control, it is often necessary to have an indivisible test-and-set instruction or group of instructions for semaphore control.

Data Transfer The main considerations for evaluating data-transfer instructions are the symmetry and the number of addressing modes available.

A block-move instruction is sometimes handy, but with good looping primitives and a variety of index registers, code can be made just as efficient without one.

Transfer of Control Look for a good set of conditionals. It is convenient to have both branch on true and branch on false statements for each of the conditions tested. One *very important* consideration is the subroutining structure of the processor, since modern structured code relys heavily on the use of subroutines. Some processors—the 1802, for example—do not have a single branch-to-subroutine instruction, and require several instructions to generate a general nestable subroutine call. Other processors—the 2650, for example—have an internal subroutine stack, and limit the number of subroutine levels that can be nested. A processor that is going to have more than a few hundred bytes of code should have an external stack with subroutine capabilities.

Machine Control The instructions in this group vary widely between machines. If interrupts are to be used, make sure that appropriate instructions are included to control the interrupts for your application. Is there only one interrupt level or many; can the interrupts be selectively disabled? If the processor is going to be running a general-purpose operating system, it is highly desirable to have a supervisor or system mode of operation where certain highly critical machine control statements are privileged to the system mode. This in conjunction with memory management makes it possible to provide a measure of security for the operating system.

Addressing Modes This section explains the commonly available addressing modes that are encountered in microprocessors. Unfortunately each manufacturer used slightly different terminology for their products. The terminology used here is a mixture of terms from different processors. Although these are all commonly accepted terms, be aware that some manufacturers will have different names for some of these modes.

Register Addressing—The data to be manipulated are in one of the processor's internal registers. This is typically a fast operation since no memory access is required to access the data, e.g., CLEAR R0.

Register Indirect or Base Register—A processor register contains the address of the data to be accessed. This is a very heavily used mode in those processors that support it. One memory access is needed to access the data. This assumes that some previous instruction has loaded the register with the address of the data. For referencing single scattered data items only once, this is a relatively inefficient mode, since another instruction is used to load the address of the data. However, it is a valuable mode for accessing the same data item or a block of data repeatedly.

Memory Indirect—A location in main memory contains the address of the data item. This is a slow but very powerful and useful addressing mode, e.g., LOAD, R0 @ADDRESS:

100	LOAD, R0 @
101	12
102	34
1234	40
1235	08
4008	76

The instruction at location 100 is a load R0 indirect via location 1234. Location 1234 contains the address of the data, 4008. Location 4008 contains the data, 76.

Absolute or Direct Addressing—The address of the data follows the instruction in memory. This is less efficient for frequently accessed data than the register indirect mode, but is better for data that is infrequently accessed.

PC Relative—This addressing mode uses a displacement from the program counter to address the data. This instruction is inherently position independent! (A position-independent program can be moved to any memory address without being relinked or reassembled in any way.) However, the range is usually limited.

Immediate Addressing—The data immediately follows the instruction, moving the number 5 to register 0.

Base Displacement—A displacement is added to a register indirect access; e.g., MOVE DO,26(A5) would move the contents of register DO to the memory location 26 bytes past the location addressed by register A5. This mode is frequently used by high-level block-structured languages, where a base register contains a pointer to a local data space for a routine and the displacement references a particular data item within the data space.

Indexing   Indexing is not an address mode itself, but is frequently applied to various addressing modes. In any indexed addressing mode the contents of an index register are added to an address calculated by some other addressing mode to get the final address. This mode is normally used for manipulating arrays of data, and is also useful in handling data structures. Auto incrementing and/or decrementing of the index registers are common and very useful features in most processors.

11-5   SUPPORT SOFTWARE

11-5a   ROM Monitors

The ROM monitor is one of the most primitive, as well as most common and readily available debugging tools. It is a simple program supplied in a ROM by the vendor of the microprocessor, and it gives the user a limited set of software debugging tools with which to develop the initial software for the system.

The ROM monitor is usually included with a single-board computer or demonstration board, and can usually be purchased separately if the designer is only buying the microprocessor components. If the ROM monitor is to be put in a board of your own design, there will be some requirements on how the system must be configured. Generally, the monitor will require a specific I/O peripheral chip addressed to some particular location, as well as some RAM configured to a specific place.

The operations generally available in the monitor are:

Examine and change memory
Display memory contents
Patch (change) memory contents
Display processor registers
Set processor registers
Single step the processor's execution
Start the processor executing at a location
Set one or more breakpoints
Load a formated object program (from cassette or paper tape)
Save memory to cassette or paper tape
Communicate with a host computer

Sometime a "one line assembler" is also available in ROM. This assembler allows the user to enter the instruction operation codes with the assembler mnemonic and enter the address and data operands by absolute numeric value. With a one-line assembler the user could, for example, enter the 8085 instruction "jump to 1000" as "JMP 1000" instead of "C30010."

11-5b   Operating System Kernel

At the time of this writing, only a few of the larger vendors are supplying operating systems for their processors. These systems are usually sold as part of a board set, or complete microcomputer, and provide extensive assistance for system software development. Other companies are also beginning to supply operating system software. Bell Labs licenses the UNIX operating system, which is written in the language C and can be installed on PDP-11 or similar computers. A look through *Electronics* and similar trade magazines will reveal a number of offerings for the new 16-bit processors. The purchased operating system kernel should be considered when purchasing a "stock" microcomputer system. You may not find exactly what you need, but you may find a system that can be customized at less expense than writing an entire operating system from scratch.

11-5c   Languages

One of the most important decisions in the design of a computer system is the choice of the language the system software will be developed in. This choice becomes increasingly important as microcomputer systems become larger and more software intensive. All microprocessor vendors have some sort of assembler available for their processors. Higher level languages (Basic, Fortran, Pascal) are less readily available, although more and more manufacturers are providing them. A number of independent software firms are also beginning to provide compilers for many chips.

Compilers and assemblers come in two varieties; one runs on the machine that it is generating code for, and the other, the cross compiler, runs on some other host computer.

Compilers and assemblers are large programs commonly requiring 16K to 128K of memory, mass storage (floppy disk or hard disk), and I/O devices. This implies that unless you are building a substantial general-purpose computer you will want to use a cross assembler or compiler. Even if you do install these tools on the new system, you will probably want to do the initial development on another host.

11-5d Application

Commercially available application software is another item to consider when choosing a processor. Programs are available for computer games, bookkeeping, engineering calculation, and just about anything you can think of. If your processor system is to serve many uses, check on the software packages that are already available. You could save much time and money compared to developing them from scratch.

11-6 ARCHITECTURE

In the selection of a processor, given that the problem has been defined, the most obvious differences will be in architecture.

While the application (power, circuit size, etc.) and needed levels of performance usually dictate a given architecture or architectures, it should be noted that selection of different processor types will greatly affect the design time. A processor with a family of compatible support chips will be a relatively easy design task as compared to say a microprogrammed bit slice processor. Keep this in mind as you decide on a particular architecture.

11-6a Single Chip

The recent proliferation of single-chip microcomputers are a testament of their acceptance by the industry. They exist in a variety of sizes, shapes, and power. Applications for these all-in-one processors range from industrial controllers to local slave processors.

The single chip computers come in three major process technologies: CMOS, NMOS, PMOS. The process used to create a given microcomputer does not seem to indicate any category of speed, voltage, or power, but the fastest seem to be NMOS and the lowest in power CMOS, though it should be noted that CMOS has made great advances in speed/power in the last few years. The majority of the processors are 4 or 8 bit and are available with a variety of on-board functions. A glance at the comparison tables shows the availability of functions that might meet your specific needs. Here are some examples of what's available: A/D converters, serial/parallel I/O ports, UV EPROM, counter timers, zero-crossing detection, high current drive, LED digit interface, touch switch interface, ability to expand, RAM/ROM off chip, analog signal processing, and almost without exception on-chip clock, RAM, and ROM.

The following are some things to consider. The on-board ROM/RAM makes these processors attractive where the parts count must be kept to a minimum. Remember that this ROM must be programmed before a designer can begin to check out a system. A solution to this has been to design a UV EPROM version of the most popular micros, thus allowing prototyping to occur before a masked version is requested.

If flexibility is a necessity, then consider the off-chip memory-expansion capability a high priority. The tradeoffs here are increased parts count for a lesser coding task of trying to compress for every byte, and subroutine call (if the stack is in RAM).

The farther your application is from a very dedicated controller, the closer you need to look at whether a microcomputer will accommodate your needs, albeit the high end micros are powerful computers with on-board RAM/ROM.

11–6b   Internal/External Stack

Speed and flexibility are the major tradeoffs here. With an internal stack the response time to an interrupt or subroutine call will be faster than that of the external stack processor. However, by placing the stack inside instead of in RAM, one limits the depth to which interrupts or subroutines can be nested. This is very apparent in high-level languages where block-structured programming makes heavy use of the stack.

11–6c   Bit-Slice Processors

Bit-slice processors are the fastest, the biggest power consumers, and usually the most costly to implement. These build-it-yourself processors are attractive to those who can't find a processor to meet their needs elsewhere. Their microprogrammability offers a flexibility to implement the application of specific instructions. The speed of execution of these instructions is much higher than that of a routine of machine code instructions on another processor.

The tradeoff for this flexibility and speed is the design time factor. In choosing a microprocessor, you have as a given that the processor will correctly execute the commands you present it. With the bit-slice processor you must debug the "processor" (ALU, sequencer, control store) and write the microcode to drive it. No easy task, but well worth the effort if the application calls for it.

11–6d   Coprocessors

Coprocessors are usually processors that have been designed to coexist with a primary processor. They have many functions: memory management, floating point, I/O processors, terminal controllers, and others. They are a big boost to the processing power of a system. The major consideration to keep in mind is that they are usually a family-type interface; that is, the Zilog Extended Processing units were designed to perform with the Zilog Z8000. The signals that the main processor and coprocessor use to communicate are very dedicated. Although the Zilog memory management unit (MMU) is really just a peripheral part, it takes advantage of the fact that the Z8000 outputs its address segment lines one full clock before the address, giving the MMU the time it needs to calculate the upper address. On the other hand, the Zilog extended processing units (EPUs) are really an extension of the processor silicon and definitely cannot be used with another processor. The Z8000 has a number of reserved instructions that it treats as NOPs (no operation). These instructions are recognized by the EPU chips as their special instructions (e.g., floating-point multiply). These co- or extended processors enhance the power of a main processor, so remember during selection that they are of a family nature.

If you are considering a larger computer configuration and wish to use these new performance-increasing parts, check their availability and compatibility with the manufacturer.

11-6e   Multiprocessing

The major concern in a multiprocessing system is the system interface itself. Processors in a multiprocessing environment need to be able to communicate in a way that prevents corruption of data in the system.

In a multiprocessing system the processors execute at the same time, which requires synchronization of the processors. A feature to look for in a processor that will aid in the development of a system such as this is a test and set instruction. That is, an indivisible read/write cycle. This allows implementation of a semaphore (a semaphore is a flag that is used to allow only one processor to enter a critical section of code or data). If this function is not available as an instruction in the processors used, it must be implemented in hardware on the system level. Another concern in a multiprocessing environment is the large task of system software development. The operating system will not be a trivial task and checkout is a complex task that must try to exercise the states (which now included combinations of states of all processors in the system) of the 2 + processors in order to ensure proper operation.

11-6f   Problem-Solving Peripherals

The following short descriptions of peripherals available to microprocessor system designers is both incomplete and brief. Chapters 9, 10, and 13 have covered these circuits in more detail and should be used as reference for these introductory comments.

Parallel/Serial I/O   The chips available for parallel and serial I/O are many, and their function is essential for all but the smallest of systems. The Motorola ACIA 6850 and the Intel USART 8251 are examples of the serial I/O chips, and the Motorola PIA 6820 and the Intel PPI 8255 are examples of the parallel I/O chips. These chips are easiest to interface within their own families.

One may use, say, a Motorola ACIA with an Intel product, but care must be taken to ensure correct timing. Unless for some exceptional reason (availability), it is advisable to stay with the family for these types of chips. Essentially what these serial/parallel I/O chips do for a system is to allow external devices (keyboards, RS-232 terminals, etc.) to be interfaced to the processor. The I/O parts have the control and data bus buffers (tristate) and data latches included on the chip. The programmable nature of the devices allows great flexibility. The functional configuration of the I/O ports is controlled by a control word in the part. System software can "reconfigure" the operation mode of the device.

With the design problem defined, the selection of a serial/parallel I/O part should not be a problem. First, consider the family parts of the processor you choose: if the right part is not available, then consider the peripheral chips of another vendor. Make sure that interface timing between processor and peripheral part can be met. In the recent past, a number of intelligent I/O processors have been introduced. These parts remove the overhead of I/O processing from the main processor, but add complexity to the I/O system. It is worth investigating the possibility of using these new parts instead of the older mainstays mentioned above. This approach may add additional expense and complexity, but the rewards in system performance could justify such a choice.

Counter/Timers   Another family of peripheral chips are the counter/timers. These chips can be used for a variety of functions, some of which include substitution of software timing loops with a hardware interrupt generated with a timer chip, a real-time clock, event counter, and programmable timing signal generation.

Memory Management Units   The memory management units (MMU) provide a logic-to-physical address translation. They also provide functions, such as virtualization, write-protect, and nonexistent memory indication. The major function of the chips is to provide multiprogramming systems with a way to easily context switch between several physical memory pages (spaces). Each user in a multiprogramming system presents a logical address to the MMU. The operating system need only adjust the contents of the MMU to map the user's logical address to a unique physical space. Larger systems designers should consider these new (in chip form) devices.

Direct Memory Access (DMA)   A direct memory access (DMA) controller is a programmable peripheral device that allows a high-speed transfer of data from a source to a destination. This transfer can be from memory to memory, memory to peripheral, or any combination of devices involved in data handling. It is all done without the main processor being involved, except for the initialization of the DMA controller. In applications where a bulk storage medium (floppy disk, hard disk, tape units, etc.) exists, a DMA controller can increase system performance dramatically.

11–6g   CRT, GPIB, RAM Controllers

Special-purpose peripheral chips are available for many functions, including those listed above. Each device will remove some overhead from the processor and minimize parts count and design time in a system that requires such support. Again a big consideration is whether these parts are within the processor family.

11–7   SYSTEM INTERFACE

As was stated earlier in the chapter, the system interface (or interconnect) will most likely be the limiting factor in the future growth of your system. It is not a major consideration in the decision of which processor one needs, but it is well worth noting the importance the system interconnect has in the life of a system.

Discrete circuitry between the processor and the system bus will usually allow a designer to interface just about any processor to any bu. This is not to say that some combinations won't be a large design task. The closer .e system bus looks like the processor bus, the easier the interfacing and synchronization tasks become. So remember your processor's requirements when specifying;

1. Synchronous/asynchronous bus
2. Memory
   a. Static/dynamic
   b. Relocatable
   c. RAM/ROM
   d. Virtual
3. Interrupts
4. Multiprocessor compatibility

11–8 DEVELOPMENT AIDS

11–8a Emulators

The emulator is probably the most powerful system debugging tool available to the microcomputer designer. An "emulator" is usually a component of a larger microprocessor development system. The emulator itself has a cable that plugs directly into the microprocessor socket of the prototype circuit to be emulated. The end of the cable acts just like the processor that is being emulated. It allows the user to monitor and control the processor better than any other single tool that is available. When considering the selection of the microprocessor, consider whether an emulator is available. It will be a great aid in hardware debugging, software debugging, and hardware/software integration.

When considering the purchase or rental of a development system with an emulator, consider the following points:

1. Can the emulator run at the processor's full speed?
2. Will the emulator allow you to use all of the processor's interrupts and traps transparently?
3. Does the development system reserve any of the processors address space for its own use, or is it all available to the user program?
4. Does the development system have logic analyzer capabilities for tracing program execution in real time? If so, can it be configured or programmed flexibly for breakpoints, limited acquisition, timing, and counting?
5. Is the overall user interface of the system acceptable?

11–8b Debug Software

Some compilers are capable of generating additional code to aid in the debugging of user programs. These compilers rely on support from the operating system of the computer the program is running on. The capabilities of debugging software vary, but they can typically provide tracing of execution flow and display of registers and memory.

11–8c Simulation

Simulators for many microprocessors are available on large host computers, and can typically be rented on a time-sharing network. A simulator allows the execution of a microprocessor program without having any of the microprocessor hardware available. The software debugging capabilities are similar to emulation, but there can be no hardware interaction, so simulation is limited to software checkout only.

11–9 SINGLE-BOARD COMPUTER VS. BUILDING YOUR OWN

As the complexity of the processor and the associated support circuitry increases, the choice of whether to build a computer or buy a single board computer (SBC) will be asked more frequently. The following sections will hopefully point out a few criteria for one to examine while making the decision.

11–9a In-House Experience

In-house experience is very important! If you are going to build a board from scratch, you need experienced people in prototyping, design, etched circuit boards (this can be farmed out), and checkout. The parts need to be purchased and stocked. Are you going to assemble it in-house? Do you have experienced technicians to get the boards working correctly? Is there test equipment available, such as scopes, logic analyzers, and prototype systems? Where is all the support software going to come from? Is someone going to write an interactive monitor (these normally come with an SBC)?

If your company is fairly large and has all the support groups to help in the task of creating a computer, then you might say yes to the above questions, but how many are you going to build? How much are you going to have to sell them for to recover your design expense?

The SBCs do have their limitations, and these must be compared to your design goals. If the SBC is incompatible with your requirements, then you may have no choice. All the sections we have dealt with previously in this chapter hold true here also. What does the SBC manufacturer offer in the way of peripheral support (these will most likely be cards that plug in the same system bus), memory, mass storage, etc.? If you choose an SBC and its associated system as an original equipment manufacturer (OEM) computer to support your needs, remember you can always design a card to fit into that system, if a specific function you need is not supported.

The choice between an SBC and building it yourself is not an easy one. Profit margin is higher on something you build yourself, but the design and development cost will be much higher. With an SBC the design time is short, and there is usually operating and support software available with the card.

More and more products are being shipped to the consumer market with OEM computers. At the same time, companies are discovering the huge overhead in developing their own computers. Sometimes large quantities and specific requirements demand that a computer be designed from scratch, but even that can be contracted out. So, as part of processor selection, deciding whether you truly need to build your own or OEM an SBC is an often overlooked option.

11–10 DESIGN OF A LABORATORY CONTROLLER

11–10a The Problem

In this example, consider a quality-control laboratory that wants to take a large volume of semiautomated test data, process that data quickly for go–no-go evaluation, and then transfer summaries of that data to a large host computer for further analysis and accounting reporting.

1. The test and measurement equipment to be operated is for the most part GPIB compatible, with the exception of one piece of equipment that requires a special parallel interface.
2. The data to be processed is a mixture of fixed-point and floating-point numbers with a precision of $4\frac{1}{2}$ digits.
3. There are approximately 50,000 floating-point calculations involved in computing the desired parameters from the collected raw data.
4. After the data is extracted and processed, it will be uploaded in bursts to a host computer for statistical analysis and accounting purposes. It is not desirable to send the raw data because of the volume of data and the response time of the host.
5. The laboratory environment is somewhat noisy, but with proper shielding it should be no problem. There are no constraints on power, and the ambient temperature is normal room temperature.

6. The system to be built is considered a prototype that may want to be expanded to multiple test stations, either by building additional units or by expanding the capabilities of this prototype.

11–10b   The Selection Process

The Hardware Interface   The processor must communicate to the outside world via a GPIB port, a special parallel interface, and two RS-232 ports (one for the host and one for the local terminal).

For the moment it is not considered necessary to have local mass storage, although it may be desired in the future.

11–10c   The Software Interface

Due to the large amount of floating-point computation to be done, it is highly desirable to have a high-level language with floating-point capability available for the software development. A PDP-11 host computer running the UNIX operating system is available with "C" Cross compilers available for the Z-80, 6800, Z-8000, PDP-11, and the 68000.

Initial development of the numerical processing algorithm on the host computer indicates that it will take approximately 30 kbytes of code. Allowing another 20 kbytes for a limited special-purpose operating system, we estimate an immediate need for 50 kbytes of code in the initial version. The 64K memory space of the Z-80 or 6800 would probably be adequate for now, but if the application were to grow, then we would be forced into a paged or overlayed system, which would be more trouble later. Therefore prime consideration will be given to the larger address space machines.

The initial version is not expected to need local mass storage. This allows a simpler operating system, since no file management is required, but what does it imply about loading the system? Will a ROM operating system be required? If the host computer can communicate over a 4800-bits per second line, then it could download at a rate of approximately 480 bytes per second. To download 64 kbytes at this rate would take about 138 s in one continuous burst, or allowing some pauses from the host, about 3 min. Since the system should need to be booted normally only once a day, this does not seem to be objectionable. Only a small ROM monitor will be required to start the system download.

One of the principal considerations for the system is the ability to expand it in the future. It may, for example, be expanded to multiple test stations, and it may be expanded to have its own general-purpose operating system locally. In that case, we should consider the possibility of that expansion now. One of the major considerations will be the operating

Processor	Operating system	Purpose	Company
LSI 11/23	RT-11	General	Digital Equipment Co.
	RSX-11M	General	
	RSX-11S	Process control	
	UNIX	General	Bell Labs
68000	VERSADOS	Real-time control	Motorola
	MSP-68000	Process control	Hemenway Assoc. Inc.
	UNIX	General	Control Systems Inc.
	IDRIS	General	Whitesmiths Ltd.
	MTOS-68K	Process and general	Industrial Programming Inc.
Z-8000	IDRIS	General	Whitesmiths Ltd.

system to use. Consider the partial list of operating systems available for various processors that appears at the bottom of page 11-20.

11–10d The Choice and Why

Communications   The communications is to be done over a standard GPIB bus and RS-232 serial lines. The GPIB controller and "talker-listener" functions are available as a set of ICs, so the communications load does not suggest any particular processor other than one that can interface with one of the available chips.

Quantity   The immediate aim is a single prototype for evaluation, with possible eventual production of 10 to 20 units. Therefore there is no great urgency to keep the parts cost to a minimum. The engineering costs could easily exceed the parts cost, so tradeoffs should be made to favor fast, easy development rather than minimum parts cost. This would indicate that a single-board computer would be a better choice than a "from scratch" layout, if the SBC meets all of our other requirements.

Address Space   The size of our application, the lack of mass storage, and our desire to avoid nonstandard paging schemes suggests that we should use one of the large address space machines.

Precision of Computation   Our precision is $4\frac{1}{2}$ decimal digits, or one part in 20,000. To handle such a number in one register would require at least a 16-bit register, so again a 16-bit processor seems desirable. The Z-80 could still be used, but at some speed penalty.

Real Time   Our application is not "real time" in that it must keep up with something beyond its control, but speed is important for test throughput. Again the choice is one of the new high-performance processors.

At this point we have narrowed the choice to one of the following:

LSI 11/23
68000
Z-8000
8086

Any one of these processors could do the job. One alternative is to consult published bench marks. For example, the April 1, 1981 issue of *Electronic Design* gives a set of bench marks comparing these four processors. These tests are somewhat ambiguous, but they seem to show a slight advantage for the 68000 overall, although the 8086 and Z-8000 both win some of the tests.

The 11/23 has the advantage of being software compatible with the large DEC minicomputers, for future expansion, but it is one of the slower of the 16-bit microprocessors.

At this point, local factors—such as tools available, familiarity of the designer with the various processors, personal preference of architecture, and last, but not least, current price—will be the deciding factor. We have chosen, for the sake of this example, to construct our laboratory system around a 68000 VERSAmodule system. This system uses the CPU card with two serial ports, four parallel ports, 64 kbytes of memory, timer, and associated control hardware. One additional board will be custom designed to hold the GPIB interface, since at the time of this writing none of the SBC packages offered for the processors considered have a GPIB board.

11–11 DESIGN OF KEYBOARD INTERFACE

11–11a The Problem

In this second example, consider a keyboard interface. We have a Z-80 processor that is used in a word processing system. The I/O consists of a floppy disk, a printer, and a terminal/keyboard. The keyboard that we are to use (it is already used in another product) has a serial interface.

The overhead to scan the keyboard would place an unneeded burden on the processor. To alleviate the problem we want to use a microcomputer to do the keyboard overhead and present the Z-80 with the ASCII character entered from the keyboard.

1. The keyboard interface is such that clock pulses are sent to the keyboard from the controller and after each pulse a serial data line is read to see whether that key is pressed; that is, after the fortieth clock pulse sent to the keyboard, the serial data line will indicate whether a particular key is pressed.
2. A table needs to be stored to allow look-up based on the clock pulse count when a true value is sent on the serial line.
3. The key debounce algorithm needs to be done in software. Other functions, such as auto repeat and learned keys, are desirable but not necessary. N-key rollover (the recognition of multiple keys pressed) must be implemented, as this is a word processing system.
4. The interface to the six LEDs on the keyboard is similar. A serial data line, along with clock line, is used to clock the data into a shift register. Once in the register (eight clocks), the data passes to a D-type latch an RC time constant later. It is at this time that the data turns on the LEDs.
5. The software for the processor will be written in assembly language, as we feel it will be approximately 1 kbyte.
6. It will be a medium-volume product, between 500 and 1000 per year.
7. There is no power problem, but the total system will have to pass company standards for static, EMI, and temperature. No specific environmental requirements exist, except for a humidity test.
8. This keyboard processor will need to communicate with the Z-80. The method is still undefined, but possibilities are shared I/O port, shared RAM (dual port), etc.
9. Considering the possible learned key function, it should be easy to add additional RAM to the processor to support this. The software structure should be designed with the idea of learned key addition.

11–11b The Selection Process

The Hardware Interface   The keyboard processor will be connected to the system bus, which will allow the Z-80 to read an I/O port containing the ASCII character calculated by the keyboard processor.

11–11c The Software Interface

Due to the relatively low-level task of scanning the keyboard and debouncing the keys, the instruction set of the processor is not of major concern.

It is desirable to have an EPROM version of the chip for prototype development. We have a microcomputer development system that supports many of the single-chip microcomputers. The selected processor should be supported on this system for software development and prototype debugging.

11–11d The Choice and Why

1. Processor Needs
    - a. Bit addressable I/O for clock pulses to keyboard and to read key data input
    - b. If possible, a Z-80/8080 bus interface
    - c. EPROM version of chip for prototyping
    - d. 1K of ROM and a minimum of 64 bytes of RAM
    - e. RAM stack pointer to assist in structured code
    - f. Expansion to outside RAM/ROM for future needs
2. Possible Selections
    - a. 8048/8748: Contains timer, 64 bytes of RAM, 1K ROM/EPROM, serial input pins, and 27 I/O lines.
    - b. 8041/8741: A special-purpose peripheral computer chip. Contains asynchronous master/slave interface, 1K ROM, 64 bytes RAM, timer, 18 programmable I/O lines, and 2 serial inputs.
    - c. 6801/68701: Contains 2K ROM, 128 bytes RAM, 3 timers, 31 I/O lines, expandable to 65 kbytes of memory, but not easily interfaced to Z-80 bus.
    - d. By consulting the selection tables you can see that there are many more chips that could possibly do the job than just the three listed here. These three are supported on our development system and used for comparison in this example only.

Which Processor?   Our choice is the 8041/8741 (as illustrated in Fig. 11-4) because of its unique master/slave interface. With the 8041 as the keyboard processor, the parts count is an unbelievable three chips, using the circuit of Fig. 11-5. The 8041

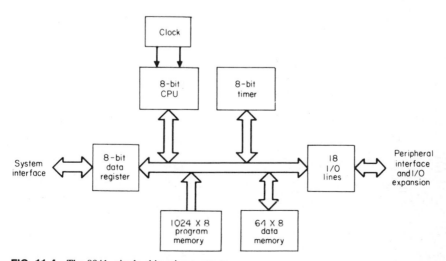

FIG. 11-4   The 8041, single-chip microcomputer.

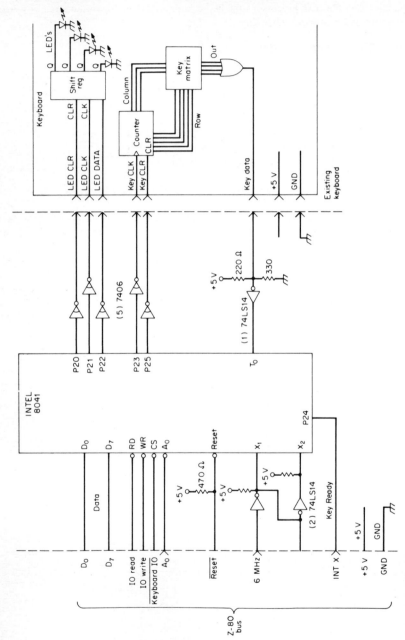

FIG. 11-5 New interface to ease system (Z-80) overhead.

11-24

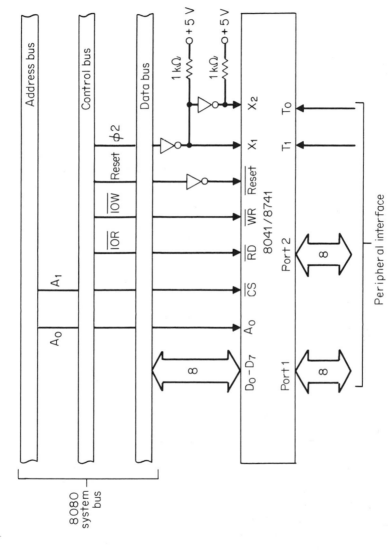

FIG. 11-6 The 8041/8741 interface to the 8080 system bus.

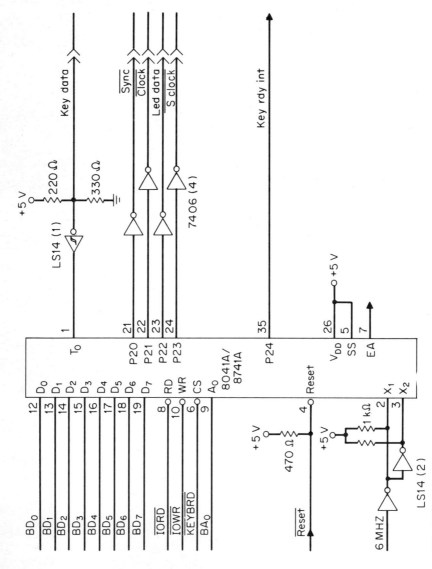

FIG. 11-7 The 8041 as a keyboard processor, where the keyboard is a serial scan type.

interfaces to the Z-80 just like a peripheral chip, as in Fig. 11-6, but it is also a complete microcomputer. With I/O lines and a timer, it fits the application perfectly. The 8041 will take care of the keyboard with four I/O lines and one serial input line. When a character needs to be sent to the system processor (Z-80), another I/O line is used to cause an interrupt to the Z-80. Although the 8041 cannot be expanded to external RAM/ROM, it was decided that the system processor can at some future time support the learn key function itself. The low parts count and manufacturing cost are primary concerns affecting this choice, along with the ease of debugging and the interconnect requirements.

Any of the other microcomputers listed or that appear in the tables would require additional circuitry to establish the communication between the system processor and the keyboard processor. With that circuitry built into the 8041, the interface is minimal (see Fig. 11-7).

The development system support is another reason to select one of the three processors listed above. It has been our experience that an emulator/assembler/debug package that is contained in a development system greatly reduces design and debug time in a hardware/firmware environment such as this.

11-12   DEVICE SELECTION CHARTS*

The tables on the following pages give information on microcomputer languages, general-purpose microprocessors, all-in-one processors, bit-slice families, and a directory of microprocessors by vendor.

---

*Tables 11-2 through 11-6 are reprinted by permission from *Electronic Design*, Vol. 28, No. 24, November 22, 1979; copyright Hayden Publishing Co., Inc., 1980.

TABLE 11–2 Summary of Microcomputer Languages

Languages	Available from	Executes on	Variables — No. of types	Variables — Name length	Math functions	Operators — Logic/bit	Operators — String	Operators — File	Array dimensions	Dynam. index
MacBasic	Analog Devices	Macsym 32	5	2	21	5	7	24	4	Yes
Ext. Basic	Data General	Nova. Eclipse	2	2	7	6	2	7[1]	Any	Yes
Bus. Basic	Data General	(μ)Nova, Eclipse	4	6	8	6	6	15	2	N/A
Basic-11	Digital Equip.	LSI-11/21,23	6	2	12	0	9	11	2	No
Basic/1000/L	Hewlett-Packard	L-Series	3	2	12	3/3	3	4	2	Yes
Basic-80	Microsoft	8080, 8085, Z80[1]	4	40	17	8	7	11	Any	No
6800 Basic	Motorola	6800. 09 EXORciser	1	2	7	0	8	5	2	No
Basic-M	Motorola	EXORset, μModule[1]	4	Any	20	4	10	Yes	2	No
6809 Basic	Motorola	6809 EXORciser[1]	6	80	25	13	11	5	2	No
Basic 1	RCA Corp.	CDP 18 S007	4	1	11	3	0	9	None	No
Basic 2	RCA Corp.	CDP 18 S007	3	1	17	4	4	9	2 (min)	No
Power Basic	Texas Inst.	TM990 modules[1]	4	5	10	4	8	4[2]	Any	Yes
12-k Basic	Wintek Corp.	6800-based	2	2	13	2	10[1]	8	8	Yes
Industr. Basic	Wintek Corp.	6800-based	1[1]	1	4	0	0	0	2	No
BCD Basic	Zilog Inc.	MCZ, ZDS	3	2	17	3	10	12	Any	Yes[1]
C-language	Whitesmiths	8080. PDP, LSI-11	10	8	6	8	Yes	Yes	5	No
C-language	Wintek Corp.	6800-based	7	8	0	10/7	0	0[1]	Any	No
C-language	Yourdon Inc.	LSI-11, PDP-11	6 min	Any[1]	0	3/3	0	0	256[2]	N/A
Interact. Cobol[1]	Data General	CS20, 30 (μNova)	N/A	30	4	0	2	5	3	Yes
Cobol 80	Microsoft	8080, 8085, Z80	6	30	5	2	5	8[1]	3	Yes
ANSI-74 Cobol	Zilog Inc.	MCZ, PDS	3	30	4	6	N/A	7	Any	Yes
Dabil	Ithaco Inc.	CompuDAS	1	160	15	6/6	3	8	Any	N/A
Forth	Ancon	8080/85/86.Z80[1]	Any	31	4	Any	N/A	N/A	Any	Yes
PolyForth	Forth Inc.	8080/86, 6800/09[1]	8[2]	128	21	18[2]	16	82[3]	Any	Yes
MP/Fortran IV	Data General	(μ)Nova. Eclipse	7	31	72[1]	9/3	0	10	128	Yes
Fortran IV	Digital Equip.	LSI-11/21, 23	7[1]	6	56	5/5	16[2]	16	7	Yes
Fortran IV	Hewlett-Packard	HP1000	5	6	58	3/3	0	5[1]	3	N/A
Fortran 80	Intel	Intellec, 8080/85[1]	4	6	38	5/5	0	8	7	No
Fortran 80[1]	Microsoft	8080. 8085. Z80	4	6	N/A	4/4	0	Yes[2]	3	No
Fortran	Motorola	MC6800, 6809[1]	2	6	9	4	0	4	N/A[2]	No
Fortran 77	Softech	8080/86, 6800/09[1]	4	6	35	3	0	7	3	Yes
Fortran	Texas Inst.	TM9900 modules[1]	6	6	12	8	0	6	3	Yes
Fortran IV[1]	Zilog Inc.	MCZ-1, PDS, ZDS	4	6	50	4	0	7	3	N/A
Jovial J73	Softech	LSI-11, Z8000[1]	7	31	0	5/5	7	0[2]	7	Yes
MDL/μ	Tektronix	8080/85, Z80, 6800/02	6	6	8	14/8	1[1]	7[1]	N/A	N/A
MPL	Motorola	6800/09 μModule[1]	11	Any	0	5/5	Yes	0[2]	3	No
BSO/Pascal	Boston Systems	6800. 8086[1]	8	132	21+	8+	0	10	Any	No
MP/Pascal	Data General	μNova. Eclipse	6	32[1]	11	13	13	19[2]	Any	No
Pascal/I	Hemenway Assoc.	M6800, Z8000	2	6	11	9/3[1]	0	0	1	No
Pascal 1000[1]	Hewlett-Packard	HP1000 M. E. F	6	150[2]	11	3/3	2	22	Any	No
Pascal 80[1]	Intel	Intellec/Isis II	10+	8	24	3/3	6	19	Any	Yes
MS-Pascal	Microsoft	8080, 8086[1]	14	19	31	16	12	18	Any	Yes[2]
Pascal	Motorola	6809. 68000[1]	10	32	6	11/4	5	18	Any	No
Pascal (UCSD)	Softech	8080/86, 6800/09[1]	8	736	17	11	6	12	Any	Yes
μP Pascal	Texas Instr.	FS, DS990[1]. TM990s	14	72	14	10	10	9	Any	No
Pascal	Whitesmiths	8080. PDP, LSI-11	6	8	11	N/A	Yes	Yes[1]	5	No
Industr. Pascal	Xycom	Z80-based[1]	5	8	11	3/3	10	10	Any	No
Pascal	Zilog Inc.	MCZ, PDS. ZDS	8+	1	14	3	2	5	Any	No
PL/M-80	Intel	8080/85. MDS800[1]	2	32	5	4	1	2[2]	1	No
PL/M-86	Intel	8086/88. MDS800[1]	5	32	5	4	17	4[2]	1	No
PLM	RCA	CDP 183007	2	31	5	4	0	0	1[1]	No
PL/65, CLS/65	Rockwell	6500 family	2	6	0	3[1]	0	0	1	Yes
PL/W	Wintek Corp.	6800[1]	3	31	5	3/3	0	0	1	No
PLZ/SYS	Zilog Inc.	MCZ, ZDS	5+	255	8	4/2	0	0	N/A	No
Strubal+	Hemenway Assoc.	M6800	3[1]	6	21	10/4	7	7	2	No

*A = Assembly; C = Cobol; F = Fortran; M = Machine; Mac = Macro assembler; P = Pascal.

Loop nesting levels	Memory access	Free-form coding	Multi state-ments	Reent-rant sub-routines	Con-cur-rency	Intermixing with other languages*	Comments
4	No	Yes	Yes	Yes	Yes[1]	None	[1]Up to 18 multitasking programs.
20	No	No	No	No[2]	Yes[1]	M	[1]No. of file types. [2]Fresh copy reach use. [3]User-implemented.
8[1]	Yes	No	No	N/A	Yes[2]	A	[1]Can be changed by programmer. [2]Interpreter is shared.
Any	No	Yes	Yes	No	No	Mac[1]	[1]Can also be intermixed with Syslib language.
256	No	Yes	No	No	No	A. F[1]	[1]Has CHAIN and INVOKE commands.
Any	Yes	Yes	Yes	No	No[2]	C. F. M. P	[1]Interpreted: 8080, Z80; compiled: all. [2]Real-time I/O supported.
8	Yes[1]	No	No	No	No	A	[1]PEEK and POKE functions.
21	Yes	No	No	No	Yes	A	[1]Compiles on 6809 EXORciser MDOS, uses run-time library.
21	Yes	No	No	No	Yes	A	[1]Compiled and interpreted.
Any	Yes	Yes	Yes	Yes	No	A. M[1]	[1]USR function used to expand language.
Any	Yes[1]	Yes	Yes	Yes	No	M	[1]PEEK and POKE functions.
10	Yes	Yes	Yes	No	No[3]	A	[1]Compiles on FS990. [2]Configured P-Basic. [3]Yes if interrupts.
N/A	No	No	Yes	No	No	None	[1]Substring operations.
20	Yes	No	Yes	No	No	A	[1]Only integers, −99999 to +99999. RAM and ROM versions.
Any	No	No	No	No	No	A. PLZ/ASM/SYS	[1]Can be redimensioned logically, IF/THEN/ELSE.
Any	Yes	Yes	Yes	Yes	Yes	M. P	
10	Yes	Yes	Yes	Yes	No	Linkable files[2]	[1]Available as external functions. [2]Real code, at run time.
Any	Yes	Yes	Yes	Yes	No	User-implemented	[1]First 8 characters significant. [2]Limited by addressing.
30	No	Yes	Yes	Yes	No	A. F. M	[1]ANSI-74 standard.
Any	No	Yes	Yes	No	No	A. F[2]	[1]Relative and independent files. [2]Compiler, assembler.
N/A	No	Yes	Yes	No	No	A. PLZ/SYS[1]	[1]Uses interpreter.
Any	Yes	Yes	Yes[1]	N/A	Yes[2]	M	[1]Maximum 160 characters per line. [2]Multiuser available.
Any	Yes	Yes	Yes	Yes	Yes	M. A	[1]Also 6502, 6800/09, 68000, TI990, LSI-11, μNova, HP21.
Any	Yes	Yes	Yes	Yes	Yes	Mac	[1]Also 1802, LSI-11, TI-990. [2]Plus user-defined. [3]Optional.
Any	Yes	Yes	No	Yes	Yes[3]	A	[1]55 arithmetic and conversion, 17 trig. [2]Multitasking.
Any	Yes	No	No	No	Yes[2]	A. Mac. Syslib	[1]Each also as array. [2]Syslib language extensions.
Any	No	No	No	No	No	A. P[2]	[1]ANSI-66. [2]Assembly and Pascal routines can be called.
Any	No	Yes	Yes	Yes	No	ASM80. PL/M-80	[1]Compiles on Intellec/Isis II, executes on 8080/85-based μCs.
Any	No	No	No	No	No	A. B. C	[1]ANSI-66. [2]Formatted/unformatted, sequential/random.
N/A	No	Yes	No	No	No	A. MPL[1]	[1]Compiles on EXORciser MDOS. [2]32k max. [3]Run-time library.
N/A	No	No	No	Yes	No	A. P	[1]Also Z80, 6502, 9900. LSI-II. μEngine (Western Digital).
N/A	Yes	Yes	No	Yes	No	A[2]	[1]Compiles on DS990, FS990. [2]User-function library.
Any	No	No	No	No	No	A. PLZ/SYS	[1]ANSI-66, without complex variables. Reads bits from ports.
Any	Yes	Yes	Yes	Yes	No	F. M	[1]Compiles on IBM 370. [2]Has target machine's I/O.
Any	No	No	No[2]	Yes	Yes	M	[1]Concatenation. [2]Need sequential line numbers.
N/A	No	Yes	Yes	No	No	A. F[3]	[1]Compiles on EXORciser. [2]MDOS utilities. [3]Efficient comp.
Any	Yes	Yes	Yes	Yes	No	A	[1]Compiles on DEC 10.
Any	Yes	Yes	Yes	Yes	Yes	A	[1]10 characters significant. [2]11 from INCLUDE option.
256	Yes	Yes	Yes	Yes	Yes	A	[1]AND, OR, NOT are bitwise.
Any	Yes	Yes	Yes	No	No	F. M	[1]Superset of Wirth Pascal. [2]All significant.
Any	Yes	Yes	Yes	Yes	No	F. M. PL/M[2]	[1]Superset of Wirth Pascal. [2]Second version.
Any	Yes	Yes	Yes	Yes	Yes	B. C. F. M. PL/M	[1]Compiles on DEC 10. [2]On "heap" only.
Any	Yes	Yes	Yes	Yes	No	A. M[2]	[1]Compiles on EXORciser. [2]Interpreter uses M-code, P-code.
Any	Yes	Yes	Yes	Yes[2]	Yes	A. F	[1]Also 6502, 68000, 9900, LSI-11, Z80, μEngine. [2]Also recursive.
31	Yes	Yes	Yes	Yes[2]	Yes	A[3]	[1]For compilation. [2]Recursive. [3]Also Component Software.
Any	Yes	Yes	Yes	Yes	Yes	M. C-language	[1]Uses language library.
Any	Yes[2]	Yes	Yes	Yes	Yes	A	[1]Compiles on any Xycom development system. [2]Pointers.
Any	Yes	Yes	Yes	Yes	No	A. PLZ/ASM. SYS	[1]First eight significant.
255	Yes	Yes	Yes	Yes	No	ASM80	[1]For compilation. [2]Port manipulation.
255	Yes	Yes	Yes	Yes	No	ASM86	[1]For compilation. [2]Port manipulation.
18	Yes	Yes	Yes	Yes	No	A	[1]Depends on memory size. [2]Includes I/O procedures for TTY.
255	Yes	Yes	Yes	Yes	No	A	
Any	Yes	Yes	Yes	Yes[2]	No	A (run time)	[1]Compiles on any 16-bit μC with Fortran. [2]Only assembly.
Any	Yes	Yes	Yes	Yes[1]	No	A. PLZ/ASM[2]	[1]Dynamic data allocation. [2]Also callable from B, C, F, P.
8	Yes[2]	Yes	No	Yes	No	A	[1]Plus user-defined types. [2]PUSH/POP construction.

TABLE 11–3 General-Purpose Microprocessors

Original source manufacturer	Processor	Process technology	Word size (data/instruction)	Direct addressing range (words)	Number of basic instructions	Maximum clock frequency (MHz)/phases	Instruction time shortest/longest[2] (μs)	TTL compatible	BCD arithmetic	On-chip interrupts/levels	Number of internal general-purpose registers
Motorola	MC14500	CMOS	1/4	0	16	1/1	1/1	Yes	No	Yes/1	1
Intel	4004	PMOS	4/8	4k	46	0.74/2	10.8/21.6	No	Yes	Yes/1	16
Intel	4040	PMOS	4/8	8k	60	0.74/2	10.8/21.6	No	Yes	Yes/1	24
NEC Microcomputers	μPD555	PMOS	4/10	1920 ×10	72	0.2/1	10/20	Yes	Yes	Yes/2	96 × 4
National Semiconductor	COP402	NMOS	4/8	1k	49	1/1	4	Yes	Yes	Yes/3	64 ×
National Semiconductor	COP402M	NMOS	4/8	1k	49	1/1	4	Yes	Yes	No	64 × 4
National Semiconductor	COP404L	NMOS	4/8	2k	49	0.25/1	16	Yes	Yes	Yes/3	128 × 4
NEC Microcomputers	μPD556	PMOS	4/8	2k	80	0.44/1	4.5/9	Yes	Yes	Yes/2	96 × 4
Panasonic	MN1498	NMOS	4/8	1k	66	0.3/1	10/20	Yes	Yes	Yes/1	64 × 4
Panasonic	MN1499	NMOS	4/8	2k	75	0.3/1	10/20	Yes	Yes	Yes/1	64 × 4
Panasonic	MN1499A	NMOS	4/8	2k	75	0.3/1	10/20	Yes	Yes	Yes/1	128 × 4
Panasonic	MN1599	NMOS	4/8	4k	125	1/1	2/4	Yes	Yes	Yes/4	256 × 4
Fairchild	2 chip F8 (3850)	NMOS	8/8	64k	69	2/1	2/13	Yes	Yes	Yes/1	64
General Instrument	8000	PMOS	8/8	1k	48	0.8/2	1.25/3.75	No	Yes	Yes/1	48
Intel	8008	PMOS	8/8	16k	48	0.8/2	12.5/37.5	No	Yes	Yes/1	6
Intel	8035/8039	NMOS	8/8	64k	96	6/1	2.5/5	Yes	Yes	Yes/1	64
Intel	8080A	NMOS	8/8	64k	78	2.6/2	1.5/3.75	Yes[3]	Yes	Yes/1	8
Intel	8085	NMOS	8/8	64k	80	5.5/1	0.8/5.2	Yes	Yes	Yes/4	8
MOS Technology	MCS-650X	NMOS	8/8	64k	56	4/1	0.5/3.5	Yes	Yes	Yes/1	0
MOS Technology	MCS-651X	NMOS	8/8	64k	56	4/2	0.5/3.5	Yes	Yes	Yes/1	0
Mostek	3874	NMOS	8/8	64k	70 +	4/1	1/6.5	Yes	Yes	Yes/4	64
Motorola	6800	NMOS	8/8	64k	72	2/2	1/2.5	Yes	Yes	Yes/1	0
Motorola	6802/6808	NMOS	8/8	64k	72	2/1	2/5	Yes	Yes	Yes/1	128/0
Motorola	6803	NMOS	8/8	64k	82	3.58/1	2/12	Yes	Yes	Yes/1	128
Motorola	6809	NMOS	8/8	64k	59	2/1	2/5	Yes	Yes	Yes/1	0
National Semiconductor	INS8060	NMOS	8/8	4k	46	4/1	5/22	Yes	Yes	Yes/1	8
National Semiconductor	NSC800	CMOS	8/8	64k	150 +	8/1	0.5/2.88	Yes	Yes	Yes/5	14
National Semiconductor	INS8040	NMOS	8/8	64k	96	11/1	1.4/2.8	Yes	Yes	Yes/1	256
National Semiconductor	INS8070	NMOS	8/8[7]	64k	74	4/1	3/1000[8]	Yes	No	Yes/2	9
NEC Microcomputers	μPD 8080A	NMOS	8/8	64k	78	2/2	1.92/8.16	Yes[3]	Yes	Yes/1	8
RCA	1802	CMOS	8/8	64k	91	6.4/1	2.5/3.75	Yes	Yes	Yes/1	16
RCA	8085AC	CMOS	8/8	64k	80	5.5/1	0.8/5.2	Yes	Yes	Yes/4	8
Signetics	2650	NMOS	8/8	32k	75	2/1	1.5/6	Yes	Yes	Yes/1	7
Signetics	8 × 300	Bi-polar	8/16	8k	NA[10]	4/1	0.25	Yes	No	No	8
Zilog	Z80	NMOS	8/8	64k	150 +	4/1	1/5.75	Yes	Yes	Yes/1	14
Intersil	6100	CMOS	12/12	4k	81	4/1	2.5/5.5	Yes	No	Yes/1	0
Toshiba	T3190	PMOS NMOS	12/12	4k	108	2.5/1	10/30	Yes	No	Yes/8	8
Advanced Micro Devices	Am29116	ECL	16/16	64k	30 +	10/1	0.1/0.2	Yes	No	No	32
Data General	mN601	NMOS	16/16	32k	42	8.33/2	1.2/29.5	Yes	No	Yes/1	4
Data General	mN602	NMOS	16/16	64k	82	8.3/2	2.4/53	Yes	No	Yes/16	4[6]
Fairchild	9440	I²L	16/16	64k[5]	42	12/1[9]	1.25/3.5	Yes	No	Yes/16	4
Fairchild	9445	I²L	16/16	64k	100	20/1[9]	0.3/5.7	Yes	No	Yes/16	4
Ferranti	F100L	Bi-polar	16/16	32k	153	14/1	1.19/14	Yes	No	Yes/1	RAM
General Instrument	CP1600/1610	NMOS	16/16	64k	87	4/2	1.6/4.8	Yes	No	Yes/1	8
Intel	8086	NMOS	16/16	1M[5]	97	5/1	0.4/37.8	Yes	Yes	Yes/1	8
Intel	8088	NMOS	16/8[1]	64k[5]	97	5/1	0.4/37.8	Yes	Yes	Yes/1	8
Motorola	MC68000	NMOS	16/16	16M[5]	61	8/1	0.5/NA[10]	Yes	Yes	Yes/1	16
National Semiconductor	INS8900	NMOS	16/16	64k	45	2/1	2.5/5	Yes	Yes	Yes/6	4
National Semiconductor	NS16008	NMOS	16/16[1]	64k[5]	78/100 +	NA[10]	NA[10]	Yes	Yes	Yes	8
National Semiconductor	NS16016	NMOS	16/16	64k	78/100 +	NA[10]	NA[10]	Yes	Yes	Yes	8
National Semiconductor	NS16032	NMOS	16/16	16M[5]	100 +	NA[10]	NA[10]	Yes	Yes	Yes	8
Panafacom	MN1610	NMOS	16/16	64k	33	2/2	2/6	Yes[3]	No	Yes/3	5
Texas Instruments	TMS9980/9981	NMOS	16/16[1]	8k	69	4/4	3.2/49.6	Yes[3]	No	Yes/4	16
Texas Instruments	TMS9985	NMOS	16/16[1]	32k	68	5/1	2.4/50	Yes	No	Yes/4	RAM
Texas Instruments	TMS/SBP9900	NMOS/ I²L	16/16	32k	69	4/4	2/31	Yes[3]	No	Yes/16[11]	16
Western Digital	WD-16	NMOS	16/16	64k	116	3.3/4	2.1/780	Yes	Yes	Yes/16	6
Western Digital	Pascal Microengine	NMOS	16/16	64k	150 +	3/4	2.4/300[8]	Yes	Yes	Yes/4	RAM
Zilog	Z8000	NMOS	16/16	48M[5]	110 +	8/1	0.75/90	Yes	Yes	Yes/1	16

1. Has 8-bit external buses and 16-bit internal buses. 2. With maximum clock. 3. Except clock lines. 4. Standard TTL or MOS circuits will suffice. 5. Range in bytes. 6. Frame Pointer too. 7. Double-precision 16-bit operations available. 8. String search. 9. Clock internally divided by 4 or 6 depending on instruction. 10. Not applicable. 11. 9980 only.

Number of stack registers	On-chip clock	DMA capability	Specialized memory & I/O circuits available	Prototyping system available	Package size (pins)	Voltages required (V)	Assembly language development system	High-level languages	Time-sharing cross software	Comments
0	Yes	No	No⁴	No	16	3 to 18	No	No	No	Needs external program counter
3 × 12	No	No	Yes	No	16	15	Yes	Yes	Yes	Superseded by 4040
7 × 12	No	No	Yes	Yes	24	15	Yes	Yes	Yes	General-purpose 4-bit μP
4	Yes	No	No	Yes	64	−10	Yes	No	Yes	ROM-less version of μPD548
RAM	Yes	No	Yes	Yes	40	4.5 to 6.3	Yes	No	Yes	The 402, 402M and 404L are ROM-less versions of COP420 and 440
RAM	Yes	No	Yes	Yes	40	4.5 to 6.3	Yes	No	Yes	Single chip μC. All three have serial I/O and event counting capability, as well
RAM	Yes	No	Yes	Yes	40	4.5 to 9.5	Yes	No	Yes	As 20 I/O lines
3	Yes	No	No	Yes	64	−10	Yes	No	Yes	ROM-less version of PD546
RAM	Yes	No	Yes	Yes	40	5	Yes	No	Yes	ROM-less version of MN1402, but 66 instructions and come in a 40-pin package
RAM	Yes	No	Yes	Yes	64	5	Yes	No	Yes	ROM-less version of MN1400
RAM	Yes	No	Yes	Yes	64	5	Yes	No	Yes	ROM-less version of MN1400, but 128 nibbles of on-chip RAM
RAM	Yes	Yes	Yes	Yes	64	5	Yes	No	Yes	ROM-less version of MN1564; chip has 12 4-bit I/O ports
RAM	Yes	Yes	Yes	Yes	40	5, 12	Yes	Yes	Yes	Usually used with program storage unit
0	No	No	Yes	Yes	40	5, −12	No	Yes	Yes	Predecessor of F8
7 × 14	No	No	Yes	Yes	18	5, −9	Yes	Yes	Yes	Predecessor of 8080, still in wide use
RAM	Yes	Yes	Yes	Yes	40	5	Yes	Yes	Yes	ROM-less versions of 8048/8049
RAM	No	Yes	Yes	Yes	40	5, 12, −5	Yes	Yes	Yes	By and large, still the most popular
RAM	Yes	Yes	Yes	Yes	40	5	Yes	Yes	Yes	8080 code compatible, has built-in clock
RAM	Yes	No	Yes	Yes	40	5	Yes	Yes	Yes	Provides 13 addressing modes
RAM	No	No	Yes	Yes	40	5	Yes	Yes	Yes	Similar to 650X but needs 2ϕ clock
RAM	Yes	No	Yes	Yes	40	5	Yes	No	Yes	ROM-less piggy-back version of 3870; accepts UV EPROMS on top
RAM	No	Yes	Yes	Yes	40	5	Yes	Yes	Yes	Available in depletion-load version
RAM	Yes	Yes	Yes	Yes	40	5	Yes	Yes	Yes	6802 has 128 × 8 on RAM; 6808 has no RAM
RAM	Yes	Yes	Yes	Yes	40	5	Yes	Yes	Yes	ROM-less version of 6801 single-chip μC
RAM	Yes	Yes	Yes	Yes	40	5	Yes	Yes	Yes	Enhanced 6800 command set
RAM	Yes	Yes	No⁴	Yes	40	5	Yes	Yes	Yes	Has handy daisy-chain capability
RAM	Yes	Yes	Yes	Yes	40	3 to 12	Yes	Yes	Yes	Executes Z80 instructions and has 8085 bus structure
RAM	Yes	Yes	Yes	Yes	40	5	Yes	Yes	Yes	ROM-less version of INS8050 single chip μC
RAM	Yes	Yes	Yes⁴	Yes	40	5	Yes	Yes	No	ROM-less version of INS8072: 64 bytes of on-chip RAM
RAM	No	Yes	Yes	Yes	40	5, 12, −5	Yes	Yes	Yes	Pin compatible but does BCD subtraction
RAM	Yes	Yes	Yes	Yes	40	3 to 12	Yes	Yes	Yes	Superseded two-chip version
RAM	Yes	Yes	Yes	Yes	40	5	Yes	Yes	Yes	CMOS equivalent to 8085A, and pin-compatible
8 × 15	No	Yes	Yes	Yes	40	5	Yes	Yes	Yes	There are 1.25 and 2-MHz versions
0	No	No	No⁴	Yes	50	5	Yes	No	Yes	Intended for high-speed controllers
RAM	No	Yes	Yes	Yes	40	5	Yes	Yes	Yes	8080 instructions are a subset
RAM	Yes	Yes	Yes	Yes	40	4 to 11	Yes	Yes	Yes	Emulates PDP-8 instruction set
RAM	Yes	Yes	Yes	Yes	36	5, −5	Yes	Yes	Yes	Has multiply and divide instructions
32	No	Yes	Yes	Yes	48	5	Yes	No	Yes	Control-oriented microprogrammable CPU, can generate CRC bits
RAM	Yes	Yes	Yes	No	40	5, 10, 14, −4.25	Yes	Yes	Yes	Emulates NOVA instruction set
RAM	No	Yes	Yes	Yes	40	3, 12, ±5	Yes	Yes	Yes	Executes the NOVA instruction set and addresses double the memory of the mN601
RAM	Yes	Yes	Yes	Yes	40	5	No	No	No	Emulates NOVA instruction set
RAM	No	Yes	Yes	Yes	40	5	Yes	Yes	Yes	Executes NOVA 3 and 4 instruction sets
RAM	No	Yes	Yes	Yes	40	5, 1.2	Yes	Yes	Yes	Can do double word operations
RAM	No	Yes	Yes	Yes	40	5, 12, −3	Yes	Yes	Yes	All internal registers can be accumulators
RAM	Yes	Yes	Yes	Yes	40	5	Yes	Yes	Yes	Has 24 addressing modes
RAM	Yes	Yes	Yes	Yes	40	5	Yes	Yes	Yes	8-bit bus version of 8086 microprocessor
RAM	No	Yes	Yes	Yes	64	5	Yes	Yes	Yes	Has 32-bit-wide internal structure
10 × 16	No	Yes	Yes	Yes	40	5	Yes	Yes	Yes	Architecture intended for data handling
RAM	No	Yes	Yes	Yes	40	5	Yes	Yes	No	8-bit bus version of dual language (8080/native) CPU, has internal 16-bit bus
RAM	No	Yes	Yes	Yes	40	5	Yes	Yes	No	Full 16-bit version, offers 8080A and native instruction sets
RAM	No	Yes	Yes	Yes	48	5	Yes	Yes	No	Expanded 16-bit version with eight 32-bit registers, six 24-bit registers and two 16-bit registers; can address 16 Mbytes
RAM	No	Yes	Yes	No	40	5, 12, −3	Yes	No	No	
RAM	Yes	Yes	Yes	No	40	5, 12, −5¹¹	Yes	Yes	Yes	The 9981 requires external clock
RAM	Yes	Yes	Yes	Yes	40	5	Yes	Yes	Yes	ROM-less version of 9940, with buses
RAM	No	Yes	Yes	No	64	5, 12, −5	Yes	Yes	Yes	Emulates 990 mini instructions
RAM	No	Yes	Yes	Yes	40	5, 12, −5	Yes	Yes	No	Very similar to DEC LSI-11
RAM	No	Yes	Yes	Yes	40	+12, ±5	Yes	Yes	Yes	Five-chip set directly executes Pascal p-code
RAM	No	Yes	Yes	Yes	40/48	5	Yes	Yes	Yes	40-pin version is the Z8002; 48-pin, the Z8001

TABLE 11–4 All-in-One Processors

Original-source manufacturer	Device	Process technology	Word size in bits (data/inst.)	On-chip RAM size	On-chip ROM/PROM size (words)	Off-chip memory expansion	Number of basic instructions	Maximum clock frequency (kHz)	On-chip clock	Instruction time shortest/longest (μs)	TTL compatible
AMI	S2000	NMOS	4/8	64 × 4	1024 × 8	No	51	1000	Yes	4.5/9	Yes
	S2150	NMOS	4/8	64 × 4	1536 × 8	No	51	1000	Yes	4.5/9	Yes
	S2200	NMOS	4/8	128 × 4	2048 × 8	Yes	59	1000	Yes	4.5/9	Yes
	S2400	NMOS	4/8	128 × 4	4096 × 8	Yes	59	1000	Yes	4.5/9	Yes
Essex International	SX-200	PMOS	4/8	64 × 4	1024 × 8	Yes	41	400	Yes	20/20	No
Hitachi	HMCS42	PMOS	4/10	32 × 4	512 × 10	No	74	780	Yes	10	No
	HMCS43/43C	PMOS/CMOS	4/10	80 × 4	1024 × 10	No	74	780/500	Yes	10	No/Yes
	HMCS44	PMOS	4/10	160 × 4	2048 × 10	No	69	780	Yes	20	No
	HMCS45	PMOS	4/10	160 × 4	2048 × 10	No	69	780	Yes	20	No
ITT Semiconductor	7150	?	?	?	N/A[1]	?	?	25	Yes	?	?
National Semiconductor	MM57109	PMOS	4/8	5 × 32	N/A[1]	Yes	70	400	No	1220/1 S	Yes
	COP410L	NMOS	4/8	32 × 4	512 × 8	Yes	40	250	Yes	16/32	Yes
	COP411L	NMOS	4/8	32 × 4	512 × 8	Yes	40	250	Yes	16/32	Yes
	COP420	NMOS	4/8	64 × 4	1024 × 8	Yes	49	1000	Yes	4/8	Yes
	COP420L	NMOS	4/8	64 × 4	1024 × 8	Yes	49	250	Yes	16/32	Yes
	COP420C	CMOS	4/8	64 × 4	1024 × 8	Yes	49	250	Yes	16/32	Yes
	COP421	NMOS	4/8	64 × 4	1024 × 8	Yes	49	1000	Yes	4/8	Yes
	COP421L	NMOS	4/8	64 × 4	1024 × 8	Yes	49	250	Yes	16/32	Yes
	COP421C	CMOS	4/8	64 × 4	1024 × 8	Yes	49	250	Yes	16/32	Yes
	COP440	NMOS	4/8	128 × 4	2048 × 8	Yes	49	1000	Yes	4/8	Yes
	COP444L	NMOS	4/8	128 × 4	2048 × 8	Yes	49	250	Yes	16/32	Yes
NEC Microcomputers	μPD548	PMOS	4/10	96 × 4	1920 × 10	Yes	72	200	No	10/20	Yes
	μPD546	PMOS	4/8	96 × 4	2000 × 8	No	80	440	Yes	4.5/9	Yes
	μPD553	PMOS	4/8	96 × 4	2000 × 8	No	80	440	Yes	4.5/9	Yes
	μPD650	CMOS	4/8	96 × 4	2000 × 8	No	80	440	Yes	4.5/9	Yes
	μPD547	PMOS	4/8	64 × 4	1000 × 8	Yes	58	440	Yes	4.5/9	Yes
	μPD547L	PMOS	4/8	64 × 4	1000 × 8	Yes	58	180	Yes	11/22	Yes
	μPD552	PMOS	4/8	64 × 4	1000 × 8	No	58	440	Yes	4.5/9	Yes
	μPD651	CMOS	4/8	64 × 4	1000 × 8	Yes	58	440	Yes	4.5/9	Yes
	μPD550	PMOS	4/8	32 × 4	640 × 8	No	58	440	Yes	4.5/9	Yes
	μPD554	PMOS	4/8	32 × 4	1000 × 8	No	58	440	Yes	4.5/9	Yes
	μPD652	CMOS	4/8	32 × 4	1000 × 8	No	58	440	Yes	4.5/9	Yes
	μPD551	PMOS	4/8	64 × 4	1000 × 8	Yes	58	440	Yes	4.5/9	Yes
Panasonic	MN1400	NMOS	4/8	64 × 4	1024 × 8	No	75	300	Yes	10/20	Yes
	MN1402	NMOS	4/8	32 × 4	768 × 8	No	57	300	Yes	10/20	Yes
	MN1403	NMOS	4/8	16 × 4	512 × 8	No	50	300	Yes	10/20	Yes
	MN1404	NMOS	4/8	16 × 4	512 × 8	No	48	300	Yes	10/20	Yes
	MN1405	NMOS	4/8	128 × 4	2048 × 8	No	75	300	Yes	10/20	Yes
	MN1430	PMOS	4/8	64 × 4	1024 × 8	No	75	200	Yes	15/30	No
	MN1432	PMOS	4/8	32 × 4	768 × 8	No	57	200	Yes	15/30	No
	MN1435	PMOS	4/8	128 × 4	2048 × 8	No	75	200	Yes	15/30	No
	MN1450	CMOS	4/8	64 × 4	1024 × 8	No	75	500	Yes	6/12	Yes
	MN1453	CMOS	4/8	16 × 4	512 × 8	No	50	500	Yes	6/12	Yes
	MN1454	CMOS	4/8	16 × 4	512 × 8	No	48	500	Yes	6/12	Yes
	MN1455	CMOS	4/8	128 × 4	2048 × 8	No	75	500	Yes	6/12	Yes
	MN1542	NMOS	4/8	152 × 4	2048 × 8	Yes	124	1000	Yes	2/4	Yes
	MN1544	NMOS	4/8	256 × 4	4096 × 8	Yes	124	1000	Yes	2/4	Yes
	MN1562	NMOS	4/8	152 × 4	2048 × 8	Yes	124	1000	Yes	2/4	Yes
	MN1564	NMOS	4/8	256 × 4	4096 × 8	Yes	124	1000	Yes	2/4	Yes
Rockwell	PPS-4	PMOS	4/8	0	0	Yes	50	200/400 Two clocks	No	5/15	No
	PPS-4/1, 4/2	PMOS	4/8	0	0	Yes	50	200/400	Yes	5/15	No
	MM77/MM77L	PMOS	4/8	96 × 4	1344 × 8	RAM only	50	100/4	Yes	10/40	Yes
	MM78/MM78L	PMOS	4/8	128 × 4	2048 × 8	RAM only	50	100/4	Yes	10/40	Yes
	MM76	PMOS	4/8	48 × 4	640 × 8	RAM only	50	100/4	Yes	10/40	Yes
	MM76/C	PMOS	4/8	48 × 4	640 × 8	RAM only	50	100/4	Yes	10/30	Yes
	MM76/E	PMOS	4/8	48 × 4	1024 × 8	RAM only	50	100/4	Yes	10/30	Yes
	MM76/L	PMOS	4/8	48 × 4	640 × 8	RAM only	50	100/4	Yes	10/30	Yes
	MM76/EL	PMOS	4/8	48 × 4	640 × 8	RAM only	50	100/4	Yes	10/30	Yes
	MM75	PMOS	4/8	48 × 4	670 × 8	RAM only	50	100/4	Yes	10/40	Yes

11-32

BCD arithmetic	On-chip interrupts/levels	Subroutine nesting levels	General-purpose internal registers	Number of I/O lines	Additional special support circuits	Package size (DIP pins)	Voltages required (V)	Prototyping system available	Assembly language programming system	High-level language programming system	Time-sharing cross software	Comments
Yes	Yes/1	RAM	RAM	29	No	40	9	No	Yes	No	No	Includes display drivers and switch interface
Yes	Yes/1	RAM	RAM	29	No	40	9	No	Yes	No	No	Expanded ROM version
Yes	Yes/3	5	RAM	29	No	40	9	Yes	Yes	Yes	Yes	Both the 2200 and 2400 include
Yes	Yes/3	5	RAM	29	No	40	9	Yes	Yes	Yes	Yes	an 8-bit a/d converter and an 8-bit d/a converter. All S2000 processors also come in an "A" version that can directly drive vacuum fluorescent displays
Yes	Yes/1	1	RAM	16	No	28	10 to 20	Yes	Yes	Yes	Yes	Has touch switch interface
Yes	No	RAM	RAM	22	No	28	−10	No	Yes	Yes	Yes	Minimal I/D version
Yes	Yes/2	RAM	RAM	32	No	42	−10/+5	Yes	Yes	Yes	Yes	Available in CMOS and PMOS versions
Yes	Yes/2	RAM	RAM	31	Yes	42	−10	Yes	Yes	Yes	Yes	Easily handles display driving
Yes	Yes/2	RAM	RAM	40	Yes	54	−10	Yes	Yes	Yes	Yes	Comes in 54 pin flat package
?	?	?	?	14	Yes	14/18/24	−15	No	No	No	No	Designed for washing machines
Yes	Yes/1	4	1	11	Yes	28	9	No	No	No	No	Has scientific calculation ability
Yes	No	2	RAM	16	Yes	24	4.5 to 9.5	Yes	Yes	No	Yes	All COPs processors include se-
Yes	No	2	RAM	15	Yes	20	4.5 to 9.5	Yes	Yes	No	Yes	rial I/O and event counting ca-
Yes	Yes/1	3	RAM	20	Yes	28	4.5 to 6.3	Yes	Yes	No	Yes	pability. Major differences be-
Yes	Yes/1	3	RAM	20	Yes	28	4.5 to 9.5	Yes	Yes	No	Yes	tween models include the I/O arrangements —input only, bi-directional, output only, etc.
Yes	Yes/1	3	RAM	20	Yes	28	2.4 to 6.3	Yes	Yes	No	Yes	I/O options include LED direct
Yes	No	3	RAM	16	Yes	24	4.5 to 6.3	Yes	Yes	No	Yes	segment drive, LED direct digit
Yes	No	3	RAM	16	Yes	24	4.5 to 9.5	Yes	Yes	No	Yes	drive, three-state push-pull,
Yes	No	3	RAM	16	Yes	24	2.4 to 6.3	Yes	Yes	No	Yes	push pull, open drain, and standard (active device to ground and a pull-up to VCC).
Yes	Yes/1	3	RAM	32	Yes	40	4.5 to 6.3	Yes	Yes	No	Yes	
Yes	Yes/1	3	RAM	20	Yes	28	4.5 to 9.5	Yes	Yes	No	Yes	
Yes	Yes/2	4	RAM	35	No	42	−10	Yes	Yes	No	Yes	Well suited for POS and ECR applications
Yes	Yes/1	3	RAM	35	No	42	−10	Yes	Yes	No	Yes	TTL-compatible I/O lines
Yes	Yes/1	3	RAM	35	No	42	−10	Yes	Yes	No	Yes	I/O handles −35-V vacuum fluorescent drive
Yes	Yes/1	3	RAM	35	No	42	+5	Yes	Yes	No	Yes	CMOS version of 546 (4% of the power)
Yes	Yes/1	1	RAM	35	No	42	−10	Yes	Yes	No	Yes	TTL-compatible I/O lines
Yes	Yes/1	1	RAM	35	No	42	−8	Yes	Yes	No	Yes	Low power version of 547 (half the current)
Yes	Yes/1	1	RAM	35	No	42	−10	Yes	Yes	No	Yes	I/O handles −35-V vacuum fluorescent drive
Yes	Yes/1	1	RAM	35	No	42	+5	Yes	Yes	No	Yes	CMOS version of 547 (4% of the power)
Yes	Yes/1	1	RAM	21	No	28	−10	Yes	Yes	No	Yes	I/O handles −35-V vacuum fluorescent drive
Yes	Yes/1	1	RAM	21	No	28	−10	Yes	Yes	No	Yes	I/O handles −35-V vacuum fluorescent drive
Yes	Yes/1	1	RAM	21	No	42	+5	Yes	Yes	No	Yes	CMOS version of 550 (5% of the power)
Yes	Yes/1	2	RAM	28	No	40	−10	Yes	Yes	No	Yes	Includes a/d converter with 2% resolution, 4% accuracy
Yes	Yes/1	2	RAM	30	No	40	5	No	Yes	Yes	Yes	Complete all-in-one controller
Yes	Yes/1	2	RAM	19	No	28	5	No	Yes	Yes	Yes	Smaller I/O version of 1400
No	Yes/1	2	RAM	13	Yes	18	5	Yes	Yes	No	Yes	All the processors in the
No	Yes/1	2	RAM	10	Yes	16	5	Yes	Yes	No	Yes	MN1400 family are available
No	Yes/2	2	2 + RAM	34	Yes	40	5	Yes	Yes	No	Yes	in at least one other technol-
No	Yes/1	2	RAM	30	Yes	40	−15	Yes	Yes	No	Yes	ogy. The 18 and 16-pin ver-
No	Yes/1	2	RAM	19	Yes	28	−15	Yes	Yes	No	Yes	sions are about the smallest
No	Yes/2	2	2 + RAM	34	Yes	40	−15	Yes	Yes	No	Yes	µCs available, although most
No	Yes/1	2	RAM	30	Yes	40	4.25 to 6	Yes	Yes	No	Yes	versions still retain at least 2/3 of the instructions available on the 1405. The CMOS versions can also be with operating voltages of up to 10 V.
No	Yes/1	2	RAM	13	Yes	18	4.25 to 6	Yes	Yes	No	Yes	
No	Yes/1	2	RAM	10	Yes	16	4.25 to 6	Yes	Yes	No	Yes	
No	Yes/2	2	2 + RAM	34	Yes	40	4.25 to 6	Yes	Yes	No	Yes	All MN1500 family processors
Yes	Yes/4	16	4 + RAM	24	Yes	40	5	Yes	Yes	No	Yes	include an 8-bit counter/timer
Yes	Yes/4	16	4 + RAM	24	Yes	40	5	Yes	Yes	No	Yes	and an 8-bit serial shift regis-
Yes	Yes/4	16	4 + RAM	48	Yes	64	5	Yes	Yes	No	Yes	ter. All I/O lines are bidirec-
Yes	Yes/4	16	4 + RAM	48	Yes	64	5	Yes	Yes	No	Yes	tional and the chips also have a power-down made to minimize power dissipation
Yes	Yes/1	2	1	12+	Yes	42	−17/+5, −12	Yes	Yes	No	Yes	Combination ROM/RAM/I/O available
Yes	No	2	1	12+	Yes	42	−17/+5, −12	Yes	Yes	No	Yes	Same as PPS-4 but has internal dk
Yes	Yes/1	2	2 + RAM	31	Yes	42	−15/+5, −10	Yes	Yes	No	Yes	I/O includes serial channel
Yes	Yes/1	2	2 + RAM	31	Yes	42	−15/+5, −10	Yes	Yes	No	Yes	Software compatible with 77
Yes	Yes/1	1	1 + RAM	31	Yes	42	−15/+5, −10	Yes	Yes	No	Yes	Primarily used for keyboard display
Yes	Yes/1	1	1 + RAM	39	Yes	52	−15/+5, −10	Yes	Yes	No	Yes	Has high-speed counter
Yes	Yes/1	1	1 + RAM	31	Yes	42	−15/+5, −10	Yes	Yes	No	Yes	Larger ROM than MM76
Yes	Yes/1	1	1 + RAM	31	Yes	40	6 to 11	Yes	Yes	No	Yes	Low power version of 76
Yes	Yes/1	1	1 + RAM	31	Yes	42	−15/+5, −10	Yes	Yes	No	Yes	Expanded ROM version
Yes	Yes/1	1	1 + RAM	22	Yes	28	−15/+5, −10	Yes	Yes	No	Yes	Reduced I/O version of 76

TABLE 11-4 (Continued)

Original-source manufacturer	Device	Process technology	Word size in bits (data/inst.)	On-chip RAM size	On-chip ROM/PROM size (words)	Off-chip memory expansion	Number of basic instructions	Maximum clock frequency (kHz)	On-chip clock	Instruction time shortest/longest (μs)	TTL compatible
Texas Instruments	TMS-1000	PMOS	4/8	64 × 4	1024 × 8	No	43	400	Yes	15/15	Yes
	TMS-1000C	CMOS	4/8	64 × 4	1024 × 8	No	43	1000	Yes	6/6	Yes
	TMS-1100	PMOS	4/8	128 × 4	2048 × 8	No	40	400	Yes	15/15	Yes
	TMS-1018	PMOS	4/8	64 × 4	1024 × 8	No	43	400	Yes	15/15	Yes
	TMS1022	PMOS	4/8	64 × 4	2048 × 8	No	43	400	Yes	15/15	Yes
	TMS1117	PMOS	4/8	128 × 4	2048 × 8	No	43	400	Yes	15/15	Yes
	TMS1121	PMOS	4/8	128 × 4	2048 × 8	No	42	400	Yes	15/15	Yes
	TMS1400	PMOS	4/8	128 × 4	4096 × 8	No	41	550	Yes	11/11	Yes
	TMS1600	PMOS	4/8	64 × 4	512 × 8	No	41	550	Yes	11/11	Yes
	TMS1700	PMOS	4/8	64 × 4	512 × 8	No	43	400	Yes	15/15	Yes
Toshiba	T3444	NMOS	4/8	16 × 8	256 × 24	Yes	3	800	Yes	3	Yes
	T3472	NMOS	4/8	16 × 4	256 × 24	Yes	67	1000	Yes	33/360	Yes
Western Digital	1872	PMOS	4/10	32 × 4	512 × 10	No	37	150	Yes	6.25/12.5	Yes
Fairchild	F38E70	NMOS	8/8	64 × 8	2048 × 8	Yes	70+	4000	Yes	1/6.5	Yes
	F3878	NMOS	8/8	64 × 8	4096 × 8	Yes	70+	4000	Yes	1/6.5	Yes
General Instrument	PIC1645	NMOS	8/12	24 × 8	256 × 12	No	30	1000	Yes	4/8	Yes
	1650	NMOS	8/12	32 × 8	512 × 12	No	30	1000	Yes	4/8	Yes
	1655	NMOS	8/12	32 × 8	512 × 12	No	30	1000	Yes	4/8	Yes
	1670	NMOS	8/12	48 × 8	1024 × 12	No	30	1000	Yes	4/8	Yes
Intel	8021	NMOS	8/8	64 × 8	1024 × 8	No	70	3000	Yes	10/20	Yes
	8022	NMOS	8/8	64 × 8	2048 × 8	No	70	3000	Yes	10/20	Yes
	8041/8741	NMOS	8/8	64 × 8	1024 × 8	Yes	90	6000	Yes	2.5/5	Yes
	8048/8748	NMOS	8/8	64 × 8	1024 × 8	Yes	96	6000	Yes	2.5/5	Yes
	8049	NMOS	8/8	128 × 8	2048 × 8	Yes	96	11000	Yes	1.4/2.8	Yes
Intersil	87C41	CMOS	8/8	64 × 8	1024 × 8	Yes	90	6000 (5V)	Yes	2.5/5	Yes
	80C48/87C48	CMOS	8/8	64 × 8	1024 × 8	Yes	96	6000 (5V)	Yes	2.5/5	Yes
Mostek	3870	NMOS	8/8	64 × 8	2048 × 8	Yes	70+	4000	Yes	1/6.5	Yes
	3872	NMOS	8/8	128 × 8	4096 × 8	Yes	70+	4000	Yes	1/6.5	Yes
	3873	NMOS	8/8	64 × 8	2048 × 8	Yes	70+	4000	Yes	1/6.5	Yes
	3876	NMOS	8/8	128 × 8	2048 × 8	Yes	70+	4000	Yes	1/6.5	Yes
Motorola	6801/68701	NMOS	8/8	128 × 8	Yes	Yes	82	3580	Yes	2/12	Yes
	6805/705	NMOS	8/8	64 × 8	1100 × 8	Yes	61	3580	Yes	2/4	Yes
	6805R2	NMOS	8/8	64 × 8	2048 × 8	Yes	61	3580	Yes	2/4	Yes
	146805	CMOS	8/8	64 × 8	1100 × 8	Yes	61	3580	Yes	2/4	Yes
National Semiconductor	INS8050	NMOS	8/8	256 × 8	4096 × 8	Yes	96	11000	Yes	1.4/2.8	Yes
	INS8072	NMOS	8/8	64 × 8	2560 × 8	Yes	74	4k	Yes	3/1000	Yes
RCA	CDP1804	CMOS/SOS	8/8	64 × 8	2048 × 8	Yes	113	8000	Yes	2/3	Yes
Rockwell	PPS-8	PMOS	8/8	0	0	Yes	100	256/4	No	4/12	No
	PPS-8/2	PMOS	8/8	0	0	Yes	100	200/4	No	5/15	No
	R6500/1	NMOS	8/8	64 × 8	2048 × 8	Yes	56	2000	Yes	1/3.5	Yes
Zilog	Z8	NMOS	8/8	144 × 8	2048 × 8	Yes	47	8000	Yes	1.5/3.75	Yes
Texas Instruments	TMS 9940E/9940M	NMOS	See Note 2	128 × 8	2048 × 8	No	68	5000	Yes	2/452	Yes
Intel	2920	NMOS	25/25	40 × 25	192 × 24	No	21	2500	Yes	0.4/0.4	Yes

1. Not applicable.   2. External 8 bits, internally 16 bits.   3. User defined.   ?Not available.

BCD arithmetic	On-chip interrupts/levels	Subroutine nesting levels	General-purpose internal registers	Number of I/O lines	Additional special support circuits	Package size (DIP pins)	Voltages required (V)	Prototyping system available	Assembly language programming system	High-level language programming system	Time-sharing cross software	Comments
Yes	No	1	2 + RAM	23/25	Yes	28/40	9 or 15	Yes	Yes	No	Yes	Aside from two supply versions, a 35-V vacuum fluorescent drive version is available (TMS 1070/1270)
Yes	No	3	2 + RAM	22/32	Yes	28/40	3 to 6	Yes	Yes	No	Yes	CMOS version of TMS-1000
Yes	No	1	2 + RAM	23/28	Yes	28/40	9 or 15	Yes	Yes	No	Yes	Also has VF drive versions (TMS 1170/1370) and is pin compatible with TMS-1000
Yes	No	N/A¹	N/A¹	N/A¹	N/A¹	28	15	N/A¹	N/A¹	N/A¹	N/A¹	Dedicated number cruncher
Yes	No	N/A¹	N/A¹	N/A¹	N/A¹	28	15	N/A¹	N/A¹	N/A¹	N/A¹	Dedicated CB PLL controller
Yes	No	N/A¹	N/A¹	N/A¹	N/A¹	28	15	N/A¹	N/A¹	N/A¹	N/A¹	Dedicated microwave oven controller
Yes	No	N/A¹	N/A¹	N/A¹	N/A¹	40	15	N/A¹	N/A¹	N/A¹	N/A¹	Dedicated appliance timer/controller
Yes	No	3	2 + RAM	19	Yes	28	9	Yes	Yes	No	Yes	VF drive version available (1470) and 15-V supply model coming
Yes	No	3	2 + RAM	32	Yes	40	9	Yes	Yes	No	Yes	VF drive version available (1670) and 15-V supply model coming
Yes	No	3	2 + RAM	21	Yes	28	9 or 15	Yes	Yes	No	Yes	Reduced ROM and I/O version of TMS-1000
3	Yes/1	8	RAM	16	No	40	5	Yes	Yes	Yes	Yes	Intended for dedicated controllers
Yes	Yes/2	8	RAM	16	Yes	42	5	Yes	Yes	Yes	Yes	Designed for keyboard/display interfacing
Yes	Yes/1	1	RAM	27	No	40	12	Yes	Yes	Yes	Yes	RAM holds BCD numbers
Yes	Yes/4	RAM	RAM	32	Yes	40	5	Yes	Yes	Yes	Yes	On-chip UV EPROM instead of mask ROM version of 3870
Yes	Yes/4	RAM	RAM	32	Yes	40	5	Yes	Yes	Yes	Yes	Similar to Mostek 3872 but has no standby RAM
Yes	Yes/1	2	RAM	4	Yes	24	5	Yes	Yes	Yes	No	Minimal I/O family members
Yes	Yes/1	2	RAM	32	Yes	40	5	Yes	Yes	Yes	No	32 programmable I/O lines
Yes	Yes/1	2	RAM	8	Yes	28	5	Yes	Yes	Yes	No	Reduced I/O version of 1650
Yes	Yes/1	2	RAM	32	Yes	40	5	Yes	Yes	Yes	No	Increased memory version
Yes	Yes/1	RAM	RAM	21	No	28	5	Yes	Yes	Yes	Yes	Minimal I/O CPU version
Yes	Yes/1	RAM	RAM	27	No	40	5	Yes	Yes	Yes	Yes	Contains two a/d converter channels
Yes	Yes/1	RAM	RAM	18	Yes	40	5	Yes	Yes	Yes	Yes	8041 has a ROM and 8741 a UV EPROM
Yes	Yes/1	8	RAM	27	Yes	40	5	Yes	Yes	Yes	Yes	8748 has UV EPROM
Yes	Yes/1	8	RAM	27	Yes	40	5	Yes	Yes	Yes	Yes	Enlarged memory version of 8048
Yes	Yes/1	8	16 + RAM	18	Yes	40	5 to 10	Yes	Yes	Yes	Yes	CMOS equivalent to Intel 8041; dissipates 50 nW
Yes	Yes/1	8	RAM	27	Yes	40	5 to 10	Yes	Yes	Yes	Yes	CMOS equivalent to Intel 8048/8748; dissipates 50 mW
Yes	Yes/4	RAM	RAM	32	Yes	40	5	Yes	Yes	Yes	Yes	Has 16-bit prog. timer
Yes	Yes/4	RAM	RAM	32	Yes	40	5	Yes	Yes	Yes	Yes	Double the memory capacity of 3870, with a standby capability on the additional 64 bytes of RAM
Yes	Yes/4	RAM	RAM	32	Yes	40	5	Yes	Yes	Yes	Yes	Some I/O lines dedicated as serial port
Yes	Yes/4	RAM	RAM	32	Yes	40	5	Yes	Yes	Yes	Yes	Same as 3870 but double the RAM
Yes	Yes/1	RAM	RAM	31	Yes	40	5	Yes	Yes	Yes	Yes	6801 has masked ROM, 701 has UV EPROM
Yes	Yes/1	0	RAM	20	Yes	28	5	Yes	Yes	Yes	Yes	Low cost 8 but μC has 8-bit timer and prescaler, 705 version has UV EPROM
Yes	Yes/1	0	RAM	20	Yes	28	5	Yes	Yes	Yes	Yes	CPU includes 8-bit a/d converter (available in mid-1980)
Yes	Yes/1	0	RAM	20	Yes	28	5	Yes	Yes	Yes	Yes	CMOS version of the 6805
Yes	Yes/1	8	RAM	27	Yes	40	5	Yes	Yes	Yes	Yes	Enlarged proprietary version of Intel 8049 processor with transparent improvements
Yes	Yes/2	RAM	RAM	0	Yes	40	5	Yes	Yes	Yes	Yes	Three-state data/address bus for user selectable I/O
Yes	Yes/1	RAM	RAM	13	Yes	40	5 to 10	Yes	Yes	Yes	Yes	Compatible with 1802 software
Yes	Yes/3	16	2	15	Yes	42	$-17/+5, -12$	Yes	Yes	No	Yes	Combination RAM/ROM/I/O support
Yes	Yes/3	16	2	15	Yes	42	$-17/+5, -12$	Yes	Yes	No	Yes	I/O chip includes clock
Yes	Yes/1	RAM	RAM	32	Yes	40	5	Yes	Yes	Yes	Yes	Single chip version 6502
Yes	Yes/6	RAM	RAM	32	Yes	40	5	Yes	Yes	Yes	Yes	Has two counter/timers and UART
Yes	Yes/4	64	RAM	16	No	40	5	Yes	Yes	Yes	Yes	Two versions available, one has a 2 k EPROM, the other 2 k ROM
N/A¹	N/A¹	0	RAM	12	No	28	5, -5	Yes	Yes	No	Yes	Analog processor accepts four analog inputs and delivers up to eight digitally processed analog outputs

TABLE 11-5 Bit-Slice Families

Original-source company	Series	Process technology	ALU part number	ALU word size (bits)	Number of ALU instructions	Can ALU do BCD arithmetic?	Maximum ALU clock rate (MHz)	General-purpose registers in ALU	ALU package size (DIP pins)	Microprogram sequencer number
Advanced Micro Devices	2900	STTL	2901A	4	16	No	16.67	16	40	2909/11
		STTL	2903	4	25	No	10	16	48	2910
Fairchild	Macrologic	STTL/ CMOS	9405/34705	4	64	No	10	8	24	9406
	F100220	ECL	F100220	8	27	Yes	50	1	68¹	F100224
Motorola	10800	ECL	10800	4	100+	Yes	20	0	48	10801
Signetics	3000	STTL	3002	2	40	No	10	11	28	3001
Texas Instruments	SBP-0400A	I²L	SBP-0400	4	512	No	5	10	40	74S482
	SBP-0401A	I²L	SBP-0401	4	512	No	5	10	40	74S482
	54/74S481	STTL or LSTTL	S481 LS481	4	24.780	No	10	0	48	74S482

1. Leadless carrier.   2. Not available.

TABLE 11-6 Directory of Microprocessors by Vendor

Manufacturer	Processor models manufactured
AEG Telefunken	Series 8000
Advanced Micro Devices	2900, 2903, 29116, Z8000,* 8048,* 8080A,* 8085*
American Microsystems	S2000, S2150, S2200, S2400, 6800,* 6802/8,* 6809,* 9900,* 9980/81,* 9985*
Data General	mN601, mN602
EFCIS	MC68000
EMM Semiconductor	R6500/1,* CP1600/10,* PIC1645, 50, 55, 70*
Fairchild Semiconductor	Macrologic (9405), F100220 (ECL), 9440, 9445, 3850, 38E70, 3878, 2900,* 3870,* 6800,* 6802*
Ferranti Ltd.	F100-L
Fujitsu	6800,* 6802,* 6809*
General Instrument	CP1600, CP1610, PIC1645, PIC1650, PIC1655, PIC1670
Harris Semiconductor	6100*
Hitachi Ltd.	MC6800, MC68000
Hughes, Semi Div.	1802
Intel	3000, 4004/4040, 8008, 8021, 8022, 8035, 8039, 8041, 8048, 8049, 8080A, 8085A, 8086, 8088, 8089, 2920
Intersil	6100, 87C41,† 87C48,† IM80C41,† IM80C48†
ITT Semi	7150, 1600*
Mitel Semiconductor	6802 (CMOS version)
MOS Technology	6502, 6503, 6504, 6505, 6506, 6507, 6512, 6513, 6514, 6515
Mostek	3870, 3872, 3874, 3876, 3850,* 8086,* Z80*
Motorola	6800, 6801/701, 6802, 6803, 6805/705, 6808, 6809, 68000, 146805, 14500, 10800, 2900,* 3850,* 3870,*
National Semiconductor	COP402, 402M, 404L, 410L, 411L, 420, 420L, 420C, 421, 421L, 421C, 440, 444L, IMP-4, IMP-8, IMP-16, INS8060, INS8900, INS8070,

*Alternate source product.
†Functionally equivalent but different technology.

Number of address bits	Maximum sequencer clock rate (MHz)	Number of sequencer commands	Sequencer stack size	Sequence package size (DIP pins)	Are parts TTL-compatible?	Voltages required (V)	Prototyping system available	Development software available	Specialized support circuits available	Comments
4	10	12	4 × 4	28/20	Yes	5	Yes	Yes	Yes	Has widest number of second sources
12	10	16	5 × 12	40	Yes	5	Yes	Yes	Yes	ALU has nine more instructions than 2901, including multiply and divide
4	10	4	16 × 4	24	Yes	5	Yes	Yes	Yes	CMOS version (34705) operates at 2 MHz
N/A²	N/A²	N/A²	N/A²	N/A²	N/A²	−4.2 to −5.7	Yes	Yes	Yes	Only 8-bit slice, sub-ns instructions
4	20	16	4 × 4	48	No	−2, −5.2	Yes	Yes	Yes	Fastest 4-bit slice available
9	10+	11	0	40	Yes	5	Yes	Yes	Yes	Only 2-bit ALU available
4	20	64	4 × 4	20	Yes	Current	Yes	No	No	Has pipeline register
4	20	64	4 × 4	20	Yes	Current	Yes	No	No	Does not have pipeline register
4	20	64	4 × 4	20	Yes	5	Yes	No	Yes	Very flexible instructions set

TABLE 11-6 (Continued)

Manufacturer	Processor models manufactured
National Semiconductor (cont'd)	INS8072, INS8040, INS8050, NS16008, NS16016, NS16032, NSC800, 29103, 8035,* 8039,* 8048,* 8049,* 8080A,* 2900*
NEC Microcomputers	μPD555, 556, 548, 546, 553, 650, 547, 547L, 552, 651, 550, 554, 652, 551, 8048,* 8080A,* 8085A,*
Panafacom	MN1600
Panasonic (Matsushita)	MN1498, 1499, 1499A, 1599, 1403, 1404, 1405, 1430, 1432, 1435, 1450, 1453, 1454, 1455, 1542, 1544, 1562, 1564
Philips	8080A*
RCA	CDP1802, 1804
Raytheon	2900*
Rockwell International	PPS-4, 4/2, MM77/77L, 78/78L, 76, 76/C, 76/E, 76/L, 76/EL 75, R6500/1, 6502,* 6503,* 6504,* 6505,* 6506,* 6507,* 6512,* 6513,* 6514,* 6515,* PPS-8, 8/2, MC68000*
SGS-ATES	Series 8000*
Sharp	SM-4, SM-5, Z80A,* Z8001,* Z8002*
Siemens	8080A,* 8085A,* 8086*
Signetics	8 × 300, 2650A, 8021,* 8035,* 8048,* 8080A,* 8085A,* 9405,* 34705*
Solid State Scientific	1802*
Synertek	Z8,* 6502,* 6503,* 6504,* 6505,* 6506,* 6507,* 6512,* 6513,* 6514,* 6515,*
Texas Instruments	74S481, SBP0400A/0410A, TMS-1000, 1000C, 1100, 1018, 1022, 1117, 1121, 1400, 1600, 1700, 1070, 1270, 1170, 1370, 1470, 1670, TMS/SBP9900, 9980/81, 9985, 9940, 8080A*
Thomson CSF/Sescosem	2900,* 6800,* 6802,* 6809*
Toshiba	T3190, 3444, 3472
Western Digital	1872, MCP1600, Pascal Microengine, INS8060,* COP4020,* 4200*
Zilog	Z8, Z80A, Z8001, Z8002

REFERENCES

Beizer, G.: *Micro-Analysis of Computer System Performance*, Van Nostrand-Reinhold Co., 1978.

Blakeslee, T. R.: *Digital Design with Standard MSI and LSI*, Wiley, 1975.

Intel: *MCS-48 Microcomputer User's Manual*, Intel Corp., 1978.

Intel: *UPI-41 User's Manual*, Intel Corp., 1978.

Madnik, S. E., and J. J. Donovan: *Operating Systems*, McGraw-Hill, 1974.

Mano, M. M.: *Computer Logic Design*, Prentice-Hall, 1972.

McNamara, J. E.: *Technical Aspects of Data Communication*, Digital Equipment Corp., 1978.

Motorola: "M68000 Family" (M68KFM), Motorola Inc., 1980.

Motorola: *MC68000 User's Manual* (MC68000UM(AD2)), Motorola Inc., 1980.

Peatman, J. B.: *Digital Hardware Design*, McGraw-Hill, 1980.

Peterson, W. W.: *Introduction to Programming Languages*, Prentice-Hall, 1974.

Smith, C. L.: *Digital Computer Process Control*, Intext Educational Publishers, 1972.

Stone, H. S.: *Introduction to Computer Architecture*, Science Research Associates, 1975.

Taub, H., and D. Schilling: *Digital Integrated Electronics*, McGraw-Hill, 1977.

Wester, J. W.: *Software Design for Microprocessors*, Texas Instruments Learning Center, 1976.

Chapter 12

OPTOELECTRONICS

William M. Otsuka   President
Optomicronix Inc.
Cupertino, Calif.
Eric G. Breeze
Earl V. Cole
Atari Corp.
Sunnyvale, Calif.

The authors were with General Instrument's Optoelectronics Division (Palo Alto, Calif.) when this chapter was written.

12-1 LED LAMPS

12–1a Introductory Theory

Light Sources   Man-made light sources in illumination and equipment applications historically could be categorized either as incandescent filament lamps, or as electric fluorescent, vapor, or neon gas lamps. Light energy produced by these lamps is spread over a wide frequency spectrum, much of it outside the band visible to the human eye. During the past decade, advances in semiconductor technology have added a new category of light source, light-emitting diodes (LEDs). These solid-state components produce light in very narrow spectrums, however (see Fig. 12-1). Table 12-1 lists measurement units used in optoelectronics to specify wavelength λ. The nanometer is the most frequently encountered unit.

Conduction Theory   According to solid-state conduction theory, during current flow across *pn* junctions, energy in the forms of photons (light) and phonons (heat) are radiated whenever injected holes or electrons are annihilated by a charge carrier of the opposite type. Quantum physics states that in crystalline solids electrons can assume only certain levels of energy, and defines energy gap as the separation between the top of the valence band and the bottom of the conduction band. This energy gap is a characteristic

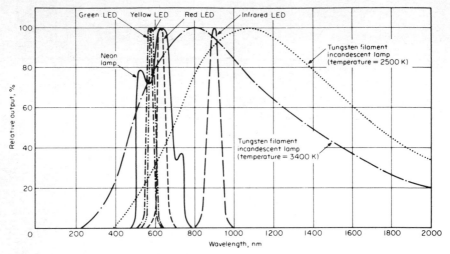

FIG. 12-1   Normalized spectrums of selected light sources.

TABLE 12-1 Measurement Units for Wavelength

Name of unit	Symbol	Value in meters	Equivalents
micron	μ or μm	1×10^{-6}	$1\ \mu m = 1000$ nm
			$1\ \mu m = 10{,}000$ Å
nanometer	nm	1×10^{-9}	1 nm $= 10$ Å
Angstrom	Å	1×10^{-10}	1 Å $= 0.1$ nm

of the semiconductor itself; magnitude (in electron volts, eV) of the energy gap determines wavelength of light that will be emitted. Out of various semiconductor materials suitable for fabrication of LEDs, considerations about useful wavelengths, conversion efficiency, and ease of doping have led manufacturers to choose gallium arsenide (GaAs), gallium phosphide (GaP), or compounds of GaAsP.

LED Characteristic Curves   Although color of light, i.e., wavelength, is determined by the material used, the amount of light produced by an LED is controlled by its driving circuit and increases rapidly as current density increases. Fig. 12-2 shows the forward current vs. forward voltage plots for several LED materials. Locations of the voltage knees on these diode curves are directly related to energy gap, with the knee for the red LED occurring at the lowest forward voltage. In this figure dynamic resistance for the red LED is 1 to 2 Ω, and for shorter wavelength materials it ranges from 7 to 15 Ω. As current density increases across the *pn* junction, more electrons and holes will be injected into the forbidden gap. Their movement causes a secondary effect that increases the number of electrons and holes available for radiative recombination,

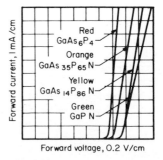

FIG. 12-2 I_F vs. V_F.

thereby raising the efficiency of the light-emission process. Fig. 12-3 shows plots of light output as a function of diode drive current for several LED materials. Note that GaP red produces light at low current densities, but also approaches saturation at a relatively low current density compared to the other materials.

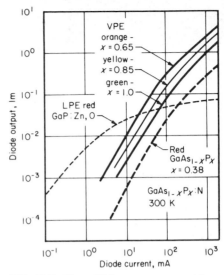

FIG. 12-3   Plots of light output vs. current density. LPE-Liquid phase epitaxial process; VPE-vapor phase epitaxial process.

Drive Circuits for High-Efficiency Output   From the slopes of the GaAsP curves in Fig. 12-3 it can be seen that a doubling of drive current leads to more than twofold increases in light output, showing that output efficiency of these LEDs improves at higher drive currents and suggesting that a pulsing drive current will give higher average light output than an equivalent steady-state current. A numerical example for the GaAsP red curve and the pulsing drive current waveform of Fig. 12-4a illustrates this gain in efficiency. Whereas a 10-mA steady-state drive current results in a light output of about 0.7 mlm, the pulsed waveform light has an average output of about 2.0 mlm (10% of about 20 mlm), as shown in Fig. 12-4b. When driving with high pulsed current, the pulse width and duty cycle will affect junction temperature. These effects are discussed in the next subsection. No flicker or other flashing sensations will be noticeable to the human eye, provided that the LEDs are pulsed at frequencies much higher than 30 Hz.

Effects of Temperature   Forward voltage drop across the LED will decrease as temperature increases, and a factor of -1.3 to -2.5 mV/°C should be anticipated. Peak wavelength of the emitted light will shift upward as temperature increases, by approximately 0.2 nm/°C or less, depending on LED material. Also, LEDs emit less light as the temperature increases with a typical negative temperature coefficient of approximately 1%/°C.

Lifetime Performance   Because they are solid-state devices, the operating lifetime of LEDs can normally be expected to exceed that of the equipment in which they are installed. However, extremely slow natural diffusion of foreign particles into the semiconductor crystalline compound, as well as other not fully understood mechanisms, cause light output to decrease somewhat over time. In the life test curves shown in Figs. 12-5 and 12-6, it can be seen that drop-off in light output is greater at higher currents, yet is typically less than 10% even after 1000 hours of continuous operation.

The lifetime of an LED is normally defined as the time at which the light output decreases to 50% of the original value. For visible LEDs, common lifetimes are quoted to be typically 100,000 hours (in excess of 11 years), under normal operating conditions.

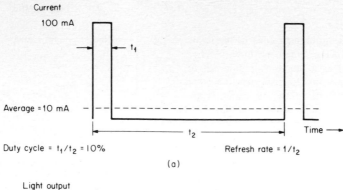

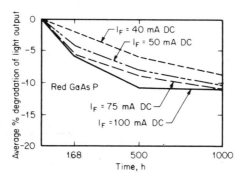

FIG. 12-4 Waveforms. (*a*) Drive current waveform; (*b*) light output waveform.

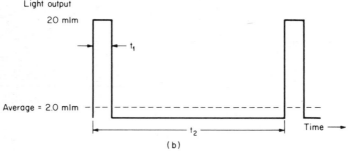

FIG. 12-5 Life test curves for sample A.

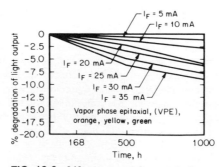

FIG. 12-6 Life test curves for sample B.

The amount of degradation is also affected by choice of operating current: lowering it lessens the degradation and extends operating lifetime.

Current Limiting   From the I_F vs. V_F curves in Fig. 12-2 it can be seen that after the knee is reached, I_F rises very rapidly with small additional increases in forward voltage V_F. A resistor should be added in series with the LED, as shown in Fig. 12-7, to provide current limiting, so that the LED is kept operating at or below a specific current level. The resistor value can be obtained from the following equation:

$$R = \frac{V_{CC} - V_F}{I_F} \qquad (12\text{-}1)$$

where V_{CC} is the supply voltage. Equation 12-1 is basic to all classes of LED applications—indicators, illuminators, bar graphs, alphanumeric displays, or optocouplers.

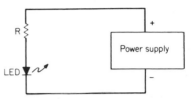

FIG. 12-7   Load resistor to limit current.

If several LEDS were connected in parallel directly (through a single common resistor) to the same regulated DC supply, the LED having the lowest V_F would "hog" most of the current, with the result that its light output would be noticeably greater than that of the remaining LEDs. To avoid this, each LED should have its own separate, series-connected current-limiting resistor.

As a rule of thumb, a luminous intensity as small as 4 to 5 mcd is sufficient for ready observation of LED output in a bright office environment with about 100 fc of illumination. For high-efficiency visible red LEDs, the required I_F is typically 10 mA, and for other colors, typically 20 mA.

Example 12–1   LED Drive Computation

An MV5752 red LED is to operate from a +5-V supply, and is to produce a light output of 15 mcd. Determine the current-limiting resistor.

Solution

Figs. 12-8 and 12-9 show the characteristic curves from the data sheet for this LED. From these curves we find that an I_F of 10 mA is needed to

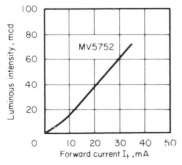

FIG. 12-8   Luminous intensity vs. I_F for MV5752.

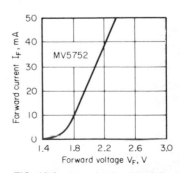

FIG. 12-9 I_F vs. V_F for MV5752.

produce 15 mcd, and that at $I_F = 10$ mA, the V_F is 1.8 V. Solving for these values we obtain

$$R = \frac{V_{CC} - V_F}{I_F} = \frac{5 - 1.8}{0.01} = \frac{3.2}{0.01} = 320 \ \Omega \qquad (12\text{-}1)$$

The closest standard value of resistor will then be 330 Ω ($\frac{1}{8}$ W rating).

12–1b   Glossary of Terms

Angstrom (symbol Å)   Measurement unit of length, equal to 10^{-10} m, and commonly used to express the wavelength of light.

Annunciator   Type of moving-point display used to convey information other than analog measurement values. A common example of this type is the backlighted numerical panel display used in elevators to indicate current floor number.

Anode   The terminal of a device that is normally biased with a positive voltage (with respect to the other terminal called the Cathode).

Area Source (also see Point Source)   A representation of a source of radiation as having both a definite position in space and nonnegligibly small dimension. Used when the viewing distance is not large with respect to the actual dimension of source.

ASCII   Abbreviation for American Standard Code for Information Interchange, a binary code that can represent letters, numerals, punctuation and special symbols, and certain control characters (such as line feed, carriage return, etc.). The full representation of this code requires seven binary bits to implement a 128-character set that includes both upper- and lowercase letters. With only six binary bits, a 64-character subset can be implemented in which letters are uppercase only.

Bandgap   In solid-state quantum theory, the energy difference (usually expressed in electron volts, eV) between two allowed bands of orbiting electrons.

Bar-Graph Display   On a bar-graph array the method of conveying information by means of a linear string of lighted elements such that *length of string* represents an analog value.

BCD (Binary-Coded Decimal)   A method of representing the numbers 0 through 9 in terms of four binary digits. BCD is sometimes called 8-4-2-1 code because the leftmost binary digit has the weight 8, the next digit has the weight 4, and so forth. The binary equivalents for decimal ten (1010) through decimal fifteen (1111) are not allowed in this code.

Bezel   A mechanical frame or rim that surrounds (and also may support) an optoelectronic device or display.

Binary Code   A code that represents all numbers in terms of only two symbols—0 and 1. For example, the first four numerals 0 through 3 would be represented in binary code as 0000, 0001, 0010, 0011.

Blanking   A circuit design technique under which a display can be kept dark for a desired period of time.

Brightness   See Luminance.

Bus   A common conductor path or set of parallel paths for transmission of signals (e.g., data bus), or energy (e.g., Power Supply Voltage Bus). In general, signals are placed on a bus from any of several sources (one at a time), and taken off the bus by one or more receivers (sometimes simultaneously). In modern electronic systems, especially microprocessor-based systems, the direction of signal flow over a bus can be unidirectional (always in one direction), or bidirectional (in one direction at certain times and in the opposite direction at other times).

Candela   Measurement unit of Luminous Intensity. Defined as one Lumen of flux radiated through a unit solid angle (Steradian).

Cathode   The terminal of a device that is normally biased negative (with respect to the other terminal called the Anode).

CIE Curve   Curve for the human eye that shows relative frequency response to light of various colors. Name of curve is derived from the organization that established it in 1931—Commission Internationale de l'Eclairage.

Common Anode   Method of connecting device terminals in which all anodes are tied together.

Contrast Ratio   The ratio of the Luminance of display elements or lamps and the Luminance of the surroundings (as a result of reflection of ambient illumination).

Cursor   Name given to the marker that points out or otherwise identifies a specific character position along a printed line of characters. On CRT displays usually a blinking underline character; on other displays an asterisk or similar character may be used.

Current Transfer Ratio   For optocoupler devices the ratio of output current to input current.

Dark Current (see also Light Current)   The current that flows through a photoconductive device in the absence of any light.

Darlington   Method of connecting two or more transistors for high current gain. The emitter of the first stage is tied directly to the base of the next stage, and collectors are tied together.

DC Plasma   Type of display that produces light through ionization of neon or other gas in a sealed glass envelope across whose terminals a high DC potential is applied.

Detector   Any device that detects or senses a change in the value of some physical quantity; for example, Luminous Intensity.

Doping   The addition of impurities to a semiconductor material in order to obtain a desired characteristic, such as producing an *n*-type or *p*-type semiconductor.

Duty Cycle   The ratio of working time to total time for an intermittently operating device.

Electroluminescence   The process of light emission, not due to heating effects alone, that results from application of an electric field to some material, usually a solid.

Electrochromic   Display built from organic or inorganic insulating solids that change color when injected with positive or negative charges.

Emittance   The power radiated per unit area of a radiating surface.

Emitter[1] On a transistor, the region from which minority carriers are injected into the base region.

Emitter[2] A source of electromagnetic wave radiation in the visible or infrared frequency spectrum.

Firmware The instructions for a computer program, or portions thereof, that are permanently stored inside the computer (usually in ROM, or read-only memory). The instructions in firmware are in contrast to instructions in software that have to be loaded into the computer separately.

Font Term originally used in printing and graphics, but adopted into optoelectronics to refer to the format of individual sections of a display (as, for example, bar-segment or dot-matrix font).

Footcandle Unit of measurement for illumination. Defined either as a surface of area 1 ft^2 upon which a luminous flux of 1 lm is uniformly distributed, or as a surface all of whose points are at a distance 1 ft from a uniform point source of 1 cd.

Footlambert Unit of measurement for Luminance (Photometric Brightness) equal to $1/\pi$ cd/ft^2, or to the uniform Luminance of a perfectly diffusing surface emitting or reflecting light at the rate of 1 lm/ft^2.

GaAsP Chemical abbreviation for gallium arsenide phosphide (semiconductor material used for the manufacture of LEDs).

GaP Chemical abbreviation for gallium phosphide (semiconductor material used for the manufacture of LEDs).

Human Factors Those characteristics of display systems that have an effect on a viewer's performance (such as viewing distance, viewing angle, character height, brightness, color, contrast ratio, font shape and sharpness, blur and flicker).

Hybrid Type of device construction method in which parts of internal circuitry are fabricated on two or more semiconductor chips that are interconnected by means of bonding wires and/or metallization deposited on the package substrate. (See also Monolithic.)

Illumination The amount of luminous flux falling upon a unit of area; usually measured in fc.

Illuminator Type of LED intended for use as high-intensity indicator, or as light source for back-panel illumination.

Incandescent Lamp or display that produces light when a metallic filament is heated white-hot in a vacuum by passing an electric current through it.

Infrared Name given to invisible radiation in region of the electromagnetic spectrum having wavelengths from 750 nm to 1000 μm.

Interactive Display Name given to category of displays used with host computer systems and having self-contained slave microprocessor and memory to perform data-update and display-refresh tasks.

Irradiance The amount of electromagnetic radiant power per unit area that flows across or onto a surface. (Also called radiant flux density.)

Isolation Resistance Optocoupler parameter that rates minimum resistance (in ohms) between input and output terminals. Always measured under a specific applied voltage.

Isolation Voltage Optocoupler parameter that rates minimum high-voltage isolation between the input and output terminals. Always specified at some very low leakage current and specified test duration.

Light Current (also see Dark Current) The current that flows in a photoconductive device in response to photons of light falling upon the device.

Light Current Sensitivity Photodetector parameter that characterizes current produced in response to light irradiance falling upon the detector. Defined as current produced in response to a stated flux per unit detector area, with measurement unit such as $\mu A/(mW \cdot cm^2)$.

Liquid Crystal Organic fluid compound whose light transmission properties can be altered from transparency to absorption by application of an electric field.

Lumen The basic photometric measurement unit of flux. One watt of radiometric flux (Radiated Output Power) at a wavelength of 555 nm is equivalent to 680 lm.

Luminous Intensity Photometric parameter defined as amount of flux radiated through a 3-dimensional solid angle. The common measurement unit is candela, where 1 cd = 1 lm/sr.

Luminance (also called Brightness and Photometric Brightness) Photometric parameter defined as luminous flux per unit solid angle per unit area of emitting surface (i.e., area source). Common measurement unit is footlambert, where 1 fL = 1 cd/ft^2.

Micron Measurement unit of length; equal to 10^6 m.

Monochromatic Light composed exclusively of a single wavelength (or very narrow range of wavelengths).

Monolithic Type of device construction method in which all parts of internal circuitry are fabricated on a single planar semiconductor chip. (See also Hybrid.)

Moving-Point Display On a bar-graph array, the method of conveying information by means of only one lighted element at a time, such that the *position* of one lighted element along a string of elements represents an analog value.

Multiplexing, Multiplexed Drive Circuit design technique for reduction of hardware under which several indicator lamps or display digits are driven on a time-shared basis from a common set of drivers, with refresh rate chosen sufficiently high to be unnoticeable to human viewers.

Nanometer Measurement unit of length; equal to 10^{-9} m, and commonly used to express wavelength of light.

Normalized A quantity whose value has been adjusted so that its full representation (100% value) is set equal to 1.

Optical Limit Switch Optocoupler device housed in a sealed package having the emitter and detector physically separated by an air-gap space. The device can perform as a limit switch when a mechanical object is inserted into the air gap and interrupts the light path, causing a change in the device output state.

Optocoupler Device consisting of light emitter and light detector, each in separate circuits electrically isolated from one another, and coupled only by means of a light path from emitter to detector.

Optoisolator   Equivalent term to optocoupler.

Peak Spectral Emission   The wavelength of greatest energy content in the total emission spectrum of the energy given off by a radiating source.

Photoconductor   Any conducting device whose output current varies according to the amount of light photons falling upon it.

Photocoupler   See optocoupler.

Photodarlington   Devices fabricated from two or more phototransistors connected in a Darlington configuration.

Photodetector   That portion of a photoconductor device that receives the light photons falling upon the device (or generated internally in the case of some types of Optocouplers).

Photodiode   Photoconductor device consisting of a diode with a light-sensitive region that responds to incoming light photons by causing variations in output current flow.

Photometric Units   System of electromagnetic radiation measurement units valid only for light wavelengths within the visible spectrum. (See also Radiometric Units.)

Photometric Brightness   See Luminance.

Photon   Theoretical physics term for the minute quantity or quantum of electromagnetic radiation having zero rest mass and energy $E = h/f$, where h is Planck's constant and f is the frequency of the radiation.

Photo SCR   Silicon rectifier (SCR) device fabricated with photosensitive gate region and housed in package having either an internal light source or provisions for external light to enter from the outside.

Phototransistor   Transistor fabricated with photosensitive gate region and housed in package having either an internal light source or provisions for external light to enter from the outside.

Photo TRIAC   Optocoupler device built by connecting two photo-SCRs "back-to-back" (i.e., with Anode of first photo-SCR tied to Cathode of second photo-SCR, and vice versa).

Point Source (also see Area Source)   An ideal representation of a source of radiation as having a definite position in space but negligibly small dimensions. It is generally valid when the viewing distance is very large with respect to the actual dimensions of the source.

PLZT Display   Display fabricated using lead lanthanium zirconate titanate material. The name is adapted from the chemical abbreviations for these elements (Pb La Zr Ti).

Radiance   (also called Radiant Sterance) Radiometric measurement unit for radiant flux per unit solid angle per unit area of emitting surface (watts per steradian per square meter).

Radiant Sterance   See Radiance.

Radiated Output Power (ROP)   The basic radiometric measurement unit of flux (in watts).

Radiation Pattern   For a light source, the angular plot of Luminous Intensity vs. angle from the main axis (in a given plane).

Radiometric Units   System of electromagnetic radiation measurement units valid for all wavelengths. (See also Photometric Units.)

Reflective Sensor   Optocoupler device housed in a sealed package having an emitter and detector positioned side-by-side in such a way that the light path from the emitter to the sensor depends upon the reflection off some physical object placed in front of the device.

Reverse Breakdown Voltage   That amount of reverse voltage applied across a device above which a further increase of voltage will cause an abrupt increase in electric current through the device.

ROP   See Radiated Output Power.

Silicon   Basic element with a chemical abbreviation, Si. (Extensively used in solid-state electronics as a semiconductor material.)

Spatial Distribution   See Radiation Pattern.

Spectral Distribution   For a radiating source, a graphical plot of energy content of the radiation vs. wavelength.

Spectral Sensitivity   For a photodetector, a graphical plot, or other representation (such as a table) that shows detector sensitivity vs. wavelength.

Steradian   The unit solid angle, i.e., the solid angle that subtends on the surface of a sphere an area equal to the radius of the sphere squared.

Ternary Compound   Any chemical compound composed of three elements.

Total Radiated Flux   Light output from a source; measured in lumens (see Lumen).

Vacuum Fluorescent   Display technology based upon vacuum tubes with phosphor-coated anodes. Heated filaments emit electrons that bombard the anodes and cause them to give off light.

Viewing Angle   In plane formed between viewer and light source, the angle between the viewer's eye and the main axis (highest Luminous Intensity axis) of the light source.

Watt   Basic measurement unit for power. One watt is defined as 1 joule of energy per second.

Wavelength (symbol λ)   Distance over which an electromagnetic wave travels during one complete cycle; mathematically $\lambda = c/f$, where c is velocity of light (in a vacuum), f is the frequency, and λ is in meters.

12–1c   Advantages of Solid-State Lamps

The low operating voltage, current, and power consumption of LEDs make them compatible with electronic drive circuits, so that interfacing is easier than with filament incandescent and electric discharge lamps. The rugged, sealed packages developed for

LEDs exhibit high resistance to mechanical shock and vibration and allow LEDs to be used in severe environmental conditions where other light sources would fail. LED fabrication from solid-state materials ensures a long operating lifetime, thereby improving overall reliability and lowering maintenance costs of the equipment in which they are installed. The range of available LED colors—from red to orange, yellow, and green—provides the design engineer with added versatility. LEDs have low inherent noise levels and also high immunity to externally generated noise. Finally, circuit response of LEDs is fast and stable, without surge currents or the prior "warm-up" period required by filament light sources.

12–1d Optical Characteristics

Lenses and Radiation Patterns Only a portion of the photons radiated by an LED emerge from its package because of various internal loss mechanisms, which include photon absorption within the semiconductor material itself, and optical reflection and refraction within the package lens. Spatial distribution of emerging light is dependent upon the optical characteristics of the lens. The diagrams in Fig. 12-10 of radiation patterns from three types of dome lenses illustrate effects of optical characteristics. Generally, the data sheet for an LED lamp will give an angular plot of its light distribution. The vertical axis with respect to the LED chip surface is represented as 0° with 100% relative output, and off-axis angles with lesser percents of output. As an example of a typical LED, see Fig. 12-11 for a plot of spatial distribution for the MV5152. The term "viewing angle" defines the off-axis angle for which light output has dropped off by 50% from what it was at 0° (on-axis). Examination of Fig. 12-11 shows that the narrowly focused MV5152 lens gives the device a viewing half-angle of 10°, or a total angle of view of 20°.

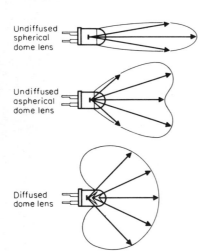

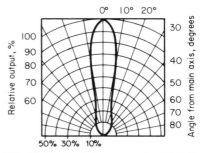

FIG. 12-10 Radiation pattern variations.   **FIG. 12-11** Radiation pattern for MV5152.

Radiometric and Photometric Measurement Units Fig. 12-12 shows plots of relative frequency response for the human eye and for silicon cell detectors (such as phototransistors). Also included in the figure is a plot of the emission spectrum of a typical infrared LED. Data for the eye come from the standard eye response established in 1931 by the Commission Internationale de l'Eclairage, commonly called the CIE curve.

An examination of these two detector curves shows that the eye's response is very

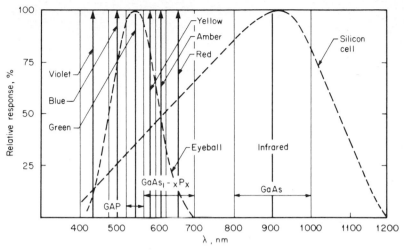

FIG. 12-12 Detector responses.

nonlinear, and that of the silicon detector is highest and comparatively uniform in the infrared region of the frequency spectrum. This difference in responses had led to use of two different systems of measurement units in optoelectronics work—radiometric units, applicable for all wavelengths of radiation, and photometric units, valid only within the visible-light spectrum.

On data sheets for infrared LEDs the light output is listed as a device parameter in a radiometric flux unit called power, P, or radiated output power, ROP, expressed as energy radiated per unit time, in watts (joules per second). This parameter includes the total energy radiated in *all* directions from the LED lens. The corresponding photometric flux unit is called the lumen. At the peak response of the human eye, which occurs at 555 nm in the green region, 1 W in radiometric units equals 680 lm in photometric units. At other visible wavelengths, a watt equals fewer than 680 lm.

Besides the flux parameter, light output from an LED can also be characterized as an intensity parameter, which is defined as the amount of flux radiated through any three-dimensional solid angle. The diagram in Fig. 12-13 defines a unit solid angle called the steradian (symbol sr). Close examination of this figure will show that the steradian can be formed at any radial distance from the light source. This therefore means that the intensity parameter is independent of the distance (along the main viewing axis) from the lens. In radiometric units the intensity is measured in watts per steradian, and in photometric units, it is measured in candelas (symbol cd), where 1 cd = 1 lm/sr.

Light sources in the foregoing discussion are presumed to be point sources, which are close approximations for individual LED lamps viewed from several feet away. However, in cases of larger rectangular and shaped lamps, illuminators, and alphanumeric displays, the visual stimulation experienced by an observer can more resemble that of an area light source rather than point source. Measurement units have therefore been developed that take into account the area of the light source. The radiometric units, collectively called radiance, are expressed in watts per steradian per square unit. The corresponding photometric units, collectively called luminance (or sometimes brightness), are expressed in lumens per steradian per square unit (i.e., candelas per square unit). Of

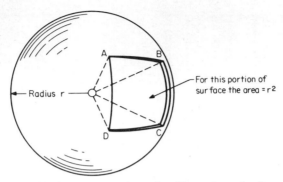

FIG. 12-13 Definition of steradian: For a sphere of radius r, a steradian is defined as the solid angle that subtends on the surface of the sphere an area equal to the square of the sphere radius (i.e., the area bounded by A, B, C, D is equal to r^2, where $AB = BC = CD = DA = r$).

these, the most frequently used is the foot lambert (candelas per square foot). Additional photometric measurement units are defined in the glossary. For a fuller treatment, refer to the chapter bibliography.

12–1e Lamp Driving Considerations

Basic LED Drive Circuits Figs. 12-14 and 12-15 show two basic drive circuits for LEDs, whether they are lamps, illuminators, segments of bar graphs or alphanumeric displays, or inputs of optocouplers. In the active low circuit, when the transistor is

FIG. 12-14 Active low drive circuit. **FIG. 12-15** Active high drive circuit.

conducting, the LED is forward biased and produces light. The value of the current-limiting resistor can be calculated as follows:

$$R = \frac{V_{res}}{I_F} = \frac{V_{CC} - (V_F + V_{CE(sat)})}{I_F} \qquad (12\text{-}2)$$

where $V_{CE(sat)}$ is the transistor saturation voltage.

In the active high circuit the LED produces light when the transistor is off (not conducting). Here the value of the resistor must meet two criteria—allow enough I_F through the LED to produce the desired light output when the transistor is off, and to ensure that the transistor is in saturation when it is conducting. When the transistor is off, the equation for R is

$$R = \frac{V_{CC} - V_F}{I_F}$$
(12-1)

When the transistor is conducting, the equation for R is

$$R = \frac{V_{CC} - V_{CE}}{I_{CE}}$$
(12-3)

where I_{CE} is the saturation current of the transistor.

Fig. 12-16 shows several TTL drive circuits, along with listings of LED on and off responses to the logical input states.

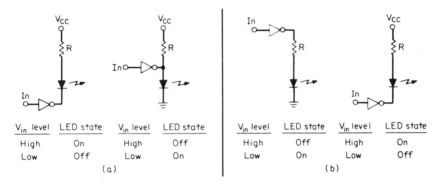

V_{in} level	LED state
High	On
Low	Off

V_{in} level	LED state
High	Off
Low	On

(a)

V_{in} level	LED state
High	Off
Low	On

V_{in} level	LED state
High	On
Low	Off

(b)

FIG. 12-16   TTL drive circuits. (*a*) Open collector output; (*b*) active pull-up output.

12–1f   LED Selection

Considerations Governing Choice of LED   In addition to a desired light output and the drive current required, choice of a specific LED for a given application can also depend on other important considerations. Among these are: package and lens combination, mounting site, LED color, ambient lighting environment, viewing angle, contrast with background that will surround the LED, and viewing distance.

Manufacturers offer LEDs in package sizes originated for incandescent lamps (both standard and short T-$\frac{3}{4}$, T-1, and T-1$\frac{3}{4}$ packages), and also supply grommets and bezels for mounting on flat panels. LEDs can also be obtained in TO-18 transistor and axial lead packages for low-profile mounting directly on printed circuit boards. Mechanical outlines and physical dimensions for several common packages are given in Fig. 12-17. Besides these conventional packages, LEDs are also available in square, rectangular, and other package outlines. Fig. 12-18 illustrates some of these variations. The shaped lamp represents an arrow that conveys direction when lighted, and the square and rectangular lamps can have legends printed directly on the image surface of their packages.

Typical LED mounting sites include interior locations on the surface or edges of printed circuit cards to show logic state or diagnostic information, exterior locations such as on front panels, or behind light filters in hand-held displays (for calculators, games, and remote control consoles). Besides its lighted or nonlighted status, an LED may be chosen to convey additional information by means of its color (e.g., red to signify danger condition, yellow for caution, green for normal). Ambient lighting may range from bright

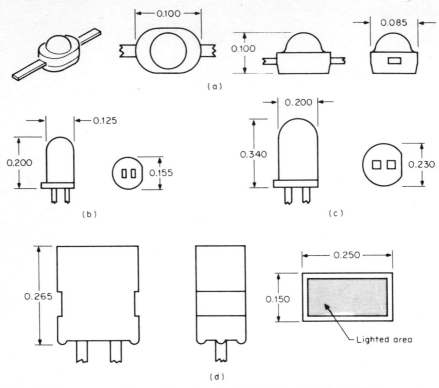

FIG. 12-17 Package outlines and dimensions. (*a*) T-¾; (*b*) T-1; (*c*) T-1¾; (*d*) 57124 rectangular.

sunlight to indoor conditions in offices and homes, and even down to subdued lighting found inside radar equipment rooms. Table 12-2 lists illumination data for several common environments.

In order for an observer to readily discern the ON or OFF status of an LED, a minimum luminous intensity must be established for any given ambient light condition. When all other factors at the mounting site (e.g., LED and lens type, viewing angle and distance, etc.) remain constant, the LED luminous intensity must be higher at higher ambient light levels. Fig. 12-19 shows an empirical plot of this subjective relationship for a MV5754. For the same LED chip and I_F, a narrowly focused lens gives high luminous intensity along the main axis, but has a small viewing angle (in the range 5° to 10°), while a lens with a wide radiation pattern gives a wide viewing angle, but causes luminous intensity to be considerably lower. The kind of background that surrounds the LED can have a strong influence on the perception by the human eye. Perceived brightness of a background depends on amount of light falling upon it, reflectivity and texture of its surface, and even its color compared with that of the LED itself. Visual perception of the LED can be improved by increasing contrast between the lighted LED and its background, through mechanical means (hoods, louvers, or recessed bezels) or by the addition of optical filters. (This is the case even though a filter somewhat attenuates the LED luminous intensity.) Viewing distance governs choice of the minimum size of the LED apparent emitting area;

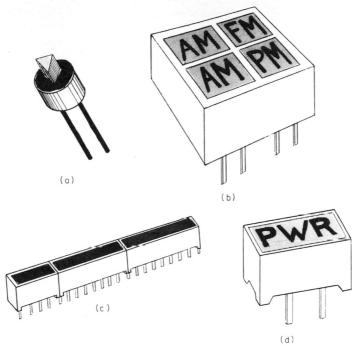

FIG. 12-18 Lamp package variations. (*a*) Shaped; (*b*) block; (*c*) bar modules; (*d*) rectangular.

subminiature packages (such as T-$\frac{3}{4}$) are adequate when the viewing distance is only a few feet, but the larger rectangular or shaped lamps may be required if the LED is to be observed from many feet away.

In some applications the various considerations bearing upon choice of an LED may at first examination lead to specifications that are unachievable by any commercially manufactured LED device. In such situations the designer then has to analyze various tradeoffs and compromises, and through that process find an available LED that at least

TABLE 12–2 Illumination Data*

Environment	Illumination (fc)
Daylight: bright clear	8500
cloudy	1500
Office: general	100
typing and computer operation	150
Manufacturing: electrical test and assembly	100
rough bench work	50
Residence: kitchen	150
study	70
general	10

*Adapted from the *Illuminating Engineering Society (IES) Handbook*, 5th ed., 1972.

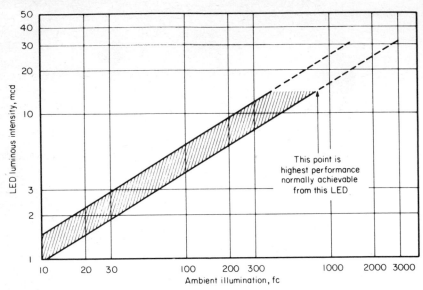

FIG. 12-19 Empirical plot for comfortable LED viewing at various ambient illumination levels. Note that data is for typical MV5754 red LED (with diffused lens), unfiltered, with grey background, viewed on-axis from a distance of approximately 3 ft.

meets the most important criteria of the application at hand. For example, make the background darker, such as black anodized aluminum, choose a higher light output device, such as a red rather than green LED because of its better output efficiency, etc.

Designer's Worksheet To facilitate the designer's task, Table 12-3 gives a concise list of considerations for proper choice of an LED. The following two design examples illustrate the use of this list.

TABLE 12–3 Designer's Worksheet

1. Type of equipment
2. Mounting site on/inside equipment
3. Mounting technique
4. Illumination environment
5. LED color
6. Background surrounding LED
7. Viewing angle
8. Viewing distance
9. Filter type (if required)
10. Other contrast improvement techniques
11. Other design criteria
12. Candidate LED types
13. Desired luminous intensity
14. I_F required at above luminous intensity
15. Drive current to be steady-state or pulsed
16. V_F corresponding to above I_F
17. Type of drive circuit
18. Value of current-limiting resistor
19. Final choice of LED type

Example 12–2 Industrial Equipment Application

The equipment is a portable laboratory instrument for benchtop use. An LED mounts on a front control panel by means of a two-piece, pop-in plastic grommet. The instrument is meant for use in a manufacturing electrical test area where an ambient illumination of 100 fc is expected. The function of the LED is to indicate power ON or OFF status of the instrument, so it can be any color except red, which is restricted by company policy to be used only for danger conditions. The background surrounding the LED is a black, nonglare, flat anodized aluminum surface. The viewing angle can be narrow and the viewing distance is only a few feet because the instrument can conveniently be moved by the operator to improve viewing.

Other design criteria are: some luminous intensity loss through a diffused lens is justified to obtain improved aesthetic appearance when lighted, and I_F to be kept to a minimum to limit power consumption. Candidate LED types are:

> MV5254: green (most desirable color—implies "go" condition)
> MV5354: yellow (possible color)
> MV5154: orange (least desirable color—too close to red)

Solution

From Fig. 12-19 for this LED family we see that, for the stated environment (manufacturing—electrical test area) with an illumination of 100 fc, the minimum luminous intensity ought to be 4 mcd. Curves of luminous intensity vs. I_F and I_F vs. V_F from the data sheet for the MV5X54 family are shown in Figs. 12-20 and 12-21. From these it can be seen that for a MV5254

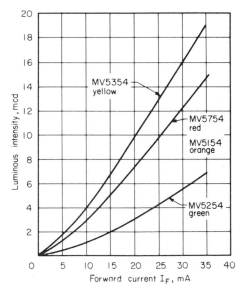

FIG. 12-20 Luminous intensity vs. I_F for the MV5X54 LED family.

green LED, a required I_F of 24 mA is needed to produce an output of 4 mcd. For this I_F the corresponding $V_F = 2.25$ V. If the LED is connected through a resistor that ties directly to the voltage source V_S, a steady-state current will keep the LED turned on whenever source power is applied. The resistor value can be calculated from

$$R = \frac{V_{res}}{I_F} = \frac{V_{CC} - V_F}{I_F}$$

$$= \frac{V_{CC} - 2.25}{0.024} \quad (12\text{-}1)$$

When V_{CC} is 5 V, R will be 115 Ω.

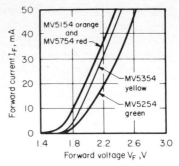

FIG. 12-21 I_F vs. V_F for the MV5X54 LED family.

Example 12–3   Consumer Product Application

The equipment is a microwave oven where the LED is driven by a temperature probe monitor and is to light when the desired temperature has been reached. The oven is intended for a home kitchen. The LED can be any size or color, and will be soldered on a printed circuit board located behind a silk-screened transparent plastic control panel just to the right of the oven door. The viewing angle is to be wide and the viewing distance can range up to approximately 10 ft. Illumination in the kitchen can range from nighttime fluorescent lighting of 150 fc to a worst-case bright summer day, with sunshine entering through windows to produce as much as 3000 fc. The LED is to be mounted behind a dark plastic filter material. A drive circuit will consist of an open collector *npn* transistor and a series-current-limiting resistor to the V_{CC} supply. A preferred LED package is a T-1$\frac{3}{4}$ and the candidate LED family is MV5X53. Another design variable is that in place of a clear plastic panel in front of the LED, the color of the panel can be matched to the LED color to improve contrast.

Solution

From the foregoing it can be seen that the most stringent lighting condition for this application is daytime when the sun is shining through the kitchen window. From Fig. 12-19 it is evident that the minimum luminous intensity required when the ambient illumination is 3000 fc will not be achievable even with MV5753, the most efficient LED in this family. Therefore use of the preferred LED will force certain compromises and tradeoffs in this application. For example, on sunshiny days a viewer would have to come closer than 10 ft and/or move from 50° off-axis angle to a position closer to the on-axis. If the LED were to be driven at its maximum I_F, the higher rate of degradation of light output might lead to an intolerable reduction of performance during the projected lifetime of the microwave oven.

Therefore the designer in this application has to consider several alternatives, namely, lower certain technical requirements of the application (such as changing the range of the ambient lighting to exclude viewing under sunlight, or making the viewing angle narrower in order to allow the use of a more narrowly focused lens that gives higher luminous intensity), provide a sophisticated pulsed current drive, improve contrast by surrounding the LED with a nonreflective black background, or even look to a different type of LED (such as the MV5752).

Design Considerations for LED Illuminators   Whereas in indicator applications an observer views the LED directly, in illuminator applications light output from the LED is *projected* onto a surface, usually the back side of a translucent, silk-screened panel or annunciator block. On data sheets for illuminators the light output is listed as luminous intensity I, and is expressed in candelas (lumens per steradian). This parameter is independent of distance from the LED point source.

After determining the distance from the LED chip to the illuminated surface, the designer can derive an illumination photometric parameter illuminance E from the listed luminous intensity I. Illuminance, also called luminous exitance and luminous incidence, is measured in lumens per square unit. For example, one footcandle of illumination is equal to one lumen falling on a 1-ft^2 surface.

When the surface is curved, as shown in Fig. 12-22, and luminous intensity I is uniform over the entire viewing angle, the following simple equation applies:

$$E = \frac{I}{d^2} \tag{12-4}$$

When the surface is flat, as shown in Fig. 12-23, the angle θ with respect to the main (vertical) axis must also be considered, and the equation for E then becomes

$$E_\theta = \frac{I \times (\cos \theta)^2}{d^2} \tag{12-5}$$

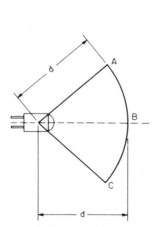

FIG. 12-22   Curved surface *ABC*.

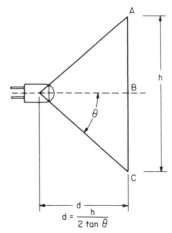

$$d = \frac{h}{2 \tan \theta}$$

FIG. 12-23   Flat surface *ABC*.

After surface size, shape, and illumination are specified, the procedure for the designer is to select an LED, decide on its mounting position with respect to the surface to be illuminated, and calculate what I_F has to be supplied by the driving circuit in order to cause the LED to produce the necessary illumination at the surface.

Example 12–4 Illuminator Design

A flat surface of dimensions 0.75 in by 0.75 in is to be lighted with an illumination of 20 fc (20 lm/ft²), and an MK9150-2 orange illuminator LED has been chosen for the light source. Inside its package the MK9150 has two LED chips connected in series, and their apparent emitting surfaces are 0.150 in below the surface of the lens. A drive circuit is to provide a steady-state I_F to light the LED.

Solution

Fig. 12-24 shows the radiation pattern for this LED. From the pattern we can see that at angles up to 30° from the main axis, the luminous intensity

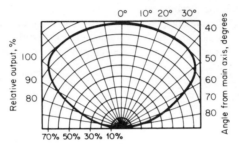

FIG. 12-24   Radiation pattern for the MK9150-2 LED.

exhibits no drop-off in value. Therefore by setting θ = 30° for this application, the position of the LED with respect to the surface to be illuminated will be as shown in Fig. 12-25.

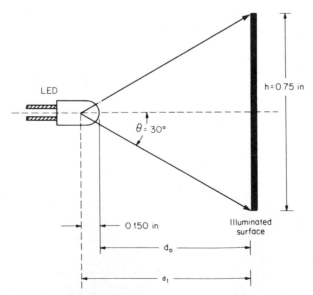

FIG. 12-25   Position of LED in illuminator design example.

Solving for d_1 and d_0 we obtain:

$$d_1 = \frac{h}{2 \tan \theta} = \frac{0.75 \text{ in}}{2 \tan 30°} = \frac{0.75 \text{ in}}{(2)(0.577)} = 0.650 \text{ in} = 0.054 \text{ ft} \tag{12-6}$$

$$d_0 = d_1 - 0.150 \text{ in} = 0.650 - 0.150 = 0.500 \text{ in}$$

Figs. 12-26 and 12-27 show curves of luminous intensity vs. I_F and I_F vs. V_F for the MK9150-2 LED. From these figures we can see that luminous

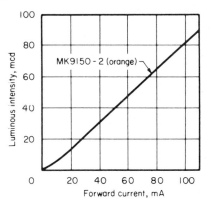

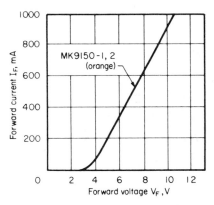

FIG. 12-26 Luminous intensity vs. I_F for the MK9150-2 LED.

FIG. 12-27 I_F vs. V_F for the MK9150-2 LED.

intensity is 80 mcd when I_F is 100 mA and V_F is 4.3 V. On the lighted surface, along the main axis the illumination E is

$$E = \frac{I}{d^2} = \frac{0.080}{(0.054)^2} = 27 \text{ fc} \tag{12-4}$$

and at 30° from the main axis the illumination is

$$E_\theta = \frac{I(\cos \theta)^2}{d^2} = \frac{(0.080)(0.866)^2}{(0.054)^2} = 20.6 \text{ fc} \tag{12-5}$$

At the extreme corners of the 0.75 in by 0.75 in surface the off-axis angle will be greater than 30°. From Fig. 12-28 we see that

$$2d_2 = \sqrt{A^2 + B^2} \tag{12-7}$$

and

$$\tan \theta = \frac{d_2}{d_1} \tag{12-8}$$

Solving for θ, we obtain

$$d_2 = \frac{\sqrt{(0.75)^2 + (0.75)^2}}{2} = 0.53 \text{ in}$$

$$\tan \theta = \frac{0.53 \text{ in}}{0.65 \text{ in}} = 0.815 \quad \text{and} \quad \theta = 39°$$

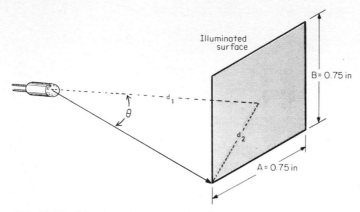

FIG. 12-28 Off-axis angle to corner of surface.

From the MK9150 radiation pattern in Fig. 12-24 we see that at 39° the luminous intensity drops off to about 98% of its on-axis value of 80 mcd. Substituting the new value of θ into Eq. 12-5, we obtain the illumination E at the extreme corners of the surface

$$E_\theta = \frac{(0.98)\ (0.080)\ (0.777)^2}{(0.054)^2} = 16.2 \text{ fc}$$

These calculations have therefore confirmed that with the top of the MK9150-2 lens positioned 0.5 in from the surface to be illuminated, and with $I_F = 100$ mA steady-state, the resulting illumination E spread over the surface will exceed 20 fc, except near the extreme corners, where it drops off to a minimum of 16.2 fc. Fig. 12-29 shows a drive circuit for this example. The value of the current-limiting resistor can be obtained from

FIG. 12-29 Drive circuit.

$$R_L = \frac{V_{res}}{I_F} = \frac{V_{CC} - (V_F + V_{CE(sat)})}{I_F} \quad (12\text{-}2)$$

Higher average illumination levels are achievable if the transistor is driven into saturation by a pulsing waveform rather than by a steady-state level.

12–2 BAR-GRAPH ARRAYS

12–2a Packages

Simple bar-graph arrays can be constructed by placing several round or rectangular, individual LED packages side-by-side. When compact size is important, or when the bar graph is to be very long, it may be more practical to use one or more packages, each containing multiple, independent LED sections. (For examples of these latter packages, see Fig. 12-30.)

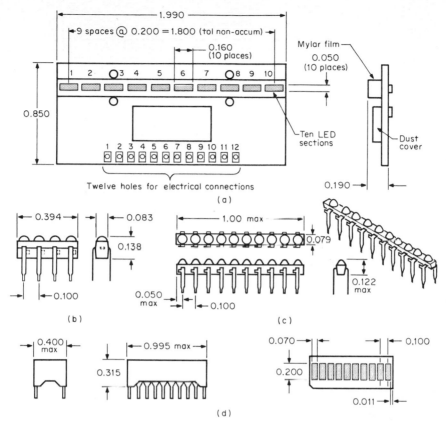

FIG. 12-30 Examples of packages with multiple LED sections. (*a*) NSM3914 module (National Semiconductor); (*b*) BPX 84 (Litronix); (*c*) TIL280 (Texas Instruments); (*d*) MV57164 (General Instrument). (*Note:* All dimensions in inches.)

12–2b Moving-Point and Bar-Graph Displays

Fig. 12-31 depicts two methods for displays of information on a bar-graph array. If only one LED lights at a time, the value of the input information is conveyed by the position of the lighted LED, and the array is called a "moving point" display. In the other method, one or more LEDs light consecutively to form a lighted string to show the input value; such an array is called a "bar graph" display.

12–2c Bar-Graph Decoding/Driving

Circuits Operating from Analog Voltage Inputs   Figs. 12-32 and 12-33 show circuits in which a number of individual LED lamps are placed in line. The portions of these circuits employ analog comparators and a voltage-divider network to perform both analog-to-digital conversion and the necessary decoding. In Fig. 12-32 the driver produces a moving-point display, and that in Fig. 12-33 produces a bar-graph display. For a linear

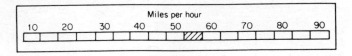

(a)

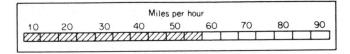

(b)

FIG. 12-31 Bar-graph array for speedometer application. (*a*) One section lights to indicate speed (moving-point display); (*b*) string of sections lights to indicate speed (bar-graph display).

indicator system all resistors in the resistor-divider network would have equal values. On the other hand, a nonlinear analog indicator system (such as inverse square, logarithmic, etc.) would require the resistors in the network to have scaled values.

In the circuit of Fig. 12-32, when V_{in} is less than the comparator voltage V_1, the section 1 LED will light, all comparator outputs go high, and all remaining LED sections remain dark. As V_{in} increases and exceeds V_1, output of comparator 1 will go low, thereby turning off the section 1 LED and allowing the section 2 LED to light. (The remaining

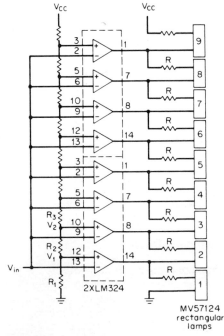

FIG. 12-32 Drive circuit for a moving-point display.

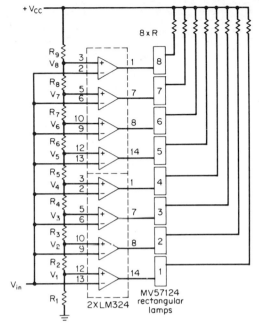

FIG. 12-33   Drive circuit for a bar-graph display.

LED sections still remain dark.) As V_{in} further increases, each higher LED section will in turn light and all other LED sections will remain dark.

In the circuit of Fig. 12-33, if V_{in} is less than V_1 (input threshold of first comparator), then *no* LED sections will light. As V_{in} increases, first the section 1 LED will light, and then, as V_{in} increases further, the section 2 LED will also light, and so on, thereby producing a string of lighted LED sections.

Integrated circuits that perform similar drive functions on analog signals to drive bar-graph displays are now available from a number of manufacturers.

To satisfy applications in which space for components is severely limited, manufacturers have developed techniques to put several LED sections into a single package or on special substrates. In such arrangements the LED sections offer a wide viewing angle, and multiple packages can be stacked on end to form bar graphs of virtually any length. Fig. 12-34 shows a schematic of a circuit built from two packages—an 18-pin and a 20-pin DIP. The LM3914 driver accepts an analog voltage input and produces individual drive lines for the 10 LED sections of the MV57164 array. Fig. 12-30 gives the dimensions and spacing of these sections. On the LM3914 the method of interconnecting pin 9, the mode-select pin, to other pins will determine how the LED sections are driven; connection to V_{CC} allows the lighting of a string of LED sections for a bar-graph display; and connection to pin 11 lights one LED section at a time for a moving-point display.

Circuits Operating from Digital Inputs   When display information is supplied in digital form, either from analog-to-digital converters or from control logic, digital ICs can be employed in the drive circuits. Fig. 12-35 shows a circuit having a TTL decoder

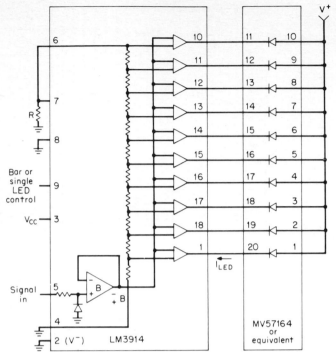

FIG. 12-34 Two-package circuit.

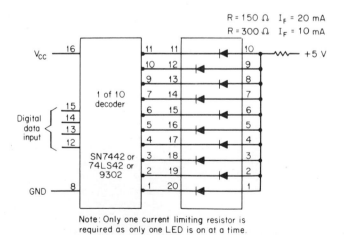

FIG. 12-35 Moving-point display circuit with digital input.

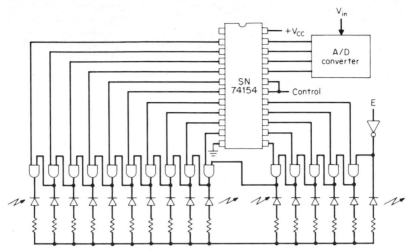

FIG. 12-36 Bar-graph display circuit with digital input.

IC that drives a moving-point display with 10 LEDs. Fig. 12-36 shows a circuit having a TTL demultiplexer IC to perform decoding and driving functions for a bar-graph display with 16 LEDs.

12–3 DISPLAYS

12–3a Introduction to Displays

Levels of Information Lamps and bar-graph arrays discussed in Secs. 12-1 and 12-2 convey binary levels of information to a viewer by means of their lighted or not lighted states, and annunciator-type information according to their locations along a string of sections. Displays, on the other hand, convey much higher levels of information because they can show additional states besides just lighted or not lighted. Each position of a digital display can show numerals 0 through 9, and each position of an alphanumeric display can show numerals and letters (in some cases the entire alphabet), plus several symbol characters.

Fonts Displays are comprised of one or a multiple of digits or characters. Each display position is, in turn, constructed from individual sections whose formats, called fonts, are arranged either as a rectangular bar segment array or a dot matrix array. Figs. 12-37 through 12-41 give examples of the most common fonts and show typical numerals, letters, and symbols that they can depict.

12–3b Types of Display Technologies

Besides the LED technology discussed so far, economic and human-factor considerations have stimulated development of several competing display technologies. To date, none of these has emerged as the most suitable technology for all application environments.

The following paragraphs summarize other technologies, and Table 12-4 compares characteristics of the five main display technologies in general use today.

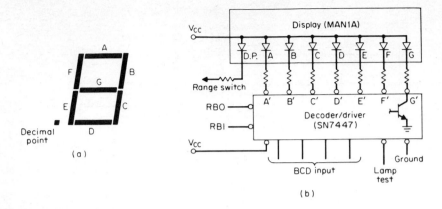

FIG. 12-37 (a)

(b)

BCD input code				Decoder/driver output states							Numeral displayed
2^3	2^2	2^1	2^0	A'	B'	C'	D'	E'	F'	G'	
O	O	O	O	O	O	O	O	O	O	1	0
O	O	O	1	1	O	O	1	1	1	1	1
O	O	1	O	O	O	1	O	O	1	O	2
O	O	1	1	O	O	O	O	1	1	O	3
O	1	O	O	1	O	O	1	1	O	O	4
O	1	O	1	O	1	O	O	1	O	O	5
O	1	1	O	1	1	O	O	O	O	O	6
O	1	1	1	O	O	O	1	1	1	1	7
1	O	O	O	O	O	O	O	O	O	O	8
1	O	O	1	O	O	O	1	1	O	O	9

"O" causes segment to light
"1" keeps segment not lighted

(c)

FIG. 12-37 Seven-segment font. (*a*) Format and identification of segments; (*b*) typical LED implementation circuit; (*c*) truth table for MAN1A and SN7447 circuits.

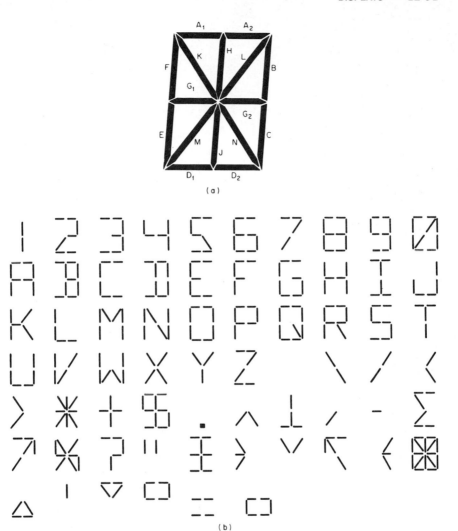

FIG. 12-38   Fourteen-segment font. (*a*) Format and identification of segments. (*Note:* Segments A and D appear as two segments each, but both halves are driven together. (*b*) Typical character set displayed.

Vacuum Fluorescent   Displays based on this technology consist essentially of vacuum tubes with phosphor-coated anodes. When current flows in the filaments, they give off electrons that bombard the anodes and cause the anodes to emit light.

DC Plasma   The principle of operation of these displays is ionization of neon gas contained in a sealed glass enclosure. When a high DC voltage is applied across the anode and cathode elements, the neon gas between them ionizes, giving off a bright orange light.

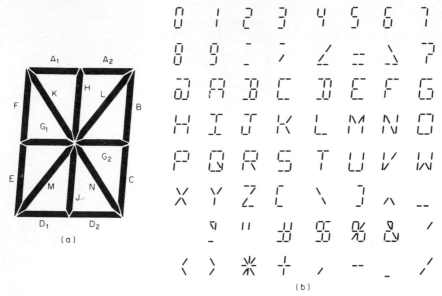

FIG. 12-39   Sixteen-segment font. (*a*) Format and identification of segments; (*b*) typical character set displayed.

Liquid Crystal   Commonly called LCDs, these displays utilize organic fluid compounds whose light transmission properties can be altered by application of an electric field (in the form of AC or DC voltages). LCDs are constructed in the form of a "sandwich" of two sheets of glass whose inner surfaces are coated with a conducting liquid-crystal fluid shaped into segments or dot patterns. Low voltages applied to one or more segments (or dots) alter the molecular structure of the liquid crystal. With no voltage applied, the liquid-crystal fluid is transparent to light (i.e., does not contrast with the surrounding glass envelope), but with voltage applied, light is absorbed by the crystal fluid and the shape of the segment (or dot) appears to the viewer to be in contrast with the surroundings.

From the foregoing, it can be seen that liquid-crystal displays are different from displays based on other technologies in that LCDs do not themselves emit light. Instead, they either transmit or absorb any light that falls on them from external light sources, usually ambient light. Their principle of operation explains why LCDs appear invisible in darkness (unless backlighting is provided), and their viewability improves as the level of ambient illumination increases. This effect is just the opposite from displays that produce light.

Much work has been expended on the development of circuits for interfacing and multiplexing LCDs to the drive electronics of the data sources. Because of the low power requirements of LCDs, they can be driven directly from complex-function IC chips without external interface circuitry.

Incandescent   Displays of incandescent technology are constructed from individual tungsten filaments (one filament for each display segment) that are sealed in some sort of glass envelope enclosure. Current applied to a filament causes it to heat up and

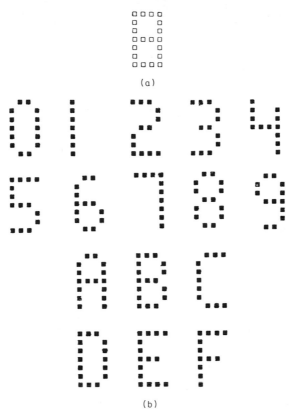

FIG. 12-40   Modified 4-by-7 dot matrix font. (*a*) Format of dots; (*b*) typical hexadecimal character set displayed.

glow, giving off white light composed of many colors. Therefore, colored filters can be placed in front of the filaments to obtain a display of a specific color. Incandescent displays are difficult to multiplex as they are resistive and pass current in both directions.

Other Display Technologies   Research conducted so far indicates only varied promise for display technologies other than those already mentioned.

Both thick- and thin-film displays based on electroluminescence phenomena (under AC and DC applied voltages) come in a variety of colors. However, these displays tend to wash out in sunlight and their lifetimes are short. Their biggest drawback, however, is the very high voltage requirement—from 200 to 600 V.

Displays based on electrochromic technology likewise exhibit residual coloration problems and have short lifetimes. Although they offer a variety of colors and operate over a wide temperature range, the designer will encounter difficulties with multiplexing of these displays.

PLZT displays take their name from the chemical abbreviation for lead lanthanium zirconate titanate. When an electric field is applied to a ceramic plate made from this material, its optical transmission property changes so that it darkens by a ratio as much

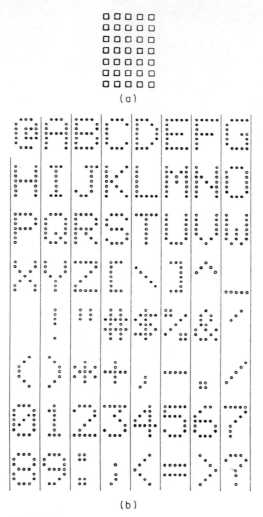

FIG. 12-41   Full 5-by-7 dot matrix font. (*a*) Format of
dots; (*b*) typical ASCII character set displayed.

as 100:1. On/off switching speed is 10–15 μs (much faster than for LCDs). Like LCDs,
these displays do not produce light; instead, their operation depends on light from external
sources.

By the beginning of the 1980s none of these latter technologies had yet made any
significant impact on the displays marketplace.

12–3c   Human Factors

As used here, the term "human factors" refers collectively to those characteristics of
display systems that have an effect on a viewer's performance. The most important of
these characteristics are: viewing distance and viewing angle, character height, brightness,

TABLE 12–4 Display Technologies Characteristics

Characteristics	LED	Vacuum fluorescent	DC plasma	Liquid crystal	Incandescent
			Technologies		
Brightness	Good to excellent	Good	Good to excellent	N/A	Excellent
Colors	Red, orange, yellow, green	Red, yellow, green, blue	Orange	External illumination	All (with filters)
Character height (in)	0.1–1.0	0.2–1.0	0.2–1.0	0.2–12	0.2–1.0
Font style	7–16 seg., dot matrix	7–16 seg., dot matrix	7–16 seg., dot matrix	7–16 seg., dot matrix	7–16 seg.
Viewing angle (degrees)	150	150	120	100	150
Operating temperature range (°C)	−40 to 85	0 to 55	0 to 55	−20 to 60	−40 to 85
Voltage	1.6 to 5 V DC	10–35 V DC 1.7–5 V AC	125–180 V DC	3–20 V AC	3–5 V DC
Power/digit	10–250 mW	20–250 mW	175–750 mW	10–200 μW	100–700 mW
Response time	50–500 ns	1–10 μs	15–500 μs	50–200 ms	10 ms
Other display functions available	Status indicator, illuminator, analog annunciator	Analog annunciator	Analog annunciator	Analog annunciator	Status indicator, illuminator
Lifetime (hours)	100,000+	50,000	50,000	50,000	1000 to 20,000
Graphics-potential	Yes	Yes	Yes	Yes	No

color, contrast ratio between display elements and the background surrounding them, font shape, sharpness, blur, and flicker.

Generally, these characteristics will determine how readily and accurately the viewer recognizes the characters produced, and the extent of fatigue a viewer may experience as duration of viewing time is increased. Experiments involving human subjects have shown that, even with the same display and identical viewing conditions, the performance is not uniform from one person to another. For this reason conclusions from human-factors studies tend to be formulated in terms of subjective guidelines and as rule-of-thumb approximations. During development of new display designs, it is therefore advisable to construct prototypes and mock-ups, and to have them evaluated under actual operating conditions by several typical viewers.

Relationship Between Character Height and Viewing Distance Fig. 12-42 defines the geometrical relationships between character height and viewing distance in terms of a height angle Φ. Studies have shown that at height angles smaller than 3.5′ (0.058°), the display characters become unrecognizable by human viewers. A height angle of 10′ (0.167°) lies within the eye's high comfort region. For a 0.5-in (0.042-ft) character height and a height angle of 10′, the viewing distance is

$$d = \frac{h}{\tan \Phi} = \frac{0.042 \text{ ft}}{\tan 0.167°} = 14 \text{ ft} \qquad (12\text{-}9)$$

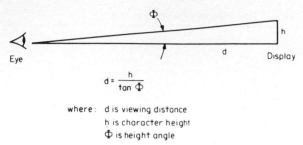

$$d = \frac{h}{\tan \Phi}$$

where: d is viewing distance
h is character height
Φ is height angle

FIG. 12-42 Relationship between viewing distance and character height.

Luminous Intensity vs. Character Height   Unlike discrete LED lamps, which can be regarded as point light sources, the comparatively larger sizes of displays cause them to be considered as area light sources. This means that as character size increases so also must the luminous intensity, in order to maintain the same brightness to the viewer's eye. Fig.12-43 shows a plot of luminous intensity vs. LED character height, where ambient illumination is held at 100 fc.

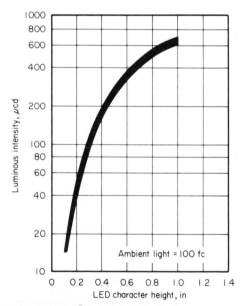

FIG. 12-43 Plot for normal viewing, luminous intensity vs. character height.

Luminous Intensity vs. Ambient Illumination   To maintain normal viewing, luminous intensity must also be increased if the ambient illumination increases. Fig. 12-44 shows two plots of this relationship, one plot for a 0.40-in LED display, and the other plot for a 0.56-in LED display.

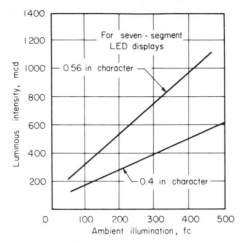

FIG. 12-44 Plots for normal viewing, luminous intensity vs. ambient illumination.

Contrast Ratio For light-emitting displays the contrast ratio between display elements and the background surrounding them is governed by luminance (brightness) of the display elements themselves and by luminance of the surroundings that arises as a result of reflection of ambient illumination. For a display brighter than its surroundings, a simple expression to determine a dimensionless contrast ratio is

$$CR = \frac{B_{display} - B_{surrounding}}{B_{surrounding}} \tag{12-10}$$

Experiments involving human viewers have shown that minimum CR for operational discrimination between display elements and surroundings would seem to be about 2:1. Military standards normally specify 10:1 as the lowest acceptable CR.

From the above expression it can be seen that when display luminance is held constant, increases in ambient illumination cause corresponding increases in luminance of the surroundings, with the result that the CR decreases. This explains, for example, why LED displays on hand calculators and wristwatches are more difficult to read when outdoors in direct sunlight than when indoors under normal office lighting conditions.

One method of maintaining visibility as ambient illumination increases is to increase the luminous intensity from the display.

Placement of an optical filter in front of a display is another method of improving viewability, particularly when ambient illumination is high. This holds true even though some luminous intensity from the display is absorbed by attenuation through the filter. With monochromatic displays, highly selective filters can be used.

Blur, "Ghosting," and Flicker These three characteristics degrade the quality of the character being displayed, thereby increasing operator fatigue and possibly also contributing to human errors during readout. The cause of blur is fuzziness at the edges of the segments or dots in the character font. "Ghosting" is an effect encountered in some multiplexed systems operating at high refresh rates. When the switching speed of digit drivers is not fast enough to guarantee complete turnoff of a preceding digit before the succeeding digit is turned on, the undesirable condition of two digits simultaneously turned on may occur for short durations.

Flicker is also an effect encountered in multiplexed systems. It occurs if refresh is too slow. To prevent flicker keep the strobing rate over 100 Hz (or even higher, if there will be rapid motion between the display and the viewer's eye).

12–3d Seven-Segment Displays

Seven-segment displays are the most commonly encountered displays because they can show all ten numerals as well as several easily recognized letters. Most displays can be driven from DC steady-state circuits, but use of multiplexing techniques can reduce the hardware requirements, especially when several displays are stacked together. To simplify descriptions of drive circuits and to keep within the space available in this chapter, only LED displays will be covered in detail.

LED Display Configurations

Single-digit LED displays are usually supplied in either of two configurations, "common anode" (CA) or "common cathode" (CC) (see Fig. 12-45). Multiple-digit arrays can also be connected on a digit-by-digit basis in a common anode or common cathode configuration (where all *a* segments are tied together, all *b* segments are

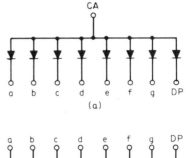

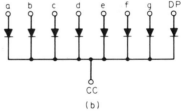

FIG. 12-45 Internal connections for LED displays. (*a*) Common anode display; (*b*) common cathode display.

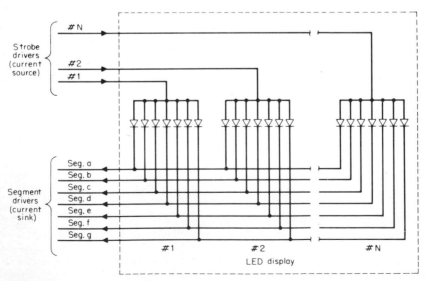

FIG. 12-46 Multiple-digit LED display having common anode configuration.

tied together, and so forth). See, for example, Fig. 12-46. Such multiple-digit arrays have to be driven from a multiplexed drive circuit.

LED Display Construction Methods   Figs. 12-47 through 12-49 depict the three most common construction methods in the manufacture of LED displays. The direct-view hybrid was the first method used. Each segment is comprised of two chips, with eight light-emitting dots per chip. Light emitted within the chip travels directly out of the top of the package to the viewer.

The monolithic, also a direct-view method, is usually used for fabrication of small displays (with character height of the order of 0.135 in) due to the restriction of the chip size/cost tradeoff. To achieve an easily readable character size, a bubble or other magnifying lens is generally placed above the display. Common applications for monolithic displays are hand-held calculators and digital watches.

Lower cost considerations led to the development of the reflector (or "light pipe") construction method (see Fig. 12-49). Here a single LED per segment is used to project digits or character sizes from 0.3 to 0.8 in. In this method a single LED is mounted at

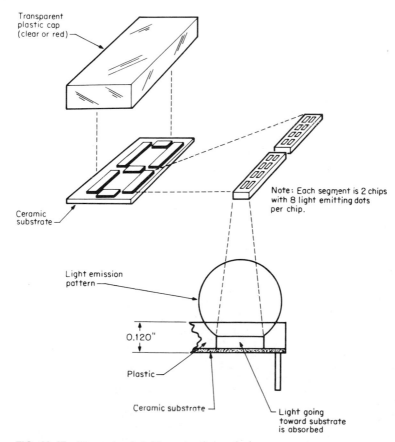

FIG. 12-47   Direct-view hybrid construction method.

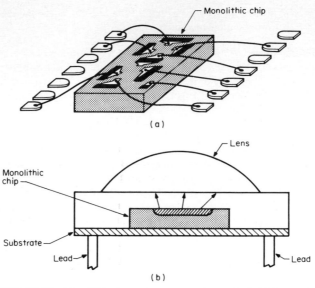

FIG. 12-48 Monolithic direct-view construction method. (*a*) Isometric view; (*b*) side view.

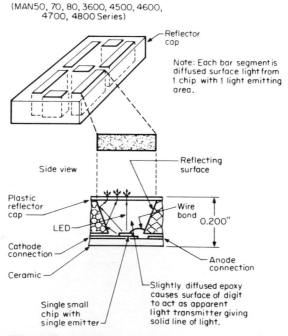

FIG. 12-49 Reflective ("light pipe") construction method.

the bottom of a shaped cavity. Light from the LED is reflected through the cavity to a diffused surface on the top of the display.

DC Drive Circuits for LEDs Fig.12-50 depicts two drive circuit hookups for common cathode displays. In the first circuit the V_{CC} power supply sources current through a current-limiting resistor to light the LED segment when the output transistor of the driver IC is not conducting (i.e., "active high" driver). The LED segment remains not lighted when the transistor conducts. In the second circuit the LED segment lights when the transistor conducts (sources current), but remains not lighted when the transistor is not conducting. Fig. 12-51 depicts an active low-drive circuit hookup for common anode displays. Here the LED segment lights when the transistor conducts (sinks current).

To summarize, the common cathode requires a segment drive circuit that can "source" current and is usually an active "high" circuit (Fig. 12-50). This cathode can be one of two types, either where the output transistor is used as a series switch to V_{CC} (Fig. 12-50b) and sources current into the LED, or where the transistor is a shunt switch and diverts current away from the LED (Fig. 12-50a). The common anode type requires a segment drive circuit that can "sink" current and is usually an active "low" type circuit (Fig. 12-51). The standard available integrated-circuit display/decoder/drivers all use similar output circuitry to the circuitry outlined in Figs. 12-50 and 12-51.

Table 12-5 lists various driver ICs available for 7-segment LED displays.

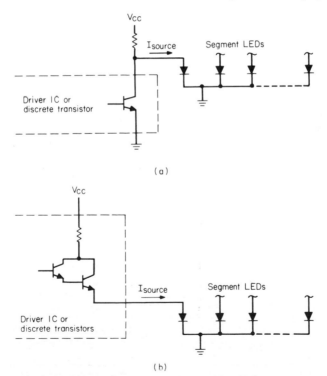

FIG. 12-50 Source drive circuits for segments in common cathode LED displays. (a) Active high driver (such as SN7449); (b) active high source driver (such as 9368).

TABLE 12–5 Seven-Segment Decoder/Driver ICs for LED Displays

Function of driver IC	Basic device number	I_{max}, mA	Manufacturer
Quad segment driver, MOS-to-LED anode	75491	50 (sink/source)	FAIR, MOT, TI
	75493	25 (source)	NS, TI
	7895/8895	19 (source)	NS
	501	40 (source)	ITT
	503	34 (source)	
	507	18 (source)	
	517	16 (source	
	518	14.5 (source)	
	522	12 (source)	
	523	10 (source)	
	491	50 (source)	
HEX digit driver, MOS-to-LED cathode	75492	250 (sink)	FAIR, MOT, NS, TI
	75494	150 (sink)	NS, TI
	8870	350 (sink)	
	8877	50 (sink)	
	8892	200 (sink)	
	500	250 (sink)	ITT
	502	200 (sink)	
	506	200 (sink)	
	510	160 (sink)	
	492	250 (sink)	
7-digit driver, MOS-to-LED cathode	75497	150 (sink)	TI
	8866	50 (sink)	NS
	546	50 (sink)	ITT
	552	500 (sink)	
	554	500 (sink)	
	556	500 (sink)	
8-digit driver, MOS-to-LED cathode	8863/8963	500 (sink)	NS
	8865	50 (sink)	
	8871	40 (sink)	
	514/525	40 (sink)	ITT
9-digit driver, MOS-to-LED cathode	526	40 (sink)	ITT
	548	60 (sink)	
	558	40 (sink)	
	75498	150 (sink)	TI
	8855	50 (sink)	NS
	8874/8876/ 8879	50 (sink)	
	8973/8974/ 8976	100 (sink)	
12-digit driver, MOS-to-LED cathode	8868	110 (sink)	NS
	8973	40 (sink)	
Segment driver, MOS-to-LED	8861 (5-Seg)	50 (source)	
	8877 (8-Seg)	14 (source)	NS

List of Manufacturers: FAIR = Fairchild Semiconductor; HAR = Harris Semiconductor; ITT = ITT Semiconductor; MOT = Motorola Semiconductor; NS = National Semiconductor; RCA = RCA Solid State Division; SIG = Signetics; SSS = Solid State Scientific; and TI = Texas Instruments.

TABLE 12–5 (*Continued*)

Function of driver IC	Basic device number	I_{out}, mA source/sink	Font: 6 and 9 with or without tails	Manufacturer
BCD to 7-segment decoder/driver Outputs: active high, internal resistive pull-up	7448	6.4 (sink) 2 (source)	W/O	FAIR, ITT, MOT, NS, SIG, TI
	74248	6.5 (sink) 2 (source)	W	TI
	74LS48	6 (sink) 2 (source)	W/O	NS, TI
	8T05	15 (sink) 4 (source)	W/O	SIG
	7856	7.5 (source)	W/O	NS
BCD to 7-segment decoder/driver Outputs: active high, current source	7857	60 (source)	W/O	NS
	7858	50 (source)	W/O	NS
BCD to 7-segment decoder/driver Outputs: Active high, open collector	7449	10 (sink)	W/O	FAIR, MOT
	74LS49	8 (sink)	W/O	NS, TI
	74249	10 (sink)	W	TI
	74LS249	8 (sink)	W	TI
BCD to 7-segment decoder/driver Outputs: active low, open collector	7447	40 (sink)	W/O	FAIR, ITT, MOT, NS, SIG, TI
	74L47	20 (sink)	W/O	TI
	74LS47	12 (sink)	W/O	NS, TI
	74247	40 (sink)	W	TI
	74LS247	24 (sink)	W	TI
	9317B	40 (sink)	W/O	FAIR
	9317C	20 (sink)	W/O	FAIR
	8T04	40 (sink)	W/O	SIG
BCD to 7-segment decoder, CMOS; Outputs: *n*-channel sink and *npn* bipolar sources	74C48	3.6 (sink) 50 (source)	W/O	HAR, NS
BCD to 7-segment latch/decoder/ driver; CMOS with bipolar outputs	4511 (14511)	25 (source)	W/O	FAIR, HAR, MOT, RCA, SSS
BCD to 7-segment decoder/driver, CMOS	14558	0.28 (sink) 0.35 (source)	W/O	MOT
BCD to 7-segment latch/decoder/ driver Outputs: constant current	9368	22 (source)	6 (W) 9 (W/O)	FAIR
	9374	18 (sink)	6 (W) 9 (W/O)	FAIR
	8673/ 8674	18 (sink)	6 (W) 9 (W/O)	NS
BCD to 7-segment latch/decoder/ driver Outputs: active low, open collector	9370	40 (sink)	6 (W) 9 (W/O)	FAIR

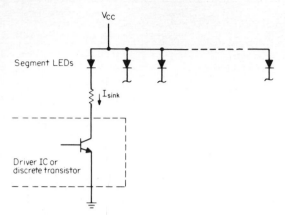

FIG. 12-51   The sink drive circuit for segments in common anode LED displays.

Example 12–5   Hexadecimal Display Drive Circuit

A single-digit display on the front panel of a laboratory instrument is to show hexadecimal numbers. Four data lines plus one strobe line are available to the display from TTL control logic inside the instrument. Incoming data will be valid for several microseconds while the strobe line goes LOW, but data lines are to be ignored while the strobe line is HIGH. Ambient illumination is 300 to 400 fc, maximum viewing distance is 10 ft, and maximum viewing angle is 100°. Design the required circuit using a green display.

Solution

For the display choose a MAN4510 (green) common anode, with a 9370 driver IC. (This driver is chosen because it decodes a usable hexadecimal character set as shown in Fig. 12-52, and because it performs a latching function to hold data as well as decoding and driving.)

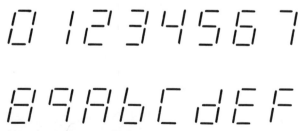

FIG. 12-52   Hexadecimal character set obtained on a 7-segment display a from 9370 decoder/driver IC.

The data sheet for the MAN4500 series of displays specifies a viewing angle of up to 150°, a value in excess of the 100° requirement in this design example. This data sheet also states that the character height is 0.40 in (0.033 ft). Choosing a height angle of 10′, substitution in Eq. 12-9 gives a value for viewing distance d that is greater than the 10 ft called for in this design example.

$$d \text{ (in feet)} = \frac{h}{\tan \theta} = \frac{0.033 \text{ ft}}{\tan 0.167°} = 11.5 \text{ ft} \qquad (12\text{-}9)$$

An examination of Fig. 12-44 shows that for high comfort viewing of a display with 0.40-in character height, the luminous intensity produced by the display should not be less than 500 μcd when ambient illumination is 400 fc.

The specification for the 9370 states that the output will sink current and is typically 20 mA (when $V_{CC} = 5$ V). The luminous intensity vs. I_F plot for the MAN4510 (see Fig. 12-53) shows a luminous intensity of 500 μcd ($I_F = 20$ mA), a value that meets the requirement in this design example.

A schematic of the circuit is shown in Fig. 12-54. The seven current-limiting resistors R_L each have the same value, which is calculated as follows. From the V_F vs. I_F plot for the MAN4510 (Fig. 12-55), we see that V_F is 2.5 V when $I_F = 20$ mA. For $V_{CC} = 5$ V, the specification for the 9370 states that V_{out} is equal to or less than 0.4 V when $I_{out} = 20$ mA. The expression for calculation for R_L in this example is

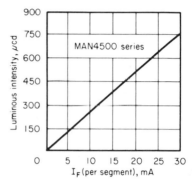

FIG. 12-53 Luminous intensity vs. I_F plot for MAN4510 display.

$$R_L = \frac{V_{CC} - V_F - V_{out}}{I_F} \qquad (12\text{-}11)$$

Setting $V_{CC} = 5$ V, the value of R_L for $I_F = 20$ mA becomes

$$R_L = \frac{5 \text{ V} - 2.5 \text{ V} - 0.4 \text{ V}}{0.02 \text{ A}} = 105 \; \Omega$$

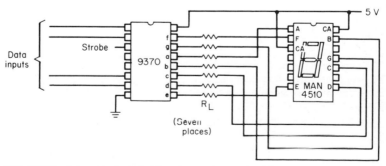

FIG. 12-54 Schematic for DC drive circuit of Example 12-5.

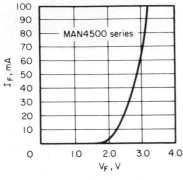

FIG. 12-55 I_F vs. V_F plot for MAN4510 display.

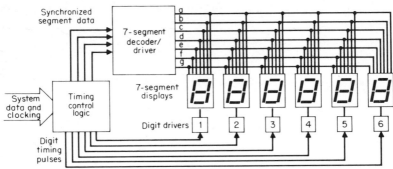

Note: Decimal point LEDs omitted for simplicity.

FIG. 12-56 Block diagram of multiplexed LED display system.

Multiplexed Drive Circuits Fig. 12-56 shows a general block diagram of a multiplexed display system. Note that only a single decoder/driver is used. For discussion purposes, a system with six digit (or character) positions has been chosen, but the same general operating principles apply to any configuration of N displays driven from common system clocking and data inputs.

The purpose of multiplexing is to time-share a single 7-segment decoder/driver among multiple digit displays, thus reducing component count and making fewer wiring connections. In a multiplexed system only *one* digit will be lighted at any one time.

The function of the timing control logic is to produce a set of sequential digit pulses, and to synchronize the segment data with these pulses (see Fig. 12-57). The multiplexing frequency f (also called refresh rate) is given by

$$f = \frac{1}{T} \quad (12\text{-}12)$$

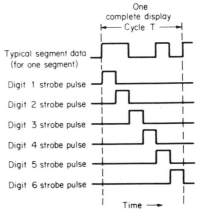

FIG. 12-57 Timing diagram for a 6-digit multiplexed display system.

where T is the time required for one complete display cycle. Here, synchronization means that data for all seven segments of digit 1 must be placed on the common segment lines at the same time that digit 1 pulse is being generated, data for all segments of digit 2 when digit 2 pulse is being generated, and so forth for the remaining digits. Although at any given moment the segment data for only one digit are applied to *all* displays simultaneously, only the display whose digit driver is being pulsed will light up; the other $(N - 1)$ displays remain unlighted because their drivers are not being pulsed.

Fig. 12-58 shows a simplified transistor drive circuit for multiplexed LED display systems. Note that this circuit has the common anode (CA) type of display, which requires that the segment drivers sink current and that the strobe (digit) drivers source current.

For the entire display system only seven current-limiting resistors are required (one per segment), because the LEDs of only *one* digit are ever able to light up at any given time during a single complete display cycle. This means that the maximum forward collector current through any segment-driver transistor will be the forward current of just *one* LED segment. But, collector current through a digit-driver transistor will be the sum of all the segment currents from the lighted (ON) segments of that digit. Therefore, digit-driver collector current can vary from zero (for "blank" character) to seven times the diode forward current (for character 8).

Because the displays are OFF for a substantial part of the display cycle, it can be seen that if the LEDs were driven at the same forward current as used under DC steady-state drive conditions, to a viewer the apparent brightness would be lower than for DC steady state. In order to obtain an equivalent apparent brightness, it is necessary to compensate by increasing the forward diode current. The ratio of ON time to OFF time for the LED is expressed as a duty cycle, defined as follows:

$$\text{Duty cycle (in percent)} = \frac{1}{N} \times 100 \qquad (12\text{-}13)$$

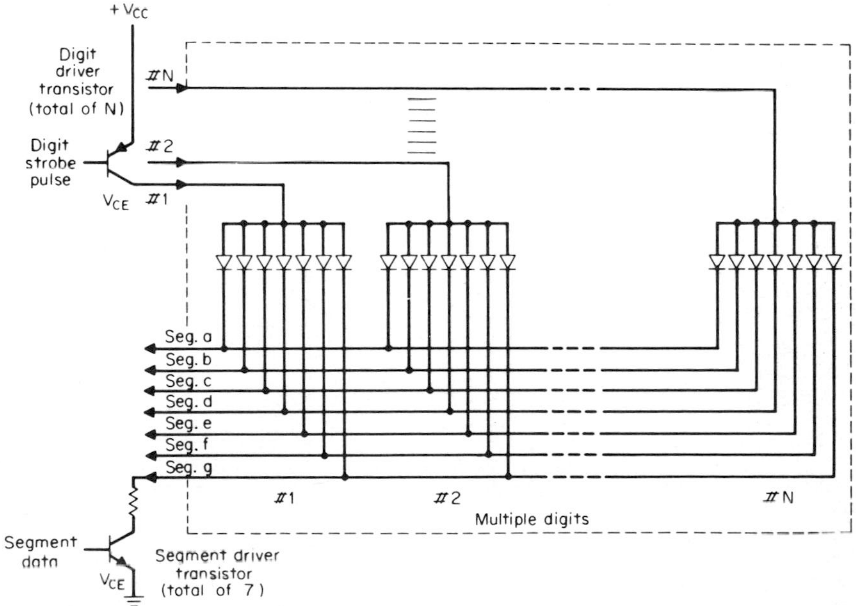

FIG. 12-58 Simplified drive circuit for multiplexed LED display systems.

From the above equation we see that the duty cycle for the 6-digit display system of Fig. 12-56 is 16.7%.

The pulsed drive current (I_P) needed to produce the same apparent brightness as the DC forward drive current (I_F) would, for the example of the 6-digit display with $I_F = 10$ mA, be calculated as follows:

$$I_P = \frac{I_{F(avg)}}{\text{duty cycle}} = \frac{10 \text{ mA}}{0.167} = 60 \text{ mA}_{peak} \tag{12-14}$$

where $I_{F(avg)} = I_{F(DC)}$.

The average forward current, $I_{F(avg)}$, through any one LED segment being driven by a maximum-width pulse as shown in Fig. 12-57 is given by

$$I_{F(avg)} = (\text{duty cycle}) (I_P) \tag{12-14}$$

Section 12-1 of this chapter pointed out that driving LEDs with current pulses results in higher efficiency light emissions than can be obtained from driving with equivalent DC currents. In multiplexed drive LED display systems this higher efficiency is utilized, but within the upper limitation imposed by the maximum peak current rating of the display. Data sheets for LED displays can specify a single absolute maximum rating for DC drive as "continuous forward current I_F," but for pulsed current drive conditions the maximum rating has to be expressed in the form of curves plotted against pulse duration or against duty cycle. Likewise, luminous intensity for pulsed drive conditions is also expressed on a curve, usually as a ratio or relative value (compared to DC conditions), plotted against pulse duration or duty cycle. The following design examples for a MAN4510 green display illustrate how to interpret these curves.

Example 12–6   Design of Four-Digit Multiplexed Display

A four-digit display system is to be built with MAN4540 displays. The multiplexing frequency (refresh rate) is to be 200 Hz. From the curve given in Fig. 12-59, determine the maximum I_p per segment, and from the curve in Fig. 12-60 determine the pulsed average luminous intensity produced from this I_p. (Assume that the maximum I_p pulsewidth will be used, as depicted in Fig. 12-57.)

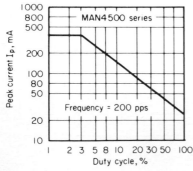

FIG. 12-59 Plot of maximum segment peak current I_P vs. duty cycle for MAN4500 series.

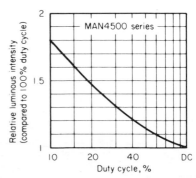

FIG. 12-60 Plot of relative luminous intensity vs. duty cycle for MAN4500 series. (*Note:* Data for this plot was obtained under condition $I_{F(avg)} = 10$ mA.)

Solution

Solving Eq. 12-13 for the duty cycle, we obtain

$$\text{Duty cycle} = \frac{1}{N} \times 100 = \frac{100}{4} = 25\%$$

From Fig. 12-59, we find that for a 25% duty cycle the maximum allowed I_p is approximately 80 mA. By substitution of this value in Eq. 12-14, we obtain the equivalent average current.

$$I_{F(avg)} = (\text{duty cycle}) \, (I_p) = (0.25) \, (80 \text{ mA}) = 20 \text{ mA}$$

From Fig. 12-53 we see that a steady-state $I_F = 20$ mA produces approximately 525 μcd. Because emission efficiency is higher under pulsed conditions, we now refer to Fig. 12-60 to find a relative luminous intensity value. From Fig. 12-60, for a 25% duty cycle, this value is seen to be 1.35 times greater than the DC luminous intensity of 525 μcd, or a pulsed average luminous intensity of 709 μcd.

Example 12–7   Design of a Ten-Digit Multiplexed Display

A 10-digit display system built with MAN4510 displays and having a multiplexing frequency of 200 Hz is to produce a desired luminous intensity of at least 500 μcd. Determine the value of I_p that will be required.

Solution

With ten digits, from the curves in Figs. 12-53, 12-59, and 12-60, we arrive at the following:

$$\text{Duty cycle} - \frac{1}{N} \times 100 = 10\% \qquad (12\text{-}13)$$

$$\text{Maximum } I_p = 160 \text{ mA (from Fig. 12-59)}$$

For 10% duty cycle, the relative luminous intensity is 1.8 (from Fig. 12-60), and the average luminous intensity is found from:

Average luminous intensity
$$= 1.8 \times \text{DC luminous intensity} = 500 \text{ μcd} \qquad (12\text{-}15)$$

From the above expression we obtain the equivalent DC luminous intensity equal to 278 μcd. And from Fig. 12-53 we see that the $I_{F(DC)}$ needed to produce 278 μcd is 10 mA. We therefore calculate I_p as follows:

$$I_p = \frac{I_{F(avg)}}{\text{duty cycle}} = \frac{0.010 \text{ A}}{0.1} = 100 \text{ mA} \qquad (12\text{-}14)$$

Example 12-8   Computing Current-Limiting Resistor

Fig. 12-61 shows a circuit for the segment and digit-driver transistors on one digit of the display system just described in Example 12-7, where $I_P = 100$ mA. Determine the value for the current-limiting resistor at a given supply voltage V_{LED}. For transistor Q_S assume a specification with a curve that shows a guaranteed $V_{CE(sat)} = 0.2$ V (at $I_C = 100$ mA), and for transistor Q_D assume a specification with a curve that shows a $V_{CE(sat)} = 0.8$ V (at $I_C = 1$ A).

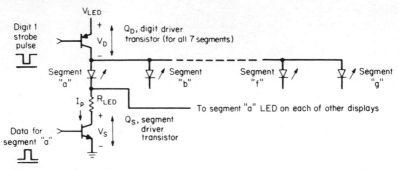

FIG. 12-61 Portion of multiplex drive circuit for digit 1 of 10-digit display system.

Solution
The value of the current-limiting resistors is calculated as follows:

$$R_{LED} = \frac{V_{LED} - V_F - V_S - V_D}{I_P} \qquad (12\text{-}16)$$

where I_p = peak segment forward current
V_F = LED forward voltage for above I_p
V_S = saturation voltage of segment-driver transistor at above I_p
V_D = saturation voltage of digit-driver transistor at collector current equal to (I_p) times (number of segments lighted)

In the above equation it can be seen that the worst-case condition for V_D occurs at 8 when all segments light (i.e., when collector current equals 7 times I_p).

From Fig. 12-55 we see that the V_F is 3.2 V when I_P is 100 mA. Upon substitution of this and the given values in the previous equation, we obtain

$$R_{LED} = \frac{V_{LED} - 3.2 \text{ V} - 0.2 \text{ V} - 0.8V}{0.1 \text{ A}} = \frac{V_{LED} - 4.2 \text{ V}}{0.1 \text{ A}} \qquad (12\text{-}16)$$

From the above equation it can be seen that if a supply voltage of 5 V is chosen for V_{LED}, then the voltage drop across resistor R_{LED} will only be 0.8 V, leaving 4.2 V (or 84% of the voltage drop) for V_F, V_S, and V_D. Under such conditions, the variations in these latter drops (as collector current through Q_D varies from two times I_p for character 1 to seven times I_p for character 8) would present noticeable variations to the viewer in the apparent brightness of the display digits as different characters are displayed. Also any slight variations in V_{LED} supply cause large variations of current.

Therefore, in this example, the V_{LED} supply voltage should be set higher than 5 V. In many items of equipment a convenient choice is the unregulated DC voltage (after rectification) that is present at the input side of the 5-V power supply for logic circuits. Typically this voltage is of the order of 6.5 V, and its amplitude variations will not be noticeable to the viewers of the display.

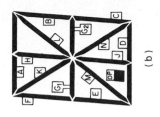

(b)

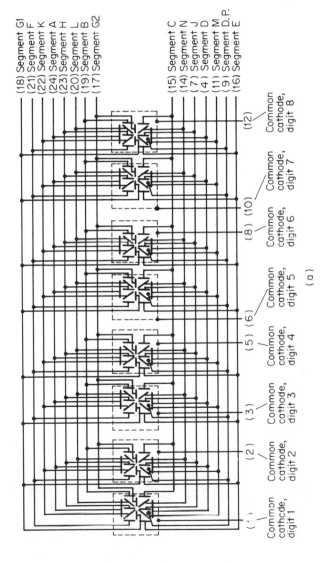

(a)

FIG. 12-62 Segment connections for MAN2815 8-digit LED display. (*a*) Internal connections and package pin designations. (*b*) Format and identification of segments for each digit position. (*Note:* Segments *A* and *D* appear as two segments each, but both halves are driven together.)

Setting V_{LED} equal to 6.5 V in this example would put 2.3 V of drop across R_{LED}, leaving 4.2 V (64%) for V_F, V_S, and V_D. Solving the equation for R_{LED}, we obtain

$$R_{LED} = \frac{6.5 \text{ V} - 4.2 \text{ V}}{0.1 \text{ A}} = \frac{2.3 \text{ V}}{0.1 \text{ A}} = 23 \ \Omega$$

12–3e   Alphanumeric Displays

Multisegment Displays   Although DC steady-state drive circuits could be used if a display had separate package pins for each of its segments, employment of multiplexing techniques will significantly reduce the circuit hardware whenever two or more displays are stacked together. Some form of multiplexing drive circuit is mandatory for multidigit displays in which corresponding segments from each digit are internally connected in parallel. For example, Fig. 12-62 shows a diagram of internal connections and package pins on the MAN2815 14-segment LED display. The MAN2815 has eight digits in a 24-pin package, with dimensions 1.39 in by 0.74 in by 0.195 in.

Fig. 12-63 shows a block diagram of a general multiplexing circuit approach for multisegment displays, and Fig. 12-64 shows a specific drive circuit design for a string of up to four MAN2815 LED displays.

Dot Matrix Arrays   Compared to bar segment displays, the greater number of independent elements on dot matrix arrays allows showing of a wider variety of characters.

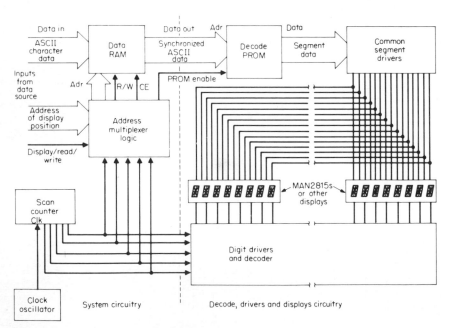

FIG. 12-63   General multiplexing approach for multisegment displays.

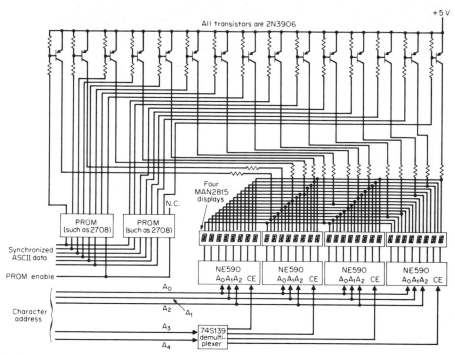

FIG. 12-64 Decode, drivers, and display portion of circuit for 32-character-wide display.

Furthermore, these characters are more attractive because their rounded shapes more closely resemble those of printed characters. (For example, compare the numerals in Fig. 12-37 with those in Fig. 12-41.)

Unlike 7-segment displays on which a separate pin connection can easily be provided to each segment, the package space limitations on dot matrix displays prohibit separate pin connections to each independent matrix element. For example, a 5-by-7 matrix would require 35 pins to access each element separately, plus one more pin for the common connection. Two such displays would thus require 72 pins, and so forth. To reduce the pin counts, manufacturers of dot matrix arrays therefore connect the elements in rows and columns, and only provide one pin per row and one pin per column. Fig. 12-65 shows a schematic for such a connection scheme for a 5-by-7 LED array having 12 pins for the array plus one more pin for a decimal point LED.

With dot matrix displays multiplex drive circuits must be employed to generate characters either by column strobing or by row strobing methods, as illustrated in Fig. 12-66. Display designers have the choice of working out their own circuits and associated timing, or else can elect to purchase display devices having built-in drive circuits. For example, the HP5082-7300 series includes a modified 4-by-7 LED dot matrix display, drivers, and latch memory, all in a single 8-pin package. With displays having built-in drives, the designer's task becomes one of integration of this display subsystem into the timing and data format generated by the other hardware inside the equipment.

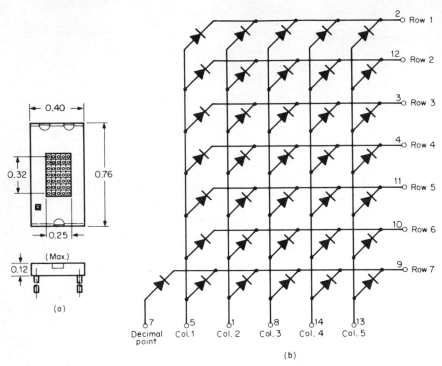

FIG. 12-65 Package dimensions and schematic for typical LED 5-by-7 matrix array (MAN2A). (*a*) Package dimensions; (*b*) schematic.

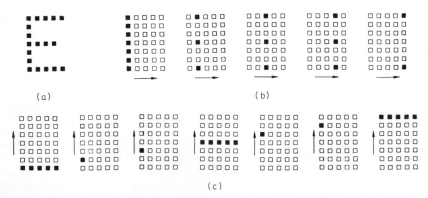

FIG. 12-66 Methods of character generation. (*a*) Final letter "E" (5-by-7 font). (*b*) Generation of "E" by column strobing method. (*c*) Generation of "E" by row strobing method.

12–3f Displays with On-Board Electronics

Definition Designer demands for higher levels of component integration have caused LED manufacturers to offer display products having drive electronics and one or more LED displays together in a single package. These complete units are called "on-board electronics" (OBE) displays, and most of them are designed to be end-stackable so that multicharacter systems can be assembled.

Packages and Construction OBE displays come in a multitude of types and sizes, ranging from single character to multiple characters, with 7-, 14-, and 16-segment or high-resolution dot matrix fonts. The on-board drive electronics can perform partial or full decode/latching/segment or dot drive, using DC or multiplexing techniques.

A basically hybrid method is employed for construction of OBE displays. The drive IC chip and one or more LED chips are first die attached on a single substrate, then the chips are interconnected by wire bonding. Depending upon the specific construction method used, the drive IC chip is positioned either on the front side of the substrate along with the LED display, or else is positioned on the rear side of the substrate.

On the smaller (and relatively expensive) "direct view" displays, a ceramic substrate is generally used, and the entire assembly (drive IC and LED chips) is encapsulated in epoxy. On some of these displays the epoxy additionally forms a cylindrical or bubble lens for the LEDs as well.

On the larger (but relatively inexpensive) multiplex displays, "reflector" light pipe techniques are usually employed. The drive IC chip, or chips, are epoxy bonded directly down onto a printed circuit board, then wire bonds are added to connect the chip pads to artwork metal traces on the board, and these traces, in turn, connect to the LEDs. After bonding is completed, a number of ways can be adopted to protect chips and bonding wire from mechanical damage. A lens reflector cap can be placed over the chips and bonding wires, a separate plastic cover cap can be glued on top, or a glob of epoxy or silicone material can simply be applied directly over the chips and bonds and allowed to solidify.

Drive Electronics Complete decoder/driver/latch electronics are incorporated on single-digit numeric OBE displays that use static DC drive. But only partial drive circuits are provided on some of the more complicated dot matrix and bar segment alphanumeric displays. On these latter displays the decoding function has to be done externally, and additional multiplex drive circuits are also required in order to arrive at a complete display system.

Typical Displays The following four types of displays are typical of those currently available from manufacturers.

> *Single-digit hexadecimal 5-×-7 dot matrix display:* Fig. 12-67 shows the block diagram for this type of direct-drive display. A number of manufacturers offer similar displays, but in different package styles. These are direct-view displays that have individual LEDs for each dot.
>
> *Three-character 5-×-7 dot matrix display:* Fig. 12-68 shows a partial-drive multiplexed display. It contains a 15-bit shift register to address its 15 columns of LEDs, but requires external drive at its 7-row anode terminals. Decoding for the display font has to be done externally, and

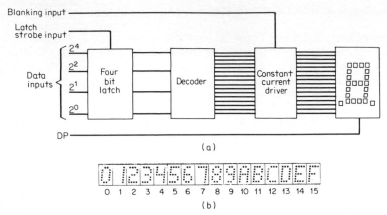

(a)

(b)

FIG. 12-67   Single-digit 5-by-7 hexadecimal display with binary input (TIL505, TIL311, HP 7300 series). (a) Block diagram; (b) character font.

drive signals have to be appropriately fed to the serial input and row anode terminals to cause the desired characters to be presented by the display. Since it is a multiplexed system, data for this display's rows and columns have to continually be updated. Fig. 12-69 shows a typical circuit to perform this updating function.

Four-character 16-segment alphanumeric displays: Fig. 12-70 shows a block diagram for a complete 4-character multiplexed display system. This type of display contains ROM for character decoding, RAM for display refresh, segment and strobe circuits, and even its own oscillator to keep the display in the multiplex mode. The display color is red, and its monolithic, direct view, 16-segment character font is approximately 0.160 in high.

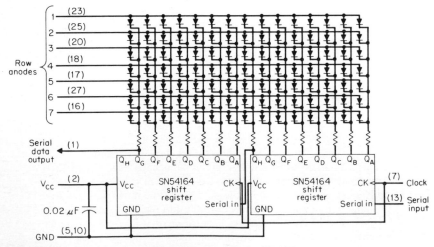

FIG. 12-68   Three-character 5-by-7 dot matrix display (TIL560).

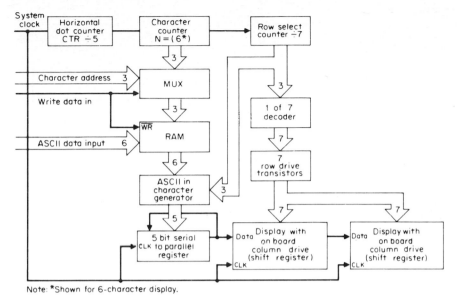

FIG. 12-69 Block diagram of system using TIL560 5-by-7 dot matrix displays.

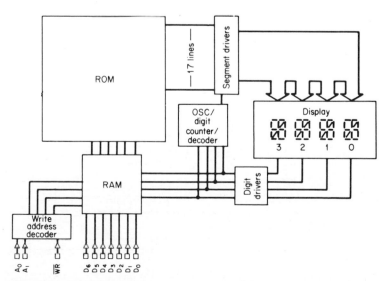

FIG. 12-70 Block diagram of 4-character multiplexed display (DL-1414/DL-1416).

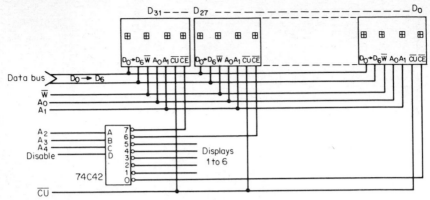

FIG. 12-71 Expanded 32-character display (DL-1414/DL-1416).

When the write enable ($\overline{\text{WR}}$) line is activated, parallel ASCII data can be written into the display asynchronously on the data lines D_0 through D_7. Fig. 12-71 shows how these displays can be expanded to form a 32-character display system.

Four-digit 7-segment drive system: Fig. 12-72 shows a block diagram for a 4-digit display system built with reflector digits $\frac{1}{2}$ in high. This display contains a DC drive circuit that consists of a 35-bit serial input register and a 35-bit parallel output register. Data for the display is serially entered into the input register, and after every thirty-fifth bit, this register's contents are automatically loaded into the output register.

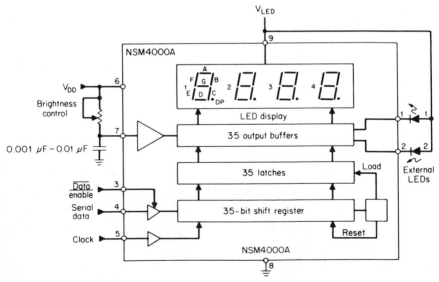

FIG. 12-72 Four-digit display module with OBE (NSM4000A).

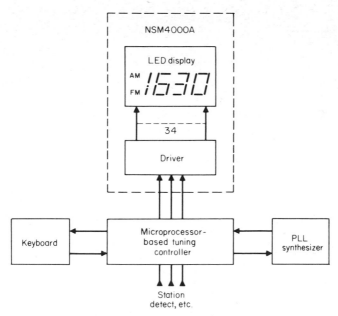

FIG. 12-73   Block diagram of typical microprocessor-driven decoding and serialization system.

An external circuit is required to decode and serialize the data into 35-bit lengths for the display. This is usually accomplished by some specialized microprocessor-based system, such as that shown in Fig. 12-73.

12–3g   Interactive Displays

Historical Development   The introduction of microprocessors during the late 1970s led to the incorporation of central processor architecture into many new areas of equipment design. By placing importance on hardware/software tradeoffs and on reduction of overhead tasks for central or ''master'' processors, this innovation spurred development of ''interactive displays.'' This is a display system where it is possible to write into the display and also read back from the display. This new display category consists of a complete system that typically is built on two printed circuit boards. One board, called the controller, has a ''slave'' microprocessor, together with its own local memory (ROM and RAM), segment/dot decoding, and circuitry for interfacing to a host computer system. The second, or display board, has several LED displays with their associated drive circuitry. Manufacturers offer interactive displays for either parallel or serial data exchange with a master processor in a host computer system, and for connection directly to TTL data busses or via communication interfaces (such as RS-232-C levels).

Available Products   Figs. 12-74 and 12-75 show two of the commercially available interactive displays. The HDSP-24XX uses 5-by-7 dot matrix displays (HDSP 2000). Display boards can be obtained in several versions having from 16 to 40 characters each,

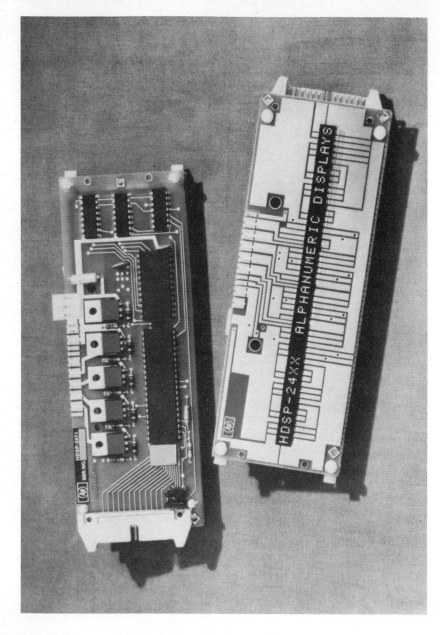

FIG. 12-74 HDSP-24XX alphanumeric display system (using HDSP-2000 5-by-7 dot matrix displays) (*Hewlett-Packard*).

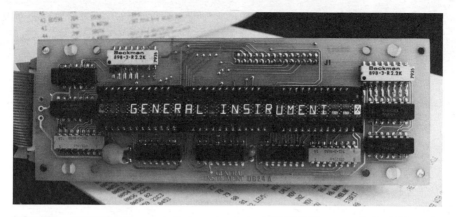

FIG. 12-75 XDS2724 alphanumeric display system (using MAN2815 14-segment displays) (*General Instruments*).

and controller boards can be obtained with decoders for 64-character ASCII code, 128-character ASCII character code, or with a socket for a PROM to accommodate a customer's own code. The XDS2724 uses 14-segment (plus decimal point) displays (MAN2815) to form a 24-character display system; versions for parallel and for serial data exchange with the host system are available. Decoding is determined by a portion of the local microprocessor's ROM on the controller board. The standard XDS2724 ROM programming operates on data in ASCII code to produce the display characters shown in Fig. 12-76.

Benefits Interactive displays simplify the designer's job, provide interim data storage to speed up the process of updating display information, and relieve the host system from fresh and digit strobing tasks. Additionally, interactive displays provide control lines back to the host system for control of apparent display brightness, and also allow the host system to retrieve display data (via the same interface used for transmission of data from host to display).

Hardware/Software Requirements Requirements vary from one manufacturer's product to another, but in general the display designer has to be concerned with the following considerations:

1. Hardware connections to the host system's microprocessor data bus and control lines, and timing of signals on these lines
2. Hardware connections to other control circuitry of host system
3. Format of data and command words to be exchanged between host and interactive display.

12–4 OPTOCOUPLERS

12–4a Optocoupler Theory

When an LED source is combined in the same package with some type of solid-state photodetector (usually silicon semiconductor structures), the resulting device is called an

ASCII Hex code	ASCII Desig-nation	Re-sponse	ASCII Hex code	ASCII Desig-nation	Character displayed	ASCII Hex code	ASCII Desig-nation	Character displayed	ASCII Hex code	ASCII Desig-nation	Character displayed
00	NULL		20	SPACE		40	@	@	60	`	`
01	SOH		21	!	!	41	A	A	61	a	a
02	STX		22	"	"	42	B	B	62	b	b
03	ETX		23	#	#	43	C	C	63	c	c
04	EOT		24	$	$	44	D	D	64	d	d
05	ENQ		25	%	%	45	E	E	65	e	e
06	ACK		26	&	&	46	F	F	66	f	f
07	BEL	(used)	27	'	'	47	G	G	67	g	g
08	BS	(used)	28	(	(	48	H	H	68	h	h
09	HT	(used)	29	)	)	49	I	I	69	i	i
0A	LF	(used)	2A	*	*	4A	J	J	6A	j	j
0B	VT		2B	+	+	4B	K	K	6B	k	k
0C	FF		2C	,	,	4C	L	L	6C	l	l
0D	CR	(used)	2D	−	--	4D	M	M	6D	m	m
0E	SO		2E	.	.	4E	N	N	6E	n	n
0F	SI		2F	/	/	4F	O	O	6F	o	o
10	DLE		30	0	0	50	P	P	70	p	p
11	DC1	(used)	31	1	1	51	Q	Q	71	q	q
12	DC2	(used)	32	2	2	52	R	R	72	r	r
13	DC3	(used)	33	3	3	53	S	S	73	s	s
14	DC4		34	4	4	54	T	T	74	t	t
15	NAK		35	5	5	55	U	U	75	u	u
16	SYN		36	6	6	56	V	V	76	v	v
17	ETB		37	7	7	57	W	W	77	w	w
18	CAN		38	8	8	58	X	X	78	x	x
19	EM		39	9	9	59	Y	Y	79	y	y
1A	SUB		3A	:	:	5A	Z	Z	7A	z	z
1B	ESC	(used)	3B	;	;	5B	[	[	7B	{	{
1C	FS		3C	<	<	5C	\	\	7C	}	|
1D	GS		3D	=	--	5D	]	]	7D	}	}
1E	RS		3E	>	>	5E	∧	∧	7E	~	~
1F	US		3F	?	?	5F	_	--	7F	DELETE	(see note)

NOTE: In response to DELETE, the XDS2724 deletes character under cursor, (see below), and causes cursor to move one position to the right.

⌗ CURSOR

FIG. 12-76 Characters displayed by XDS2724 in response to ASCII data from the host computer system.

optocoupler or sometimes optoisolator. Light from the LED, generally an infrared emitter, reaches the detector through a transparent medium, such as a plastic light pipe or, in some packages, an air gap. This arrangement of components produces a device that allows optical coupling of signals between two separate electronic circuits, even though they are electrically isolated from one another. Devices can attain isolation ratings of 2000 to 3750 V and higher, depending on package design.

Fig. 12-77 shows a typical optoisolator, while Fig. 12-78 shows a diagram of the phototransistor detector. When forward current (I_F) flows through the input LED, the light or photons it produces fall on the transistor's photosensitive base region and generate

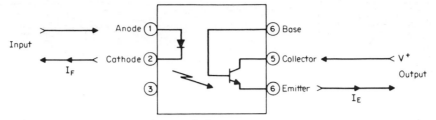

FIG. 12-77 Typical phototransistor optoisolator.

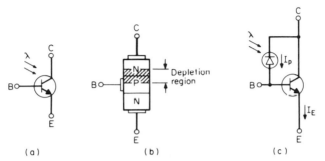

FIG. 12-78 Phototransistor detector. (*a*) Symbol; (*b*) structure; (*c*) equivalent circuit.

photocurrent (I_p). The resulting emitter current (I_E) is equal to the transistor's current gain H_{FE} multiplied by I_p. (The ratio of I_E to I_F is called the current transfer ratio.)

Fig. 12-79 shows a diagram of a photodarlington transistor detector. Here photon current is generated primarily in the first stage, and is labeled I_{p_1} in the figure. The detector's output current, the second-stage emitter current I_{E_2}, is equal to the product of I_p and the current gains of the two transistors, i.e., $I_{E_2} = I_{p_1} \times H_{FE_1} \times H_{FE_2}$. For a given I_F through the input LED, this detector therefore will produce a larger output current than will the single-stage detector described previously. Another version of a Darlington transistor detector is shown in Fig. 12-80. In this split configuration the base connection to the final stage is made available as V_{B_2}, so that an external resistor can be inserted.

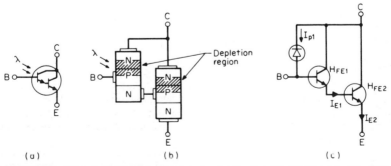

FIG. 12-79 Photodarlington transistor detector. (*a*) Symbol; (*b*) structure; (*c*) equivalent circuit.

This provides a way to adjust gain-bandwidth and improve noise margin in TTL circuitry. Optocouplers with this type of detector, such as a 6N139, can operate with input I_F as low as 0.5 mA and yet have output currents that are compatible with driving TTL circuits.

Fig. 12-81 shows a photo-SCR detector. Here photogenerated current in the base collector of the *npn* transistor will cause this transistor to conduct. This will now

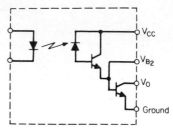

FIG. 12-80   Split photodarlington transistor detector.

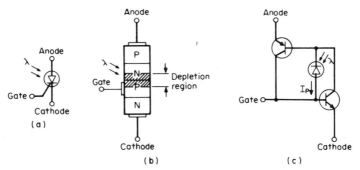

FIG. 12-81   Photo-SCR detector. (*a*) Symbol; (*b*) structure; (*c*) equivalent circuit.

turn on the *pnp* transistor, which in turn will sustain conduction of the *npn* transistor (latch-up), even with photocurrent removed.

Fig. 12-82 shows a detector in which photons from the input LED fall upon a photodiode that connects to an amplifier. In turn the amplifier drives a Schmitt trigger and logic gate whose standard open collector output offers normal current sinking capability to drive logic circuits. Optocouplers built with this detector, such as the MCL611, typically require an I_F of only 10 mA and will respond to data rates up to 1 MHz.

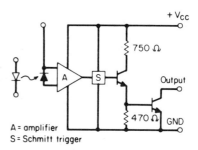

A = amplifier
S = Schmitt trigger

FIG. 12-82   Amplifier/logic gate detector.

12–4b   Types of Optocouplers

Figs. 12-83 through 12-85 show typical optocoupler packages and corresponding device schematics. For punched paper tape and card readers and other applications where mechanical constraints prohibit the use of these standard packages, optocouplers can be constructed as a special case, using LED sources and silicon detectors in separate packages, as depicted in Fig. 12-86.

Detector portions of optocouplers range in complexity from single-photodiode phototransistors and photo-SCRs on up to digital, amplifier, and switching circuits. Along with electrical isolation, the functions performed by optocouplers on input signals can include linear amplification, digital coupling, or control of power circuits, such as TRIAC

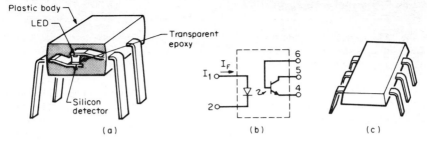

FIG. 12-83 Typical package with internal light-transmission path (MCT2). (*a*) Cutaway view of package; (*b*) schematic; (*c*) standard mini-DIP package.

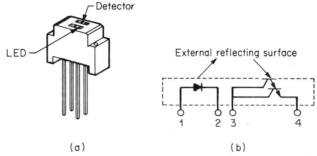

FIG. 12-84 Typical package with reflective light path (MCA7 reflective object sensor). (*a*) External view; (*b*) schematic.

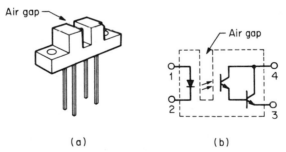

FIG. 12-85 Typical package with air-gap light path (MCA8 slotted optical limit switch). (*a*) External view; (*b*) schematic.

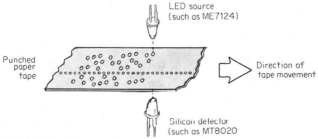

FIG. 12-86 Example of optocoupler constructed from separate emitter and detector packages for optical tape- or card-reader application.

drives. Optocouplers provide isolation interfaces for linear circuits, for digital logic-to-logic and logic-to-power circuits, and even for power-to-logic circuits, such as monitors of AC power sources.

Advantages Optocouplers offer several other advantages in addition to electrical isolation of circuits; they often also isolate people from high voltages. Optocouplers eliminate power-supply ground loops, and other interference from control circuits into loads, and (unlike transformers) prevent reflections of transients and noise spikes from loads back into the driving circuits. When used to replace electromechanical components, such as relays and limit switches, optocouplers eliminate contact bounce, operate faster, exhibit much higher reliability, and do not require mechanical adjustments.

12–4c Optocoupler Parameters

Radiated Output Power, Radiant Intensity, and Light Current Sensitivity (for sources and detectors built in separate packages)   Radiant flux or radiated output power (ROP), measured in watts, is used to specify output from infrared LEDs. Data sheets generally define it as the total flux radiated at its specified wavelength. Radiant intensity is also used to measure infrared output from an LED. It is measured in watts per steradian (w/st) and is defined as the amount of flux through a unit solid angle.

For silicon transistors the parameter light current sensitivity (symbol S_{ceo}) characterizes current produced in response to light irradiance falling upon the detector. S_{ceo} is defined as the current produced for a given flux per unit area, and data sheets usually list it as microamps produced per milliwatt per square centimeter of irradiance. The source of light is usually an infrared emitter, with a stated peak wavelength, and/or a tungsten source operating at a specified color temperature.

Dark Current [symbol I_{ceo} or $I_{ceo}(DARK)$]   This parameter is defined as the output current that flows in the absence of light from the source. For optocoupler packages with internal light transmission paths this means detector output current without input current to the LED emitter. Called leakage current on some data sheets, this current arises from internal charge carrier mechanisms and has to be taken into account during worst-case design analysis. In optocoupler applications having external light transmission paths (e.g., reflective sensors or slotted optical limit switches), during worst-case analysis of the circuit, one should measure detector output current empirically under actual ambient light conditions with the LED light source turned off. This current should be small enough to sustain the circuit in an off condition.

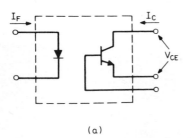
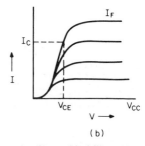

(a)   (b)

FIG. 12-87   Curves for current transfer ratio. (*a*) Schematic; (*b*) graphical illustration.

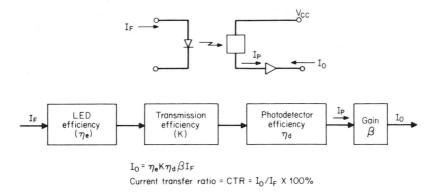

$$I_O = \eta_e K \eta_d \beta I_F$$

Current transfer ratio = CTR = I_O/I_F X 100%

Notes: 1. LED efficiency is a function of I_F, temperature and age.
2. Photodetector efficiency is a function of dark current, temperature, photo-sensitive area and bias.
3. Gain is a function of the bias and temperature.

FIG. 12-88 Detailed definition of current transfer ratio parameter.

Current Transfer Ratio The ratio of an optocoupler's output current to the LED input current is called the current transfer ratio (symbol CTR and usually expressed in percent). Data sheets list CTR at given input currents I_F and known output conditions (such as V_{CE} at specified voltages). Refer to Fig. 12-87 for an example of typical data sheet curves, and Fig. 12-88 for a more detailed definition of this parameter.

Isolation Voltage and Resistance These parameters serve as ratings of an optocoupler's input-to-output electrical isolation capability. Isolation voltage, symbol V_{iso}, is defined as a high potential that, if applied between the shorted input LED terminals and the shorted output terminals, would neither damage the device nor produce current flow (in excess of some very small value specified under the test conditions). Some data sheets list two different V_{iso} ratings, one called surge isolation with a very short application time (e.g., 1 s), and another called steady-state isolation with a longer application time (typically, 1 min). Data sheets may also include an isolation resistance rating, in ohms, measured at a lower V_{iso} (say 500 V).

Switching and Response Delay Times

These parameters describe dynamic output response of the optocoupler to pulse current waveforms applied to its input LED and outputted from the detector. Data sheets, in addition to listing test conditions for these parameters, also give characteristics of the input waveform and show schematics of the test circuits. Fig. 12-89 shows the test circuit for a typical optocoupler, the 4N35, and Fig. 12-90 shows waveforms on which the following four parameters are defined: delay time

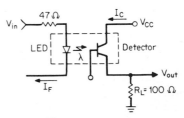

FIG. 12-89 Test circuit.

(t_d), rise time (t_r), storage time (t_s), and fall time (t_f). On some data sheets the sum of t_d and t_r is referred to as t_{on}, and the sum of t_s and t_f is referred to as t_{off}.

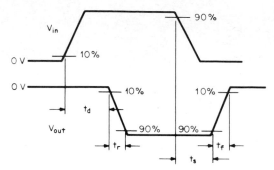

FIG. 12-90   Input and output waveforms.

12–4d   Circuit Applications

Logic-to-Logic Interface with 6N138, 6N139   Figs. 12-91 and 12-92 show inverting and noninverting optocoupler circuits for electrically isolated logic-to-logic interfaces. In these circuits the supply voltage and type of logic driver (CMOS or bipolar)

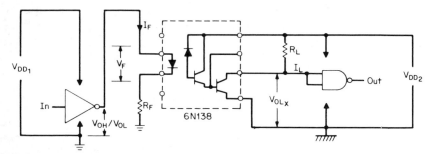

FIG. 12-91   Inverting logic-to-logic interface circuit.

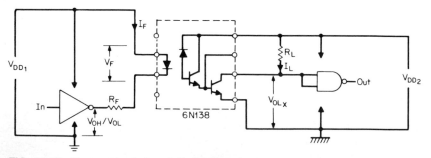

FIG. 12-92   Noninverting logic-to-logic interface circuit.

at the input side of the circuit can be different from those on the output side. For the inverting circuit the equation for calculating the value of R_F is

$$R_F = \frac{V_{OH_1} - V_F}{I_F}$$ (12-17)

TABLE 12–6 Calculated Values of R_F and R_L for LED Drivers and Receivers

Logic type	Supply voltage	Circuit function	R_F (Ω)	R_L (Ω)*
CMOS	5 V	Noninverting	2000	1000
		Inverting	510	—
	10 V	Noninverting	5100	2200
		Inverting	4700	—
74 XX	5 V	Noninverting	2200	750
		Inverting	180	—
74LXX	5 V	Noninverting	1800	1000
		Inverting	100	—
74SXX	5 V	Noninverting	2000	1000
		Inverting	360	—
74LSXX, 74HXX	5 V	Noninverting	2000	560
		Inverting	180	—

*The value of R_L is the same whether for an inverting or noninverting circuit.

and for the noninverting circuit is

$$R_F = \frac{V_{DD_1} - V_F - V_{OL_1}}{I_F} \qquad (12\text{-}18)$$

where V_{DD_1} = is input circuit supply voltage
V_{OH_1} and V_{OL_1} = driver's logic "1" and "0" voltages, respectively
V_F and I_F = the forward voltage and forward current of the LED inside the 6N138

For both circuits the equation for calculating the value of R_L is

$$R_L = \frac{V_{DD_2} - V_{OL_X}}{I_L} \qquad (12\text{-}19)$$

where V_{DD_2} = output circuit supply voltage

I_L = portion of output current that flows through R_L
V_{OL_X} = saturation voltage of 6N138 output transistor when its collector current is the sum of I_L plus input current of logic receiver

Table 12-6 lists calculated R_F and R_L values for various combinations of common driver and receiver types.

Logic-to-Logic Interface with MCL601, MCL611   Fig. 12-93 shows a schematic for an optocoupler having what can be called a logic gate detector. Photons fall upon a high-speed photodiode that connects to an amplifier. After amplification, the detected signal drives a Schmitt trigger that improves noise immunity by providing a threshold and hysteresis. At its output side this detector has a standard open collector circuit that offers normal current sinking capability. The additional stages in the detector

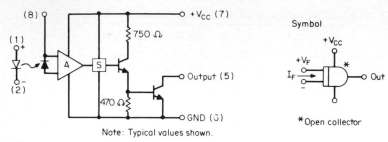

FIG. 12-93   Schematic of MCL601 and MCL611.

make MCL601 and MCL611 optocouplers attractive for line receiver applications over twisted wire lines at data rates up to 1 MHz.

Fig. 12-94 shows circuits that allow the optocoupler's input LED to be brought into conduction from either logic ON (pull-up load) drivers or logic OFF (pull-down load) drivers. Here inputs to the optocoupler are not specified as HI or LOW as with inputs to logic gates; instead inputs are specified as ON or OFF, according to whether or not current flows through the input LED of the optocoupler. Thus the input may be ON for logic drive HI (pull-up load system, as in Fig. 12-94a) or for logic drive LOW (pull-down load system as in Fig. 12-94b, for open collector devices). As a convenience of notation, reference will be made to a pull-down type of load input connected as shown in Fig. 12-94b. For this latter connection a logic LOW means ON, and a logical HI means OFF.

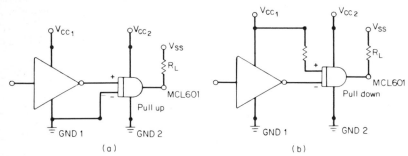

FIG. 12-94   Connection to drivers. (a) For logic "HI" (pull-up load) driver (logic source). (b) For logic "LOW" (pull-down load) driver (logic sink).

Fig. 12-95 shows a circuit for interfacing a twisted pair line. For a line termination an external diode has been added to the input of the optocoupler.

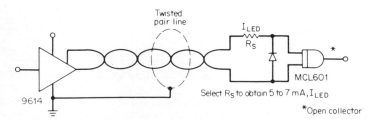

FIG. 12-95   Circuit for interface over twisted line pair.

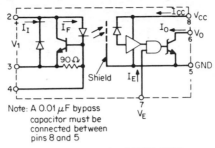

Note: A 0.01 μF bypass
capacitor must be
connected between
pins 8 and 5

FIG. 12-96 Schematic of HCPL-2602.

Logic-to-Logic Interface with HCPL-2602   Fig. 12-96 shows an optocoupler having a built-in line termination. This feature makes the HCPL-2602 attractive for use in high noise environments where immunity to differential noise and common mode rejection are important. Besides supplying I_F for the LED, the line-termination portion of the circuit clamps the transmission line voltage and regulates current flow through the LED. An input current of 5 mA to the optocoupler will produce an output that will sink a TTL gate fan-out of 8, with typical propagation delay from input to output of 45 ns. The optocoupler itself is capable of data rates up to 10 Mb/s, but the actual speed of operation may be limited by characteristics of the transmission line.

Figs. 12-97 through 12-99 show three line driving circuits. The last of these, the polarity reversing circuit, uses two 2602s and a flip-flop to provide optimum noise rejection as well as balanced time delays. The input LED of the first 2602 conducts on one polarity of the drive signal, and the input LED of the second 2602 conducts on the other polarity.

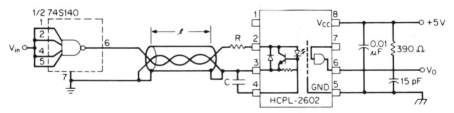

FIG. 12-97 Polarity nonreversing drive circuit.

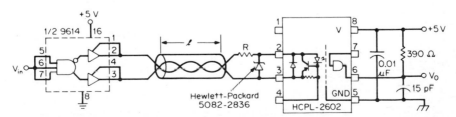

FIG. 12-98 Polarity reversing, single-ended drive circuit.

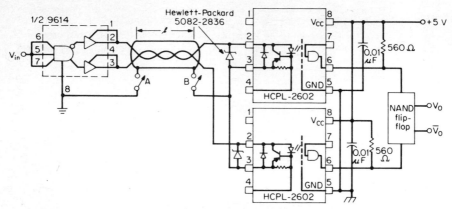

FIG. 12-99 Polarity reversing, split-phase drive circuit.

Linear Circuit Interfaces Fig. 12-100 shows the schematic for a simple opto-coupler having a symmetrical bilateral silicon photodetector. Optocouplers in this H11F family perform similarly to ideal isolated field-effect transistors (FETs) for distortion-free control of low-level AC and DC analog signals. On-state AC small signal resistance

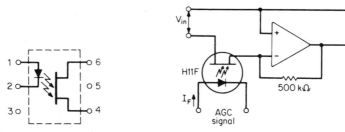

FIG. 12-100 Schematic of H11F opto-coupler.

FIG. 12-101 Circuit with optocoupler used as variable resistor.

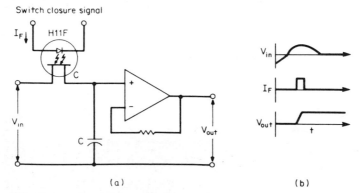

FIG. 12-102 Circuit with optocoupler used as analog switch (*a*) Schematic; (*b*) waveforms.

across the optocoupler's two output terminals is under 500 Ω and is highly linear, within 99% or better. When used as analog switches, these optocouplers have extremely low offset voltage, capability for 60-V peak-to-peak signals, and switching speeds under 15 μs for turn-on and turn-off.

Fig. 12-101 shows an H11F optocoupler connected to act as a variable resistor at the input side of an operational amplifier. This circuit provides 70 dB of stable gain control when an electrically isolated automatic gain-control signal drives the input LED with I_F ranging from 0 to 30 mA.

Fig. 12-102 shows an H11F optocoupler used as an analog switch in a sample-and-hold circuit. The optocoupler output "closes" (i.e., goes to a low on-resistance state) when I_F from an isolated control drive circuit flows through the LED. Accuracy and range in this circuit are improved over those of circuits using conventional FET switches because the H11F has no charge injection from the control signal. For this optocoupler circuit the V_{in} can be of either polarity with a magnitude of up to 30 V.

Logic-to-Power Interfaces   Optocouplers with SCR detectors find wide application wherever power circuits with high AC voltage loads are to be switched from logic-driven control circuits. In Fig. 12-103 an MCS2 optocoupler allows half-wave conduction through a resistive load as long as the logic gate supplies I_F to the input LED. Resistor R_{GK} improves noise immunity and thermal stability at high temperatures.

Fig. 12-104 shows a circuit that uses an MCS6200 optocoupler for bidirectional control of a small DC motor's rotation. Note that the MCS6200 detector has two SCRs, with a separate LED for each. The state of the circuit's input logic flip-flop determines

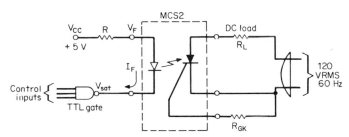

FIG. 12-103   Optocoupler circuit for half-wave conduction.

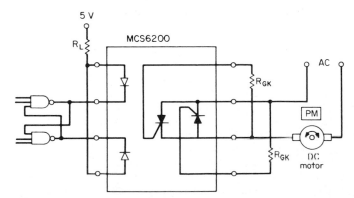

FIG. 12-104   Optocoupler circuit for bidirectional motor control.

which input LED will be supplied with I_F. Photons from the conducting LED will turn on the appropriate SCR, thereby providing a current flow path to the motor.

Fig. 12-105 shows a schematic for the MOC3030 and MOC3031 optocouplers. Their detector functions as a zero-voltage-crossing, bilateral TRIAC driver. A typical application is for a logic-to-power interface to drive an external high-power TRIAC that connects in-series with an AC load, as shown in Fig. 12-106. The 39 Ω resistor and 0.01-μF capacitor in the load portion of the circuit are added to reduce dV/dt to stop false triggering of the TRIAC; for some types of TRIAC devices and loads, these passive components may not be necessary.

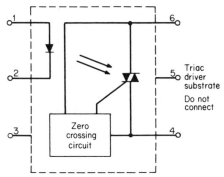

FIG. 12-105 Schematic for MOC3030 and MOC3031 optocouplers.

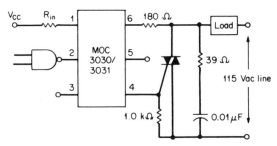

FIG. 12-106 Optocoupler circuit to switch external TRIAC.

Power-to-Logic Interface   Many control applications have requirements for logic circuitry to monitor the ON or OFF status of AC power lines and other AC sources. Fig. 12-107 shows both detailed and simplified schematics of the MID400 optocoupler that can perform this interface directly. At their input side the back-to-back LEDs can be connected through a series dropping resistor directly to the AC source. At the output side pin 8 connects to V_{CC} (up to 7 V), and an open collector TTL output is provided at pin 6. This logic output remains LOW as long as AC input current to the LEDs exceeds 4 mA, rms. The switching time of the detector amplifier is intentionally designed to be slow, so as to prevent changes at the MID400 output during the zero-crossing period occurring every half-cycle of the AC input waveform. The slow switching time allows the MID400 logic output to remain LOW if input current to the LEDs is absent over a few milliseconds.

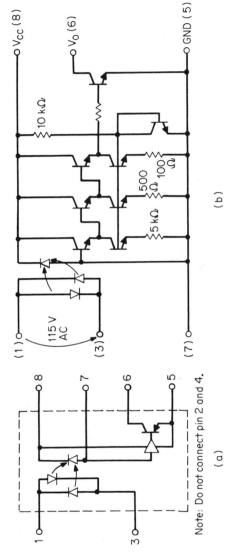

Note: Do not connect pin 2 and 4.

(a)

(b)

FIG. 12-107 MID400 schematics. (a) Simplified schematic; (b) detailed schematic.

12-75

If logic drive from the MID400 requires adjustable delays for either turn-on or turn-off, some additional components should be added, as shown in Fig. 12-108. Input and output waveforms achievable with this circuit are illustrated in Fig. 12-109. Addition of the 555 also increases the output drive capability of the circuit.

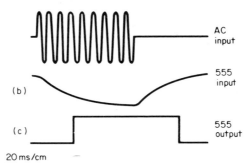

FIG. 12-108 The MID400 circuit with adjustable delay.

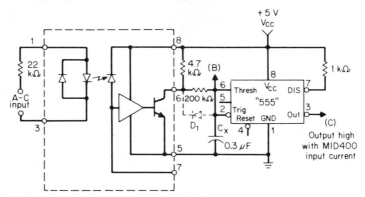

FIG. 12-109 Turn-on and turn-off delay.

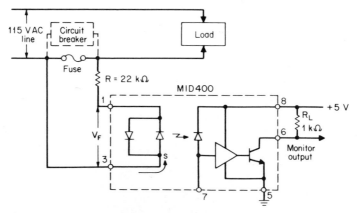

FIG. 12-110 Monitor circuit with MID400.

One simple but important class of application for the MID400 is monitoring of a fuse or circuit breaker on an AC power line, as shown in Fig. 12-110. Here the MID400 logic output stays off (HIGH) as long as the fuse remains intact or the circuit breaker is closed, but goes LOW if the fuse blows or circuit breaker trips. This, of course, only holds true provided that $+5$ V is maintained at pin 8; the monitor output will be erroneous if the $+5$-V supply to the MID400 fails.

12-5 DEVICE SELECTION CHARTS

The tables on the following pages present information concerning alphanumeric displays, Leds, miscellaneous indicators, and optocoupler devices.

TABLE 12–7 Alphanumeric Displays

Digit outline	Device no.	Color	Description	Brightness or luminous intensity (per seg. min)	Applications
	MAN1A	Red	0.270-in; common anode; LHDP; direct view	74 μcd @ 10 mA	Instruments, test equipment, office machines, computers
	MAN101A	Red	0.270-in; common anode; polarity/overflow; direct view	74 μcd @ 10 mA	
	MAN2A	Red	0.320-in; X-Y 35 diode, alphanumeric; direct view	125 μcd @ 10 mA	Business machines, calculators, computers, industrial control equipment
	MAN3610A MAN51A MAN71A MAN81A	Orange Green Red Yellow	0.3-in; common anode; RHDP	510 μcd @ 10 mA 125 μcd @ 10 mA 125 μcd @ 10 mA 320 μcd @ 10 mA	
	MAN3640A MAN54A MAN74A MAN84A	Orange Green Red Yellow	0.3-in; common cathode; RHDP	510 μcd @ 10 mA 125 μcd @ 10 mA 125 μcd @ 10 mA 320 μcd @ 10 mA	Instruments, test equipment, computers, office machines, computers, automobiles, clocks, radios, communication equipment, calculators, CB radios

Display	Part No.	Color	Description	Brightness	Applications
	MAN3630A MAN53A MAN73A MAN83A	Orange Green Red Yellow	0.3-in; common anode; polarity and overflow	510 μcd @ 10 mA 125 μcd @ 10 mA 125 μcd @ 10 mA 320 μcd @ 10 mA	
	MAN6610 MAN6640	Orange Orange	0.560-in; common anode; RHDP; 2-digit 0.560-in; common cathode; RHDP; 2-digit	510 μcd @ 10 mA	POS terminals, computers, instruments, test equipment, clocks, radios, TV channel indicators
	MAN6710 MAN6740	Red Red	0.560-in; common anode; RHDP; 2-digit 0.560-in; common cathode; RHDP; 2-digit	125 μcd @ 10 mA	
	MAN6630 MAN6650	Orange Orange	0.560-in; common anode; RHDP; 1½-digit 0.560-in; common cathode; RHDP; 1½-digit	510 μcd @ 10 mA	
	MAN6730 MAN6750	Red Red	0.560-in; common anode; RHDP; 1½-digit 0.560-in; common cathode; RHDP; 1½-digit	125 μcd @ 10 mA	
	MAN6660 MAN6680	Orange Orange	0.560-in; common anode; RHDP 0.560-in; common cathode; RHDP	510 μcd @ 10 mA	

TABLE 12-7 *(Continued)*

Digit outline	Device no.	Color	Description	Brightness or luminous intensity (per seg. min)	Applications
	MAN6760	Red	0.560-in; common anode; RHDP	125 μcd @ 10 mA	POS terminals, computers, instruments, test equipment, clocks, radios, TV channel indicators
	MAN6780	Red	0.560-in; common cathode; RHDP		
	MAN8610	Orange	0.800-in; common anode; RHDP	600 μcd @ 10 mA	
	MAN8630	Orange	0.800-in; common anode; RHDP; ±1 overflow		
	MAN8640	Orange	0.800-in; common cathode; RHDP		
	MAN8650	Orange	0.800-in; common cathode; RHDP; ±1 overflow		
	MAN2815	Red	0.135-in; common anode; 14-segment alphanumeric; 8 characters	60 μcd @ 2.5 mA (average current)	Compact computers, test equipment, desktop calculators, communications equipment, verification systems

TABLE 12-8 Leds

LED outline	Device no.	Viewed color/lens color or effect	Luminous intensity (typ.) @ forward current	Typical viewing angle	Maximum power	Maximum DC current	Forward voltage (typ.)	Applications
	MV5053	Red/flooded	1.6 mcd @ 20 mA	80°			1.70 V	Instruments, printed circuit board indicators, board-mounted panel display, different lens effect and viewing angles. MV5020 series offers leads with standoffs for assembly ease. General-purpose indicators
	MV5054-1 MV5054-2 MV5054-3	Red	2.0 mcd @ 10 mA 3.0 mcd @ 10 mA 4.0 mcd @ 10 mA	40°	180 mW	100 mA	1.80 V	
	MV5152 MV5153 MV5154	Orange	40.0 mcd @ 20 mA 6.0 mcd @ 20 mA 10.0 mcd @ 20 mA	28° 65° 24°			2.00 V	Computers, general-purpose indicators, instruments, test systems, mini- and microprocessors, process-controlled industrial systems, sorting machines, assembly equipment, vending machines, telephone equipment, backlight panels. High-intensity indicators in four colors
	MV5252 MV5253 MV5254	Green	15.0 mcd @ 20 mA 3.5 mcd @ 20 mA 3.0 mcd @ 20 mA	28° 65° 24°			2.20 V	
	MV5352 MV5353 MV5354	Yellow	45.0 mcd @ 20 mA 8.0 mcd @ 20 mA 10.0 mcd @ 20 mA	28° 65° 24°	105 mW	35 mA	2.10 V	
	MV5752 MV5753 MV5754	High-efficiency red	40.0 mcd @ 20 mA 9.0 mcd @ 20 mA 10.0 mcd @ 20 mA	28° 65° 24°			2.00 V	
	MV51642	Orange	3.5 mcd @ 10 mA	90°			2.00 V	Portable equipment, general-purpose indicators and matrix panel displays, test equipment and systems, sorting machines, vending machines. High-intensity indicators in four colors
	MV52642	Green	4.0 mcd @ 20 mA	90°			2.20 V	
	MV53642	Yellow	4.5 mcd @ 10 mA	90°			2.10 V	
	MV57642	High-efficiency red	3.5 mcd @ 10 mA	90°			2.00 V	

TABLE 12–9 Miscellaneous Indicators

Package outline	Device no.	Viewed color/ lens color or effect	Luminous intensity (typ.) @ forward current	Typical viewing angle	Maximum power	Maximum DC current	Forward voltage (typ.)	Applications
Rectangular*	MV52124 MV53124 MV57124	Green Yellow High-efficiency red	3.0 mcd @ 20 mA 4.0 mcd @ 20 mA 4.0 mcd @ 20 mA	100°	105 mW	35 mA	2.00 V	Legend backlight, panel indicator, bar graph, display button. Mounting hardware available
Bar graph	MV57164	High-efficiency red	1.0 mcd @ 10 mA (per segment)	Wide angle	750 mW	300 mA (30 mA per segment)	2.50 V	Analog measurement, audio instruments, meters, gauges
0.5″ rectangular*	MV57173	High-efficiency red	10 mcd @ 20 mA	Wide angle	200 mW	35 mA	2.00 V	Panel indicators, backlight legends, light arrays

*Pop-ins available.

TABLE 12–10 Optocoupler Devices

Package outline	Device no.	Output configuration	Emitter voltage (max)	Detector				Minimum DC surge-isolation voltage	Typical operating speed or bandwidth	Applications
				Minimum output voltage (BV_{ceo})	Type h_{FE}	Maximum $V_{CE(sat)}$	Minimum current transfer ratio			
Transistors										
	MCT2E	Transistor	1.5 V @ 20 mA	30 V	250	0.4 V @ 2 mA	20%	3550 V	150 kHz	AC line/digital logic isolator, logic isolator, line receiver, cable receiver, relay monitor, power-supply monitor, UL recognized
	MCT271 MCT272 MCT273 MCT274	Transistor	1.5 V @ 20 mA	30 V	420 500 280 360	0.4 V @ 2 mA	45–90% 75–150% 125–250% 225–400%	3550 V	7 μs 10 μs 20 μs 25 μs	Switching networks, power-supply regulators, digital logic inputs, microcircuit inputs, appliance sensor systems, appliance controls, UL recognized
	MCT275	Transistor	1.5 V @ 20 mA	80 V	170	0.4 V @ 2 mA	70–210%	3550 V	7 μs	Telecommunications, high-voltage industrial control, relay driver, telephone, UL recognized
	MCT276 MCT277	Transistor	1.5 V @ 20 mA	30 V	90 420	0.4 V @ 2 mA	15–60% 100%–up	3550 V 2500 V	3.5 μs 15 μs	Data processing, microprocessor input, high-speed digital logic, UL recognized
	4N25	Transistor	1.5 V @ 50 mA	30 V	250	0.5 V @ 2 mA	20%	2500 V	300 kHz	Low-cost products for logic isolator, telecommunications, line/cable receiver, high-frequency feedback and control system, monitoring circuits

TABLE 12–10 (*Continued*)

	Part	Type								Application
	4N35	Transistor	1.5 V @ 10 mA	30 V	100	0.3 V @ 5 mA	100%	3550 V	150 kHz	Low current, low-power products for industrial control and consumer, monitoring circuits, line receiver, UL recognized
	MCT6	Transistor pair	1.5 V @ 20 mA	30 V	—	0.4 V @ 2 mA	20%	1500 V	150 kHz	Data line isolation, telephone signal coupling, line/cable receiver, mobile equipment
	MCT4	Transistor	1.5 V @ 40 mA	30 V	—	0.5 V @ 2 mA	15%	1000 V	300 kHz	Logic isolation, line or cable receiver for high hermeticity
Darlingtons	MCA231 MCA255	Darlington transistor	1.5 V @ 20 mA	30 V 55 V	50,000 25,000	1.2 V @ 50 mA 1.0 V @ 50 mA	200% 100%	3550 V	10 kHz	High current, low capacitance, and fast switching products for reed relay, pulse transformer, multiple contact control applications. Telecommunication, remote control logic isolation and alarm monitoring circuits, AC line/logic coupling
	4N29 4N32	Darlington transistor	1.5 V @ 50 mA	30 V	5000	1.0 V @ 2 mA 1.0 V @ 2 mA	100% 500%	2500 V 2500 V	30 kHz	Low-capacitance medium-speed products for data isolation, logic conversion, line/cable receiver, monitoring circuits, or mechanical feedback controls.
	6N138 (MCC670)	Split-Darlington	1.7 V @ 1.6 mA	7 V	—	$0.4 \text{ V} @ I_f = 1.6 \text{ mA}$ $I_0 = 4.8 \text{ mA}$ $V_{cc} = 4.5 \text{ V}$	300%	3000 V	t_{PHL} @ 10 μs t_{PHL} @ 35 μs	CMOS logic interface, telephone ring detector, low input TTL interface, power-supply isolation, UL recognized
	6N139 (MCC671)			18 V	—	$0.4 \text{ V} @ I_f = 5 \text{ mA}$ $I_0 = 15 \text{ mA}$ $V_{cc} = 4.5 \text{ V}$	400%		t_{PHL} @ 1 μs t_{PHL} @ 7 μs	

SCRs

Package outline	Device no.	Output configuration	Emitter voltage (max.)	Detector					Minimum isolation voltage	Applications
				V_{GT} (max.)	On-voltage (max.)	Holding current (max.)	I_{FT} (max.)	Blocking voltage		
	MCS2400	SCR	1.5 V @ 20 mA	1 V	1.3 V @ 100 mA	0.5 mA	14 mA	400 V	3550 V	Lower power ICs to AC line isolation, relay functions, latches for DC circuits, home appliances, consumer and industrial control logic. UL recognized
	MCS6201	Bidirectional SCRs	1.5 V @ 20 mA	1 V	1.3 V @ 100 mA	2 mA	14 mA	200 V	2500 V	AC power control, TRIAC triggering, AC motor control, power-supply polarity control, appliance control, logic interface

Logic gates

Package outline	Device no.	Output configuration	Emitter Voltage (max.)	Detector					Minimum DC surge isolation voltage	Minimum operating frequency	Applications
				ΔI_f (typ.)	I_{OH_L} (max.)	V_{OL} (max.)	I_{CC} (typ.)				
	MCL601	Open-collector logic gate	1.5 V @ 20 mA	1 mA	200 µA	0.4 V @ 16mA	20 mA	2000 V	0.1 MHz	Digital logic isolator, DC voltage sensor, pulse shaping, level shifting, logical level conversion
	MCL611			5 mA					1.0 MHz	

12-85

TABLE 12–10 (*Continued*)

AC line monitor

Package outline	Device no.	Output configuration	Emitter voltage (max.)	Detector				Minimum DC surge isolation voltage	Switching times T_{on}, T_{off} (typ.)	Applications
				On-state rms input current (min.)	Off-state rms input current (max.)	V_{OL} (max.)	I_{OH} (max.)			
	MID400	Open-collector logic gate	1.5 V @ 30 mA	4.0 mA	0.15 mA	0.4 V	100 µA	3550 V	1.0 ms	Monitors AC "line-down" conditions, "closed loop" interface between electromechanical elements and microprocessors. Time-delay isolation switch

Sensors

Package outline	Device no.	Output configuration	Maximum emitter voltage	Detector				Typical bandwidth	Applications
				Minimum BV_{ceo}	Typical dark current	Maximum $V_{CE(sat)}$	Minimum current transfer ratio		
	MCT8	Slotted limit switch, transistor	1.5 V @ 20 mA	30 V	5 nA @ 10 V	0.4 V @ 50 µA	1%	150 kHz	Tape reader, mark sensor, end-of-tape detector, end-of-film detector, metal processing equipment, length measurement, coded disk detection, edge sensor, textile
	MCA8	Slotted limit switch, Darlington		30 V	5 nA @ 5 V	1.0 V @ 2 mA	15%	0.8 kHz	processing equipment, fluid volume and velocity control, level detector, object sensor, strobing light control, stroboscope
	MCA7	Reflective sensor, Darlington	1.5 V @ 20 mA	30 V	5 nA @ 5 V	—	0.1%	0.8 kHz	Object sensing, end-of-tape detection, length measurement, industrial processing equipment.

12-86

Emitters

Package outline	Device no.	Radiated power (typ.)	On-axis irradiance or intensity (typ.)	Maximum forward voltage	Maximum DC current	Maximum power	On/off delay (typ.)	Applications
	ME61	550 μW	250 mW/cm^2	1.5 V @ 50 mA	50 mA	75 mW	10 ns	Card readers, encoders, alarm and sector systems, level indicator, end-of-tape detection
	ME7024	1.0 mW	81.2 mW/st	1.5 V @ 20 mA	100 mA	150 mW	100 ns	
	ME7124	3.0 mW	243.6 mW/st	1.= V @ 50 mA	100 mA	150 mW	500 ns	
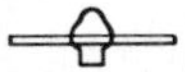	ME7161	3.0 mW	—	1.8 V @ 50 mA	50 mA	75 mW	500 ns	

Detectors

Package outline	Device no.	Sensitivity, μA/(mW·cm^2) (typ)	$V_{CE(sat)}$ (max.)	Maximum DC current	Minimum BV_{ceo}	Dark current (typ.)	Band-width (typ.)	Applications
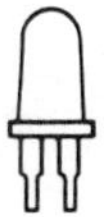	MT8020	350	0.2 V @ 1.6 mA	40 mA	30 V	1.5 nA	300 kHz	Optical switching, intrusion alarm, process control, tape and card reader, level controls, character recognition

REFERENCES

Bylander, E. G.: *Electronic Displays.* McGraw-Hill, New York, 1979.

Chappell, A., ed.: *Optoelectronics: Theory and Practice.* McGraw-Hill, New York, 1978.

Gage, S. et al.: *Optoelectronics Applications Manual.* McGraw-Hill, New York, 1977.

Kaufman, J. E., ed.: *IES Handbook,* Illumination Engineering Society, New York, 1972.

Meister, D. et al.: *Guide to Human Engineering Design for Visual Displays.* Produced for the U.S. Navy by the Bunker-Ramo Corporation, Canoga Park, Calif., August 1969.

Chapter 13

LSI PERIPHERAL DEVICES

Brian Cayton Marketing Manager
Standard Microsystems Corp.
Hauppauge, N.Y.

13–1 THE NEED FOR LSI PERIPHERALS

The microprocessor revolution signified a major change in the direction of system design. Systems, instead of being built with hard-wired logic consisting of hundreds or thousands of packages, could be built with the combination of a microprocessor and software.

The reasons for the success of this revolution were many, but they revolved around two key points. The number of components could be greatly reduced and changes could, in theory at least, be made easily via software changes.

It was further envisioned that the microprocessor itself, with the aid of some small amounts of external logic, would control the microprocessor's peripherals. Typically, these included:

Keyboards
CRTs
Displays
Disks
Printers
Serial I/O
Parallel I/O
Tape Drives

The challenge has been to have the microprocessor control all of these peripherals and still have time to run the programs that are its prime responsibility. A second consideration with respect to having a microprocessor control peripherals is that in production systems changes just aren't as easy as envisioned. A system may be as locked in by software as hardware.

Finally, many of the required control functions are beyond the capability of all but the high-speed bit-slice microprocessors. Use of these as controllers is, in general, economically unsound when compared to even SSI/MSI implementations. Many of the LSI controllers of today, such as the UART, are architecturally the implementation of standardized SSI/MSI circuits that originated concurrent with, and even prior to, the microprocessor.

As a result, the microprocessor revolution spawned not a revolution but an evolution: the peripheral evolution. This evolution represents the microprocessor's role changing from that of performing the control tasks to supervising them.

The peripheral controllers themselves may be programmable and may be faster and more complex than the microprocessor.

13–1a   Glossary of Terms

The terms below pertain to this chapter and are defined as follows.

Analog Data   An electrical representation of information in which the data has an exact proportional relationship to the actual information.

ASCII   This is an acronym for American Standard Code for Information Interchange. ASCII is a digital code adopted as a standard to facilitate the interchange of data among various types of data processing and data communications equipment.

Asynchronous Data Communications   The most popular form of data exchange from computers to and from remote terminals. Asynchronous data transmission is a method of transferring data where the timing of character placement on connecting communication lines is not critical. Each transferred character is preceded by a start bit and followed by a stop bit, permitting the interval between characters to vary.

Attributes   As applied to video displays, attributes are those special variations that enable words or characters to be emphasized or set apart. Some popular attributes include blink, underline, reduced intensity, and reversed video.

Baud   The unit of signaling speed in data communications. It is equal to the number of signal events per second and is used as a measure of serial data flow. In the binary case, the baud rate is equal to the number of bits per second.

Baud Rate Generator   An LSI integrated circuit used to generate the receiver and transmitter clocks for data communications devices.

BISYNC Communications   "Binary synchronous communications" is one form of synchronous data communications in which a variety of special control characters are used to synchronize the receiver to the transmitter.

Bit   Short for binary digit, the bit is the basic data element of digital computers and digital communications. A bit may have a value of zero or one.

Bus   A path over which digital information is transferred from any of several sources to any of several destinations. The sources and destinations may be inside or outside of the computer.

Character Generator   Used with most types of displays and matrix printers, a character generator (ROM) converts binary data representing a character address into the physical shape of the character.

CRC   CRC is cyclical redundancy check, a method for testing serial data transmitted over data communication links for errors. It consists of processing the data mathematically by dividing it by a specified polynomial. CRC-16, CRC-CCITT are two of the polynomials.

CRT   Short for cathode-ray tube. CRTs are the picture tubes used in televisions and computer terminals.

Data Communications   Data communications is a very broad term referring to the communication of digital data from one point to another.

Digital Data   An electrical representation of information in which the information is encoded as strings of binary elements or bits. All modern computers utilize digital data. In addition, telephone systems are increasingly making use of the conversion of analog voice information into digital data for improved communication efficiency.

Floppy Disk   A type of computer memory in which a flexible plastic disk is covered with a magnetic coating on which data is stored.

Graphics   A term used with CRT displays to describe the presentation of physical shapes, such as maps or graphs, on a computer terminal.

Keyboard Encoder   An integrated circuit that converts mechanical switch closures to digital codes.

Kilobit   A thousand bits.

LSI   Large-scale integration. LSI is generally used to define integrated circuits that contain over 100 equivalent logic elements within one chip.

Microprocessor   A microprocessor is a single chip that has all of the control elements of a computer. The addition of memory elements for program storage converts a microprocessor into a microcomputer, literally a computer on a chip. Hand-held calculators are built around simple versions of such microcomputers.

MOS   Stands for metal-oxide semiconductor, a semiconductor integrated-circuit technology that utilizes a thin layer of oxide as an insulator between the metal controlling element, the "gate," and the semiconductor body of the transistor "switch." MOS technology allows the relatively inexpensive processing of highly complicated transistor arrays, allowing such functions as calculator chips, microprocessors, RAMs, ROMs, and UARTs to be economically implemented.

MSI   Medium-scale integration. MSI is generally used to define integrated circuits that contain 10 to 100 equivalent logic elements within one chip.

Multiplexing   Multiplexing is a technique whereby a communications link is divided into two or more channels.

N-MOS   N channel MOS. Current flow in N-channel MOS transistors is caused by the movement of electrons, which are negative charges. N-MOS technology allows fabrication of high-density, high-performance, low-cost integrated circuits.

Peripheral   A general term designating various kinds of electronic equipment that operate in combination or conjunction with a computer, but are not physically part of the computer. Peripheral devices typically display computer data, store data from the computer and return the data to the computer on demand, prepare data for human use, or acquire data from a source and convert it to a form usable by a computer.

P-MOS   P-channel MOS. Current flow in P-channel MOS transistors is caused by the movement of holes, which are positive charges. P-MOS is the oldest form of MOS technology and is rarely used for new designs.

PROM   Programmable read-only memory. PROMs are programmable nonvolatile memories. Once programmed, they remember data without power. PROMs may be programmed by a customer as opposed to ROMs, which must be programmed during manufacture.

RAM   Random access memory. RAMs are typically used for "scratchpad" or temporary data storage in computer or microcomputer systems. Electrical data can be read into and then read out of a random access memory. RAMs will generally lose their data if power is removed.

ROM   Read-only memory. Typically used for storage of programs or fixed data, ROMs, unlike RAMs, will retain data without power. ROMs are programmed during manufacture and, unlike EPROMs, cannot be programmed by the user.

SDLC   SDLC is a communication line discipline associated with the IBM Systems Network Architecture (SNA) that offers a number of advantages to users of data networks. SDLC is IBM's newest line discipline, and initiates, controls, checks, and terminates information exchanges or communication lines. It is designed for full duplex operation, simultaneously sending and receiving data.

Shift Register   A serial form of computer memory in which data is "shifted" through the chip (or device) from input to output.

Silicon Gate   A form of MOS technology in which the gate, or control element, of the transistor is formed by highly doped polycrystalline silicon. Silicon-gate technology offers the user the advantage that the gate electrodes are generally self-aligned to the source and drain electrodes, thereby reducing parasitic capacitances and improving circuit speed.

SSI   Small-scale integration. SSI is generally used to define integrated circuits that contain less than 10 equivalent logic elements within one chip.

Synchronous Data Communications   A method of transferring serial binary data between computer systems or between a computer system and a peripheral device; binary data is transmitted at a fixed rate, with the transmitter and receiver synchronized. Synchronized characters are located at the beginning of each message or block of data to synchronize the flow.

UART   Universal asynchronous receiver/transmitter. An LSI integrated circuit used for asynchronous data communications.

USART   Universal synchronous-asynchronous receiver/transmitter. An LSI integrated circuit used for synchronous or asynchronous data communications.

USRT or USYNRT   Universal synchronous receiver/transmitter. An LSI integrated circuit used for synchronous data communications.

VDAC   CRT Video Display Attributes Controller. An LSI integrated circuit (produced by Standard Microsystems) that provides the character generator, graphics circuitry, and attributes circuitry for a CRT display all on one chip.

VTAC CRT Video Timer and Controller. An LSI integrated circuit (produced by Standard Microsystems) that provides, on one chip, all of the timing and control signals for a CRT display.

VLSI Very large-scale integration. A relatively new term used to describe the next higher level of integrated-circuit density after LSI. VLSI is generally used to define integrated circuits that contain 1000 equivalent logic elements or more on one chip.

13-2 DATA COMMUNICATION CIRCUITS

Data communications is a very broad term defining the transmission (usually electronic) of data from one point to another. Covered within the full sense of this term are all of the elements involved with the transmission of data. These elements include:

Channels
Modems
Serial interfaces
Communication processors
Data-link configuration
Information codes
Protocols

To ensure that two or more stations may coherently pass data between one another, a set of rules must be established and followed by both. These rules, or protocols, define the electrical, physical, and functional characteristics of the communication link.

To organize the various sets of protocols established by the various standards committees, a hierarchy of levels have been established.

Level 1 contains the physical, electrical, and functional interchange to establish the physical link between the terminal equipment. Examples of level-1 standards are RS-232C and RS449.

Level 2 contains the data-link controls, such as BISYNC (IBM), SDLC (IBM), DDCMP (DEC), ADCCP (ANSI). Key features of these data-link control protocols are shown in Table 13-1. The key point to note is that these controls pertain to a single data link.

TABLE 13-1 Synchronous Protocol Features

Feature	Protocol				
	DDCMP	BISYNC	SDLC	ADCCP	HDLC
Synchronous/asynchronous	Yes	No	No	No	No
Full duplex	Yes	No	Yes	Yes	Yes
Half duplex	Yes	Yes	Yes	Yes	Yes
Data transparency	Count	Character stuffing	Bit stuffing	Bit stuffing	Bit stuffing
Error detection	CRC-16	CRC-16	CRC-CCITT	CRC-CCITT	CRC-CCITT
CRC calculation	Text and header	Text only	Text and header	Text and header	Text and header
Point to point	Yes	Yes	Yes	Yes	Yes
Multipoint	Yes	Yes	Yes	Yes	Yes

Level 3 establishes the path controls, or the format and control procedures for connections in a network. These controls include routing and traffic control. An example of level 3 is the packet switching protocol, X.25 (CCITT).

Level 4 is the system control. Examples are DECNET (DEC) and SNA (IBM's system network architecture).

Level 5 and up are not as clearly defined. ANSI established in one committee a five-level hierarchy in which level 5 is user control. In another ANSI committee, a six-level hierarchy has been established. The International Standardization Organization (ISO) has established a seven level protocol.

It is the purpose of this section just to cover those functions directly pertaining to LSI peripherals, in essence covered by levels 1 and 2.

The most popular format for data transmission is asynchronous. This format specifies that each word (usually between five and eight data bits) be preceded by a "start" bit or a "space," defined as a logic "0" and followed by one or more stop bits or "marks" defined as logic "1"s. Between words, the line marks, that is, stop bits are transmitted continuously. Because each individual character is framed by these start and stop bits, the receiver is resynchronized each character time, allowing unequal intervals between characters. The character format is illustrated in Fig. 13-1.

FIG. 13-1 A synchronous data format.

The data is transmitted with the least significant bit (LSB) first, and the most significant bit (MSB) last. The last data bit may be followed by a parity bit, which is optional, and may be odd or even.

The length of the data word, the parity, and the minimum number of stop bits allowed between words (1, 1½, or 2) may vary from system to system. Also, the actual bit transmission rate, or "baud" rate, may vary. Therefore, a communication device to interface a microprocessor to an asynchronous data link must be programmable.

13–2a The UART

One of the first, and to date the most popular, asynchronous receiver/transmitter device is the universal asynchronous receiver/transmitter (UART). This device combines a separate receiver and transmitter each with its own independent data bus, serial port, and clock. The control logic, which sets the number of data bits and whether parity is odd or even (or present), is shared by the receiver and transmitter. This architecture allows full duplex operation at different data rates. The logic controlling the number of stop bits affects only the transmitter. This is because the receiver only looks for one stop bit. It is the transmitter that needs to know the minimum number of stop bits to insert.

The block diagram for the transmitter is shown in Fig. 13-2a. Note that it is designed for "handshaking" operation with the host microprocessor. The two major building blocks are a transmitter holding or buffer register and the parallel to serial shift register.

A clock, usually with a rate of 16 times the desired baud rate, is applied. Initially, since the transmitter holding register is empty, the transmitter buffer empty output is high, signifying to the microprocessor that it is ready to receive data. Data is transmitted when the data strobe input is pulsed low. The leading edge of the data strobe loads the

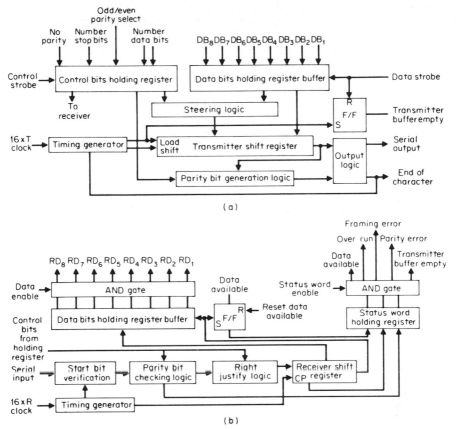

FIG. 13-2   UART block diagram. (*a*) UART transmitter block diagram. (*b*) UART receiver block diagram.

data from the bus into the holding register. The trailing edge causes the parallel transfer of data into the shift register, and the actual serial shifting out of data.

Since the holding register is free as soon as the parallel transfer occurs, even if the actual serial shifting of data is not complete, the transmitter buffer empty output goes high, indicating that a new word may be loaded. This time, if the serial shift register is not free, the data is held until the serial shift is complete; the transfer of the new word into the shift register is then allowed to take place.

The UART may be used in either an interrupt or polling configuration with a microprocessor. In the former, the transmitter buffer empty output sends an interrupt to the microprocessor, which services it by loading in new data and providing a data strobe. If the microprocessor does not provide a new data strobe, the UART keeps marking the line (transmitting stop bits).

Like the transmitter, the receiver also has a shift register and holding register as its main blocks. Data is shifted serially into the serial-to-parallel shift register. When an entire word is assembled, it is transferred to the holding register, freeing the shift register

TABLE 13–2 UARTS

		+5 V only?	1½ SB?	Maximum baud rate (kbaud)	Pull-ups?	Enhanced dist marg?	Int Osc?	Process
SMC	COM2017		X	25	X			P
	COM2017H		X	40	X			P
	COM2502			25	X			P
	COM2502H			40	X			P
	COM8017	X	X	40	X			N
	COM8018	X	X	40	X	X		N
	COM8502	X	X	40	X			N
	COM1893	X	X	40		X		N
AMI	S1883		X	12	X			P
G.I.	AY-5-1013			40	X			P
	AY 3-1014	X	X	30				N
	AY 3-1015	X	X	30	X	X		N
Harris	HM 6402	X	X	125				C
	HM 6403	X	X	125			X	C
Intersil	IM 6402	X	X	125				C
	IM 6403	X	X	125			X	C
SSS	SCP 1854	X	X	100				C
T.I.	TMS 6011			12	X			P
W.D.	TR 1602		X	40	X			P
	TR 1863	X	X	62		X		N

to receive the next word. When the word is available on the data bus, the data available output is raised. Three error-status bits are made available via the status register.

The first, parity error, indicates that a parity bit of the wrong polarity has been received. A framing error indicates that a 0 stop bit has been received. Finally, an overrun error indicates that the data available has not been reset. Since the reset data available input is normally strobed when data is read, an overrun will normally indicate that a new word has been received, overwriting the previous word held in the holding register.

The ease of asynchronous communications derives from the fact that both the transmitter and receiver may have separate clocks. To overcome clock frequency drift, jitter, and to provide noise immunity, a clock of a fixed multiple of the data rate, usually 16, is employed. When the receiver detects a start bit via a 1-to-0 transition at the serial input, it starts counting clock pulses. After eight pulses (nominally the center of the bit), the serial input is sampled. If a 0 is detected, it is assumed that a valid start bit has been detected. If a 1 is detected, it is assumed that the first transition was noise, and the receiver is reset. Once a valid start bit is detected, the data is sampled every 16 clock pulses.

Some of the newer UARTs utilize clocks of 32 or 64 times the incoming data rate to increase the accuracy of the center sample, thus increasing data distortion margins.

Table 13-2 provides a UART selection chart.

13–2b ACIAs

Less general purpose than the UART are the bus-oriented communication devices developed by some microprocessor manufacturers. These devices may simplify the timing and hardware requirements imposed by the microprocessor by having their data bus described from the outset to match a specific microprocessor.

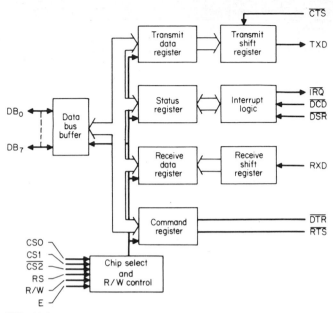

FIG. 13-3   Block diagram of the 6850 ACIA.

An example of such a device is the asynchronous communication interface adapter (ACIA). A typical device is illustrated in the block diagram of Fig. 13-3. The ACIA (6850) is designed for direct connection to the 6800 microprocessor bus. The transmitter and receiver data busses are multiplexed on chip for direct connection to the 6800 data bus. Control and status are likewise multiplexed onto the data bus.

The ACIA looks like two bidirectional memory locations to the microprocessor. One is the transmitter/receiver, the second, control/status. Register selection is via one address bit to register select (RS) and the read/write (R/$\overline{\text{W}}$) control signal.

R/$\overline{\text{W}}$	RS	Register selection
0	0	Control register
1	0	Status register
0	1	Receive data register
1	1	Transmitter data register

Additionally, three chip selects are provided for ACIA selection.

The ACIA is designed for interrupt-driven operation with $\overline{\text{IRQ}}$ output (interrupt request signaling) anytime the ACIA requires servicing.

In addition to the functions of the UART, the ACIA includes the RS232 modem control signals of Request To Send (RTS), Data Carrier Detect (DCD), and Clear To Send (CTS).

TABLE 13–3 Baud-Rate Generators

	SMC COM5016	SMC COM8116	SMC COM8136	SMC COM8146	SMC COM8046	National MM 5307	Fairchild F 4702	Harris H 6405
On-chip oscillator	Yes	Yes	Yes	Yes	Yes	Yes	Yes	Yes
Frequency control latch	Yes	Yes	Yes	Yes	Yes	No	No	No
f_x output	No	No	No	No	Yes	Yes	Yes	Yes
$f_x/4$ output	No	No	Yes	Yes	Yes	No	Yes	Yes
$f_o/16$ output	No	No	No	No	Yes	No	No	No
Process	PMOS	NMOS	PMOS	NMOS	NMOS	PMOS	CMOS	CMOS
Power	+5, +12	+5	+5	+5	+5	+5 − 12	+5	+5
Package	18	18	18	14	16	14	16	16
Full duplex	Yes	Yes	Yes	No	No	No	No	No

13–2c   Baud-Rate Generators

The devices described so far have a major requirement in common: they need a clock source. Although the clock can be generated by a crystal oscillator and divider logic, the different clock rates, or baud rates, make the design of these divider chains difficult. Several LSI clock generators called "baud-rate generators" greatly simplify this requirement by providing, from one crystal source, the various clock frequencies desired (see Table 13-3).

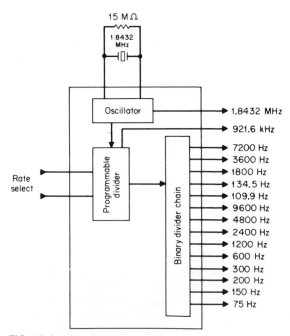

FIG. 13-4   The MC14411 baud-rate generator.

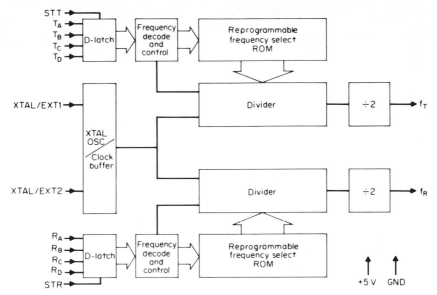

FIG. 13-5 The COM8116 dual baud-rate generator.

The 14411-type baud-rate generator simultaneously provides all the standard baud rates. External decoding logic is then used to select the specific baud rate desired. A block diagram of this device is shown in Fig. 13-4.

Programmable baud-rate generators just provide one output, but this output is selected via a 4- or 5-bit address. This eliminates the external logic and allows direct microprocessor selection of the baud rate. Programmable baud-rate generators are available with two independently selectable clocks, allowing full duplex operation at different transmit and receive frequencies. Two examples of programmable baud-rate generators are the COM8126 single baud-rate generator and the COM8116 dual baud-rate generator by Standard Microsystems. A block diagram of the COM8116 is shown in Fig. 13-5.

Example 13–1   Design of Full Duplex Asynchronous Microprocessor Interface

A full duplex asynchronous microprocessor interface is required. Baud rates as well as data formats must be selectable by the microprocessor.

Solution
A COM8017 UART is used with a COM8116 baud-rate generator. This part selection allows the use of a single $+5$-V supply and selection of 16 different receive/transmit frequencies. As shown in Fig. 13-6, the receiver and transmitter data busses are connected together, along with the inputs and outputs of the status register. Register selection and the various operating signals are provided by address selection. The receiver and transmitter baud rates are programmed by the data bus. If the baud rates are fixed, or expected to be seldom changed, the inputs to the baud-rate generator may be connected to mechanical switches rather than to the data bus.

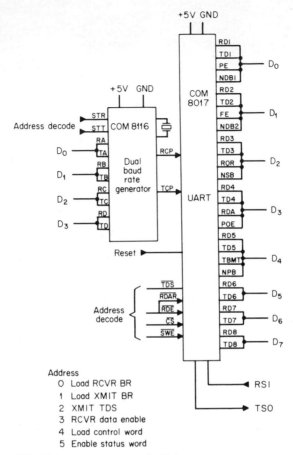

FIG. 13-6 Circuit of Example 13-1.

13-2d Synchronous Data Interfacing

Although simple to implement, asynchronous protocols are inefficient due to overhead imposed by start and stop bits for every word. For an ASCII word of 7 data bits, three extra bits are required, start, stop, and parity. Although this is not a major handicap at low data rates, high-speed communication over typically overburdened data links require higher efficiencies. These are provided by synchronous protocols.

With synchronous protocols, the receiver derives its clock from the incoming data. This clock is at the same frequency as the incoming data.

Synchronous data communication protocols can be broken down into two main categories, character controlled and bit oriented.

In character-controlled protocols, specific character patterns have meaning with regard to framing message blocks, identifying addresses, establishing communication links, terminating links, etc. In bit-oriented protocols a limited number of bit sequences have data-link control meaning. It is the position in a frame that determines the function of a word.

The most popular character-controlled protocol (CCP) is IBM's BISYNC protocol. In BISYNC, the first portion, or "field" is defined as the "header." The header is introduced by a bit sequence called start of header (SOH) and contains the address of the receiver, the message type (control or data), acknowledgment of the previous message, and so on. The header is terminated by the start of text (STX) character. The text may be of variable length or omitted if the message is a control message. End of text (ETX) terminates the text and introduces the block check character (BCC).

A major deficiency of a CCP is that is it not naturally transparent. If a block of numeric data contains a bit sequence the same as, for example, ETX, the receiver will begin looking for a check character. Specific control procedures must therefore be established to provide message transparency.

Bit-oriented protocols are much more straightforward. A flag character consisting of a zero, 6-ones, and another zero always delimit the beginning and end of a message. The first character field after the flag is an address, the next, a control field, then an information field, and finally a frame check sequence. The last character transmitted is, once again, a flag.

In IBM's SDLC, the control and address are each 8 bits long.

Actual information, as transmitted, will never contain the flag sequence of 6-ones, due to the control feature called "bit stuffing." Whenever five successive ones are encountered in the data by the transmitter protocol controller, a zero is inserted. The receiver always "strips" a zero encountered after 5-ones to maintain message integrity.

USRTs   The earliest synchronous receiver/transmitter ICs were basically UARTs modified for CCP synchronous operation. Examples are the SMC COM2601 and AMI S2350. The sync word is stored in a specific register. Receipt of a sync bit sequence initiates word reception in the receiver. The transmitter sends data from either the data buffer or the sync register, depending on command. These parts are not communication controllers, but rather synchronous interfaces.

USARTs   More recent than the UART or USRT, the universal synchronous/asynchronous receiver transmitter (USART) combines a bus-oriented UART and USRT on one chip. In addition, modem controls are typically added.

The 8251 (Intel, SMC, NEC, AMD) USART is shown in Fig. 13-7. It may be broken into seven blocks: receiver control, receiver shift register, transmitter control, transmitter buffer, data-bus buffer, read/write control, and modem control. By combining the transmitter and receiver parallel data ports and the format register programming all on the one 8-bit data bus, the 40 pins required by both the UART and USRT are reduced to 28 pins.

Data control and status are communicated to the microprocessor via the data bus and stored internally in five registers: transmit data, receive data, mode, control, and status. An unusual architectural feature of the 8251A is that, other than a chip select, the only address bus connection is the control/data pin. It is the sequence of operation that tells the 8251A whether a control word is actually a mode select, command sync, or status.

In the asynchronous mode, operation is much the same as in the UART, except that the handshaking is via the status register rather than by dedicated pins. The first control byte received after a reset is the mode control, which specifies whether operation is

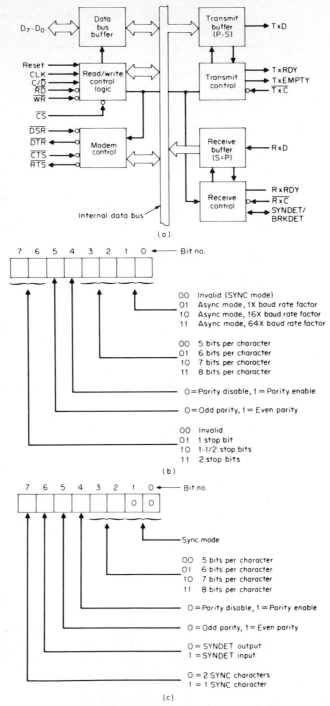

FIG. 13-7 The 8251A USART. (*a*) Block diagram; (*b*) async mode-control word; (*c*) sync mode-control word.

synchronous or asynchronous, and if synchronous, the number of bits per word, parity, and the number of stop bits.

The synchronous mode, like the asynchronous, is determined by the mode-control byte. Here the byte determines the sync detect output polarity and the number of sync characters required to establish synchronization instead of the no longer needed number of stop bits.

Synchronous operation is somewhat more complex than asynchronous. In transmission, the serial output remains "high" (marking) until the USART receives the first character (usually a sync character) from the microprocessor. After a command instruction has enabled the transmitter and clear to send (CTS) goes low, the first character is serially transmitted at the clock frequency.

Once transmission has started, synchronous data protocols require that the serial data stream continue at the clock rate or sync will be lost. If a data character is not provided by the microprocessor before the USART transmit buffer becomes empty, the sync character(s) loaded directly following the mode instruction will be automatically inserted in the data stream. The sync character(s) are inserted to fill the line and maintain synchronization until the new data characters are available for transmission. If the USART becomes empty and must send the sync character(s), the transmitter buffer empty output is raised to signal the processor that the buffer is empty and sync characters are being transmitted.

In synchronous reception, character synchronization can be either external or internal. If the internal sync mode has been selected and the enter hunt (EH) bit has been set by a command instruction, the receiver goes into the hunt mode.

Incoming data on the serial input is sampled and the contents of the receive buffer is compared with the first sync character after each bit has been loaded until a match is found. If two sync characters have been programmed, the next receive character is also compared. When the (two contiguous) sync character(s) programmed have been detected, the USART leaves the hunt mode and is in character synchronization.

If external sync has been specified in the mode instruction, a "one" applied to the sync detect (input) for at least one clock cycle will synchronize the USART.

Parity and overrun errors are treated the same way in the synchronous as in the asynchronous mode. If not in hunt, parity will continue to be checked, even if the receiver is not enabled. Framing errors do not apply in the synchronous format.

The processor may command the receiver to enter the hunt mode with a command instruction that sets EH if synchronization is lost. Under this condition, the received-data register will be cleared to all "ones."

Similar in many regards to the 8251A, the 2651 PCI (programmable communications interface) and the 2661 EPCI (enhanced programmable communications interface) combine a USART with a baud-rate generator. Transparent BISYNC operation is added, governed by the ASCII character DLE. If the control characters DLE and STX (start of text) are received in sequence, all subsequent bit patterns, including control characters, are treated as data until the next DLE is received. The block diagram of the 2651 is shown in Fig. 13-8.

The data-bus, transmitter, receiver, and control registers are very similar to those of the 8251A. The modem control register is also very similar, but adds the modem control data carrier detect (DCD). The most important addition is the on-chip baud-rate generator.

The 2651 and 2661 are essentially pin compatible. The 2661 has some software improvements and less critical data-bus timing.

A device selection chart for USARTs is given in Table 13-4.

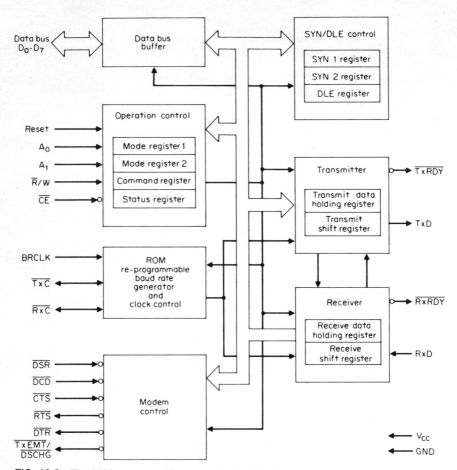

FIG. 13-8 The 2651 programmable communications interface.

TABLE 13–4 USARTs

	SMC COM1671	W.D. UC 1671	SMC, Intel 8251A	SMC, SIG 2651	SMC, SIG 2661	Zilog S10
Number of pins	40	40	28	28	28	40
Power supplies	3	3	1	1	1	1
Bus oriented?	No	No	No	Yes	Yes	No
Baud-rate generator?	No	No	No	Yes	Yes	Z80 only
Maximum data rate	1 Mbaud	1 Mbaud	64 kbaud	800 kbaud	800 kbaud	880 kbaud
Self-test modes	Local	Local	None	Local/ Remote	Local/ Remote	None
Modem controls	6	6	4	5	5	4
Transparent mode?	Yes	Yes	No	Yes	Yes	No
Auto echo?	Yes	Yes	No	Yes	Yes	No
CRC generation?	No	No	No	No	No	Yes
Full duplex channels	1	1	1	1	1	2
External JAM SYNC	No	No	No	No	Yes	No
Programmable address?	Yes	Yes	No	No	No	No

13-16

Example 13-2   Using the 8251 USART with the 8085 Microprocessor

Design an asynchronous data port for an 8085-based terminal system. The receiver and transmitter will operate at the same baud rate, which will be in the range of 110 to 19,200 baud. The serial port will operate with the RS-232C protocol.

Solution

An 8251 is selected for the R/T function. It directly interfaces with the 8085 data bus and allows the later addition of synchronous operation without hardware changes. A COM8046 baud-rate generator is picked to supply the baud rates, which will be programmable via the microprocessor.

The reference clock output of the COM8046 baud-rate generator also provides a convenient source for the timer clock of the 8155 RAM I/O chip, as well as the master clock of the 8251A. The circuit is shown in Fig. 13-9.

The 8155 provides the system RAM and two I/O ports. One port is used to input keyboard data, the second port is utilized to latch the baud-rate selection for the COM8046 baud-rate generator.

The COM8251A receives the baud-rate clock from the programmable f_o output of the COM8046 and receives its master clock from the reference output f_x.

The modem control signals and the serial I/O go to the RS-232 connector through a 1488 RS-232 driver and a 1489 RS-232 receiver.

The control/data selection is made by address A_0. Chip select is derived by the 74LS5138 3-to-8 decoder by address bits A12, A13, A14, and A15.

In this system address 9000H (hex) is decoded as the I/O register for data transmission and reception, and address 9001H is a mode/command instruction.

13-2e   Multiprotocol Data Communication Devices

The data communication devices described so far have one property in common: they are primarily interface devices. The multiprotocol synchronous receiver/transmitters tend more toward the category of data-link controllers, and present a much higher degree of complexity.

Like the USARTs and USRTs, the primary function of the multiprotocol receiver/transmitter is serial-parallel and parallel-serial conversion, relieving the microprocessor of the demands of high-speed interfacing.

In addition to this basic task, these devices add the bit stuffing and insertion required by SDLC/HDLC, flag detection and transmission, CRC generation and checking, and idle generation and recognition. The Zilog "SIO" is currently the only device to include asynchronous operation.

Table 13-5 lists the key characteristics of the popular multiprotocol receiver/transmitters. The COM5025 and the 2652 are pin compatible, except for pin 1. On the COM5025, pin 1 is a +12-V supply; on the 2652, it is a chip select.

These devices are on the order of complexity of a microprocessor, and to design using them is an extensive hardware/software exercise. Rather than spend the better portion of this chapter with a detailed explanation of these parts, it is recommended that the designer refer to the manufacturers' data sheets.

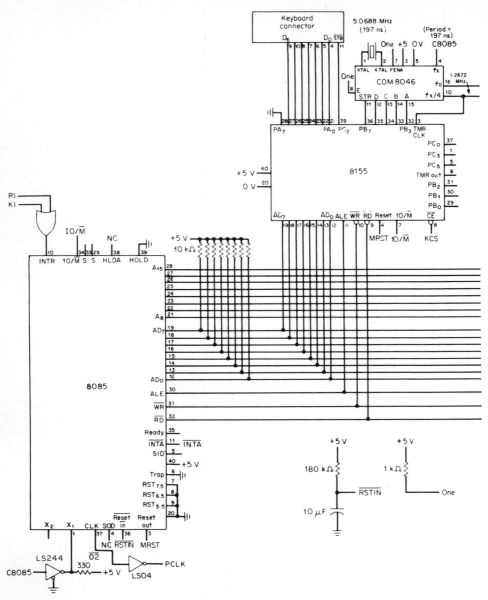

FIG. 13-9 The 8251A USART design application of Example 13-2.

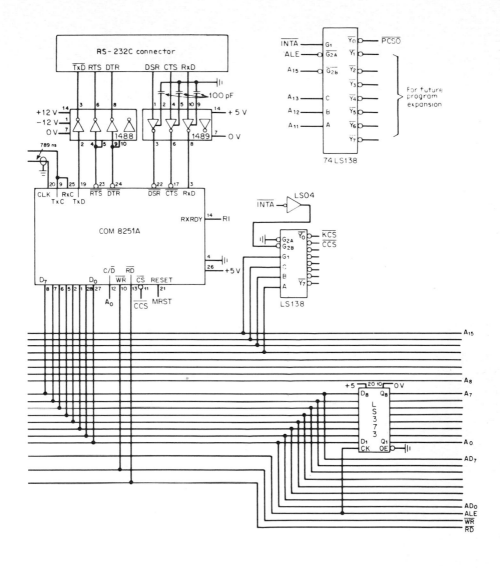

FIG. 13-9 (*Continued*)

TABLE 13-5 Multiprotocol Receiver/Transmitters

	SMC COM5025	Signetics 2652	Zilog S10	Fairchild 3846	Motorola 6854	Intel 8273	Western Digital 1933
Maximum data rate	1.5 Mbaud	1–2 Mbaud	0.55–0.88 Mbaud	1 Mbaud	0.6–1 Mbaud	64 kbaud	1 Mbaud
Package pins	40	40	40	40	28	40	40
8- or 16-bit bus?	Yes	Yes	No	Yes	No	No	No
Character length variable	Yes	Yes	Yes	Yes	Yes	No	Yes
System clock required?	No	No	Yes	No	Yes	No	No
Separate REC/XMIT interrupts?	Yes	Yes	No	Yes	No	Yes	No
Receiver FIFO	0	0	2	0	2	0	0
XMIT FIFO	0	0	0	0	2	0	0
Loopback mode	Yes	Yes	No	Yes	No	Yes	Yes
Multiprotocol	Yes	Yes	Yes	Yes	No	No	No
BISYNC CRC	*	*	Yes	Yes	No	No	No
Secondary address recognition	Yes	Yes	Yes	Yes	No	Yes	Yes
Global address recognition	Yes	Yes	Yes	Yes	No	Yes	Yes
Extended address/control	Yes	No	No	Yes	Yes	No	Yes
Residual character handling	Yes	Yes	Yes	Yes	No	Yes	Yes
NRZI coding	No	No	No	No	Yes	Yes	Yes
Digital phase lockup	No	No	No	No	No	Yes	Yes
Short frame rejection?	Yes	Yes	No	No	Yes	Yes	No
Second source	Yes	Yes	No	No	No	No	No
+12 V required?	Yes	No	No	No	No	No	No

*Calculates CRC, but not in strict accordance to the BISYNC protocol.

13–3 CRT CONTROLLERS

13–3a The CRT System

Most of today's video display systems utilize an intensity-modulated raster scan, as shown in Fig. 13-10. In this type of display, the CRT's electron beam is scanned across and down the screen of the CRT in a zigzag fashion to form a raster. Since the horizontal scan frequency is typically 15.720 kHz or more, as compared with 50 or 60 Hz for the vertical, the screen is "painted" with a series of horizontal lines. The number of lines for the system described is the horizontal frequency divided by the vertical frequency of 60 Hz, or 262 lines.

Data is displayed in a dot matrix, as illustrated in Fig. 13-11, in which the beam's intensity is increased to form a visible dot, or decreased for an "undot." The dot matrix is typically 5 × 7 (5 dots across, 7 dots high) or 7 × 9. Since a CRT scan goes across an entire line, this means that the top row of dots of all the characters on one line is scanned, then the next row, etc. The entire screen is scanned in one-sixtieth of a second, and then the procedure begins again.

The comparative slowness of the human eye, coupled with the persistance of the CRT phosphors, integrates this so the display is viewed as a complete page. Nevertheless, since the data fades in a matter of milliseconds, the entire page must be stored in memory and readdressed continually by the CRT for page refresh.

It should be obvious that some means is necessary to convert data words in a standard code, such as ASCII, into the actual shape of the character. This function is performed by a character generator ROM.

The following functions are therefore required for control of a CRT system:

1. Memory Addressing for page refreshing
2. Generation of screen blanking during retrace
3. Generation of vertical and horizontal synchronization pulses
4. Scan line addressing to the character generator

One other function may or may not be provided by a CRT controller. Let's look at it.

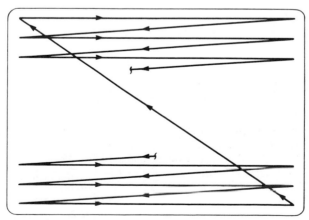

FIG. 13-10 Raster scan.

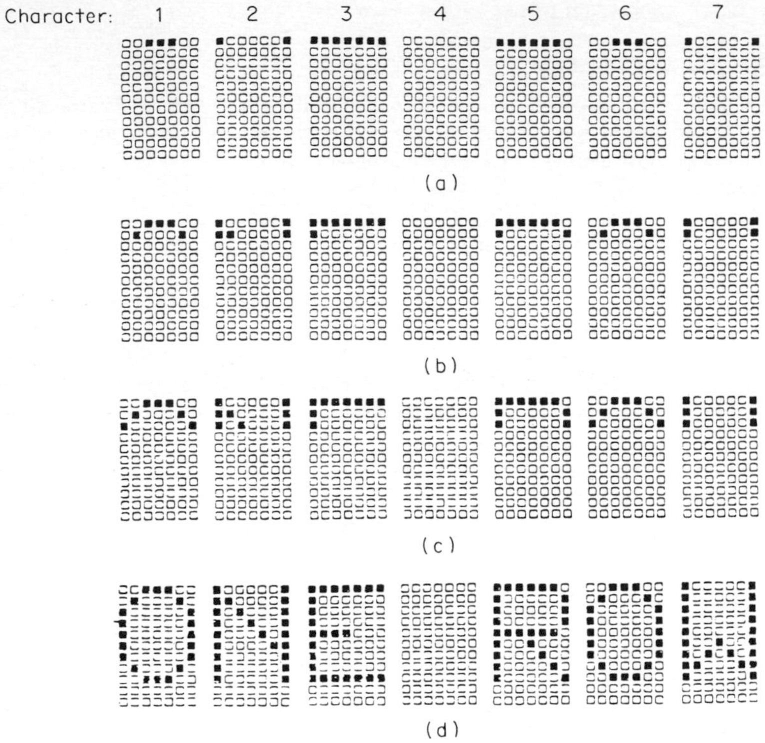

FIG. 13-11 Dot matrix display. (*a*) First line of character row; (*b*) second line of character row; (*c*) third line of character row; (*d*) ninth line of character row.

The continuous scanning of the screen must not be interrupted or synchronization will be lost. Since the scan is continuous, data must be constantly provided so that data is replaced in each character slot before it fades. Therefore, the CRT demands continuous access to the refresh memory. How then does the system grab memory addressing in order to move data into the desired character location? There are several possible methods, each with its own advantages and disadvantages.

The simplest way, typical of "glass teletypes," is to allow the CRT via the CRT controller (CRTC) to completely control RAM addressing. The cursor signal is then used to indicate the block in time so that the page memory address from the CRTC is the same as the page memory location to be filled. Although this circuit is very simple (Fig. 13-12), it is limited in speed. It can take 16.6 ms for the raster to come around to a given position. This results in a practical upper data entry speed of 600 baud for asynchronous data transmission with eight data bits.

To speed things up, the most obvious solution is to allow the system controller (typically a microprocessor) to seize the memory whenever it has data to load into the RAM. This, however, results in flashing of the CRT screen. If the flashing is not objectionable in the specific application, then this scheme is satisfactory.

If data writing is infrequent, it is possible to blank the display during writing. This is only very slightly more complex than allowing flashing. Infrequent short blanks are

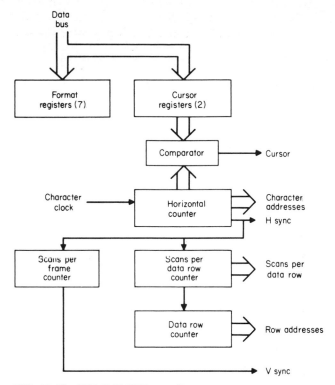

FIG. 13-12   CRT 5027 CRT controller.

practically invisible. If much data is to be written, the blanking is more obvious and may be objectionable.

There are, however, periods during which the display is intentionally blanked, regardless of incoming data. These periods are the retrace intervals when the display is blanked to hide the retrace lines. Typically, 20–25% of each horizontal line period is blanked for retrace, and 5–10% of the vertical period (or field).

By allowing the system controller page memory access during these intervals, data can be written into memory without any visible effect. This implies that the microprocessor must yield priority to the CRT controller anytime the microprocessor wants to access the page memory during the active display period. This pause is easily accomplished by putting the microprocessor into a "wait" state until the next retrace cycle, although it does somewhat slow the microprocessor's throughout.

For still higher data rates, a cycle-sharing approach may be used, in which half of the microprocessor cycle time is given to the microprocessor and the other half to the CRT. For the 6500 and 6800 microprocessors, for example, addresses are provided during Ø1 and data are examined during Ø2. If the CRT addresses the RAM during Ø2, and reads data during Ø1, the CRT and microprocessor are mutually transparent, that is, they do not interfere with each other. This system does have the drawback that RAMs of twice the speed are required than would be otherwise.

Other techniques using additional hardware to DMA data from microprocessor memory into row buffers may also be used. With row buffering, the CRT controller does not address the actual page memory, but rather a small row memory or buffer. During the blank scan line that appears between data rows, the microprocessor allows transfer of data from the page memory to the row memory. For the remaining scan lines of the data row, the CRT controller reads data from the row buffer.

Which method is best? The answer depends on the system requirements.

If data is to be written infrequently, only one or two characters at a time, and cost is the prime consideration, simply blanking the screen may be totally acceptable. However, if blocks of data are to be moved, the screen will be blanked for unacceptably long periods.

At the opposite end of the spectrum with regard to throughout, the cycle-sharing technique allows no interruption of the microprocessor and no interruption of the display. This method does present the highest hardware overhead. The other two methods of controlling memory access are the most popular, the utilization of the row buffer and the utilization of the retrace interval.

The row-buffer method places a constant overhead burden onto the microprocessor whether or not new data is being displayed. This is because the microprocessor has to service the CRT buffer every data row. But this overhead does not go up when a large block of data is being manipulated.

The placement of data only during retrace allows the processor and display to be independent, except when data is being moved. In applications where data is updated infrequently, and cost is a major factor, this method is the most popular.

Attributes   In addition to displaying data, it is frequently desirable to highlight characters or words. This function is performed by "attributes." Attributes include:

Blink   A character or field of characters blinks
Reverse video   A character or field of characters appears dark on a light
   background, instead of light dots on a dark background
Underline   An underline appears under the character or characters
Half-intensity   The characters appear at a reduced intensity
Strikethrough   A line appears through the undesired character
Blank   Selected words are blanked out. This attribute is often used for
   "protected fields" of data for restricted viewing.

How are these attributes inserted into the video data? The two most popular methods of adding attributes to a video stream are "embedded attributes" (also called "field attributes") and "invisible attributes."

Embedded attributes take advantage of the fact that ASCII uses 7-bit words, while most popular microprocessors utilize an 8-bit bus. When the eighth bit is low, the seven lower order bits define a character. When the eighth bit is high, the seven lower order bits define an attribute for all subsequent characters. Another attribute "embedded" in the data stream may be used to turn off the attribute or add a new one. Each embedded attribute therefore affects a field of characters. Since a word of memory is lost from viewing each time a word is an attribute, each attribute change is accompanied by a blank or space in the display. Therefore, the minimum field for an attribute is a full word, preceded and followed by a space.

"Invisible" or "character at a time" attributes can be generated by using a memory more than 8 bits wide. Each character (7 bits) is thereby accompanied by its own attribute information (5–8 bits). This information is easily generated with 12- or 16-bit microprocessors. Alternatively, attribute data can be decoded from a data stream and placed in the lower order or higher order portions of the data word.

Let's look at a common pitfall in system design with embedded attributes. An attribute field, for example, reverse video, is turned on at the middle of one data row and turned off at the middle of the next data row. Attribute words are inserted accordingly:

```
N XXXXX     RV YYYYYYYYYYYYYYYYYYYYYYYYYY     Data row 1
YYYYY  N   XXXXXXXXXXXXXXXXXXXXXXXXXXXXXX     Data row 2
```

where N defines the start of the normal character field and RV defines the start of the reverse-video character field. The first data row appears exactly as desired. However, the top scan line of the second data row appears with reverse video at the left; all subsequent scan lines of data row 2 appear as normal video all the way across.

In this example, the designer forgot that each data row appears as a series of scan lines. The N embedded in the second data row cleared the attribute, which then stayed cleared when the second scan line began.

The two most popular techniques for avoiding this pitfall are:

1. Automatically clear all attributes at the beginning of a data row
2. Latch the attribute that appears at the last scan line of a data row and reinsert it at the beginning of the next data row

Graphics It is frequently desirable to draw pictures of graphs on a video display as well as to display alphanumeric data. This is called "graphic display." The three most popular forms of graphics are "dot," "wide," and "thin" or "line" graphics.

In "dot graphics" the screen is regarded not as, for example, 16 data rows of 64 characters, but as 256 rows of 256 dots. Each dot, in turn, is a bit of memory. Dot graphics allow high resolution, but it is difficult to mix them with alphanumeric data.

"Wide graphics" break each character block into six or eight "pixels." Each pixel is controlled by a memory bit as represented in Fig. 13-13a. Since this block can be either a standard character, generated by a character generator, or a graphic, it is relatively simple to add to a display.

"Thin graphics" are used for drawing thin lines—for business forms, for example. Here the display controller provides a selection of lines, each controlled by a memory bit (see Fig. 13-13b).

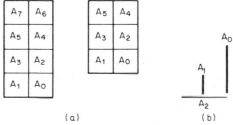

(a)   (b)

FIG. 13-13 Graphics. (a) Wide graphics; (b) thin graphics.

Description of the CRT Controller The first widely available CRT controller, and at this time the most popular, is the 5027/5037 VTAC (Video Timer and Controller), first introduced by Standard Microsystems and now supplied by several other vendors. The 5037 is identical to the 5027, with the exception of having some additional modes of operation in the interlace mode.

The 5027/5037 CRT controller provides the character addressing in a row-column format. That is, each character is addressed as an X-Y coordinate. This allows treatment of characters the same in hardware and software. The alternative method is linear addressing in which each character is treated as a sequential number starting from character 1. In X-Y addressing the first character of the second data row of an 80-column display would be addressed by $Y = 2, X = 1$. In a linear address it would be character 81.

Registers are loaded from the microprocessor or a separate PROM for complete format control. This includes such characteristics as the number of rows, the number of characters per data row, sync widths, blanking time, margins, and character block size.

A feature of the 5027/5037 is that it also can "scroll," that is, shift the lines vertically up or down via a simple command. Scrolling is used to allow entry of data beyond one page by shifting all the displayed lines up and allowing entry of new data on the bottom row as follows:

Before scroll	After scroll
Row 1	Row 2
Row 2	Row 3
⋮	⋮
Row 23	Row 24
Row 24	Row 1 (new data)

The 6845 CRT controller is quite similar to the CRT 5027/5037 with the key difference that it provides linear binary RAM addressing instead of X-Y addressing. Like the 5027/5037, the 6845 has programmable character size and programmable screen format. An enhanced version of the 6845, called the 6545, was recently introduced. It is essentially a 6845 with the capability of X-Y or linear addressing.

The 8275 CRTC is a radical departure from the controllers already described in that it is designed for row-buffer operation only. The block diagram is shown in Fig. 13-14. The 8275 is designed to fit between a DMA controller and the CRT, and to handshake with both the microprocessor and the DMA controller. The 8275 does not provide as much format flexibility as the other CRTCs described, although it does provide on-chip decoding of visual attributes.

Originally designed in France by Thomson-CSF, with Standard Microsystems in the United States as a second source, the 96364 CRT controller is designed for an absolute minimum parts count and a minimum cost system, both at the expense of limited flexibility. The screen format is fixed at 64 characters per row, with 16 data rows.

The 96364 provides the CRT controls already mentioned, but adds on-chip cursor control. An off-chip decoder or PROM converts ASCII control codes to 96364 control codes. The 96364 provides four-direction cursor movement and various other functions, such as cursor home, carriage return, line feed, erase page, and erase line. Additionally the 96364 will place data presented to it directly into memory during horizontal retrace intervals. The circuit for a 17-chip terminal is shown in Fig. 13-15.

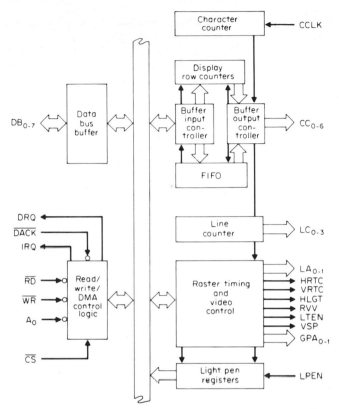

FIG. 13-14 The 8275 CRTC block diagram.

Another nonprogrammable CRT controller is the 8350, which is provided by National Semiconductor (see Fig. 13-16). Unlike the 96364, the 8350 may be mask programmed during manufacturing for different formats. Also, unlike the other CRT controllers, the 8350 incorporates the high-speed dot oscillator and counter on-chip. A microprocessor is required for character entry control.

A CRT controller selection chart is give in Table 13–6.

13–3b Character Generators

The conversion of ASCII data from the video RAM of a display system requires:

1. A ROM to convert the ASCII into character-shape information
2. A shift register to serialize the parallel output ROM data
3. Synchronizing latches
4. Attributes and graphics circuitry

A character generator IC consists as a minimum of a ROM and may not contain the other elements.

The 2513 (G.I., Signetics) character generator consists of a 2560-bit ROM with five

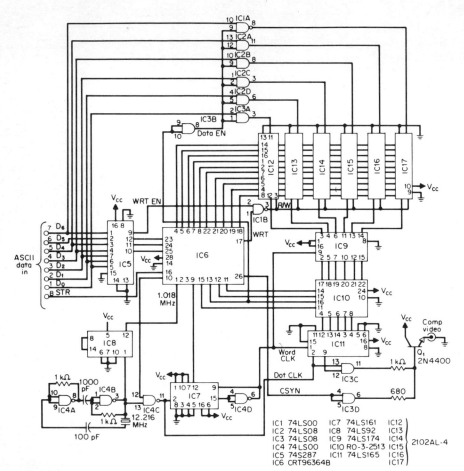

FIG. 13-15 Application of the 96364 CRT controller.

outputs. This generator provides 64 characters in a 5 × 7 format display. It is low in cost and is available in upper- and lowercase units. Note that it takes two 2513s to provide a full 128-character generator. However, for low end terminals, a 64-character, uppercase-only display is often satisfactory.

The 6670 (Motorola) character generator is very similar, except that it provides 128 characters. An upgrade of the 6670, the 66700 character generator, provides 128 7 × 9 characters and provides an internal shift to provide ''descenders'' where lowercase characters such as ''g'' ''y'' descend below the line.

A character generator may contain the video shift register in addition to the ROM. Two examples of character generators incorporating video shift registers are the 8678 (National) and the SMC CRT 7004. The 8678 provides 64 7 × 9 characters, while the CRT 7004 provides 128 7 × 11 characters. The 7 × 11 format is somewhat unusual in that it allows for 7 × 9 characters plus descenders.

The newest character generators pack latches, attributes circuitry, and graphics on the chip in addition to the character generator. An example is the CRT 8002 (SMC,

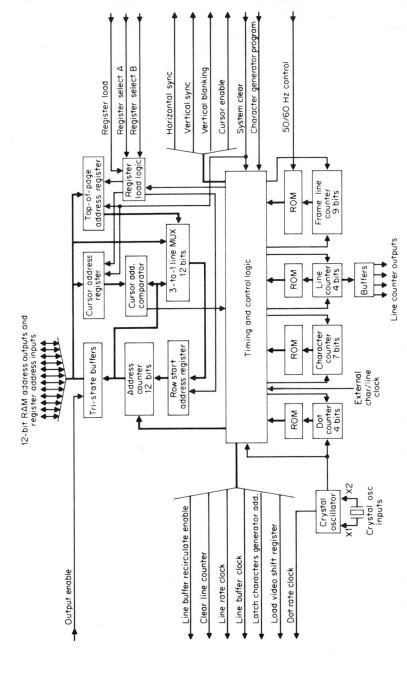

FIG. 13-16 The 8350 CRTC block diagram.

TABLE 13–6 CRT Controllers

	SMC CRT 5027 CRT 5037	Mot. 6845	Natural DP 8350	Intel 8275
Programmable:				
Character/data row	Yes	Yes	Yes	Yes
Data row/frame	Yes	Yes	Yes	Yes
Raster scans/data row	Yes	Yes	Yes	Yes
Raster scans/frame	Yes	Yes	Yes	Yes
Front porch	Yes	Yes	Yes	No
Sync width	Yes	Yes	Yes	Yes
Back porch	Yes	Yes	Yes	No
Sync delay	Yes	Yes	Yes	No
Blank delay	Yes	No	No	No
Cursor delay	Yes	No	No	No
Vertical data position	Yes	Yes	Yes	No
Interlaced/noninterlaced	Yes	Yes	No	No
50-Hz/60 Hz	Yes	Yes	Yes	Yes
Programmed via:				
Processor data bus	Yes	Yes	No	Yes
External ROM	Yes	No	No	No
Mask option	Yes	No	Yes	No
Scrolling:				
Single line	Yes	Yes	Yes	Yes
Multiline	Yes	Yes	Yes	Yes
Page mode	Yes	No	Yes	Yes
Window mode	No	Yes	Yes	Yes
TTL compatible	Yes	Yes	Yes	Yes
Addressing mode:				
Row/column	Yes	No	No	No
Linear	No	Yes	Yes	Yes
Speed:				
Character counter	4 MHz	2.5 MHz	2.5 MHz	3.125 MHz
Dot counter	N/A	N/A	25 MHz	N/A
Visual attributes	No	No	No	Yes
DMA control required	No	No	No	Yes
On-chip row buffers	No	No	No	Yes
Light pen register	No	Yes	No	Yes
Dynamic RAM compatible	Yes	No	Yes	N/A
Power	+5, +12	+5	+5	+5
Package	40 pin	40 pin	40 pin	40 pin

SSS), which includes thin and wide graphics, reverse video, blank, blink, underline, and strike-through on the chip, as well as 128 7×11 characters. A simplified block diagram is shown in Fig. 13-17.

Note that all incoming data is resynchronized by the load/shift input, which controls whether the 8002 is inputting data or shifting it out. The attribute enable input allows operating the 8002 with a 16-bit-wide memory for character attributes and an 8-bit memory for embedded attributes.

Example 13–3 Design of Video Controller

Design a video controller for a word-editing terminal. The maximum data rate is 19.2 kbaud.

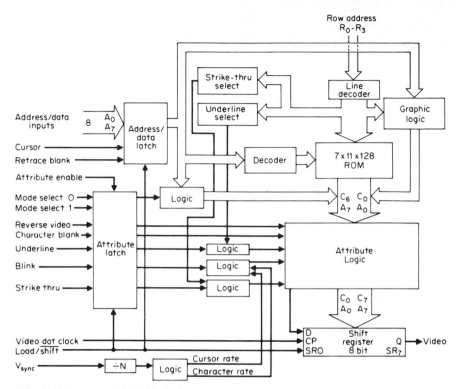

FIG. 13-17   The CRT 8002 block diagram.

Solution

A very large number of design decisions are required for a terminal design, all relating to a tradeoff between cost and capability. The first decision is the display format. In this application, a display of 80 characters per data row with 24 data rows is selected, as it is the most popular configuration for medium-capability terminals. Attributes and graphics are desired to optimize character shape and readability, and a 7×9 character font with lowercase descenders is selected over a 5×7 font.

The following calculations can be made:

1. Vertical sync frequency = 60 Hz (the line frequency).
2. Each character is seven dots wide. To allow for intercharacter spacing, a nine-dot-wide character block is chosen, allowing for a space on either side of each character.
3. Allowing for left- and right-hand screen margins, the number of character times per data row is picked as 25% more than the number of characters, or $1.25 \times 80 = 100$.
4. The number of scan lines per data row is 9 (based on the 7×9 character) + 2 (for the descending "tails" on the lowercase characters) + 1 for interrow spacing, or 12 in all. The number of active scan

lines is therefore $24 \times 12 = 288$. In addition, a margin of 20 scan lines for retrace results in 308 scan lines total.

5. The horizontal frequency = 60 Hz $\times 308 = 18,400$ Hz.
6. The character clock rate = 1840 kHz $\times 100$ character times per row = $1,840,000$ Hz.
7. The dot clock = 1840 kHz $\times 9$ dots per character = $16,560$ kHz.

As the data rate of 19.2 kbaud is well within the capability of the memory contention technique of character entry during horizontal blanking, this method is selected as the best tradeoff of capabililty vs. cost. When the microprocessor tries to address an area of video RAM, the Ready line of the 8085 is lowered, putting the microprocessor into a wait state by extending the memory cycle until the next horizontal retrace.

The 5037 is selected as the CRT controller. It is teamed with an 8002 character generator to minimize the parts count.

To keep the memory 8 bits wide to reduce cost, the attributes are embedded. Software will be used to clear the attribute at the start of each scan line. Since the 8002 has an internal attribute latch, the embedded attribute is entered or deleted via its attribute enable input.

Attributes are recognized by the most significant bit of the data word; when high, a word is recognized as an attribute.

The complete circuit is shown in Fig. 13-18.

13–4 FLOPPY DISK CONTROLLERS

The flexible, or floppy, disk is one of the most practical ways to add additional memory to a microprocessor-based system, combining random access with relatively high storage density.

The floppy disk itself is a flexible Mylar-coated disk with a magnetic oxide, and protected by a thin jacket. The jacket has a radial slot to allow a read/write head to contact the rotating disk. The disk is formatted into tracks and sectors as shown in Fig. 13-19.

Disks are available in 2 sizes: 8-in "full size" disks and $5\frac{1}{4}$-in "minifloppies." Given these two sizes, there are 2 formats, "hard sector" and "soft sector."

In the hard sector disk, 32 sector holes are punched in the diskette. These holes are used to define the sector boundaries. In the soft sector diskette, there are no sector holes, but only one index hole. The number of sectors is user defined and separated by recording gaps and headers.

Although hard-sectored disks allow more recorded data because recording space is not taken up by gap and header, soft sectoring allows positive sector identification each time the sector heading is read. The soft-sectored format has achieved greater acceptance as a standard than has the hard-sectored format.

To date, most disk systems utilize recording on one side of the disk (single sided) with one bit location every 4 μs (single density). To increase the available storage on a diskette, double-density recording (one bit per 2 μs) is becoming more popular, and double-sided recording is starting to overcome some early mechanical problems. The system essentials remain the same, however. For simplicity, the discussion will be based upon a single-density system.

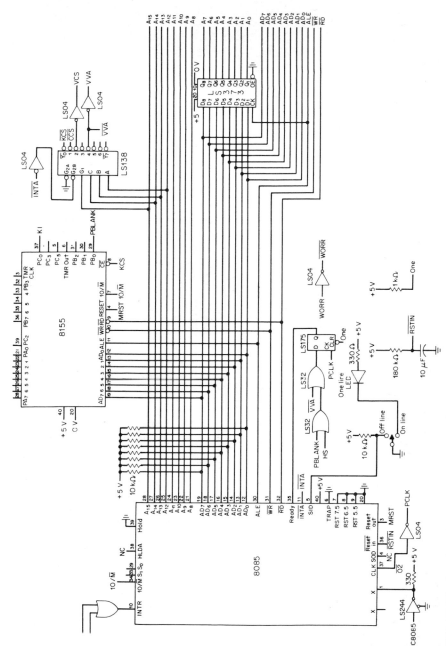

FIG. 13-18 Complete video controller.

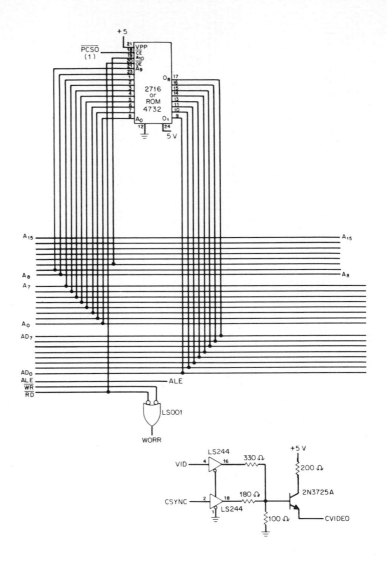

FIG. 13-18 (*Continued*)

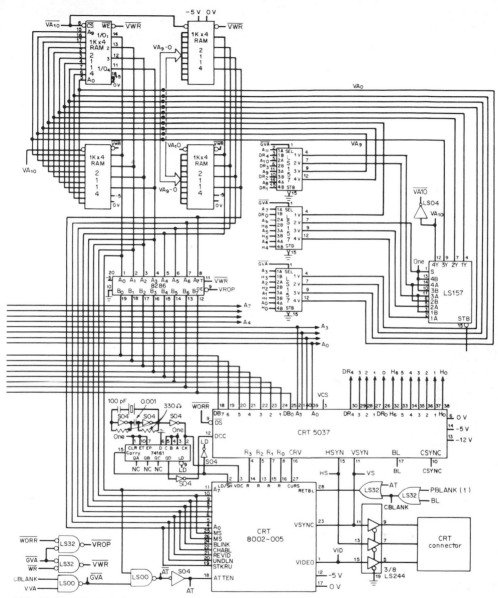

FIG. 13-18 (*Continued*)

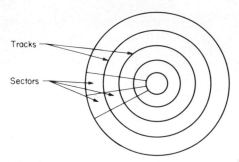

FIG. 13-19 Floppy disk.

The sequence of writing to or reading from a disk is typically as follows:

1. *Motor On*   After the disk is rotating, 1 s is allowed for the speed to stabilize.
2. *Select Direction/Step*   The proper track is selected by selecting the direction to move the head, then stepping it as required, one track position at a time. A write gate is used to prevent accidental writing while moving the head.
3. *Head Load*   When the proper track is reached, the head is lowered onto the disk.
4. *Read/Write Sector*   With the head lowered, the controller proceeds to read the ID fields, which separate the sectors, until the proper sector is reached.

For a disk to be interchangeable between systems, a standard format must be used. Both the IBM 3740 single-density and the IBM System 34 double-density formats have achieved a de facto standard status.

If a speed variation occurs when rewriting a data record within a sector, that record may become physically larger and extend past the previous ending. To avoid having data errors due to this phenomonon, "gaps" are provided to separate records, ID fields from data fields, and the index from the data. Four types of gaps are used in the IBM compatible formats:

     Gap 1   The "index gap," gap 1, separates the index from the first record identification on each track. Gap 1 is defined as 16 bytes of hexadecimal "FF" followed by 4 bytes of "00."

     Gap 2   The "ID" gap, gap 2, separates each record identification field from its data field. Initially, it consists of 6 bytes of hexadecimal "F" followed by 4 bytes of "00." Gap 2 is allowed to vary in length when the file is updated to allow for disk-speed variations.

     Gap 3   The "data gap," gap 3, terminates a record with 17 bytes of "FF" and 4 bytes of "00."

     Gap 4   Gap 4 is called the "free index gap," and is used just prior to the index hole.

13–4a   The 1791 Floppy Disk Formatter/Controller

The basic block diagram of a floppy disk controller using the 1791 controller (Western Digital, Synertek, Standard Microsystems) appears in Fig. 13-20. The device's pins are split into two groups, the computer interface and the floppy disk interface.

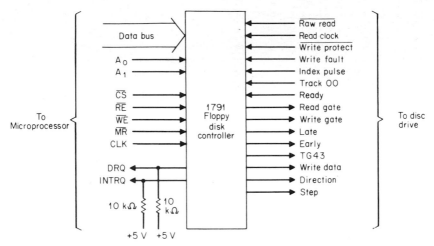

FIG. 13-20 The 1791 floppy disk formatter/controller.

On the microprocessor side, an 8-bit data bus is used for transfer of status, control, and data. The type of word is programmed via A_0 and A_1, the register select lines, as shown in Table 13-7.

The data request and interrupt request outputs provide the handshaking for an interrupt-driven microprocessor system.

The floppy disk interface is more complex, but it is oriented around the standard inputs and outputs of commercial disk drives. First, let's discuss the inputs.

Raw read and *read clock* are, respectively, the read data and clock extracted from the disk by a data separator circuit.

Write protect is used to inhibit writing onto a disk that has had its write protect tab removed.

Write fault signifies a disk drive problem that is preventing proper writing onto the disk.

Index pulse signifies the index mark.

Track 00 informs the floppy disk controller that the head is over track 00.

Ready signals the controller that the disk is rotating at full speed.

The outputs from the controller are:

Read gate signals that the field of zero's at the end of a gap have been received.

TABLE 13–7 Register Select Programming

A_1	A_0	Read mode	Write mode
0	0	Status register	Command register
0	1	Track register	Track register
1	0	Sector register	Sector register
1	1	Data register	Data register

Write gate tells the controller that a write is to be performed.

Late and *early* are for "write precompensation." They tell the disk drive how to allow for the skew that occurs in a magnetic recording medium with different data patterns. This skew occurs on the inside tracks of a disk where the bits are packed closer together (tracks 43 to 76).

TG 43 is used in conjunction with write and early to turn on the precompensation circuit when tracks 43 to 76 are being written.

Write data is the serial data to the disk.

Direction and *step* provide the head step and direction control.

This section provides an outline of the functions provided by a floppy disk controller. The actual design of a system is a rather complex undertaking that is heavily dependent on the drive selected and the system requirements. The designer is referred to application notes provided by both the LSI and drive manufacturers.

13–5   KEYBOARD ENCODERS

In its basic form, a keyboard consists of a matrix of push-button switches. For an MPU system to utilize a keyboard, the following functions are required:

1. Switch closure detection
2. Switch debounce
3. Keyswitch encoding
4. Roll over

These four functions are provided by the LSI keyboard encoder.

Switch closure detection is most often provided by the key identification scheme of keyboard scanning. With this method each key is identified as an X-Y coordinate, as shown in Fig. 13-21.

To scan the keyboard, the encoder presents a series of "X outputs" and has a set of "Y inputs." These outputs and inputs are sequentially scanned, walking a logic 1 through the X outputs and looking for it on each Y input. When a particular Y input is examined and is found to have a logic 1 present, the state of the X and Y counters is also examined. A logic 1 on Y_2 when X_2 is high indicates that the "S" key is depressed.

To allow for the fact that mechanical keys are noisy and may bounce a large number of times after key closure or release, most encoders wait a period of time longer than the expected bounce period and retest the key closure. If the closure is still present, a valid key closure is considered present and data is outputted.

A key design question is "What does a designer want to have happen if one key is already depressed when a second closure occurs?" This problem of multiple closures is called "roll over."

FIG. 13-21   Simplified keyboard encoder matrix.

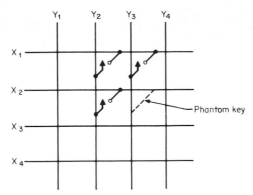

FIG. 13-22   Phantom closures.

Two methods of roll-over control are popular in alphanumeric keyboards.

Two-key roll over inhibits detection of a second closure while one key is depressed. Most two-key roll-over methods will output data from a second key if it is still held down when the first is released. Two-key roll over is easily implemented by stopping the X-Y scan while a key closure is present, then resuming the scan when the key is released.

N-key roll over allows encoding of a second key closure, or for that matter, a third, fourth, etc., while a key is depressed. The scan continues even though a key is depressed. Depressed key locations are stored in an internal memory.

N-key roll over is preferable where high typing rates are expected. However, since an error in N-key roll over is most often one incorrect data entry, whereas two-key roll

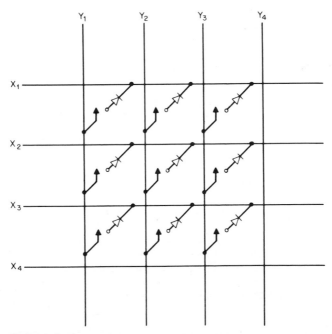

FIG. 13-23   Preventing phantom closures with diodes.

TABLE 13–8 Keyboard Encoders

	SMC KR 2376	SMC KR 3600	National MM 5740	G.I. AY-5-3600
Number of keys	88	90	90	90
Static protected	Yes	Yes	Yes	Yes
Input/output TTL	Yes	Yes	Yes	Yes
N-key roll over	No	Yes*	Yes*	Yes*
N-key lockout	Yes	Yes*	Yes*	Yes*
On-chip oscillator	Yes	Yes	No	Yes
Debounce circuit	Yes	Yes	Yes	Yes
Number of data bits	9	10	9	10
Data outputs buffered	No	Yes	Yes	Yes
Data complement control	Yes	Yes	No	Yes
Pulsed strobe	Yes	Yes*	Yes*	Yes*
Level strobe	No	Yes*	Yes*	Yes*
Repeat key function	No	No	Yes	No
Shift-lock indicator	No	No	Yes	No
Power	+5, −12	+5, −12	+5, −12	+5, −12

*Programmable.

over generally produces missed keys, two-key roll over is often preferred when large amounts of numeric data are to be entered. A missing digit is more noticeable than an incorrect one.

Since an N-key roll-over system allows three or more keys to be depressed simultaneously, precautions must be taken to prevent "phantom" closures, as illustrated in Fig. 13-22. Simultaneous closure of keys 1, 2, and 3 creates a phantom key between X_2 and Y_3 via the circuit created by the connection of X_1 to Y_2 to X_2 to Y_3. This problem is usually solved by placing diodes in series with the keys, as shown in Fig. 13-23.

Refer to Table 13-8 for device selections.

Example 13–4 Keyboard Encoder Design

Design a 76 ASCII encoded keyboard with the layout shown in Fig. 13-24. Two-key roll over is acceptable, thus saving the cost of the diodes. The key switches are specified as having a maximum bounce of 4 ms.

Solution

A KR3600-ST ASCII encoder is selected. It has programmable two-key or N-key roll over via pin 4.

A 50-kHz nominal key scan clock is selected as the center of the minimum and maximum scan rates.

Consulting the data sheet, a 0.02-μF capacitor from pin 31 to ground will result in a 6-ms strobe delay, which is more than adequate to compensate for the key bounce. The completed design is illustrated in Fig. 13-25.

A variation of the KR3600 or AY-5-3600 keyboard encoder is the "Pro." It is programmed to produce not ASCII, but a binary sequential output that can drive a PROM. This allows simple changing of a key switch function without expensive keyboard change or encoder custom programming. This device is shown in Fig. 13-26.

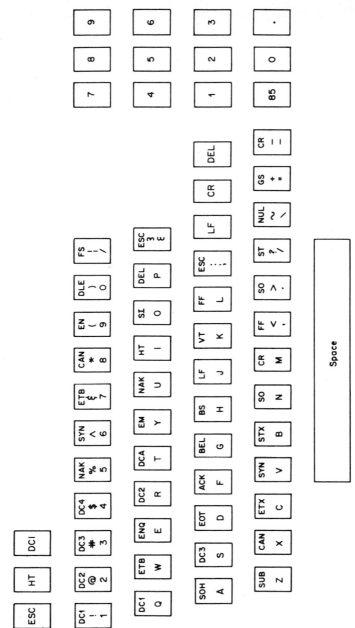

FIG. 13-24 Keyboard layout of Example 13-4.

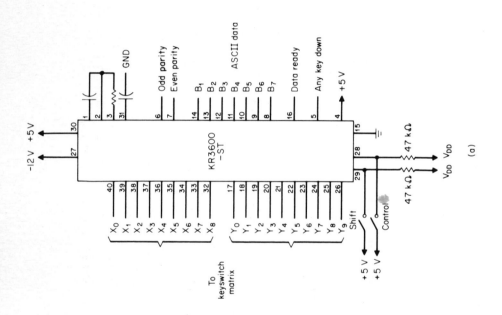

(a)

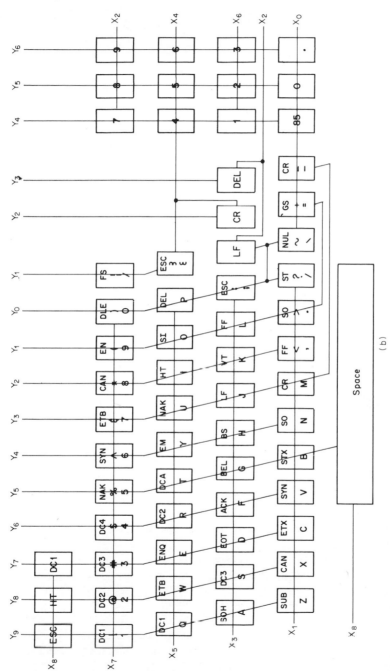

FIG. 13-25 Keyboard encoder of Example 13-4. (*a*) ASCII encoder; (*b*) keyboard.

(b)

13-43

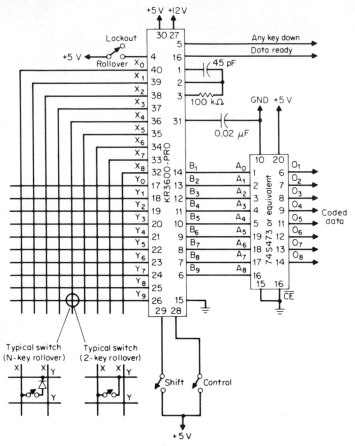

FIG. 13-26   The KR3600-Pro keyboard encoder.

REFERENCES

Cayton, B.: "Interlaced Video Displays Utilize Low Cost CRT Controllers," EDN, Sept. 5, 1979, pp. 207–212.

Herman, M.: "Video Terminal Strips Down to 26 Chips," *Electronic Design,* May 24, 1980.

Intel: *Intel Component Data Catalog.*

Lancaster, D.: *TV Typewriter Cookbook,* Howard W. Sams, Indianapolis, Ind., 1976.

Lesea, A., and R. Zaks: "Floppy Disk Controller Design Must Start With the Basics," EDN, May 20, 1978, pp. 129–137.

————: *Microprocessor Interfacing Techniques,* Sybex, 1979.

Motorola, *M6800 Microprocessor Application Manual.*

National Semiconductor, "DP8350 Series Programmable CRT Controllers," Santa Clara, Calif.

Standard Microsystems, *Standard Microsystems Data Catalog,* Hauppauge, N.Y.

Weissberger, A.J.: *Data Communication Handbook,* Signetics Corp.

Western Digital, *Western Digital Product Handbook.*

Chapter 14

INTERFACE CIRCUITS

Carroll Smith Applications Engineer
Texas Instruments Corp.
Dallas, Tex.

14–1 INTRODUCTION

The widespread use of integrated logic circuits, such as TTL, MOS, and CMOS components, in microprocessor and microcomputer systems, has created an increasing need for interface devices. These devices take the low voltage/low current levels from computer output ports or other low-level circuits and interface these levels to higher voltage/higher current peripheral hardware, such as displays, relays, lamps, and printers.

These interface devices can be placed in various categories depending upon their general use. These categories are:

1. Peripheral Drivers
2. Line Circuits
3. Display Drivers
4. Memory Interface

14–1a Peripheral Drivers

Interface circuits could be made from discrete transistors to drive high-voltage/high-current peripheral hardware. Additional components would usually be required, such as current-limiting resistors or clamp diodes in the case of inductive loads. Peripheral drivers are

TABLE 14-1 Peripheral Driver Selection Guide

Peripheral drivers with logic gates

Switching voltage	Maximum output current	Typical delay time	Drivers per package	Internal clamp diodes	Logic gate function			
					AND	NAND	OR	NOR
15 V	300 mA	15 ns	2	—	SN75430 SN75431	SN75432	SN75433	SN75434
20 V	300 mA	21 ns	2	—	SN75450B SN75451B	SN75452B	SN75453B	SN75454B
30 V	300 mA	33 ns	2	—	SN75460 SN75461	SN75462	SN75463	SN75464
30 V	500 mA	33 ns	2	—	SN75401	SN75402	SN75403	SN75404
35 V	700 mA	300 ns	4	Yes		SN75437		
50 V	350 mA	300 ns	2	Yes	SN75446	SN75447	SN75448	SN75449
55 V	300 mA	33 ns 100 ns	2	— Yes	SN75470 SN75471 SN75476	SN75472 SN75477	SN75473 SN75478	SN75474 SN75479
55 V	500 mA	33 ns 100 ns	2	— Yes	SN75411 SN75416	SN75412 SN75417	SN75413 SN75418	SN75414 SN75419

Peripheral drivers without logic gates

Switching voltage	Maximum output current	Typical delay time	Drivers per package	Internal clamp diodes	Device types			
35 V	1.5 A	500 ns	4	Yes Yes —	ULN2064 SN75064 ULN2074	ULN2066 SN75066 SN75074	ULN2068 SN75068 UDN2841	UDN2845
35 V	500 mA	1 μs	7	Yes	ULN2001A	ULN2002A	ULN2003A	ULN2004A
50 V	1.5 A	500 ns	4	Yes Yes —	ULN2065 SN75065 ULN2075	ULN2067 SN75067 SN75075	ULN2069 SN75069	
60 V	500 mA	130 ns	7	Yes	SN75466	SN75467	SN75468	SN75469

integrated circuits that usually contain two or more high-voltage/high-current output transistors with all the necessary current-limiting and bias resistors that allow the device to be connected directly to TTL, MOS, or CMOS logic levels with a minimum of external components. In addition, some of the drivers include internal logic gates (AND, NAND, OR, NOR) that allow logic functions to be performed with the interface circuit.

A summary of typical peripheral driver devices is shown in Table 14-1.

14–1b   Line Circuits

Line circuits are a class of interface devices designed to transfer digital information signals over various transmission lines. Line circuits consist of line drivers, line receivers, and transceivers. The devices may be used for general-purpose applications or designed for some industry standard data-transmission configuration. Typical standards are RS-232C, RS-422A, RS-423A IEEE-488, IBM 360/370, and the new proposed EIA RS-485.

A typical computer terminal may have one or more external peripheral subsystems, such as a printer, etc. It is necessary to communicate with these peripherals, usually over relatively short distances. Single-ended drivers and receivers are normally used in these applications. RS-232C is the EIA industry standard developed for single-ended transmission over short distances at slow data rates.

Single-ended data transmission can be used for longer distances. EIA standard RS-423A extends the maximum distance to 4000 ft at up to 1 kbits/s or up to 300 ft at 100 kbits/s.

It also provides waveshaping dependent on data rate and wire length to control reflections and radiated emission or crosstalk. Another improvement of the RS-423A is the requirement for high-impedance outputs, when the power is off, to avoid loading the transmission line.

One problem with single-ended data transmission at long distances and high data rates is that it becomes difficult to distinguish between a valid data signal and unwanted signals introduced by external environments, such as ground shifts and noise. For longer distances and higher bit rates, differential data transmission can be used to reject unwanted ground and noise signals.

Unwanted signals appear as common mode signals and are rejected by the differential line receiver input and differential driver output.

Selection of the proper line driver and receiver is a function of a number of variables. These are:

1. Line length
2. Line characteristic impedance
3. Data rate
4. Single-ended or differential requirements
5. Logic level, i.e., standard TTL, RS-232C, etc.

Table 14-2 summarizes the various specifications adopted by EIA for data transmission, and Table 14-3 summarizes the line circuits available from TI designed to meet the EIA specifications or general-purpose applications. A description of these different line circuits and the existing industry standards are discussed in Sec. 14-3. Numerous examples of typical applications are used to help the designer make the correct selection for his or her system.

TABLE 14-2 EIA Data Transmission Specification Summary

Specification	RS-232C	RS-423A	RS-422A	RS-485/P.N.1360
Mode of operation	Single ended	Single ended	Differential	Differential
Number of drivers and receivers allowed on one line	1 Driver/ 1 Receiver	1 Driver/ 10 Receivers	1 Driver/ 10 Receivers	32 Drivers/ 32 Receivers
Maximum cable length	50 ft	4000 ft	4000 ft	4000 ft
Maximum data rate	20 kbits/s	100 kbits/s	10 Mbits/s	10 Mbits/s
Maximum voltage applied to driver output	±25 V	$+6$ V	-0.25 to 6 V	-7 to 12 V
Driver output signal				
Loaded	±5 V	±3.6 V	±2 V	±1.5 V
Unloaded	±15 V	±6 V	±5 V	±5 V
Driver load	3 to 7 kΩ	450 Ω min	100 Ω	54 Ω
Maximum driver output current (high-impedance state)				
Power on				±100 μA
Power off	max 1300 Ω	±100 μA	±100 μA	±100 μA
Output slew rate	30 V/μs max	Controls provided	—	—
Receiver input voltage range	±15 V	±12 V	-7 V to $+7$ V	-7 V to $+12$ V
Receiver input sensitivity	±3 V	±200 mV	±200 mV	±200 mV
Receiver input resistance	3 to 7 kΩ	4 kΩ min	4 kΩ min	12 kΩ min

TABLE 14–3 Interface Devices Available for Line Circuits

	Drivers				Receivers	
Output	Drivers per package	Type numbers	Input		Receivers per package	Type numbers
EIA Standard RS-422A						
Differential	2	SN55158, SN75158 SN75159 UA9638C	Differential		2	SN75157 µA9637AC
	4	AM26LS31C MC3487 SN55151, SN75151 SN55153, SN75153 SN75172 SN75174			4	AM26LS32AC AM26LS33AC MC3486 SN75173 SN75175
EIA Committee TR30.1 Draft PN1360 (April 1980)						
Differential	4	SN75172 SN75174	Differential		4	SN75173 SN75175
EIA Standard RS-423A						
Single ended	2	SN75156 µA9636AC	Singled ended		2	SN75157 µA9637AC
	4	SN75186 SN75187			4	AM26LS32AC AM26LS33AC MC3486 SN75173 SN75175

TABLE 14-3 (Continued)

	Drivers			Receivers	
Output	Drivers per package	Type numbers	Input	Receivers per package	Type numbers
EIA Standard RS-232C					
Single ended	2	SN75150, SN75156, µA9636AC	Single ended	2	SN75152
Single ended	4	SN75186	Single ended	4	SN75154, SN75189, SN75189A
IBM 360/370					
Single ended	2	SN75123, SN75126, SN75130	Single ended	3	SN75124
				7	SN75125, SN75127
				8	SN75128, SN75129
General purpose					
Single ended	2	SN55121, SN75121, SN75361A, SN55450B, SN75450B, SN55451B, SN75451B	Single ended	2	SN55122, SN75122, SN55140, SN75140, SN55141, SN75141, SN55142A, SN75142A, SN55143A, SN75143A
Single ended	4	DS7831, DS8831, DS7832, DS8831	Differential	2	SN55107A, SN75107A, SN55107B, SN75107B, SN55108A, SN75108A, SN55108B, SN75108B, SN55115, SN75115, SN55182, SN75182, SN75207, SN75207B, SN75208, SN75208B
Differential	2	DS7831, DS8831, DS7832, DS8831, SN55109A, SN75109A, SN55110A, SN75110A, SN55113, SN75113, SN55114, SN75114, SN55183, SN75183, SN55450B, SN75450B, SN75112			

14-6

14–1c   Display Drivers

There are five main types of displays in use today and each type requires its own unique set of drive conditions. In most cases, the display driver interfaces between a microcomputer chip, the output ports of a microprocessor, or other digital logic systems, and supplies the necessary current and voltage levels required by the display technology in use.

The five common types of displays are:

1.   Visible light-emitting diodes (VLED's)
2.   Vacuum fluorescent displays
3.   DC plasma or gas discharge
4.   AC plasma
5.   Liquid crystal (LCD)

Table 14-4 summarizes the basic differences between drive requirements for these ·displays. The numbers given are only relative since the actual drive depends on many factors, such as character height, number of digits, multiplex or direct drive, and numeric or alphanumeric.

Typical applications illustrating devices and techniques to interface displays are discussed in Sec. 14-4.

TABLE 14–4 Display Technologies

Display	Voltage	Current	Comments
VLED	Low 1.8–3.0 V	High 1–100 mA	Requires high current for large displays Easy to drive (TTL compatible)
VF	High 30–60 V	Low <1 mA	Easy to drive (requires 50–100-mA filament current)
DC plasma	Very high 100 V	Low <1mA	Mature technology: History of reliability problems
AC plasma	Very high 100 V	Low AC driven	New technology: Used for large-area display panels
LCD	Low 3–5 V	Low <1 μA	Difficult to drive (requires AC voltage) Lack of standard interface circuits

14–1d   Sense Amplifiers

Core memories are still used in some applications where a nonvolatile memory is required after power is removed. Sense amplifiers detect bipolar differential input signals from the core memory and provide interface circuitry between the core memory and the logic section. Low-level pulses originating in the memory are transformed into logic levels compatible with TTL logic levels.

14–1e   Core Memory Drivers

Core memories cannot be driven directly by MOS or TTL logic devices. Interface circuits are needed to supply the higher voltage and current that are required by the magnetic cores. Drivers that source as well as sink current are also required. A typical core driver will source or sink 600 mA and have two operating voltage levels, 5 V for TTL logic gates and typically 24 V for the output drivers. The integrated logic gates and/or decoding functions can be used to minimize package count or possibly reduce some of the overhead on the computer and external hardware.

14–1f MOS Memory Drivers

MOS memory drivers are essentially voltage- and current-level shifters that interface between TTL and MOS logic levels. Many MOS memories are TTL voltage compatible, but require high currents due to high fan-out when many chips are addressed. Some MOS memories can present a high capacitance load, so high peak currents are required from the driver.

A device such as the TMS 4050 has TTL-compatible inputs, except for the chip-enable input. This input requires a voltage swing from +0.6 V to +12 V for proper operation so the TTL-to-MOS-level interface conversion can easily be performed.

14–2 PERIPHERAL DRIVER APPLICATIONS

Peripheral drivers are integrated circuits that can be used to interface between TTL, MOS, CMOS logic levels and higher voltage, higher current components, such as lamps, relays, or motors.

The devices usually include two or more independent sections with integrated resistors to perform level shifting. Some devices include logic gates (AND, NAND, OR, NOR) and clamp diodes to protect the output transistors from back emf transients when switching inductive loads. Maximum output current varies from 300 mA to 1.5 A. The lower current devices have high-speed switching capability with typical delay times of 15 to 30 ns.

A summary of available drivers was shown in Table 14-1. The choice of which driver to use depends upon the particular application. Design examples will be used to illustrate how some of the various choices could be made.

14–2a Lamp Drivers

Example 14–1 Design of a Lamp Driver

Design a circuit to monitor a high TTL logic level and indicate the presence of that voltage. If the logic level goes low, a #47 incandescent lamp will be latched on and remain on until the system is reset.

Solution

A #47 incandescent lamp is rated at 6.3 V and 150 mA. Most of the drivers listed in Table 14-1 would work in this application. The device rated for the lowest voltage (15 V) and current (300 mA) will be selected due to cost. Also the addition of logic gates incorporated into the driver will minimize the external components required. An SN75432 dual NAND driver is a good choice for the application. The schematic shown in Fig. 14-1 shows how one dual driver can serve as a latch as well as to interface the incandescent bulb.

Some incandescent lamps have enough inductance to cause destructive reverse voltage transients during switching. A clamp diode can be placed from the output transistor collector to the positive voltage rail to eliminate the voltage spikes across the output transistor. The clamp diode is included in many of the peripheral driver circuits, but not in the SN75432. If this extra protection is desired or required, an external clamp diode could be added to the circuit as shown in Fig. 14-1.

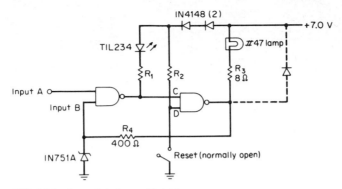

FIG. 14-1 Lamp interface with latch.

A green VLED, the TIL234, serves to indicate the presence of the high TTL level applied to input A. Resistor R_1 limits the current through the TIL234 to 25 mA when input A is high and the output of the first NAND gate is low. This low output, also applied to input C of the second NAND gate, holds its output high and the lamp will be off.

When input A goes low, the output of the first NAND gate will go high, turning off the TIL234. Since both inputs to the second NAND gate are now high, its output goes low, turning on the lamp. At the same time, the low input now on input B latches the output of the first NAND gate high. It will remain latched high until the system is reset, even if input A returns to logic level high.

The system is reset by bringing input D low. If a high level is now present on input A, the output of the first NAND gate will go low, turning the TIL234 on and the lamp off. The green TIL234 now indicates that the system has been reset and is ready to operate again when input A goes low.

The SN75432 operates with a V_{CC} of 5 V. This operating voltage is obtained from the 7-V supply by dropping the voltage through two 1N4148 diodes. A separate $+5$-V supply could be used if desired, or the entire system could be operated from $+5$ V with some reduction in brightness of the #47 lamp.

One problem with incandescent lamps is that the peak turn-on surge current of a cold lamp may be as much as 10 times the normal hot on-current. The initial peak surge turn-on current of the #47 lamp could be as high as 1.5 A. This application example illustrates one way to limit the surge current so that a smaller, lower cost peripheral driver can be used without peak currents exceeding the device capabilities and causing device deterioration. Resistor R_3 (8 Ω) will limit the peak current to approximately 500 mA during the start-up period. For the short time (200 ms) required for the filament resistance to increase, the power dissipation capability of the SN75432 will not be exceeded.

The supply voltage is increased to 7-V DC to compensate for the voltage drop across R_3 at the normal operating current of the lamp. To prevent excessive voltage on input B when the second NAND gate is off, a 5.1-V zener is placed across the gate. Resistor R_4 (400 Ω) limits the current into the zener to approximately 5 mA. This current will flow through the #47 lamp but will not be sufficient to cause the lamp to glow. The lamp driver circuit can be modified to make use of this zener current to reduce the peak inrush

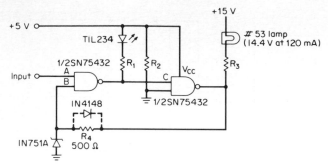

FIG. 14-2 Modified lamp interface circuit with latch.

current. This modification is shown in Fig. 14-2. This basic circuit could be used for higher voltage lamps, but the peak current must be limited to keep the SN75432 within its current and power dissipation capability. Fig. 14-2 shows how the circuit could be changed to drive a #53 lamp rated at 14.4 V and 120 mA. A separate +5-V supply is used for the V_{CC} rail and R_4 increased to 500 Ω to limit the current through the 5-V zener to 20 mA.

This circuit eliminates the high peak inrush current of a cold incandescent lamp. The zener current of approximately 20 mA through the lamp serves as "keep alive" current. This current is not high enough for the lamp to be seen as "on," yet the low current heats the filament enough so that the peak current due to a cold filament is significantly reduced. The current can be adjusted by means of R_4 for optimum keep-alive current. The circuit is easily adapted for higher voltage lamps. The value of R_4 has to be increased to a higher resistance to control the zener and keep-alive current through the lamp. A 1N4148 diode, as shown in Fig. 14-2, can be placed across R_4 to ensure a logic level low at input B when the lamp is on, if R_4 exceeds 500 Ω.

Higher voltage lamps require higher voltage capability in the output transistors. Dual NAND drivers, such as the SN75472, are available with a switching voltage capability to 55 V.

Another way to decrease the high surge current of an incandescent lamp is to use two switches. One section of the dual switch has a current-limiting resistor and turns on initially. After a short time delay (approximately 200 ms) the second section of the driver turns on without a current-limiting resistor and the lamp is at full brilliance.

Example 14–2 Delayed Turn-On

Design a circuit featuring an incandescent lamp driver with delayed turn-on.

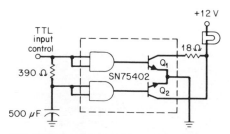

FIG. 14-3 Incandescent lamp—delayed turn-on.

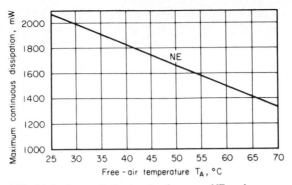

FIG. 14-4   Power dissipation derating curve NE package.

Solution

A possible circuit is shown in Fig. 14-3. An SN75402 NAND gate was chosen for this application. This device is packaged in the 2075-mW NE 14-pin DIP that allows the current capability to be increased to 500 mA. Both sections of the device can operate at maximum ratings at the same time, as long as the package power-dissipation ratings are not exceeded.

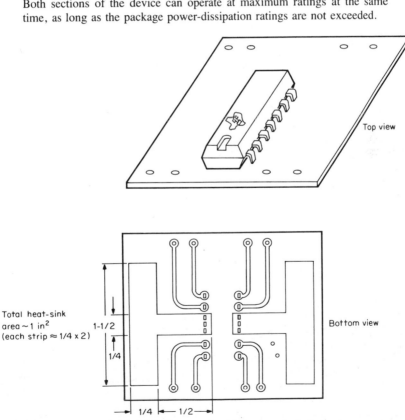

FIG. 14-5   Typical PCB layout for NE package. (Dimensions are in inches.)

Figure 14-4 shows the power dissipation derating curve for the NE package. To increase the power-handling capability of this package, large copper areas can be provided on the PCB. A typical layout is illustrated in Fig. 14-5. Another method of increasing the power-handling capability of the NE package is to use a heat sink clipped or soldered to the ground/substrate pins.

Also, the final equipment packaging must be considered. There is little benefit to using large areas of copper on the PCB or external heat sinks if the PCB is placed inside a closed box with no ventilation, especially if there are other heat-producing components in the box, such as transformers, lamps, etc. It is vital to monitor and control the temperature rise inside the cabinet to ensure long-term reliability.

Our applications example is based on a GE 1815 incandescent lamp. However, any lamp could be used, if the appropriate voltage and current rated device were selected from Table 14-1.

14–2b   Relay Drivers

Where a great number of peripheral drivers are required in a system, a device with low power dissipation would offer an advantage. A new series of drivers is now available, the members of which have only 55 mW of power dissipation vs. the 350 mW required for most drivers. The series is as follows:

SN75446   Dual driver with "AND" enable
SN75447   Dual driver with "NAND" enable
SN75448   Dual driver with "OR" enable
SN75449   Dual driver with "NOR" enable

These devices have TTL-compatible inputs, 70-V minimum off-state voltage, 50-V minimum latch-up voltage, 300-ns typical delay time, and have integrated clamp diodes built into the chip. The maximum output current is 350 mA for each section.

A typical design example using these devices could be as hammer drivers in a printer application. A typical application could require 100 or more drivers. A driver with low power dissipation would save considerable power, create less heat, and provide higher system reliability.

Example 14–3   Design of Hammer Driver

Design an interface circuit to drive 128 electromechanical relays (printer hammers) requiring 24-V DC at 100 mA each.

Solution

An SN75447 will be used. This is a dual "NAND enabled" driver with integrated clamp diodes for transient protection. The NAND function was chosen so that a high input signal will turn on the driver and a low input will turn the driver off. For compatibility with system logic, the designer could alternately choose the AND, NOR, or OR functions as desired.

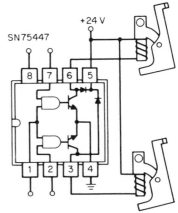

FIG. 14-6   Dual-hammer driver.

The 128 hammers will require 64 SN75447 devices. Each device requires +5-V DC for V_{CC} at a maximum current of 18 mA. Total I_{CC} current
for all 64 drivers would be 1.152 A maximum.

Fig. 14-6 shows a typical circuit for each device driving two hammers.
If the enable input (pin 1) is not used, it should be tied to the V_{CC} rail.

14–2c   Current Sensing

The low input current required by the SN75446 series can also be used to
an advantage when driven by an optoisolator. This is illustrated by the
following example.

Example 14–4   Current-Sensing Alarm Driver

Suppose you need to sense the presence of current in a circuit and activate
an alarm—relay, horn, etc.—if the current has ceased. In addition, the alarm
circuit needs to be isolated from the current source, and it should stay
activated until the circuit is manually reset.

Solution

Fig. 14-7 shows a circuit with very few components that could satisfy the
requirements. Isolation is achieved with a TIL154 low-cost optoisolator.
Isolation voltage is specified as 2500-V rms, and the device has a UL listing.

Since this circuit requires current to flow through the LED emitter
continuously, degradation effects in the LED emitter can be minimized by
keeping its forward current very low. By making use of the high sensitivity

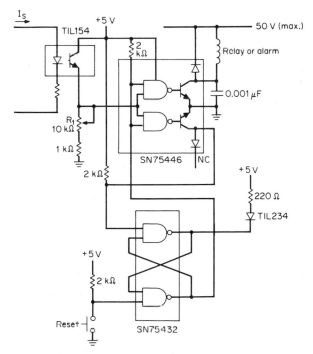

FIG. 14-7 Current monitor.

of the SN75446, the forward current through the LED can be reduced to several milliamperes, thus essentially eliminating degradation effects in the coupler.

The current being sensed is passed through the LED emitter. The photocurrent produced holds the two inputs to the SN75446 high and the output is held off. If the LED current drops below the amount required to keep the input high, both output sections of the SN75446 will go low.

The latch, using a dual SN75432 as described in an earlier example, will hold the alarm on until reset. A TIL234 green VLED indicates that the system is reset and current is flowing through the LED in the optoisolator. Sensitivity of the circuit can be adjusted with R_1.

If currents being monitored are higher than 4 or 5 mA, a shunt could be used to divide part of the current into the LED emitter.

Note that the load on the SN75446 can go as high as 50 V and 350 mA in this circuit.

In some applications involving the switching of inductive loads, the fast rise time and high-voltage transient occurring during turn-off can force the output transistor into a secondary breakdown condition. In such cases the collector voltage reaches V_{CC_2} within a few nanoseconds. To prevent undesired breakdown, the collector voltage slew rate should be reduced to 1 V/ns or less. This gives the gate sufficient time to provide a low base-to-ground impedance before the collector voltage is extremely high, and collector-to-emitter breakdown is prevented. A 0.001-μF capacitor from the collector of the output transistor to ground is usually adequate to accomplish this function.

14–2d   Lamp Flashers

The following design example illustrates how a high-current lamp flasher can be built using a handful of components. This high-current flasher can be used as part of an alarm system or annunicator panel.

Example 14–5   Lamp Flasher Design

Design a low-frequency lamp flasher circuit.
Solution
While the voltage levels of the SN75446 series are TTL logic levels, the input currents are comparable to many CMOS devices. Fig. 14-8 shows how

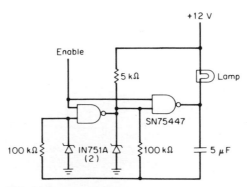

FIG. 14-8   Lamp flasher.

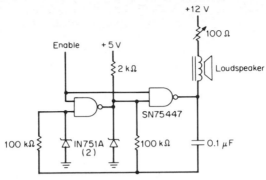

FIG. 14-9 High-current audio oscillator.

the SN75447 can be used to make a low-frequency oscillator that will flash a lamp at approximately a 50% duty cycle. Note that the resistance and capacitance values are comparable with CMOS circuits, while the output can switch up to 50 V and 35 mA.

The resistance and capacitance values can be changed to increase the oscillator frequency into the audio range to drive a high-power loud speaker. This circuit, shown in Fig. 14-9, could be used in conjunction with the lamp flasher as part of an alarm system. The common enable pin can be used on both the flasher circuit and the audio circuit to enable both circuits simultaneously.

14–2e   Load Isolation

Example 14–6   Electrical Isolation of Loads

Design a circuit to drive and electrically isolate a 50-V 300-mA load from a 4-bit CMOS microcomputer, such as the TMS 1000C.

Solution

The "R" output ports from the CMOS TMS 1000C can source or sink several milliamperes of current. While electrical isolation can be achieved with an optoisolator, the limited current available from the CMOS microcomputer usually dictates use of a high CTR (current transfer ratio) isolator at a forward IR emitter current of 1 or 2 mA, such as the 4N47. These isolators are quite expensive.

Alternate approaches use discrete transistors to increase the current drive to the isolator. A lower cost isolator can then be used, but the additional components required to achieve the current gain add to the system cost. By using the SN75446 series of peripheral drivers that require very low switching current, i.e., 10 µA maximum, a low-cost isolator, such as the TIL111, can be used. With 2 mA of current into the IR emitter of the TIL111, sufficient current transfer can be achieved in its phototransistor to switch the SN75446 series without any additional components.

The interface components for a SN75447 driver are shown in Fig. 14-10. Any device from the SN75446 series could be used depending on the logic desired.

The R outputs ports of the TMS 1000C can either source or sink current.

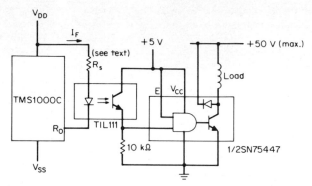

FIG. 14-10   TMS1000C interface (R_o line sink).

Fig. 14-10 shows the TMS 1000C sinking the IR emitter current with the R_o port.

The data sheet specifications on the TMS 1000C define a ''low'' voltage of 1.8 V at 2.9 mA. The voltage drop across the IR emitter is approximately 1.2 V. The resistor in series with the IR emitter is calculated as follows:

$$R_s = \frac{V_{CC} - V_{IR} - V_{OL}}{I_F} \qquad (14\text{-}1)$$

where V_{CC} = supply voltage = 5 V

    V_{IR} = forward voltage of IR emitter = 1.2 V

    V_{OL} = TMS 1000C low-level output voltage = 1.8 V

    I_F = forward IR emitter current = 2 mA

$$R_S = \frac{5 - 1.2 - 1.8}{0.002} = 1000 \ \Omega$$

When R_o is pulled low, the SN75447 input will be pulled high by the TIL111 phototransistor and the SN75447 output transistor will sink current through the load. The same logic could be accomplished by sourcing current

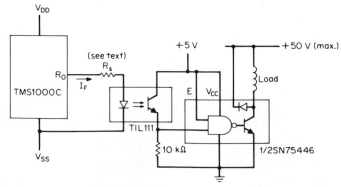

FIG. 14-11   TMS1000C interface (R_o line source).

through the IR emitter and using the AND SN75446. Fig. 14-11 shows the IR emitter sourced by current from R_o. The value of R_s is calculated in the following manner.

The phototransistor in the TIL111 has to supply a current of only 250 μA minimum to swing the voltage across the 10 kΩ resistor above 2.4 volts and supply the required input current of 10 μA for a "high" logic input level into the SN75446. Therefore

$$R_s = \frac{V_{OH} - V_{IR}}{I_F} \tag{14-2}$$

where V_{OH} = TMS 1000C high-level output voltage = 3.1 V
$\quad V_{IR}$ = forward drop of IR emitter = 1.2 V
$\quad I_F$ = forward current thru IR emitter = 2 mA

$$R_s = \frac{3.1 - 1.2}{0.002} = 950 \ \Omega$$

This basic circuit can be used with other members of the TMS 1000 family, such as the 15-V PMOS TMS 1000, the 9-V TMS 1000 NLL, TMS 1100 NLL, etc., to provide isolation and level shifting.

14–2f Level Shifting

Example 14-7 illustrates a means of level shifting from PMOS to TTL logic levels.

Example 14–7   Design of PMOS-to-TTL Level Shifter

Design an interface circuit to level shift from a 9-V 4K TMS 1400NLL to TTL from one R output line and drive a 50-V/300-mA load from another R output line.

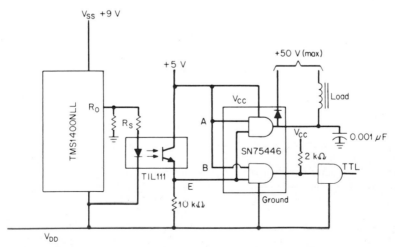

FIG. 14-12   A PMOS-to-TTL level shift.

Solution

The schematic in Fig. 14-12 shows a circuit using a TIL111 driving an SN75446. One section of the SN75446 drives the high voltage/high current load and the other section interfaces to the TTL circuits.

The TMS 1400NLL R lines are capable of higher current than the TMS 1000C discussed in Example 14-6. However, degradation effects can be minimized in the TIL111, if the current is limited to 5 mA or less. With the high input sensitivity of the SN75446, 5 mA into the IR emitter of the TIL111 is more than sufficient to switch the peripheral driver.

R_s is calculated as described in Example 14-6 with the following change:

$$R_s = \frac{V_{OH} - V_{IR}}{I_F} \tag{14-3}$$

where V_{OH} = high-level output of R line = 7.3 V for the TMS 1400NLL

V_{IR} = forward voltage drop of the IR emitter = 1.2 V

I_F = forward current through the IR emitter = 5 mA

$$R_s = \frac{7.3 - 1.2}{0.005} = 1220 \ \Omega$$

Note that the PMOS devices can only source current from the R lines and that a pull-down resistor is required on the R lines to ensure a low output when the R line is off.

14–2g   Driving High-Current Loads

The TMS 1000C can interface directly with the SN75446 series without the optoisolator if isolation is not required. The following section will illustrate this interface circuit and discuss how the peripheral drivers can be paralleled for higher current capability.

Example 14–8   High Current Load Driver

Drive a 50-V/600-mA load from the R output ports of a TMS 1000C.

Solution

Fig. 14-13 shows a typical circuit. The R output ports can either source or sink current (active high or active low); however, the low logic level is not TTL compatible. This is not a problem if an external 10-kΩ pull-down resistor is used on the input to the SN75446 to ensure a logic low.

Under conditions of logic output high, the V_{OH} will be a minimum of 3.1 V. Input current to the SN75446 will be a maximum of 40 μA with all inputs tied together.

When the R_o output goes low, the minimum V_{OL} is 1.8 V. This voltage will not ensure a low logic level into the SN75446 gate unless an external pull-down resistor of 10 kΩ is used. Under low conditions the R_o output will only sink current and the 10-k Ω pull-down resistor will pull the input gate of the SN75446 to logic low.

Note that two inputs to the SN75446 could be driven from other R lines of the TMS 1000C to make use of the AND logic integrated into the

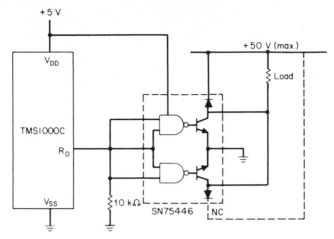

FIG. 14-13 TMS 1000C high-current interface. (*Note:* Clamp diode connected as shown for inductive load.)

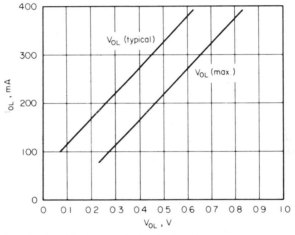

FIG. 14-14 The V_{OL} vs. I_{OL} for the SN75446.

SN75446. Separate 10-kΩ pull-down resistors would be required on each R line. The outputs cannot be paralleled in this case, and each output will require an independent load.

Since both output transistors in the example are connected in parallel, the limiting factor on maximum current capability is the package power dissipation. A secondary factor is the distribution of load current between the outputs. Since both output transistors are in the same monolithic silicon chip, the V_{OL}'s will be closely matched. It is good engineering practice, however, to derate the output current by 10 to 20% as a safety margin to account for any slight unbalance in the current distribution.

Fig. 14-14 shows the typical and maximum V_{OL} as a function of output current. For the required current rating of 600 mA, each transistor will conduct approximately 300

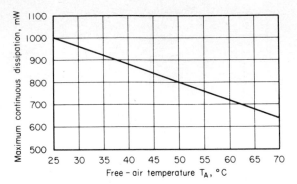

FIG. 14-15 Power-dissipation derating curve SN75446 (P Package).

mA, and the maximum V_{OL} at 300 mA can be found from the figure as approximately 0.65 V.

Collector-to-emitter power dissipation will be 0.65 times 0.6 A, or 390 mW. In addition, the V_{CC} rail must supply approximately 25 mA maximum for the logic circuitry. This represents another 125 mW, or total package power dissipation of approximately 515 mW.

The power-dissipation derating curve for the SN75446 (P DIP Package) is shown in Fig. 14-15. Five hundred fifteen milliwatts is well under the maximum device capability of 70°C. The curves in Figs. 14-14 and 14-15 can be used to calculate power dissipation at higher current levels if desired. Note that other conditions also affect the package

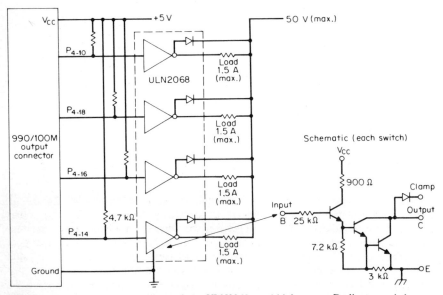

FIG. 14-16 High-current interface using a ULN2068 quad high-current Darlington switch.

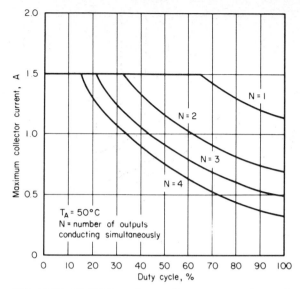

FIG. 14-17 ULN2068 maximum collector current vs. duty cycle
($T_a = 50°C$).

temperature rise. If the SN75446 is placed close to a hot component, such as a power transformer, the package temperature could be exceeded unless adequate cooling is provided inside the system enclosure.

Example 14–9 Quad Driver for 500-mA Relays

Design an interface circuit to interface between the output ports of a TM990/100M-1 microprocessor board and four 24-V 500-mA relays. These output ports are typical of most NMOS microprocessors available on today's market, and can usually drive at least one standard TTL gate.

Solution
The TM990/100M-1 microcomputer board has TTL compatible output ports available on connector P4. Four of the ports can be used to drive a ULN2068 quad high-current Darlington switch. The ULN2068 can satisfy the high-voltage/high-current relay drive requirements. A circuit is shown in Fig. 14-16. The internal clamp diodes in the ULN2068 are used to protect the output transistors from reverse transients produced by switching the inductive relay loads.

The only thing critical about the design of Example 14-9 is the power dissipation in the Darlington drivers. The ULN2068 is packaged in a 16-pin NE dual-in-line plastic package with a total power dissipation of 2075 mW at 25°C free air temperature. The power-dissipation derating curve for this package can be determined from Fig. 14-4.
Figs. 14-17 and 14-18 show the maximum collector current vs. duty cycle at $T_a = 50°C$ and $T_a = 70°C$, respectively. Pins 4, 5, 12, and 13 are connected to ground as well as to the internal heat sink. This heat sink can be used to an advantage to decrease the operating temperature of the device as described in Example 14-2.

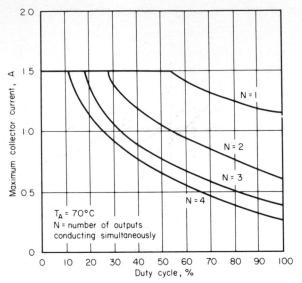

FIG. 14-18 ULN2068 maximum collector current vs. duty cycle ($T_a = 70°C$).

Example 14–10 Device Protection

Design a circuit to provide protection against SN75447 device failure if IC power is removed while load power is still applied.

Solution

If V_{CC_1}, which normally supplies power to the logic gate circuit, is removed while V_{CC_2} is still applied, the output transistor's blocking voltage capability can be significantly reduced. This is due to the increase in base-terminating impedance of the output transistor. This lower breakdown voltage can result in high leakage current with excess power dissipation and unwanted current flowing through the load and can result in device failure.

Fig. 14-19 shows a practical method of applying V_{CC_1} from V_{CC_2} if the regulated V_{CC_1} supply voltage is inadvertently removed from the circuit. A zener breakdown voltage of 4 to 5 V less than V_{CC_2} will establish a V_{CC_1} voltage that will provide the necessary bias voltage to keep the output transistor at its full blocking capability. If, for example, V_{CC_2} is 24 V, an 1N5249B 19-V, 5% zener could be used to provide a 5-V supply level for V_{CC_1}. The example shown in Fig. 14-19 uses an SN75447 peripheral driver in a typical circuit, and the comments are applicable to any driver. However, power dissipation in the zener must be taken into account if higher voltage devices and/or devices with a significantly higher I_{CC} current are used. It may be necessary to use

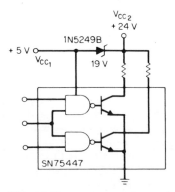

FIG. 14-19 V_{CC_1} protection with zener.

higher power rated zeners or stack several zeners in series to keep each zener within its power-dissipation rating.

14–2h Balanced Line Drivers

While line circuits will be discussed in Sec. 14-3, many of the peripheral drivers can be used as line drivers. The following example shows a typical circuit.

Example 14–11 Driving Twisted Pairs

Drive a 140 Ω twisted pair line from TTL logic levels.
Solution
Fig. 14-20 shows an SN75450B used as a balanced line driver. Note that the emitter of one of the output transistors and the substrate must be connected to the most negative DC voltage in the system to keep the substrate diodes reverse biased and provide isolation between the various circuit functions in the chip.

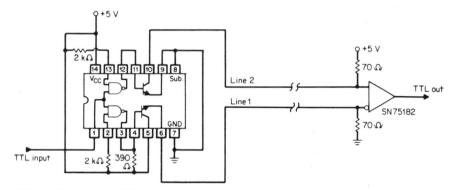

FIG. 14-20   An SN75450B used as a line driver.

Having the separate substrate connection allows one of the output transistors to be used as an emitter follower and current source and the other output transistor to be used as a current sink.

The other TTL inputs, pin 2 and pin 13, are available to perform additional logic functions, if desired, otherwise they are tied to V_{CC} through the 2-kΩ pull-up resistors.

Another example of the SN75450B application as a line driver is shown in the next example.

Example 14–12 Negative-Voltage Line Driver

Some line circuit specifications require voltage swings below ground. One special military standard, for example, swings from a 0- to a -12-V level. Design a circuit to meet this requirement.
Solution
Fig. 14-21 shows how the SN75450B could be used to interface standard TTL levels to negative levels by making use of the external substrate connection. Zener diodes are used to perform level shifting to drive the output transistors.

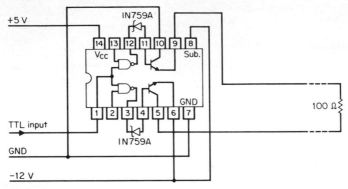

FIG. 14-21   TTL interface to negative logic levels.

14–2i   Stepper Motor to TTL Conversion

Example 14–13   Stepping Motor Driver

Design a circuit to interface a two-phase stepping motor to TTL levels.

The stepping motor has the following characteristics:

> Operating voltage, 16–18 V (nom. 17 V)
> Two-phase pulsewidth = 8 ms
> Advance cycle time = 16 ms
> Motor current = 400 mA

These characteristics are typical of small stepping motors used for paper advance on printers, etc. The motor is phase sensitive and contains two separate windings. The input waveform steps the motor once for each 16-ms paper advance cycle. This cycle consists of ϕ_1 and ϕ_2 input pulses, as shown in Fig. 14-22.

Solution

The TL376C is a monolithic bipolar three-channel stepper-motor controller designed for this type of application. The three output transistors can source or sink up to 500 mA.

The circuit diagram of a motor driver using the TL376C is shown in Fig. 14-23. Direction of rotation is controlled by the phase shift of the input TTL-level drive pulse.

Input voltage requirements for the TL376C are specified as $(V_{CC}/2) + 0.8$ V for high level and $(V_{CC}/2) - 0.2$ V for low level. An SN75461 is used to provide level shifting from the input TTL to the logic levels required to operate the TL376C.

The TL376C is encapsulated in the 2075-mW NE package previously discussed. Care must be exercised to provide power dissipation with large copper areas on the PCB or by means of a heat sink, as several outputs will always be on at the same time when driving the motor.

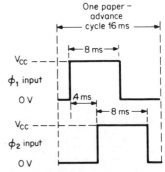

FIG. 14-22   Stepper motor input voltage waveforms.

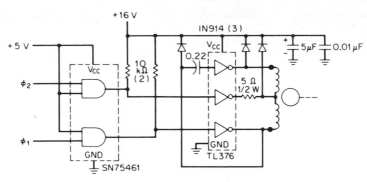

FIG. 14-23   Stepper motor driver interface.

While most of the peripheral drivers are designed for general-purpose applications and can be used in a variety of circuits, there are a few specialized devices designed for some specific applications.

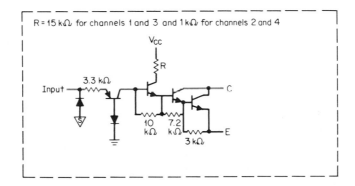

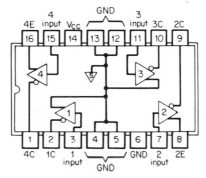

$\overline{S}$ = substrate

FIG. 14-24   Schematic and pin-out of UDN2845.

Two examples of these specialized devices are the UDN2841 and the UDN2845. Both devices are quad high-voltage (50-V), high-current (1.5-A), with TTL-compatible inputs. They are encapsulated in the 2-W NE power package.

The UDN2841 is intended for current sink applications with the load connected to ground and the device sinking to the negative supply. The UDN2845 is a sink and source combination for use in bipolar switching applications where both ends of the load are floating.

A schematic of each driver in the quad is shown in Fig. 14-24.

Use of the UDN2845 as a motor driver will be illustrated in the following example.

Example 14–14   TTL-to-DC Motor Interface

Interface a TTL-level source to a DC motor with capability for forward and reverse operation.

Solution

The UDN2845 quad has two sections designed to source current and two sections designed to sink current. A standard H configuration, as shown in Fig. 14-25, could be used to drive the DC motor. When the input is high and the enable input is high, the motor will reverse. If the enable is low, the motor will stop.

The UDN2845 is designed to source and sink current with negative power supplies. Collectors of transistors 2 and 4 are connected to ground and source current out of their emitters. Emitters of transistors 1 and 3 are connected to the negative supply and sink current to the negative rail.

A positive 5 V on a separate V_{CC} pin provides proper level shifting to operate all four drivers with TTL-level inputs above ground.

Since two sections of the device will always be on at the same time, the total current capability will be limited by the 2075-mW power dissipation of the NE package. Good heat-sink techniques must be used on pins 4, 5,

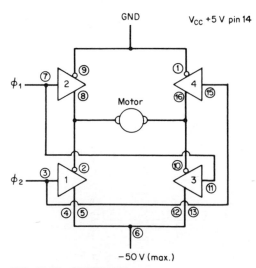

FIG. 14-25   UDN2845 H motor driver.

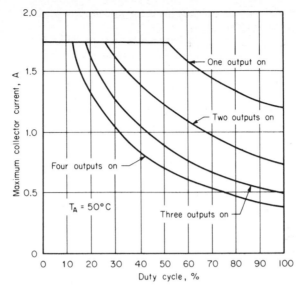

FIG. 14-26 Maximum collector current vs. duty cycle for UDN2845.

12, and 13 with large copper areas on the board or an external metal heat sink.

One of the curves shown on the UDN2845 data sheet is reproduced in Fig. 14-26. This curve shows the maximum collector current vs. duty cycle when two outputs are conducting simultaneously at 50°C free air temperature.

Additional curves and information can be found on the UDN2845 data sheet.

14–3 LINE CIRCUITS

System designers are constantly faced with the problem of transmitting digital data from one system to another. The increasing use of computers in a wide variety of industries has increased the need for data transmission between the components making up the system.

Many systems use a central computer with a number of remote terminals. These remote terminals may be located at distances of a few feet away or possibly thousands of feet. Even a small independent system will usually have one or more pieces of peripheral hardware, such as a printer located at some, usually short, distance away.

The need for data-transmission standards became apparent as the industry matured. At first, each system manufacturer established its own standards for its own products. This forced the systems engineer to either stay with the same manufacturer for the central processor and all external hardware, or fabricate what could be a complex interface circuit to convert from one system to another.

By adopting industry standards, the various manufacturers have allowed the systems engineer to choose the standard that matches the application, then mix and match computers and peripheral hardware that use the chosen standard (regardless of manufacturer) to achieve the best overall solution.

In general, these standards are defined by the state-of-the-art technology capabilities of semiconductor line drivers and receivers. As technology has improved, new standards have been written, extending the capabilities of the system. Also system requirements for higher data rates, due to improved computer speed and power, have spurred new product development for line circuits with improved characteristics meeting these system requirements. The line circuits all have a common purpose, to transfer, without error, digital information over greater distances and/or to more devices, at lower cost, than would otherwise be possible with standard discrete logic circuits.

It is very important for a designer of peripheral computer hardware to be intimately familiar with the various data-transmission standards. With the right choice, the hardware can be more versatile by interfacing into many standard systems, and have immediate customer acceptance. On the other hand, a nonstandard interface designed into a piece of peripheral hardware could cause many problems with a customer's system design, and the product would probably meet an early death.

Key considerations in selecting a data-transmission standard are line length, bit rate, environment, noise, etc., and whether or not the system will have to interface with other existing or future systems.

14–3a EIA Standards

Fig. 14-27 summarizes four EIA standards in use today. One of these standards, RS-485, is still in committee and has not been released in its final approved form as of this writing.

To illustrate typical applications using these standards, a complete system using local and remote terminals will be designed. A block diagram of this proposed system is shown in Fig. 14–28.

The individual interfaces making up the system will be used as examples to show how possible components are selected on the basis of system requirements.

Digital signals can be transmitted either single ended or balanced over various lines or cables. These can include coaxial cables, twisted-pair lines, flat or ribbon cables, or discrete wires. The twisted-pair line for balanced operation has several advantages over other transmission lines:

1. Cancellation of noise due to the alternating polarity of the magnetic circuits provided by adjacent twists of the line.
2. Both wires in the pair are equally affected by electrostatically coupled noise, resulting in a net common-mode signal with respect to the ground return.
3. Voltage differences between parts of the system appear as common-mode signals and can be rejected by the receiver.
4. Characteristic impedance is fairly uniform, making the line easy to terminate, usually with a 100 Ω resistor at the receiver. Table 14-5 shows characteristics of typical twisted-pair lines available from several vendors.
5. Twisted-pair lines are low cost with long life and are mechanically very rugged.

RS-232C The RS-232 was introduced in 1962 and is now widely used in the industry. This standard was developed for single-ended data transmission over short distances and slow data rates. Basic requirements for the RS-232C transmitter and receiver are summarized below.

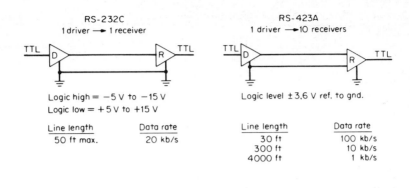

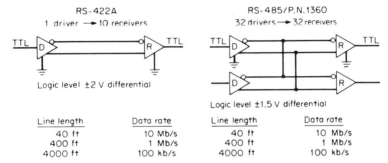

FIG. 14-27 EIA data-transmission standards.

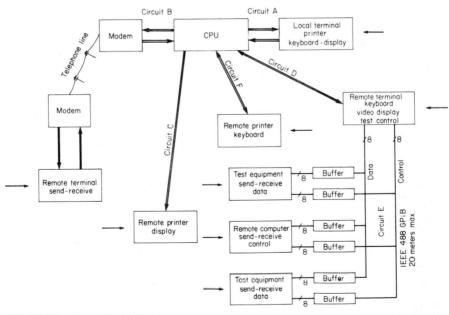

FIG. 14-28 System block diagram.

TABLE 14–5 Characteristics of Twisted-Pair Lines

Description	Impedance (Ω)	Wire	Capacitance per foot (pF)	Manufacturer	Manufacturer number
Shielded twisted pair	100	#20 (7 × #28)	15	Belden	8227
Twisted pair, vinyl insulation	110	#22 solid	12	Belden	8481
Twisted pair, plastic jacket	110	#22 solid	25	Alpha	1793
Twisted pair, plastic insulation	100	#20 (10 × #30)	15	Alpha	1918
Twisted pair, vinyl insulation	100	#22 solid	15	Belden	8795

RS-232C Driver Requirements

1. Output must withstand an open circuit or short circuit to ground from either supply, or any other conductor in the cable.
2. Power-off impedance $>300\ \Omega$.
3. Maximum open-circuit voltage ± 25 V.
4. Maximum short-circuit output current 500 mA.
5. Absolute value of the output drive, into a 3000- to 7000-Ω load must be >5 V and <15 V.
6. Output rise and fall times, within the transitional limits of plus and minus 3 V, shall not exceed 1 ms.
7. Output slew rate shall not exceed 30 V/μs.
8. Maximum data rate is 20,000 bits/s.

TABLE 14–6 RS-232C Connector Pin Definitions

25-pin Connector pin number	EIA RS-232 name	Description
1	AA	Protective ground
2	BA	Data transmitted from terminal
3	BB	Data received from modem
4	CA	Request to send
5	CB	Clear to send
6	CC	Data set ready
7	AB	Signal ground
8	CF	Carrier detector
9–14	—	Undefined
15	DB	Transmitted bit clock, internal
16	—	Undefined
17	DD	Received bit clock
18 and 19	—	Undefined
20	CD	Data terminal ready
21	—	Undefined
22	CE	Ring indicator
23	—	Undefined
24	DA	Transmitted bit clock, external
25	—	Undefined

RS-232C Receiver Requirements

1. Input impedance $>3000\ \Omega$ and $<7000\ \Omega$.
2. Maximum effective shunt capacitance from receiver input and connecting cable <2500 pF with no inductive reactive component.
3. Open-circuit input voltage <2.0 V.
4. Maximum data rate is 20,000 bits/s.
5. Input voltage limits ±25 V.

To facilitate interconnection of the equipment using the RS-232C, pin-out of a 25-pin D connector has been standardized for the RS-232C interface. Most systems only require part of the defined signals, so it is necessary to check the hardware thoroughly before making the connection. The RS-232C connector pin definitions are shown in Table 14-6.

Example 14–15   RS-232C Communication Link

Interface a central processor (CPU) to a local terminal that includes a printer, video display, and keyboard. This terminal will be located on the same table with the CPU and a cable length of 10 ft will connect the parts together. The data rate desired is 9600 baud. This portion of the system is shown as circuit A in Fig. 14-28.

A typical CPU could be the TM990/100M-1 board. This board uses the TMS9900 16-bit microprocessor chip with TIBUG monitor available in EPROMs. Also included on the board are 256 16-bit words of random access memory (RAM) to hold program data that can be expanded to 512 words, if desired. All address, data, and control lines are brought to the board connectors so the board can be expanded to use the entire capability of the 16-bit TMS9900 in larger systems.

A serial I/O port is jumper selectable as either RS-232C or a 20-mA TTY interface. Sixteen parallel I/O ports are handled by a TMS901 on the board. These outputs are TTL compatible and capable of one standard TTL load.

Solution

The RS-232C serial I/O port on the TM990/100M-1 board will be used to interface the local terminal. A circuit diagram of the interface is shown in Fig. 14-29. The RS-232C is a logical choice for this interface due to the low baud rate and short distance required.

The TMS9902 on the CPU board is a programmable system I/O controller with TTL-compatible outputs. As shown in Fig. 14-29, the TTL output of the TMS9902 drives an SN75188 quad RS-232C line driver. These drivers are also installed on the CPU board, and the RS-232C-level signals are available on the 25-pin standard EIA connector mounted on the board. A similar 25-pin connector is also used on the terminal. As shown in Fig. 14-29, a quad SN75189A receiver and a quad SN75188 driver are used inside the terminal to interface the RS-232C levels to standard TTL logic levels.

The SN75188 is a quad line driver that meets the RS-232C requirements. Its input is TTL compatible and has a current-limited output of 10 mA at the RS-232C logic levels. The SN75189A is a quad receiver that accepts the RS-232C logic levels on its input and

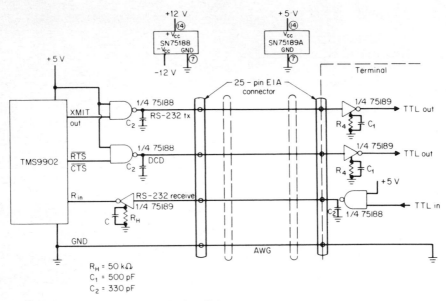

FIG. 14-29   The RS-232 communications link.

delivers TTL-compatible levels on its output. These TTL levels are used to communicate with devices inside the terminal.

The slew rate for the driver is specified as >30 V/μs in the RS-232C standard. Some devices, such as the SN75188, require an external capacitance, either in the transmission line or external to the chip. The 330-pF capacitor shown in Fig. 14-29 on the output of each driver serves this purpose. Other integrated-circuit drivers, such as the μA9636, can be programmed with an external resistor for the correct slew rate.

Resistor R_4 is used to set the hysteresis in the SN75189A for improved noise immunity. The 500-pF capacitor across the resistor also helps improve the noise immunity by filtering out short-duration noise pulses. Note that the data transfer is unidirectional, i.e., a driver in the CPU talks to a receiver in the terminal over one line and a driver in the terminal talks to a receiver in the CPU over another line. The third line shown in Fig. 14-29 carries additional control signals from the CPU to the terminal.

Other devices designed for the RS-232C that could have been used in this application are shown in Table 14-7.

TABLE 14–7 RS-232C Device Selection Guide

Driver	Number per package	Receiver	Number per package	Comments
SN75150	Dual	SN75152	Dual	Also meets MIL-STD-188C
SN75156	Dual			
μA9636	Dual	SN75154	Quad	
SN75186	Quad			
SN75188	Quad	SN75189	Quad	
		SN75189A	Quad	Improved hysteresis over SN75189

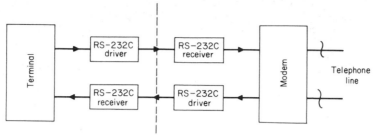

FIG. 14-30   Terminal/modem interface.

Example 14–16   Modem Interface

Design an interface system to allow bidirectional data transmission from the CPU to a remote terminal located several thousand miles away. Data rates of up to 9600 baud are required.

Solution

Standard long-distance telephone lines can be used to transfer bidirectional data at low data rates. This is illustrated in circuit B of Fig. 14-28 where a modem is used to connect to the telephone at the CPU and at the remote terminal. Most of the modems commercially available today use RS-232C interfaces, as the data rates are low and they are usually close to the CPU or terminal. The interface between the CPU to the modem and the remote modem to the terminal would be essentially the same as described in Example 14-15. A block diagram of this system is shown in Fig. 14-30.

RS-422A/423A   The RS-423A, proposed in 1965, applies to single-ended applications similar to RS-232C; however, the data rate and line length were significantly increased. The standard also features controlled waveshaping, dependent upon cable length and data rate, to control reflections and radiated emission or crosstalk.

TABLE 14–8 Comparison of RS-423A and RS-422A

Specification	RS-423A	RS-422A
Mode of operation	Single-ended driver differential receiver	Differential driver and receiver
Maximum data rate over 4000 ft	100 kbaud/s	10 Mbaud/s
Maximum voltage applied to driver output	±6 V	−0.25 to +6 V
Driver output signal Loaded Unloaded	±3.6 V ±6.0 V (reference to ground)	±2.0 V ±5.0 V (differential)
Driver load	450 Ω minimum	100 Ω minimum
Receiver input voltage range	±12 V	−7 to +7 V
Supply voltage	Positive and negative	+5 V only

The increase in line length and data rate capability of RS-423A is due primarily to the use of very sensitive receivers with differential inputs. The use of sensitive differential receivers significantly improved the performance of RS-423A over RS-232C. By going to a differential output from the driver *and* differential input on the receiver, further improvement in performance can be achieved. The differential configuration nullifies effects of any ground shifts and noise signals that appear as common-mode voltages on the driver outputs and receiver inputs.

EIA Standard RS-422A applies to the differential mode of operation. This standard allows data rates up to 10 Mbaud for line lengths of 40 ft and data rates of 100 kbaud for line lengths to 4000 ft.

A comparison of RS-422A and RS-423A is shown in Table 14-8.

Drivers designed to meet RS-422A are capable of transmitting a 2-V differential signal to a twisted-pair line terminated in 100 Ω. The receivers are capable of detecting a ±200-mV differential signal in the presence of a common-mode signal from −7 to +7 V. Both RS-423A and RS-422A allow one transmitter and ten receivers on the line at the same time.

Example 14-17   Application of RS-422A Interface

An interface is required for the long line transferring data to the remote printer and display shown in circuit C of Fig. 14-28. This circuit requires unidirectional transfer of data from the CPU to the remote hardware.

Solution

For this application, we choose RS-422A. A schematic diagram of the system is shown in Fig. 14-31.

As with the RS-232C output to the modem, one of the TTL-compatible data lines in the TMS9902 controller chip can be used to drive the interface to the remote printer.

We will use RS-422A for this application, although RS-423A could have been used. The advantage of RS-422A over RS-423A, in addition to higher performance, is the need for only +5-V power. Drivers for RS-423A require both positive and negative power supplies. A dual μA9638C driver is used to interface the TTL output from the TMS9902 to the twisted-pair cable. A dual μA9637AC receiver is used to interface the twisted pair to the TTL input of the printer and display. In RS-422A (and RS-423A) one driver is capable of communicating with up to ten receivers. With RS-422A the data rate can be up to 100 kbaud over line lengths to 4000 ft, compared to 1.0 kbaud for RS-423A over the same distance.

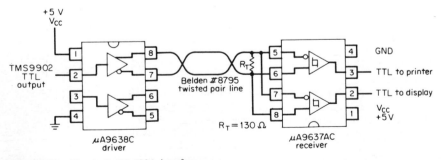

FIG. 14-31 Circuit C, RS-422A interface.

TABLE 14–9 Devices Available for RS-422A

Transmitters	Receivers
SN75158 (dual)	SN75157 (dual)
SN75159 (dual)	
μA9638C (dual)	μA9637AC (dual)
AM26LS31C (quad)	AM26LS32AC (quad)
	AM26LS33AC (quad)
MC3487 (quad)	MC3486 (quad)
SN75151	
SN75153	
SN75172 (quad)	SN75173 (quad)
SN75174 (quad)	SN75175 (quad)

Other devices that could have been used for this application are shown in Table 14-9.

RS-485   A new standard is under consideration that overcomes some of the limitations of RS-423A in regard to multiple transmitters and receivers on the same line. Initially this standard was called P.N. 1360, but its designation was recently changed to P.N. 1488. However, Texas Instruments and most data books still refer to the standard as P.N. 1360. The specification in the finalized form will be assigned an RS-485 number. The key features of RS-485 compared to RS-422A are:

- Driver common-mode output voltage -0.25 to $+6$ V in RS-422A extended to -7 to $+12$ V with power on or off.

- Receiver common-mode input voltage of -7 to $+7$ V in RS 422A is extended to -12 to $+12$ V.

- Receiver input impedance increased from 4 kΩ minimum to 12 kΩ minimum.

- RS-422A and RS-423A only allows one transmitter and ten receivers on the same line. RS-485 allows up to 32 transmitters and 32 receivers on the same line.

- Devices satisfying RS-485 requirements will have built-in protection circuitry to prevent failure if two transmitters are turned on at the same time. This feature is called contention protection.

Example 14–18   Using RS-485

Circuit D of Fig. 14-28 is also a long line, but differs from Circuit C in that bidirectional communication is required. Design a circuit to transfer data from the remote terminal back to the CPU for processing.

Solution

Fig. 14-32 is a schematic of circuit D using new devices designed for RS-485. While only two transmitters and receivers are used in circuit D, additional transmitters and receivers could be added anywhere along the line. The line should be terminated at both ends with a resistor having the same value as the characteristic impedance of the line.

If additional transmitters and receivers are placed on the line, the connection to the main line should be as short a distance as possible to minimize unwanted line reflections.

As shown in Fig. 14-32, an SN75176 transceiver is used at each end of the line. This device internally connects the driver outputs to the receiver inputs, and provides complementary enables for bidirectional data communication.

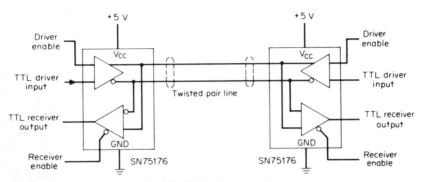

FIG. 14-32   Circuit D, RS-485 (P.N. 1360) bidirectional interface.

The SN75176 is designed to operate with a single $+5$-V supply and maintain a high-impedance output over a common-mode range from -7 to $+12$ V, with power supply on or off. The transmitter has a maximum delay time of 50 ns and rise and fall times less than 80 ns. This allows data rates up to 4 Mbaud. The total power consumption is only 162 mW with the transmitter enabled and 137 mW with receiver enabled.

The SN75176 features transmitter contention protection through the use of both positive and negative current limiting and thermal shut down. The sink current is limited to 500 mA out to 12-V common mode, and the source current is limited to -250 mA to -7-V common mode. For additional protection, a thermal sensing circuit causes the device to go into a high-impedance state whenever the chip temperature exceeds approximately 150°C.

The SN75176 receiver is similar to existing RS-422A differential receivers except for a higher input impedance and extended common-mode range. The receiver features ± 200-mV sensitivity over a -12 to $+12$-V common-mode range, 12-kΩ minimum input impedance, and 35-ns maximum propagation delay. Excellent noise margin is realized by a 50-mV hysteresis on the input.

Instead of a transceiver, separate transmitters and receivers are available, if desired. The transmitter in the SN75176 is equivalent to one section of an SN75172 quad differential driver and the receiver in the SN75176 is equivalent to one section of an SN75173 quad differential receiver. An alternate pair of devices with the same electrical characteristics but slightly different enable logic, is the SN75174 quad driver and the SN75175 quad receiver. Fig. 14-33 shows the pin-out and internal logic configuration of these four devices.

Note that these differential receivers can be used for RS-423A, RS-422A, or the newer RS-485 interface circuits. The transmitters, however, can only be used for RS-422A or RS-485 and not for RS-423A, since the outputs do not swing negative with respect to ground as required by RS-423A.

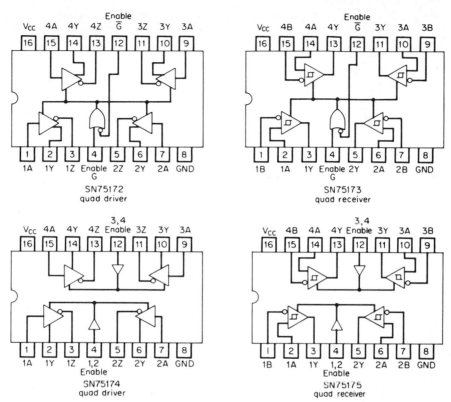

FIG. 14-33   P.N. 1360 drivers and receivers.

Example 14–19   Using the SN75177 and SN75188 Devices as Repeaters

Suppose the remote terminal shown in circuit D of Fig. 14-32 were 6000 feet from the CPU. This is beyond the capabilities of RS-423A, RS-422A, or RS-485. A modem could be used in conjunction with telephone lines, but the data rate would have to be significantly reduced. Devise an alternate solution.

Solution

The SN75177 and SN75188 transceivers are designed to be used as repeaters to extend the maximum cable distance and maintain the high data rate of RS-485. The enable inputs on the two devices are complementary so that the pair can be used for bidirectional communication. A schematic of circuit D using these repeaters for a 6000-ft line is shown in Fig. 14-34.

14–3b   The IEEE 488 GPIB Bus

Because of the large number of manufacturers building programmable instruments that must be easily and economically interfaced, there is a need for a standard industrywide instrumentation bus. For this reason the general-purpose interface bus (GPIB) defined by IEEE Standard 488 has standardized the interface system used to interconnect program-

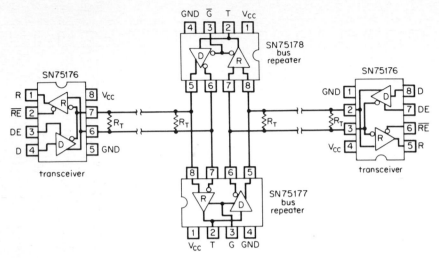

FIG. 14-34 Long line application with repeaters.

mable and nonprogrammable instruments, computers, and peripheral hardware necessary to build a complete instrumentation system.

This allows a user to purchase instruments from many different manufacturers and then connect them together using off-the-shelf cables. The IEEE 488 GPIB uses a 16-line bidirectional bus that carries data at rates of up to 1 Mbaud (1 Mbyte/s) on eight lines with handshaking and bus management signals on the other eight lines. Up to 15 instruments can be tied together with a maximum line length of 20 m (65.6 ft).

The block diagram shown in Fig. 14-28 includes programmable remote instrumentation. Components required to implement a typical programmable design will be described in the following example.

Example 14–20   IEEE 488 GPIB Bus

The RS-485 interface allows bidirectional communication over circuit D between a remote terminal and the central CPU. Several pieces of programmable equipment that need to be controlled by the terminal are located close to this remote terminal. This part of the system is shown as circuit E in our hypothetical system. Interconnect these devices.

Solution
IEEE Standard 488 GPIB will be used to interface the remote terminal and the peripheral programmable equipment. The GPIB uses a 16-line bidirectional bus that can tie together up to 15 instruments over a maximum line length of 20 m. The resulting circuit is shown in Fig. 14-35.

The Texas Instruments SN75160A octal-bus data transceiver is designed to provide the interface between the GPIB (bus) and the bus controller. The transceiver's outputs have a built-in termination network, required by the IEEE 488 Standard, that presents a high impedance to the bus when power is removed from the device. Each receiver has a minimum of 400-mV hysterisis for additional noise margin.

The SN75160A is an improved version of the original SN75160. It has lower power and faster speeds than the SN75160, as shown in Fig. 14-36.

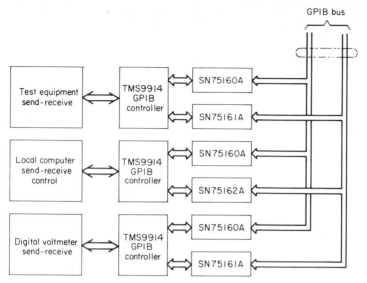

FIG. 14-35   GPIB interface.

A functional block diagram of the SN75160A and its pin-out is shown in Fig. 14-37. The direction of data flow is controlled by the talk enable (TE) input. All eight channels are simultaneously in the receive mode when TE is low, and data is received from the bus and transferred to the bus controller. When TE is in the high state, all eight channels go to the transmit mode, and data will be transferred onto the bus. Each driver features a totem-pole output that can actively drive the bus high or low to give the fastest data rates possible.

Eight lines of the IEEE 488 bus are used for data transfer and eight lines are used for control. The SN75160A octal transceiver was designed to implement the eight-line bus data bus.

The eight control lines are handled by either an SN75161A or an SN75162A. The SN75161A includes with the TE and direction-control inputs the necessary logic to enable each channel in the correct direction for the exchange of bus-management and handshaking signals. Three of the channels (NDAC, NRED, and SRQ) have open collector driver outputs, as required by IEEE 488. These lines are always used in a wired-OR configuration. The other five channels have totem-pole outputs.

The SN75162A offers an alternate method of implementing the eight-line control bus. It is identical to the SN75161A, except that the direction of the REN and IFC channels is controlled by a separate input called the system controller (SC). With this additional flexibility, control of the entire bus system may be transferred from one instrument to another in multiple controller systems.

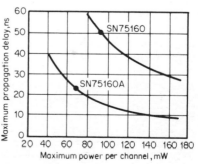

FIG.   14-36   Speed/power   comparison, SN75160 vs. SN75160A.

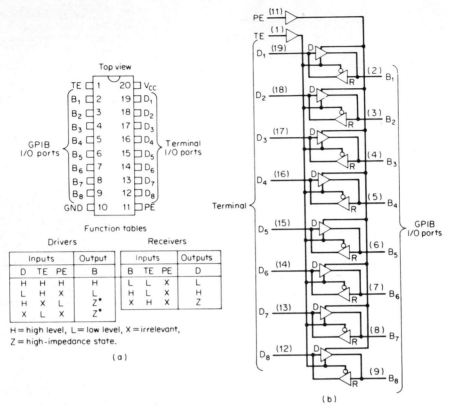

FIG. 14-37 The SN75160A. (*a*) Pin-out and truth table; (*b*) functional diagram.

Fig. 14-35 shows our system with three pieces of programmable hardware communicating through the IEEE bus to the remote terminal.

A TMS9914 GPIB controller device performs the interface function between the 16-wire bus and the microprocessor located inside the programmable equipment. The TMS9914 relieves the microprocessor of the task of maintaining the IEEE protocol. By utilizing the interrupt capabilities of the device, the bus does not have to be continually polled, and fast responses to changes in the interface configuration can be achieved. A bus controller, such as the TMS9914, is a very simple way of interfacing the IEEE bus, because it is tailored to the bus and requires no extra logic devices or complicated printed circuit board layout.

Communication between the microprocessor and the TMS9914 is carried out via memory-mapped registers. There are 13 registers within the TMS9914, six read and seven write. They are used for both passing control data and getting status information from the microprocessors.

The three least significant address lines from the microprocessor are connected to the register select lines RS_0, RS_1, and RS_2 and determine the particular register selected. The high-order address lines are decoded by external logic (see Fig. 14-38) to cause the CE input to the TMS9914 to be pulled low when any one of eight consecutive addresses are selected. Thus the internal registers appear to be situated at eight consecutive locations

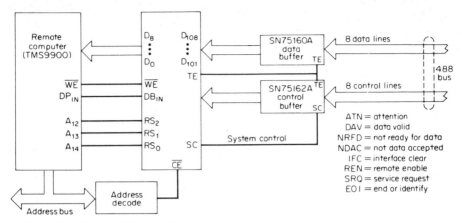

FIG. 14-38   Typical GPIB controller.

within the microprocessor address space. Reading or writing to these locations transfers information between the TMS9914 and the microprocessor. Note that reading and writing to the same location will not access the same register within the TMS9914 since they are either read-only or write-only registers. For example, a read operation with $RS_2 - RS_0 = 011$ gives the current status of the GPIB interface control lines, whereas a write to this location loads the auxillary command register.

Each SN75160A or SN75161A/SN75162A on the bus interface is given a 5-bit address enabling it to be addressed as a talker or a listener. This address is both read by the microprocessor and written into the address register as part of the initialization procedure. The TMS9914 responds by causing a My Address (MA) interrupt and entering the required addressed state when this address is detected on the GPIB data lines.

Additional information on the TMS9914 can be found in the *TMS9914 GPIB Adapter Preliminary Data Manual* available from Texas Instruments.

14-3c   Coax Lines

Coax transmission lines provide good isolation from radiated noise and crosstalk. Although generally used for single-ended transmission, dual coax cables may be employed for

TABLE 14-10 Characteristics of Typical Coaxial Lines

Type	Wire	Nominal impedance (Ω)	Nominal capacitance per foot (pF)	Attenuation per 100 ft $f \approx 10$ MHz (dB)
RG-58A/U	#21 (19 × #33)	50	29.5	1.6
RG-59B/U	0.023 Solid copper	75	20.5	1.1
RG-63B/U	0.025 Solid copper	125	10	0.6
RG-22B/U (dual)	Two 7 × 0.0152	95	16	1.6

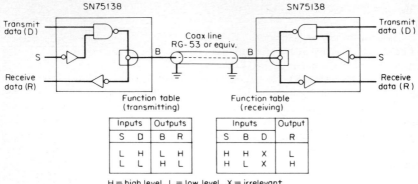

H = high level, L = low level, X = irrelevant

FIG. 14-39   Circuit D, coax line.

balanced-line transmission. Table 14-10 shows characteristics of typical coax lines used for data transmission.

Example 14–21   Coax Line Operation

Design a coax cable interface for circuit F of Fig. 14-28 that requires communication with a remote printer and keyboard located 100 ft away.

Solution

Since the remote printer and keyboard will require bidirectional communication, a transceiver, such as the SN75138, would be a good choice for this application.

The SN75138 quad bus transceivers are designed for two-way data transmission over single-ended transmission lines. The high current drivers (100 mA) are open collector and capable of driving 50 Ω coax lines.

The receiver input is internally connected to the driver output, and has a high impedance to minimize loading of the transmission line.

The receivers also feature a threshold of 2.3 V (typical), providing a wider noise margin than would be possible with a receiver with the usual TTL threshold. Receiver operation is not affected when the strobe turns off all drivers.

A schematic of circuit F is shown in Fig. 14-39. Only one channel of the SN75138 is used in this example. Three other sections are available for additional circuits, if needed. Also because of the high driver output current and the high receiver input impedance, a very large number (typically hundreds) of transceivers may be connected to a single data bus. The devices are designed to minimize loading of the data bus when the power-supply voltage is zero.

14–4   DISPLAY DRIVERS

14–4a   VLED Displays

Table 14-4 summarized basic differences between drive requirements for the five most common types of displays. The reader should refer to Chap. 12 for a detailed discussion of the optical characteristics of these displays, since the emphasis of this section is on interfacing.

The first display technology shown in Table 14-4 is visible light-emitting diodes. (VLEDs). These displays are probably the easiest to interface since they require low voltage and relatively moderate current.

VLEDs are diodes and behave as diodes in a circuit. They pass current when forward biased and have low leakage current when reverse biased below their breakdown voltage.

The diodes are made from gallium arsenide phosphide (GaAsP) instead of the germanium or silicon normally used for diodes, transistors, and integrated circuits. The wavelength of the light emitted is dependent upon the ratio of arsenic to phosphorous in the material. The amount of light emitted is directly proportional to the forward current through the diode.

This light-emission dependence on forward current means that the VLED should be operated from a constant current source to control the brightness. Any variation in forward current results in a variation in light output from the VLED. This is particularly important where numeric or alphanumeric characters are involved and the emitters are adjacent to each other.

The VLED displays may be categorized as follows:

1. *Discrete VLED Lamps*   These devices may be used as pilot lamps, indicator lights, trouble status indicators, etc. Their big advantages are compatibility with standard TTL levels, long life, low heat generation, attractive appearance, and, in general, they are just easy to use.

    As mentioned earlier the devices do need to be operated from a quasi-constant current supply for brightness uniformity. From TTL levels, the quasi-constant current can be obtained by a current-limiting resistor in series with the VLED from the positive $+5$-V rail. The low current required by the VLED (10 to 20 mA) and the low voltage drop across the VLED (typically 1.8 V) gives excellent results with the simple resistor circuit.

2. *VLED Displays*   Individual light-emitting diodes can be arranged in arrays of diodes to form different patterns. One of the most popular is the seven-segment numeric format. Another is a linear configuration to make a bar graph type of display. The individual diodes can be arranged in a dot matrix (usually 5×7) for presentation of alphanumeric information. A low-cost alphanumeric display can be made with 14 or 16 segments that make a very readable and easy-to-drive display. In all these arrangements, the same basic rule applies; i.e., each VLED diode needs a relative constant current to maintain uniform brightness.

A multidigit display is made up of several individual seven-segment characters. The VLEDs that make up the seven segments can be connected in a common anode or common cathode configuration. The common anode connection requires individual segment current to be sunk from each separate cathode and total digit current to be sourced into the common anode. The common cathode connection requires individual segment current to be sourced into each segment and total digit current sunk from the common digit cathode.

If the display consists of only a few digits, the segments can be driven by continuous DC current. Separate connections to each digit's segments are required. Several decoder/driver IC chips are available to drive the individual segments. The SN7447 or SN74247 is designed to drive seven individual segments of a common anode display and is capable of sinking up to 40-mA segment current. The MC14411 decoder/driver will source up to 25 mA into a common cathode display.

These IC chips all require separate external resistors in series with each segment to limit the current and provide the constant current for uniform brightness. For a multidigit display with say 10 characters, 80 segment connections and 10 digit connections would be required.

The number of interconnections to the display can be significantly reduced by multiplexing the display. Multiplex operation requires connecting all corresponding segments together and turning on the digits sequentially. Segment data is loaded into the segment lines at the appropriate time to give the correct information on the display. While only one digit is on at a time, the multiplex rate is high enough that, to the eye, the display appears to be on continuously. A 10-digit display now only requires eight segment lines (seven segments plus a decimal) and ten digit lines, resulting in eighteen connections instead of the ninety connections required for direct drive. Multiplex operation does have one disadvantage. The peak segment current must be increased in proportion to the reduced duty cycle to keep the same average current so the apparent brightness looks the same to the eye.

Small "calculator" displays are usually made from a monolithic red-emitting GaAsP chip. The seven segments and decimal point are formed on the monolithic chip using photomask and diffusion techniques common to silicon integrated-circuit technology. Due to cost considerations, the bar is limited in size, restricting the digit height from 0.050 to 0.100 in. A plastic magnifying lens ($1.5\times$ to $2.5\times$) is used to increase the digit height at the expense of viewing angle. These displays are called "calculator" displays because of their wide use in VLED calculators. Although most calculators today use LCD displays, the small VLED displays still find wide use in small hand-held consumer devices, such as games and toys.

Many calculator chips provide source and sink capability in their I/O ports and are designed to drive the small multidigit displays without external drivers. Very popular microcomputer chips used in hand-held electronic games are from the TMS1000 family. These devices will only source current out of their output ports. A digit driver is required to sink the display cathode current, whereas the anodes of the small displays can usually be driven out of the output ports without an external segment driver.

Example 14–22   Microcomputer to Display Interface

Interface a TMS1000NLL microcomputer to a six-digit TIL-393-6 display.
Solution
The TMS1000NLL is a 4-bit microcomputer that has 1024 bytes of ROM, 64×4 bits of RAM, eleven individually addressed R output ports, and eight parallel latched "Q" data output ports. The device is designed to operate from a standard 9-V "transistor"-type dry cell.

The TIL393-6 display has a common cathode and seven anode segments plus decimal point formed on a monolithic chip of GaAsP. Each segment requires a peak drive current of approximately 10 mA if multiplexed at one-sixth duty cycle. This current can easily be handled by the latched Q data lines on the microcomputer. The cathode current will be equal to the sum of all segment currents, or approximately 80 mA. (Seven segments plus decimal are on at the same time for worst-case conditions.)

An SN75492 digit driver can be used to sink the digit current. This driver has six Darlington high-gain transistors with current-limiting input resistors in the same 14-pin DIP package. The inputs are designed to interface

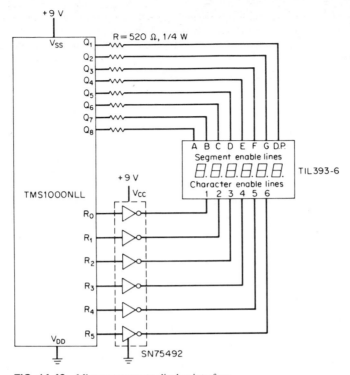

FIG. 14-40 Microcomputer to display interface.

directly into the TMS1000 family, although the device will function equally well from standard TTL levels.

Fig. 14-40 shows how the interface circuit could be implemented with the SN75492. Initialization and oscillator components are not shown on the schematic since they are irrelevant to this example.

Display segments are driven from the Q lines through segment resistors R. These resistors are selected to limit the current to 10 mA, and are calculated as follows:

$$R = \frac{V_{OH} - V_{VLED} - V_{SAT}}{I_{PEAK}} = \frac{8.0 - 1.8 - 1.0}{0.010} = 520 \ \Omega \qquad (14\text{-}4)$$

where V_{OH} = high-level output voltage of the TMS1000
V_{VLED} = forward voltage drop of display segment chip
V_{SAT} = saturation voltage drop of SN75492

Selection of display current is dependent upon many factors, including desired appearance, battery life, and viewing conditions, i.e., bright ambient light or darkened room. This example used a peak current of 10 mA through each segment of the display and will result in a very bright display. If battery life is a major consideration, the display current can be reduced by increasing the series segment resistor value. Peak currents on the TIL393-6 can be reduced to 6 mA and still have an acceptable appearing display, unless a poorly designed filter results in excessive light loss.

Most applications will have a diffused red plastic filter over the display for improved contrast ratio. This filter, particularly if poorly designed, can lose considerable light that must be made up by increased segment current.

If VLED displays are used with larger digit height, such as a 0.3-in TIL313 or 0.5-in TIL322, the TMS1000NLL family cannot source enough current to drive the segments directly, and it will be necessary to use a segment driver as well as the digit driver. The SN75491 is a popular quad segment driver. This device will source up to 50 mA and has inputs compatible with the PMOS TMS1000NLL.

Fig. 14-41 shows how the TMS1000NLL interface circuit can be modified to drive six TIL322 displays. At one-sixth duty cycle and 48-mA peak current, the average segment current will be 8 mA. This segment current will result in a very attractive looking display. However, we are no longer talking about operating the system from a 9-V transistor dry cell. Instead, AC line power converted to 9-V DC through suitable transformers, rectifiers, and regulators or high-capacity rechargable batteries would be used for system power. Worst-case maximum digit current is now 48 mA times 8 with all segments on at the same time, or 384 mA. This exceeds the ratings of the SN75492 digit driver.

The ULN2004 is a device that can be used in place of the SN75492. This device is classified as a peripheral driver, and data sheets may be found in the *Peripheral Driver*

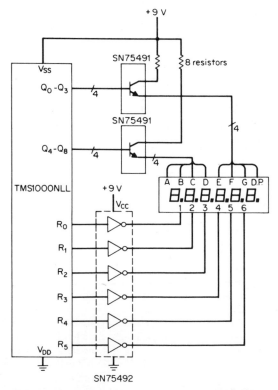

FIG. 14-41   Segment and digit drive for VLED displays.

Data Book from Texas Instruments. It is an excellent VLED digit driver with seven 500-mA rated transistors in the same package. The ULN2004 has inputs designed for interfacing with PMOS and other devices with TTL, CMOS, or high-voltage PMOS logic levels.

Linear arrays of VLEDs are replacing analog meters in many applications, particularly stereo amplifiers and tuners. The VLEDs form a bar graph type of display and can "dress up" a front panel with flashing lights.

Devices are available to drive the VLED bar-graph displays; however, they are usually not classified as VLED drivers. Instead, they are called analog level detectors and listed as special Functions in the Texas Instruments *Linear Control Circuits Data Book*.

Example 14–23   Bar-Graph Display

Design a drive circuit for a bar-graph display that will indicate zero balance for a stereo system using 10 VLEDs on each side of center. The first four VLEDs on each side are green light emitters, the next four VLEDs are yellow emitters, and the last two VLEDs are red emitters.

Solution

Two LM3914 analog level detectors can be used to accomplish this function. Each device will drive up to 10 VLEDs at a constant current programmed by an external resistor. Individual current-limiting resistors are not required in series with each VLED.

The LM3914 has 10 comparators and a reference voltage network to detect the level of a signal at the analog input. The output of each comparator drives one of the output stages. The first comparator turns on the first LED driver when the input level reaches approximately 125 mV. Each 125-mV increment in input voltage level turns on another VLED output driver until all 10 outputs are on with approximately 1.25-V input.

A circuit diagram for the display and driver is shown in Fig. 14-42. The unique thing about the LM3914 is the programmable constant current drivers in the output. Resistor R_1 sets the output current and the circuit will compensate for variations in VLED forward voltage drop. This is a real

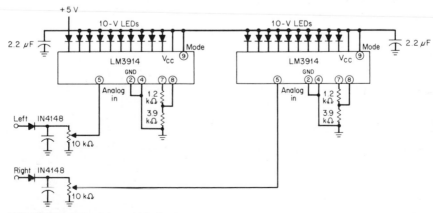

FIG. 14-42 Bar-graph stereo indicator.

advantage when VLEDs of different colors are mixed in the same system. If an external series resistor were used in each segment line to control the current, different values of resistors would have to be chosen to compensate for the difference in VLED forward voltage drop. The forward voltage drop of the green VLEDs may be as high as 3.0 V, compared to 1.8 V for the red VLEDs. This difference, when operated at a 5-V or lower supply voltage, can make a noticeable difference in segment brightness if the same resistor is used to limit the current. The LM3914 has approximately the same current through each output driver, independent (within limits) of the VLED forward voltage drop.

The analog input and discrete VLED driver output stages form a crude analog-to-digital converter. The input analog voltage can be scaled either by active amplifiers or attenuators for a 2-V input as full-scale indication. (VLED number 10 turned on.)

The analog input voltage for the two channels can be picked up at a number of circuit points in a typical stereo amplifier.

Another unique feature of the LM3914 is that control via pin 9 provides either a bar graph type of display or moving dot (only one output driver on at a time). There are numerous applications where this is a preferred display instead of a continuous bar graph.

14–4b   Vacuum Fluorescent Displays

The next display technology mentioned in Table 14-4 is vacuum fluorescent (VF). This technology makes use of light emission from a fluorescent material in a vacuum tube when struck by electrons. Originally, the displays were made with each seven-segment numeric display in its own vacuum tube with a separate grid and filament in the tube. Now the technology has improved to where a single glass structure can contain a multidigit 7-segment, 14-segment, X-Y dot array, or custom-designed special-character display.

Each digit of the VF multidigit display has its own internal grid used for multiplexing. Depending upon the character height and number of digits, there will be one or more filaments in front of the grid for the source of electrons. The structure is very similar to a triode vacuum tube with a filament, grid, and anode. A positive voltage on the anodes and grids accelerates electrons emitted from the filament and visible light is emitted from the anode areas struck by electrons. When the grids are grounded or negative, the electron current is shut off and the entire character is dark. The grids are sequentially turned on, one at a time, to multiplex the display, and decoded anode data is applied for the correct character information. Multiplex rates are high enough that the display does not flicker. This technique is very similar to the multiplex operation of VLED multidigit displays, as we previously discussed.

Operating voltages for the VF display are moderately high, typically 24 to 30 V; however, well within low-cost semiconductor process technology.

Example 14–24   Vacuum Fluorescent Display Driver

Design an interface circuit to drive a 12-digit 10-segment VF display. The 10 segments include 7 numeric segments, a decimal point, and 2 custom anode designs on several digits used to display special symbols.

Solution

This interface problem will be solved with two SN75512 display drivers.

The SN75512 can be used to drive segments of the VF display or to

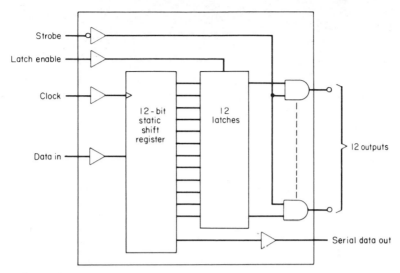

FIG. 14-43 The SN75512 functional diagram.

scan the digit lines. While designed originally for driving VF X-Y matrix-type displays, this example will show the device's versatility by driving a 10-segment multidigit VF array. One SN75512 is used to decode the segment lines and another SN75512 multiplexes the grid lines.

A functional block diagram of the SN75512 is shown in Fig. 14-43. The device has a serial data port that accepts a 12-bit serial data word at rates up to 1.0 MHz. The data is converted to parallel data with a 12-bit serial-to-parallel shift register on the low-to-high clock transition.

The 12-bit parallel data is loaded into 12 output latches when the latch enable input is high. Each totem-pole output is capable of 60-V operation

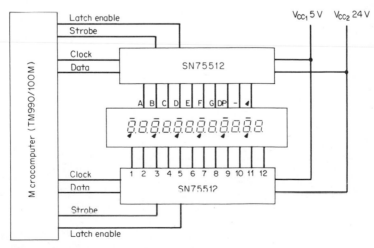

FIG. 14-44 Vacuum fluorescent drive.

and sourcing up to 25 mA. This capability will drive most VF displays that are available on the market. The active low strobe input enables all the outputs. Serial data output from the shift register may be used to cascade drivers for driving large X-Y display arrays. The serial data output are not affected by the latch enable or strobe inputs.

A schematic of the VF interface circuit is shown in Fig. 14-44. A 10-bit serial data word into driver 1 will select the segment line to be driven. Driver 2 multiplexes the grids. The two clocks must be synchronized so that the correct segment information is displayed on the proper digit.

14–4c   AC Plasma Displays

This technology is of fairly recent origin and is being pursued by a number of laboratories at the present time. Interest stems from the need for a flat panel large-screen display that can compete with cathode-ray tubes for computer terminal applications.

The AC plasma display is an X-Y matrix sandwiched between two flat pieces of glass that have a precisely controlled separation. The X and Y electrodes are deposited on the internal surfaces of the glass plates and covered by an insulating dielectric. The two plates are sealed around their outside edges. After evacuation, the space between the glass is backfilled with a gas under low pressure. The backfilling and final sealing are similar to construction of a DC plasma display; however, the electrodes in a DC plasma display are immersed in the media gas, whereas the electrodes of the AC plasma are isolated by a dielectric layer. Fig. 14-45 illustrates the construction of the display pane. Since the electrodes of the AC plasma display are isolated by the dielectric medium, DC operation is not possible. When an AC voltage exceeding the breakdown voltage of the gas is applied between the X electrode and the Y axis, the gas is ionized and the discharge that occurs emits a visible spot of light at the intersection of the selected X and Y row and column. By addressing each intersection point, information can be displayed by the glowing spots of emitted light. Once the discharge is initiated, the display element can be maintained active without additional addressing. The data retention prop-

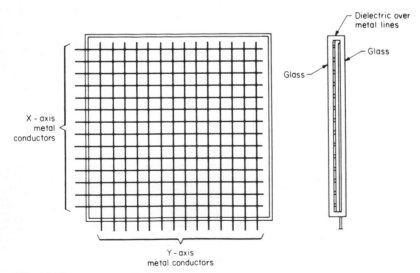

FIG. 14-45   AC plasma display.

erty of the AC plasma display eliminates the necessity of a memory map for simple information displays. The voltage required to maintain the active discharge is less than the original voltage required to ionize and initiate the initial breakdown. A cell (defined as the intersection between the X electrode and Y electrode) that is glowing can be maintained indefinitely by the reduced AC voltage. This period of time in which no panel information is being altered is called the "sustain" mode.

Since the sustaining electrode voltage is less than the potential required to fire the cell, this voltage can be applied to the rows and columns continuously. If a cell has not been previously fired, the sustaining voltage will have no effect on the cell. To turn on a cell, a pulse is superimposed on the sustain voltage at a specific X-Y location. Once fired, the cell will emit light until the voltage across the cell is reduced below its sustain-voltage level.

To turn off a cell location, an erase pulse is addressed to the specific cell. This pulse has sufficient amplitude and duration to counterbalance the sustaining voltage and extinguish the plasma discharge at that specific X-Y location.

Let us now discuss the design of an interface circuit to drive a 256 × 256 AC plasma display panel. This interface will use SN75502A and SN75503A AC plasma drivers from Texas Instruments. These devices were designed specifically for AC plasma display applications. They are made with a BIDFET process that combines several process technologies on the same silicon chip. The three process technologies are bipolar, DMOS, and CMOS. Prior to the development of the BIDFET process, the high-voltage require-

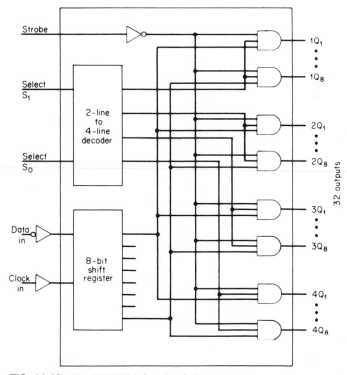

FIG. 14-46   The SN75502A functional diagram.

ments of the plasma panel prevented the use of integrated-circuit interface circuits, so many discrete components were required to accomplish this function. This resulted in a very expensive interface circuit that significantly added to the total system cost.

The output drive transistors are totem-pole structures formed by DMOS transistors (double-diffused MOS) with clamp diodes to both ground and V_{CC}, and to provide 100-V output voltage swing capability. Inputs are diode-clamped *pnp* structures using standard bipolar processing and they have TTL levels.

Each SN75502A drives 32 X-axis lines, and the SN75503A drives 32 Y-axis lines. The 256 × 256 AC plasma display panel will require eight X-axis drivers and eight Y-axis drivers.

The SN75502A performs the select operation along the X axis. TTL-compatible serial data is loaded into the data input port at a maximum data rate of 4.0 MHz. The device has an 8-bit shift register and a 2-to-4-line decoder that steers the serial 8-bit data string to one of four groups of eight outputs. A functional block diagram of the SN75502A is shown in Fig. 14-46. Select lines S_0 and S_1 select one of the four groups of eight outputs. The outputs are activated by the strobe input. When the strobe input is driven low, the selected 8-bit output group reflects the inverted data input string, while the other 24 outputs remain low. Outputs that are switched high form the positive select pulse.

The first component of the sustain waveform, the base pulse, is generated external to the X-axis drivers and is applied to all electrodes addressed along the X axis.

The SN75503A Y-axis driver also has 32 totem-pole structures made with the DMOS process to obtain a 100-V output voltage swing capability. The functional block diagram is shown in Fig. 14-47. These devices each have a 32-bit shift register with indefinite latch capability when the clock input is either high or low. Information at the data input, meeting the setup time requirements, is transferred into the shift register on the positive-going edge of the clock signal. Each shift register output drives its respective Q output through two gates controlled by the strobe and sustain inputs, respectively. The strobe

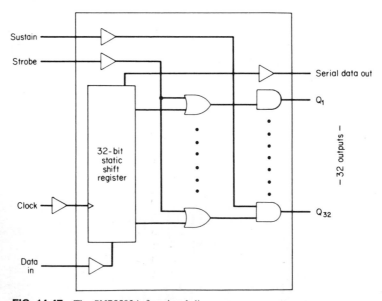

FIG. 14-47   The SN75503A functional diagram.

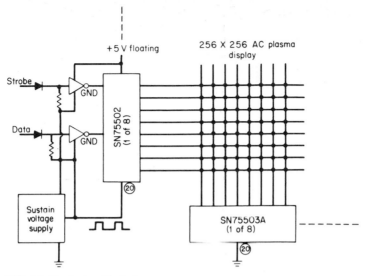

FIG. 14-48   System block diagram.

input controls the outputs for writing and erase functions, while the sustain input controls the output for system sustaining along the Y axis. The serial-data output can be used to cascade shift registers.

Unlike the SN75502A, the SN75503A can operate on all 32 of its outputs at one time. For this reason, the SN75502A is used to scan the electrodes along which the information is written. Selection of one specific bank of eight drivers out of the four banks available is determined by the select inputs S_0 and S_1. When selected, the state of the eight outputs of the section is determined by the data stored in the 8-bit shift register shown in Fig. 14-46. Data is shifted serially into the shift register on the positive transition of the clock. The maximum data rate is 4 MHz, and the strobe input must be held to a logic 1 level during data entry.

Logic 0's loaded into the shift register input select the outputs that will switch high when the data is loaded and the strobe input pulled low.

To reduce the output voltage swing requirements of the drivers, one or both drivers can be floated on a base voltage waveform generated by an external source. In this example, only the SN75502A X-axis driver will be floated and the SN75503A will be used on the ground rail.

Fig. 14-48 shows how pin 20 (ground) of all the SN75502A drivers can be tied to the sustain power supply and the entire driver floated at this power-supply level. The sustain power supply shown in Fig. 14-48 produces a voltage pulse that swings from 0 to $+115$ V with a frequency of 20 to 50 kHz. The X-axis drivers require $+5$-V DC and $+85$ V DC for operation. These supplies must also float with the sustain voltage level.

When the SN75502A outputs are low, the outputs (and X-axis lines) will follow the sustain voltage waveform. When one or more outputs go high, the voltage applied to the X-axis line will be 85 V above the sustain base voltage.

The voltage applied to the Y axis will be 0 or ground when the outputs are low and $+85$ V when the SN75503A goes high. Separate $+5$- and $+85$-V supplies are required for these devices, and are referenced to system ground. The voltage swing on the Y axis

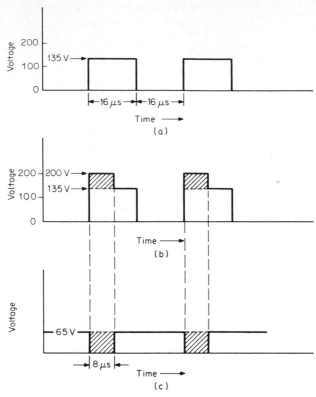

FIG. 14-49   Sustain waveform. (*a*) Bulk sustain waveform; (*b*) composite sustain waveform; (*c*) *Y*-axis sustain waveform.

is added to the base sustain voltage on the *X* axis to give a net voltage swing across the cell.

Waveforms of the base sustain voltage and the *Y*-axis voltage are illustrated in Fig. 14-49. When the SN75503A *Y*-axis driver swings from +85 V to ground, the 85 V is added to the bulk sustain voltage of 115 V, resulting in a voltage drop of 200 V across the cell. Fig. 14-49*b* shows the composite sustain waveform applied to all cells in the matrix. This voltage is sufficient to sustain all cells, since it is effectively placed across all cells in the matrix at the same time.

While a glowing cell can be sustained by the 200-V peak voltage, a voltage of 265 V is required to initiate the glow discharge. To write to a specific cell, the *X*-axis driver is turned on and the *Y*-axis driver is turned off at the same time. This produces a net peak voltage of 265 V across the cell, and the cell will break down into the conduction region. Waveforms of the write operation are shown in Fig. 14-50.

Once fired, the sustain voltage will maintain conduction until the cell is extinguished by another selected pulse called the erase pulse. This pulse is produced the same way as the write pulse. The *X*-axis driver is turned on and the *Y*-axis driver turned off at a specific address in the matrix to extinguish the pulse at that location.

Since the *X*-axis drivers are floating up and down on the sustain voltage pulse, it is necessary to isolate the input lines, since they are driven by standard TTL levels. When

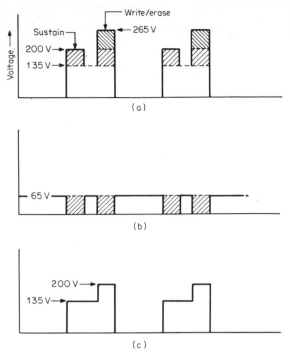

FIG. 14-50 Write/erase waveform. (*a*) Composite sustain/write/erase waveform; (*b*) *Y*-axis sustain/write/erase waveform; (*c*) *X*-axis write/erase waveform.

the sustain voltage is at ground, the data can be entered as standard TTL levels with no problem. When the SN75502A floats to the peak sustain voltage level, the input diode D_1 effectively isolates the driver from the rest of the system.

Since all inputs will be low due to the floating pull-down resistors, it is necessary to add an inverter on each input, as shown in Fig. 14-48. In the schematic an SN74LS04 is shown to be compatible with a total TTL system.

The SN75500A and SN75501A are drivers functionally the same as the SN75502A and SN75503A, except for MOS-compatible inputs. With these devices, CMOS inverters on the input would be more compatible with system design and would consume less total power.

Alternate isolation procedures could use optoisolators or transformer coupling; however, the diode isolation is effective and low in cost.

The purpose of this example is to illustrate a typical interface to the AC plasma display with the SN75502A and SN7550A drivers. The software and additional hardware for the complete system has not been discussed.

Additional systems application assistance is available from Texas Instruments, if required.

14—4d DC Plasma Displays

The DC plasma or gas discharge display is a mature display technology that has been available for many years.

This display is characterized by a bright orange light emitted from the display and produced when a mixture of neon/argon gas is electrically excited by a high voltage in the order of 120 to 170 V. While the appearance is very attractive and manufacturing cost competitive with most multidigit displays, the displays have suffered from several problems.

1. Historically, the DC plasma displays have had a reputation for poor reliability. The problems stem primarily from contact failure where a connection is made from a deposited metal layer on the glass to the external driver circuits.
2. The displays require high-voltage DC for operation. Integrated-circuit drivers have not always been easy to manufacture with breakdown voltages in excess of 100 V, and the cost of discrete driver circuits has not been competitive with VLED drive circuits.

With the recent introduction of high-voltage DMOS processing, integrated-circuit drivers are now inexpensive and available. Also, improvements in contruction techniques have eliminated most of the contact reliability problems, so DC plasma continues to be popular for large-area multidigit displays.

Example 14–25 DC Plasma Display Driver

Design an interface circuit to decode BCD intput data and drive a $4\frac{1}{2}$-digit DC plasma display for a digital panel meter.

Solution

The SN75584A is a popular decoder/driver capable of driving most DC plasma multidigit displays. Fig. 14-51 shows the functional block diagram of this device. The segment output drive transistors are made with DMOS technology for 100-V capability. Output current is programmable from 0.1

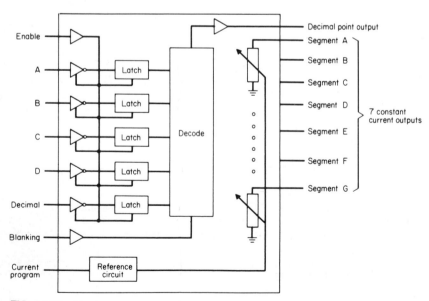

FIG. 14-51 The SN75584A functional diagram.

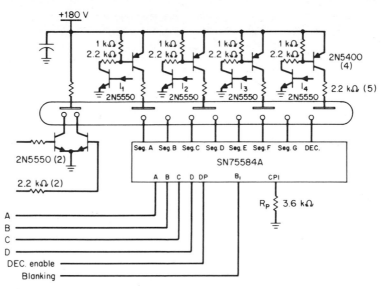

FIG. 14-52 DC plasma drive circuit.

to 4 mA with one external resistor. The device has *pnp* inputs compatible with TTL, PMOS, or CMOS logic levels and is designed to decode four lines of BCD data plus a decimal point while the enable input is at a low level.

The device decodes BCD digits 0 to 9 and blanks the display with input data greater than BCD 9.

The display can be directly driven using a separate SN75584A for each digit. For the $4\frac{1}{2}$-digit display, four SN75584A devices would be used and the plus and minus signs would be switched with separate *npn* transistors.

If the display is multiplexed, only one SN75584A would be required, but separate anode drivers would be needed. Fig. 14-52 shows the complete multiplexed interface circuit between TTL levels and the $4\frac{1}{2}$-digit DC plasma display.

Multiplexing is accomplished with the four 2N5400 *pnp* transistors. These transistors are turned on sequentially in synchronization with the correct BCD data applied to the input of the SN75584A. As shown on the schematic, four additional *npn* 2N5400 transistors are required to switch the *pnp*'s from standard TTL logic. Input from a multiplex circuit will be applied to the bases of the 2N5550s. This signal may be obtained from separate TTL logic or a microcomputer.

The plus and minus signs are handled separately from the multiplex circuit, since they cannot be decoded and driven by the SN75584A, and treating them separately simplifies both software and hardware.

When setting up the digit drive signals to multiplex the display, a small delay time should be introduced between characters to allow the gas plasma tube to recover. During this dead time, the segments should be blanked to ensure glow extention.

Multiplex rates will depend upon the manufacturer's specification for the particular tube used. Some tubes require a separate keep-alive cell to start the first scan character. Refer to recommended circuits usually supplied with the various gas plasma tubes.

The SN75584A supplies a programmed constant current into the segments. The constant current driver eliminates some problems, such as slow-segment turn-on associated with other drive techniques.

The segment current is programmed with resistor R_p connected between the CPI pin and ground. The value of this resistor in kilohms is determined by $3.6/I_0$, where I_0 is in milliamperes.

The 3.6-kΩ resistor shown in the schematic forces a 1.0-mA constant current in each segment. The values of R_p and the constant current should be determined from the manufacturer's recommended operating conditions shown on the tube's data sheet.

REFERENCES

Texas Instruments: *The Peripheral Driver Data Book,* Dallas, Tex., 1981.

Texas Instruments: *The Line Driver and Line Receiver Data Book,* Dallas, Tex., 1981.

Texas Instruments: *The Display Driver Data Book,* Dallas, Tex., 1977.

Texas Instruments: "AC Plasma Display," Application Rep., Bulletin SCA-204, Dallas, Tex.

Index

ABOUT THE EDITOR

Arthur B. Williams is Vice President of Engineering, Research, and Development at Coherent Communications Systems Corporation. Formerly Head of Advanced Techniques of the Telesignal Division, Singer Corporation, he has also worked as a senior circuit design engineer at Stelma, Inc., and a filter design engineer at Burnell and Company. Mr. Williams is the author of numerous technical design articles for trade publications and the IEEE Circuit Theory Transactions, as well as two books: *Electronic Filter Design Handbook* (McGraw-Hill) and *Active Filter Design* (Artech House). He holds a patent on a digital frequency discriminator.